Flora Australiensis

A Description of the Plants of the Australian Territory

VOLUME 6: THYMELEAE TO DIOSCORIDEAE

GEORGE BENTHAM
FERDINAND VON MUELLER

CAMBRIDGE
UNIVERSITY PRESS

CAMBRIDGE UNIVERSITY PRESS

Cambridge, New York, Melbourne, Madrid, Cape Town,
Singapore, São Paolo, Delhi, Tokyo, Mexico City

Published in the United States of America by Cambridge University Press, New York

www.cambridge.org
Information on this title: www.cambridge.org/9781108037433

This edition first published 1873
This digitally printed version 2011

ISBN 978-1-108-03743-3 Paperback

FLORA AUSTRALIENSIS.

FLORA AUSTRALIENSIS:

A DESCRIPTION

OF THE

PLANTS OF THE AUSTRALIAN TERRITORY.

BY

GEORGE BENTHAM, F.R.S.,

ASSISTED BY

BARON FERDINAND VON MUELLER, C.M.G., F.R.S.,

GOVERNMENT BOTANIST, MELBOURNE, VICTORIA.

VOL. VI.

THYMELEÆ TO DIOSCORIDEÆ.

PUBLISHED UNDER THE AUTHORITY OF THE SEVERAL GOVERNMENTS
OF THE AUSTRALIAN COLONIES.

LONDON:
L. REEVE & CO., 5, HENRIETTA STREET, COVENT GARDEN.

1873.

LONDON:
SAVILL, EDWARDS AND CO., PRINTERS, CHANDOS STREET
COVENT GARDEN.

CONTENTS.

CONSPECTUS OF THE ORDERS CONTAINED IN THE SIXTH VOLUME.

CLASS I. DICOTYLEDONS.

SUBCLASS III. MONOCHLAMYDEÆ.

(Continued from Vol V.)

*** *Ovary free but enclosed within the base of the perianth, 1- rarely 2-celled, with 1 ovule in each cell. Style simple but not oblique. Flowers mostly hermaphrodite. Leaves entire.*

CV. THYMELEÆ. Ovule pendulous. Hairs silky or spreading. Bark stringy.

CVI. ELÆAGNACEÆ. Ovule erect. Perianth contracted over the otherwise free ovary. Indumentum scurfy.

**** *Flowers strictly unisexual, very rarely polygamous. Perianth present, at least in one of the sexes, usually small and sometimes double. Ovary superior.*

CVII. NEPENTHACEÆ. Ovary 3- or 4-celled, with many ovules in each cell. Seed albuminous. Scandent shrubs. Leaves terminating in pitchers.

CVIII. EUPHORBIACEÆ. Ovary 3-celled, rarely 1- 2- or several-celled, with 1 or 2 pendulous ovules in each cell, and as many styles or stigmatic branches as cells. Albumen usually copious.

CIX. URTICEÆ. Ovary 1-celled, with 1 ovule, and 1 or 2 oblique styles or unilateral stigmas. Albumen usually scanty. Stamens opposite the perianth lobes.

***** *Perianth none besides small bracts.*

CX. CASUARINEÆ. Bracts 2 or 4 in decussate pairs. Trees or shrubs with jointed stems and branches, leafless except a whorl of teeth at each joint. Male flowers in catkins, females in cones.

CXI. PIPERITÆ. Bracts 1 under each flower. Herbs, shrubs or trees, with articulate branches and flat leaves. Flowers in spikes, racemes or heads, hermaphrodite or unisexual.

****** *Ovary inferior.*

CXII. ARISTOLOCHIACEÆ. Flowers hermaphrodite. Perianth with an oblique or valvately lobed limb. Stamens round the base of the style. Ovary 3- or 6-celled, with several ovules in each cell. Herbs or climbers.

CXIII. CUPULIFERÆ. Flowers unisexual. Perianth small, various. Ovary 1- to 6-celled, with 1 or 2 ovules in each cell. Nuts seated on or enclosed in an involucre. Trees or shrubs.

CXIV. SANTALACEÆ. Flowers usually hermaphrodite, often minute. Perianth-lobes valvate. Stamens opposite the lobes. Ovary with few pendulous ovules rarely conspicuous till after fecundation. Drupe indehiscent. Shrubs, herbs or rarely trees.

CXV. BALANOPHOREÆ. Fleshy scapigerous leafless root-parasites. Flowers unisexual, small, in dense terminal heads or spikes. Perianth-lobes valvate. Stamens 3. Ovary 1- or 2-celled, with 1 pendulous ovule in each cell.

SUBCLASS IV. GYMNOSPERMÆ.

Flowers strictly unisexual, without perianths. Anther-cells 2 or more sessile on the scale-like connectives or scales of a catkin or cone. Ovules in the axils or upon the scales of a cone or a fleshy cup or receptacle without any ovary.

CXVI. CONIFERÆ. Branching trees or shrubs, with needle-like or rigid entire leaves or scales. Stamens in catkins with scale-like connectives, bearing 2 or few anther-cells.

CXVII. CYCADÆ. Trunks woody, usually simple, with a palm-like crown of large pinnate leaves. Anther-cells numerous on the concealed under side of the scales of a large cone.

CLASS II. MONOCOTYLEDONS.

Stem not distinguishable into pith, wood and bark, but when perennial, consisting of bundles of fibres irregularly imbedded in cellular tissue, with a firmly adherent rind outside. Seeds with one cotyledon, the embryo undivided, the young stems developed from a sheath-like cavity on one side. Leaves radical or alternate and parallel veined or rarely pinnate and crowning the undivided stem.

* *Ovary inferior.*

CXVIII. HYDROCHARIDEÆ. Aquatic plants with regular mostly unisexual flowers. No albumen.

CXIX. SCITAMINEÆ. Flowers irregular, one or more or all but one of the stamens reduced to staminodia or wanting. Seeds albuminous.

CXX. ORCHIDEÆ. Flowers irregular, only one or two of the stamens perfect and inserted on the style. Seeds minute, with a homogeneous embryo. Inflorescence centripetal.

CXXI. BURMANNIACEÆ. Flowers regular. Anthers nearly sessile, the cells separated. Seeds minute, with a homogeneous embryo. Inflorescence centrifugal. Leaves laterally flattened or reduced to scales.

CXXII. IRIDEÆ. Flowers regular or nearly so. Anthers opening outwards. Seeds albuminous. Inflorescence centrifugal. Leaves laterally flattened or terete.

CXXIII. AMARYLLIDEÆ. Flowers regular or nearly so. Anthers opening inwards. Placentas axile. Seeds albuminous. Inflorescence centripetal. Leaves chiefly radical, veinlets when present transverse.

CXXIV. TACCACEÆ. Flowers regular or nearly so. Anthers opening inwards. Placentas parietal. Seeds albuminous. Inflorescence centripetal (umbellate). Leaves radical, large, often divided.

CXXV. DIOSCORIDEÆ. Flowers small, unisexual, regular. Anthers opening inwards. Seeds albuminous. Stems usually twining. Leaves alternate, veinlets reticulate.

FLORA AUSTRALIENSIS.

ORDER CV. THYMELEÆ.

Flowers hermaphrodite or rarely diœcious. Perianth simple, tubular or campanulate, 4-lobed or in genera not Australian 5-lobed, usually regular, the lobes imbricate in the bud, with the addition in many genera not Australian of small scales, alternating with the lobes at their base. Stamens either 2 only, or as many as the lobes of the perianth or twice as many; filaments inserted in the throat or within the tube; anthers with 2 parallel cells opening longitudinally. Ovary free within the base of the perianth, 1- or rarely 2-celled, with 1 or rarely 2 or 3 pendulous anatropous ovules. Style simple, with a terminal entire capitate or truncate stigma. Fruit an indehiscent nut drupe or berry, or rarely a 2-valved capsule. Seed with or without albumen; embryo straight, with a superior radicle.—Shrubs trees or rarely herbs, with a stringy bark. Leaves alternate or opposite, always simple and entire. Flowers in terminal or axillary clusters heads umbels racemes or spikes, rarely solitary, often surrounded by an involucre of 4 or more bracts, differing more or less from the stem-leaves.

A considerable Order, widely distributed over most parts of the globe.

Stamens 2. Perianth-lobes 4 . . . . 1. PIMELEA.
Stamens 4, alternating with the perianth-lobes. Densely tufted prostrate shrub . . . . 2. DRAPETES.
Stamens twice as many as perianth-lobes. Shrubs or small trees.
Ovary 1-celled, with 1 ovule. Hypogynous scales 4, free or united in pairs . . . . 3. WIKSTRÖMIA.
Ovary 2-celled, with 1 ovule in each cell. Hypogynous scales united in a short cup . . . . 4. PHALERIA.

1. PIMELEA, Banks and Soland.

(Thecanthes, *Wikstr.;* Gymnococca, *Fisch. et Meyer;* Heterolæna *and* Calyptrostegia, *C. A. Mey.;* Macrostegia, *Turcz.*).

Perianth tubular, with a spreading or rarely erect 4-lobed limb, without scales but often slightly thickened or folded round the throat. Stamens 2, inserted in the throat opposite the 2 outer perianth-lobes.

No hypogynous scales. Ovary 1-celled, with 1 pendulous ovule. Style elongated, attached to one side of the ovary immediately below the apex. Fruit a small drupe, with a membranous or succulent epicarp, the endocarp nut-like, crustaceous, often hooked at the top. Seed pendulous with a membranous testa; albumen scanty or copious; cotyledons broad or narrow, rather thick, longer than the radicle.—Shrubs undershrubs or herbs. Leaves opposite or alternate. Inflorescence varied within the limits of the Order, but never umbellate. Perianth white pink or yellowish, often silky-villous.

The genus is limited to Australasia, comprising, besides the Australian species, one of which is also from New Zealand, nine others confined to New Zealand.

SECT. 1. **Thecanthes.**—*Involucral bracts united into a 4-lobed cup. Perianth-tube glabrous, not circumsciss.—Glabrous annuals.*

Involucral lobes very broad, shorter than the entire part, usually marked with forked veins . . . . 1. *P. punicea.*
Involucral lobes reaching to about the middle, with the midrib alone prominent.
  Involucral lobes very broad. Filaments twice as long as the perianth-lobes. Flowers white . . . . 2. *P. concreta.*
  Involucral lobes acute. Filaments much shorter than the perianth-lobes. Flowers red . . . . 3. *P. cornucopiæ.*
Involucral lobes reaching nearly to the base, several-nerved and longer than the flowers. Filaments much shorter than the perianth-lobes. Flowers red . . . . 4. *P. sanguinea.*

SECT. 2. **Eupimelea.**—*Involucral bracts free, like the stem-leaves or rather broader. Perianth-tube silky-villous, not circumsciss.—Prostrate or much branched shrubs, with flat or concave leaves usually opposite. Tasmanian or mountain species.*

Stem and leaves glabrous. Involucral bracts rather broader than the stem-leaves.
  Leaves under ½ in. long. Perianth-tube 2 to 3 lines . . . 5. *P. alpina.*
  Leaves mostly above ¾ in. Perianth-tube about ½ in. . . . 6. *P. longifolia.*
Stem and under-side of the leaves silky-villous, upper side glabrous. Involucral bracts like the stem-leaves . . . . 7. *P. cinerea.*
Stem and leaves on both sides densely silvery-silky. Involucral bracts like the stem-leaves . . . . 8. *P. Milligani.*

(43. *P. leptostachya* has the perianth not circumsciss, but the flowers in spikes, without bracts and the leaves alternate.)

SECT. 3. **Heterolæna.**—*Involucral bracts free, much broader than the leaves. Perianth-tube not circumsciss—Shrubs. Leaves opposite, glabrous, flat (not concave) or with the margins recurved or revolute. Species all Western.*

Leaves oblong or lanceolate, acute or mucronate, ¾ to above 1 in. long. Perianth-tube above ½ in. long, with long spreading hairs, the upper part with short appressed hairs . . . . . . 9. *P. spectabilis.*
Leaves of *P. spectabilis* but shorter and less acute. Perianth-tube with only a few scattered hairs above the long spreading ones . . . . 10. *P. Lehmanniana.*
Leaves flatter shorter broader and more obtuse. Involucral bracts often with coloured margins. Perianth-tube 4 to 6 lines long . . . . 11. *P. hispida.*
Leaves oblong-linear or lanceolate, mostly acute, the margins much recurved. Perianth-tube under ½ in. long . . . . . 12. *P. rosea.*
Leaves ovate or oblong, obtuse, under ½ in. long, the margins much recurved. Perianth-tube under ½ in. long . . . . . 13. *P. ferruginea.*

Leaves narrow, 2 to 3 lines long, the margins much recurved.
Perianth-tube 2 to 2½ lines long . . . . 14. *P. brachyphylla.*

(16. *P. brevifoliæ*, with small concave leaves seems also to have the perianth-tube scarcely circumsciss.)

SECT. 4. **Calyptrostegia.**—*Flowers hermaphrodite or in some specimens female by abortion. Perianth-tube after flowering (except in* P. leptostachya) *circumsciss above the ovary, leaving the lower portion only persistent round the fruit. Anthers with a narrow connective, the cells very distinct and after they open placed back to back.*

SUBSECT. 1. **Calyptridium.**—*Flower-heads terminal, with 4–6 broad persistent involucral bracts.—Shrubs (or one species a hard annual?) with opposite leaves.*

* *Western species.—Leaves flat or concave, glabrous as well as the branches. Cotyledons usually narrow.*

Involucral bracts all glabrous.
Perianth perfectly glabrous . . . . 15. *P. sylvestris.*
Perianth more or less hairy.
Leaves penniveined, elliptical or lanceolate. Hairs of the perianth all spreading, few only in the upper part . . . 16. *P. brevifolia.*
Leaves narrow without lateral veins. Hairs of the perianth short and appressed in the upper part, with or without long spreading ones lower down . . . . 17. *P. Maxwelli.*
Inner involucral bracts silky-villous inside.
Leaves linear or lanceolate. Short persistent portion of the perianth very densely hispid, deciduous portion villous with appressed hairs.
Bracts scarcely acuminate, much shorter than the perianth. Leaves mostly linear . . . . 18. *P. angustifolia.*
Bracts herbaceous, acuminate, nearly as long as the perianth-tube. Leaves mostly lanceolate . . . . 19. *P. nervosa.*
Leaves ovate oblong or broadly lanceolate. Flower-heads nodding.
Perianth silky-villous throughout, hairs of the lower part above the ovary often longer but scarcely spreading.
Bracts scarcely acuminate . . . . 20. *P. sulphurea.*
Bracts herbaceous, acuminate. Flower-heads large. Short persistent portion of the perianth very densely villous . . . . 21. *P. floribunda.*
Perianth with long spreading hairs in the lower part. Bracts large, obtuse, thin and coloured, but not concealing the flowers . . . . 22. *P. suaveolens.*
Perianth glabrous in the lower part, the long narrow-linear lobes hairy. Bracts very large, obtuse, coloured, completely enclosing the flowers . . . . 23. *P. physodes.*

** *Eastern species.—Leaves more or less concave, glabrous as well as the stem. Cotyledons usually broad.*

Perianth nearly glabrous at the base, hairy upwards. Hairs of the receptacle very long . . . . 24. *P. glauca.*
Perianth hairy throughout. Hairs of the receptacle short.
Leaves narrow, acute, very concave. Involucral bracts acuminate . . . . 25. *P. colorans.*
Leaves mostly oblong, obtuse. Bracts scarcely acuminate.
Leaves with 1 or 2 prominent marginal or submarginal veins underneath. Flower-heads erect. Involucral bracts with a prominent midrib . . . . 26. *P. collina.*
Leaves without prominent marginal veins. Bracts large and thin . . . . 27. *P. spathulata.*

*** *Eastern species.—Glabrous silky-hairy or tomentose. Leaves flat or with the margins more or less recurved. Cotyledons usually broad.*

Branches and leaves glabrous.
 Leaves narrow, under 1 in. long . . . . 28. *P. linifolia.*
 Leaves oblong or broad, above 1 in. long . . . . 29. *P. ligustrina.*
Branches silky-hairy. Leaves glabrous or loosely silky-hairy . 30. *P. humilis.*
Branches and underside of the leaves silvery-silky . . . . . 31. *P. sericea.*
Branches and underside of the leaves white-tomentose . . . . 32. *P. nivea.*

Subsect. 2. **Phyllolæna.**—*Flower-heads with numerous involucral bracts not broader than the leaves.—Western species.*

Leaves mostly alternate, glabrous or loosely silky-villous. Perianth-lobes short . . . . 33. *P. imbricata.*
Leaves mostly opposite, villous. Perianth very hispid, the lobes longer than the tube . . . . 34. *P. villifera.*

Subsect. 3. **Choristachys.**—*Flowers in clusters spikes or racemes, without involucres, or the bracts not broader than the leaves and very deciduous. Leaves flat or with slightly recurved margins.—Eastern or tropical species.*

Leaves opposite.
 Flower-clusters mostly axillary, small. Fruit usually succulent . . . . 35. *P. drupacea.*
 Flowers large, red, in a terminal dense spike . . . . . 36. *P. hæmatostachya.*
 Flowers very small, in terminal clusters or spikes. Plant glabrous.
  Leaves mostly oblong. Flower-clusters lengthening into spikes or racemes . . . . 37. *P. spicata.*
  Leaves mostly ovate or ovate-lanceolate. Flower-clusters not lengthening . . . . 38. *P. filiformis.*
Leaves mostly alternate, silky-villous.
  Leaves rather broad, 1½–3 in. long. Flowers shortly spicate. Perianth 3–4 lines long . . . . 39. *P. latifolia.*
  Leaves linear. Perianth not 2 lines long.
   Fruiting spike shortly capitate . . . . 40. *P. simplex.*
   Fruiting spike long and interrupted.
    Perianth-hairs silky appressed . . . . 41. *P. sericostachya.*
    Perianth-hairs rigid and spreading . . . . 42. *P. trichostachya.*
Leaves alternate, glabrous or slightly silky. Fruiting spike long and interrupted. Perianth small, not circumsciss . . . . 43. *P. leptostachya.*

Sect. 5. **Malistachys.**—*Flowers (small) strictly diœcious. Male perianth with a slender tube. Anthers with a narrow connective, the cells very distinct and after they are open placed back to back. Ovary abortive. Female perianth-tube after flowering circumsciss above the ovary, the lower portion persistent round the somewhat succulent fruit and almost adnate to it. Leaves silky-villous or hairy.—Western species.*

Flower-clusters all sessile and axillary . . . . 44. *P. argentea.*
Flower-clusters terminal or on axillary peduncles . . . . . 45. *P. clavata.*

Sect. 6. **Dithalamia.**—*Flowers (small) strictly diœcious. Male perianth with a slender tube; anthers with a narrow connective, the cells very distinct, and after they are open placed back to back; ovary abortive or rudimentary. Female perianth wholly persistent with small lobes divided to the ovary, or rarely with a short tube and tardily circumsciss. Fruit not at all, or slightly succulent. Leaves opposite, flat, or nearly so.*

Flower-clusters all axillary.
 Lateral veins of the leaves very diverging. Male perianth-tube 1 to 1½ lines long . . . . 46. *P. axiflora.*
 Lateral veins nearly parallel to the midrib. Male perianth-tube 4 to 5 lines long . . . . 47. *P. leptospermoides.*

Flower-clusters terminal, or in the forks.
  Leaves linear-lanceolate, mostly ½ to 1 in. long.
    Flowers more or less silky hairy . . . . 48. *P. microcephala.*
    Flowers quite glabrous . . . . 49. *P. pauciflora.*
  Leaves oblong, with recurved margins, 2 to 4 lines long . . 50. *P. elachantha.*
  Leaves small, ovate, coriaceous, more or less concave.
    Diffuse or very much branched low shrubs. Flowers glabrous, or sparingly ciliate.
      Flowers mostly solitary, upper leaves and perianths ciliate with a few long hairs . . . . 51. *P. pygmæa.*
      Flowers clustered, quite glabrous as well as the leaves . 52. *P. serpyllifolia.*
    Erect, shortly dichotomous shrubs. Flowers silky-villous.
      Leaves mostly obtuse. Female perianth-tube not produced above the ovary . . . . 53. *P. flava.*
      Leaves mostly acute. Female perianth-tube produced above the ovary and sometimes tardily circumsciss . . 54. *P. petrophila.*

SECT. 7. **Epallage.**—*Flowers hermaphrodite or more or less diœcious. Perianth-tube usually circumsciss after flowering, leaving the lower portion persistent round the fruit. Anthers rather flat, with a broad dorsal connective, the cells closely parallel on the inner face, the whole anther usually rolled back after flowering. Flowers in clusters or heads, rarely solitary, or in dense oblong spikes.*

Flowers strictly diœcious. Leaves alternate, softly silky-villous.
  Flowers solitary in the upper axils. Female perianth shortly and equally silky-villous . . . . 55. *P. Bowmanni.*
  Flowers in clusters. Female perianth with the persistent portion clothed with very long spreading hairs. Male perianth shortly and equally silky-villous . . . . . . 56. *P. ammocharis.*
Flowers hermaphrodite, or on some specimens female.
  Softly villous plants. Flowers small, bracts 2 or rarely 4, unequal and deciduous.
    Hairs appressed. Leaves mostly oblong, rarely ½ in. long . 57. *P. curviflora.*
    Hairs spreading. Leaves ovate, distinctly petiolate, under ½ in. long . . . . 58. *P. hirsuta.*
    Hairs scarcely spreading. Leaves ovate or oblong, ½ to 1½ in. long. Flowers rather larger . . . . 59. *P. altior.*
  Softly villous, or rarely nearly glabrous plants, involucral bracts several, not much broader than the leaves.
    Leaves flat, the midrib scarcely conspicuous. Bracts usually numerous.
      Filaments shorter than the corolla . . . . 60. *P. octophylla.*
      Filaments longer than the corolla . . . . 61. *P. petræa.*
    Leaves erect, concave, with the midrib prominent underneath. Bracts usually about 6.
      Leaves small (under ¼ in.) oblong, rather broad . . . 62. *P. phylicoides.*
      Leaves ¼ to near ½ in. long, narrow oblong . . . . . 63. *P. Eyrei.*
      Leaves linear, mostly about ½ in. long . . . . 64. *P. longiflora.*
  Stem and leaves glabrous, leaves concave. Bracts (4 to 6) much broader than the leaves.
    Leaves narrow, coriaceous, acute. Perianth circumsciss above the ovary.
      Flower-heads usually nodding. Perianth equally silky-villous . . . . 65. *P. stricta.*
      Flower-heads usually erect. Perianth with long more spreading hairs in the lower part . . . . 66. *P. Preissii.*
    Leaves broad, 1 to 1½ in. long. Flowers at length spicate. Bracts very deciduous. Perianth not circumsciss . . . 67. *P. Holroydi.*

P. grandiflora, Don. Hort. Cantab., and *P. prinifolia Nois.* quoted in Steudel's Nomenclator, are garden names which cannot now be identified.

SECT. 1. THECANTHES.—Involucral bracts united into a 4-lobed cup. Flowers hermaphrodite. Perianth glabrous, the tube not circumsciss.—Glabrous annuals, with opposite or alternate leaves.

1. **P. punicea,** *R. Br. Prod.* 359. An erect glabrous slightly branched annual of ½ to 1 ft. Leaves mostly opposite, lanceolate, very acute or mucronate, about 1 in. long. Flower-heads on a rather long erect terminal peduncle; thickened at the end. Involucre broadly turbinate, 6 to 8 lines diameter, divided to below the middle into 4 broad obtuse lobes, marked with forked veins, the two outer ones often slightly dilated and overlapping the others at the base. Flowers red, much exserted. Perianths on very short conical pedicels within the involucre at or near its base, the tube about 3 lines long, the lobes about 1 line. Filaments about half the length of the lobes; anthers oblong, with a narrow connective. Epicarp membranous. Seed with a scanty albumen and broad cotyledons.—Meissn. in DC. Prod. xiv. 497; Endl. Iconogr. t. 11. *Thecanthes punicea,* Wikstr. in Trans. R. Acad. Stockh. 1818, 272.

N. Australia. Arnhem N. and S. Bays, *R. Brown;* Arnhem land, *M'Kinlay;* Sims' Island, *A. Cunningham.* In R. Brown's specimens the leaves are slightly mucronate and the forked veins of the involucre are not so prominent as in the others

Var. *breviloba,* F. Muell. Involucres ¾ in. diam., the flowers exceedingly numerous. —Sandstone tableland, Upper Victoria river and Hooker's and Sturt's Creeks, *F. Mueller;* Purdie's Ponds, *M'Douall Stuart.*

2. **P. concreta,** *F. Muell. Fragm.* v. 73. A glabrous annual, with the habit and foliage of *P. punicea,* but the peduncle in the only specimen seen shorter than the last leaves and the flowers white. Involucre broad, divided to about the middle into 4 very broad obtuse or scarcely acuminate lobes, veinless except the slightly conspicuous midrib. Perianths much exserted, the tube nearly 3 lines long, the lobes short and obtuse, scarcely ¾ line long. Filaments at least twice as long as the lobes, with small oblong anthers.

N. Australia. Camden Harbour, N. W. Coast (*Herb. F. Mueller*).

3. **P. cornucopiæ,** *Vahl. Enum.* i. 305. An erect glabrous rather stiff annual of about 1 ft. Leaves alternate or the lower ones opposite, sessile or nearly so, lanceolate or oblong-linear, obtuse or nearly acute, mostly ¾ to 1½ in. long. Flower-heads on a terminal peduncle. Involucre turbinate with a long tapering base, about ½ in. diameter, divided to near the middle into broad acute lobes, with the midribs alone conspicuous. Flowers numerous, usually whitish, on short flattened pedicels within the involucre at or near its base. Perianths scarcely protruding beyond the involucral lobes, the slender tube about 2 lines long, circumsciss after flowering shortly above the ovary, the lobes small and obtuse. Filaments very short; anthers ovate, with a narrow connective. Epicarp membranous. Seed with a scanty albumen and broad cotyledons.—R. Br. Prod. 359; Meissn. in DC. Prod. xiv. 496; *Thecanthes cornucopiæ,* Wikstr. in Trans. R. Acad. Stockh. 1818, 271; *Calyptrostegia cornucopiæ,* Endl. Gen. Pl. Suppl. iv. part 2, 60.

Queensland. Endeavour river, *Banks and Solander, A. Cunningham;* Port Curtis, Keppel Bay, Shoalwater Bay, Broad Sound, *R. Brown;* Cape York, *Daemel;* Port Denison, *Fitzalan;* Burdekin river, *Bowman;* Rockhampton and Rockingham Bay, *Thozet* and others.

4. **P. sanguinea,** *F. Muell. Fragm.* i. 84. A glabrous annual, at first simple, but soon branching from the base into numerous decumbent or ascending simple or slightly branched stems, seldom exceeding 6 in. Leaves more crowded than in the allied species, oblong-linear or lanceolate, obtuse or rather acute, $\frac{1}{2}$ to $\frac{3}{4}$ or rarely 1 in. long. Flower-heads shortly pedunculate or almost sessile above the last leaves. Involucre rather broad, divided nearly to the base into ovate acute lobes of about $\frac{1}{2}$ in., the midrib prominent and a few faint lateral veins at the base. Perianths red, much shorter than the involucre, the tube not 2 lines long, the lobes scarcely above $\frac{1}{2}$ line and obtuse. Stamens shorter than the lobes, with the short anthers of *P. cornucopiæ*.

Queensland. Upper Roper river, *F. Mueller;* Cape river, *Bowman.*

A specimen from alluvial flats, Mount King, Glenelg district, *Martin*, referred by F. Mueller, Fragm., vii. 3, to *P. sanguinea*, with the evidently red flowers of that species, has the habit and involucres of *P. cornucopiæ;* but it is insufficient to determine absolutely its affinities.

Sect. 2. Eupimelea.—Involucral bracts free, like the stem leaves, or rather broader. Flowers hermaphrodite, the perianth-tube not circumsciss, silky-villous. Seeds, where known, with a scanty albumen, and broad cotyledons —Prostrate or much-branched shrubs, with flat or concave leaves, usually opposite.

To this section belong the several New Zealand species of *Pimelea.*

5. **P. alpina,** *F. Muell.; Meissn. in DC. Prod.* xiv. 511. A low much-branched prostrate or shortly erect shrub, glabrous or nearly so, except the inflorescence; leaves mostly opposite, rather crowded on the flowering branches, oblong or oblong-lanceolate, rather obtuse, under $\frac{1}{2}$ in. long. Flower-heads small, sessile within the last leaves, with 4 to 6 bracts or floral leaves rather broader than the stem-leaves. Receptacle, shortly villous. Flowers hermaphrodite, but in some measure dimorphous, in some specimens the perianth-tube slender, nearly 3 lines long, and the styles short; in others, the tube not two lines. and the styles long, but perfect anthers and ovaries in both. Perianth more or less hairy, the tube not circumsciss, the lobes about 1 line long. Filaments very short; anthers with a rather broad connective, but when open the cells placed back to back or nearly so.

Victoria. Munyong mountains, Albert Range, mounts Wellington and Latrobe, Bawbaw and Cobberas mountains at an elevation of 4000–6000 ft., *F. Mueller.*

6. **P. longifolia,** *Banks and Sol.; Meissn. in DC. Prod.* xiv. 516. An erect much-branched shrub of 2 ft., glabrous except the inflorescence, and often a minute tuft of hairs on the apex of each leaf. Leaves opposite, sessile, crowded, lanceolate, flat or nearly so, green or glaucous, $\frac{3}{4}$ to 1 in. long in Moore's specimens; but often much larger in New Zealand ones. Flower-heads terminal, sessile with a few involucral

bracts rather broader and shorter than the stem-leaves, but passing into them and rather shorter or longer than the flower. Receptacle very villous. Perianth white, silky-villous, the tube about ½ in. long, not circumsciss after flowering, the lobes about 2 lines long. Filaments about half as long as the lobes; anthers oblong, with a narrow connective, the cells when open placed back to back. Fruit not seen in Moore's specimens; in the New Zealand ones, dry with a membranous epicarp.—Hook. f. Handb. N. Zeal. Fl. 242; *P. congesta,* Moore and Muell. in F. Muell. Fragm. viii. 9.

N. S. Wales. Lord Howe's island, *C. Moore.*—The species is frequent in New Zealand, where the leaves are usually much largėr and not so pale coloured when dry, but in some specimens for instance from Dusky Bay, *Hector*, they are very much like those of the Lord Howe's island specimens, the involucral bracts, although sometimes but little different from the stem-leaves, are in other New Zealand specimens quite as distinct as in those from Lord Howe's island.

7. **P. cinerea,** *R. Br. Prod.* 361. A straggling shrub of 3 to 6 ft. resembling *P. drupacea* in habit and inflorescence, but the branches, the underside of the leaves and the perianths more densely clothed with long rather spreading silky hairs. Leaves mostly opposite, oblong or elliptical, flat, glabrous on the upper side, rarely above 1 in. long. Flower-heads small, terminal or in the old axils, with 4 to 6 bracts or floral leaves similar to the stem-leaves, but smaller. Perianth small, the lobes nearly 1 line long, but the constricted portion of the tube above the ovary not above ¼ line, the whole persistent when in fruit, and the base becoming then somewhat succulent. Stamens much shorter than the lobes, the filaments longer than the anthers; connective narrow. Seeds with a rather copious albumen, but the cotyledons broad.—Meissn. in DC. Prod. xiv. 509; *P. Gunnii,* Hook. f. Fl. Tasm. i. 332.

Tasmania. Mount Wellington (Table mountain), *R. Brown, A. Cunningham, Gunn;* South Port, *C. Stuart.*

F. Muell. Fragm vii. 8, unites this with *P. drupacea,* but the perianth is very different.

8. **P. Milligani,** *Meissn. in DC. Prod.* xiv. 509. A bushy densely branched shrub, the leafy branches and both sides of the leaves densely clothed with silvery-silky hairs, the older branches denuded, showing the scars of the fallen leaves. Leaves opposite, crowded and imbricate, ovate or oblong-elliptical, flat, thick and soft, not exceeding ½ in. Flower-heads terminal, sessile, the floral leaves not different from the others. Perianth villous, the tube 1½ to 2 lines long, or rather more, persistent or perhaps sometimes tardily circumsciss, the lobes above 1 line long. Filaments short; anthers oblong, with a rather broad connective, but the cells quite distinct, and at length placed back to back. Fruit not seen.

Tasmania. Mount Sorrell, at an elevation of 3000 ft., *Milligan.*

SECT. 3. HETEROLÆNA.—Involucral bracts free, much broader than the stem-leaves. Flowers hermaphrodite, the perianth-tube not circumsciss. Anthers with a narrow connective. Seeds where known with a copious albumen and narrow cotyledons.—Shrubs. Leaves opposite, glabrous, flat (not concave) or with the margins recurved or revolute.

9. **P. spectabilis,** *Lindl. Bot. Reg.* 1841, *t.* 33. An erect shrub, attaining sometimes 3 or 4 ft., glabrous except the flowers. Leaves mostly opposite, rather crowded, linear-oblong or lanceolate, flat or the margins slightly recurved, $\frac{3}{4}$ to $1\frac{1}{2}$ in. long, the uppermost often shorter and broader. Flower-heads usually large, globular, surrounded by 4 to 6 ovate or ovate-lanceolate bracts, often coloured on the margins, from half as long to nearly as long as the flowers. Perianths more or less tinged with pink, on short hirsute pedicels. The tube slender, $\frac{1}{2}$ to $\frac{3}{4}$ in. long, not circumsciss, slightly hairy round the ovary, from thence to above the middle hirsute with very long spreading hairs, the upper portion and the narrow lobes with short appressed silky hairs, the whole either persisting on the fruit, or opening at the base and falling off altogether. Filaments usually nearly as long as the lobes; anthers oblong, the connective narrow, the cells when open placed back to back. Epicarp membranous. Seed with a copious albumen, the embryo nearly cylindrical, the cotyledons narrow, longer than the radicle.—Meissn. in DC. Prod. xiv. 504; Bot. Mag. t. 3950; *Heterolæna spectabilis*, Fisch. et Mey. Ind. Sem. Hort. Petrop. x. 46; C. A. Mey. in Bull. Acad. Petrop. iv. (1845), 74; *Pimelea Verschaffeltii*, Morren in Ann. Soc. Agri. Gand, iii. 584, t. 166 (*Meissn.*).

W. Australia. *Drummond, 3rd coll. n.* 283, 284, *5th coll. n.* 425; Kalgan and Blackwood rivers, *Oldfield;* Franklin river and Brockman's Brook, *Maxwell.*

Var. *distans.* Leaves in distant pairs, less acute and flowers smaller.—King George's Sound, *McLean.*

10. **P. Lehmanniana,** *Meissn. in Pl. Preiss.* i. 603, ii. 270, *and in DC. Prod.* xiv. 504. A shrub attaining 2 or 3 ft., glabrous except the flowers, closely resembling some forms of *P. spectabilis*, but the leaves broader and shorter ($\frac{1}{2}$ to $\frac{2}{3}$ in. long and rather rigid), the hairs of the lower portion of the perianth-tube more rigid and not so long, and the upper portion sprinkled only with a few often rather long rigid hairs, or glabrous, and not silky-villous; the flowers are also said to be white, or of a yellowish tint and not pink. The perianth-tube appears to break off sometimes above the long-haired portion (not circumsciss above the ovary as in *Calyptrostegia*), and the anthers are much smaller than in *P. spectabilis.* I have not succeeded in finding any ripe seeds with a perfect embryo.—*Calyptrostegia Lehmanniana*, Endl. Gen. Pl. Suppl. iv. part 2, 61.

W. Australia. King George's Sound and neighbouring districts, *Baxter, Drummond, Preiss, n.* 1271, *Oldfield, Maxwell, R. Mueller.*

Var. ? *ligustrinoides.* Leaves oblong, $\frac{3}{4}$ to 1 in. long. Flower-heads often nodding, with very large involucral bracts.—Swan river, *Drummond, 1st coll.* This form closely resembles *P. ligustrina* in foliage, but in the latter species the hairs of the perianth are much shorter and very caducous, and the perianth-tube is decidedly circumsciss immediately above the ovary.

11. **P. hispida,** *R. Br. Prod.* 360. An erect shrub, attaining 2 to 4 ft., with slender branches, glabrous except the inflorescence. Leaves opposite, varying from ovate and under ½ in. to oblong or oblong-lanceolate and ¾ in. long or rather more, sessile or scarcely petiolate, flat or the margins scarcely recurved. Flower-heads terminal, globular, with 4 broadly ovate bracts shorter than the flowers, usually with a narrow coloured edge and silky-villous inside. Receptacle hispid. Perianth-tube 4 to 7 lines long, not circumsciss above the ovary, but sometimes breaking off near the top, copiously hispid at and below the middle with long white spreading hairs, shorter appressed hairs being mixed with the long ones above the middle and on the narrow lobes. Filaments nearly as long as the lobes; anthers oblong with a narrow connective. The fruit seems to vary, short and nearly sessile or longer and stipitate within the perianth, the epicarp membranous. Seed with copious albumen, the embryo not seen perfect.—Meissn. in DC. Prod. xiv. 503; Bot. Mag. t. 3459; Bot. Reg. t. 1578; Lodd. Bot. Cab. t. 1966; *Heterolœna hispida,* C. A. Mey. in Bull. Acad. Petrop. iv. (1845) 73.

W. Australia. King George's Sound and adjoining districts, *R. Brown, A. Cunningham, F. Mueller* and others.

P. lanata, R. Br. Prod. 360, Meissn. in Pl. Preiss. i. 604, and in DC. Prod. xiv. 500, *Calyptrostegia lanata,* Endl. Gen. Pl. Suppl. iv. part 2, 61, appears to be a slight variety of *P. hispida* with the leaves usually but not always shorter and broader, and the smaller perianth breaking off more readily above the middle of the tube.

12. **P. rosea,** *R. Br. Prod.* 360. A much-branched shrub, closely allied to *P. ferruginea,* to which F. Mueller reduces it as a variety. It is of taller growth and more slender. Leaves opposite, linear or linear-oblong, often above ½ in. long, with the recurved or revolute margins of *P. ferruginea.* Flowers larger than in that species, but of a similar structure, pink or white. Involucral bracts 4, broad and membranous, but more acuminate than in *P. ferruginea* and often slightly hairy inside or ciliate on the margins.—Meissn. in Pl. Preiss. i. 602, and in DC. Prod. xiv. 503; Bot. Mag. t. 1458, and possibly Lodd. Bot. Cab. t. 88; *P. Hendersoni,* Grah. in Bot. Mag. t. 3721; *Heterolœna rosea,* C. A. Mey. in Bull. Acad. Petrop. iv. (1845) 73 and *H. Hendersoni,* C. A. Mey. l.c. 74.

W. Australia. Swan River, *Fraser, Drummond, 1st coll. n.* 550, *Preiss, n.* 1262, 1267, 1276; King George's Sound or adjoining districts, *R. Brown, Fraser, Drummond,* (*2nd coll.?*) *n.* 110, 166, *F. Mueller;* Sabina river, *Walcot.*

13. **P. ferruginea,** *Labill. Pl. Nov. Holl.* i. 10, t. 5. A stunted much-branched shrub, glabrous except the inflorescence, from under 1 ft. to about 2 ft. high. Leaves opposite, usually crowded, sessile, ovate or oblong, obtuse or scarcely mucronate, rather firm, with recurved or revolute margins, from about ¼ in. long when broad to ½ in. or rather more when narrow on luxuriant shoots. Flower-heads terminal, globular. Involucre of 4 orbicular membranous and coloured bracts, quite glabrous, shorter than the flowers, with often an outer pair more like the stem-leaves. Perianth-tube about 4 lines long, not circumsciss

after flowering, hispid with long spreading hairs from near the top of the ovary to above the middle, the upper portion and the narrow lobes silky with appressed hairs. Filaments usually as long as the lobes; anthers oblong, with a narrow connective, the cells when open placed back to back. Fruit short within the persistent perianth, not acuminate; epicarp membranous. Seed with copious albumen, the cotyledons narrow, longer than the radicle.—*P. decussata*, R. Br. Prod. 360; Meissn. in Pl. Preiss. i. 602, ii. 270, and in DC. Prod. xiv. 502; Sweet, Fl. Austral. t. 8; Lodd. Bot. Cab. t. 1283; Maund, Botanist, t. 136; *P. diosmifolia*, Lodd. Bot. Cab. t. 1708; *Heterolæna decussata*, C. A. Mey. in Bull. Acad. Petrop. iv. (1845) 73.

W. Australia. King George's Sound and adjoining districts, *R. Brown, A. Cunningham, Drummond*, 3*rd coll. n.* 286, *Preiss, n.* 1272 and many others; extending to Cape Arid, *Maxwell*.

14. **P. brachyphylla,** *Benth.* An erect much-branched shrub, from under 1 ft. to about 2 ft. high, glabrous except the inflorescence. Leaves opposite, rather crowded, sessile, oblong or oblong-linear, with strongly recurved margins, from under 2 lines to about 3 lines long. Flower-heads small, globular, with an involucre of 4 to 6 broadly ovate bracts, shorter than the flowers, the inner ones slightly silky-hairy inside. Receptacle hirsute. Flowers hermaphrodite or in some specimens female. Perianth-tube about 2 to 2½ lines long in the hermaphrodite flowers, shorter in the female, not circumsciss, more or less hirsute with spreading hairs. Filaments in the hermaphrodite flowers nearly as long as the lobes; anthers ovate, with a narrow connective, the cells when open placed back to back; in the female flowers the filaments short with small empty anthers and the style longer. Fruit small, ovoid, not beaked. Seed not seen quite ripe.—*P. brevifolia*, Meissn. in DC. Prod. xiv. 497 not of R. Br.

W. Australia. King George's Sound and adjoining districts, *R. Brown, Drummond*, 5*th coll. n.* 429, *F. Mueller;* dense thickets N. of Israelite Bay, *Maxwell*.

SECT. 4. CALYPTROSTEGIA.—Involucral bracts free, various in size or number, sometimes very deciduous, rarely entirely deficient. Flowers hermaphrodite or in some specimens female by abortion. Perianth-tube after flowering circumsciss above the ovary, leaving the lower portion only persistent round the fruit. Anthers with a narrow connective, the cells when open placed back to back.

SUBSECT. 1. CALYPTRIDIUM.—Flower-heads terminal, with 4 to 6 broad persistent involucral bracts. Leaves opposite.

15. **P. sylvestris,** *R. Br. Prod.* 361. A shrub of 2 to 3 ft. perfectly glabrous except the receptacle. Leaves opposite, oblong or lanceolate, mostly ½ to ¾ in. long, more or less concave. Flower-heads globular; involucral bracts 4 to 6, ovate-lanceolate, shorter than or as long as the flowers. Receptacle and very short pedicels hirsute. Perianth perfectly glabrous, the tube slender, 3 to 4 lines long, circumsciss above the ovary. Filaments as long as the lobes; anthers narrow-oblong, with a

narrow connective. Epicarp membranous. Seed with a copious albumen, the embryo nearly cylindrical, with narrow cotyledons.—Meissn. in Pl. Preiss. i. 605, and in DC. Prod. xiv. 506; Bot. Mag. t. 3276; Bot. Reg. t. 1582; Lodd. Bot. Cab. t. 1965; *Calyptrostegia sylvestris*, C. A. Mey. in Bull. Acad. Petrop. iv. (1845) 74; *Pimelea graciliflora*, Hook. Bot. Mag. t. 3288 (with rather broader leaves), Meissn. ll. cc.; *Calyptrostegia graciliflora*, Endl. Gen. Pl. Suppl. iv. part 2, 61.

W. Australia. King George's Sound, *R. Brown*, *Preiss*, *n.* 1270, *Baxter;* Swan river, *Drummond*, 1*st coll.*, *n.* 551, *Preiss*, *n.* 1274; also in *Drummond*, 2*nd coll.* 165, and 3*rd coll. n.* 289.

Var. *æruginosa*, F. Muell. Involucral bracts larger and more obtuse and the whole plant drying blue like the var. *tinctoria* of *P. suaveolens*. *P. æruginosa*, F. Muell. Fragm. vii. 2.

W. Australia. *Drummond.*

16. **P. brevifolia,** *R. Br. Prod.* 359, *not of Meissn.* A small branching shrub, apparently 6 in. to 1 ft. high, glabrous except the flowers. Leaves opposite, sessile, elliptical oblong or lanceolate, somewhat concave, under ½ in. long, usually rather rigid and distinctly penniveined underneath when dry. Flower-heads terminal, globular; involucral bracts 4 to 6, broadly ovate, shorter than the flowers and glabrous inside. Receptacle hirsute. Perianth-tube more or less hirsute with spreading hairs, the upper part often glabrous and always without appressed hairs, 3 to 4 lines long, tardily or perhaps not at all circumsciss, the lobes about 1 line long. Filaments nearly as long as the lobes; anthers ovate, with a narrow connective, fruit only seen very young.—*P. modesta*, Meissn. in Pl. Preiss. ii. 268, and in DC. Prod. xiv. 502; *Calyptrostegia brevifolia*, C. A. Mey. in Bull. Acad. Petrop. iv. (1845) 74, as to Brown's synonym.

W. Australia. King George's Sound, *R. Brown*, also in *Drummond*, 4*th coll. n.* 238.

Var. *angustifolia.* Leaves narrower, but broader than in *P. Maxwelli*, and distinctly penniveined.—Cape Arid, *Maxwell.*

Var. *membranacea.* Leaves and bracts much thinner, the latter orbicular membranous and coloured.—West Australia, *Drummond* (*Herb. F. Mueller*).

Although it seems doubtful whether the perianth-tube breaks off above the ovary after flowering, the foliage and habit indicate the place of this species in *Calyptrostegia* rather than in *Heterolæna*. Brown's species was described from a single specimen with smaller leaves and flowers than in Drummond's plant, but it seems to be the same species, approaching nearest to the var. *angustifolia.*

17. **P. Maxwelli,** *F. Muell. Herb.* A shrub with erect slender branches from under 1 ft. to 1½ ft. high, glabrous except the inflorescence. Leaves opposite, linear or oblong-lanceolate, under ½ in. long, flat or scarcely concave, with the midrib very prominent underneath. Flower-heads small, globular; involucral bracts 4 to 6, ovate, acute or shortly acuminate, shorter than the flowers and glabrous or the inner ones very slightly silky-hairy inside. Perianths slender, the tube 3 to 4 lines long; tardily circumsciss a little way above the ovary and silky-hairy, the hairs of the lower portion longer and more spreading, on the upper portion shorter and more appressed, the lobes narrow,

about 1 line long. Filaments more than half as long as the lobes; anthers small, with a narrow connective. Fruiting portion of the perianth acuminate above the fruit, which however I have not seen ripe.—*P. angustifolia var.? canescens*, Meissn. in DC. Prod. xiv. 499.

W. Australia. Gordon and Kalgan rivers, *Oldfield;* eastward of King George's Sound, *Baxter;* M Callum's Inlet, Esperance Bay, *Maxwell;* also with the perianths almost glabrous in the upper portion, Gardiner river and Cape Arid, *Maxwell.*

Some specimens have female perianths only, with abortive anthers and long styles; in others all the perianths are hermaphrodite, but both forms of flower appear to be fertile.

18. **P. angustifolia,** *R. Br. Prod.* 360. An erect rather slender not much branched shrub of 1 to 2 ft., glabrous except the inflorescence. Leaves opposite, linear or linear-lanceolate, concave, usually erect and about ½ in. long, but sometimes ¾ in. Flower-heads globular; involucral bracts 4, broadly ovate, shorter than the flowers, scarcely acuminate, silky-villous inside. Perianths white or pink, the tube 4 to 5 lines long, circumsciss above the ovary, loosely but densely silky-villous, the hairs not much longer in the persistent than in the deciduous portion, the lobes about 1½ lines long. Filaments from half as long to nearly as long as the lobes; anthers with a rather broad connective, but the cells quite distinct and placed back to back after they are open. Fruit short, but not seen ripe.—Meissn. in Pl. Preiss. ii. 269, and in DC. Prod. xiv. 499 (excluding the vars. α and ϑ); *Calyptrostegia angustifolia*, C. A. Mey. in Bull. Acad. Petrop. iv. (1845) 74.

W. Australia. King George's Sound and adjoining districts, *R. Brown, F. Bauer, Baxter, A. Cunningham, Drummond, 3rd coll. n.* 287, and many others; also between Moore and Murchison rivers, *Drummond, 6th coll. n.* 215. Brown's specimen in the Banksian herbarium has the leaves rather longer and the persistent portion of the perianths more densely and longer silky-villous than the others. R. Brown's own specimen is a poor one with smaller flowers than usual.

19. **P. nervosa,** *Meissn. in Pl. Preiss.* ii. 269, and in *DC. Prod.* xiv. 500. An erect slightly-branched shrub of 1 to 2 ft., glabrous except the inflorescence. Leaves opposite, linear-lanceolate, concave, ½ to 1 in. long, the lateral veins often but not always conspicuous underneath when dry. Flower-heads globular, usually larger than in *P. angustifolia*, smaller than in *P. floribunda;* involucral bracts 4, broadly ovate, acuminate, rigidly herbaceous, often veined outside, silky-villous inside. Perianth-tube slender, usually 5 to 6 lines long, circumsciss above the ovary, the short persistent portion very densely covered with silvery-silky spreading hairs, the deciduous portion with shorter appressed hairs. Filaments nearly as long as the perianth-lobes; anthers with a narrow connective. Fruit short, not acuminate. Seed with a copious albumen, and nearly terete embryo, the cotyledons narrow, twice as long as the radicle.—*Calyptrostegia nervosa*, Walp. Ann. iii. 324.

W. Australia, *Drummond;* S. Coast, *Baxter;* Lake Sapphire, *Harper;* Stirling Range, *F. Mueller* (with smaller flowers); M'Callum's Inlet and eastward to Cape Le Grand, *Maxwell.*

20. **P. sulphurea,** *Meissn. in Bot. Zeit.* 1848, 396, *and in DC. Prod.* xiv. 506. An erect slightly branched shrub of 1 to 2 ft., glabrous except the inflorescence, often assuming a bluish colour in drying. Leaves opposite, ovate or ovate-oblong, 2 to 4 lines or rarely ½ in. long, rather thick, flat or concave, the uppermost pair usually rather larger. Flower-heads globular, nodding; involucral bracts 6 or sometimes 8, ovate or ovate-oblong, very obtuse, thinner and larger than the stem-leaves, the inner ones ciliate on the margins and silky-hairy inside. Perianth-tube 3 to 4 lines long, circumsciss above the ovary, silky-villous, the hairs of the lower portion longer and looser but scarcely spreading, lobes narrow, above 1 line long. Filaments nearly as long as the lobes; anther-connective not very narrow, but the cells placed back to back when open. Fruit not seen ripe.

W. Australia. Swan river, *Drummond, 1st coll. n.* 549, *Preiss, n.* 1278; Vasse and Canning rivers, *Oldfield;* between Swan river and King George's Sound, *Harvey.*

Var. ? *macrocephala.* Leaves narrow, flower-heads larger, the inner bracts above ½ in. long thin and much ciliate.—Blackwood river, *Oldfield.* Perhaps referable rather to *P. suaveolens.*

21. **P. floribunda,** *Meissn. in DC. Prod.* xiv. 505. Erect, slightly branched and glabrous except the inflorescence, 1 to 1½ ft. high and apparently annual, although hard and almost woody at the base. Leaves opposite, sessile or contracted into a very short petiole, ovate-oblong or broadly lanceolate, mostly obtuse, rather thick, flat or concave. Flower-heads terminal, globular, rather large, usually nodding; involucre of 4 or rarely 6 very broadly ovate herbaceous bracts, sometimes nearly as long as the flowers, the inner ones silky-hairy inside; receptacle villous. Flowers hermaphrodite. Perianth-tube slender, 5 to 6 lines long, circumsciss after flowering immediately above the ovary, the lower persistent portion not acuminate when in fruit and very densely clothed with silvery-silky somewhat spreading hairs, the deciduous portion silky with closely appressed hairs, the lobes about 1½ lines long. Filaments nearly as long as the lobes; anthers narrow oblong, the connective not very narrow, but the open cells placed back to back.

W. Australia. Between Moore and Murchison rivers, *Drummond, 6th coll. n.* 214; Champion Bay, *Oldfield, C. Gray.*

22. **P. suaveolens,** *Meissn. in Pl. Preiss.* i. 603, and in *DC. Prod.* xiv. 504. An erect shrub, sometimes branching at the base only and under 1 ft., but attaining 2 or 3 ft. when more branched, glabrous except the flowers. Leaves opposite, from ovate-lanceolate to oblong-linear, more or less concave, mostly ½ to 1 in. long. Flower-heads globular; involucral bracts 4 to 8, broad, often as long as the flowers, usually ciliate, the inmost pair silky-hairy inside. Receptacle and pedicels villous. Perianths yellow when fresh, the tube slender, varying from 5 to 7 lines long, circumsciss above the ovary after flowering, the persistent portion acuminate and loosely hairy when in fruit, the deciduous portion clothed at the base with loose spreading hairs, which are rare and often wanting, the lobes, about 2 lines long, have a few appressed hairs.

Anthers with a rather broad connective, but the cells when open placed back to back. Fruit acuminate, rather long, the epicarp membranous. Seed with a copious albumen and narrow cotyledons.—*Calyptrostegia suaveolens*, Endl. Gen. Suppl. iv. part 2, 61; *P. macrocephala*, Hook. Bot. Mag. t. 4543, copied into Lem. Jard. Fleur. t. 76; Meissn. in DC. Prod. xiv. 504; *Calyptrostegia macrocephala*, Walp. Ann. iii. 324; *Calyptrostegia Drummondii*, Turcz. in Bull. Soc. Imp. Nat. Mosc. 1852, ii. 178.

W. Australia. Swan river, *Drummond*, 1*st coll. n.* 548, *Preiss*, *n.* 1268; Geographe Bay, *Oldfield; King George's Sound, *Harvey*, *Oldfield*, *Maxwell*, also *Drummond*, 5*th coll. n.* 426.

Var. *tinctoria*. Leaves usually somewhat shorter and broader turning bluish in drying. *P. tinctoria*, Meissn. in Pl. Preiss. i. 603, and in DC. Prod. xiv. 501; *Calyptrostegia tinctoria*, Endl. Gen. Suppl. iv. part 2, 61.—*Drummond* 4*th coll. n.* 249, 5*th coll. n.* 427, *and Suppl. n.* 83. *P. Menkeana*, Lehm.; Meissn. in Pl. Preiss. i. 604, and in DC. Prod. xiv. 503, or *Calyptrostegia Menkeana*, Endl. Gen. Suppl. iv. part 2, 61 (*Drummond*, *n.* 427) only differs from the ordinary var. *tinctoria* in the smaller flowers.

23. **P. physodes,** *Hook. Ic. Pl.* t. 865. An erect shrub of about 3 ft., glabrous except the flowers. Leaves opposite, rather crowded, sessile, oval-oblong, acute or obtuse, rather thick, concave, ½ to ¾ or rarely 1 in. long. Flower-heads terminal, nodding, completely enclosed in a large ovoid-globular involucre of about 6 broad obtuse red and yellow membranous bracts, the outer ones about 1 in. the innermost often 2 in. long. Receptacle villous. Perianth-tube about 5 lines long, circumsciss above the ovary, the lower persistent part glabrous, the deciduous portion about as long, but hairy as well as the narrow-linear lobes, which are about 4 lines long. Stamens much longer, inserted at the orifice of the tube, with a slightly prominent transverse fold in the throat, starting from each side of the base of each filament; anthers narrow, the cells when open placed back to back. Fruit narrow, about 3 lines long; epicarp membranous; endocarp minutely rugose. Seed with a copious albumen and narrow cotyledons.—Meissn. in DC. Prod. xiv. 504; *Macrostegia erubescens*, Turcz. in Bull. Soc. Imp. Nat. Mosc. 1852, ii. 178.

W. Australia. E. Mount Barren, *Maxwell*, and probably the same locality, *Drummond*, 5*th coll. n.* 424, *and Suppl. n.* 84. Notwithstanding the different shape of the perianth and the extraordinary development of the involucre, there seems no reason to separate this plant from the genus which as a whole is so very well defined.

24. **P. glauca,** *R. Br. Prod.* 360. An erect much-branched shrub, from ½ to 1½ ft. high, glabrous except the inflorescence. Leaves opposite, from ovate to oblong-lanceolate or almost linear, sometimes all under ½ in., sometimes ½ to ¾ in. long, or even longer, flat or concave, with the midrib prominent underneath. Flower-heads globular; involucral bracts usually 4, ovate or ovate-lanceolate, shorter than the perianth, sometimes not much, sometimes considerably broader than the stem-leaves, the inner ones ciliate on the margin and more or less silky-hairy inside. Receptacle densely covered with long hairs. Perianth-tube 4 to 5 lines long, circumsciss after flowering considerably above the ovary, the persistent portion rarely glabrous or sprinkled only with

hairs, the deciduous portion silky-villous, the lobes about 1½ lines long, Filaments usually half the length of the lobes; anthers oblong, with a narrow connective. Fruit sessile or nearly so within the perianth, the epicarp membranous. Seed not seen perfectly ripe, but the cotyledons appear to be rather broad.—Meissn. in DC. Prod. xiv. 501; Rudge in Trans. Linn. Soc. x. 286, t. 13; Hook. f. Fl. Tasm. i. 334; Lodd. Bot. Cab. t. 1611; *P. humilis*, Lindl. Bot. Reg. t. 1268 not of R. Br.; *P. intermedia*, Lindl. Bot. Reg. t. 1439; Maund, Botanist, v. t. 243; Meissn. in DC. Prod. xiv. 501; *Calyptrostegia glauca* and *C. intermedia*, C. A. Mey. in Bull. Acad. Petrop. iv. (1845) 74; *P. campicola*, A. Cunn. Herb.; *P. linifolia* var. ? *subnervosa*, Meissn. in DC. Prod. xiv. 498, (at least as to Cunningham's plant); *P. Preissii*, Schlecht. in Linnæa, xx. 581, not of Meissn.; *P. Schlechtendahliana*, Meissn. in Bot. Zeit. 1848, 394, and in DC. Prod. xiv. 500; *Calyptrostegia Schlechtendahliana*, Walp. Ann. iii. 324; *P. myrtifolia*, Schlecht. l.c. 582.

Queensland. Plains of the Condamine, *Leichhardt;* Darling Downs, *Law;* Dee river, *Bowman;* Rockhampton, *Dallachy;* Burnet river, *Haly.*

N. S. Wales. Port Jackson to the Blue Mountains, *R. Brown* and others; Illawarra, Bathurst, Liverpool Plains, *A. Cunningham;* New England, *C. Stuart.*

Victoria. Port Phillip, *R. Brown;* Murray and Ovens rivers, *F. Mueller;* Wimmera, *Dallachy;* Glenelg river and Wendu Vale, *Robertson;* mouth of the Glenelg river, *Allitt.*

Tasmania. Derwent river, *R. Brown;* common in light sandy soil, especially on the N. coast, *J. D. Hooker.*

S. Australia. Spencer's Gulf, *R. Brown, Warburton;* Lofty Range, Guichen Bay, Samuda, *F. Mueller.*

This species is sometimes confounded with the *P. linifolia*, the most ready distinction is in the persistent base of the perianth being nearly glabrous but half concealed by the long hairs of the receptacle. The leaves are also usually concave when dry, but this distinction cannot always be safely relied on.

25. **P. colorans,** *A. Cunn.; Meissn. in DC. Prod.* xiv. 499. An erect shrub of 2 or 3 ft., glabrous except the inflorescence. Leaves opposite, petiolate, linear or linear-lanceolate, tapering at both ends, concave, ½ to 1 in. long. Flower-heads globular, usually nodding. Involucral bracts 4, broadly ovate, 1-nerved, coloured, shortly acuminate, with sometimes a third outer pair passing into the form of the stem-leaves. Receptacle shortly villous. Perianth like that of *P. linifolia*, silky-villous, but the lower persistent portion much less so, and often bearing only a few hairs as in *P. glauca.*

N. S. Wales. Macquarrie river, *A. Cunningham.*

26. **P. collina,** *R. Br. Prod.* 359. An erect shrub, glabrous except the inflorescence, closely resembling *P. linifolia*, but the leaves are slightly concave with the lateral veins more prominent underneath, running into a marginal nerve when the leaves are narrow, more pinnate when the leaves are broader, the involucral bracts are also usually more rigid and prominently veined. It differs from the narrow leaved forms of *P. glauca* in the involucral bracts being glabrous inside as well as out, the short hairs to the receptacle, and the perianth usually villous from the base.—Meissn. in DC. Prod. xiv. 497; *P. marginata*,

Meissn. l.c.; *P. colorans*, Mitch. Trop. Austr. 362 not of A. Cunn.; *P. Mitchelli*, Meissn. in DC. Prod. xiv. 506.

Queensland. Port Bowen, *R. Brown;* Marañoa river, *Mitchell;* dividing range between Flinders and Burdekin rivers, *Thozet.*

N. S. Wales. Wellington valley, *A. Cunningham;* New England, *C. Stuart.*

Var. *latifolia.* Leaves shorter and broader.

N. S. Wales. Hastings river, *Butler.*

Victoria. Mount M'Ivor and Mount Ida, *F. Mueller;* Creswick, *Whan.*

P. Cunninghamii, Meissn. in DC. Prod. xiv. 498, from Liverpool plains, appears to be a variety of *P. collina* with the marginal nerves of the leaves less conspicuous and the persistent base of the perianth nearly glabrous after flowering, but without the acumination of *P. glauca*, and the hairs of the receptacle are short, and the bracts glabrous inside and rather thin. In long series of specimens from various localities some may be met with which seem absolutely to connect *P. glauca*, *P. collina* and *P. linifolia.*

27. **P. spathulata,** *Labill. Pl. Nov. Holl.* i. 9, t. 4. A much-branched shrub, attaining 2 or 3 ft., glabrous except the inflorescence, resembling in many respects *P. linifolia* but never allied to *P. glauca,* and apparently distinct from both. Leaves opposite, linear linear-oblong or sometimes linear-spathulate, flat or somewhat concave, the margins often thickened and nerve-like, but never recurved. Flower-heads nodding, the involucral bracts ovate, obtuse or acute, often coloured, glabrous inside as well as out. Perianths of *P. linifolia*, but much less hairy at the base, with a few long hairs, the hairs of the receptacle short as in *P. linifolia.*—*P. cernua,* R. Br. Prod. 359; Meissn. in DC. Prod. xiv. 497; Hook. f. Fl. Tasm. i. 333; Knowl. and West. Fl. Cab. t. 72; *P. nutans,* Meissn. in Linnæa, xxvi. 348, and in DC. Prod. xiv. 498; *P. Lindleyana*, Meissn. in DC. Prod. xiv. 499 (partly); *Calyptrostegia spathulata* and *C. cernua,* C. A. Mey. in Bull. Acad. Petrop. iv. (1845) 74.

N. S. Wales. Mount Caley, *A. Cunningham;* collected also by *Mitchell* in his expedition of 1835.

Tasmania. Port Dalrymple, *R. Brown;* common in dry sandy tracts throughout the island, *J. D. Hooker.*

S. Australia. St. Vincent's Gulf, *Blandowski.*

28. **P. linifolia,** *Sm. Bot. N. Holl.* 31, t. 11. An erect shrub, from under 1 ft. to 2 or 3 ft. high, glabrous except the inflorescence, the branches usually slender and virgate. Leaves opposite, on very short petioles, linear or oblong, passing also into linear-spathulate or linear-lanceolate, from under $\frac{1}{2}$ in. to about 1 in. long, nearly flat, the margins in the dry state usually slightly recurved and rather convex than concave. Flower-heads terminal, globular, erect; involucral bracts 4, ovate or ovate-lanceolate, often nearly as long as the flowers, glabrous inside as well as out, rather thin, the midrib not very prominent. Receptacle shortly villous. Perianth-tube 4 to 6 lines long, silky-villous from the base, circumsciss rather above the ovary, the lobes 1 to $1\frac{1}{2}$ lines long. Filaments usually about half the length of the lobes; anthers oblong, with a narrow connective, the cells when open placed back to back. Fruiting base of the perianth not acuminate; epicarp membranous. Seed oblong; albumen scanty; cotyledons ovate-oblong, flat but rather thick.—R. Br. Prod. 359; Meissn. in DC. Prod. xiv. 497;

Hook. f. Fl. Tasm. i. 334; Bonpl. Jard. Malm. t. 31; Bot. Mag. t. 891; Lodd. Bot. Cab. t. 1668; *P. filamentosa*, Rudge in Trans. Linn. Soc. x. 287, t. 14; *P. involucrata*, Herb. Banks; *Passerina involucrata*, Thunb. Cat. Mus. Nat. Acad. Ups. xiii. 106; *Calyptrostegia linifolia*, C. A. Mey. in Bull. Acad. Petrop. iv. (1845) 74; *P. paludosa*, R. Br. Prod. 360, Meissn. in DC. Prod. xiv. 499; *P. collina*, A. Cunn. Herb. (partly); *P. rigida*, Meissn. in DC. Prod. xiv. 502; *P. linoides*, A. Cunn. in Field, N. S. Wales, 326; *Calyptrostegia linoides*, Endl. Gen. Suppl. iv. part 2, 61; *P. Lindleyana*, Meissn. in DC. Prod. xiv. 499 (partly).

Queensland. Brisbane river, Moreton Bay, *A. Cunningham* and many others; Rockingham Bay, *Dallachy*.

N. S. Wales. Port Jackson to the Blue Mountains, *R. Brown*, *Sieber*, *n.* 206, and *Fl. Mixt. n.* 476; southward to Illawarra, *A. Cunningham*, and Twofold Bay, *F. Mueller;* northward to Hastings, Macleay and Clarence rivers, *Butler;* New England, *C. Stuart*.

Tasmania. Port Dalrymple, *R. Brown;* W. bank of the Tamar, *C. Stuart;* also collected by *Gunn*, *n.* 3, *Milligan*, *n.* 396, and *Archer*.

Var.? *Andersoni*, Meissn. Leaves flat or slightly concave. Involucral bracts somewhat silky-villous inside. Perianth much less hairy at the base and the hairs of the involucre rather longer, apparently connecting *P. linifolia* with *P. glauca*. Blue Mountains, *Sieber*, *n.* 207, *A. Cunningham*.

The *P. linifolia* and the four preceding species are sometimes not very easy to distinguish, although in their ordinary forms they appear to be separated by well-marked characters.

29. **P. ligustrina,** *Labill. Pl. Nov. Holl.* i. 9, t. 3. An erect shrub, attaining in the ordinary form 5 or 6 ft., glabrous except the inflorescence or the young shoots slightly silky-hairy. Leaves opposite, from ovate to oblong or elliptical, rather thin, 1-nerved and more or less distinctly penniveined, 1 to $1\frac{1}{2}$ in. long or on luxuriant shoots twice that size. Flower heads rather large, globular; involucral bracts in the typical form 4 or rarely 5 or 6, very broad, as long as the perianth-tubes, glabrous or nearly so. Flowers hermaphrodite or in some specimens female with shorter perianths, but in both cases apparently fertile. Perianth-tube fully 5 lines long in the hermaphrodite flowers, shorter in the females, in both more or less silky-hairy and circumsciss above the ovary after flowering, the lobes about $1\frac{1}{2}$ lines long. Filaments in the hermaphrodite flowers nearly as long as the lobes, the anthers oblong, with a narrow connective; in the females the anthers small empty on short filaments and the style longer. Fruit acuminate, the beak longer in the hermaphrodite than in the female specimens. Epicarp membranous. Seed and embryo as in *P. linifolia.*—R. Br. Prod. 360; Meissn. in DC. Prod. xiv. 505; Hook. f. Fl. Tasm. i. 333; *Calyptrostegia ligustrina*, C. A. Mey. in Bull. Acad. Petrop. iv. (1845) 74; *P. elata*, F. Muell. First Gen. Rep. 17; Meissn. in Linnæa, xxvi. 349, and in DC. Prod. xiv. 505.

N. S. Wales. Blue Mountains, *R. Cunningham*, *Woolls*, *Miss Atkinson*.

Victoria. Wilson's Promontory, Hardinger Range, Cobberas mountains, *F. Mueller;* Mount Buller, *Soues;* mouth of the Glenelg, *Allitt*.

Tasmania. Port Dalrymple, *R. Brown;* common in dense humid forests, chiefly in the northern parts of the island, *J. D. Hooker;* King's Island, *McGowan*.

S. Australia. Rivoli Bay, *F. Mueller*.

Var. *hypericina.* A shrub attaining 8 to 10 feet, with the foliage and general characters of the typical form, but the involucral bracts usually 6 to 8, much shorter than the flowers and silky-pubescent or hoary, the flowers rather smaller and more slender. *P. ligustrina*, Bot. Reg. t. 1827; *P. hypericina*, A. Cunn. in Bot. Mag. t. 3330, Meissn. in DC. Prod. xiv. 505; *Calyptrostegia hypericina*, C. A. Mey. in Bull. Acad. Petrop. iv. (1845) 74; *P. elegans*, F. Moore in Illustr. Horticole, viii. t. 295.

N. S. Wales. Hastings and Clarence rivers, *C. Moore;* Sydney woods, Paris Exhibition 1855, n. 207, *M'Arthur;* Illawarra, *A. Cunningham*, *Shepherd;* Mittagong, *Woolls.*

Victoria. Dandenong Ranges, *F. Mueller;* Upper Genoa river, *Stevenson.*

Var.? *macrostegia.* Leaves oblong. Flower-heads larger than usual, nodding, with the broad involucral bracts as long as the flowers. Perianth-hairs spreading, rigid and exceedingly caducous, leaving the persistent base quite glabrous, the upper part often glabrous from the first, the whole plant resembling the Swan river form of *P. Lehmanniana.* Perhaps a distinct species but cannot be sufficiently characterized without more satisfactory specimens.

S. Australia. Sandy scrub, Kangaroo island, *Waterhouse.*

30. **P. humilis,** *R. Br. Prod.* 361. A small shrub, branching from the base, more or less silky-villous at least on the stems, the foliage often becoming glabrous, from under 6 in. to 1 or even 1½ ft. high. Leaves opposite, rather crowded, sessile, oblong or almost ovate-oblong, obtuse or rarely almost acute, ¼ to near ½ in. long, flat or slightly convex or concave, the midrib prominent underneath, and the lateral veins sometimes conspicuous. Flower-heads globular, the involucral bracts 4 or sometimes 6, ovate or broadly ovate-oblong, usually obtuse, slightly silky-villous inside, either nearly as long as the perianth-tubes or shorter. Perianths silky-villous, the tube 4 to 5 lines long, circumsciss after flowering considerably above the ovary, the lobes about 1½ lines long. Filaments usually about half as long as the ovary; anthers oblong, with a narrow connective, the cells when opened placed back to back. Fruit shortly stipitate within the perianth, but not seen ripe.—Meissn. in DC. Prod. xiv. 502; Hook. f. Fl. Tasm. i. 334.

Victoria. Wimmera, *Dallachy;* Mount Ararat, *Green;* Little river, *F. Mueller;* Burra-Burra, *Hinteracker;* Wendu vale, *Robertson;* Portland, *Allitt.*

Tasmania. Port Dalrymple, *R. Brown;* abundant on dry hills and pastures throughout the island, *J. D. Hooker.*

S. Australia. Mount Lofty range, *F. Mueller;* near Adelaide, *F. Mueller*, *Blandowski.*

31. **P. sericea,** *R. Br. Prod.* 361. A bushy shrub of 1 to 2 ft. resembling *P. nivea*, but readily known by the dense indumentum of the branches and underside of the leaves consisting of silvery-silky hairs. Leaves opposite, crowded, ovate or oblong, under ½ in. long, glabrous above, flat or the margins recurved (concave when fresh?). Flower-heads terminal; involucral bracts 4 to 6, rather larger than the stem-leaves. Perianth like that of *P. nivea*, the tube usually 4 or 5 lines long, circumsciss above the ovary after flowering, the lobes at least 1½ lines long. Filaments rather longer than the anthers, connective narrow, the cells placed back to back when open. In some specimens the flowers are all female with smaller perianths and abortive anthers.—Meissn. in DC. Prod. xiv. 509; Hook. f. Fl. Tasm. i. 333; *P. lanata*, Hensl. in Maund, Botanist, ii. t. 61 not of R. Br.

Tasmania. Mount Wellington (Table mountain), *R. Brown;* summits of all the mountains at an elevation of 3000–4000 ft., *J. D. Hooker.*—Henslow describes the leaves as concave which they may be when fresh; in the dried state the margins are usually distinctly recurved.

32. **P. nivea,** *Labill. Pl. Nov. Holl.* i. 10, t. 6. An erect bushy or straggling shrub, the branches and underside of the leaves white with a close dense more or less crisped tomentum, accompanied sometimes by a few silky hairs but never densely silky as in *P. sericea.* Leaves opposite, ovate or orbicular, rarely broadly elliptical-oblong, under ½ in. long, rather thick, glabrous above, with recurved margins. Flower-heads globular, terminal; involucral bracts 4 to 6, rather larger than the stem-leaves. Perianths white or pink, tomentose or silky, the tube usually 4 to 5 lines long, circumsciss immediately above the ovary after flowering, the lobes 1½ lines long. Filaments nearly as long as the lobes, the anthers with a narrow connective, the cells when open placed back to back; in some specimens the perianths are smaller with abortive anthers, but the flowers fertile in both forms. Fruit small, scarcely acuminate, the epicarp membranous.—R. Br. Prod. 361; Meissn. in DC. Prod. xiv. 509; Hook. f. Fl. Tasm. i. 332; Knowl. and Westc. Fl. Cab. t. 9; *P. incana,* R. Br. l.c.; Meissn. in DC. Prod. xiv. 509; Bot. Reg. 1838, t. 24: Maund, Botanist, t. 147; *Heterolæna nivea* and *H. incana,* C. A. Mey. in Bull. Acad. Petrop. iv. (1845) 73.

Tasmania. Port Dalrymple and Mount Wellington (Table Mountain), *R. Brown;* abundant throughout the island, ascending to the summits of the Western Mountains, at an elevation of 3500 ft., *J. D. Hooker.*

C. A. Meyer places this species in his *Heterolæna,* characterized chiefly by the perianth remaining entire when in fruit, but I find it always circumsciss in *P. nivea* as the fruit ripens. I have not been able to find any perfectly ripe seeds so as to verify the embryonal character, which however certainly is the same in *Heterolæna* as in some species at least of the *Calyptridium* section of *Calyptrostegia.*

Subsect. 2. Phyllolæna.—Flower-heads terminal, with numerous involucral bracts not at all or scarcely broader than the leaves.

33. **P. imbricata,** *R. Br. Prod.* 361. A small erect much-branched shrub, from under 6 in. to 1½ ft. high, sometimes clothed from the base with long silky hairs, sometimes with all the leaves except the uppermost glabrous. Leaves usually crowded, either mostly alternate or nearly all opposite, from oblong-lanceolate and under ¼ in. long to linear and above ½ in., the upper ones almost always ciliate with long hairs, all flat or slightly concave. Flower-heads globular; involucral bracts 8 or more, scarcely broader than the stem-leaves and much shorter than the flowers. Perianth-tube hirsute with long spreading hairs, usually 2½ to 3 lines long, but sometimes rather longer, circumsciss above the ovary after flowering, the upper portion and the narrow lobes silky with appressed hairs. Filaments nearly as long as the lobes; anthers ovoid or oblong with a narrow connective, the cells when open placed back to back. Ripe seed not seen.—Meissn. in Pl. Preiss. i. 605, ii. 270, and in DC. Prod. xiv. 507.

W. Australia. King George's Sound and neighbouring districts, *R. Brown*, and many others.

A variable species of which there are 3 principal forms:—

1. *Baxteri*, Meissn. Densely branched with small glabrous leaves, flowers usually pink.—King George's Sound, *R. Brown*, *Baxter*, *Preiss*, *n.* 1273, &c.; Fitzgerald and Plantagenet ranges, *Maxwell*.

2. *gracillima*, Meissn. Taller with slender branches, short narrow glabrous leaves, smaller flower-heads and even the bracts sometimes nearly glabrous.—*P. microcephala*, Meissn. in Pl. Preiss. i. 606 not of R. Brown; *Drummond*, 1*st coll. n.* 552.

3. *piligera*. Low or tall. Leaves narrow, often ½ in. long, all or mostly clothed with long loose silky hairs. Flower-heads and flowers rather large, flowers usually white.—*P. pilibunda*, A. Cunn. Herb.; *P. nana*, Grah. in Edinb. New Phil. Journ. xxix. 174; Bot. Mag. t. 3833; Meissn. in Pl. Preiss. i. 606, ii. 272, and in DC. Prod. xiv. 508; *P. crinita*, Lindl. in Bot. Reg. 1838, Misc. 59; Meissn. in DC. Prod. xiv. 507; *Calyptrostegia nana*, Endl. Gen. Suppl. iv. part 2, 61. Swan river, *Drummond*, 1*st coll. n.* 553; *Preiss*, *n.* 1275, 1277; Port Gregory and Vasse river, *Oldfield*; Phillips river, *Maxwell*.

Drummond's 4th coll. n. 236, appears to be intermediate between the first and the third forms.

34. **P. villifera,** *Meissn. in Pl. Preiss.* ii. 271, *and in DC. Prod.* xiv. 508. An erect rather coarse branching shrub, villous with rather stiff hairs. Leaves mostly opposite, sessile, oblong-lanceolate, flat, thick but soft, 4 to 6 or 8 lines long. Flower-heads globular, densely hirsute; involucral bracts 8 or more, similar to the stem-leaves or rather more lanceolate, obtuse, often as long as the flowers. Perianths hirsute with rather long hairs, altogether scarcely above 3 lines long, circumsciss above the ovary after flowering, the lobes at least as long as the tube. Filaments about as long as the lobes; anthers with a rather broad connective, but the cells when open placed back to back.—*Calyptrostegia villifera*, Walp. Ann. iii. 324.

W. Australia. *Drummond*, 4*th coll. n.* 239.—This is certainly allied to *P. imbricata*, of which some varieties have small perianths, but the shape is different, besides that in *P. villifera* they are equally hispid on the lobes and the tube.

Subsect. 3. Choristachys.—Flowers in clusters spikes or racemes, without involucres or the bracts not broader than the leaves and very deciduous. Leaves flat or with slightly recurved margins. Flowers small, except in *P. hæmatostachya*. Seeds, where known, with scanty albumen and broad cotyledons.

35. **P. drupacea,** *Labill. Pl. Nov. Holl.* i. 10, t. 7. A straggling shrub, attaining 6 to 8 ft. but often much lower, the branches more or less silky-hairy. Leaves all opposite, from ovate to oblong-elliptical or oblong-linear, obtuse or mucronulate, ¾ to 2 in. long, glabrous above, pale and often slightly silky-hairy underneath, flat or with recurved margins. Flower-heads terminal on the young shoots, but mostly appearing axillary and sessile from the extreme shortness of the flowering branches, subtended by 2 bracts similar to the stem-leaves but smaller. Flowers small, not numerous, white tinged with pink, on small turbinate hirsute pedicels. Perianth silky-hairy, the tube about 2 lines long, circumsciss above the ovary after flowering, the lobes about 1 line long. Filaments very short;

anthers with a narrow connective, the cells when open placed back to back. Fruit a drupe, enclosed in the membranous persistent base of the perianth, red or black when ripe, the epicarp more or less succulent, the endocarp smooth and shining. Seed with the albumen not copious and broad cotyledons.—R. Br. Prod. 361; Meissn. in DC. Prod. xiv. 515; Hook. f. Fl. Tasm. i. 331; Sweet Fl. Austral. t. 52; Lodd. Bot. Cab. t. 540; *Gymnococca drupacea*, Fisch. et Mey. Ind. Sem. Hort. Petrop. x. (1845) 46.

Victoria. Sealer's Cove, *F. Mueller.*
Tasmania. Derwent river and Port Dalrymple, *R. Brown;* abundant in humid forests in a rich soil, *J. D. Hooker.*

P. umbratica, A. Cunn. Herb.; Meissn. in DC. Prodr. xiv. 510, from Logan Vale, as far as the specimens go, is undistinguishable from *P drupacea*, except in the rather smaller flowers. The station is however so far distant that more perfect specimens may possibly show it to be distinct.

36. **P. hæmatostachya,** *F. Muell. Fragm.* i. 84. An erect perennial of 1 to 2 ft. perfectly glabrous except the inflorescence, somewhat glaucous, not much branched. Leaves opposite, sessile or nearly so, oblong or lanceolate, obtuse, mostly 1 to 1½ in. long. Flower-heads pedunculate above the last stem-leaves, large and hirsute, at first globular, but soon lengthening into a dense spike of 2 to 4 in. Involucral bracts 4 to 8, very deciduous so as to be seen only on the very young head, linear, membranous and hairy. Flowers of a blood-red colour, numerous and crowded. Perianth slightly silky-hairy, the slender tube above ½ in. long, circumsciss above the ovary after flowering, the lobes about 2 lines long. Filaments at least as long as the lobes; anthers oblong, with a narrow connective, the cells when open placed back to back. Fruit not succulent, the persistent base of the perianth very hairy, the epicarp membranous. Seed with scanty albumen and broad cotyledons.

Queensland. Burdekin river and Peak Downs, *F. Mueller;* Dawson river, *Leichhardt;* Rockingham and Edgecombe Bays, *Dallachy;* Port Denison, *Fitzalan;* Nerkool Creek and Bowen river, *Bowman;* Suttor river, *Sutherland;* Rockhampton, *O'Shanesy, Thozet.*

37. **P. spicata,** *R. Br. Prod.* 362. A small much-branched shrub, usually glabrous, with slender wiry branches. Leaves opposite, very shortly petiolate, oblong-elliptical, from under ½ in. to nearly 1 in. long. Flowers very small, in heads at first short and shortly pedunculate above the last leaves, but soon lengthening out into more or less interrupted spikes or rather racemes, from under ½ in. to nearly 1 in. long, without involucral bracts, the slender rhachis short pedicels and perianths all quite glabrous. Perianth-tube slender, about 1½ lines long, circumsciss about the middle after flowering, the lobes scarcely half as long. Filaments short; anthers with a narrow connective, the cells when open placed back to back. Fruiting base of the perianth about 2 lines long, bottle-shaped; fruit acuminate, the epicarp membranous, tipped when young with a few long hairs.—Meissn. in DC.

Prod. xiv. 514; Rudge in Trans. Linn. Soc. x. 288, t. 14, f. 2; *Calyptrostegia spicata,* Endl. Gen. Pl. Suppl. iv. part 2, 61.

N. S. Wales. Port Jackson and neighbouring districts, *R. Brown, A. & R. Cunningham,* and others.

P. spiculigera, F. Muell. Herb. from near Lake Muir, in W. Australia, *J. R. Muir,* is evidently very nearly allied to *P. spicata,* but the specimen is wholly insufficient to characterize it either as a variety or distinct species.

38. **P. filiformis,** *Hook. f. in Hook. Lond. Journ.* vi. 280, *and in Fl. Tasm.* i. 331, t. 95. A diffuse or prostrate shrub, glabrous or nearly so, with slender almost filiform branches, closely allied to *P. spicata,* but with the leaves usually smaller and broader, and the flower-heads do not lengthen into a spike. Perianths glabrous or sprinkled with a few hairs, of the size of those of *P. spicata,* but the lobes longer in proportion to the tube, and the filaments rather longer.—Meissn. in DC. Prod. xiv. 514.

Tasmania. Apparently rare, although found in abundance in one spot near Penguete, Launceston, *Lawrence, Gunn, Archer.*

39. **P. latifolia,** *R. Br. Prod.* 362. A spreading silky-hairy shrub or undershrub of 1 to 2 ft. Leaves alternate, elliptical or lanceolate, acute, tapering into a very short petiole, 1½ to 3 in. long, glabrous above, silky-hairy underneath. Flower heads more or less lengthened into short spikes, shortly pedunculate above the last leaves, without involucral bracts, becoming lateral or in the forks of the branches by the elongation of the lateral shoots, the rhachis of the fruiting spike often ½ in. long. Perianth-tube about 2 lines long, circumsciss about the middle after flowering, leaving a bottle-shaped fruiting base as in *P. spicata,* the lobes rather short and obtuse. Filaments short; anthers with a narrow connective, the cells when open placed back to back. Fruit shortly acuminate, epicarp membranous, endocarp more hooked at the end than in many species.—Meissn. in DC. Prod. xiv. 514.

Queensland. Cumberland island, *R. Brown;* Port Denison, *Fitzalan;* Rockhampton, *Thozet;* Head of the Dee river, *Bowman;* Mount Elliott and Mount Mueller, *Dallachy.*

40. **P. simplex,** *F. Muell. in Linnæa,* xxv. 443. An erect slender branching annual of about 1 ft., glabrous except the inflorescence and much resembling *P. trichostachya* (with which F. Mueller proposes to unite it) in habit and foliage, but with a different inflorescence and flowers. Leaves alternate, linear, mostly about ¼ in. long. Flower-heads small, depressed-globular, surrounded when young by 2 to 4 deciduous involucral bracts like the stem-leaves but smaller, and remaining dense and compact after flowering, the hirsute rhachis not exceeding 2 lines. Flowers smaller than in *P. trichostachya,* the perianth-hairs shorter and less spreading, the tube much shorter above the ovary, but similarly circumsciss after flowering and the fruit the same.—Meissn. in Linnæa, xxvi. 350, and in DC. Prod. xiv. 511.

Victoria. Murray desert, *Dallachy.*
S. Australia. Cudnaka and N. of Lake Gairdner, *F. Mueller.*

41. **P. sericostachya,** *F. Muell. Fragm.* iv. 162. Apparently shrubby at the base, with erect branching stems, all under 1 ft. in the specimens seen but said to attain 1½ ft., the whole plant clothed with long silvery-silky hairs which soon disappear from the upper side of the leaves. Leaves alternate, lanceolate or oblong-linear, ½ to 1 in. long. Flowers very small, in heads compact when in very young bud, but soon lengthening into interrupted spikes often several inches long, always pedunculate above the last leaf, without involucral bracts. Perianth shortly silky-hairy, the tube slender, about 2 lines long, circumsciss above the ovary after flowering, the lobes about ½ line. Anthers abortive in all the specimens seen, but F. Mueller describes them as nearly sessile in bisexual flowers. In fruit the persistent base of the perianth enlarges to 3 lines; fruit acuminate, the epicarp membranous, the endocarp scarcely hooked.

Queensland. Sellham river, *Bowman.*

Var. *parvifolia.* Leaves mostly about 2 lines long.
N. S. Wales. Lachlan river, *Frazer.*

42. **P. trichostachya,** *Lindl. in Mitch. Trop. Austr.* 355. An erect slender branching annual of about 1 ft., glabrous and glaucous except the inflorescence. Leaves alternate, linear, concave, ¼ to ½ in. long. Flower-heads at first short, but soon lengthening into a slender interrupted hairy spike or raceme of 1 to 2 in., without any involucral bracts different from the stem-leaves, the flowers small "yellow" and very shortly pedicellate. Perianth not above 2 lines long, the tube circumsciss above the ovary after flowering, the persistent base clothed with long spreading hairs; the lobes not above ½ line long. Filaments short; anthers oblong with a narrow connective, the cells when open placed back to back. Fruit acuminate, the epicarp membranous.—Meissn. in DC. Prod. xiv. 514; *Calyptrostegia trichostachya,* Walp. Ann. iii. 325.

Queensland. On the Maranoa river, *Mitchell;* Bokhara Creek, *Leichhardt;* Darling Downs, *Law.*
N. S. Wales. Murray river near the junction with the Murrumbidgee, *F. Mueller.*

43. **P. leptostachya,** *Benth.* A slender undershrub (or annual with a hard base?), much branched, with erect or ascending stems of ½ to 1 ft., glabrous or sprinkled with a few silky hairs especially on the inflorescence and flowers. Leaves alternate, lanceolate or oblong-linear, ½ to 1 in. long. Flowers small, in loose spikes scarcely forming a head when very young, and lengthening to 1 or 1½ in., without involucral bracts. Perianth about 2½ lines long, sprinkled with appressed hairs, not circumsciss after flowering, the lobes rather long, filaments short; anthers with a narrow connective, the cells placed back to back when open. Fruiting perianth about 3 lines long, bottle-shaped at the base, contracted into a short neck terminating in the persistent lobes. Fruit acuminate, the epicarp membranous, hairy at the apex, the endocarp tubercular-rugose, not hooked.

Queensland. Herbert's Creek and Rockhampton, *Bowman.* The species is evidently allied to *P. trichostachya* and *P. sericostachya*, but is nearly glabrous and differs from the whole section in the persistent apex of the perianth, whilst other characters prevent the placing it in *Eupimelea*. It is very different in the shape of the perianth as well as in the alternate leaves from *P. spicata*, to which it is inadvertently referred by F. Mueller, Fragm. viii. 9.

SECT. 5. MALISTACHYS, *C. A. Mey.*—Involucral bracts free, usually small or little different from the stem-leaves. Flowers (small) strictly diœcious. Male perianth with a slender tube; anthers with a narrow connective, the cells very distinct, and when open placed back to back; ovary abortive. Female perianth-tube circumsciss above the ovary after flowering, the lower portion persistent round the somewhat succulent fruit and almost adnate to it. Leaves silky-villous or hairy.

44. **P. argentea,** *R. Br. Prod.* 362. An erect shrub with usually virgate branches, attaining 5 or 6 ft., more or less clothed with soft silky hairs, appressed or somewhat woolly on the branches, and when abundant giving the plant a silvery-white appearance. Leaves opposite or occasionally alternate, sessile, those below the inflorescence oblong or lanceolate, flat or concave, often above 1 in. long, smaller on side branches; the floral ones usually shorter and broader, varying however from very like the stem-leaves and in distant pairs along the long leafy branches, to short and broadly ovate closely approximate or imbricate in leafy spikes. Flowers diœcious, very small and often numerous, in axillary clusters shorter than the floral leaves. Males usually on slender pedicels; perianth-tube glabrous, filiform, 1 to 1½ lines long, the lobes short and broad, bearing a few hairs. Anthers often nearly sessile and sometimes only one perfect, the connective narrow. Female perianths sessile, silky-hairy, the tube shorter than in the males but circumsciss above the ovary after flowering, the lobes almost acute, the anthers abortive or rudimentary. Style exserted, the terminal stigma hispid with long hairs. Fruiting base of the perianth ovoid, hispid, about 1 line long, the membranous or scarcely thickened epicarp coming off readily with it, but not connate with it as stated by C. A. Meyer, leaving the crustaceous endocarp (not the seed) minutely tubercular-rugose.—Meissn. in Pl. Preiss. i. 607; *P. argentea*, *P. vestita*, *P. Shuttleworthiana*, and *P. myriantha*, Meissn. in DC. Prod. xiv. 513; *Calyptrostegia argentea*, C. A. Mey. in Bull. Acad. Petrop. iv. (1845) 74.

W. Australia. Goose Island Bay, *R. Brown;* Middle Island, S. Coast, *A. Cunningham;* Phillip's river, Lake Leven, Gardner river, *Maxwell;* N. of Stirling range, *F. Mueller;* thence to Swan river, *Drummond 1st coll. n.* 730, 731, *Preiss, n.* 1264, 1265; northward to Murchison river and south-westward to Vasse river, *Oldfield.*

I have been unable to sort into varieties the supposed species distinguished by Meissner, notwithstanding the very different aspects they sometimes assume, the opposite and alternate, long and short, narrow and broad, distant and close leaves, as well as the degrees of indumentum pass very variously into each other. I find the plant constantly diœcious. The supposed "filiform persistent base" of the male perianth appears to be in fact the pedicel; the rudimentary ovary, when appreciable, is within the deciduous tube.

45. **P. clavata,** *Labill. Pl. Nov. Holl.* i. 11. An erect-growing shrub, ranging from 6 or 8 ft. to twice that height, the foliage slightly silky-

hairy and usually of a pale colour, or the slender branches bearing more spreading hairs. Leaves opposite, lanceolate, acute, tapering into a short petiole, flat or with recurved margins, mostly $\frac{3}{4}$ to 1 in. long, but varying from $\frac{1}{2}$ to $1\frac{1}{2}$ in., usually glabrous above, slightly silky-hairy underneath. Flowers small, strictly diœcious, in little terminal heads, with two or three small very deciduous bracts, and the leaves of the short axillary flowering branches being also often few and deciduous, the heads, when advanced, appear to be on axillary leafless peduncles. Perianths more or less silky-hairy, the males with a filiform tube $1\frac{1}{2}$ to nearly 2 lines long, the lobes ovate and obtuse. Anthers about as long as the filaments, ovate with a narrow connective. Female perianth more hairy than the male, the tube very shortly produced above the ovary but circumsciss after flowering, the anthers abortive; stigma large but scarcely penicillate. Fruiting base of the perianth about 2 lines long; fruit not acuminate, the epicarp thick and somewhat fleshy, the endocarp crustaceous, but thinner than in most species. Seed with a scanty albumen and broadly oblong cotyledons.—R. Br. Prod. 361; Meissn. in DC. Prod. xiv. 510; *P. viridula,* Lindb. in Finsk. Vet. Soc. Forhandl. ix. 60 (*F. Muell.*).

W. Australia. King George's Sound and on the coast to the eastward, *R. Brown, A. Cunningham,* and many others; Fitzgerald river, *Maxwell;* Donnelly river, *T. C. Carey;* Warren river, *Walcott.*

SECT. 6. DITHALAMIA.—Involucral bracts either none or few and not very different from the stem-leaves. Flowers (small) strictly diœcious. Male perianth with a slender tube; anthers with a narrow connective, the cells very distinct and when open placed back to back, the ovary abortive or rudimentary. Female perianth wholly persistent, with small lobes divided to the ovary, or rarely with a short tube tardily circumsciss. Fruit not succulent. Leaves opposite or alternate, flat or nearly so.

46. **P. axiflora,** *F. Muell. First Gen. Rep.* 17 (*and Pl. Vict. t.* 76 *ined.*). A glabrous shrub of several feet, with virgate rather slender branches. Leaves opposite, linear or linear-lanceolate, acute, tapering into a short petiole, 1 to 2 or even 3 in. long, membranous, flat or with recurved margins. Flowers small, in little axillary sessile heads with 2 to 4 small involucral bracts, the males numerous, the females fewer in the head. Perianth glabrous or slightly silky-hairy, the male tube 1 to $1\frac{1}{2}$ lines long, the lobes much shorter; anthers on very short filaments, with a narrow connective, the cells when open placed back to back. Female perianth scarcely projecting above the ovary, with very short lobes, somewhat enlarged round the fruit which is about 2 lines long; epicarp membranous.—Meissn. in Linnæa, xxvi. 345, and in DC. Prod. xiv. 514.

N. S. Wales. Clyde river, *C. Moore;* Twofold Bay, *F. Mueller;* Cape Howe, *C. Walter.*

Victoria. Dandenong ranges, Ben Nevis, Mount Macedon, *F. Mueller.*

Var.? *alpina,* F. Mueller. A low shrub with very small ovate coriaceous leaves, usually under $\frac{1}{2}$ in. long. Mount Latrobe and Mount Hotham, at an elevation of 5000 ft.,

F. Mueller.—The aspect of these specimens is so totally different, that one would suppose them to belong to a distinct species, but some specimens from Mount Barelley, with the foliage nearly of *P. drupacea*, seem to connect the two forms.

47. ? **P. leptospermoides,** *F. Muell. Fragm.* vii. 2. An erect shrub of 1 to 3 ft., slightly silky-hairy, the foliage of a pale almost glaucous hue. Leaves mostly alternate, oblong, ¾ to 1½ in. long, glabrous above, with 2 or 3 veins on each side of the midrib and nearly parallel to it prominent underneath. Flowers in terminal and axillary sessile clusters, without involucral bracts different from the stem-leaves, those in the specimens seen all males. Perianth silky-hairy, with a slender tube 4 to 5 lines long and narrow lobes about 1 line. Anthers oblong, nearly sessile, the connective rather broad, but the cells quite distinct and at length placed almost back to back. Ovary apparently abortive.

Queensland. Cawarra, *Thozet.*—Until the fertile flowers have been observed, the affinities of this species must remain uncertain.

48. **P. microcephala,** *R. Br. Prod.* 361. A much-branched spreading shrub, usually quite glabrous except the flowers, the branches rigid but slender. Leaves opposite, linear-lanceolate, acute or obtuse, mostly ¼ to 1 in. long, or on some luxuriant shoots narrow and 1½ in. long, flat or concave. Flower-heads small, terminal, with 2 to 4 involucral bracts rather shorter and broader than the stem-leaves but variable. Flowers strictly diœcious. Male perianth more or less silky-hairy, the tube very slender, fully 2 lines long, the lobes about one-third as long. Filaments very short; anthers with a narrow connective, the cells when open placed back to back. Female perianth very short and villous, the small erect lobes just protruding above the ovary, usually enlarged with the fruit, which is then 2 lines long and sometimes protrudes beyond the lobes. Epicarp membranous. Seed with a scanty albumen and broad cotyledons.—Meissn. in DC. Prod. xiv. 515; *P. distinctissima*, F. Muell. First Gen. Rep. 17; *Calyptrostegia microcephala*, Endl. Gen. Pl. Suppl. iv. part 2, 61.

Queensland. Curriwillighie, *Dalton;* Peak Downs, *Burkitt.*
N. S. Wales. Peel's range, *A. Cunningham;* Murray and Darling desert to the Barrier range, *Victorian and other Expeditions.*
Victoria. Avoca and Murray rivers, *F. Mueller;* N.W. districts, *L. Morton.*
S. Australia. Kangaroo Island and Petrel Bay, *R. Brown* (the specimens very imperfect, the flowers fallen away); Murray desert, *F. Mueller;* Lake Gillies, *Burkitt;* Enola harbour, *Forrest.*
W. Australia. Murchison river, *Oldfield*, and probably the same locality, *Drummond;* Dick Hartog's island, *Milne.*

49. **P. pauciflora,** *R. Br. Prod.* 360. A much-branched glabrous shrub, attaining sometimes 8 or 10 ft., but often much lower, closely allied to and much resembling luxuriant specimens of *P. microcephala*, but usually more slender, with linear-lanceolate leaves. Flowers few in the head, and the perianth and receptacle quite glabrous, the male perianth not quite so slender as in *P. microcephala*, but the diœcious character and the structure of the flowers and fruits quite the same. The fruit is said to be red when ripe, but it does not appear to be at

all succulent.—Meissn. in DC. Prod. xiv. 515; Hook. f. Fl. Tasm. i. 335; Lodd. Bot. Cab. t. 179.

Queensland. Dividing range, Moreton Bay, *A. Cunningham;* Mount Lindsay, *W. Hill;* Warwick, *Beckler;* Wide Bay, *Bidwill.*

N. S. Wales. N.E. of Lachlan river, *A. Cunningham;* M'Leay Bellinger and Richmond rivers, *C. Moore;* M'Leay river, *Beckler;* New England, *C. Stuart.*

Victoria. Avon river, Gipps' Land, *F. Mueller;* Fitzroy river, *Robertson.*

Tasmania. Port Dalrymple, *R. Brown;* in rich soil by the banks of rivers but not common, *J. D. Hooker.*

Some of the specimens appear at first sight to have the axillary inflorescence of *P. axiflora*, but the short axillary flowering branches are more developed with larger bracts, and the branches all end in flower-heads.

50. **P. elachantha,** *F. Muell. First Gen. Rep.* 17, *and Fragm.* vii. 6 (*excl. syn. A. Cunn.*). A low shrub with very numerous slender branches, more or less hoary with appressed hairs. Leaves opposite, ovate lanceolate or narrow-oblong, mostly 2 to 4 lines long, usually glabrous or nearly so. Flowers diœcious, minute, in little sessile heads either terminal or in the forks, with 2 to 4 involucral bracts similar to the leaves, the rhachis and very short pedicels hirsute. Male perianth silky-hairy, rarely above 1 line long, the lobes about as long as the tube. Filaments short; anthers with a narrow connective, the cells when open placed back to back. Female perianth oblong, about $\frac{3}{4}$ line long at the time of flowering, ovoid and 1 line long when in fruit, with 4 minute dark-coloured lobes, slightly prominent above the ovary.—*P. Hewardiana*, Meissn. in Linnæa, xxvi. 346, and in DC. Prod. xiv. 511.

Victoria. Mount Arapiles, Wimmera, *Dallachy*, and probably the same locality, *Mitchell;* Bacchus Marsh, *F. Mueller;* near Portland, *Allitt.*

The New England specimens referred here by F. Mueller appear to me to belong to *P. curvifolia*. *P. umbratica*, A. Cunn., is either a form of *P. drupacea* or some species closely allied to it.

51. **P. pygmæa,** *F. Muell. in Linnæa*, xxvi. 346. A dwarf prostrate much-branched shrub, forming densely matted almost moss-like patches or tufts of several inches diameter, glabrous except a few long cilia towards the ends of the upper leaves. Leaves mostly opposite, usually crowded, ovate, coriaceous, concave, 1 to $1\frac{1}{2}$ or rarely nearly 2 lines long. Flowers diœcious, solitary in the upper axils, usually with a pair of bracts similar to the stem-leaves. Perianth glabrous, tube of the males 1 to $1\frac{1}{2}$ lines long tapering at the base, the lobes under 1 line. Anthers shorter than the filaments, with a narrow connective, the cells when open placed back to back; ovary present but abortive. Female perianth scarcely $1\frac{1}{2}$ lines long, the tube ovoid, the lobes about as long as the tube, divided to the ovary, persistent membranous and brittle round the fruit. Fruit not 1 line long, the epicarp membranous. Seed with a scanty albumen, the cotyledons (not seen perfect) apparently broad.—Meissn. in DC. Prod. xiv. 511; Hook. f. Fl. Tasm. i. 335.

Tasmania. Summits of the Western Mountains at an elevation of 4000 ft., *Gunn, Archer.*—The flowers described by F. Mueller are the males, those described by *J. D. Hooker* the females.

52. **P. serpyllifolia,** *R. Br. Prod.* 360. A low rigid densely branched and leafy shrub, usually glabrous. Leaves opposite or scattered, often crowded, ovate obovate or oblong, 2 to 3 lines long, coriaceous and somewhat concave. Flowers yellowish, very small, diœcious, in terminal heads, sessile within the last leaves or involucral bracts not very different from the stem-leaves, the whole inflorescence glabrous. Male perianth rather above 1 line long, the ovate lobes longer than the rather broad tube. Filaments short; anthers with a narrow connective, the cells when open placed back to back. Female perianths smaller than the males at the time of flowering, but enlarging to above 1 line when in fruit, the small erect lobes much shorter than the tube and persistent. Fruit with the epicarp apparently somewhat succulent.—Meissn. in DC. Prod. xiv. 511; *M. cluytioides,* Meissn. in Pl. Preiss. ii. 271, and in DC. Prod. xiv. 511; *Calyptrostegia cluytioides,* Walp. Ann. iii. 324.

Victoria, *Mitchell;* Murray desert and Wilson's Promontory, *F. Mueller;* Port Phillip, *R. Brown, Harvey;* Wendu Vale, *Robertson.*

Tasmania. Flinders island, *Milligan;* King's island, *M'Gowan.*

S. Australia. Memory Cove, Port Lincoln, and St. Vincent's Gulf, *R. Brown;* Tumby Bay, *Wilhelmi;* Encounter and Holdfast Bays, *F. Mueller;* Spencer's Gulf, *Warburton.*

W. Australia. Towards the Great Bight, *Baxter, Maxwell.*

53. **P. flava,** *R. Br. Prod.* 361. An erect shrub, with opposite or forked usually virgate branches, slightly silky-hairy or nearly glabrous, the whole plant often turning blue-green in drying. Leaves opposite, obovate oblong or almost orbicular, flat, more or less coriaceous, glabrous, prominently 1-nerved or rarely showing also the lateral veins, mostly 2 to 4 lines but sometimes nearly ½ in. long. Flower-heads terminal, with 4 involucral bracts, larger and broader than the stem leaves, either rather longer or shorter than the flowers. Flowers white or yellowish, diœcious. Male perianth-tube about 1½ lines long, the lobes about half as long. Filaments short; anthers with a narrow connective and the cells very distinct, but often somewhat turned inwards. Female perianth about 1½ lines long when in flower, 2 lines when in fruit, divided to the ovary into 4 very small rounded lobes.—Meissn. in DC. Prod. xiv. 510; Hook. f. Fl. Tasm. i. 333; *Calyptrostegia flava,* Endl. Gen. Pl. Suppl. iv. part 2, 61.

N. S. Wales. Euryalean scrub, *A. Cunningham.*

Victoria. Wimmera and Murray rivers, *Dallachy;* Grampians, *F. Mueller;* Skipton, *Whan;* near Portland, *Allitt.*

Tasmania. Port Dalrymple, *R. Brown;* in poor clay soil but not common, *J. D. Hooker;* Hobarton, George town and on the east coast, *Gunn.*

S. Australia. Memory Cove, *R. Brown;* Murray river to St. Vincent's Gulf, *F. Mueller, Behr;* Kangaroo island, *Waterhouse.*

W. Australia. Coast opposite Middle island, *R. Brown.*

Var. *diosmifolia,* Meissn. Branches rather more spreading, leaves more ovate and rigid, often smaller. *P. diosmifolia,* A. Cunn. Herb.; *P. dichotoma,* Schlecht. in Linnæa, xx. 581; *P. parvifolia,* Meissn. in Linnæa, xxvi. 345.—As observed by Meissner this is rather a form dependent on the arid localities, than a distinct variety., it includes most of the above mentioned desert specimens.

54. **P. petrophila,** *F. Muell. in Linnæa,* xxv. 442. A shrub of 1 to 2 ft., with erect dichotomous slightly hairy branches, the foliage glabrous. Leaves opposite, oblong-lanceolate, from under ½ in. to nearly ¾ in. long, flat or concave, with the midrib prominent underneath and the lateral veins often conspicuous. Flower-heads small, terminal or in the forks, with 4 involucral bracts similar to the leaves or rather broader. Receptacle villous. Flowers diœcious, the perianths silky-villous. Males with a slender tube about 2 lines long, the lobes about 1 line. Filaments short; anthers with a narrow connective, the cells when open placed back to back. Female perianth about 2 lines long altogether at the time of flowering, enlarging to about 3 lines when in fruit, very shortly produced above the ovary and sometimes tardily circumsciss, but usually wholly persistent round the fruit. Epicarp membranous, endocarp acuminate.—Meissn. in Linnæa, xxvi. 347, and in DC. Prod. xiv. 500.

S. Australia. Flinders range, *F. Mueller.*

SECT. 7. EPALLAGE, *C. A. Mey.*—Involucral bracts free, few or numerous, like the stem-leaves or broad. Flowers hermaphrodite or diœcious the perianth-tube usually circumsciss above the ovary after flowering, rarely wholly persistent. Anthers rather flat, with a broad dorsal connective, the cells closely parallel on the inner face, the whole anthers usually rolled back after flowering.

55. ? **P. Bowmanni,** *F. Muell. Herb.* An erect shrub, softly silky-hairy all over. Leaves alternate, crowded, sessile, oblong-lanceolate, 1-nerved, ½ to ¾ in. long, silky on both sides. Flowers solitary in the upper axils. Perianth-tubes rather slender, shortly and equally silky-villous, about 2 lines long, circumsciss above the ovary after flowering, the lobes more than half as long. Anthers in one specimen examined all abortive, in another specimen with longer perianth-lobes they were oblong, with a broad dorsal connective, the cells parallel on the inner face but more prominent and distinct than in most *Epallages.* Fruit not seen ripe.

Queensland. Broad Sound, *Bowman.* Evidently a very distinct species, but requiring further examination of more advanced specimens.

56. **P. ammocharis,** *F. Muell. in Hook. Kew Journ.* ix. 24, *and Fragm.* vii. 5. A shrub of 2 or 3 ft., usually much branched, the foliage densely silky-villous with soft silvery hairs. Leaves alternate, sessile, crowded or imbricate, oblong or elliptical, 3 or 4 lines long. Flower-heads depressed-globular, often nodding, closely sessile, surrounded by numerous bracts not differing from the stem-leaves, shorter than the male flowers but often longer than the female. Flowers diœcious. Male perianth-tube slender, silky-villous, 3 to fully 4 lines long, the lobes 1 to 1½ lines. Filaments short; anthers with a broad dorsal connective, the cells closely parallel on the inner face. Female perianth with a shorter tube, slender at the time of flowering but covered with silky hairs almost as long as the tube itself and spreading like a pappus on

the fruiting perianth, which is entirely nerveless, the lobes shorter than in the males and the anthers abortive.—Meissn. in DC. Prod. xiv. 507.

N. Australia. Upper Victoria river and Sturt's Creek, *F. Mueller;* twenty miles south of Port Nichol, N.W. coast, *Maitland.* The latter specimens are considered as a var. *Maitlandi,* by F. Mueller, I can find no difference except that the flowers are larger. The specimens from Roebuck Bay, *Martin,* quoted by F. Mueller, are doubtful, being in leaf only.

57. **P. curviflora,** *R. Br. Prod.* 362. A shrub or undershrub much branched especially near the base, with wiry ascending or erect branches, attaining 1 to 2 ft. more or less silky-hairy. Leaves alternate or here and there opposite, varying in breadth, rarely ½ in. long and often not ¼ in., glabrous above, sprinkled or clothed with appressed hairs underneath. Flowers small, hermaphrodite or female, in little heads really terminal but mostly appearing axillary from the shortness of the flowering branches, the involucral bracts few, usually 2 only, small and unequal. Perianth silky-hairy, the tube slender, usually curved below the middle, 2 to 2½ lines long, circumsciss about the middle after flowering, the upper portion falling off very early, leaving the style shortly protruding from the persistent base, the lobes not above half as long as the tube, the two inner ones sometimes smaller than the outer ones. Filaments very short; anthers with a very broad convex dorsal connective, the cells closely parallel on the inner face. Fruiting base of the perianth somewhat curved and contracted into a neck, the fruit itself shortly acuminate; epicarp membranous, endocarp not conspicuously hooked at the top. Seed apparently with a scanty albumen and broad cotyledons, but not seen very perfect.—Meissn. in DC. Prod. xiv. 512; Rudge in Trans. Linn. Soc. x. 285, t. 13; *Calyptrostegia curviflora,* C. A. Mey. in Bull. Acad. Petrop. iv. (1845) 74; *P. gracilis,* R. Br. Prod. 362; Meissn. in DC. Prod. xiv. 512; Hook. f. Fl. Tasm. i. 331; *Calyptrostegia gracilis,* Endl. Gen. Suppl. iv. part 2, 61; *P. congesta,* A. Cunn. Herb. non R. Cunn.; *P. thymifolia,* Presl. Bot. Bem. 107; *P. Muelleri,* Meissn. in Linnæa, xxvi. 351, and in DC. Prod. xiv. 512.

N. S. Wales. Port Jackson to the Blue Mountains, *R. Brown, Sieber, n.* 205 *and Fl. Mixt. n.* 476; Macquarrie and Lachlan rivers, *A. Cunningham;* Mudgee, *Woolls;* Wilson's Peak, *Leichhardt;* New England, *C. Stuart, C. Moore.*

Victoria. Port Phillip, *Gunn;* Portland, *Allitt;* Wendu vale, *Robertson.*

Tasmania. Port Dalrymple, *R. Brown;* common by the banks of streams, &c., *J. D. Hooker.*

S. Australia. Holdfast Bay, Lynedoch valley, *F. Mueller, Behr.*

Var. *sericea.* More silky, the leaves often hairy on both sides. Flowers sometimes rather larger, sometimes rather smaller and all female on some specimens both of this and the following variety. *P. propinqua,* A. Cunn., Meissn. in DC. Prod. xiv. 512; N. of Bathurst, *A. Cunningham;* New England, *C. Stuart;* Mount M'Ivor and Ararat *F. Mueller;* Creswick, *Whan.*

Var. *pedunculata.* More luxuriant, the leaves often ½ to 1 in. long. Flower-heads sometimes almost sessile in the axils as in the typical forms, but often on slender lateral or axillary peduncles (or flowering branches) ½ to 1 in., flowers rather larger.

Queensland. Warwick, *Beckler;* Darling Downs, *Law.*

N. S. Wales. New England, *C. Stuart,* Clarence river, *Beckler;* Castlereagh river, *C. Moore.*

Var. *micrantha.* Densely branched, silky-hairy, with shorter more canescent leaves. Flowers very small and apparently sterile in the specimens examined. *P. micrantha*, F. Muell.; Meissn. in Linnæa, xxyi. 351, and in DC. Prod. xiv. 512. Murray and Darling Desert and St. Vincent's Gulf, *F. Mueller.*

Var. *alpina*, F. Muell. Diffuse much branched and less silky, with short broad leaves often opposite. Mount Kosciusko, Cobberas and Munyong mountains at an elevation of 4500 ft., *F. Mueller.*

58. **P. hirsuta,** *Meissn. in DC. Prod.* xiv. 513. A much-branched shrub of 1 to 2 ft., the branches hirsute as well as the foliage, the perianths with silky but more or less spreading hairs. Leaves mostly alternate, ovate or elliptical, very shortly petiolate, ¼ to nearly ½ in. long. Flowers very few together, in terminal or apparently axillary clusters, with only 1 or 2 small bracts similar to the stem-leaves. Perianth-tube slender, 1½ to 2 lines long, circumsciss above the ovary after flowering, leaving a bottle-shaped persistent base as in *P. curviflora*, but quite straight; lobes rather narrow. Filaments very short; anthers with a broad dorsal connective, the cells closely parallel on the inner face. Fruit acuminate, the epicarp membranous, the endocarp with a distinctly hooked beak as in *P. latifolia.*—*P. congesta*, R. Cunn. in several herb., not of A. Cunn.; *P. villifera*, A. Cunn. Herb.

N. S. Wales. Blue Mountains, *A. and R. Cunningham, Leichhardt, Miss Atkinson.*

P. ovalifolia, Meissn. in DC. Prod. xiv. 502, from Port Jackson (or Blue Mountains?), Gaudichaud, which I have not seen, must, from the character given, be the same as this species.

59. **P. altior,** *F. Muell. Fragm.* i. 84. A much-branched spreading shrub, attaining 5 or 6 ft., hirsute with short spreading hairs. Leaves opposite, shortly petiolate, from broadly ovate or orbicular and under ½ in. long, to oblong elliptical and 1½ to 2 in. long, flat, shortly hirsute on both sides. Flower-heads terminal or in the forks, with 2 to 4 involucral bracts similar to the stem-leaves but deciduous. Flowers not numerous. Perianth hirsute with more or less spreading hairs, the tube slender, about 3 lines long, circumsciss after flowering shortly above the ovary, the lobes about 1 line long. Filaments very short; anthers large, oblong, with a broad dorsal connective, the cells closely parallel on the inner face. Fruit about 2 lines long, the epicarp thinly membranous. Seed with a scanty albumen and ovate cotyledons.

Queensland. Brisbane river, Moreton Bay, *F. Mueller;* Archer's Creek, *Leichhardt.*

N. S. Wales. Hastings river, *Frazer, C. Moore;* Macleay and Clarence rivers, *Beckler;* Tweed river, *C. Moore;* Richmond river, *Henderson;* New England, *C. Stuart.*

Leichhardt's and Henderson's specimens are remarkably luxuriant with long narrow leaves, and appear at first sight very different from the original ones of F. Mueller, with short almost orbicular leaves, but there are many intermediate ones.

60. **P. octophylla,** *R. Br. Prod.* 361. A low shrub, with erect virgate branches, clothed with a soft silky wool which soon wears off. Leaves alternate or occasionally opposite, sessile, oblong, erect, ½ to 1 in. long, rather coriaceous, with a few veins besides the midrib often prominent underneath, covered with long soft hairs or soon becoming

glabrous. Flower-heads terminal, rather large, with an involucre of 8 or more bracts, smaller than, but otherwise similar to the stem-leaves, and much shorter than the flowers, the rhachis and perianths very villous with long somewhat spreading hairs. Perianth-tube 6 to 7 lines long, circumsciss immediately above the ovary after flowering, the lobes about 1 line long. Filaments rather short; anthers oblong, with a broad dorsal connective, the cells closely parallel on the inner face. Fruit small, with a membranous epicarp. Seed with a scanty albumen and broad cotyledons.—Meissn. in DC. Prod. xiv. 508; *P. Behrii* and *P. viminea*, Schlecht. in Linnæa, xx. 583.

Victoria. Port Philip, *R. Brown*, *F. Mueller;* Grampians *F. Mueller;* Wimmera, *Dallachy;* Portland, *Robertson*, *Allitt.*

S. Australia. Murray river to St. Vincent's and Spencer's Gulfs, *F. Mueller*, *Behr*, *Wilhelmi*, *Blandowski*, and others; Kangaroo island, *Waterhouse.*

W. Australia ? There is a specimen in Herb. F. Mueller labelled "Freemantle, W. Australia," but the collector is not named, and there may be some mistake.

61. **P. petræa,** *Meissn. in Linnæa,* xxvi. 347, *and in DC. Prod.* xiv. 508. A small shrub, with much of the habit of *P. octophylla.* Leaves scattered, crowded, oblong-linear, softer than in *P. octophylla*, more or less silky-hairy, from under ½ in. to about ¾ in. long. Flower-heads globular, with very numerous densely packed heads, surrounded by 20 or more involucral bracts similar to the stem-leaves and nearly as long as the perianth-tubes. Perianth villous with spreading hairs, the tube 4 to 5 lines long, circumsciss immediately above the ovary after flowering, the lobes about 1 line long. Filaments nearly as long as the lobes; anthers small, with a broad dorsal connective, the cells closely parallel on the inner face. Style long.

S. Australia. Cudnaka, *F. Mueller;* near Salt Creek, *Behr;* Gawler ranges, *Sullivan;* Lake Gillies, *Burkitt.*

F. Mueller, Fragm. vii. 6, unites this with *P. octophylla*, but the involucres, the long filaments, the short anthers, besides the foliage, appear to be constant.

62. **P. phylicoides,** *Meissn. in Pl. Preiss.* ii. 271, *and in DC. Prod.* xiv. 507. An erect shrub, from under 1 ft. to near 2 ft. high, villous but usually less so than *P. octophylla*, with shorter, more rigid hairs. Leaves alternate or here and there opposite, oblong or almost ovate, mostly 1½ to 3 lines long, sessile, coriaceous, more or less concave with the midrib prominent underneath. Flower-heads very much smaller than in *P. octophylla;* involucral bracts about 6 to 8, rather larger and broader than the stem-leaves but shorter than the flowers. Perianth silky-hairy, the tube slender, 2½ to 3½ lines long, circumsciss immediately above the ovary after flowering, the lobes less than half as long as the tube. Filaments short, anthers ovate, with a broad dorsal connective, the cells closely parallel on the inner face. Ovary crowned by a few hairs.—*Calyptrostegia phylicoides*, Walp. Ann. iii. 324.

Victoria. Murray desert and Wimmera, *Dallachy;* Port Philip and Point Lonsdale, *F. Mueller*, *Harvey;* Portland, *Allitt;* near Mount Thank, *Robertson.*

S. Australia. Arid stony places, St. Vincent's Gulf, Mount Torrens, *F. Mueller;* Encounter Bay, *Whittaker;* Cancarara, *Schulzen.*

63. **P. Eyrei,** *F. Muell. Fragm.* v. 109. An erect slender shrub of 2 to 4 ft., more or less silky-hairy and sometimes almost silvery. Leaves opposite, sessile, usually erect, oblong or oblong-linear, concave, $\frac{1}{4}$ to nearly $\frac{1}{2}$ in. long. Flower-heads terminal, with 4 to 6 involucral bracts rather broader than the stem-leaves, but otherwise similar. Perianths white, closely silky, the slender tube 4 to 5 lines long, circumsciss above the short ovary after flowering, the lobes narrow, about 2 lines long. Filaments very short; anthers oblong, with a broad convex dorsal connective, the cells scarcely distinct, closely parallel on the inner face.

W. Australia. Sandy plains, Eyre's Ranges, Phillips and Fitzgerald rivers, *Maxwell;* wet places, Mount Barker, *Oldfield.*

64. **P. longiflora,** *R. Br. Prod.* 361. An erect slender shrub, from 1 to 4 ft. high, loosely sprinkled with soft hairs or the foliage at length glabrous. Leaves mostly alternate, linear, concave, from $\frac{1}{4}$ to above $\frac{1}{2}$ in. long and usually erect. Flower-heads globular, with about 5 to 8 lanceolate involucral bracts shorter than the flowers. Receptacle densely hispid with long hairs. Perianth silky with appressed hairs, the tube 4 to 6 lines long, slender, circumsciss above the ovary after flowering. Filaments very short, anthers broadly oblong, with a broad dorsal connective, the cells closely parallel on the inner face. Fruit not seen ripe.—Meissn. in Pl. Preiss. i. 606; ii. 271, and in DC. Prod. xiv. 507; Bot. Mag. t. 3281; *Calyptrostegia longiflora,* Endl. Gen. Suppl. iv. part ii. 61.

W. Australia. King George's Sound and adjoining districts, *R. Brown, A. Cunningham,* and many others; Cape Riche, *Preiss, n.* 1263, and others.

Var. *latifolia.* Leaves oblong, mostly opposite, the involucral bracts almost ovate-lanceolate in some specimens, narrow in others.—*Calyptrostegia villosa,* Turcz. in Bull. Soc. Imp. Nat. Mosc. 1852, ii. 178; *P. villosa,* Meissn. in DC. Prod. xiv. 508.—W. Australia, *Drummond, 5th coll. n.* 428.

65. **P. stricta,** *Meissn. in Linnæa,* xxvi. 348 *and in DC. Prod.* xiv. 501. A loosely branched rather slender shrub of 2 or 3 ft., glabrous except the inflorescence. Leaves opposite, lanceolate or linear-lanceolate, contracted into a short petiole, flat or concave, from under $\frac{1}{2}$ in. to about $\frac{3}{4}$ in. long. Flower-heads globular, usually nodding; involucral bracts 4, broadly ovate, acuminate, herbaceous, often as long as the perianth-tubes, silky-villous inside and more concave and closely appressed than in most species. Perianth silky-villous, the tube about 5 lines long, circumsciss considerably above the ovary after flowering, the lobes scarcely $1\frac{1}{2}$ lines long. Anthers nearly sessile, broadly oblong, with a broad dorsal connective, the cells closely parallel on the inner face. Fruit stipitate within the base of the perianth, oblong, the epicarp membranous. Seed with a rather copious albumen, the cotyledons oblong, slightly broader than the radicle.

Victoria. Wimmera, *Dallachy, L. Morton;* Wendu Vale, *Robertson.*
Tasmania. Swanport, *Story.*
S. Australia. Near Adelaide, *F. Mueller;* Salt Creek, *Oswald;* Kangaroo island, *Waterhouse.*

This species is very closely allied to the Western *P. Preissii*, and may perhaps prove to be a variety only.

66. **P. Preissii,** *Meissn. in Pl. Preiss.* i. 601, *and in DC. Prod.* xiv. 500. An erect slender shrub of 1 to 2 ft., glabrous except the inflorescence. Leaves opposite, linear-lanceolate or oblong-linear, slightly concave, ¼ to above ½ in. long. Flower-heads globular, involucral bracts 4 to 6, broadly ovate or ovate-lanceolate, often shortly acuminate, shorter than the flowers, the inner ones slightly ciliate and silky-villous inside. Perianth silky-villous, with longer but scarcely spreading hairs on the lower portion, the tube 4 to 5 or rarely 6 lines long, circumsciss above the ovary after flowering, the lobes about 2 lines long. Filaments very short; anthers oblong, with a broad dorsal connective, the cells closely parallel on the inner face. Fruiting base of the perianth acuminate, the fruit not seen ripe.—*P. Neypergiana*, Hortul., according to Dcne. Rev. Hortic. ser. 4, i. (1852) 80.

W. Australia. Swan river, *Drummond, 1st coll. n.* 554; *Preiss, n.* 1266; Harvey river, *Oldfield.*

67. **P. Holroydi,** *F. Muell. Fragm.* vi. 159, t. 59. An erect perennial or shrub, glabrous and glaucous except the inflorescence as in *P. hæmatostachya*. Leaves opposite, sessile or nearly so, from almost orbicular to oval-oblong, 1 to 1½ in. long. Flower-heads at first depressed globular, but lengthening into an ovoid or oblong spike; involucre of 4 to 6 bracts, glabrous and glaucous like the stem-leaves, but shorter than the flowers, very broad and falling off before the fruit ripens. Perianth villous with long spreading silky hairs, which also cover the rhachis; the tube about ½ in. long, swelling round the fruit at the base after flowering but not circumsciss, the lobes about 1½ lines long. Filaments nearly as long as the lobes; anthers ovate or oblong, with a rather broad dorsal connective and recurved when old, the cells parallel on the inner face, but more distinct than in most *Epallages*. Ovary tipped with a few long hairs. Fruit acuminate, the style less lateral than in most species, the epicarp membranous. Seed with a scanty albumen and broad cotyledons.

N. Australia. Gorges of the Hammersley Range, N. W. Coast, *C. Harper. (Herb. F. Mueller.)*—This is a remarkable species, approaching *P. hæmatostachya* in foliage and inflorescence, but with the broad involucral bracts and persistent perianth of *Heterolæna* and the anthers of *Epallage.*

2. DRAPETES, Lam.

(Kelleria, *Endl.;* Daphnobryon, *Meissn.*)

Perianth tubular or almost campanulate, the limb 4-lobed, more or less spreading, with 1 or 2 scales opposite each lobe in the throat. Stamens 4, inserted in the throat, alternating with the lobes. No hypogynous scales. Ovary 1-celled, with a pendulous ovule. Style elongated, slightly lateral. Fruit a small drupe, with a thinly fleshy epicarp, the endocarp nut-like, crustaceous. Seed pendulous, without albumen; cotyledons ovate, thick.—Small prostrate densely tufted shrubs. Leaves alternate, small, imbricate, concave. Flowers in small terminal heads, sessile within the last leaves which are not at all or scarcely different from the lower ones.

Besides the Australian species, which is endemic, the genus comprises two from New Zealand, one from Borneo, and one from Antarctic America. Endlicher and others have proposed limiting the genus to that species in which the perianth is circumsciss above the ovary, leaving the lower portion persistent round the fruit, as in the section *Calyptrostegia* of *Pimelea*, whilst in all the other species the perianth remains entire, and is usually cast off as the fruit enlarges; but this difference is no more than what is admitted as sectional only in *Pimelea*, and the genus as a whole is as well marked in habit as in the stamens alternating with the lobes, not opposite to them as in *Pimelea*.

1. **D. tasmanica,** *Hook. f. in Hook. Kew Journ.* v. 299, t. 7, *and Fl. Tasm.* i. 330. The prostrate intricately-branched tufts of this little plant extend often to 1 ft. or more in diam., and appear sometimes somewhat silky from the cilia of the upper leaves. Leaves imbricate on the flowering branches, often somewhat distant on slender barren shoots, linear, concave, keeled, 1 to 1½ lines long, the younger ones ciliate especially towards the ends, the older ones glabrous or nearly so. Flowers 4 to 6 together, sessile within the last leaves and not exceeding them, on a densely hispid receptacle (or on exceedingly short pedicels). Perianth hairy about 2 lines long, the lobes as long as the tube with two small gland-like scales opposite each lobe, the whole cast off unbroken from the fruit. Filaments usually about half as long as the perianth-lobes; anthers opening in 2 cells, placed back to back, with a narrow connective. Fruit ovate, slightly compressed, about 1½ lines long.—*Daphnobryon tasmanicum*, Meissn. in DC. Prod. xiv. 566.

Victoria. Munyong mountains and Mount Kosciusko, at an elevation of 6000 ft. very rare, *F. Mueller*.

Tasmania. Summits of the Western mountains, abundant, *Gunn*.

Very closely allied to the New Zealand *D. Dieffenbachii*, which has however rather larger less hairy perianths, with the lobes rarely above half as long as the tube. The pairs of scales of the perianth-throat are also in Dieffenbach's specimen confluent into a single entire or notched scale opposite each lobe, which induced Meissner to distinguish it generically. In another New Zealand specimen, however, gathered by Colenso, I find the scales distinct in pairs as in the Australian plant.

3. WIKSTRŒMIA, Endl.

Perianth tubular, with a spreading 4-lobed limb without scales in the throat. Stamens 8, the anthers sessile, those opposite the perianth-lobes inserted in the throat, the alternate ones in the tube. Hypogynous scales 4, free or more or less united in pairs. Ovary with 1 pendulous ovule; style very short. Fruit a berry-like drupe, the epicarp succulent sometimes thin, the endocarp coriaceous or crustaceous. Seed without albumen.—Shrubs or trees. Leaves opposite or rarely here and there alternate. Flowers in short terminal or axillary racemes spikes or heads, without involucral bracts.

The genus extends over a great part of tropical Asia and the islands of the Archipelago and the Pacific. The only Australian species has a wide range over the area of the genus.

1. **W. indica,** *C. A. Mey. in Bull. Acad. Sc. Petersb.* i. (1843) 357. A shrub, sometimes low and spreading, sometimes almost arborescent, glabrous or the slender branches slightly silky-hairy. Leaves from ovate and obtuse to ovate-lanceolate and acute or oblong-lanceolate and tapering at both ends, rarely above 2 in. long and sometimes all under 1 in., usually rather thin and glabrous. Flowers few together, very shortly pedicellate in small terminal heads sometimes growing out into short spikes, the common peduncle usually under 4 lines long, erect or slightly recurved. Perianth of a greenish yellow, glabrous or sprinkled with a few hairs, the tube scarcely 3 lines long, the lobes about 1 line. Hypogynous scales 4, small and narrow, approximate in opposite pairs and sometimes the two connate at the base. Drupe red, about $\frac{1}{4}$ in. diam., the endocarp rather hard.—Meissn. in DC. Prod. xiv. 543; *Daphne indica*, Linn. Sp. Pl. i. 511; R. Br. Prod. 362; Hook. and Arn. Bot. Beech. t. 15; F. Muell. Fragm. vii. 1; *W. fœtida*, A. Gray in Seem. Journ. Bot. iii. 302; Seem. Fl. Vit. 207; *W. Shuttleworthii*, Meissn. in Denkschr. Regensb. Bot. Ges. iii. 287; *W. Shuttleworthiana*, Meissn. in DC. Prod. xiv. 544; *W. viridiflora*, Meissn. in Denkschr. Regensb. Bot. Ges. iii. 286 and in DC. Prod. xiv. 546; Benth. Fl. Hongk. 297.

N. Australia. Arnhem N. bay, *R. Brown;* Cleveland Bay, N. W. coast, *A. Cunningham.*

Queensland. Shoal Bay passage, *R. Brown;* Port Denison, *Fitzalan;* Edgecombe and Rockingham bays, *Dallachy;* Rockhampton, *Thozet;* Logan river, *A. Cunningham, Frazer;* Burnett and Brisbane rivers, *F. Mueller.*

N. S. Wales. Port Jackson to the Blue Mountains, *R. Brown, A. & R. Cunningham,* and others; northward to Hastings and Clarence rivers, *Beckler;* New England, *C. Stuart,* and numerous stations in *Leichhardt's* collection; southward to Illawarra, *Herb. F. Mueller* (without the collector's name).

The species appears to be also in the Indian Archipelago, in S. China, Sikkim, and the islands of the N. and S. Pacific. It is, however, not always easy to determine the limits to be assigned to it. The character derived from the perfect freedom or the union in pairs of the hypogynous scales appears to be of little or no value.

4. PHALERIA, Jack.

(Drymispermum, *Reinw.;* Leucosmia, *Benth.*)

Perianth tubular, with a spreading 4-lobed rarely 5 or 6-lobed limb, without scales in the throat. Stamens twice as many as the perianth-lobes, in 2 rows, those opposite the lobes inserted in the throat, the alternate ones in the tube. Hypogynous scales united in a short sinuate or lobed cup. Ovary 2-celled, with 1 pendulous ovule in each cell; style terminal, elongated. Fruit a drupe, with a succulent epicarp, the endocarp coriaceous or hard, 2-celled or 1-celled by abortion. Seed one in each cell without albumen, cotyledons thick and hemispherical.—Shrubs or trees. Leaves opposite, petiolate, larger than in most *Thymeleæ.* Flowers white or yellowish, several together in lateral or terminal sessile or pedunculate heads, surrounded by an involucre of about 4 bracts much shorter than the perianth-tube. Perianth longer than in most *Thymeleæ*, white or yellowish. Stamens usually as long as or longer than the perianth-lobes, but variable in this respect in different individuals of the same species.

The genus is dispersed over the Eastern Archipelago, Southern Asia, and the islands of the North and South Pacific. The species are difficult to discriminate, especially from the few specimens in herbaria. Jack's name, published in the Malayan Miscellany, has six years' precedence over Reinwardt's.

Flower-heads terminal. Perianth-tube glabrous. 5–6 lines long . . . 1. *P. Blumei.*
Flower-heads terminal. Perianth-tube loosely pubescent, 7–8 lines long . . . 2. *P. Neumanni.*
Flower-heads mostly axillary or lateral. Perianth-tube glabrous, $1\frac{1}{4}$–$1\frac{1}{2}$ in. long . . . 3. *P. clerodendron.*

1. **P. Blumei,** *Benth.* var. *latifolia.* A bushy glabrous shrub. Leaves oblong-elliptical or oval-oblong, shortly acuminate, more or less contracted at the base but the petiole distinct, mostly about 6 in. long and 2 broad, but varying from 4 to 8 in. and broader or narrower in proportion. Flower-heads mostly terminal, on a very short common peduncle. Involucral bracts 4 to 6, oblong or obovate, obtuse, about 3 to 4 lines long. Flowers usually numerous. Perianth-tube glabrous, 5 to 6 lines long, the lobes obtuse, pubescent inside especially towards the end, usually 4 but varying occasionally to 5 or even 6. Drupe shortly acuminate, when 2-celled 4 or 5 lines in diameter and furrowed outside opposite the partition, when 1-celled more ovoid and acuminate; epicarp succulent but not very thick; endocarp coriaceous. —*Drymispermum Blumei*, Dcne.; Meissn. in DC. Prod. xiv. 604.

Queensland. Cape York, *M'Gillivray, W. Hill, Daemel.*—Also in Java and Sumatra.

There is great confusion between this and *P.* (*Drymispermum*) *laurifolia*, Dcne., although distinctly characterized by Decaisne, Meissner and Miquel as having the perianth glabrous outside in the one, pubescent in the other. The Hookerian herbarium has two specimens, cultivated at different times in the Kew Gardens. The one figured in Bot. Mag. t. 5787 as *P laurifolia*, has the perianth glabrous, and would therefore belong to *P. Blumei*, whilst the description refers rather to the other specimen with pubescent perianths which appears to be the same as the Javanese *D. longifolium*, Teysm., and is perhaps also identical with Decaisne's Timor *P. laurifolia*. The Australian specimens have the leaves rather broader and less tapering than is usual in the Javanese specimens of either species.

2. **P. Neumanni,** *F. Muell.* A bushy shrub, glabrous except the flowers. Leaves oblong-elliptical, shortly acuminate, tapering below the middle, much narrowed at the base and narrowly decurrent along the short petiole, 4 to 8 in. long, $1\frac{1}{2}$ to $2\frac{1}{2}$ in. broad. Flower-heads terminal, sessile or nearly so. Involucral bracts 4 to 6, rather broad, obtuse. Flowers numerous in the head. Perianth hoary-pubescent outside, the tube 7 to 8 lines long; lobes usually 5 but sometimes 4 only, about 2 lines long, obtuse. Fruit rather larger than in *P. Blumei* and more frequently 1-seeded by abortion.—*Drymispermum Neumanni*, F. Muell. Fragm. v. 26.

Queensland. Rockingham Bay and Herbert river, *Dallachy.*—Probably not distinct from the true *P. laurifolia* from Timor, of which, however, I have seen no specimen for comparison.

3. **P. clerodendron,** *F. Muell.* An arborescent shrub, scarcely 12 ft. high, quite glabrous. Leaves elliptical-oblong, much acuminate,

tapering at the base but distinctly petiolate, 6 to 8 in. long. Flower-heads axillary or lateral, nearly sessile. Perianth glabrous outside, the tube 1¼ to 1½ in. long, the lobes usually slightly pubescent inside, about 3 lines long. Involucre and fruits not seen.—*Drymispermum clerodendron*, F. Muell. Fragm. vii. 1.

Queensland. Rockingham Bay, *Dallachy*. This species comes very near to *D. Cumingii*, Meissn., from the Philippine islands, but the flowers are still larger.

ORDER CVI. ELÆAGNACEÆ.

Flowers hermaphrodite or diœcious. Perianth tubular, free but persistent at the base and contracted above the ovary, the upper portion deciduous, 2- or 4-lobed, or in male flowers the perianth divided to the base. Stamens equal to and alternate with the perianth-lobes or twice as many, inserted at the mouth of the tube or at the base of the perianth; anthers versatile, with two parallel cells opening longitudinally. Ovary 1-celled, with 1 erect anatropous ovule. Fruit indehiscent, consisting of the persistent enlarged and usually succulent base of the perianth lined by the thin pericarp. Seed with a membranous or rather thick testa; albumen none or very thin; embryo straight, with a short inferior radicle.—Trees or shrubs, sometimes climbing, more or less covered with a scurfy or silvery indumentum consisting of stellate or peltate scales. Leaves alternate, entire. Flowers usually yellow or greenish, in axillary clusters cymes or short spikes, with 1 bract often very deciduous under each flower.

A small Order, chiefly Asiatic and European, with a very few American, northern, or tropical species; the only Australian genus has nearly the same area as the Order.

1. ELÆAGNUS, Linn.

Flowers hermaphrodite. Perianth 4-lobed. Stamens 4, inserted at the orifice of the tube. Style elongated, recurved at the top, with a lateral stigma.

This, the principal genus of the Order, has several Asiatic or European and one North American species; the only Australian species is one widely distributed over Ceylon E. India, and the Eastern Archipelago.

1. **E. latifolia,** *Linn.; Schlecht. in DC. Prod.* xiv. 610, *and in Linnæa*, xxx. 347. A shrub sometimes climbing over the tallest trees, sometimes erect and middle-sized or with weak straggling stems, covered except the upper surface of the leaves with the scurfy scales of the Order. Leaves petiolate, ovate ovate-lanceolate or elliptical, usually acuminate but sometimes obtuse, 2 to 4 in. long or larger on barren branches. Flowers several together in axillary spikes or clusters often not exceeding the petioles, with the bracts all small and deciduous, but sometimes longer with 1 or 2 of the lower bracts leafy and persistent. Perianths shortly pedicellate, the lower persistent portion narrow at the time of flowering, scarcely above 1 line long, closed at the top round the style by a disk-like annular prominence, the free part of the tube

much broader, ovoid-campanulate, about 2 lines long, slightly contracted at the top, the limb spreading to 3 or 3½ lines diameter. Filament very short. Ovary glabrous. Fruiting base of the perianth about ¾ in. long, slightly furrowed when dry. Seed about ½ in. long.—Wight, Ic. t. 1856.

Queensland. Rockingham Bay, *Dallachy;* Pioneer river, *Vernet.*—The species is common in Ceylon and in various parts of India and the Archipelago. It should include several of the Indian species enumerated by Schlechtendahl, and referred here by Thwaites, Enum. Ceyl. Pl. 252, though perhaps not the true *E. arborea* of Roxburgh. In Australia the indumentum is either all silvery-white or more or less mixed with or covered by ferruginous scales. The stature is very variable, and the size of the flowers also in a less degree. The Australian specimens agree perfectly with many of the Cingalese ones; I have observed no spines on any of them. The fruit is said to be edible.

ORDER CVII. NEPENTHACEÆ.

Flowers diœcious. Male perianth of 4 rarely 3 sepal-like segments, imbricate in the bud. Stamens 4 to 16, the filaments united in a central column; anthers united in a head, in 1 or 2 rows, 2-celled, the cells opening outwards in longitudinal slits. Female perianth as in the males, or rarely the segments united at the base. Ovary 4- rarely 3-angled, with as many cells as angles, the carpels opposite the perianth-segments; stigma sessile, with as many lobes as ovary-cells, the lobes entire or bifid. Ovules very numerous in each cell, attached to a placenta inserted on the dissepiment, ascending and anatropous. Capsules coriaceous, opening loculicidally in 4 rarely 3 valves. Seeds very numerous, imbricated upwards; testa membranous, produced at each end into a capillary point or tail. Embryo straight, in the axis of a fleshy albumen; cotyledons linear; radicle short, inferior.—Shrubs or undershrubs with herbaceous branches, glabrous or sparingly hairy, replete with spiral vessels, the stems prostrate sarmentose or climbing by means of tendrils terminating the leaves. Leaves alternate, elongated, without stipules, the base or the winged petiole clasping the stem, marked with a few nearly parallel longitudinal veins and numerous transverse veinlets, the midrib usually produced either into a tendril or into a pendulous pitcher provided with an operculum or lid, the margin of the orifice thickened into a peristome, and having usually an external spur at the back. Flowers small, green, in terminal racemes or panicles.

The Order, limited to the single genus *Nepenthes*, is spread over Southern Asia and the islands of the Indian and South Pacific Oceans, but most abundant in the Indian Archipelago. The only Australian species, described as endemic, is as yet too imperfectly known for accurate determination.

1. NEPENTHES, Linn.

Characters and distribution of the Order.

1. **N. Kennedyi,** *F. Muell. Fragm.* v. 154. Stems or branches rather stout, glabrous or when young pubescent with stellate hairs.

Leaves on long petioles, linear-lanceolate or oblong, membranous, 1 ft. long or more including the petiole, $1\frac{1}{2}$ to $2\frac{1}{2}$ in. wide, with 5 to 8 longitudinal nerves on each side of the midrib; pitchers 3 to 4 in. long, nearly cylindrical, slightly inflated below the middle, without any crest on the anterior ribs, the orifice narrow, scarcely produced into a neck, the posterior spur stout, the peristome narrow, with numerous transverse veins; operculum elliptical, with numerous minute glands on the inner surface. Flowers and fruits unknown.—Hook. f. in DC. Prod. xvii. 98.

Queensland. Cape York, *Jardine;* Cape Sidmouth, *C. Moore*. The species may perhaps prove to be a variety only of the widely-spread *N. phyllamphora*, Willd., from which it differs in the much more slender peristome of the pitcher.

ORDER CVIII. EUPHORBIACEÆ.

Flowers always unisexual. Perianth either simple and calyx-like or almost petal-like, usually small, or double with 4 or 5 petals alternating with the calyx-lobes, or sometimes entirely wanting in one or both sexes. Stamens various. Ovary superior, consisting of 3 or sometimes 2, or more than 3, united or 1-celled, or rarely 2-celled carpels, very rarely reduced to a single one. Styles as many as carpels, free or more or less united, entire or divided, the stigmatic surface usually lining their inner face. Ovules 1 or 2 in each carpel, pendulous from the inner angle of the cells, the funicle usually thickened into a cellular mass often termed an *obturator*. Fruit either capsular, separating into as many 2-valved cocci as carpels, leaving a persistent axis, or more rarely succulent and indehiscent with the endocarp consisting of as many indehiscent nuts or cocci as carpels or cells. Seed laterally attached at or above the middle, with or without an arillus or caruncle. Embryo straight, with flat cotyledons and a superior radicle, in a fleshy albumen, or very rarely the cotyledons fleshy, with little or no albumen.—Trees shrubs or herbs, often abounding in milky juice, exceedingly various in habit. Leaves alternate or opposite, rarely divided or compound, usually with stipules. Inflorescence very varied. Flowers usually small.

A very large Order, most abundant within the tropics, both in the New and the Old World, gradually diminishing in numbers in more temperate regions, and very few ascending into alpine or cold climates. Of the 37 Australian genera, three are generally distributed over the greater part of the area of the Order; six range over the tropical or subtropical regions of both the New and the Old World; twelve more are also tropical or subtropical, but confined to the Old World; and five of these appear to be limited to Asia, including the Archipelago and some of the South Sea Islands; one genus, *Sebastiania*, is a tropical American one, represented by a single species in the Old World; and one, *Baloghia*, is only known, out of Australia, in New Caledonia and Norfolk island; the remaining fourteen genera are endemic in Australia, including amongst them the whole tribe of *Stenolobeæ*.

TRIBE 1. **Euphorbieæ.**—*Involucre calyx-like, including several male flowers, each of a single stamen without any perianth, and one central female one, a single pedicellate pistil without any or rarely with a perianth; the whole flower-head resembling a single flower. Embryo with broad cotyledons and a narrow radicle.*

No perianth under the ovary, the Australian species leafy herbs or undershrubs . . . . 1. EUPHORBIA.

Perianth 4- or 6-lobed under the ovary. Shrubs with opposite virgate branches, leafless at the time of flowering . . . . . 2. CALYCOPEPLUS.

TRIBE 2. **Stenolobeæ.**—*Flowers distinct, both sexes with a perianth. Embryo linear, the cotyledons not at all or scarcely broader than the radicle. Shrubs often heathlike, with entire coriaceous leaves, or rarely herbs with small membranous leaves.* (See the observations p. 54.)

Ovules 2 in each cell.
- Anther-cells opening in terminal pores. Styles 2-fid. Capsule depressed-globular. Flowers (white) in head-like racemes forming terminal corymbs . . . . 3. PORANTHERA.
- Anther-cells opening longitudinally. Styles usually 2-lobed. Capsule depressed-globular. Flowers axillary . . . . (Tribe PHYLLANTHEÆ.)
- Anther-cells opening longitudinally. Ovary and capsule 3-lobed, the lobes ending in simple styles.
  - Stamens free or scarcely united at the base. Flower-clusters mostly axillary.
    - Capsule 3-celled. Leaves in alternate threes . . . . 4. MICRANTHEUM.
    - Capsule 1-celled and 1-seeded. Leaves solitary . . . 5. PSEUDANTHUS.
  - Stamens, all or at least the inner ones, united in a central column. Flower-clusters terminal . . . . 6. STACHYSTEMON.

Ovules 1 in each cell.
- Stamens indefinite, usually numerous.
  - Stigma peltate, entire or 3-lobed. Petals usually small. Flowers axillary . . . . 7. BEYERIA.
  - Styles 2-fid. Petals usually longer than the calyx. Flowers mostly terminal . . . . 8. RICINOCARPUS.
  - Styles 2- to 4-fid. Calyx petal-like but no petals. Capsule usually 1-celled, 1-seeded. Flowers axillary . . . . 9. BERTYA.
  - (24. CROTON, besides the cotyledons, differs in the stamens inflected, &c.)
- Stamens definite, twice as many as petals or calyx-lobes or fewer.
  - Petals present. Style-branches fringed. Capsule without appendages. Flower-cymes terminal . . . . 10. MONOTAXIS.
  - No petals. Style-branches entire. Capsule with 6 erect toothlike appendages. Flowers in dense sessile clusters in the axils or at leafless nodes . . . . 11. AMPEREA.

TRIBE 3. **Antidesmeæ.**—*Flowers distinct, both sexes with a perianth. Embryo with broad cotyledons and a narrow radicle. Trees or shrubs, the flowers small in catkin-like spikes or racemes. Ovary usually reduced to a single cell with 2 ovules. Styles* 3.

Stamens 2 to 5. Ovary 1-celled. Styles 3, small. Drupe 1-seeded 12. ANTIDESMA.

TRIBE 4. **Phyllantheæ.**—*Flowers distinct, both sexes with a perianth. Embryo with broad cotyledons and a narrow radicle. Ovules* 2 *in each cell. Flowers in axillary clusters or solitary.*

Calyx- or perianth-lobes imbricate in the bud.
- Petals present at least in the males, sometimes small and glandlike.
  - Stamens alternating with the petals. Rudimentary pistil in the males. Herbs or undershrubs . . . . 13. ANDRACHNE.
  - Stamens near the centre of a broad concave disk round a 3-fid abortive pistil. Trees or shrubs . . . . 14. ACTEPHILA.
- No distinct petals. Perianth of 4 to 6 calyx-like or petal-like lobes or segments, all similar or the inner ones rather larger.
  - Stamens 4 or more, central, free, without any rudimentary pistil. Styles linear, undivided. *Leaves opposite* . . 15. DISSILIARIA.

Stamens indefinite in a central column, without any rudimentary pistil. Stigmas *large flat*, usually lobed . . . 16. PETALOSTIGMA.
Stamens 2 to 5, central, free or united, without any rudimentary style. Styles linear or short.
Perianth-lobes or segments erect or spreading. Styles usually lobed . . . . 17. PHYLLANTHUS.
Perianth turbinate, the male flat-topped, the small orifice closed by minute lobes, the female open, the lobes minute or obsolete. Styles usually entire . . . . 18. BREYNIA.
Stamens 4 or 5, surrounding a 2-fid or 3-fid abortive pistil. Styles 2-fid . . . . 19. SECURINEGA.
Stamens 5, 6 or more, surrounding a broad central disk.
Ovary 3-celled. Styles 3, clavate or broad . . . . . 20. NEORŒPERA.
Ovary 1-celled. Style 1, flat reniform or orbicular . . 21. HEMICYCLIA.
Calyx-lobes valvate in the bud. Petals present. Stamens 5, on a central column which terminates in an abortive pistil.
Ovary usually 2-celled. Drupe with 2 indehiscent pyrenes. Veinlets of the leaves transverse between the primary veins 22. BRIEDELIA.
Ovary usually 3-celled. Capsule 3-dymous, separating into 2-valved cocci. Veinlets of the leaves reticulate between the primary veins . . . . 23. CLEISTANTHUS.

TRIBE 5. **Crotoneæ.**—*Flowers distinct, both sexes with a perianth, sometimes minute in the males. Embryo with broad cotyledons and a narrow radicle. Ovules* 1 *in each cell. Flowers, at least the males, in spikes racemes or panicles. Stamens usually indefinite.*

Calyx-lobes or segments valvate. Petals present.
Stamens inflected at the end in the bud. Anther-cells parallel, adnate. Styles 2-fid or 4-fid . . . . 24. CROTON.
Anthers erect in the bud, the cells parallel, adnate. Styles bifid. Flowers paniculate . . . . 25. ALEURITES.
Calyx-lobes or segments (at least in the males) valvate. No petals.
Anther-cells distinct, erect straight and parallel. Styles entire 26. CLAOXYLON.
Anther-cells distinct, linear, wavy or tortuose. Styles divided into capillary branches . . . . 27. ACALYPHA.
Anthers erect, the cells adnate and parallel. Styles 2-fid, the inner surface fringed or with much-raised papillæ . . . 28. ADRIANA.
Anther-cells parallel attached above the base or the anthers versatile.
Stamens usually 8 or fewer. Styles entire or 2-fid, not fringed. Trees or shrubs . . . . 29. ALCHORNEA.
Stamens few or many. Styles entire, not fringed. Twiners or (in species not Australian) erect herbs . . . . . . 30. TRAGIA.
Stamens usually numerous. Styles undivided, usually fringed or very papillose. Trees or shrubs . . . . . . 31. MALLOTUS.
Anthers 4-lobed, opening in 4 valves or longitudinally in 2 valves. Styles undivided, fringed or not. Trees or shrubs 32. MACARANGA.
Calyx-lobes or segments imbricate. Petals present. Stamens indefinite (above 6), central. Styles undivided. Leaves entire, coriaceous.
Anther-cells placed back to back, confluent at the apex . . 33. CODIÆUM.
Anther-cells parallel and distinct, opening outwards . . . 34. BALOGHIA.
Calyx small and open or minute. No petals Stamens 2 to 6, exserted. Styles undivided.
Calyx usually dividing into 2 or 3 broad lobes. Seeds carunculate.
Ovary 2-celled. Stamens 6 or fewer. Capsule didymous, compressed, tardily dehiscent on the margins . . . . 35. CARUMBIUM.

Ovary 3-celled. Stamens 3 or 2. Capsule separating into 3 2-valved cocci . . . . 36. SEBASTIANIA.
Calyx minute. Stamens 3 or 2. Seeds without any caruncuIus 37. EXCÆCARIA.

TRIBE 1. EUPHORBIEÆ.—Involucre resembling a calyx, toothed or lobed, including several male flowers, each of a single stamen without any perianth, and one central female flower, a single pedicellate pistil without any or rarely with a perianth, the whole flower-head resembling a single flower. Embryo with broad cotyledons and a narrow radicle.

1. EUPHORBIA, Linn.

(Anisophyllum, *Haw.*)

Flower-heads resembling single flowers. Involucre small, cup-shaped, with 4 or 5 small teeth alternating with and often concealed by as many horizontal prominent glands, which are sometimes expanded into or bordered by petal-like appendages. Within are about 10 to 12 male flowers consisting each of a single stamen with an articulated filament, and usually intermixed with or surrounded by thin membranous bracts, and in the centre a single female flower consisting of a stipitate 3-celled ovary protruding from the involucre, with 1 pendulous ovule in each cell. Style 3-cleft, the branches (or distinct styles) entire or 2-lobed. Capsule separating into 3 2-valved cocci.—Herbs or shrubs abounding in milky juice often very acrid. Stem-leaves entire or denticulate, in the majority of non-Australian but in very few Australian species alternate without stipules, the flowering-branches umbellate, dichotomous with usually opposite leaves; the majority however of the Australian species belong to a section with herbaceous dichotomous stems with the leaves opposite from the base and small interpetiolar stipules, and some species not Australian are succulent leafless shrubs.

A very large genus dispersed over nearly the whole world. Of the eighteen Australian species one is a common tropical weed, three are maritime plants extending more or less over the coasts of the Indian Archipelago, southern Asia and the Pacific islands, the remaining fourteen appear to be endemic, although one or two may be nearly allied to E. Indian species.

SECT. 1. **Anisophyllum.** *Leaves all opposite with small interpetiolar stipules. Involucral glands usually bordered by a petal-like appendage (except* E. atoto *and* E. pilulifera). *Seeds without any caruncuIus.*

* *Flower-heads in small distinct terminal cymes with the floral leaves or bracts as long as or longer than the involucres. Capsules and whole plant glabrous.*

Seeds smooth. Appendages of the involucral glands scarcely conspicuous . . . . 1. *E. atoto.*
Seeds rugose. Appendages conspicuously petal-like, entire.
Leaves ovate-orbicular. Cymes rather dense. Involucres about 1 line long . . . . 2. *E. Sparmanni.*
Lower leaves ovate, upper ones lanceolate or linear. Cymes loose with filiform peduncles. Involucres ½ line long . . 3. *E. Mitchelliana.*
(See also 14. *E. Macgillivrayi*, with much larger involucral appendages.)
Seeds rugose. Appendages more or less lobed . . . . . . 12. *E. myrtoides.*

** *Flower-heads solitary or two together, terminal or in the upper axils. Stems hairy or pubescent.*

Stem and leaves pubescent or shortly hirsute.
Involucres 1½ lines long, the glands with a broad palmately lobed appendage . . . . 4. *E. schizolepis.*
Involucres not ½ line long, the glands with a very small usually lobed appendage . . . . 5. *E. Schultzii.*
Stems slender, sprinkled with long spreading hairs. Involucres under 1 line, the glands with a narrow entire white border . 6. *E. Armstrongiana*

*** *Flower-heads solitary in the upper or in nearly all the axils or forks.*

Stem and leaves pubescent or hirsute.
Involucres 1½ lines long, the glands with a broad palmately lobed appendage . . . . 4. *E. schizolepis.*
Involucres about ½ line long, the gland-appendages usually lobed. Stems much branched, usually 6 in. to 1 ft. . . . 7. *E. australis.*
Involucres about 1 line long, the gland-appendages entire. Dwarf plant with rigid stems of 1–2 in. . . . . . 8. *E. Muelleri.*
Whole plant quite glabrous.
Involucres about 1 line long. Dwarf plant with rigid stems of 1–2 in. . . . . 8. *E. Muelleri.*
Involucres ½–¾ line long.
Diffuse or prostrate branching stems forming at length a perennial rhizome. Gland-appendages usually entire.
Styles notched only at the end . . . . 9. *E. Drummondii.*
Styles slender bifid . . . . 10. *E. alsinæflora.*
Annual, with erect and slender stems. Gland-appendages usually entire . . . . 11. *E. Wheeleri.*
Annual, diffuse and much branched. Gland-appendages usually lobed . . . . 12. *E. myrtoides.*

**** *Flower-heads numerous, in dense or rather loose terminal or axillary cymes, the floral leaves, except sometimes the lowest pair, reduced to small bracts shorter than the involucres.*

Dwarf glabrous perennial. Leaves ovate or oblong. Gland-appendages large and very white . . . . 13. *E. micradenia.*
Pubescent perennial with erect or ascending stems. Leaves ovate or oblong. Gland-appendages large and very white . 14. *E. Macgillivrayi.*
Glabrous erect or procumbent annual. Leaves linear. Gland-appendages rather large . . . . 15. *E. serrulata.*
Pubescent perennial with erect or ascending stems, the ultimate branches filiform and cymes loose. Gland-appendages large and very white . . . . 16. *E. filipes.*
Hirsute annual. Leaves ovate or lanceolate. Flower-heads minute, very numerous in dense headlike axillary cymes. Involucral glands without appendages . . . . 17. *E. pilulifera.*

SECT. 2. **Eremophila.**—*Leaves opposite or the lower ones and sometimes those of lateral branches alternate. Stipules very minute or none. Involucral glands without appendages. Seeds carunculate.*

Stems erect, dichotomous. Leaves linear. Flower-heads solitary in the axils . . . . 18. *E. eremophila.*

The section *Tithymalus*, with the stem-leaves below the flowering branches alternate, the inflorescence dichotomous, the primary branches forming an umbel, the floral leaves opposite without stipules, and the involucral glands without petal-like appendages, has no endemic representative in Australia; but one of the common European weeds, *E peplus*, Linn., Boiss. in DC. Prod. xv. ii. 141, a glabrous annual with the umbel usually of about 3 rays and the involucral glands crescent-shaped, the capsule glabrous and smooth, the seeds pitted, is said to be now common in cultivated ground in New South Wales and West Australia, and probably in other colonies.

Euphorbia Brownii, Baill. Adans. vi. 290, was described from a specimen without flowers, believed to have been brought by Baudin's Expedition from the West coast of Australia, and which Baillon referred to *Euphorbia* from some general resemblance to his *E. Cleopatra* from New Caledonia. It remains however very uncertain whether it is a *Euphorbia* at all, and the station, like others attached to plants of the Baudin Expedition, is very little to be relied upon.

SECT. 1. ANISOPHYLLUM, *Roxb.*—Herbs, either annual or with a perennial base, usually much branched and often prostrate. Leaves all opposite, usually oblique at the base, with small interpetiolar stipules. Glands of the involucre usually (but not always) bordered by a white petal-like appendage or margin. Seeds without any carunculus.

The species of this section run very much one into another, and are difficult to define. It is possible, therefore, that some of the following, founded upon a small number of specimens may prove to be varieties only.

1. **E. atoto,** *Forst.*; *Boiss. in DC. Prod.* xv. ii. 12. A glabrous diffuse or procumbent perennial of 1 to 1½ ft., the primary stems thick and hard, the branches more slender and sometimes dichotomous. Leaves opposite, shortly petiolate, broadly oblong or rarely narrow, obtuse or mucronulate, more or less cordate and usually unequally so at the base, rather thick, 1 to 1¼ in. long. Stipules usually fringed when old. Flower-heads in small dichotomous cymes in the upper axils, scarcely exceeding the leaves and forming a terminal leafy corymb. Bracts or floral leaves oblong, about as long as the involucres or rather longer. Involucres shortly pedicellate, nearly 1 line long, the glands transversely oblong, with very narrow scarcely distinct borders. Capsules glabrous. Seeds smooth.—Baill. Adans. vi. 282; *E. oraria*, F. Muell. in Herb. Kew.; *E. levis*, Poir.; Boiss. in DC. Prod. xv. ii. 13.

N. Australia. Water island, Montague Sound, N.W. Coast, *A. Cunningham*; Port Essington, *Armstrong*; La Grange Bay, *Hughan*; Port Darwin, *Schultz*, n. 601; Gulf of Carpentaria, *R. Brown.*

Queensland. Sandy Cape, *R. Brown*; Port Curtis, *M'Gillivray*; Isles off Cape Flattery and Moreton island, *F. Mueller*; Sir C. Hardy's island, *Henne*; Rockingham Bay, *Dallachy.*

A sea-coast plant, found also on the coasts of E. India, the Archipelago, and the Pacific islands.

2. **E. Sparmanni,** *Boiss. Cent. Euph.* 5. A glabrous and glaucous perennial, with a hard base and diffuse or divaricately branched stems attaining 1 to 1½ ft. Leaves opposite, shortly petiolate, ovate-orbicular, oblique and often unequally cordate at the base, ½ to 1 in. long. Stipules small, ciliate-toothed or lobed. Flower-heads in small cymes either terminal or in the upper axils, with the bracts or floral leaves mostly exceeding the involucres. Involucres pedicellate, about 1 line long, the orifice ciliate inside, the glands bordered by a petal-like appendage nearly as broad as the gland itself. Capsule glabrous, rather larger than in *E. atoto.* Seeds rather prominently angled and trans-

versely rugose between the angles.—*E. ramosissima*, Boiss. in DC. Prod. xv. ii. 14; Baill. Adans. vi. 283, not of Hook. and Arn.

N. S. Wales. E. coast, *R. Brown;* Manly Beach, *Woolls;* also *Sieber, n.* 632 (*Baillon*). A specimen from Lord Howe's island, *C. Moore*, appears also to belong to it.

The species is also on Pitcairn's Island in the Pacific. It sometimes nearly resembles *E. atoto*, but differs in the short broad leaves, the appendages to the involucral glands, and the slightly rugose seeds. The true *E. ramosissima*, Hook. and Arn., more common on the Pacific islands, is Boissier's *E. Chamissonis.*

3. **E. Mitchelliana,** *Boiss. in DC. Prod.* xv. ii. 25. A perfectly glabrous perennial, attaining 1 to 1½ ft., the rhizome at length woody, the stems erect or diffuse, slender, dichotomous, the ultimate branches filiform. Leaves opposite, shortly petiolate, the lower ones ovate and small, the upper oblong or linear, entire, ½ to 1 in. long, oblique or unequally cordate at the base. Flower-heads very small, not numerous, in loose terminal dichotomous cymes, the pedicels filiform, the floral leaves or bracts linear, mostly as long as or longer than the flower-heads. Involucre about ½ line long, the glands bordered by a petal-like obovate or orbicular appendage, varying from ¼ to ½ line in breadth. Capsule glabrous. Seeds transversely rugose.

N. Australia. Sweers island, *Henne.*

Queensland. Port Bowen, Keppel Bay, Northumberland islands, *R. Brown;* Sandy Beach, Lizard island, *A. Cunningham;* Port Curtis, *M'Gillivray;* Port Denison, *Fitzalan;* Rockingham Bay, *Dallachy;* Belyando river, *Mitchell;* Bowen river, *Bowman;* Sutton river, *Thozet;* Rockhampton, *O'Shanesy;* N. Kennedy district, *Daintree.*

Var. *glauca.* Leaves rather more coriaceous. Flower-heads very small, in irregular somewhat elongated leafy cymes. Gulf of Carpentaria, *R. Brown, Londesborough;* Nichol Bay, *Gregory's Expedition;* King's Sound and Collier Bay, *Chapman.*

Var. *stenophylla.* Leaves linear, 1–1½ in. long or the lower ones shorter and linear-lanceolate.—Port Darwin, *Schultz, n.* 38, 505, 549 *and* 854.

4. **E. schizolepis,** *F. Muell.; Boiss. in DC. Prod.* xv. ii. 20. An annual, but with hard, often woody-looking stems of about 1 ft. high, dichotomously branched and more or less pubescent with crisped hairs. Leaves opposite, almost sessile, ovate or ovate-oblong, acute, oblique and unequally cordate at the base, ½ to 1 in. long, sprinkled on both sides with short hairs. Flower-heads shortly pedicellate in the upper axils, forming sometimes terminal leafy cymes. Involucre about 1½ lines long, the glands peltate with a broad spreading palmately lobed or fringed appendage, not so white as in most species. Capsule and seeds not seen.

N. Australia. Upper Victoria river, *F. Mueller.*

Var.? *glabra;* perfectly glabrous, with the gland-appendages much less lobed; perhaps a distinct species.—Gulf of Carpentaria, *F. Mueller.*

5. **E. Schultzii,** *Benth. sp. n.* A slender annual of 6 in. to 1 ft., apparently erect when young, but soon much-branched and diffuse, with pubescent or shortly hirsute stems and foliage. Leaves opposite, nearly sessile, ovate or oblong, serrate, very oblique or semicordate at the

base, the lower ones 4 to 8 lines long, those of the lateral branches half that size. Flower-heads very small, solitary or 2 together, terminating short leafy branches in the upper axils. Involucre not ½ line long, hirsute, the glands small with a very narrow usually lobed petal-like margin sometimes scarcely conspicuous. Capsule small, hirsute, with prominent angles. Seeds rugose.

N. Australia. Port Darwin, *Schultz, n.* 15, 237, 844 *and* 879.—A very poor specimen from Camden harbour, in Herb. F. Mueller; may belong to the same species.

6. **E. Armstrongiana,** *Boiss. in DC. Prod.* xv. ii. 47. An annual of 1 ft. or more, with long slender ascending loosely dichotomous branches, more or less sprinkled with long spreading hairs. Leaves opposite, in distant pairs, ovate or orbicular, very obtuse, entire, very oblique at the base, mostly under ½ in. long. Stipules entire or with short subulate lobes. Flower-heads terminal, solitary or 2 together, on very short pedicels, the last pair of leaves as long as the involucre. Involucre under 1 line long, glabrous, the glands broad with a narrow white border. Capsule above 1 line long, glabrous. Seeds marked by deep transverse furrows.

N. Australia. Port Essington, *Armstrong;* Port Darwin, *Schultz, n.* 22.

7. **E. australis,** *Boiss. Cent. Euph.* 15 *and in DC. Prod.* xv. ii. 36 Apparently a perennial, forming at length a hard woody rhizome, the stem prostrate, much branched, often 1 ft. long, more or less villous as well as the foliage. Leaves opposite, nearly sessile, ovate-oblong or nearly orbicular, obtuse, mostly serrate, very oblique at the base, 3 or 4 lines long, those of the flowering branches much crowded. Flower-heads solitary in the upper axils, the short flowering branches usually crowded at the ends of the principal ones. Involucres and capsules small as in *E. Drummondii*, but more or less hirsute and the petal-like appendages of the glands more or less lobed, varying from white to red.—Baill. Adans. vi. 283; *E. vaccaria*, Baill. l.c. 286.

N. Australia. Victoria river, *F. Mueller;* Nichol Bay, *M. Brown;* Gulf of Carpentaria, *R. Brown, Gregory's Expedition.*
Queensland, *Bowman;* Gilbert river, *F. Mueller.*
N. S. Wales. Near the Barrier range, *Beckler.*
S. Australia. Near Morunda, *Beckler;* in the interior, *McDoual Stuart;* Lake Gillies, *Burkitt.*

Var. *erythrantha.* Appendages of the involucral glands very red. *E. erythrantha,* F. Muell. Fragm. ii. 152; Baill. Adans. vi. 284. To this variety belong more especially Beckler and Burkitt's specimens from the desert interior, but in some others the glands assume a reddish tint.

8. **E. Muelleri,** *Boiss. in DC. Prod.* xv. ii. 27. A dwarf plant with a thick woody rhizome and procumbent or ascending rather rigid stems of 1 to 2 in., densely pubescent with short hairs or nearly glabrous. Leaves opposite, nearly sessile, rather crowded, ovate orbicular or the upper ones oblong, rather thick, entire, 2 to 5 lines long, very oblique at the base, and the lower ones sometimes broader than long. Flower-heads larger than in *E. australis* and *E. Drummondii*, solitary and pedicellate

in the upper axils, forming almost a terminal leafy cyme. Involucre about 1 line long, glabrous or pubescent. Capsule 1½ lines long, glabrous or shortly hairy. Styles short. Seeds irregularly rugose.

N. Australia, *F. Mueller;* Port Darwin, *Schultz, n.* 439, 485.

9. **E. Drummondii,** *Boiss. Cent. Euph.* 14 *and in DC. Prod.* xv. ii. 36. A prostrate or diffuse much-branched plant, closely resembling the European *E. chamæsgee*, Linn., and when flowering the first year easily confounded with it, but always quite glabrous and forming at length a perennial thick or woody rhizome. Leaves opposite, orbicular ovate or oblong, obtuse or notched, entire or serrulate, very oblique at the base, 2 to 4 lines long, firmer than in *E. chamæsgee.* Stipules entire fringed or lobed. Flower-heads very small, shortly pedicellate in the upper axils. Involucres about ½ line long, the glands with a narrow white border entire or nearly so. Capsule under 1 line long, glabrous. Styles notched only, varying from almost none to nearly as long as in *E. chamæsgee.* Seeds rugose.—*E. chamæsgee, E. Ferdinandi, E. Drummondii* and *E. Dallachyana,* Baill. Adans. vi. 284, 285.

N. Australia. Victoria river, *F. Mueller;* Gulf of Carpentaria, *Lanesborough.*
Queensland. Broad Sound and Thirsty Sound, *R. Brown;* Brisbane river, *F. Mueller;* Rockhampton, *Dallachy,* and others; Curriwillighie, *Dalton.*
N. S. Wales. Port Jackson, *R. Brown;* Lachlan river, *A. Cunningham;* Darling river, *Mitchell;* and thence to the Barrier range, *Victorian and other Expeditions;* New England, *C. Stuart.*
Victoria. Snowy river, *F. Mueller;* Wimmera, *Dallachy.*
Tasmania. Swan port, *Story.*
S. Australia. Spencer's Gulf, *R. Brown;* Murray river and Capunda, *F. Mueller;* Gawler ranges, *Sullivan.*
W. Australia. Swan river, *Fraser, Drummond,* 1*st coll. n.* 670; Murchison river, *Oldfield.*

Var.? *rubescens.* Rather less prostrate than the ordinary form and apparently annual, very much branched, with the branches often angular. Inflorescence flowers and fruits as in the common *E. Drummondii,* except that the styles are rather longer. Perhaps a distinct species.—Dirk Hartog's Island, *A. Cunningham.*

10. **E. alsinæflora,** *Baill. Adans.* vi. 288. A glabrous and glaucous perennial, with a hard knotted base or rhizome, and much-branched ascending or decumbent stems, under 6 in. high in the specimens seen. Leaves opposite, very shortly petiolate, ovate or oblong, entire or obscurely serrulate, oblique or semi-cordate at the base, mostly 4 to 6 lines long. Stipules fringed or divided. Flower-heads solitary in the upper axils, very shortly pedicellate, not forming distinct cymes. Involucre rather above ½ line long, the petal-like appendages of the glands rather broad, entire crenate or sinuate. Styles rather slender and bifid. Fruit not seen.

N. Australia. Mount King, Glenelg river, N. W. Coast, *Martin.*
Queensland? A specimen from Warwick, *Beckler,* appears to be the same species.

11. **E. Wheeleri,** *Baill. Adans.* vi. 286. A perfectly glabrous plant, apparently annual, branching from the base into erect or ascending slender stems of 6 in. to 1 ft. Leaves opposite, in rather distant pairs, oval-oblong, entire or slightly serrulate, very oblique at the base, rarely

above ½ in. long. Stipules entire or lobed. Flower-heads solitary in the forks and upper axils, on very short pedicels. Involucre scarcely above ½ line long, the glands with a petal-like entire or slightly lobed appendage scarcely broader than the gland itself. Capsule above 1 line long, on a long stipes. Seeds deeply rugose.—*E. divaricata*, A. Cunn. Herb.

N. Australia. Greville island, Regent's river and Montague Sound, N. W. Coast, *A. Cunningham;* King's Sound, *Hughan.*
S. Australia. Between Stokes' Range and Cooper's Creek, *Wheeler.*

This plant appears to be quite distinct from *E. Drummondii* in habit as well as in the capsule nearly twice as large. It may, however, prove to be the first year's state of *E. alsinæflora.*

12. **E. myrtoides,** *Boiss. in DC. Prod.* xv. ii. 15. A diffuse or spreading much-branched annual, with the stem at length hard and almost woody at the base so as to appear perennial. Leaves opposite, shortly petiolate, obliquely ovate, very obtuse, entire or minutely denticulate, very unequal and often semicordate at the base, rather coriaceous but very unlike those of a Myrtle, rarely above ½ in. long. Stipules scarcely prominent. Flower-heads shortly pedicellate in the upper axils, forming sometimes rather dense dichotomous leafy cymes, the floral leaves longer than the involucres. Involucres smaller than in *E. atoto*, but considerably larger than in *E. Drummondii*, the glands with a narrow petal-like border usually denticulate or lobed. Capsule about 1¼ lines long, the cocci more distinct than in the allied species, and prominently keeled. Seeds rugose.—*E. Sharkoensis*, Baill. Adans. vi. 287.

N. Australia. N.W. Coast, Dampier's Archipelago, *A. Cunningham;* Despard's island, *Bynoe.*
W. Australia. Useless Harbour, Sharks' Bay, *R. Brown;* Port Gregory, *Oldfield.*

13. **E. micradenia,** *Boiss. in DC. Prod.* xv. ii. 27. A dwarf perennial with a woody rhizome, closely allied to *E. Macgillivrayi* and *E. serrulata*, differing from the former in its perfectly glabrous stem and foliage, and from the latter in its perennial rhizome, shorter and broader leaves, and in the involucral glands very small within the petal-like appendages, which are about 1 line diameter. Stems in the specimens seen rarely above 6 in. long. Leaves mostly semicordate, ovate or the upper ones oblong-lanceolate, obtuse, more or less serrulate, ½ to 1 in. long. Involucres inflorescence and fruit those of the more petaloid specimens of *E. Macgillivrayi*, of which it may be a variety.

N. Australia. Islands of the Gulf of Carpentaria, *R. Brown;* Port Darwin, *Schultz, n.* 545.
Queensland. Albany island, *F. Mueller.*

14. **E. Macgillivrayi,** *Boiss. in DC. Prod.* xv. ii. 26. A perennial but often flowering the first year so as to appear annual, the rhizome at length hard and woody. Stems erect or ascending, from under 6 in. to above 1 ft. high, more or less pubescent as well as the foliage or at length nearly glabrous. Leaves opposite, petiolate, the lower ones usually ovate-oblong and under ½ in. long, the upper ones narrow-

oblong often above 1 in., obtuse, serrulate, oblique at the base or semicordate. Stipules often lobed or fringed. Flower-heads small, crowded in rather dense shortly pedunculate axillary or terminal cymes, the floral leaves reduced to small bracts or the lower pair only developed and leafy. Involucre $\frac{1}{2}$ to $\frac{3}{4}$ line long, the bracts bordered by white petal-like entire appendages, varying from $\frac{1}{2}$ to 1 line in diameter. Capsule glabrous, above 1 line long. Seeds rugose.

Queensland. Thirsty Sound, *R. Brown;* Port Molle and Gould island, *M'Gillivray;* Brisbane river, *F. Mueller;* Port Denison, *Fitzalan;* Rockhampton, *Dallachy, Thozet;* Walloon, *Bowman;* Suttor river, *Thozet.*

N. S. Wales. Clarence river, *Beckler;* New England, *C. Stuart.*

15. **E. serrulata,** *Reinw.; Boiss. in DC. Prod.* xv. ii. 25. A glabrous annual, the stems often hard at the base, erect or procumbent, sometimes exceeding 1 ft. Leaves opposite, very shortly petiolate, linear, obtuse, rather firm, more or less denticulate, oblique and sometimes slightly cordate at the base, $\frac{3}{4}$ to $1\frac{1}{2}$ in. long. Flower-heads in rather dense shortly pedunculate cymes in the upper axils, much like those of *E. Macgillivrayi,* the bracts small, the petal-like appendages of the involucral glands usually smaller than in that species.

N. Australia. Port Essington, *Armstrong;* Port Darwin, *Schultz, n.* 234; Escape Cliffs, *Hulse.*

Queensland. Keppel Bay and Thirsty Sound, *R. Brown;* Rockhampton, *O'Shanesy* (the latter specimen imperfect and doubtful).

This species extends over the sea-coasts of the Archipelago, from Timor to S. China.

16. **E. filipes,** *Benth.* A perennial allied to *E. Mitchelliana* and *E. Macgillivrayi,* with the habit and loose slender inflorescence of the former and the pubescent stems and foliage of the latter. Stems slender, erect or ascending, 6 in. to 1 ft. high, the ultimate branches filiform. Leaves opposite, oblong, the lower ones short and rather crowded, the upper ones narrow and distant. Flower-heads in rather loose irregular cymes, terminal or in the upper axils, the common peduncle as well as the petals filiform, the lower pair of floral leaves as long as the involucre, the upper ones reduced to small bracts. Involucres and capsules as small as in *E. Mitchelliana,* but the petal-like appendages to the glands broadly obovate or orbicular, the involucre sometimes hairy, the capsule glabrous. Seeds rugose.

N. Australia. Islands of the Gulf of Carpentaria, *R. Brown, Henne;* Fitzmaurice river, *F. Mueller.*

17. **E. pilulifera,** *Linn.; Boiss. in DC. Prod.* xv. ii. 21. A prostrate or ascending branched annual, attaining 1 to 2 ft., the branches hirsute with spreading hairs and often rufescent. Leaves opposite, shortly petiolate, from ovate to ovate-lanceolate or oblong, $\frac{3}{4}$ to $1\frac{1}{2}$ in. long, rather obtuse, usually denticulate, very oblique and narrow or semicordate at the base. Stipules small, linear, inserted on a transverse raised line. Flower-heads minute and numerous, crowded in head-like cymes on short peduncles in one axil of each pair of leaves or terminating the branches. Involucre about $\frac{1}{3}$ line long, the glands

small and entire, without the petal-like appendages of the preceding species. Capsule ½ to ¾ line diameter, more or less hairy. Seeds slightly rugose.

Queensland. Rockingham Bay, *Dallachy;* common about Rockhampton, *Dallachy, Bowman*, and others.—A common tropical weed of cultivation both in the New and the Old World.

SECT. 2. EREMOPHILA, *Boiss.*—Herbs. Leaves opposite or the lower ones and sometimes those of luxuriant branches alternate. Stipules very minute or obsolete. Involucral glands without petal-like appendages. Seeds carunculate.

18. **E. eremophila,** *A. Cunn. in Mitch. Trop. Austr.* 348. An erect glabrous hard annual or perhaps perennial, usually dichotomous and from 6 in. to 1 ft. high, but in some situations taller and more slender. Lower leaves and sometimes a few on lateral branches alternate, all the others opposite, petiolate, linear or rarely linear-lanceolate or oblong, more or less remotely serrulate or sometimes quite entire, ½ to above 1 in. long, leaving when fallen off a gland-like scar and sometimes a second one immediately above, the stipules often scarcely perceptible or quite wanting. Flower-heads solitary in one axil only of the pair of leaves, on a short pedicel. Involucre scarcely 1 line long, the glands (5 or 4) broad, reniform, without appendages. Bracts within the involucre few and short. Capsule about 2 lines long, glabrous and smooth, the stipes rather long. Seeds granular-rugose, with a rather large variously shaped caruncle.—Boiss. in DC. Prod. xv. ii. 70 and Euph. Ic. t. 43; *E. deserticola*, F. Muell. in Linnæa, xxv. 440.

N. Australia. Groote island, Gulf of Carpentaria, *R. Brown;* Greville island, Regent's river N.W. Coast and Goulburn island, *A. Cunningham;* Victoria river, *F. Mueller* (with large leaves often alternate).

Queensland. Broad Sound and Shoalwater Bay, *R. Brown;* Port Curtis, *M'Gillivray;* Three isles, Barrier reef Passage, and islands of Moreton Bay, *F. Mueller;* Rockingham and Edgecombe Bays, *Dallachy;* Rockhampton, *Bowman* and others; Warrego river, *Mitchell.*

N. S. Wales. Murray and Darling rivers and thence to the Barrier Range, *F. Mueller, Victorian and other Expeditions*; Clarence river, *Beckler;* New England, *C. Stuart.*

Victoria. Murray river, *F. Mueller.*

S. Australia. Spencer's Gulf, *R. Brown;* Flinders' range, Cudnaka, Akaba, &c., *F. Mueller;* Mount Searle, *Warburton;* Gawler range, *Sullivan;* Lake Gillies, *Burkitt;* Cooper's Creek, *Howitt's Expedition* (with thick succulent stems and small leaves).

W. Australia. Murchison river, *Oldfield, Drummond, 6th coll. n.* 88; Dirk Hartog's island, *A. Cunningham.*

2. CALYCOPEPLUS, Planch.

Flower-heads resembling single flowers. Involucre campanulate or open, 4-lobed, with or without small glands between the lobes. Male flowers collected in 4 clusters of 4 or more within the involucre and opposite its lobes, each one subtended and more or less embraced by a bract, the outer 1 or 2 much enlarged and enclosing the cluster; each flower consisting of an articulated filament (or pedicel with a short filament) without any perianth and a single anther, with 2 parallel cells opening longitudinally. Female flower solitary in the centre of the head, pedicellate. Perianth 4- or 6-lobed. Ovary sessile within the perianth, 3-celled, with 1 ovule in each cell. Style 3-cleft, the branches (or separate styles) entire or 2-lobed. Capsule separating into 3 2-valved cocci. Seeds smooth, carunculate.—Shrubs (or undershrubs) with opposite virgate branches usually leafless at the time of flowering. Leaves opposite or sometimes whorled, narrow, entire, very deciduous. Stipules very minute or obsolete. Flower-heads small, axillary or terminal.

The genus is limited to West Australia. It is united by Boissier with *Euphorbia*, but being confirmed by a second species showing the distinctive characters still more prominently, I have followed Baillon in retaining it, although not on the grounds advanced by him, for he appears to consider the flower-heads as single flowers in *Euphorbia* and as heads of flowers in *Calycopeplus*, a difference in which I am quite unable to concur.

Branches terete. Lobes of the involucre and of the female perianth wholly green or scarcely bordered . . . . 1. *C. ephedroides.*
Branches flat or angular. Lobes of the involucre and of the female perianth with broad white margins . . . . 2. *C. marginatus.*

1. **C. ephedroides,** *Planch. in Bull. Soc. Bot. Fr.* viii. 31. An erect glabrous shrub of several feet, with virgate ephedra-like terete branches, usually leafless at the time of flowering. Leaves on the young shoots opposite or in whorls of 3 or 4, petiolate, linear or lanceolate, entire, 1 to 1½ in. long, leaving after falling a persistent gland-like base, described sometimes as an adnate stipule. Flower-heads almost sessile, solitary or 2 together in each axil of the pair or whorl of leaves, within 2 to 4 broad short scale-like bracts. Involucre broadly campanulate, about 1½ lines diameter, with 4 rounded lobes, the margins of the 2 outer ones often dilated and whitish, 4 glands alternating with them very small and stipitate. Clusters of small flowers usually enclosed in one very broad bract, all the other bracts much smaller. Perianth of the female flowers very short, with 6 broad equal lobes. Styles free almost or quite from the base. Capsule glabrous.—*Euphorbia paucifolia*, Klotzsch in Pl. Preiss. i. 174; Boiss. in DC. Prod. xv. ii. 175, and Euph. Ic. t. 120; *Calycopeplus paucifolius*, Baill. Adans. vi. 319.

W. Australia. Swan river, *Drummond*, 1*st coll. n.* 669: Canning river, *Preiss.*

2. **C. marginatus,** *Benth.* Branches opposite, rigid, virgate and almost leafless as in *C. ephedroides*, but either flattened or angular as in *Amperea spartioides.* Leaves very few, small, linear, very deciduous. Stipules minute. Flower-heads pedunculate, solitary in the forks or at the ends of the branches or on short axillary branches bearing a single pair of bracts or small leaves. Involucre deeply divided into 4 broadly ovate lobes not 1 line long, with broad white petal-like margins and without the intervening glands of *C. ephedroides.* Clusters of male flowers like those of *C. ephedroides*, but very short, not exceeding the involucre and at least 2 of the outer bracts broad and enclosing the cluster one within

the other. Pedicel of the female flower short, the perianth deeply divided into 4 broad white-bordered lobes like those of the involucre but longer, quite as long as the ovary at the time of flowering, persisting under the fruit, and the edges usually crenulate. Capsule ovoid, glabrous, nearly 3 lines long. Styles united in a short slender column at the base, spreading upwards, acute and entire. Seeds smooth, caruncular.

W. Australia. Towards Cape Riche, *Drummond, 5th coll. n.* 213.

Ephedra arborea, F. Muell. Syst. Arr. Pl. Carpentaria, 14 (App. Journ. Lanesborough Exped.), referred to under doubtful species by Parlatore in DC. Prod. xvi. ii. 359, from the foot of Newcastle Range in North Australia, appears to me most likely to be a third species of *Calycopeplus* with the sessile flower-heads of *C. ephedroides* and the angular branches of *C. marginatus*. The flower-heads in the specimens are, however, all so deformed by grubs that their true structure cannot be ascertained.

TRIBE 2. STENOLOBEÆ.—Flowers distinct (not enclosed in a calyx-like involucre), both sexes with a perianth. Embryo linear, the cotyledons not at all or scarcely broader than the radicle. Shrubs often heath-like, with entire coriaceous leaves or rarely herbs with small membranous leaves.

The principal character which distinguishes the two principal divisions of this small exclusively Australian tribe, from the tribes of *Phyllantheæ* and *Crotoneæ* respectively, that derived from the embryo, is probably constant, although rarely to be ascertained in herbarium specimens. The tribe is, however, natural in character and well-marked geographically. A few species of *Phyllanthus* may approach *Pseudanthus* in habit, but readily distinguished by the ovary and capsule, and some species of *Beyeria* which have been confounded with *Croton* are as easily known by their stamens and style. The other genera have no representatives of their habit either in *Phyllantheæ* or *Crotoneæ*.

3. PORANTHERA, Rudge.

Flowers monœcious. Male fl.: Calyx petal-like, deeply divided into 5 segments imbricate in the bud. Petals 5, small or occasionally deficient, with a small gland at the base of each. Stamens 5, opposite the calyx segments; anther-cells completely divided, forming 4 distinct cells opening in terminal pores either quite distinct or at length confluent into 2. Rudimentary ovary of 3 small clavate or membranous bodies. Female fl.: Calyx and petals of the males. Stamens 0. Ovary broad, flat or concave at the top, 6-lobed, 3-celled, with 2 ovules in each cell. Styles 3, divided to the base into 2 linear branches. Capsule depressed globular, opening in 3 loculicidal valves or separating into 3 2-valved cocci, the valves usually separating more readily than the cocci, the whole falling away leaving a persistent clavate axis. Seeds strongly reticulate. Embryo terete, curved, the cotyledons not broader than the radicle.—Herbs either annual or after the first year becoming suffrutescent. Leaves alternate, membranous, entire, narrow or small. Stipules small, acuminate. Flowers small, white, in very short dense racemes almost contracted into heads, each one in the axil of a floral leaf or bract, the head-like racemes solitary or more frequently several in a terminal leafy corymb.

The genus is limited to Australia.

Leaves linear, sessile or shortly contracted at the base. Plants usually erect and becoming woody at the base.
Plant rarely above 6–8 in. high. Leaves with much revolute margins. Flower-heads solitary or in close corymbs.
Stipules jagged or toothed. Eastern plant . . . . 1. *P. ericifolia.*
Stipules entire. Western or S. Australian plant . . . . . 2. *P. ericoides.*
Plants mostly 1 ft. high or more. Leaf-margins less revolute. Flower-heads loosely corymbose.
Plant rather slender, rarely much above 1 ft. high. Western species . . . . 3. *P. Huegelii.*
Plant 1–3 ft. high. Eastern species . . . . 4. *P. corymbosa.*
Leaves obovate or linear-spathulate contracted into a long petiole.
Diffuse slender annual with small flowers . . . . 5. *P. microphylla.*

1. **P. ericifolia,** *Rudge in Trans. Linn. Soc.* x. 302, t. 22. A small undershrub, the erect stems usually under 6 in. and rarely nearly 1 ft. high, often minutely scabrous-pubescent. Leaves crowded, sessile, linear, with revolute margins, $\frac{1}{4}$ to $\frac{1}{2}$ in. long.. Stipules scarious, lanceolate, more or less lobed or jagged. Flowers small and numerous, the head-like racemes pedunculate, forming a dense terminal leafy corymb, the peduncles usually shorter than the subtending leaves, each flower pedicellate in the axil of a linear-spathulate bract. Male calyx under 1 line long, the segments whitish with dark streaks. Petals narrow, not half so long as the calyx. Female flowers fewer than the males, at the base of the raceme, the calyx and petals similar but rather smaller. Capsule about 1 line diameter.—Muell. Arg. in DC. Prod. xv. ii. 191.

N. S. Wales. Port Jackson, *R. Brown*, *Sieber*, *n.* 118, *Woolls*, and others.

2. **P. ericoides,** *Klotzsch in Pl. Preiss.* ii. 232. A small undershrub closely resembling *P. ericifolia*, with the same heath-like crowded leaves, and compact inflorescence sometimes reduced to a single head-like raceme at the end of the branch, but the stipules appear to be constantly quite entire. It is either quite glabrous or very slightly scabrous-pubescent.—Muell. Arg. in DC. Prod. xv. ii. 191; *P. piceoides*, Kl. in Pl. Preiss. ii. 232; *P. glauca*, Kl. l.c. 231, Muell. Arg. l.c. 192 (with rather more developed inflorescence); *P. arbuscula*, Sond. in Linnæa, xxviii. 567; *P. cicatricosa*, F. Muell. (ined.?)

S. Australia. Encounter Bay, *F. Mueller*.

W. Australia. King George's Sound to Swan river, *Baxter*, *Drummond*, 1*st coll. n.* 674, *Preiss*, *n.* 1227 *and* 2044, *Oldfield*, and others, and eastward to Cape Arid, *Maxwell*, the eastern specimens as well as the S. Australian showing generally a more developed corymb (*P. glauca*), but many even among these with the compact raceme solitary or nearly so.

3. **P. Huegelii,** *Klotzsch in Pl. Preiss.* ii. 231. Very near to the taller looser-flowered specimens of *P. ericoides* but more slender, usually taller and less woody at the base, some specimens coming very near to the smaller forms of the eastern *P. corymbosa*. Leaves more acute than in *P. ericoides*, less crowded, and not so much revolute. Flowers in a looser less corymbose panicle, the lower or all the branches much longer than the subtending leaves. Flowers and fruits the same as in that species.—Muell. Arg. in DC. Prod. xv. ii. 192.

W. Australia, *Drummond;* King George's Sound, *Preiss*, *n.* 2047, *F. Mueller*, and others; Blackwood and Kalgan rivers, *Oldfield;* eastward along the coast from Cape Arid to Cape Paisley, *Maxwell.*

4. **P. corymbosa,** *Brongn. in Duperr. Voy. Coq. Bot.* 219, t. 50, f. A. A much larger and more shrubby plant than the preceding species, the stems from 1 to 3 ft. high, not much branched. Leaves linear or sometimes linear-lanceolate, usually obtuse, $\frac{3}{4}$ to $1\frac{1}{2}$ or even 2 in. long, crowded in the lower part, more distant in the upper part of the stem, sometimes quite flat but the margins more frequently narrowly revolute. Stipules entire. Head-like racemes forming a terminal corymb, with the lateral branches much longer than the subtending leaves. Capsule rather larger and less lobed than in the preceding species.—Muell. Arg. in DC. Prod. xv. ii. 192; *P. linarioides,* Sieb. Pl. exs.; Baill. Etud. Euph. t. 25, f. 1 to 9; *P. arbuscula,* Sieb. Pl. exs. (the young inflorescence not yet fully developed); *P. ericifolia,* Hueg. Bot. Archiv. t. 8 (raised from Port Jackson seeds and not the western *P. Huegelii*).

N. S. Wales. Port Jackson to the Blue Mountains, *R. Brown, A. Cunningham, Sieber, n.* 116, 117, and many others; Illawarra, *A. Cunningham, Shepherd;* near Mount Imlay, Twofold Bay, *F. Mueller* (smaller specimens coming very near to *P. Huegelii*).

5. **P. microphylla,** *Brongn. in Duperr. Voy. Coq. Bot.* 218, t. 50 B. A low diffuse glabrous slender annual, sometimes becoming hard at the base, the branches ascending to from 3 to 6 in. Leaves from linear-spathulate to obovate, obtuse, tapering into a rather long petiole, flat or with the margins slightly recurved, $\frac{1}{4}$ to $\frac{1}{2}$ in. long or sometimes longer when narrow or scarcely 2 lines long when obovate. Flower-heads in small very leafy corymbs, the outer floral leaves usually exceeding the flowers. Flowers smaller than in the other species and sometimes minute, but their structure the same, the petals very variable, sometimes fully half as long as the calyx, sometimes some or all much smaller or entirely deficient. After the flowers have fallen the rhachis is often much elongated.—Muell. Arg. in DC. Prod. xv. ii. 193; Klotzsch in Pl. Preiss. ii. 230; Hook. f. Fl. Tasm. i. 343; *P. Drummondii,* Klotzsch, l.c. 231.

N. Australia. Port Darwin, *Schultz, n.* 54.

Queensland. Sandy Cape and Keppel Bay, *R. Brown;* Port Curtis, *M'Gillivray;* Endeavour river, *A. Cunningham;* Brisbane river, *Prentice.*

N. S. Wales. Port Jackson to the Blue Mountains, *R. Brown, A. & R. Cunningham,* and others; Bathurst, *A. Cunningham;* New England, *C. Stuart;* Hastings and Macleay rivers, *Beckler.*

Victoria. Melbourne, *Adamson;* Buffalo Range, *F. Mueller;* Wendu Vale, *Robertson;* Portland, *Allitt;* Wimmera, *Dallachy.*

Tasmania. A common weed from the sea-coast to the tops of the mountains, *J. D. Hooker.*

S. Australia. Mount Gambier, St. Vincent's and Spencer's Gulfs, Torrens river, *F. Mueller,* and others.

W. Australia. King George's Sound, *Preiss, n.* 2045, *F. Mueller;* Swan river, *Drummond,* 1*st. coll. n.* 675, *Preiss, n.* 2048; eastward to Cape Arid, *Maxwell.*

4. MICRANTHEUM, Desf.

(Caletia, *Baill.*)

Flowers usually monœcious. Male fl.: Perianth of 6 petal-like segments, the three inner ones often rather larger. No internal disk. Stamens either 3 opposite the outer perianth-segments, or 6 opposite all the segments, or 8 or 9 more irregularly inserted anthers with 2 parallel cells opening longitudinally. Rudimentary ovary small and lobed. Female fl.: Perianth of the males. Ovary 3-lobed and 3-celled, the lobes alternating with the inner perianth-segments and terminating in simple styles. Ovules 2 in each cell. Capsule 3-celled. Seeds 2 (or 1 by abortion) in each cell, oblong, smooth, carunculate. Embryo linear, straight, the cotyledons twice as large as and not much broader than the radicle.—Much branched heath-like shrubs. Leaves on very short petioles, small, narrow, entire, coriaceous, in alternate threes on each side of the stem, supposed to be the 3 leaflets of a compound leaf, but without any common petiole. Flowers small, solitary or few together in the upper axils, the males on short pedicels, the females usually sessile.

The genus is limited to Australia.

Stamens 3 . . . . 1. *M. ericoides.*
Stamens 6 . . . . 2. *M. hexandrum.*

1. **M. ericoides,** *Desf. in Mem. Mus. Par.* iv. 253, t. 14. A heath-like shrub of 1 to 2 ft., the branches and sometimes the foliage hirsute. Leaves or leaflets in threes, linear, flat or with slightly recurved margins, about 3 or rarely 4 lines long, glabrous and smooth or scabrous-pubescent. Flowers in the axils of floral leaves which are often solitary or only 2 together, the males on pedicels shorter than the leaves. Perianth-segments ovate or oval-oblong, about $\frac{3}{4}$ line long, the inner ones flat, the outer concave and smaller. Rudimentary ovary usually 3-lobed. Female flowers larger, the inner perianth-segments 2 lines long. Capsule smooth, about 3 lines long.—Muell. Arg. in DC. Prod. xv. ii. 195; *M. boroniaceum,* F. Muell. Fragm. i. 32.

Queensland. Port Bowen, *R. Brown;* Burnett river, *F. Mueller.*
N. S. Wales. Port Jackson, *R. Brown,* and others; Dogwood Creek, *Leichhardt.*

2. **M. hexandrum,** *Hook. f. in Hook. Lond. Journ.* vi. 283, *Fl. Tasm.* i. 342. An erect shrub, attaining in beds of rivers 8 to 10 ft., quite glabrous or the branches shortly pubescent. Leaves in threes, linear or oblong, $\frac{1}{4}$ to $\frac{1}{2}$ in. long and less spreading than in *M. ericoides.* Flowers larger than in that species, often one in the axil of each of the 3 floral leaves. Male perianth with the inner segments $1\frac{1}{2}$ lines long, the outer ones shorter. Stamens 6 to 9. Rudimentary ovary 3- or more-lobed. Female perianth 2 lines long. Ovary with 3 thick diverging lobes or styles, alternating with the inner perianth-segments as in *M. ericoides.*—*Caletia micrantheoides,* Baill. Etud. Euph. 554, t. 26, f. 1 to 18, Adans. vi. 326; *C. hexandra,* Muell. Arg. in DC. Prod. xv. ii. 194.

N. S. Wales. Cox's river, *A. Cunningham;* George river, *Macarthur;* Camden and Berrima, *Woolls;* Illawarra, *Shepherd.*

Victoria. Genoa river and Buffalo range, *F. Mueller.*

Tasmania. Port Dalrymple, *R. Brown;* moist shady ravines near Launceston, *Laurence, Gunn,* and others ; South Esk river, *Archer.*

This species has been very unnaturally distinguished as a genus from *M. ericoides* on account of a supposed difference in the position of the rudimentary ovary in the male flowers, which is slightly affected by the difference in the number of stamens. The relation of the carpels of the female flowers to the perianth-segments is the same in both species.

5. PSEUDANTHUS, Sieb.

Flowers monœcious. Male fl.: Perianth petal-like or rather rigid, of 6 segments nearly equal or one of the inner ones deficient or replaced by a long filament. Stamens 3, 6 or more (as many as 20), free or very shortly united at the base with the small rudimentary ovary. Anthers with 2 separate cells opening outwards in 2 valves. Female fl.: Perianth as in the males. Ovary 2- or 3-lobed, with thick diverging stigmatic lobes, 2- or 3-celled when very young with 2 ovules in each cell, but the dissepiments very early obliterated and all the ovules but one abortive. Capsule oblong, 1-celled, 1-seeded, opening in 4 or 6 valves. Seed oblong, smooth, carunculate. Embryo linear, the cotyledons longer but scarcely broader than the radicle.—Heathlike shrubs. Leaves opposite or alternate, small, coriaceous, obtuse, with thick margins and the midrib prominent underneath. Stipules small, subulate or with a broad base. Flowers small, sessile or shortly pedicellate in the upper axils, the males often several together, the females more sessile, solitary and alone or with one or more males.

The genus is endemic in Australia. It is generally described as having the stamens united in a central column without any rudimentary ovary. This I have only found to be the case in *P. polyandrus,* which I have therefore transferred to *Stachystemon.* In the typical *P. pimeleoides* the stamens are certainly free as figured by Endlicher, and I have always found in that as in all other true species of *Pseudanthus,* a central 3-lobed or sometimes 2-lobed rudimentary ovary. From *Micrantheum* and the typical *Caletia,* which I have as above restored to *Micrantheum, Pseudanthus* is readily distinguished by its solitary leaves, and by the capsule constantly 1-celled and 1-seeded by abortion. The stamens vary in number in both genera.

Stamens 6. Eastern species.
  Male perianth-segments linear, 5–6 lines long . . . . . 1. *P. pimeleoides.*
  Male perianth-segments ovate or oblong-lanceolate, 1 line long or less.
    Leaves mostly ovate, 1–2 lines long. Perianth 1 line long, inner stamens nearly as long.
      Stamens 6 . . . . 2. *P. ovalifolius.*
      Stamens 3 . . . . 3. *P. micranthus.*
    Leaves mostly ovate or orbicular, 1–1½ lines long. Perianth ½ line on a still shorter pedicel. Stamens very short . 4. *P. divaricatissimus.*
    Leaves mostly oblong-linear, 2–4 lines long. Perianth ½ line on a pedicel longer than itself. Stamens short . 5. *P. orientalis.*
Stamens 9–20. Western species.
  Male perianth pedicellate, the segments (3–6) all similar and about ½ line long. Branches minutely pubescent . . . 6. *P. virgatus.*
  Male perianth sessile, 5 of the segments ½ line long, the sixth filiform, red and 2–3 lines. Plant glabrous . . . . . 7. *P. nematophorus.*

1. **P. pimeleoides,** *Sieb. in Spreng. Syst. Cur. Post.* 25. An erect much-branched glabrous shrub 1 to 2 ft. high. Leaves alternate, scarcely petiolate, lanceolate or linear, acute, mostly 4 to 6 lines long, smooth. Male flowers clustered at the ends of the branches, shortly pedicellate and very conspicuous from their coloured yellowish linear perianth-segments 4 to 5 or even 6 lines long and exceeding the upper leaves. Stamens 6, closely clustered round a small central rudiment of the ovary, which is sometimes slightly raised and shortly adnate to one of the filaments, but the whole are generally quite free as figured by Endlicher, and I have never seen them united in a column as represented by Baillon. Female flowers few and inconspicuous. Perianth-segments lanceolate, acute, jagged on the margin, rigid, about $1\frac{1}{4}$ lines long, the ovate ones rather smaller. Ovary narrow, 3-lobed, the dissepiments very imperfect at the time of flowering and only one ovule fertilised. Capsule oblong, acute, $2\frac{1}{2}$ to 3 lines long, smooth, 1-seeded. —Muell. Arg. in DC. Prod. xv. ii. 196; Endl. Atakta, 11, t. 11, the analysis copied in Flora, 1832, ii. t. 4; Baill. Etud. Euph. t. 25, f. 16.

Queensland. *Burdekin Expedition;* Whitsunday island, *C. Moore;* Repulse Bay, *A. Cunningham.*

N. S. Wales. Port Jackson to the Blue Mountains, *Sieber, n.* 292 *and Fl. Mixt. n.* 528, and others; Shoalhaven, *C. Moore;* Illawarra, *Shepherd.*

2. **P. ovalifolius,** *F. Muell. in Trans. Phil. Inst. Vict.* ii. 66. A densely branched rigid low spreading shrub, quite glabrous or with a slight scabrous pubescence on the angles of the branches and midrib of the leaves. Leaves scattered, occasionally opposite but mostly alternate, on exceedingly short petioles, mostly ovate but varying from orbicular and 1 line diameter to oblong and 2 lines long, very obtuse, rigid, concave or complicate and often recurved at the end. Male flowers very shortly pedicellate in the upper axils; perianth white, the segments not very unequal, about 1 line long. Stamens 6 round a minute rudiment of the ovary, the three inner ones nearly as long as the perianth, the outer ones short and one sometimes very short. Female flowers sessile, the perianth-segments about $\frac{1}{2}$ line long, red with white margins. Ovary narrow. Fruit not seen.—*Caletia ovalifolia,* Muell. Arg. in Linnæa, xxxiv. 55 and in DC. Prod. xv. ii. 194; Baill. Adans. vi. 327.

Victoria. Mount Zero, Grampians, *Wilhelmi.*
Tasmania. Flinders Island, *Milligan.*

3. **P. micranthus,** *Benth.* A small rigid much-branched glabrous shrub, the branches slender, scarcely angular. Leaves very shortly petiolate, ovate, coriaceous, complicate, 1 to 2 lines long. Stipules very minute. Male flowers 2 or 3 together on turbinate pedicels of $\frac{1}{4}$ to $\frac{1}{2}$ line, within a very few small brown bracts. Perianth-segments about $\frac{1}{2}$ line long, the 3 inner ones rather larger and more petal-like than the outer. Stamens 3, nearly as long as the perianth and alternating with the inner segments. Rudimentary ovary very small, with 3 lobes alternating with the stamens. Female perianth-segments 5 in the

flowers examined, nearly 1 line long when full-grown. Ovary narrow, 2-celled or nearly 3-celled, tapering into as many long divergent styles or stigmatic lobes. Capsule obliquely ovoid, about 2 lines long, with a single seed.

S. Australia. Near Adelaide, *Whittaker*.

4. **P. divaricatissimus,** *Benth.* A depressed densely branched or divaricate shrub, quite glabrous or minutely scabrous-pubescent. Leaves scattered, mostly alternate, ovate or oblong, obtuse, 1 to 2 lines long, recurved at the end but scarcely concave. Flowers very small in the upper axils, the males on pedicels not longer than the perianth. Perianth-segments ovate, obtuse, about ½ line long, the inner ones rather larger than the outer. Stamens 6, all short, the inner filaments half as long as the perianth, the outer much shorter, the anthers rather large. Capsule oblong, 1-seeded, larger than in *P. orientalis.*—*Caletia divaricatissima,* Muell. Arg. in Linnæa, xxxii. 79 and in DC. Prod. xv. ii. 194.

N. S. Wales. King's Tableland, Blue Mountains, at an elevation of 3000 ft., *A. Cunningham;* Blue Mountains, *R. Cunningham.*

Var. *orbiculare.* Leaves orbicular, about 1 line diameter.

Victoria. Summits of the rocky mountains on the M'Alister river, Mount Macedon and granite rocks on the Yowaka river, *F. Mueller.*

5. **P. orientalis,** *F. Muell. Fragm.* ii. 14. A low densely branched glabrous shrub. Leaves scattered, mostly alternate, often crowded, oblong-linear or linear spathulate, 2 to 4 lines long, obtuse, thick, concave but often recurved at the end. Flowers in the upper axils very much smaller than in *P. ovalifolia,* the males on pedicels of nearly 1 line. Perianth "yellow," scarcely ½ line long, the segments nearly equal. Stamens 6, the inner ones shorter than the perianth, the outer ones very short. Female flowers sessile, sometimes in the same axil as the males, the perianth rather larger. Ovary still narrower than in *P. ovalifolia.* Capsule narrow-oblong, about 2 lines long, 1-seeded by abortion.—Muell. Arg. in DC. Prod. xv. ii. 197; *Caletia orientalis,* Baill. Adans. vi. 327; *C. linearis,* Muell. Arg. in Linnæa, xxxii. 79 and in DC. Prod. xv. ii. 194; Baill. Adans. vi. 327.

N. S. Wales. Sandhills near Port Jackson and Botany Bay, *A. and R. Cunningham, Mossman, F. Mueller;* Tweed river, *C. Moore;* also in *Leichhardt's* collection.

6. **P. virgatus,** *Muell. Arg. in Linnæa,* xxxiv. 56 *and in DC. Prod.* xv. ii. 197. A low shrub, the branches sometimes diffuse or straggling and 1 to 1½ ft. long, sometimes more erect and shorter, more or less pubescent or sprinkled with very short spreading hairs. Leaves opposite or on a few branches alternate, ovate oblong or almost linear, from 1½ lines long when broad to 3 or even 4 lines when narrow, obtuse. Flowers solitary or clustered, the males on pedicels of about ½ to 1 line. Perianth-segments 3 to 6, most frequently 4 and often unequal; ovate, obtuse or acuminate, entire or denticulate, rarely above ½ line long. Stamens varying from about 8 to 15, or perhaps more, clustered but always with a small rudiment of the ovary in the centre,

the inner ones or nearly all as long as the perianth, and all free or very shortly connate at the base. Female flowers sessile and solitary, alone or with the males. Capsule ovoid, 1-seeded, nearly 3 lines long when ripe.—*Chrysostemon virgatus*, Klotzsch in Pl. Preiss. ii. 232; *Pseudanthus occidentalis*, F. Muell. Fragm. i. 107; *Chorizotheca micrantheoides*, Muell. Arg. in Linnæa, xxxii. 76.

W. Australia. *Drummond*, 1*st coll. n.* 725; 5*th coll. n.* 222; near Bakewell, *Preiss*, *n.* 1230; Kalgan river, *Oldfield*, *F. Mueller*; Fitzgerald and Gardner rivers, *Maxwell*.

P. nitidus, Muell. Arg. in DC. Prod. xv. ii. 197, described from a specimen in Herb. DC., said to be from King George's Sound, *Cumming* (probably for *A. Cunningham*, as Cumming never was there), is unknown to me. The description agrees with *P. virgatus*, except that there are said to be scattered glands on the receptacle alternating with the stamens.

7. **P. nematophorus,** *F. Muell. Fragm.* ii. 14. A shrub of 1 to 2 ft., with the aspect of the larger narrow-leaved specimens of *P. virgatus*, but quite glabrous and more virgate in habit. Leaves opposite or alternate, linear or narrow-oblong, $\frac{1}{4}$ to $\frac{1}{2}$ in. long. Flowers minute, sessile and clustered in the upper axils or at the ends of the branches, the males usually 3 to 7 together; outer perianth-segments 3, broadly ovate, scarcely $\frac{1}{2}$ line long, the inner rather smaller and 2 only, the place of the third occupied by a red filament 2 to 3 lines long, which F. Mueller describes as a staminode and Mueller Arg. as a gland of the disk, but which may be rather an altered segment, having a stamen closely opposite to it like the 2 normal inner segments. Stamens 15 to 20 or even more, scarcely exceeding the perianth, with short thick filaments, the rudimentary ovary in the centre minute and 2- or 3-lobed. Female flower more closely sessile, and one only in the centre of the cluster, more tardily developed than the males and sometimes none in the cluster; perianth-segments narrow, acute, usually 4 only. Ovary 2-lobed. Capsule oblong, rather acute, about 2 lines long, 4-valved, 1-seeded.—Muell. Arg. in DC. Prod. xv. ii. 197.

W. Australia. Murchison river, *Oldfield*, *Drummond*, 6*th coll. n.* 89.

6. STACHYSTEMON, Planch.

Flowers monœcious. Male fl.: Perianth rather rigid, of 6 segments nearly equal and entire. Stamens indefinite, united in a central column, the anthers sessile on irregular protuberances or borne on distinct filaments, each with 2 separate distinct parallel cells, opening outwards in 2 valves. No rudimentary ovary. Female fl.: Perianth-segments broader and thinner than in the males and usually fringed. Ovary 2- or rarely 3-lobed, with thick diverging stigmatic lobes, 2- or rarely 3-celled when very young, but the dissepiments early obliterated and all the ovules but one abortive. Capsule (where known) oblong, 1-seeded. Seed oblong, smooth, caruncalate.—Heath-like shrubs. Leaves alternate, small, coriaceous, obtuse with the midrib or keel prominent underneath. Stipules very small. Flowers sessile or shortly pedicellate, clustered at the ends of the branches, usually several males with 1 to 3 females. Staminal column sometimes very long and cylindrical, and always exceeding the perianth.

The genus is endemic in Australia and closely allied to *Pseudanthus*, differing only in the stamens united in a column without any central rudiment of the ovary.

Stamens 10–25, the outer filaments almost free from the column, the inner or upper ones very short . . . . 1. *S. polyandrus.*
Stamens very numerous, all sessile on a long cylindrical column.
Leaves oblong, under ¼ in. long. Anthers closely packed on a column of less than ½ in. . . . . 2. *S. brachyphyllus.*
Leaves linear, mostly ½ in. long. Anthers not dense, on a column of ¾ in. or more . . . . 3. *S. vermicularis.*

1. **S. polyandrus,** *Benth.* A glabrous shrub, the stems in all the specimens seen erect, slightly branched, 6 to 8 in. high, arising several together from a woody stock. Leaves very shortly petiolate or almost sessile, oblong or linear, erect or slightly spreading, concave and keeled, rarely above ¼ in. long and the upper ones shorter. Flowers in terminal clusters, sessile or nearly so, usually 6 to 8 males with 1, 2 or more females. Male perianth "yellow" about 1 line long. Andrœcium from a little longer than, to twice as long as the perianth, consisting of from 10 to 25 stamens, of which the outer ones have their filaments free almost to the base, the inner ones more or less united in a column, and sometimes 2 or 3 anthers sessile or nearly so at the top of the column. Female perianth about ¾ line long, the segments broad thin and fringed on the margin. Ovary 2-celled. Capsule not seen.—*Pseudanthus polyandrus,* F. Muell. Fragm. ii. 153; Muell. Arg. in DC. Prod. xv. ii. 196; *P. chryseus,* Muell. Arg. in Flora, 1864, 486 and in DC. Prod. l.c.

W. Australia. *Drummond, 5th coll. n.* 221; Oldfield river, *Maxwell.*—This species connects in some measure *Stachystemon* with *Pseudanthus*, but I find the stamens always really united, without any trace of the rudimentary ovary, and have therefore transferred it to *Stachystemon.*

2. **S. brachyphyllus,** *Muell. Arg. in Linnæa,* xxxii. 76, *and in DC. Prod.* xv. ii. 198. A glabrous shrub, with erect stems of about 1 ft., closely resembling *S. polyandrus* except in the staminal column. Leaves oblong, under ¼ in. long. Flowers in terminal clusters, often as many females as males, the former closely sessile, the latter on pedicels of nearly 1 line. Male perianth of 6 lanceolate acute rather thick entire segments. Anthers exceedingly numerous, sessile, and closely packed, from the base to the end, on irregular gland-like protuberances of a cylindrical column of about 4 lines, the 2 cells of each anther quite separate, having the appearance of distinct 1-celled anthers. Female perianth-segments fully 1 line long, broad, acute or acuminate, much imbricate with fringed margins. Ovary with 2 long recurved stigmatic lobes. Capsules ovoid, acute, nearly 3 lines long, with a single seed.

W. Australia. *Drummond, 4th coll. n.* 95.—This species had been well distinguished in the Hookerian herbarium by Planchon, with the manuscript name of *S. brevifolius,* which Mueller Arg. in common fairness ought to have adopted.

3. **S. vermicularis,** *Planch. in Hook. Lond. Journ.* iv. 471, t. 15. An erect glabrous shrub, with virgate stems of about 1 ft. Leaves linear, acute or obtuse, not so thick as in the preceding species and 4 to 8 lines

long. Inflorescence of *S. brachyphyllus*, but the males usually more numerous on pedicels of 1 to 1½ lines, the females few and sessile. Male perianth of 6 very narrow acute segments, scarcely 1 line long. Staminal column linear, ¾ to 1 in. long, the anthers sessile on protuberances not so closely packed as in *S. brachyphyllus*. Female perianth-segments broader than in the males, acute, fringed, 1 or 2 of the inner ones sometimes reduced to slender filaments. Ovary deeply divided into 2 long lobes stigmatic inside in their upper half. Capsule not seen.—Muell. Arg. in DC. Prod. xv. ii. 198.

W. Australia. Swan river, *Drummond, 2nd coll. n.* 234, *Clarke.*

7. BEYERIA, Miq.

(Calyptrostigma, *Klotzsch;* Beyeriopsis, *Muell. Arg.*)

Flowers monœcious or rarely diœcious. Male fl.: Calyx of 5 rarely 4 broad segments, imbricate, concave and more or less petal-like. Petals as many, small or rarely exceeding the calyx, or fewer or more. Glands as many as petals and alternating with (or rarely opposite to) them. Stamens numerous, with very short filaments, crowded on a hemispherical receptacle, without any central rudimental ovary; anthers with 2 distinct parallel cells opening outwards longitudinally in 2 valves. Female fl.: Calyx-segments thicker and narrower than in the males, often enlarging after flowering. Ovary 3-celled, entire, with 1 ovule in each cell; stigma sessile, broad, entire or 3-lobed, peltate and flat or more or less calyptriform and almost conical, more rarely deeply 3-lobed. Capsule 3-celled, 3-seeded, or rarely oblique and 1-seeded by abortion. Seeds oblong, smooth, carunculate. Embryo narrow-linear, the cotyledons longer but scarcely broader than the radicle.—Shrubs often more or less glutinous, with alternate leaves, usually narrow, with recurved or revolute margins, white underneath with a close stellate pubescence. Flowers small, axillary, the males solitary or in clusters of 2 or 3, rarely racemose, the females solitary.

The genus is limited to Australia.

SECT. 1. **Eubeyeria.**—*Anthers twice as long as broad, adnate to an entire or scarcely lobed connective. Stigma entire or scarcely lobed.—Eastern species (one species also Western).*

Leaves oblong lanceolate or broadly linear, flat or with recurved margins.
 Leaves mostly 1–2 in. long. Capsule glabrous . . . . . . 1. *B. viscosa.*
 Leaves mostly 1–2 in. long. Capsule hairy . . . . 2. *B. lasiocarpa.*
 Leaves under 1 in., usually narrow. Capsule glabrous . . . 3. *B. opaca.*
Leaves narrow-linear almost terete with a hooked point . . . 4. *B. uncinata.*

SECT. 2. **Beyeriopsis.**—*Anthers short, the cells quite distinct, either adnate to a deeply 2-lobed connective or partially free with the connective more entire. Stigma entire or scarcely lobed.—Species all Western.*

Leaves ovate or lanceolate, rounded or cordate at the base. Petals glabrous or scarcely tomentose inside.
 Leaves broadly ovate, ½–1 in. long. Male pedicels slender, pubescent, longer than the perianth . . . . 5. *B. latifolia.*

Leaves ovate, not exceeding ½ in. Male pedicels very short thick and hirsute . . . . 6. *B. cygnorum.*
Leaves mostly lanceolate, the margins much revolute. Male pedicels slender, glabrous, longer than the perianth . . . 7. *B. cinerea.*
Leaves broadly ovate. Petals bearded inside . . . . 8. *B. cyanescens.*
Leaves narrow, sessile or tapering into a short petiole. Petals usually hirsute inside, at least at the base.
Leaves oblong, with revolute margins. Petals broad, rather large. Capsule equally 3-seeded . . . . 9. *B. lepidopetala.*
Leaves all linear, much revolute, scarcely showing any white undersurface.
Petals orbicular. Capsule oblique, 1-seeded.
Ovary and capsule 2-horned on the fertile side . . . . 10. *B. similis.*
Ovary and capsule without appendages . . . . 11. *B. brevifolia.*
Petals small and irregular, often wanting. Capsule equally 3-seeded . . . . 12. *B. Drummondii.*

SECT. 3. **Oxygyne.**—*Anthers of* Beyeriopsis. *Stigma deeply 3-lobed. Male flowers in a loose raceme.—Tropical species.*

Leaves ovate or lanceolate, tapering at the base, 1–1½ in. long . 13. *B. tristigma.*

SECT. 1. EUBEYERIA.—Anthers twice as long as broad, adnate to an entire or scarcely lobed connective. Species all strictly Eastern except the common *B. viscosa,* which extends also into W. Australia.

The differences in the anthers of the two sections, although as observed by Baillon not marked enough to justify the maintenance of distinct genera, when some species such as *B. viscosa* and *B. Drummondii* are otherwise so nearly allied, is yet very easily observed when the two are compared.

1. **B. viscosa,** *Miq. in Ann. Sc. Nat. ser.* 3, i. 350, t. 15. A tall shrub or tree, the flowering and fruiting branches usually viscid. Leaves from oval-oblong to oblong-lanceolate or broadly linear, mostly obtuse, tapering into a petiole, the margins often recurved, glabrous above, pale or white-tomentose underneath, usually from 1 to above 2 in. in length. Flowers axillary or lateral, on recurved pedicels of ¼ to nearly ½ in., the females solitary, the males often 2 or 3 together. Calyx-segments broad, coloured, about 2 lines long. Petals small. Anthers (about ½ line long) twice as long as broad, the cells wholly adnate to the entire connective. Female calyx smaller than the male, the thickened pedicel appearing to form part of it, the segments narrow with frequently small glands alternating with them. Ovary nearly globular, the large calyptriform sessile stigma closely appressed when young, raised and flat or concave when fully out. Capsule ovoid-globular, about 4 lines long, hard and glutinous, 3-celled and 3-seeded.—Muell. Arg. in DC. Prod. xv. ii. 202; *Croton viscosum,* Labill. Pl. Nov. Holl. ii. 72, t. 222; *Calyptrostigma viscosum* and *C. oblongifolium,* Klotzsch in Pl. Preiss. i. 176; *Beyeria oblongifolia,* Hook. f. Fl. Tasm. i. 339.

Queensland. Mount Flinders and Cape Porteous, *A. Cunningham;* Moreton Bay, *F. Mueller;* sources of Cape river, *Bowman;* Mount Wheeler, *Sutherland.*

N. S. Wales. Blue Mountains, *R. Cunningham;* Liverpool plains, *A. Cunningham;* Moore river, *Mitchell;* New England, *C. Stuart;* between the Lachlan and Bogan rivers, *L. Morton.*

Tasmania, *Labillardière;* Port Dalrymple and Derwent river, *R. Brown;* Abundant in shady places especially on the rocky banks of rivers, *J. D. Hooker.*

W. Australia. Swan river, *Drummond, Harvey;* Rottenest island, *Preiss, n.* 2387; Sharks Bay. *M. Brown.*

Var. *latifolia*, with broad but not long leaves, W. Australia, *Drummond, 5th coll. n.* 217. The leaves in the species vary exceedingly in size in breadth and in the whiteness of the under surface. In the Western specimens they are generally but not always broader than in the Eastern ones; in those from Tasmania and from the Blue Mountains they are large and long, and the flowers rather large; in the interior of N. S. Wales and in Queensland both leaves and flowers are smaller.

2. **B. lasiocarpa,** *F. Muell.; Muell. Arg. in Linnæa,* xxxiv. 59, *and in DC. Prod.* xv. ii. 201. A tall almost arborescent shrub, closely resembling the larger broader-leaved forms of *B. viscosa* and recently regarded by F. Mueller as a variety only, but originally distinguished by him as well as by Mueller Arg. and by Baillon as a species by the larger fruiting perianth, the capsule densely hirsute with hyaline hairs, and the larger stigma slightly raised on a central prominence of the capsule.—Baill. Adans. vi. 307.

N. S. Wales. Twofold Bay, *F. Mueller;* also a form with less hairy capsules intermediate in several respects between the typical *B. viscosa* and the *B. lasiocarpa* from New England, *C. Stuart, Leichhardt;* Hastings river, *Beckler;* but the specimens are scarcely sufficient to determine whether the two should be retained as species or varieties.

3. **B. opaca,** *F. Muell. in Trans. Phil. Soc. Vict.* i. 16, *and in Hook. Kew Journ.* viii. 210. An erect shrub of 1 to 2 ft., closely allied to *B. viscosa*, of which F. Mueller considers it a variety, with more slender stems and smaller leaves and flowers. Leaves rarely above 1 in. long and mostly ½ to ¾ in., oblong or almost linear rarely ovate, very obtuse, with revolute margins or nearly flat, pale or very white underneath. Flowers of *B. viscosa*, but smaller in every respect, and the anthers rather shorter in proportion to their breadth.—Sond. in Linnæa, xxviii. 565. and *B. ledifolia*, Sond. l.c. but not Klotzsch's plant; *B. Backhousii*, Hook. f. Fl. Tasm. i. 339; *Hemistemma? Leschenaultii*, DC. Syst. Veg. i. 444; *B. Leschenaultii*, Baill. Adans. vi. 307.

N. S. Wales. Euryalean scrub, *Fraser.*

Victoria. Sea-coast, Port Fairy, *Gunn;* Cape Otway, *F. Mueller;* Murray river, *Wilhelmi.*

Tasmania. Islands of Bass's Straits, *R. Brown, Gunn, Backhouse, Milligan, M'Gowan.*

S. Australia. Memory Cove, Port Lincoln and Kangaroo island, *R. Brown;* Murray river to Lakes Tyrrell and Albert, Point Nepean and Guichen Bay, *F. Mueller;* Rivoli Bay, *Robertson, F. Mueller;* Port Lincoln, *Wilhelmi.*

Var. *linearis.* Leaves narrow-linear, truncate or emarginate, 1–1½ in. long, with revolute margins.—Alps on the Macalister, *F. Mueller;* near Adelaide, *Blandowski.*

4. **B. uncinata,** *F. Muell.; Baill. Adans.* vi. 306. An erect viscid shrub, with slender slightly canescent branches. Leaves almost sessile, linear-subulate, recurved or hooked at the point, the margins so closely revolute as to be almost terete with a groove underneath, ½ to ¾ in. long. Male flowers unknown. Female flowers like those of *B. opaca*, but much smaller. Baillon further distinguishes it by the thinner calyx-segments, more free from the ovary, and by the very caducous stigma, distinctions however scarcely warranted by the specimens.

S. Australia? Murray desert, *F. Mueller.*

SECT. 2. BEYERIOPSIS.—Anthers short, the cells quite distinct, either adnate to a deeply 2-lobed connective, or partially free with the connective more entire. Species all Western.

5. **B. latifolia,** *Baill. Adans.* vi. 304. A more or less viscid shrub of 6 to 8 ft., the branches usually hoary with a stellate pubescence. Leaves broadly ovate, obtuse, flat or with recurved margins, sometimes almost cordate at the base, glabrous above, very densely and softly tomentose underneath, $\frac{1}{2}$ to 1 in. long, on petioles of 1 to 2 lines. Flowers monœcious, both sexes solitary in our specimens, on slender pedicels of $\frac{1}{2}$ in. or more. Calyx-segments about $\frac{3}{4}$ line long, broad, very concave and often slightly jagged. Petals broad and nearly as long as the calyx, glabrous or scarcely tomentose inside, alternating with the glands. Stamens very numerous, the anther-cells quite distinct, attached by the centre, with minute appendages at the end.—*Beyeriopsis latifolia*, Muell. Arg. in DC. Prod. xv. ii. 200.

W. Australia. Towards Cape Riche, *Drummond, 5th coll. n.* 216; Point Henry, *Oldfield.*

6. **B. cygnorum,** *Baill. Adans.* vi. 309. A much-branched more or less pubescent shrub. Leaves ovate or ovate-lanceolate, obtuse, rounded or slightly cordate at the base, the margins much revolute, 3 to 6 lines long, on petioles of 1 to 2 lines. Flowers apparently diœcious, axillary, the males solitary or several together on exceedingly short thick hispid pedicels. Calyx-segments about $\frac{3}{4}$ line long, very concave, almost saccate, hispid outside near the base. Petals shorter than the calyx and quite glabrous. Stamens numerous, the anthers short, with distinct cells. Female flowers solitary on pedicels rather longer than those of the males, the calyx quite glabrous. Petals small. Ovary tapering at the top, the peltate stigma not very large.—*Beyeriopsis cygnorum*, Muell. Arg. in DC. Prod. xv. ii. 199.

W. Australia. Between Moore and Murchison rivers, *Drummond, 6th coll. n.* 85.

7. **B. cinerea,** *Baill. Adans.* vi. 309. A much-branched erect shrub, more or less scabrous-pubescent or shortly hispid and possibly glutinous. Leaves lanceolate or ovate-lanceolate, with the margins often so much revolute as to appear linear, rather acute, rounded or almost cordate at the base, 3 to 5 lines long, on petioles of about 1 line. Flowers in the specimens seen all males, solitary or 2 or 3 together on filiform pedicels rather longer than the petioles. Calyx-segments orbicular, glabrous, about $\frac{3}{4}$ line long, each with a very prominent dorsal keel or protuberance near the base. Petals rather shorter than the calyx, orbicular, glabrous. Stamens numerous, anthers short, the cells distinct, tipped with small appendages, those of 3 central stamens rather larger and not tipped.—*Beyeriopsis cinerea*, Muell. Arg. in DC. Prod. xv. ii. 200.

W. Australia. Swan river, *Drummond, 1st coll. n.* 724.—This species at first sight much resembles *B. cygnorum*, but the flowers are different.

8. **B. cyanescens,** *Benth.* Branches short, spreading, grey but glabrous. Leaves crowded, broadly ovate or ovate-lanceolate; the

margins much revolute, ¼ in. long, stellate-pubescent when young but becoming glabrous above, white-tomentose underneath. Pedicels of the male flowers slender. Petals rigid, bearded inside.—*Beyeriopsis cyanescens*, Muell. Arg. in DC. Prod. xv. ii. 200.

W. or S. Australia. "Iles Steriles" (Recherche Archipelago?), *Herb. Mus. Par.* (*Muell. Arg.*). I have not seen this plant, and am not sure which of the islands off the south or west coast were provisionally designated under the above name, which appears never to have been published.

9. **B. lepidopetala,** *F. Muell. Fragm.* i. 230. A viscid shrub of several feet, the branches sprinkled with a scabrous stellate pubescence. Leaves on very short petioles, oblong or linear, ½ to 1 in. long, rather thick, obtuse with revolute margins, glabrous above, white-tomentose underneath. Flowers monœcious, both males and females solitary on rather slender pedicels of ¼ to ½ in., the female pedicel as in most species thickening upwards after flowering. Male calyx-segments nearly orbicular, about ¾ line diameter, much imbricate. Petals short, broad, hairy inside, alternating with 5 prominent disk-glands. Stamens numerous; anther-cells short, distinct, each with a small terminal appendage. Female calyx-segments rather narrower and more rigid than in the males; petals larger and fringed but yet shorter than the calyx. Stigma broadly but distinctly 3-lobed. Capsule ovoid-globular, nearly 3 lines long.—*Beyeriopsis lepidopetala*, Muell. Arg. in DC. Prod. xv. ii. 200.

W. Australia. Murchison river, *Oldfield.*

10. **B. similis,** *Baill. Adans.* vi. 309. An erect shrub, with virgate branches, 1 to 1½ ft. high, minutely viscid-pubescent or nearly glabrous. Leaves sessile or nearly so, linear, obtuse, with revolute margins, mostly ¾ to 1 in. long. Flowers monœcious, the males often 2 together in the axils on pedicels of about 1 line, the females solitary on pedicels at first very short but lengthening under the fruit to 5 or 6 lines. Male calyx-segments orbicular, very concave, almost saccate, about ¾ line long. Petals shorter, with a short tuft of hairs inside at the base. Anther-cells adnate to a connective deeply lobed in the outer ones, less so in the inner ones. Female calyx-segments thicker than in the males, the petals small and irregular. Ovary with 2 ovate or horn-like appendages on one side, which enlarges much after flowering, throwing the rather small peltate stigma quite obliquely to the other side. Capsule ovoid, oblique, 2½ to 3 lines long, usually 1-seeded on the horned side, the other cells remaining unenlarged.—*Beyeriopsis similis*, Muell. Arg. in DC. Prod. xv. ii. 200.

W. Australia. Between Moore and Murchison rivers, *Drummond, 6th coll. n.* 86.

11. **B. brevifolia,** *Baill. Adans.* vi. 309. A branching heath-like shrub, glabrous but apparently viscid. Leaves linear, obtuse, with revolute margins, tapering into a short petiole, ¼ to ½ in. or rarely longer. Flowers monœcious, both males and females on slender glabrous pedicels, often ½ in. long in the females, shorter in the males. Flowers of *B. similis* although in one of the two males examined I could not

find the petals, the ovary and capsule similarly oblique and one-seeded by abortion, but without the two horns or appendages of that species, and the stigma larger and broadly conical.—*Beyeriopsis brevifolia,* Muell. Arg. in DC. Prod. xv. ii. 201.

W. Australia. Towards Cape Riche, *Drummond, 5th coll. n.* 215.—The species very closely resembles both *B. similis* and *B. Drummondii.*

12. **B. Drummondii,** *Muell. Arg. in Linnæa,* xxxiv. 58, *and in DC. Prod.* xv. ii. 201. An erect slender slightly viscid shrub of 2 to 3 ft., the branches often flattened. Leaves narrow-linear, with revolute margins, resembling those of the linear variety of *B. opaca.* Flowers smaller than in that species, apparently diœcious, both sexes on slender pedicels of about 1 line, lengthening under the fruit to 2 lines. Male calyx-segments 5 or sometimes only 4, 1 to 1¼ lines long. Petals very small and glabrous. Disk continuous, irregularly crenate-lobed. Anthers scarcely longer than broad, the cells quite distinct and attached in the centre only. Female calyx-segments rigid, scarcely 1 line long. Capsule globular or slightly ovoid, about ¼ in. long but variable in size. —*Calyptrostigma ledifolium,* Klotzsch in Pl. Preiss. i. 176.

W. Australia. *Drummond, 5th coll. n.* 214, 220, *and* (*Suppl.?*) *n.* 13; Middle Mount Barren and Fitzgerald river, *Maxwell.*

SECT. 3. OXYGYNE. *F. Muell.*—Anthers of *Beyeriopis.* Stigma deeply divided into 3 narrow lobes. Male flowers in a loose raceme.

13. **B. tristigma,** *F. Muell. Fragm.* vi. 181. A small viscid shrub, glabrous except the underside of the leaves. Leaves resembling those of the common Olive, oval-elliptical or oblong-lanceolate, tapering into a short petiole, the margins recurved or revolute, 1 to 1½ in. long. Flowers monœcious, the males very small, few in a loose raceme of about 1 in. at the end of the branches, the pedicels 2 to 4 lines long. Calyx-segments 5, not ½ line long. Petals three times as long, ovate, obtuse, fringed inside at the base with a tuft of hairs. Stamens very numerous on a hirsute convex or hemispherical receptacle; anther-cells short, quite distinct. Female flowers on a pedicel of ½ in. or longer when in fruit, either solitary at the end of the branches or 1 or 2 at the base of the male raceme, smaller than the males at the time of flowering, but the calyx-segments enlarging under the fruit to nearly 1 line. Ovary capsule and seed of the genus, but the stigma divided to the base or nearly so into 3 narrow flat recurved lobes.

Queensland. Hinchinbrook island, Rockingham Bay, *Dallachy.* The racemose male flowers, the petals much longer than the calyx, and the divided stigma or style bring this species near to *Ricinocarpus,* but the stamens are entirely those of *Beyeria,* and the stigmatic lobes are closely recurved as in that genus, to which on the whole it appears to be the nearest related.

8. RICINOCARPUS, Desf.

(Rœperia, *Spreng.*)

Flowers monœcious. Male fl.: Calyx deeply divided into 4 to 6 usually 5 lobes or segments. Petals as many as calyx-lobes and usually longer, rarely deficient. Glands as many as petals and alternating with them. Stamens numerous, united in a central column without any rudimentary ovary; filaments shortly free; anthers reflexed, with 2 parallel cells opening outwards longitudinally in 2 valves. Female fl.: Calyx and petals of the males, very deciduous or rarely persistent. Ovary 3-celled, with 1 ovule in each cell. Styles 3, shortly united at the base, deeply divided into 2 branches. Capsule separating into 3 2-valved cocci. Seeds oblong, smooth, caruncular. Embryo (where known) linear, straight, the cotyledons longer but scarcely broader than the radicle.—Shrubs either glabrous or stellate-tomentose. Leaves alternate, entire, linear oblong or lanceolate, the margins recurved or revolute, usually pale white or tomentose underneath, without stipules. Flowers solitary or clustered, or the males rarely racemose, terminal or rarely apparently axillary from the reduction of the flowering branch, the females either alone or surrounded by or by the side of the male cluster or raceme. Pedicels usually subtended by small scale-like bracts, and often bearing a pair of bracteoles.

The genus is limited to Australia.

Glabrous plants with linear leaves. Flowers solitary or clustered.—Species all Western except *R. pinifolius.*
Bracteoles deciduous or none. Male calyx divided scarcely below the middle.
Female calyx deciduous. Capsule more or less tuberculate or echinate.
Capsule obtuse, as broad as long, densely muricate. Style-branches nearly terete, spreading or recurved. Eastern species . . . . 1. *R. pinifolius.*
Capsule obtuse, nearly twice as long as broad, slightly tuberculate. Styles of *R. pinifolius.*—Western species 2. *R. tuberculatus.*
Capsule acuminate, strongly tuberculate. Style-branches flattened and incurved . . . . 3. *R. cyanescens.*
Female calyx persistent and much enlarged . . . . . 4. *R. psilocladus.*
Bracteoles persistent. Male calyx divided nearly to the base, the female persistent but not much enlarged. Capsule quite smooth. . . . . 5. *R. glaucus.*
Glabrous plant with lanceolate leaves. Glands stipitate, adnate to the calyx segments.—Doubtful Tasmanian species . . . 6. *R. major.*
Branches and calyx tomentose. Flowers solitary or clustered. —Eastern or Northern species.
Leaves linear, the margins much revolute, under 1 in. long. Petals longer than the calyx. Ovary muricate . . . . 7. *R. Bowmanni.*
Leaves linear, the margins revolute, 1 to 2½ in. long. Petals as long as the calyx. Ovary tomentose . . . . 8. *R. ledifolius.*
Leaves linear, the margins much revolute, 1½ to 3 in. long. Petals much shorter than the calyx. Ovary tomentose . 9. *R. rosmarinifolius.*
Leaves lanceolate, flat, hoary on both sides, 1½ to 3 in. long. Petals much shorter than the calyx. Ovary tomentose . 10. *R. marginatus.*
Leaves narrow-oblong, 2 to 2½ in. long, the margins somewhat revolute . . . . 11. *R. speciosus.*
Male flowers several in a raceme.
Branches and inflorescence tomentose. Petals longer than the calyx. Ovary densely hirsute . . . . 12. *R. trichophorus.*
Branches and inflorescence glabrous, glutinous. Petals none. Ovary glabrous. Styles shortly 2-lobed . . . . 13. *R. muricatus.*

1. **R. pinifolius,** *Desf. in Mem. Mus. Par.* iii. 459, t. 22. An erect glabrous shrub of 2 to 3 ft. Leaves rather crowded, linear, mucronate or almost obtuse, with the margins revolute to the midrib, $\frac{3}{4}$ to $1\frac{1}{2}$ in. long, contracted into a short petiole. Flowers in a terminal cluster, usually 1 female with 3 to 6 males, but sometimes either the female or the males deficient, the pedicels $\frac{1}{2}$ to $\frac{3}{4}$ in. long said to bear a pair of bracteoles above the middle, but so deciduous that I have but very rarely seen any traces of them even when the bud is young, each pedicel embraced at the base to the subtending bract. Male calyx $1\frac{1}{2}$ to 2 lines long, divided to the middle or rather lower into 4 to 6 lobes. Petals white, usually about $\frac{1}{2}$ in. long. Female calyx more deeply divided and falling away very early. Capsule nearly globular, very obtuse, densely muricate, about $\frac{1}{2}$ in. long.—Muell. Arg. in DC. Prod. xv. ii. 205; Baill. Etud. Euph. t. 12, f. 39 to 44, and Adans. vi. 294; Hook. f. Fl. Tasm. i. 338; Endl. Iconogr. t. 124; *Rœperia pinifolia,* Spreng. Syst. iii. 147; *Echinosphæra rosmarinoides,* Sieb. Pl. Exs.; *Ricinocarpus sidæformis,* F. Muell. in several Herb., quoted by Baill. Etud. Euph. 344 as *R. sidæfolius.*

Queensland. Moreton island, *A. Cunningham, M'Gillivray, F. Mueller.*

N. S. Wales. Port Jackson, *R. Brown, Sieber, n.* 293, *Woolls,* and others; Currejong, Blue Mountains, *Miss Atkinson;* northward to Richmond and Clarence rivers, *Henderson;* southward to Illawarra, *A. Cunningham.*

Victoria. Port Philip and Melbourne, *Gunn, Adamson, F. Mueller,* and others; Wilson's Promontory and Brighton to Cape Howe, *F. Mueller* and others.

Tasmania. Abundant on sandhills near the sea on the north coast, *Gunn* and others.

2. **R. tuberculatus,** *Muell. Arg. in Linnæa,* xxxiv. 60, *and in DC. Prod.* xv. ii. 205. An erect glabrous shrub closely resembling *R. glaucus* in foliage, but evidently taller and more robust and more nearly allied to *R. pinifolius.* Leaves linear, with much revolute margins, usually $\frac{1}{2}$ to 1 in. long. Male flowers often 6 or more together in terminal clusters with usually one female. Male pedicels $\frac{1}{2}$ to 1 in. long, without bracteoles at the time of flowering. Calyx about $1\frac{1}{2}$ lines long, divided to the middle or scarcely lower into 5 lobes. Petals more than twice as long. Female pedicels shorter than the males, the calyx more deeply divided and falling away very early. Capsule about $\frac{1}{2}$ in. long, nearly twice as long as broad, obtuse, obscurely tuberculate.

W. Australia. Lucky Bay, *R. Brown, Drummond, 4th coll. n.* 84.

3. **R. cyanescens,** *Muell. Arg. in Linnæa,* xxxiv. 60, *and in DC. Prod.* xv. ii. 205. An erect glabrous much-branched shrub of 4 to 10 ft., the foliage assuming often a bluish tint in the dried state. Leaves linear, from under $\frac{1}{2}$ to near 1 in. long, rather thick, more spreading and the margins much less revolute than in *R. glaucus* and *R. tuberculatus.* Flowers often solitary, the males sometimes 2 to 4 together; pedicels rather longer than the leaves, without bracteoles at the time of flowering. Calyx scarcely $1\frac{1}{2}$ lines long, divided to about the middle. Petals nearly $\frac{1}{2}$ in. long, villous inside at the base. Glands glabrous. Female calyx rather larger than the male and more deeply divided, falling away

very early. Ovary contracted at the top into a very short neck, crowned by the deeply divided styles, which are somewhat flattened, spreading from the base and then incurved. Capsule ½ in. long and nearly as broad, acuminate, tuberculate.

W. Australia, *Drummond,* 4*th coll.* 86 *in part, and Suppl. n.* 15; sandy places along the coast from Esperance Bay to Cape Paisley and Port Malcolm, *Maxwell.*

This species is united with *R. glaucus* by Baillon, Adans. vi. 295; but besides the different foliage and the absence of bracteoles, the calyx and above all the ovary styles and capsule are very different.

4. **R. psilocladus,** *Benth.* A glabrous or scabrous shrub, resembling some forms of *R. glaucus,* but more rigid, the branches somewhat flexuose. Leaves linear, with much revolute margins, ¾ to 1½ in. long. Flowers solitary or the males 2 together terminating leafy branches, on pedicels of 3 to 4 lines without bracts at the time of flowering. Male calyx fully 2 lines long, divided to about the middle into 5 lobes. Petals twice as long. Female flower not seen, but the persistent fruiting calyx evidently enlarged with coriaceous lobes nearly 4 lines long. Capsule not seen except the persistent axis which is about the length of the calyx.—*Bertya gummifera β psiloclada,* Muell. Arg. in DC. Prod. xv. ii. 210, quoted by Baillon Adans. vi. 299, as *Bertya psiloclada.*

W. Australia, *Drummond,* 2*nd* or 3*rd coll. n.* 153. Mueller Arg. must have determined this originally from a very bad specimen, for the one in the Hookerian herbarium, which he has himself identified, although not good, is yet sufficient to show that it is a *Ricinocarpus,* and not a *Bertya.*

5. **R. glaucus,** *Endl. in Hueg. Enum.* 18. An erect glabrous shrub of 1 to 2 ft. Leaves very shortly petiolate, linear, with the margins revolute to the midrib, thicker than in *R. pinifolius* and usually shorter (½ to 1 in. long) but very much like those of a few forms of that species. Flowers often solitary, but usually 2 to 4 males together, with or without a female, on pedicels of ½ to 1 in., each subtended by a concave bract and bearing below the middle a pair of scale-like lanceolate bracteoles, which are often above 1 line long and usually persistent at the time of flowering. Male calyx about 1½ lines long, divided nearly to the base into rather acute segments, ciliate inside. Petals white, rather narrow, 4 to 5 lines long. Glands surrounded by a tuft of long hairs at the base of the petals. Female flowers either without petals or losing them very early. Ovary perfectly glabrous but surrounded at the base by hairs which appear all to proceed from the disk. Styles divided to the base into reflexed branches. Capsule glabrous and smooth, obtuse, about 4 lines long and nearly as broad, surrounded by the persistent but scarcely enlarged calyx.—Klotzsch in Pl. Preiss. ii. 229; Muell. Arg. in DC. Prod. xv. ii. 205; *R. undulatus,* Lehm. in Pl. Preiss. ii. 370.

W. Australia. Swan river, *Drummond,* 1*st coll., Preiss, n.* 2016, 2017, 2031, *Oldfield;* Champion Bay, *Oldfield;* King George's Sound, *R. Brown, Huegel;* Stirling range, *F. Mueller;* towards Cape Riche, *Drummond,* 5*th coll. n.* 220; Cape Riche and S.W. Bay, *Maxwell.*

6. ? **R. major,** *Muell. Arg. in Linnæa,* xxxiv. 59, *and in DC. Prod.* xv. ii. 204. A glabrous shrub. Leaves almost sessile, linear-spathulate or lanceolate, the margins slightly recurved, obtuse or apiculate, contracted at the base, above 1 in. long and 3 to 4 lines broad, those of the lateral branches smaller. Male flowers unknown. Female flowers on very short pedicels. Calyx nearly 3 lines long, persistent. Petals about as long, acute. Glands stipitate, adhering to the calyx-segments. Ovary glabrous and smooth; styles shortly united at the base in a narrow column.

Tasmania. *Verreaux in herb. DC.* I have not seen this plant, which must be very unlike any other *Ricinocarpus* in foliage, and the male flowers being unknown, the genus must be very uncertain. Perhaps also the station may be erroneous, for I am not aware that Verreaux visited any but well known parts of Tasmania.

7. **R. Bowmanni,** *F. Muell. Fragm.* i. 181. An erect bushy shrub, usually small, the branches and inflorescence tomentose. Leaves almost sessile, linear, the margins usually recurved to the midrib, smooth or scabrous-tuberculate, from under ½ in. to nearly 1 in. long. Male flowers "pink," in terminal clusters of 3 to 6, the pedicels about as long as the leaves. Calyx densely tomentose, the segments obtuse, about 2 lines long. Petals not twice as long. Glands flat, mostly 2-lobed and hairy. Female flowers solitary, alone or in the male cluster. Capsule 4 to 5 lines long, hirsute with rather long hairs which at length wear off. Styles divided to the base.—Muell. Arg. in DC. Prod. xv. ii. 206; *R. puberulus,* Baill. Etud. Euph. 344, name only, referred by Muell. Arg. without doubt to *R. Bowmanni,* but described by Baill. Adans. vi. 295, from a specimen with male flowers only, as doubtfully distinct.

Queensland. Upper Maranoa river, *Mitchell.*
N. S. Wales. Lower Macquarrie river, *Bowman;* desert north of Arbuthnot range, *A. Cunningham, Fraser;* also in *Leichhardt's* collection, with remarkably tuberculate, almost muricate leaves.

8. **R. ledifolius,** *F. Muell. Fragm.* i. 76. A shrub attaining 8 to 10 ft., the branches and inflorescence tomentose. Leaves shortly petiolate, oblong-linear or lanceolate, with recurved margins, white-tomentose underneath, 1 to 2 in. long. Male flowers in terminal clusters of 3 to 5, the pedicels 3 to 5 lines long, occasionally bearing 2 flowers. Calyx tomentose, about 2 lines long, the segments very obtuse. Petals about as long as the calyx. Glands more or less united in a crenulate ring. Female flowers solitary, alone or with the males, on a rather longer and stouter pedicel, and rather larger. Capsules about 2 lines long, densely stellate-tomentose.—Muell. Arg. in DC. Prod. xv. ii. 206; Baill. Adans. vi. 294.

Queensland. Burdekin river, *F. Mueller;* Darling Downs, *Dallachy;* Rockhampton, *O'Shanesy;* Herbert's Creek, *Bowman.*

9. **R. rosmarinifolius,** *Benth.* A slender twiggy shrub, the branches inflorescence and underside of the leaves hoary with a minute tomentum. Leaves narrow-linear, the margins much revolute, 1½ to

above 3 in. long, the upper or outer surface becoming quite glabrous, tapering into a very short petiole. Flowers clustered at the ends of the branches, the males small, on short pedicels, but not seen fully out. Calyx tomentose outside. Petals small, glabrous. Female flowers 1 or 2 in the male clusters, on pedicels at first short but lengthening to nearly 1 in. Calyx-segments above 1 line long, much narrower and more deeply divided than in *R. ledifolius.* Ovary tomentose, tapering into 3 bifid styles. Fruit not seen.—*Croton rosmarinifolium,* A. Cunn. Herb.

N. Australia. Montague and York Sounds, N.W. coast, *A. Cunningham.*

10. **R. marginatus,** *Benth.* A tall shrub, the branches inflorescence and both sides of the leaves hoary with a very short and close but soft tomentum. Leaves very shortly petiolate, lanceolate, rather obtuse, quite flat, but the margins slightly thickened and nerve-like, the primary veins very divergent from the midrib and prominent underneath, 1½ to nearly 3 in. long. Flowers in terminal clusters, the males rather numerous on pedicels of 1 to 2 lines, the females few in the same cluster, the pedicels lengthening to nearly 1 in. Male calyx-segments 5, about 1 line long, tomentose outside. Petals considerably shorter, glabrous, obovate. Female calyx rather larger. Ovary tomentose, tapering into 3 shortly bifid styles. Fruit not seen.—*Croton marginatum,* A. Cunn. Herb.

N. Australia. York Sound, N W. coast, *A. Cunningham.*

11. **R. speciosus,** *Muell. Arg. in DC. Prod.* xv. ii. 204. This species, only known from a fragmentary specimen in the Hookerian Herbarium in bud only, is evidently closely allied to *R. ledifolius,* but much larger in all its parts, with a more copious rufous tomentum on the branch and inflorescence. Leaves oblong-lanceolate, 2 to 2½ in. long, ½ in. broad. Inflorescence the same as in *R. ledifolius* but the buds larger.

N. S. Wales. Wilson river, Port Macquarrie, *Backhouse.*

12. **R. trichophorus,** *Muell. Arg. in Linnæa,* xxxiv. 60, *and in DC. Prod.* xv. ii. 206. A shrub, apparently tall, the branches and inflorescence tomentose, rusty when young, at length white. Leaves linear, the margins much revolute, 1 to 2 in. long, tomentose when young, becoming glabrous above when old. Male flowers 4 to 6 in a loose raceme of 1 in. or more, the pedicels about ¼ in. long, each in the axil of a small tomentose bract, without bracteoles. Calyx about 3 lines long, deeply divided into lanceolate acuminate lobes. Petals narrow, not twice as long. Female flower solitary at the base of the male raceme on a pedicel of 3 or 4 lines, the calyx rather larger than in the male. Ovary densely hairy. Capsule obtuse, about 5 lines long and broad, hirsute.

W. Australia. Towards Cape Riche, *Drummond, 5th coll. n.* 219.

13. **R. muricatus,** *Muell. Arg. in Linnæa,* xxxiv. 61, *and in DC. Prod.* xv. ii. 207. An erect shrub, glabrous except the underside of

the leaves, but glutinous. Leaves nearly sessile, linear, the margins closely revolute, $\frac{3}{4}$ to $1\frac{1}{2}$ in. long. Flowers small, the males rather numerous in a raceme of about 1 in. or rather longer. Pedicels 1 to nearly 2 lines long, each in the axil of a small subulate bract, without bracteoles. Calyx above 1 line long, divided nearly to the base into obtuse ciliate-fringed lobes. Petals none. Stamens about as long as the calyx. Female flowers solitary at the base of the male raceme on a pedicel of $\frac{1}{4}$ to $\frac{1}{2}$ in. Calyx-segments narrow more acute than in the males, persistent under the fruit but not enlarged. Ovary glabrous. Styles long, divided to about $\frac{1}{3}$ into 2 branches. Capsules about 4 lines long and broad, tuberculate and glutinous.

W. Australia. Towards Cape Riche, *Drummond, 5th coll. n.* 218 *and Suppl. n.* 85.

9. BERTYA, *Planch.*

Flowers monœcious. Male fl.: Perianth (calyx?) deeply divided into 5 petal-like segments, without inner petals or glands. Stamens numerous, united in a central column without any rudimentary ovary; filaments shortly free, spreading or recurved; anthers with 2 parallel cells opening longitudinally and outwards in 2 valves. Female fl.: Perianth of the males, but the segments usually smaller and narrower, and sometimes much enlarged round the fruit. Ovary 3-celled, with 1 ovule in each cell, but usually only one of the three fertilized. Styles 3, free or shortly united at the base, each one more or less deeply divided into 2 to 4 (usually 3) branches. Capsule ovoid or oblong, obtuse, or acute, usually 1-celled and 1-seeded by abortion. Seed oblong, smooth, carunculate; embryo (where known) linear, straight, the cotyledons longer but scarcely broader than the radicle.—Shrubs often glutinous, more or less stellate-tomentose, or glabrous. Leaves alternate, without stipules, the margins recurved or revolute, rarely flat, glabrous above when full-grown, tomentose or white underneath. Flowers axillary, solitary or few together, pedicellate or almost sessile, with 3 to 8 small bracts on the pedicel, either persistent and imitating a calyx (but imbricate and not uniseriate) or deciduous.

The genus is limited to Australia.

Leaves with revolute or recurved margins. Flowers sessile or on very short pedicels. Bracts 5 to 8, persistent.
  Leaves narrow, revolute to the midrib. Ovary glabrous, tapering at the top
    Fruiting perianth much enlarged.
      Young shoots densely hirsute . . . . 1. *B. gummifera.*
      Whole plant glabrous or nearly so . . . . 2. *B. pinifolia.*
    Fruiting perianth not enlarged. Plant glabrous . . . . . 3. *B. Cunninghamii.*
  Leaves narrow, revolute to the midrib. Ovary densely villous, tapering at the top. Fruiting perianth scarcely enlarged, or much shorter than the capsule.
    Flowers pedicellate, the perianth-segments scarcely above 1 line . . . . 4. *B. rosmarinifolia.*
    Flowers sessile or nearly so, the perianth-segments nearly 2 lines long . . . . 5. *B. Mitchelli.*

Leaves with the margins less revolute showing the underside.
Ovary obtuse.
Leaves narrow, 1 to 2 in. long. Fruiting perianth much enlarged. Ovary densely villous . . . . 6. *B. oleæfolia.*
Leaves ovate or orbicular, 2 to 4 lines long, almost bullate. Fruiting perianth slightly enlarged. Ovary stellate-tomentose . . . . 7. *B. rotundifolia.*
Leaves flat or with recurved margins. Flowers pedicellate. Bracts usually few, very deciduous or none.
Leaves narrow, 1 to 2 in. long. Pedicels 2 to 3 lines. Ovary villous . . . . 8. *B. pedicellata.*
Leaves ovate or oblong, about ½ in. long. Pedicels slender, 4 to 8 lines long. Ovary glabrous . . . . 9. *B. pomaderroides.*

1. **B. gummifera,** *Planch. in Hook. Lond. Journ.* iv. 473. An erect viscid shrub of several feet, resembling some of the narrower leaved specimens of *B. oleæfolia,* but more villous with a looser indumentum and the ovary different. Leaves linear, with the margins revolute to the midrib, 1 to nearly 2 in. long, thinly scabrous-tomentose outside. Flowers almost sessile, with 5 to 8 imbricate bracts as in *B. oleæfolia.* Male perianth about 3 lines long, glabrous, contracted into a very short turbinate pedicel above the bracts. Ovary quite glabrous, tapering at the top into the styles, which are united in a column of about 1 line, then spreading, each one divided into 2 or 3 branches more slender than in *B. oleæfolia.* Female perianth much enlarged after flowering, attaining ½ in. in length and enclosing the capsule.—Muell. Arg. in DC. Prod. xv. ii. 210.

N. S. Wales. Hunter's river and perhaps Croker's Range, *A. Cunningham;* but there is some confusion in his labels of this and of *B. rosmarinifolia* which he had regarded as one species.

2. **B. pinifolia,** *Planch. in Hook. Lond. Journ.* iv. 473. A tall shrub, glabrous or nearly so and apparently viscid. Leaves narrow-linear, sessile or nearly so, with the margins much revolute, mostly 1 to 1½ in. long. Flowers almost sessile, much larger than in *B. Cunninghamii.* Bracts thick, unequal, the outer ones linear, the inner ones mostly acuminate from a broad base. Male perianth-segments nearly 2 lines long. Ovary quite glabrous, tapering into a neck or united base of the styles of nearly 1 line, the styles divided to below the middle usually into 3 branches. Capsule enclosed in the enlarged perianth which, although apparently not full-grown in any of our specimens, is already above 3 lines long, oblong and obtuse.—Muell. Arg. in DC. Prod. xv. ii. 211.

Queensland. Brisbane river, *Fraser.*

3. **B. Cunninghamii,** *Planch. in Hook. Lond. Journ.* iv. 473. A shrub attaining 6 ft., with long rather slender viscid branches, otherwise glabrous. Leaves narrow-linear, with much revolute margins, mostly about ½ in. long, tapering at the base. Flowers small, solitary, on pedicels varying from ½ to 1 line. Bracts small, thick, narrow, very unequal, at or near the top of the pedicel. Perianth-segments broad, obtuse, 1 to 1½ lines long. Ovary ovoid, quite glabrous, contracted at

the top. Styles very shortly united or almost free, deeply divided into 2 or 3 branches. Fruiting perianth not enlarged. Capsule ovoid-oblong, 2 to 3 lines long.—Muell. Arg. in DC. Prod. xv. ii. 211.

N. S. Wales. Lachlan river, and frequent in the N.W. interior, *A. Cunningham;* New England, *C. Stuart.*

Victoria. Snowy river, *F. Mueller.*

4. **B. rosmarinifolia,** *Planch. in Hook. Lond. Journ.* iv. 473. A handsome bushy heath-like shrub, attaining 6 to 8 ft., the young branches and foliage clothed with a short close stellate tomentum wearing off with age. Leaves sessile or nearly so, linear, with the margins much revolute, mostly about ½ in. and rarely nearly 1 in. long. Flowers small, mostly solitary, on a peduncle of ½ to nearly 1 line below the bracts, which are small thick nearly equal about 6 in number and assuming more the appearance of a calyx than in any other species. Perianth-segments rather above 1 line, the female not much enlarged after flowering. Ovary densely villous. Styles divided to the base into 2 or 3 branches. Capsule ovoid-oblong, stellate-hirsute, 3 to 4 lines long and usually at least half as long again as the perianth-segments.—Muell. Arg. in DC. Prod. xv. ii. 210; Hook. f. Fl. Tasm. i. 339; *Croton rosmarinifolium,* A. Cunn. in Field, N.S. Wales, 355; *Ricinocarpus tasmanicus,* Sond. in Linnæa, xxviii. 562; *Bertya tasmanica,* Muell. Arg. in Linnæa, xxxiv. 63, and in DC. Prod. xv. ii. 211.

N. S. Wales. Cox's river, *A. Cunningham;* head of Macleay river, *C. Moore;* New England, *C. Stuart.*

Tasmania. Abundant on Nile rivulet and South Esk river, *Gunn;* Great Swanport, *Backhouse.*—I can see no difference between the Tasmanian and the N.S. Wales plants.

5. **B. Mitchelli,** *Muell. Arg. in Linnæa,* xxxiv. 63, *and in DC. Prod.* xv. ii. 210. An erect shrub of several feet, the young branches and foliage closely stellate-tomentose, becoming at length glabrous, nearly allied to *B. rosmarinifolia,* but stouter and more rigid. Leaves linear, with the margins much revolute, thick and coriaceous, ½ to 1 in. long. Flowers larger and more sessile than in *B. rosmarinifolia,* with 5 to 8 bracts. Perianth-segments rather broad in the males, narrower in the females, nearly 2 lines long, scarcely enlarged under the fruit. Ovary and styles of *B. rosmarinifolia,* of which this may prove to be a variety.—*Ricinocarpus Mitchelli,* Sond. in Linnæa, xxviii. 563.

N. S. Wales. Castlereagh river, *C. Moore;* Darling river, *Herb. F. Mueller;* Murray river, Mitta-Mitta, Lake Koorong, *F. Mueller;* Wimmera, *Dallachy.*

6. **B. oleæfolia,** *Planch. in Hook. Lond. Journ.* iv. 473. An erect shrub of 3 to 4 ft., the branches densely stellate-tomentose. Leaves very shortly petiolate, linear or oblong-lanceolate, with recurved margins, coriaceous, 1 to 2 in. long and sometimes above 2 lines broad, thinly scabrous, tomentose above, densely white-tomentose underneath. Flowers more or less diœcious, solitary in the axils and almost sessile, with 5 to 8 unequal bracts, the inner ones nearly 2 lines long. Male perianth with 5 oval-oblong lobes 2½ to 3 lines long, contracted into a

very short turbinate pedicel within the bracts. Female perianth-lobes narrower and more acute than the males. Ovary obtuse, densely hirsute. Styles free from the base and divided nearly to the base into 3 or 4 branches. Capsule enclosed in the enlarged perianth but not seen ripe.—Muell. Arg. in DC. Prod. xv. ii. 209.

Queensland. Shattered gullies, Mantuan Downs, *Mitchell.*
N. S. Wales. Wellington valley, *A. Cunningham.*

7. **B. rotundifolia,** *F. Muell. Fragm.* iv. 34. A rigid much-branched probably low shrub, the branches densely stellate-tomentose. Leaves shortly petiolate, ovate or orbicular, coriaceous, convex and almost bullate, 2 to 4 lines long, becoming glabrous above, white-tomentose underneath. Male flowers not seen. Female flowers nearly sessile, surrounded by a few minute tomentose bracts. Fruiting calyx about 1½ lines long, with membranous glabrous ovate-oblong segments. Capsule ovoid, 3 lines long, obtuse, retaining a scattered stellate tomentum. Styles 2- or 3-branched, united in a short column at the base.—Muell. Arg. in DC. Prod. xv. ii. 209.

S. Australia. Kangaroo island, *Waterhouse;* the specimens in leaf only, except one with a single attached capsule in Herb. F. Mueller.

8. **B. pedicellata,** *F. Muell. Fragm.* iv. 143. An erect shrub, with the habit and foliage nearly of *B. oleæfolia,* but the white tomentum very close and soon disappearing from the branches. Leaves linear or oblanceolate, with recurved margins, tapering into a short petiole, glabrous above, white-tomentose underneath, 1½ to 2 or even 2½ in. long. Flowers solitary or 2 or 3 together on a short common peduncle, the lower bracts closely complicate, ovate-lanceolate, acuminate, 3 to 4 lines long and very deciduous, smaller ones on the pedicels very deciduous or none. Perianths glabrous. Males only seen loose, with oblong petal-like segments 2 lines long, the staminal column and anthers entirely those of the genus. Female perianths on pedicels of 1 to 2 lines, with 3 or 4 small linear very deciduous bracts, the segments linear, acuminate, nearly 1½ lines long. Styles free from the base, 2- or 3-branched. Ovary tomentose-villous. Capsule narrow-ovoid, rather acute, 3 to 4 lines long, the surrounding perianth not at all enlarged in some specimens, somewhat longer and broader than when in flower in others.—Baill. Adans. vi. 298.

Queensland. Rockhampton, *Thozet.*

9. **B. pomaderroides,** *F. Muell. Fragm.* iv. 34. A much more slender spreading shrub than any other of the genus, the branches sometimes filiform, slightly and minutely stellate-tomentose or soon glabrous. Leaves shortly petiolate, ovate or oblong, very obtuse, rather thin, flat or with slightly recurved margins, glabrous above, white-tomentose underneath, from under ½ in. to fully 2 in. long. Pedicels of both sexes axillary, solitary, slender, 4 to 8 lines long, with 2 or 3 very deciduous bracts near the top, or sometimes perhaps without any. Male perianth-segments petal like, oblong, 1½ lines long, the staminal

column rather longer. Female perianth-segments subulate-acuminate, scarcely above 1 line long and not enlarged under the fruit. Ovary glabrous, tapering at the top. Styles united in a short column at the base, each one 2- or 3-branched. Capsule (not yet ripe) narrow, acute, 4 to 5 lines long.—Muell. Arg. in DC. Prod. xv. ii. 209; Baill. Adans. vi. 298; *B. oblongifolia*, Muell. Arg. in Flora, 1864, 471, and in DC. Prod. xv. ii. 209.

N. S. Wales. In the interior, *C. Stuart;* Bents Basin, near Port Jackson, *Woolls.*

10. MONOTAXIS, Brongn.

(Hippocrepandra, *Muell. Arg.*)

Flowers monœcious. Male fl.: Calyx of 4 or 5 imbricate spreading usually petal-like segments. Petals as many as calyx-segments and shorter or longer. Stamens twice as many as petals or fewer, the filaments distinct or very shortly united at the base, without any rudiment of the ovary; anthers with 2 distinct small almost globular cells, separated by a curved thick connective. Female fl.: Calyx and petals of the males. Ovary 3-celled, with 1 ovule in each cell. Styles 3, each one deeply divided into 2 fringed branches. Capsule globular or tridymous, without appendages, separating into 3 2-valved cocci. Seeds ovate or oblong, smooth, carunculate. Embryo linear, straight or slightly curved, the cotyledons much longer but scarcely broader than the radicle.—Herbs or undershrubs, usually small and glabrous. Leaves alternate, entire, flat or with recurved margins. Stipules very small. Flowers small, in dense head-like cymes, sessile or shortly pedunculate in the forks or at the ends of the branches between the last leaves, the flowers more or less pedicellate within the cymes, the males usually numerous, the females single in the centre or few. Bracts usually several, small and scalelike, subtending the pedicels or the outer ones empty.

The genus is endemic in Australia.

SECT. 1. **Eumonotaxis.**—*Flowers mostly 4-merous, calyx-segments almost valvate. Petals shorter than the calyx.*

Eastern species. — Flower heads pedunculate. Calyx-segments obtuse or scarcely acute.
 Erect annual, slightly branched. Leaves 1 to 2 in. long. Female flowers several in the head . . . . 1. *M. macrophylla.*
 Stems numerous from a thick perennial stock. Leaves under ½ in. Female flowers solitary in each head . . . . 2. *M. linifolia.*
Western species.—Flower-heads sessile. Calyx segments acuminate . . . . 3. *M. occidentalis.*

SECT. 2. **Hippocrepandra.**—*Flowers usually 5-merous; calyx segments distinctly imbricate. Petals longer than the calyx.—All Western species with sessile flower-heads.*

Leaves lanceolate, sessile
 Stems thick, rigid, about 1 ft. high. Leaves ½ to 1 in. . . . 4. *M. lurida.*
 Stems slender, about ½ ft. high. Leaves under ½ in. . . . 5. *M. megacarpa.*

Leaves linear-lanceolate, not above ½ in., tapering into a short petiole. Stems slender . . . . 6. *M. gracilis.*
Leaves narrow-linear, the margins closely revolute, not above ½ in. Stems slender. Stipules subulate, persistent . . . . 7. *M. grandiflora.*

SECT. 1. EUMONOTAXIS.—Flowers mostly 4-merous. Calyx-segments almost valvate. Petals shorter than the calyx.

1. **M. macrophylla,** *Benth.* An erect glabrous slightly branched annual of about 1 ft. Leaves opposite or alternate, on rather long petioles, oblong or ovate-oblong, obtuse, entire, thin and flat, 1 to 2 in. long. Flower-heads (or dense cymes) rather larger than in *M. linifolia* shortly pedunculate above the last leaves, containing several female flowers intermixed with or surrounded by numerous males. Male calyx of 4 very slightly imbricate petal-like segments of about ¾ line. Petals minute. Stamens 7 or 8. Ovary of the females only seen in very young bud, the styles then short and involute, and of the capsules the specimens only show the persistent axis, about 1 line long, from which the cocci have fallen away.

Queensland (or N. S. Wales ?). Summit of Mount Danger near Moreton Bay, *A. Cunningham.*—The plant has much of the aspect of some forms of *Euphorbia geniculata.*

2. **M. linifolia,** *Brongn. in Ann. Sc. Nat. ser.* 1, xxix. 387 *and in Duperr. Voy. Coq. Bot.* 224, *t.* 49 B. A small glabrous undershrub, with a thick woody stock or rhizome, and numerous herbaceous wiry ascending or diffuse stems of 6 in. to 1 ft. Leaves not numerous, opposite or alternate, the lower ones or those of the barren stems small, ovate or cuneate and sometimes 2- or 3-toothed, the others quite entire, oblong lanceolate or linear, usually acute and tapering into a short petiole, green on both sides and rarely above ½ in. long, without any or with minute tooth-like stipules. Flower-heads (or dense cymes) shortly pedunculate above the last leaves, consisting of about a dozen males each on a pedicel of ½ to ¾ line surrounding a single almost sessile female. Bracts minute and scale-like round each pedicel and a few ovate empty ones forming a sort of involucre. Male calyx of 4 or very rarely 5 ovate obtuse or scarcely acute segments of about ½ line, very slightly imbricate or sometimes perhaps quite valvate in the bud. Petals much shorter, broadly cordate. Stamens nearly as long as the calyx, usually 8 but sometimes 7 only. Female flower rather larger than the males. Styles 3, divided to the base into 2 fringed branches. Capsule glabrous, about ¼ line long, 3-celled.—Muell. Arg. in DC. Prod. xv. ii. 212; Baill. Adans. vi. 291; *M. tridentata,* Endl. Atakta, 8, t. 8.

N. S. Wales. Port Jackson, *R. Brown, A. and R. Cunningham, F. Mueller,* and many others.

Occasionally when the head has no female flower the rhachis grows out forming a short irregular raceme. The arrangement of the flowers in this species shows an approach to that of *Euphorbia.*

3. **M. occidentalis,** *Endl. in Hueg. Enum.* 19. An undershrub with the habit sometimes nearly of *M. linifolia,* but usually smaller, more

branched, and sometimes appearing annual (flowering the first year ?). Leaves ovate oblong or linear-cuneate, acute or almost obtuse, entire, 1 to 4 lines long, tapering into a short petiole. Flower-heads or cymes sessile in the forks or within the last leaves, with usually rather fewer flowers than in *M. linifolia*. Male calyx-segments 4 as in that species and scarcely imbricate, but always acuminate. Female flowers and fruits of *M. linifolia*.—Klotzsch in Pl. Preiss. ii. 229; *M. cuneifolia*, Klotzsch l.c. i. 176; *M. linifolia* var. *occidentalis*, Muell. Arg. in Linnæa, xxxiv. 63 and in DC. Prod. xv. ii. 212; Baill. Adans. vi. 292; *M. porantheroides*, F. Muell. in several Herb.

W. Australia. King George's Sound and adjoining districts, *Huegel*, *Oldfield*, *F. Mueller*; Swan river, *Preiss*, *n.* 1222.

SECT. 2. HIPPOCREPANDRA.—Flowers usually pentamerous. Calyx-segments distinctly imbricate. Petals longer than the calyx.

4. **M. lurida,** *Benth.* Stems, from a thick woody base or rhizome, erect, simple or slightly branched below the inflorescence and there often almost umbellate, much more rigid and thicker than in any species except *M. megacarpa*, about 1 ft. high. Leaves sessile, oblong-lanceolate, with recurved or revolute margins, thicker than in the other species, ½ to 1 in. long. Flower-heads or dense cymes closely sessile in the forks or within the last leaves. Male calyx-segments 5, about ½ line long, imbricate. Petals unguiculate, with a broadly cordate-ovate lamina at least 1 line long. Stamens 10 or 9, on slender filaments. Female calyx-segments obtuse. Petals rather smaller than in the males, contracted at the base but not unguiculate. Styles deeply divided into 2 fringed lobes. Capsule globular, about 2 lines long.—*Hippocrepandra lurida*, Muell. Arg. in Linnæa, xxxiv. 61 and in DC. Prod. xv. ii. 207 (erroneously referred by Baillon to *M. megacarpa*); *M. Oldfieldii*, Baill. Adans. vi. 293.

W. Australia. Murchison river, *Oldfield*, *Drummond*, *6th coll. n.* 37.—The plant has something of the aspect of some forms of *Euphorbia eremophila*.

5. **M. megacarpa,** *F. Muell. Fragm.* iv. 143. Stems, from a perennial base, erect or ascending, simple or slightly branched below the inflorescence, and there sometimes umbellate as in *M. lurida*, but much more slender and mostly 6 to 8 in. high. Leaves sessile, lanceolate or linear, with recurved margins, under ½ in. long. Flower-heads or clusters closely sessile, usually 2 or 3 females with about twice as many males, all on slender pedicels longer than in *M. lurida*, but the whole structure and the peculiar form of the petals precisely as in that species of which this may possibly prove a variety, however different its aspect.—Baill. Adans. vi. 293, but not the synonym adduced.

W. Australia. Murchison river, *Oldfield*, and a single specimen from *Drummond* in Herb. F. Mueller; Boxvale, 50 miles E. from York, *Miss Wells*.

6. **M. gracilis,** *Baill. Adans.* vi. 293. An erect branching slender undershrub of ½ to 1 ft. Leaves linear-lanceolate, contracted into a very short petiole, the margins more or less recurved, rarely above ½ in.

long. Flower-heads or clusters sessile within the last leaves, consisting of 1 to 3 females and about twice as many males, on short filiform pedicels. Male calyx-segments 5, rather acute, scarcely above ½ line long. Petals more than twice as long, contracted into a short broad claw and obscurely auriculate above the claw. Stamens 10 or fewer by abortion. Female calyx-segments more acuminate than in the males. Petals less distinctly unguiculate. Capsule nearly globular, deeply 3-furrowed.—*Hippocrepandra gracilis*, Muell. Arg. in Linnæa xxxiv. 62, and in DC. Prod. xv. ii. 207.

W. Australia, *Drummond*, 3*rd coll. n.* 18.

7. **M. grandiflora,** *Endl. in Hueg. Enum.* 19. A small undershrub, with a woody base, and numerous erect or ascending slender wiry stems, rarely above 6 in. high. Leaves narrow-linear, rather rigid, mucronate, with closely revolute margins, contracted into a very short petiole, rarely above ½ in. long. Stipules more prominent than in any other species, almost setiform, often 1 line long. Flower-heads or clusters closely sessile in the forks or between the last leaves, consisting sometimes of several males only, sometimes of 1 or 2 females with or without a few males. Males on slender pedicels. Calyx-segments 5, acute, nearly 1 line long. Petals about half as long again, shortly unguiculate and biauriculate above the claw. Female flowers on short thick pedicels. Calyx-segments acuminate. Petals narrower than in the males, almost acute, tapering at the base, but without any distinct claw.—Klotzsch in Pl. Preiss. ii. 230; *M. ericoides*, Klotzsch in Pl. Preiss. i. 177; *Hippocrepandra ericoides*, Muell. Arg. in Linnæa xxxiv. 62 and in DC. Prod. xv. ii. 208.

W. Australia. Swan river, *Drummond*, 1*st coll. and* 3*rd coll. n.* 19, *Preiss*, *n.* 1218; Hay district, *Preiss*, *n.* 2142; N. of Stirling Range, *F. Mueller*.

M. bracteata, Nees in Pl. Preiss. ii. 230 (*Hippocrepandra Neesiana*, Muell. Arg. in Linnæa xxxiv. 62, and in DC. Prod. xv. ii. 208, *Monotaxis Neesiana*, Baill. Adans. vi. 293), is founded on bad specimens of this plant from near York, *Preiss*, *n.* 1219, with shorter leaves than usual, but all belonging to lateral branches, the longer ones of the main stems having all fallen away.

11. AMPEREA, A. Juss.

Flowers monœcious or diœcious. Male fl.: Perianth campanulate, somewhat petal-like, 3- to 5-lobed, without inner petals. Stamens twice as many as perianth-lobes or fewer, the filaments free or shortly united at the base, without any rudimentary ovary and sometimes surrounded by as many small glands as perianth-lobes; anthers 2-celled or 1 or more of the outer ones 1-celled, the cells distinct, globular or ovoid, parallel, opening longitudinally in 2 valves, the connective usually tipped with a small gland. Female fl.: Perianth more deeply divided than the males into 5 rarely 4 rather rigid lobes, persistent but scarcely enlarged under the fruit. Ovary 3-celled, with 1 ovule in each cell. Styles 3, more or less deeply divided into 2 branches. Capsule ovoid, crowned by a ring of 6 erect tooth-like appendages, each on the back of one of the valves, separating into 3 2-valved cocci. Seeds ovoid-oblong, smooth, carunculate. Embryo, where known, linear, slightly curved, the cotyledons longer but scarcely broader than the radicle.—Perennials or undershrubs with a hard often woody base or rhizome, the stems erect or decumbent, usually rigid, sometimes almost or quite leafless. Leaves when present alternate, linear, either entire with closely revolute margins, or flat and then sometimes toothed. Stipules small brown and scarious. Flowers very small in small axillary closely sessile tufts, surrounded by scarious bracts, the males usually numerous, the females few or solitary, all on very short pedicels or almost sessile. Capsule small.

The genus is endemic in Australia, and the species all Western except *A. spartioides.*

Male flowers 3-merous.
  Stems slender, diffuse. Leaves flat, oblong-lanceolate, petiolate . 1. *A. protensa.*
  Stems twining, almost leafless . . . . 2. *A. volubilis.*
Male flowers 4–5-merous.
  Leaves linear with revolute margins.
    Stems terete. Flowers very numerous, in dense tufts with deeply subulate fringed bracts.
      Leaves distant. Stems slender but rigid . . . . 3. *A. micrantha.*
      Leaves crowded, rigid. Stems thick, ascending. Stipules and bracts very conspicuous . . . . 4. *A. conferta.*
    Stems angular or compressed . . . . 5. *A. ericoides.*
  Stems leafless or nearly so when in flower, erect, rigid, compressed or 3 angled . . . . 6. *A. spartioides.*

A. subnuda, Nees in Pl. Preiss. ii. 229, is *Gyrostemon brachystigma*, F. Muell. above v. 146.

1. **A. protensa,** *Nees in Pl. Preiss.* ii. 229. Stems, from a perennial base, numerous, slender, diffuse, somewhat compressed but scarcely angular, 6 in. to 1 ft. long. Leaves oblong-lanceolate, acute, contracted into a petiole, flat or nearly so, mostly about $\frac{1}{2}$ in. long. Stipules broad, deeply fringed. Male perianths pedicellate, about $\frac{3}{4}$ line long, divided to about the middle into 3 almost valvate lobes. Stamens usually 5, nearly as long as the perianth, 1 or 2 outer ones with only 1 cell to the anthers, the other anthers 2-celled. Female flowers sessile, usually 1 in the cluster surrounded by a few males. Perianth divided to the base into 5 segments. Styles divided to the base but all connected by a ring or falling off together. Capsule about 1 line long, the dorsal teeth or protuberances less prominent than in other species and very obtuse.—Muell. Arg. in DC. Prod. xv. ii. 213.

W. Australia, *Drummond, 4th coll. n.* 85; Swan River, *Preiss, n.* 1214 (*partly*).

2. **A. volubilis,** *F. Muell. Herb.* Stems elongated, twining, terete or nearly so, leafless or nearly so at the time of flowering. Leaves very few on some of the young shoots, linear, rigid, with much revolute margins, $\frac{1}{4}$ to $\frac{1}{2}$ in. long, but often abortive from the first. Stipules small, ovate or lanceolate, mostly entire. Male flowers in dense clusters, the outer bracts ovate mostly entire, the inner ones very small. Perianth deeply 3-lobed. Stamens 6 or fewer, the outer ones often, but not always, with 1-celled anthers. Female flowers solitary in our

specimen. Perianth deeply 4- or 5-lobed, the lobes rigid and acute. Capsule about 1 line long, the tooth-like appendages small and acute.

W. Australia, *Drummond, 4th coll n.* 87 *and 5th coll. Suppl. n.* 60.

3. **A. micrantha,** *Benth.* Stems from a perennial base several, usually simple, from under 6 in. to above 1 ft. high, rather slender but rigid, terete or nearly so. Leaves linear, with revolute margins, about $\frac{1}{2}$ in. long, less rigid than in *A. ericoides.* Stipules ovate or lanceolate, deeply fringed with setiform lobes. Flowers very small and numerous, crowded in all the axils in small dense tufts of brown fringed bracts, all males in our specimens. Pedicels very short. Perianth broadly campanulate, scarcely $\frac{3}{4}$ line diameter, 5- or rarely 4-lobed. Stamens varying from 2 to 5. Female flowers not seen.

W. Australia, *Drummond, 3rd coll. n.* 19; also in Herb. F. Muell. unnumbered or with the n. 223, but not that number of other collections where it is given to *Phyllanthus scaber;* Busselton, *Priess.*

4. **A. conferta,** *Benth.* Stems from a woody base, apparently decumbent, thick and hard, terete, 8 to 10 in. long in our specimens. Leaves crowded and clustered in the axils, linear with revolute margins, coriaceous, mostly $\frac{1}{2}$ in. long or rather more. Stipules broad, brown, fringed. Flowers all female in our specimens, sessile and almost concealed in dense tufts of brown scarious setaceously fringed bracts. Calyx or perianth deeply divided into 4 or 5 unequal lanceolate mucronate or aristate segments. Ovary crowned by 6 long acute dorsal appendages. Styles 3, very shortly united at the base, curved reflexed and protuding between the appendages, shortly bifid at the end. Capsule scarcely more than 1 line long, the erect appendages nearly as long as the capsule itself.

W. Australia, *Drummond, n.* 29.

5. **A. ericoides,** *A. Juss. Tent. Euph.* 112, t. 10, *according to Muell. Arg. in DC. Prod.* xv. ii. 214. Stems, from a hard often woody base or rhizome, several, ascending or erect, rigid, flattened or triquetrous. Leaves sessile or nearly so, linear, mucronate-acute, with revolute margins, rigid, mostly about $\frac{1}{2}$ in. or the lower ones nearly 1 in. long. Stipules subulate or dilated at the base, entire or slightly lobed, often minute or deciduous. Flower-clusters usually consisting of 1 female with 3 or 4 males on exceedingly short pedicels, surrounded by small brown scarious bracts. Male perianth about $\frac{3}{4}$ line long, divided to the middle into 4 or 5 lobes, often of a bluish tint in the bud. Stamens usually 8, one or two of the outer ones with 1-celled anthers. Glands small at the base of the very short staminal column. Female perianth-segments usually 5, deeper and more acute than in the males. Capsule rather more than 1 line long, the dorsal appendages short broad and obtuse. Styles deeply divided.—*A. rosmarinifolia,* Klotzsch in Pl. Preiss. i. 176.

W. Australia. King George's Sound and neighbouring districts, *R. Brown, A. Cunningham, Preiss, n.* 1225, *Drummond, 2nd coll. n.* 233, *and 3rd coll. n.* 206, and many others.

6. **A. spartioides,** *Brongn. in Duperr. Voy. Coq.* 226, t. 49 A. Stems, from a hard woody base or rhizome, erect, 1 to 2 ft. high, rigid, flat or 3-angled, often above 1 line or even 2 lines broad, usually leafless at the time of flowering. Leaves few only on the young stems or in the lower portion, cuneate-oblong, often toothed, contracted into a short petiole, $\frac{1}{2}$ or sometimes 1 in. long, the floral ones when present few and very much smaller, linear and entire. Stipules small, deeply fringed or lobed. Flowers nearly sessile in clusters at the nodes, the males often numerous, the females solitary, either alone or surrounded by a few males. Bracts small, broad, mostly fringed. Male perianth nearly 1 line long, broadly campanulate, divided to the middle into 4 rarely 5 lobes. Stamens 8 rarely 10, all with 2-celled anthers. Female perianth more deeply 5-lobed. Ovary crowned by 6 acute dorsal teeth or appendages. Styles rather short, more or less bifid.—Muell. Arg. in DC. Prod. xv. ii. 214; A. Rich. Sert. Astrol. 53, t. 20; *A. cuneiformis*, F. Muell. Herb. and in Baill. Etud. Euph. 455; *Leptomeria xiphoclada*, Sieb. in Spreng. Syst. Cur. Post. 109.

N. S. Wales. Port Jackson or Blue Mountains, *R. Brown*, *Sieber*, *n.* 135, *and Fl. Mixt. n.* 524, and many others; New England, *C. Stuart.*

Victoria. Gipps' Land, *F. Mueller.*

Tasmania. Abundant in poor sandy soil, *J. D. Hooker.*

S. Australia. Rivoli Bay, *F. Mueller;* Corner Inlet, *Wilhelmi.*

TRIBE 3. ANTIDESMEÆ.—Flowers distinct, both sexes with a perianth. Ovary usually reduced to a single cell, with 2 ovules. Styles 3. Embryo with broad cotyledons and a narrow radicle.—Trees or shrubs, with small flowers in catkin-like spikes or racemes, both sexes with a perianth.

This tribe, if it includes the small and scarcely sufficiently known Madagascar genus *Thecacoris*, is chiefly distinguished from *Phyllantheæ* by habit and inflorescence, and is therefore reduced by Mueller Arg. to a subtribe of that tribe; on the other hand the 1-celled (1-carpellary) ovary with 3 styles, had caused it formerly to be considered as forming the distinct Order of *Antidesmeæ* or *Stilagineæ*.

12. ANTIDESMA, Linn.

Flowers diœcious, the males in dense or interrupted spikes or catkins, the females in spikes or racemes. Male fl. very small: Perianth of 3 to 5 segments, slightly imbricate in the bud. Stamens 2 to 5, opposite the perianth-segments, round a central rudimentary ovary. Anthers 2-celled, the cells separated by and terminating the thick more or less 2-lobed connective. Glands alternating with the stamens and often concrete with the rudiment of the ovary in a depressed lobed mass. Female fl.: Perianth of the male. Glands more distinct and usually flattened. Ovary 1-celled or rarely 2-celled (in the very young state 3-celled ?) with 2 ovules in each cell. Styles 3, very short, usually 2-lobed. Fruit a small more or less oblique drupe. Seed usually 1 only, without any arillus or caruncula. Cotyledons broad.—Trees or shrubs. Flowers small, the spikes solitary or several together in the upper axils or forming a terminal panicle.

The genus is spread over the tropical and subtropical regions of the Old World. Among the Australian species two only are recognised as identical with widely dispersed Asiatic species, the remaining five here described, as well as two or three insufficiently known, may be endemic in Australia, but the characters by which the species are distinguished in this difficult genus are as yet very uncertain and for the most part minute.

Flowers inside, glands and rudimentary ovary in the males, ovary in the females, pubescent or hirsute. Spikes paniculate.
Female flowers nearly sessile. Perianth-segments ovate acute. Leaves rounded at the end . . . . 1. *A. Ghæsembilla.*
Female flowers pedicellate. Perianth-segments very broad and obtuse. Leaves often obtusely acuminate . . . . 2. *A. Dallachyanum.*
Flowers in the interior perfectly glabrous.
Spikes or racemes mostly paniculate. Flowers nearly sessile 3. *A. Bunius.*
Spikes or racemes mostly simple. Female flowers pedicellate.
Leaves ½ to 1 in. long, ovate or obovate. Spikes ½ to 1 in. Perianth-segments broad. Fruit small . . . . 4. *A. parvifolium.*
Leaves ovate or elliptical, 1 to 2 inches long. Female perianth-segments narrow-lanceolate. Fruit small . . 5. *A. Schultzii.*
Leaves oblong-lanceolate or elliptical, 2 to 4 in. long. Female perianth shortly and broadly 4-lobed. Fruit small, black 6. *A. erostre.*
Leaves oblong or elliptical, mostly sinuate, 2 to 4 in. long. Female perianth shortly and broadly 4-lobed. Fruit rather large . . . . 7. *A. sinuatum.*

1. **A. Ghæsembilla,** *Gærtn. Fruct.* i. 189, t. 39. A shrub or small tree, the young branches foliage and inflorescence more or less pubescent or tomentose, the full-grown leaves often glabrous. Leaves on very short petioles, broadly ovate obovate or nearly orbicular, very obtuse, rounded or contracted at the base, rather thin, but often shining above, 1½ to 2 or rarely 3 in. long. Male spikes dense in most Indian specimens less so in the Australian ones, 1 to 2 in. long, the females shorter and looser, both solitary or more frequently several in a terminal panicle. Male flowers sessile or nearly so; perianth deeply divided into 5 or rarely 4 ovate ciliate segments not ½ line long. Stamens varying from 3 to 5, the filaments at least 1 line long. Glands bread, hirsute. Female flowers on very short thick pedicels. Ovary when young pubescent or hirsute, but usually becoming glabrous as it enlarges. Styles short, united at the base, spreading upwards, shortly 2-lobed. Drupes "purple," not above 3 lines long, usually obliquely ovoid and 1-seeded, but said to be occasionally didymous and 2-seeded.—Muell. Arg. in DC. Prod. xv. ii. 251; *A. paniculatum*, Roxb.; Wight Ic. t. 820, and other synonyms adduced by Muell. Arg. l.c.

N. Australia. Careening Bay, N.W. coast, *A. Cunningham;* Point Pearce, Victoria and Fitzmaurice rivers, *F. Mueller;* Port Darwin, *Schultz, n.* 694, 748.—This species is widely diffused over East India and the Archipelago from Ceylon to S. China.

2. **A. Dallachyanum,** *Baill. Adans.* vi. 337. A shrub or small tree, closely allied to *A. Ghæsembilla*, the young shoots rather less pubescent and the adult foliage often almost glabrous. Leaves from ovate to lanceolate-elliptical, obtuse or shortly and obtusely acuminate, 2 to 4 in. long on petioles of 2 to 4 lines. Male spikes pubescent, mostly paniculate, but sometimes solitary, more slender than in *A. Ghæsembilla*

and more or less interrupted. Perianths sessile, deeply divided into 4 or 5 broad concave hirsute segments ½ line long. Stamens 3 to 5, usually 4. Rudimentary ovary rather large, hirsute as well as the glands. Female flower racemose, the pedicels ½ to 1 line long. Perianth-segments very broad, ciliate and hirsute as well as the glands and ovary. Styles broad, 2-lobed. Fruit obovoid "white and acid," twice as large as in *A. Ghæsembilla.*

Queensland. Rockingham Bay, *Dallachy.*
N. Australia ? Some male specimens from Port Essington, *Armstrong,* appear to belong to this species, but with smaller leaves.

3. **A. Bunius,** *Spreng.; Muell. Arg. in DC. Prod.* xv. ii. 262. A tree, usually quite glabrous. Leaves oblong, obtuse, acute or shortly acuminate, 4 to 5 in. long or sometimes more, somewhat coriaceous and shining, on petioles of 3 or 4 lines. Male spikes slender, interrupted, 4 to 6 in. long. Stamens usually 3 or 4. Female racemes much shorter, with the flowers nearly sessile. Fruits about 3 lines long, on pedicels of 1 to 1½ lines long.—Wight Ic. t. 819.

Queensland. Rockingham Bay, *Dallachy.*—The species extends over the Indian Archipelago to the Philippines and South China. The Australian specimens are imperfect, but appear to belong to it.

There are also among Dallachy's Rockingham Bay collections two other species or varieties of *Antidesma* with the large glabrous leaves of *A. Bunius;* 1. with apparently short racemes of "large very acid fruits, black when ripe;" and 2. with broader leaves and apparently short racemes of "large acid fruits, white when ripe." The latter may possibly be a glabrous variety of *A. Dallachyana,* the former a large-fruited variety of *A. Bunius;* but this cannot be determined without complete specimens of both sexes.

4. **A. parvifolium,** *F. Muell. Fragm.* iv. 86. A bushy shrub of about 4 ft., the young shoots and inflorescence very slightly pubescent, the adult foliage glabrous. Leaves ovate obovate or orbicular, very obtuse, not exceeding 1 in. and mostly smaller, on very short petioles. Male spikes axillary, interrupted, slender, ½ to 1 in. long. Perianth divided to the middle into 4 broad membranous glabrous lobes. Stamens 2 to 4. Glands and rudimentary ovary thick and glabrous. Female racemes usually shorter than the males, the flowers very shortly pedicellate. Perianth smaller than in the males, the lobes ovate, glabrous as well as the glands and ovary. Fruit "red," not above 2 lines long.

Queensland. Port Denison, *Fitzalan, Dallachy.*

5. **A. Schultzii,** *Benth.* A shrub of 8 to 15 ft. the young branches and inflorescence slightly pubescent, the adult foliage glabrous. Leaves ovate obovate or elliptical-oblong, obtuse or scarcely acuminate, thin, 1 to 2 in. long, on very short petioles. Male flowers unknown. Female racemes solitary or 2 together in the axils, slender, under 1 in. long, the flowers shortly pedicellate in the axils of small bracts. Perianth deeply divided into 5 narrow-lanceolate acute ciliate segments, about ½ line long. Glands very broad, almost petal-like, truncate and often crenate, more than half as long as the perianth and perfectly

glabrous as well as the ovary. Styles short, broad, deeply lobed. Fruit apparently small, but not seen ripe.

N. Australia. Port Darwin, *Schultz, n.* 610, 743.

6. **A. erostre,** *F. Muell. Herb.* A shrub, perfectly glabrous in the specimens seen. Leaves oblong-lanceolate or elliptical, often acuminate, 2 to 4 in. long, on a short petiole. Male flowers unknown. Fruiting racemes axillary, slender, 1 to 1½ in. long, quite glabrous. Pedicels about 1 line long. Perianth shortly 4-lobed. Fruit small, "black," glabrous. Styles very short.

Queensland. Rockingham Bay, *Dallachy.*—Perhaps a form of the E. Indian *A. diandrum*, Roth, with which, however, it cannot be properly compared until the male flowers are known.

7. **A. sinuatum,** *Benth.* A tree of about 30 ft., glabrous or the young shoots slightly pubescent. Leaves oblong or elliptical, entire or deeply sinuate, 2 to 4 in. long, on petioles of 3 to 6 lines. Male flowers unknown. Female fruiting racemes solitary, 1 to 3 in. long. Pedicels 1 to nearly 2 lines. Perianth glabrous or nearly so, broadly urceolate, shortly lobed. Fruits compressed, 3 to 4 lines long. Styles very short.

Queensland. Rockingham Bay, *Dallachy.*

TRIBE 4. PHYLLANTHEÆ.—Flowers distinct (not enclosed in a calyx-like involucre), both sexes with a perianth. Embryo with broad cotyledons and a narrow radicle. Ovules 2 in each cell.—Trees, shrubs, or herbs. Leaves entire or rarely crenate, often coriaceous or if membranous usually small. Flowers in axillary clusters, or solitary.

This tribe differs from *Crotoneæ* in the ovules always in pairs, and generally in habit and inflorescence. From the biovulate *Stenolobeæ* the embryo affords perhaps the only constant distinction, although there are no two genera belonging to the two tribes which have not several other characters to separate them.

13. ANDRACHNE, Linn.

Flowers monœcious, in axillary clusters or the females solitary. Male fl.: Calyx more or less deeply divided into 5 or 6 lobes or segments. Petals as many as calyx-lobes and shorter than them. Glands as many as petals and opposite to them. Stamens as many as petals and alternate with them; anther-cells distinct, parallel, opening longitudinally in 2 valves. Rudimentary ovary in the centre very small. Female fl.: Calyx usually larger than in the males. Petals minute or none. Ovary 3-celled, with 2 ovules in each cell. Styles 3, more or less deeply divided into 2 entire branches. Capsule separating into 3 2-valved cocci. Seeds curved, 3-angular, rugose, not caruncu-late. Embryo curved, with broad cotyledons.—Herbs or undershrubs, with procumbent ascending or erect branching stems. Leaves alternate, petiolate, entire, usually small. Flowers very small, pedicellate in the axils, the females solitary, with or without a few males in the same axil.

The genus contains but few species, dispersed over the temperate and subtropical regions of both the New and the Old Worlds. The only one in Australia is also in Timor and in the Eastern Archipelago. The habit is often that of a *Phyllanthus*, from which the genus differs in the presence of petals and of a central rudimentary ovary in the male flowers.

1. **A. Decaisnei,** *Benth.* Apparently annual, but the stems hard and woody-looking at the base, much branched, decumbent, attaining 1 to 2 or even 3 ft., the whole plant softly villous. Leaves broadly obovate or obovate-oblong, $\frac{1}{2}$ to $\frac{3}{4}$ in. long, on rather long petioles. Male flowers 2 or 3 together on very short pedicels. Calyx-segments 5, lanceolate, acute, spreading, about $\frac{1}{2}$ line long. Petals narrow, nearly as long as the calyx. Female flowers solitary in the same axils as the males, on pedicels attaining 1 line when in fruit. Calyx-segments under the fruit broadly ovate, fully 1 line long, the base of the calyx contracted into a distinct stipes. Styles divided to the base into 2 branches. Capsule depressed, orbicular, villous, about 2 lines diameter.—*A. fruticosa*, Dcne. according to Muell. Arg. in DC. Prod. xv. ii. 235, not of Linn.

N. Australia. Islands off the North coast, *R. Brown;* Victoria river, *F. Mueller;* and if the synonymy is correct, also in Timor and Java. The true *Andrachne fruticosa* of Linnæus, to which Decaisne had referred the Timor plant, is now transferred to *Breynia*, and Mueller Arg. had therefore retained the specific name *fruticosa* as Decaisne's, for the Timor plant, which I have not seen, but is according to him a true *Andrachne.* If, however, the Australian one is really the same, the name, besides the confusion it occasions with the Linnæan plant, is totally inapplicable as it is certainly not shrubby.

Queensland. Near Peak Downs, *Bowman.*

Var. *orbicularis.* Leaves smaller, orbicular. Styles divided to the middle only. Petals smaller. Perhaps a distinct species.

W. Australia. Port Walcot, *C. Harper.*

14. ACTEPHILA, Blume.

Flowers monœcious, in axillary clusters or solitary. Male fl.: Perianth of 5 or 6 segments spreading out flat, the inner ones rather larger than the outer. Petals (or petal-like glands?) small, as many as perianth-segments, and alternating with them or none. Stamens 3 to 6, inserted near the centre of a broad concave disk and surrounding a 2- or 3-fid style, the ovary abortive. Anthers with 2 cells at first parallel and turned inwards, at length usually divaricate, opening longitudinally in 2 valves. Female fl.: Perianth and disk of the males. Ovary 3-celled with 2 ovules in each cell. Styles 3, shortly 2-fid, or entire, free or united at the base. Fruit separating into 3 2-valved cocci, the pericarp hard. Seeds large, with a membranous brittle testa; albumen very scanty or none. Embryo curved, the cotyledons very thick fleshy and folded one over the other or much contortuplicate.—Trees or shrubs. Leaves alternate, petiolate, entire, usually large, coriaceous when full-grown. Flowers rather small, often several females as well as males in the same cluster, all pedicellate. Capsule globular, usually large and smooth.

The genus comprises few species, dispersed over tropical Asia; the Australian ones are, however, all endemic.

Leaves on petioles of $\frac{1}{4}$ to 1 in.
  Petals present (very small). Stamens 5 or 4. Leaves oblong-lanceolate or narrow-elliptical, tapering into the petiole.
    Pedicels $\frac{1}{2}$ to 1 in. long . . . . 1. *A. grandifolia.*
    Pedicels 1 to 3 lines long . . . . 2. *A. Mooreana.*
  No petals. Stamens 5. Leaves obovate or broadly elliptical . 3. *A. latifolia.*
  No petals. Stamens 3. Leaves ovate-lanceolate, rounded at the base, on rather long petioles . . . . 4. *A. petiolaris.*
Leaves sessile or nearly so, oblong, cordate at the base . . . . 5. *A. sessilifolia.*

1. **A. grandifolia,** *Baill. Adans.* vi. 330, 360, t. 10. A glabrous tree or shrub of various heights. Leaves petiolate, oblong-lanceolate or elliptical, shortly and obtusely acuminate, tapering towards the base, coriaceous, smooth, 5 to 8 in. long, on a petiole varying from under $\frac{1}{2}$ in. to $\frac{3}{4}$ in. Flowers of both sexes in the same cluster and in the specimens seen rather more females than males, on pedicels of $\frac{3}{4}$ to 1 in. Male perianth spreading to a diameter of 3 lines, the inner segments larger than the outer. Petals? 5, very small, obovate, inserted under the raised margin of the broad disk. Stamens 4 or 5, rather shorter than the perianth. Female perianth rather larger than the male. Styles 3, free and shortly united at the base, more or less bifid. Fruit fully $\frac{1}{2}$ in. diameter, epicarp splitting at the sutures before the endocarp separates as in *Dissiliaria.* Cotyledons described by Baillon as large and very much contorted.—*Lithoxylon grandifolium,* Muell. Arg in Linnæa xxxiv. 65 and in DC. Prod. xv. ii. 232.

N. S. Wales. Richmond river, *Fawcett, Henderson.*

2. **A. Mooreana,** *Baill. Adans.* vi. 330, 366. A tree or shrub closely resembling *A. grandifolia.* Leaves oblong-lanceolate, rather obtuse, entire or irregularly sinuate, tapering at the base into a petiole of about 5 lines, finely veined, about $2\frac{1}{2}$ in. long. Pedicels of the male flowers not above 1 line long, of the females about 3 lines, the structure of the flowers otherwise as in *A. grandifolia* (Baillon).

N. S. Wales. Mount Lindsay, *C. Moore.*—I have not seen any specimen.

3. **A. latifolia,** *Benth.* A glabrous shrub or tree. Leaves obovate or broadly elliptical, sometimes nearly orbicular, obtuse or shortly and obtusely acuminate, rounded or tapering at the base, firmly coriaceous, penniveined as in *A. grandifolia,* but the reticulate veinlets obscure, 3 to 6 in. long, on a petiole varying from under $\frac{1}{2}$ in. to near 1 in. long. Pedicels slender, $\frac{1}{2}$ to 1 in. long. Perianth-segments 5 or 6, petal-like with glandular ends, about 2 lines long in the males, rather longer in the females. No petals. Stamens 5. Styles 3, united at the base, rather short, spreading, slightly dilated and notched at the end but not branched. Fruit not seen.

Queensland. Cape York, *Daemel;* Rockingham Bay, *Dallachy.*

4. **A. petiolaris,** *Benth.* A glabrous shrub or tree. Leaves ovate-lanceolate or ovate-acuminate, rounded at the base, thinner than in

A. grandifolia, 3 to 4 in. long, on a petiole of 1 in. or more. Male perianth-segments, 1 line long, petal-like with dark streaks. Petals none. Disk, stamens and rudimentary ovary of *A. grandifolia,* except that there are only 3 stamens. Female flowers of that species but rather smaller and no petals. Disk and ovary the same. Styles rather more united at the base.

Queensland. Rockingham Bay, *Dallachy.*

5. **A. sessilifolia,** *Benth.* A glabrous shrub of 4 to 6 ft. Leaves oblong, obtuse, entire, slightly cordate at the base and sessile or nearly so, thinly coriaceous, 2 to 3 in. long. Male flowers not seen. Female pedicels slender, about 1 in. long. Perianth deeply 6-lobed, about 2 lines diameter, with very small petals (or petal-like lobes of the disk ?). Styles 3, united at the base, rather short, spreading, undivided. Capsule globular, coriaceous, about ½ in. diameter, slightly scabrous-punctate.

Queensland. Caves mountains, five miles west of Morinisi, *Thozet.*—There seems to be but little doubt that this is an *Actephila,* although I have only seen a single female flower in Herb. F. Mueller, the other specimens being in fruit only.

15. DISSILIARIA, F. Muell.

Flowers monœcious (or diœcious ?) in axillary clusters or solitary (appearing terminal when the terminal bud is not developed). Male fl.: Perianth of 4 to 6 segments, imbricate in the bud, the inner ones rather larger and more petal-like. Glands none. Stamens few or many, inserted on a central receptacle without any rudimentary ovary; filaments very short. Anthers dorsally attached, the cells parallel, opening longitudinally in 2 valves. Female fl.: Perianth of 3 or 4 segments. Disk shortly cup-shaped or annular. Ovary 3-celled, with 2 ovules in each cell. Styles linear, spreading, undivided, free or very shortly connate at the base. Capsule more or less tridymous, dividing into 2-valved cocci, the pericarp thick and hard, the epicarp usually separating from the endocarp in each valve. Seeds without any caruncului. Albumen copious. Colyledons flat and broad.—Trees or shrubs. Leaves opposite, undivided, entire or crenulate, penniveined and slightly reticulate. Male flowers very small, on short pedicels, females few together or solitary. Fruits usually solitary.

The genus is endemic in Australia; it is, however, as yet insufficiently known, the male flowers having been seen only in *D. tricornis,* and the three species may not therefore be strictly congeners. They all, however, differ from all other Australian *Phyllantheæ* by their opposite leaves.

Capsule 8 or 9 lines diameter, the thick corky exocarp deeply furrowed on the back of the cocci . . . . 1. *D. baloghioides.*
Capsule 4 or 5 lines diameter, the corky exocarp slightly furrowed on the back of the cocci . . . . 2. *D. Muelleri.*
Capsule about 4 lines diameter, truncate on the top, the cocci not furrowed on the back, with a short conical point or horn on the top 3. *D. tricornis.*

1. **D. baloghioides,** *F. Muell.; Baill. Adans.* vii. 359. A glabrous tree. Leaves ovate oblong or elliptical, entire, coriaceous, smooth and

shining, 2 to 3 in. long, on a petiole of about $\frac{1}{4}$ in. Female flowers (which I have not seen) 3-merous (*Baillon*). Fruiting pedicels 2 to 4 lines long, thick, solitary, apparently terminal. Capsule 8 to 9 lines diameter, minutely tomentose, with a thick corky exocarp separating from the hard endocarp, marked outside with narrow furrows between the cocci and a broad deep furrow along the line of dehiscence on the back of each coccus. Ovules 2 in each cell but only one usually enlarged. Ripe seeds not seen.

Queensland. Shady woods, Moreton Bay, *A. Cunningham*, also *Leichhardt*, and Pine river, *Fitzalan;* but I have only seen Cunningham's specimens.

2. **D. Muelleri,** *Baill. Adans.* vii. 359, t. 1. A glabrous shrub or tree. Leaves broadly ovate, sometimes almost cordate, obtuse, irregularly crenulate, 1 to 3 in. long and broad, on a petiole of about $\frac{1}{4}$ in. Female flowers in short cymes apparently terminal, the fruiting pedicels $\frac{1}{4}$ to $\frac{1}{2}$ in. long. Calyx of 3 or 4 broad segments, imbricate in the bud. Petals rather longer than the calyx. Disk short, crenulate, close round the base of the ovary. Capsule globular, 4 to 5 lines diameter, slightly tridymous and furrowed on the backs of the cocci, the thick corky exocarp separating from the endocarp as in *D. baloghioides.*

Queensland. Rockhampton, *Thozet, Dallachy.*—I have not seen the specimens described by Baillon, but some male specimens from *O'Shanesy* may belong to the same species although the leaves are more obovate and shortly tapering at the base. In these the perianth-segments are 6, about $\frac{3}{4}$ line long, and the stamens numerous, but shaped entirely as in *D. tricornis.*

3. **D. tricornis,** *Benth.* A shrub, with the young branches pubescent, the adult foliage nearly glabrous. Leaves ovate elliptical or almost lanceolate, very obtuse, crenulate, rounded or contracted at the base, coriaceous, shining and veined above, pale with the midrib pubescent underneath, 1 to $1\frac{1}{2}$ in. long, on a petiole of $\frac{1}{4}$ in. or less. Flowers in sessile clusters in the upper axils or appearing terminal from the non-development of the terminal bud, the males numerous, on slender pedicels of 1 to 3 lines, the females 1 or 2 in each axil on shorter thick pedicels. Bracts small, villous. Male perianth glabrous, of 4 rarely 6 broad segments, $\frac{1}{2}$ to $\frac{3}{4}$ line long, somewhat petal-like, especially the inner ones which are rather longer. Stamens 4, rarely 6 or more, on a hairy receptacle, longer than the perianth. Female perianth of 3 rather acute or mucronulate glabrous segments, and 3 smaller inner lanceolate or linear ones alternating with them. Disk very hairy, but not prominent. Ovary very short, tapering into 3 distinct styles much thickened at the base as in *Pseudanthus.* Capsules on pedicels of 2 to 4 lines, pubescent, about 4 lines diameter, tridymous, flat-topped, hard, with a small conical point or horn on the top of each coccus but no furrow on the back. Seeds only ripening one of the two ovules of each cell, ovoid, without any caruncula.

N. Australia. Port Essington, *A. Cunningham*, and (probably the same plant but not in fruit) *Armstrong.*

Queensland. Rockingham Bay, *Dallachy.*

16. PETALOSTIGMA, F. Muell.

(Hylococcus, *R. Br.*)

Flowers monœcious, ? in axillary clusters or the females solitary. Male fl.: Perianth of 4 to 6 imbricate calyx-like segments. Glands none. Stamens indefinite, united in a central column without any rudimentary ovary, the filaments shortly free; anthers adnate, with parallel cells opening longitudinally in 2 valves. Female fl.: Perianth as in the males, but the segments narrower and very deciduous. Ovary 4-celled or sometimes 3-celled, with 2 ovules in each cell. Styles 4 or 3, expanded into large flat almost petal-like stigmatic branches. Fruit globular or almost ovoid, with a fleshy exocarp and a hard endocarp, separating into 4 or 3 2-valved and spuriously 2-celled cocci. Seeds oblong, slightly compressed, with a small caruncuius (sometimes wanting ?).—A tree. Leaves alternate, entire. Stipules minute or none. Inflorescence of *Phyllanthus*, but the flowers larger than is usual in that genus.

The genus consists of a single species, endemic in Australia. The flowers are said to be monœcious, but probably on different branches or expanding at different times, for our specimens are all unisexual.

1. **P. quadriloculare,** *F. Muell. in Hook. Kew Journ.* ix. 17. A small or moderate-sized tree, the branches and underside of the leaves closely silky or more loosely tomentose. Leaves shortly petiolate, in the typical form ovate or sometimes almost orbicular, very obtuse or almost acute, ½ to 1½ in. long, becoming glabrous above when old. Male flowers several together, on very short pedicels. Perianth-segments orbicular or broadly obovate, silky-pubescent or villous, varying from scarcely more than 1 line to nearly 2 lines long. Staminal column villous with long hairs, the free part of the filaments glabrous. Anthers somewhat incurved, the glabrous or hairy connective shortly projected beyond the cells. Branches of the styles cuneate, more or less undulate and crenate. Fruit orange-coloured, often ½ in. diameter. Seeds slightly compressed, smooth.—Muell. Arg. in DC. Prod. xv. ii. 273; *Hylococcus sericeus*, R. Br. in Bauer Ic. ined.; Mitch. Trop. Aust. 389; *Petalostigma triloculare*, Muell. Arg. l.c. 274; *P. Australianum*, Baill. Adans. vii. 356, t. 2.

N. Australia. N.W. Coast, *A. Cunningham;* islands of the Gulf of Carpentaria, *Henne;* Arnheim's Land, *F. Mueller;* in the interior, lat. 20°, *M'Douall Stuart;* Port Darwin, *Schultz, n.* 94, 298, 299, 447, 449.

Queensland. Broad Sound, *R. Brown;* Endeavour river and Moreton Bay, *A. Cunningham;* Moore river, *Mitchell;* Rockhampton, Port Denison, Edgecombe Bay, *Dallachy*, and others; Albany island, *W. Hill.*

N. S. Wales. Clarence river, *Beckler, C. Moore;* N. S. Wales woods, *London Exhibition*, 1862, *n.* 90.

Var. *glabrescens.* Leaves elliptical-lanceolate, 1½ to nearly 3 in. long, becoming nearly glabrous, and the ovary and fruit much less villous than in the typical form or only shortly tomentose or sometimes quite glabrous.—Moreton Bay, *W. Hill;* Clarence river, *C. Moore; London Exhibition*, 1862, *n.* 91; Cape Sidmouth, *Curdie.*

The species varies exceedingly in the shape of the leaves, the size of the flowers and the indumentum. The more glabrous forms have generally smaller flowers ; both occur with 3-locular as well as 4-locular ovaries and fruits. It is possible, however, that there may be two species distinguished by the shape of the style-branches in conjunction with the size of the flowers; but the specimens I have seen, numerous and various as they are, have been insufficient for determining the point; the great majority having either male flowers only, or fruits only from which the styles have fallen away,

17. PHYLLANTHUS, Linn.

(Glochidion, *Forst.*; Kirganelia, *A. Juss.*; Synostemon, *F. Muell.*; Reidia, *Wight.*)

Flowers monœcious or rarely diœcious, in axillary clusters or solitary. Male fl.: Perianth-segments (or rarely lobes) 6 or rarely 5, or in some sections 4, more or less distinctly in 2 rows, all similar and often petal-like, or the inner ones rather larger. Glands when present usually distinct, but wanting in some sections. Stamens 3 or rarely 2 or 5 or more, the filaments united in a central column or free; anthers with 2 distinct parallel or divergent cells, opening longitudinally or almost transversely in 2 valves. Female fl.: Perianth-segments usually 6 or 5, narrower or more herbaceous than in the males, and sometimes enlarged after flowering. Ovary 3- or more-celled, with 2 ovules in each cell. Styles free or united, more or less deeply 2-lobed or rarely entire. Capsule separating into 2-valved cocci or loculicidally dehiscent. Seeds triangular in the cross section, the inner edge straight; the back semicircular, without any caruncalus. Cotyledons broad.—Herbs, shrubs or trees. Leaves alternate, entire, often distichous so as to give the branchlets the appearance of pinnate leaves. Stipules small, usually brown and scarious, or thick, or edged with white, usually persistent. Flowers small, the males most frequently clustered, but usually few together, the females solitary in the same or in different axils.

The genus is numerous in species dispersed over the warmer regions both of the New and the Old World. Of the forty-four Australian species, I have been only able to identify four with widely-spread Asiatic ones, the others appear to be all endemic; but, especially in the section *Glochidion*, the species are distinguished by characters so vague, that it is possible that some may be still referred to Indian or Archipelagan types.

SECT. 1. **Glochidion.**—*Trees or shrubs the leaves often large. Stamens* 3 *or more, the anthers erect and sessile on a central column, free or more or less connate, the cells parallel with the connective projecting beyond them. Ovary* 3- *or more-celled. Styles short, thick, erect, often connate at the base. No glands or disk in either sex.*

Ovary and capsule slightly furrowed, 5- to 7-celled . . . 1. *P Ferdinandi.*
Ovary and capsule deeply lobed, 3-celled . . . . 2. *P. lobocarpus.*

SECT. 2. **Synostemon.**—*Undershrubs shrubs or perennial herbs. Leaves small. Stamens* 3, *the anthers adnate to a central column, the cells parallel. Ovary* 3-*celled. Styles distinct or connate at the base. No glands or disk in either sex.*

Male perianth-segments narrow, erect, herbaceous or rigid, united or free. Hoary or glaucous undershrubs or rarely small shrubs.
  Styles thick but free. Stems from a woody base decumbent or ascending. Leaves rather rigid, glaucous.
    Leaves sessile, the lower ones cordate, the upper ones ovate . . . . 3. *P. ditassoides.*

Leaves linear.
Leaves ½ to 1 in. long. Male perianth-segments erect but free . . . . 4. *P. Adami.*
Leaves under ½ in. long. Male perianth-segments connate nearly to the apex . . . . 5. *P. thesioides.*
Styles more or less connate or very short. Stems branching, ascending or erect.
Male perianth-segments united to the middle.
Undershrub. Leaves linear, distant . . . . . . 6. *P. hirtellus.*
Small shrub. Leaves small obcordate or emarginate, clustered at the nodes . . . . 7. *P. rigens.*
Male perianth-segments free. Leaves ovate or obovate, sessile. Branches much compressed or angular . . 8. *P. ochrophyllus*
Male flower unknown. Leaves ovate, short, rigid. Branches nearly terete, hoary-tomentose . . . . . 9. *P. rigidulus.*
Male perianth-segments ovate. Erect much-branched almost leafless undershrub . . . . 10. *P. ramosissimus.*
Male perianth-segments small, spreading.
Stems short, leafy. Stipules very spreading. Anthers short, round the dilated summit of the column . . . 11. *P. rhytidospermus.*
Erect branching shrubs. Anthers oblong, occupying nearly the whole column.
Leaves petiolate, thin. Pedicels filiform . . . . . 12. *P. albiflorus.*
Leaves almost sessile, coriaceous. Pedicels very short.
Leaves obovate-orbicular, glabrous, 3 to 4 lines long . 13. *P. crassifolius.*
Leaves ovate or oblong, hoary, 1 to 3 lines long . . 14. *P. elachophyllus.*

SECT. 3. **Kirganelia.**—*Shrubs, often large. Leaves distichous, usually petiolate. Stamens* 5, *the filaments usually connate at the base. Glands present. Ovary and capsule* 3- *or more-celled, usually fleshy or succulent. Styles distinct or connate at the base, short in the Australian species.*

Leaves obovate-oblong. Flowers on filiform pedicels of 2 to 3 lines. Capsule globular, 3 lines diameter . . . . . . 15. *P. Novæ-Hollandiæ.*
Leaves ovate or orbicular. Capsule depressed-globular, about 2 lines diameter. Styles exceedingly short.
Pedicels filiform, under 2 lines, but mostly longer than the perianth . . . . 16. *P. reticulatus.*
Pedicels mostly shorter than the perianth . . . . . . 17. *P. baccatus.*

SECT. 4. **Paraphyllanthus.**—*Shrubs or herbs of varied habit. Stamens* 3, *the filaments free or more or less united. Anthers free, the cells parallel, opening longitudinally. Glands present. Ovary* 3-*celled. Styles free. Capsule dry.*

* *Filaments more or less united.*

Leaves narrow-oblong or linear-lanceolate.
Annuals or perennials with virgate branches and distichous leaves.
Flowers nearly sessile. Capsule tuberculate . . . . 18. *P. Urinaria.*
Pedicels 1 to 3 lines long. Ovary tuberculate. Capsule smooth . . . . 19. *P. trachygyne.*
Flowers nearly sessile. Ovary and capsule smooth . . 20. *P. maderaspatanus.*
Erect bushy shrub. Leaves rigid, under ½ in. long. Flowers nearly sessile . . . . 21. *P. Mitchelli.*
Leaves obovate-oblong. Pedicels filiform, 2 to 4 lines long.
Perianth scarcely enlarged under the fruit . . . . . 22. *P. Gasstrœmii.*
Fruiting perianth much enlarged, as long as the capsule . 23. *P. Dallachyanus.*

(See also 29. *P. grandisepalus*, which has the filaments sometimes united at the base.)

** *Filaments free.*

† *Fruiting perianth much enlarged, usually as long as or longer than the capsule.*

Leaves ovate or lanceolate, acute . . . . 24. *P. subcrenulatus.*
Leaves obovate or oblong, obtuse or mucronate.
Male pedicels filiform, 2 to 4 lines long. Western species 25. *P. calycinus.*
Male pedicels 1 line long or less. Eastern or Northern species.
Capsule glabrous.
Glabrous or glaucous plant. Stems decumbent from a woody base . . . . 26. *P. flagellaris.*
Glabrous plant not glaucous. Shrubby with elongated slender branches. Seeds smooth . . . . 27. *P. similis.*
Glabrous. Branchlets very numerous, ½ in. long . . 28. *P. microcladus.*
Glabrous or glaucous. Shrubby and branched. Seeds striate . . . . 29. *P. grandisepalus.*
Minutely hoary-tomentose. Fruiting perianth less enlarged . . . . 30. *P. Carpentariæ.*

(See also 23. *P. Dallachyanus*, in which the filaments are sometimes almost free.)

Capsule pubescent or hairy.
Hoary-tomentose plant, shrubby and branched. Capsule pubescent . . . . 31. *P. Fuernrohrii.*
Tomentose or villous plant, shrubby and branched. Capsule sprinkled with hairs . . . . 32. *P. hebecarpus.*

†† *Fruiting perianth shorter than the capsule.*

Low diffuse annuals. Leaves oblong, obtuse.
Stipules minute. Capsule depressed-globular, scarcely 2 lines diameter . . . . 33. *P. lacunarius.*
Stipules spreading. Capsule globular, 3 lines diameter . 34. *P. trachyspermus.*
Undershrubs or shrubs. Leaves small, coriaceous, not distichous. Fruits almost sessile.
Low diffuse glabrous undershrub . . . . 35. *P. australis.*
Erect shrub, with virgate branches more or less pubescent. Eastern species . . . . 36. *P. thymoides.*
Erect bushy glabrous shrub. Western species . . . . 37. *P. scaber.*
Shrubs with the leaves distichous on the young branches.
Leaves small, coriaceous, rigid. Pedicels very short.
Leaves ovate, about 2 lines long, hoary as well as the perianth . . . . 38. *P. indigoferoides.*
Leaves oblong, about 3 lines long, glabrous as well as the perianth . . . . 39. *P. aridus.*
Leaves about ½ in. long, membranous. Fruiting pedicels filiform, 3 to 6 lines long . . . . 40. *P. Gunnii.*

(See also 29. *P. grandisepalus* and 30. *P. Carpentariæ*, in which the perianth is sometimes less enlarged.)

SECT. 5. **Euphyllanthus.**—*Trees shrubs or herbs of varied habit. Leaves usually distichous. Stamens* 3, *the filaments free or more or less united; anthers free, the cells short, more or less diverging or opening transversely, and often separated by a broad connective. Glands present. Ovary* 3-*celled. Styles free. Capsule dry. Flowers usually minute.*

(*P. Niruri*, Linn., a common tropical weed belonging to this section, has not yet been observed in Australia.)

Stems annual or from a perennial base, slightly branched, rather rigid, ½ to 1 ft. high. Leaves oblong . . . . . 41. *P. simplex.*

Stems from a woody base, numerous, filiform, under 6 in. Leaves obovate or orbicular . . . . 42. *P. filicaulis.*
Stems annual, filiform, branching, ½ to 1½ ft. long. Flowers very minute . . . . 43. *P. minutiflorus.*

SECT. 6. **Reidia.**—*Trees or shrubs with distichous leaves. Male perianth of* 4 *segments. Stamens* 2, *the filaments united in a central column, anthers with separate parallel cells having the appearance of* 4-*celled anthers. Female perianth of* 4 *to* 6 *segments. Styles* 3.

Branchlets with distichous leaves, several at the ends of the branches with the appearance of pinnate leaves . . . . 44. *P. Armstrongii.*

SECT. 1. GLOCHIDION.—Trees or shrubs, the leaves often large. Stamens 3 or 4 (or in species not Australian more), the anthers erect and sessile on a central column, free or more or less connate, the cells parallel, with the connective projecting beyond them. Ovary 3- or more-celled. Styles short, thick, erect, often connate at the base. No glands or disk in either sex.—Genus Glochidion, *Forst.*, Bradleia, *Gærtn.*

The species of this section are numerous in tropical Asia and Africa, and exceedingly difficult to characterize well from dried specimens, and it is very possible that some of the forms enumerated below may prove identical with some of those from the Archipelago, although I have been unable to match them precisely with any of our specimens.

1. **P. Ferdinandi,** *Muell. Arg. in Flora* 1865, 379, *and in DC. Prod.* xv. ii. 300. A small tree quite glabrous in the typical form except the ovary. Leaves shortly petiolate, elliptical or ovate-lanceolate, more or less acuminate, often obliquely contracted at the base, usually somewhat coriaceous and shining on the upper side, 2 to 4 in. long. Flower-clusters in the typical form sessile in the axils, the pedicels 2 to 3 lines long. Male perianth-segments nearly 2 lines long in the few specimens where I have seen it fully out, but usually much smaller. Anthers 3 or rarely 4, linear, erect on a very short central column, with the connective very shortly produced above the parallel cells. No glands within the perianth in either sex. Female perianth smaller than in the male, with narrower segments. Ovary pubescent, scarcely contracted into 5 to 7 short erect thick more or less united styles. Capsules glabrous or nearly so, orbicular, 5- to 7-celled, and much depressed in the centre in the typical form, slightly furrowed between the cells, about ½ in. diameter when fully ripe.

N. Australia. Islands of the Gulf of Carpentaria, *R. Brown.*
Queensland. Rockingham Bay, *W. Hill, Dallachy;* Fitzroy island, *M'Gillivray;* Wide Bay, *C. Moore.*
N. S. Wales. Port Jackson, *R. Brown* and others.

The following forms may possibly prove to be distinct species:—
Var.? *minor.* Leaves smaller. Pedicels shorter. Styles longer and more slender.—Bremer river, *A. Cunningham;* New England, *C. Stuart.*
Var.? *supra-axillaris.* Leaves of the typical form or larger and more coriaceous. Flower-clusters very shortly pedunculate and inserted shortly above the axils. Capsule 4- to 6-celled, not more than 4 lines diameter and less depressed in the centre, quite glabrous in most specimens, but pubescent in a few.—Rockingham Bay, *Dallachy;* Rockhampton, *Thozet, O'Shanesy.*

Var.? *mollis.* Branches, foliage, and flowers softly pubescent. Anther-column very short.—Rockingham Bay, *Dallachy.* A sub-variety from the same locality has the leaves narrow and not 2 in. long.

2. **P. lobocarpus,** *Benth.* A small tree, the young branches minutely pubescent. Leaves shortly petiolate, oblong-lanceolate or elliptical, obtuse or almost acute, usually oblique at the base, 2 to 3 in. long, green above, very pale or white underneath and rather thin. Stipules very small. Flowers diœcious, the males 2 or 3 together, on recurved pedicels of about 1 line, the females solitary, on pedicels lengthening to 2 lines. Male perianth-segments about $\frac{3}{4}$ line long, obtuse and concave, the 3 inner ones rather longer than the 3 outer. Anthers 3, erect and connivent on a very short column, the cells parallel, with the connective much produced beyond them. No glands within the perianth in either sex. Female perianth rather smaller than the male, and slightly pubescent. Ovary depressed, almost 3-partite, the carpels deeply 2-lobed. Styles 3, short and thick, erect in the central depression, stigmatic inside. Capsule about 4 lines diameter, much depressed, more or less deeply divided into 6 or fewer lobes according to the number of seeds perfected. Seeds "orange-red when ripe," but not quite ripe in the specimens seen.

Queensland. Rockhampton, *O'Shanesy;* Nerkool Creek, *Bowman.*

SECT. 2. SYNOSTEMON.—Undershrubs shrubs or perennial herbs. Leaves usually small. Stamens 3, the anthers adnate to a central column, with parallel cells opening longitudinally. Ovary 3-celled. Styles distinct or connate at the base, usually spreading. No glands or disk in either sex.

The section is limited to Australia.

3. **P. ditassoides,** *Muell. Arg. in Flora* 1864, 487, *and in DC. Prod.* xv. ii. 326. An undershrub with a short thick woody base and wiry slightly branched ascending stems of $\frac{1}{2}$ to 1 ft., somewhat angular, glaucous, sprinkled with a few short hairs as well as the foliage which turns blackish in drying. Leaves sessile, cordate-ovate or almost orbicular, obtuse or almost acute, coriaceous, $\frac{1}{2}$ to 1 in. long, the upper ones gradually smaller ovate-lanceolate or lanceolate. Flowers according to Mueller Arg. monœcious, in our specimens diœcious, the males solitary, or 2 or 3 together on pedicels of 2 to 3 lines. Perianth-segments narrow, erect, about $1\frac{1}{2}$ lines long. Anthers connate and twice as long as the column below. Female flowers solitary, on very short pedicels. Perianth-segments broader than in the males and spreading. Ovary depressed, 3-celled, pubescent. Styles diverging from the base, thick, 2-lobed. Capsule not seen.

N. Australia. South Goulburn island, *A. Cunningham;* Port Essington, *Armstrong.*

4. **P. Adami,** *Muell. Arg. in DC. Prod.* xv. ii. 327 (♀). Stems from a thick woody base, ascending or erect, simple or branched, from a few inches to above 1 ft. high, more or less angular, glaucous as well as the

foliage. Leaves linear, acute or obtuse, rigid, rather thick, ½ to 1 in. long. Flowers diœcious, the males few together, almost sessile. Perianth-segments narrow, erect, about ¼ line long. Anthers connate, about twice as long as the column below. Female flowers on pedicels of 1 to 2 lines. Perianth-segments spreading, at first of the size of the males, but growing out to above 2 lines. Styles thick but rather long and bifid. Capsule ovoid, almost acute, glabrous and glaucous, 3-celled. —*Synostemon glaucus*, F. Muell. Fragm. i. 33; *Phyllanthus bossiæoides*, A. Cunn. Herb.; *P. stenocladus*, Muell. Arg. in Flora 1864, 536, and in DC. Prod. xv. ii. 327 (♂).

N. Australia. Port Keats, N.W. coast, *A. Cunningham;* M'Adam range, Point Pearce, Providence hill, *F. Mueller;* Port Darwin, *Schultz, n.* 460; Port Essington, *Armstrong.*

5. **P. thesioides,** *Benth.* Probably herbaceous, quite glabrous, the stems slender but rigid, much branched, erect or ascending, the specimens seen 4 to 8 in. long. Leaves linear linear-cuneate or the lower ones oblong-spathulate, obtuse, rather thick, flat or with recurved margins, 2 to near 6 lines long. Flowers apparently diœcious, solitary on pedicels of 1 to 1½ lines. Male perianth tubular, 2 lines long, narrow, slightly dilated upwards, with 6 very short broad thick rounded and inflexed lobes, the 2 or 3 inner ones still smaller than the outer ones. Anthers connate, occupying rather more than half the length of the staminal column, the connective scarcely projecting beyond the cells. Female flowers much smaller, the perianth-segments short and spreading. No disk. Ovary glabrous, 3-celled, with 3 short thick free styles, spreading at the end but not lobed.

Queensland. Near Brisbane, but very local, *C. Prentice* (female specimen).
N. S. Wales. Lachlan river, *L. Moreton* (male specimen).

I describe this from two single specimens, which, though of different sexes and from different stations, appear to me to belong to one species, allied to *P. Adami*, but different as well in foliage as in the male perianth and the female styles. The characters will have, however, to be verified from further specimens.

6. **P. hirtellus,** *Muell. Arg. in DC. Prod.* xv. ii. 326. Apparently herbaceous, or perhaps an undershrub, the stems in the specimens seen not above 6 in. long, slender, minutely pubescent as well as the foliage. Leaves distichous but rather distant, linear or oblong, mostly somewhat cuneate, ¼ to ½ in. long. Stipules very minute. Flowers mostly solitary, the males on pedicels of 1 to 1½ lines. Perianth tubular, nearly 2 lines long, narrow, slightly dilated upwards, with 6 broad lobes not ⅓ as long as the tube, 2 of the inner ones smaller than the others. Anthers connate, occupying about half the length of the staminal column, the connective very slightly projecting beyond the cells. Female flowers on pedicels of 1 to 2 lines or sometimes 3 lines when in fruit. Perianth divided to the base into ovate very obtuse segments of nearly ¾ line. Styles united at the base, shortly free, broad and spreading at the top. Capsule ovoid, pubescent, 4 or 5 lines long.—*Synostemon hirtellus*, F. Muell. Fragm. iii. 89.

Queensland. Rockhampton, Connors river, Walloon, *Bowman.*

7. **P. rigens,** *Muell. Arg. in Flora* 1864, 513, *and in DC. Prod.* xv. ii. 325. Probably a small shrub, with rigid terete branches quite glabrous and perhaps spinescent. Leaves very small, clustered at the nodes, obcordate or cuneate and emarginate, rather thick, 1 to 3 lines long. Male flowers "sessile. Perianth-segments connate high up, ovate, obtuse; anthers connate, with the connective projecting beyond the cells" (*F. Mueller*). Female flowers on slender pedicels much longer than the leaves. Perianth-segments linear, rigid, rather above 1 line long. Ovary glabrous, contracted at the top. Styles 3, very short, spreading, rather broad, entire or notched. Capsule glabrous, 3 to 4 lines long.—*Synostemon rigens*, F. Muell. Fragm. ii. 153.

N. S. Wales. Upper Darling river, *Bowman;* Mutanic range, *Beckler;* both single small specimens in an imperfect state in Herb. F. Mueller, but very different from any other species known to me, although with something of the habit of *P. thymoides.*

8. **P. ochrophyllus,** *Benth.* Stems from a woody base, ascending or erect, under 1 ft. high, the branches angular or flattened, hoary or glaucous as well as the foliage, but scarcely tomentose. Leaves almost sessile, ovate or obovate, acute or obtuse, ½ to ¾ in. long. Flowers apparently diœcious, both sexes solitary in the axils, pendulous, on pedicels at first of about 1 line, but in the females growing out to about 2 lines. Perianth narrow, the segments erect, nearly 2 lines long. No glands. Anthers 3, erect, more or less connate, twice as long as the column below, tipped with the small projecting connectives. Female perianth rather smaller than the male, but enlarged after flowering. Styles 3, erect, connivent or more or less connate, shortly divided at the end. Capsule globular, 4 to 5 lines diameter. Seeds smooth, at least when unripe.

N. Australia. Port Darwin, *Schultz, n.* 428 (males) *and* 489 (females).

9. **P. rigidulus,** *F. Muell.; Muell. Arg. in Linnæa* xxxiv. 72, *and in DC. Prod.* xv. ii. 370. Stems erect, rigid, 1 to 2 ft. high, hard and woody at the base, with virgate branches, hoary as well as the foliage with a minute tomentum. Leaves almost sessile, ovate or broadly oblong, mucronate-acute or the lower ones broader and obtuse, rather rigid, under ½ in. long. Stipules small, brown. Flowers solitary, almost sessile, all females in our specimens. Perianth cylindrical, 1¼ lines long, tomentose outside, divided nearly to the base into narrow rigid segments. No gland or disk. Ovary tomentose. Styles erect, connivent and more or less connate, shortly 2-lobed at the end. Capsule (which I have not seen) globular, ashy-tomentose. (*Muell. Arg.*)

N. Australia. Gulf of Carpentaria, *F. Mueller.*

10. **P. ramosissimus,** *Muell. Arg. in Linnæa* xxxiv. 70, *and in DC. Prod.* xv. ii. 326. A slender wiry rigid much-branched undershrub of about 1 ft., leafless or nearly so at the time of flowering. Leaves few and only on the very young branches, linear, thick, almost terete, 2 to 4 lines long. Stipules small but persistent. Male flowers sessile

or nearly so, but not yet fully out in the specimens. Perianth-segments ovate, about ½ line long. No glands. Anthers 3, connate, occupying nearly the whole of the column. Female flowers on pedicels varying from 2 to 6 lines, solitary or 2 together. Perianth narrow, about 1 line long. Ovary glabrous. Styles free, thick, dilated and shortly 2-lobed at the end. Capsule ovoid, 3-celled, 3 to 4 lines long.—*Synostemon ramosissimus,* F. Muell. Fragm. i. 33.

Queensland. Mackenzie range, *F. Mueller.*
N. S. Wales. Between the Darling river and Cooper's Creek, *Beckler.*

11. **P. rhytidospermus,** *F. Muell.; Muell. Arg. in Linnæa* xxxiv. 70, *and in DC. Prod.* xv. ii. 327. Stems from a woody base, decumbent or ascending, 6 to 8 in. high, the branches angular, glabrous and glaucous as well as the foliage. Leaves sessile, somewhat distichous, oblong or broadly linear, acute or mucronulate, 3 to 5 lines long. Stipules persistent, subulate and conspicuously spreading, often 1 line long. Flowers monœcious, very small, nearly sessile. Male perianth-segments ovate, spreading, petal-like, about ⅓ line long. No glands. Staminal column very short, anthers 3, very short, adnate round the dilated end and projecting slightly above it, forming a disk of ½ line diameter. Female perianth-segments linear or lanceolate, acute, rather rigid, 1 line long. Ovary glabrous. Styles short, erect or scarcely spreading at the top, very shortly lobed. Capsule 3-celled, ovoid, glabrous, about 3 lines long.

N. Australia. Depôt Creek, Upper Victoria river, *F. Mueller.*

12. **P. albiflorus,** *F. Muell.; Muell. Arg. in Linnæa* xxxiv. 70, *and in DC. Prod.* xv. ii. 326. A much-branched glabrous shrub, varying from 3 or 4 ft. to twice that height, the branches terete or slightly angular, the smaller ones slender but rigid. Leaves obovate-oblong to narrow-cuneate, very obtuse, sometimes mucronulate, contracted into a very short petiole, membranous, glaucous underneath, rarely above ½ in. long. Stipules small, brown. Flowers solitary, on filiform pedicels of 4 to 5 lines, surrounded by small brown scale-like bracts, the female pedicels sometimes much longer. Perianth-segments nearly equal, ovate, rather above 1 line long. No glands. Anthers 3, connate, occupying nearly the whole of the central column. Ovary globular, 3- or sometimes 4-celled, glabrous. Styles distinct, recurved, cuneate and emarginate at the end. Capsule depressed-globular, nearly 4 lines diameter.

Queensland. Brisbane river, Moreton Bay, *F. Mueller, C. Stuart;* Rockhampton, *Dallachy* and several others.

13. **P. crassifolius,** *Muell. Arg. in Flora* 1864, 513, *and in DC. Prod.* xv. ii. 325. A rigid glabrous divaricately branched shrub, sometimes low and spreading, but sometimes attaining 3 or 4 ft. Leaves distichous, obovate or orbicular, rigidly coriaceous, scabrous-puncticulate, 3 to 4 lines long, all tipped in our specimens with a black gland. Stipules small and gland-like. Flowers very small, few in the axils, on very short pedicels, surrounded by small broad black bracts. Male perianth-segments ovate, about ¾ line long, the inner 3 rather longer than the

outer ones. No glands. Anthers 3, connate, occupying nearly the whole of the central column, the connectives shortly produced beyond the cells. Female perianth-segments about 1 line long, the outer ones thick and scabrous outside. Ovary glabrous. Styles 3, erect or incurved, deeply bifid. Capsule not seen.

W. Australia. Sharks Bay, *Milne;* Murchison river, *Oldfield.*—The original specimens examined by Mueller Arg. were Milne's. Oldfield's Murchison river ones, which have numerous female flowers, were only seen by him after his character was drawn up.

14. **P. elachophyllus,** *F. Muell. Herb.* A bushy shrub, with numerous rigid branchlets, hoary as well as the foliage with a minute almost papillose pubescence. Leaves distichous, ovate or oblong, rigidly coriaceous, 1 to 3 lines long. Stipules almost gland-like. Male flowers very small, few in the axil, on exceedingly short pedicels. Perianth-segments ovate, rather thick, about $\frac{1}{4}$ line long. No glands. Anthers 3, very short, connate in a ring round the dilated apex of the column as in *P. trachyspermus.* Female perianth-segments narrow, fully $\frac{3}{4}$ line long under the fruit. Styles short, erect, distinct, very shortly bifid. Capsule 3-celled, globular, glabrous, 3 lines diameter.

Queensland. Newcastle range, *F. Mueller;* Einasleigh river, *Daintree.*

SECT. 3. KIRGANELIA.—Shrubs, often large. Leaves distichous. Stamens 5, the filaments usually connate at the base. Glands present, ovary and capsule 3- or more-celled, usually fleshy or succulent. Styles distinct or connate at the base.

15. **P. Novæ-Hollandiæ,** *Muell. Arg. in DC. Prod.* xv. ii. 346. A small spreading glabrous shrub. Leaves distichous, petiolate, oblong-obovate, very obtuse, often acute at the base, membranous, 4 to 8 lines long. Flowers probably diœcious, the males in clusters of 2 to 4, on filiform pedicels of 2 to 3 lines, the females solitary, with thicker pedicels. Male perianth-segments 5, broad, obtuse, petal-like, about 1 line long. Glands distinct. Stamens 5, the 3 inner filaments connate, the 2 outer free or nearly so; anthers quite distinct, the cells parallel. Female flowers none in our specimens. Ovary according to Mueller Arg. 5-celled. Styles 5, bifid and recurved. Capsule somewhat fleshy, globular, 2 lines diameter.—*P. uberiflorus,* Baill. Adans. vi. 343.

Queensland. Port Denison, Edgecombe Bay, *Dallachy.*

16. **P. reticulatus,** *Poir.* var. glaber, *Muell. Arg. in DC. Prod.* xv. ii. 345. A shrub of several feet, quite glabrous (the typical Asiatic form slightly pubescent), the branches slender. Leaves distichous, petiolate, ovate-obovate or orbicular, very obtuse or rarely almost acute, mostly under 1 in. long. Flowers small, in axillary clusters usually of 3 or 4 males and a single female, all on filiform pedicels rarely exceeding 2 lines. Male perianth-segments 5, broad, very obtuse, $\frac{1}{2}$ line long or 1 or 2 outer ones shorter. Stamens usually 5, the 2 or 3 inner ones with their filaments more or less united, the outer ones free or nearly so. Glands present but variable, sometimes scarcely conspicuous. Female

perianth rather larger than the male. Ovary usually with about 8 cells, but varying from 6 to 12. Capsule depressed-globular, succulent when young but at length nearly dry and furrowed between the seeds, about 2 lines diameter.—*Anisonema eglandulosum*, Dcne. Herb. Tim. Descr. 154, and other synonyms given by Muell. Arg.

N. Australia.? I have not seen any Australian specimens, but the plant is given as Australian by Mueller Arg. on the faith of a specimen in Herb. Franqueville, said to be from Tasmania, *Expedition of the Astrolabe*. The special localities, however, of the Australian plants of the early Expeditions are often falsely noted in French herbaria, and if really Australian the specimen in question was most probably from the north coast opposite Timor. The above description is therefore taken chiefly from A. Cunningham's specimens gathered at Coepang in Timor.

17. **P. baccatus,** *F. Muell. Herb.* A large spreading or diffuse glabrous shrub. Leaves distichous, shortly petiolate, ovate broadly elliptical obovate or almost orbicular, varying from ½ to 1 in., and here and there some old leaves at the ends of the shoots twice that size, or in other specimens nearly all 1 to 2 in. long. Flowers in axillary clusters, usually 1 female sessile and a few males on pedicels of 1 to 2 lines, rarely the female also pedicellate; the short flowering branches often lose their leaves early, and then appear like leafless racemes or spikes of 1 to 2 in. Male perianth-segments 5, broad, obtuse, about ¾ line long or the outer ones smaller. Glands 5. Stamens 5, all free from the base; anther-cells parallel. Female perianth more rigid than in the males and often deciduous. Ovary broad, the exceedingly short styles almost or quite concealed in a central depression. Fruit much depressed, nearly 3 lines diameter, much more succulent than that of *P. reticulatus*.—*P. Novæ-Hollandiæ*, Baill. Adans. vi. 343, not of Muell. Arg.

N. Australia. Vansittart Bay and Greville island, Regent's river, N.W. coast, *A. Cunningham*; Victoria river, *F. Mueller*; Port Darwin, *Schultz*, *n.* 860.

Some specimens in Herb. R. Brown from Prince of Wales island may be a variety of this species with smaller narrow leaves and the flowers, as far as I can find, all males, the plant either diœcious or the females not yet developed.

SECT. 4. PARAPHYLLANTHUS, *Muell. Arg.*—Herbs or shrubs of varied habit. Stamens 3, the filaments free or more or less united; anthers free, the cells parallel, opening longitudinally. Glands present. Ovary 3-celled. Styles free. Capsule dry.

18. **P. Urinaria,** *Linn.*; *Muell. Arg. in DC. Prod.* xv. ii. 364. An erect ascending or procumbent glabrous annual or perennial of 1 to 2 ft., with angular stems and numerous slender branchlets resembling pinnate leaves. Leaves distichous, narrow-oblong, nearly sessile, often all under ¼ in., but sometimes nearly ½ in. long on the main stem. Stipules small, often bordered with white. Flowers minute, nearly sessile, the females solitary with or without 2 or 3 males, all turned to the lower side of the branch away from the leaves. Male perianth of 6 ovate or obovate segments, about ¼ line long. Anthers 3, distinct, erect on a column nearly as long as themselves, the cells parallel. Glands globular. Female perianth-segments narrower and more rigid than the males, about ½ line long. Ovary 3-celled. Styles free, spreading,

dilated and 2-lobed at the end. Capsule depressed-globular, scarcely furrowed, scaly-tuberculate or almost muricate. Seeds more or less distinctly marked with transverse ridges or rows of tubercles.—*P. echinatus*, A. Cunn. Herb.

N. Australia. South Goulburn island, *A. Cunningham;* Port Darwin, *Schultz*, *n.* 85, 203.—A common weed in tropical Asia and Eastern Africa.

19. **P. trachygyne,** *Benth.* Stems from a perennial base, decumbent or erect, simple or slightly branched, often compressed, 1 to 1½ ft. long, the whole plant glabrous and rather glaucous. Leaves very shortly petiolate or almost sessile, oblong-lanceolate or linear, acute or obtuse, ½ to 1 in. long. Stipules minute, brown. Flowers apparently diœcious. Males clustered, the filiform pedicels about 2 lines long in some specimens, not above 1 line in others. Perianth-segments 6, nearly equal, petal-like, nearly 1 line long. Glands globular. Stamens 3, filaments shortly united at the base; anther-cells parallel. Female flowers solitary or 2 together, with short pedicels. Perianth-segments narrow, not ¾ line long. Disk with 6 broad thin lobes. Ovary densely verrucose. Styles 3, bifid, recurved and closely appressed to the ovary. Capsules depressed-globular, smooth, the warts disappearing as the ovary enlarges, not 2 lines diameter.

N. Australia. Port Darwin, *Schultz*, *n.* 112 (males with short pedicels); *n.* 660 (females with flattened stems and obtuse leaves); *n.* 668 (females with rather acute leaves); *and* 788 (males with longer pedicels and acute leaves).

20. **P. maderaspatanus,** *Linn.; Muell. Arg. in DC. Prod.* xv. ii. 362 var. *angustifolius.* An erect simple or branched rather rigid annual (or perennial?) of 1 to 1½ ft., the branches slender, virgate, somewhat angular. Leaves distichous, oblong-linear or cuneate, obtuse or mucronate, contracted towards the base but sessile or nearly so, ½ to 1 in. long. Stipules usually bordered with white. Flowers very small, usually 1 female with or without 2 or 3 males in each axil, the pedicels about ½ line long. Male perianth-segments 6, obovate, about ¼ line long. Glands minute. Anthers 3, distinct, erect on the top of a short column, the cells parallel. Female perianth about twice the size of the male. Ovary 3-celled. Styles distinct, spreading, dilated and very shortly 2-lobed at the end. Capsule depressed, 3-furrowed, about 1½ lines diameter. Seeds elegantly marked on the back with minute tubercles arranged in 10 to 12 longitudinal rows.—Wight Ic. t. 1895; *P. brachypodus*, F. Muell. in several herb.

N. Australia. Intercourse island, Dampier's Archipelago, *A. Cunningham;* Port Walcot, N.W. coast, *Harper;* Upper Victoria and Fitzmaurice rivers, *F. Mueller;* Port Darwin, *Schultz*, *n.* 877.

Queensland. Rockhampton, *Bowman*, *O'Shanesy;* Charlesville, *Giles;* Peak Downs, *Burkitt.*

The species is common in the tropical and subtropical regions of the Old World. The shape of the leaves in the Australian specimens is very nearly that figured by Wight, although much narrower than in the commoner forms of the species.

21. **P. Mitchelli,** *Benth.* An erect bushy shrub, much resembling some of the narrow-leaved less pubescent forms of *P. thymoides*, with

which it is united by Mueller Arg., but the plant is quite glabrous, the male flowers larger on shorter pedicels, with the segments more united at the base, and the filaments in all the specimens I have examined united to above the middle. Leaves not distichous, narrow-cuneate, rigid, complicate, 1½ to 3 lines long. Flowers probably diœcious, all males in the specimens seen.—*Micrantheum triandrum*, Hook. in Mitch. Trop. Austr. 342; *P. triandrus*, Muell. Arg. in DC. Prod. xv. ii. 195 (among the species excluded from *Micrantheum*) a name reserved p. 299 for the *Kirganelia triandra*, Blanco; *P. thymoides* var. Muell. Arg. l.c. 372.

Queensland. Pyramid depôt, *Mitchell.*

22. **P. Gasstrœmii,** *Muell. Arg. in DC. Prod.* xv. ii. 358. An erect glabrous shrub or undershrub of 1 to 2 ft. the branches often compressed, the smaller ones slender, with the general aspect of *P. Gunnii.* Leaves very shortly petiolate, obovate-oblong, very obtuse, contracted at the base, membranous, from under ½ in. to nearly ¾ in. long. Stipules brown. Flowers very small, monœcious or almost diœcious, the males in clusters of 3 or 4, the females solitary. Male perianth-segments usually 6, about ½ line long, very obtuse and petal-like with dark centres. Glands small. Anthers 3, distinct, ovate, rather large, erect on a short slender column, the cells parallel, minutely tipped with the projecting connective. Female perianth-segments larger and more acute than in the males, but not enlarged after flowering as in *P. Dallachyanus*, greenish, bordered with white. Ovary 3-celled. Styles longer than the ovary, linear, bifid according to Mueller Arg., entire in the flowers examined. Capsule depressed-globular, glabrous, smooth, about 2 lines diameter.—*P. indigoferoides*, A. Cunn. Herb.

Queensland. Burnet river, *F. Mueller.*

N. S. Wales. Port Jackson (Cabramatta), *Woolls;* Hunter's river, *M'Arthur;* Hastings and Macleay rivers, *Beckler;* New England, *C. Stuart;* near Liverpool and Illawarra, *A. Cunningham.*

23. **P. Dallachyanus,** *Benth.* A glabrous shrub, the young branches often flattened, the smaller branchlets slender, 3 to 5 in. long. Leaves distichous, obovate broadly oblong or almost orbicular. Flowers monœcious, the males clustered few together on filiform pedicels of about 2 lines, the females solitary in a very few of the upper axils and on some branches none at all, the pedicels either not longer than the males or sometimes ½ in. long. Male perianth-segments ovate, petal-like, obtuse or mucronate, ¾ to nearly 1 line long. Stamens 3, the filaments united to about the middle; anther-cells parallel, but separated by a broad thick connective. Female perianth larger, the segments ovate, enlarging round the fruit to about 3 lines. Disk with a broad free margin. Styles 3, thick, diverging, 2-lobed to about the middle.

Queensland. Rockingham Bay, *Dallachy.*—With the foliage nearly of *P. Gunnii*, this has the filaments more or less united as in the preceding species, and the enlarged fruiting perianth of *P. grandisepalus* and its allies.

24. **P. subcrenulatus,** *F. Muell. Fragm.* i. 108. A glabrous branching shrub, of 1 to 1½ ft., the branches acutely angular. Leaves sessile

or nearly so, ovate-lanceolate, more acute than in almost any other species, rounded at the base, rather rigid, with a prominent midrib, 4 to 8 lines long in the N. S. Wales specimens, larger and thinner in the Queensland ones. Stipules brown. Flowers monœcious, solitary or very few together, on pedicels of ½ to 1½ lines. Male perianth of 6 ovate petal-like segments, nearly ¾ line long. Stamens 3, the filaments distinct from the base; anther-cells parallel, but distinct and almost stipitate. Glands conspicuous. Female perianth longer than the male, the segments more herbaceous, bordered with white, and growing out to 1½ lines or more under the fruit. Styles 3, not long but slender and divided to about the middle into 2 branches. Capsule globular, glabrous, smooth, fully 3 lines diameter. Seeds slightly striate longitudinally.—Muell. Arg. in DC. Prod. xv. ii. 368.

Queensland. Upper Brisbane river, *F. Mueller;* Rockhampton, *O'Shanesy.*
N. S. Wales. New England, *C. Stuart;* Clarence river, *Beckler;* St. Aubins, Invermein, *Backhouse.*

25. **P. calycinus,** *Labill. Pl. Nov. Holl.* ii. 75, t. 225. A glabrous shrub of 1 to 2 ft. Leaves oblong-cuneate, very obtuse, contracted at the base but scarcely petiolate, ¼ to ½ in. or when very luxuriant nearly ¾ in. long. Stipules small, brown or more or less white. Flowers monœcious, on pedicels of 2 to 4 lines, the females solitary with or without 2 or 3 males in the same axils, the fruiting pedicels lengthening to ½ in. or more and thickened towards the end. Male perianth-segments 6 or sometimes 5, petal-like, pink bordered with white, ovate-oblong, about 1½ lines long. Glands prominent and broad. Stamens 3, filaments free from the base; anther-cells distinct but parallel. Female perianth larger than in the males, the segments broadly ovate and after flowering lengthening out to 3 lines or even more. Disk with a free undulate margin. Ovary 3-celled. Styles free, or very shortly united at the base, somewhat spreading, shortly 2-lobed. Capsule globular, slightly depressed. Seeds in some specimens marked with 8 to 10 longitudinal striæ or slightly raised ribs, in others smooth or obscurely 2- or 3-ribbed.—Kl. in Pl. Preiss. i. 179; Muell. Arg. in DC. Prod. xv. ii. 371; *P. cygnorum,* Endl. in Hueg. Enum. 19; Muell. Arg. l.c.; *P. pulchellus,* Endl. l.c.; *P. Preissianus,* Kl. in Pl. Preiss. i. 179; *P. pimeleoides,* A. DC. Not. Pl. Rar. Jard. Gen. ix. 15; Lehm. in. Pl. Preiss. ii. 230.

S. Australia. Port Lincoln, *Wilhelmi;* Spencer's Gulf, *Warburton.*
W. Australia. Swan river, *Drummond,* 1*st coll.*, *Preiss, n.* 1212; Murchison river, *Oldfield;* Champion Bay, *C. Grey;* Carnac island, *Preiss, n.* 1213; King George's Sound and adjoining districts, *Harvey, F. Mueller, Oldfield.*—The above all not far from the coast, but also Blackwood river, 90 miles from the sea, *Oldfield.*

Var. *parviflora.* Flowers much smaller, the female perianth scarcely 2 lines long after flowering.—W. Australia, *Burgess.*

Two species are usually distinguished, according as the seeds are striate or smooth, but the striæ are often not apparent till the seed is quite ripe, and amongst the very numerous specimens in herbaria very few have ripe seed, and the two forms are otherwise absolutely undistinguishable.

26. **P. flagellaris,** *Benth.* An undershrub slightly hoary or at length glabrous, with a woody base and procumbent simple or slightly branched stems of 6 in. to above 1 ft., more or less flattened. Leaves distichous, very shortly petiolate, obovate, from nearly orbicular to oblong, rarely above ½ in. long. Stipules minute. Flowers apparently diœcious, the males 2 to 4 together on pedicels of about 1 line, the females solitary on pedicels attaining 2 lines when in fruit. Male perianth-segments petal-like, oblong, nearly 1 line long when fully out and often contracted and thickened at the base. Glands large. Stamens 3, the filaments free; anther-cells parallel. Fruiting perianth-segments enlarged to 2 lines. Capsule depressed-globular, 2½ lines diameter, glabrous. Styles free, shortly bifid. Seeds smooth.

N. Australia. Goulburn islands, *A. Cunningham.*

27. **P. similis,** *Muell. Arg. in Linnæa* xxxiv. 71, *and in DC. Prod.* xv. ii. 369. A glabrous shrub, the branches elongated, slender, slightly compressed. Leaves distichous, shortly petiolate, obovate-oblong, membranous, mostly about ½ in. long. Flowers monœcious, shortly pedicellate, the males in clusters of 3 to 6, the females solitary. Male perianth-segments about ½ line long, petal-like. Glands large. Stamens 3, the filaments free; anther-cells parallel. Female perianth-segments somewhat enlarged after flowering, the inner ones bordered with white and as long as the capsule, the outer ones smaller. Ovary glabrous. Styles free, spreading, divided to about the middle into 2 branches. Capsule depressed-globular, not 2 lines diameter. Seeds smooth.

Queensland. Moreton Bay, *F. Mueller;* Demon Creek, *C. Stuart.*

28. **P. microcladus,** *Muell. Arg. in Linnæa* xxxiv. 71, *and in DC. Prod.* xv. ii. 369. A densely branched shrub, glabrous or nearly so, the main branches rather stout, the ultimate branchlets very numerous, solitary or clustered, filiform, about ½ in. long, mostly with 3 leaves each. Leaves scarcely petiolate, cuneate or obovate-spathulate, obtuse, rigid, glabrous. 4 to 7 lines long. Stipules minute. Flowers monœcious, mostly solitary, the males on very short pedicels, the females on longer ones, attaining ½ to ¾ in. when in fruit. Male perianth-segments ovate. Glands prominent. Stamens 3, the filaments free; anther-cells parallel. Female perianth when in fruit about 2 lines diameter. Glands connate into a lobed cup. Ovary glabrous. Styles deeply divided. Capsule depressed-globular, about 2 lines diameter.

Queensland. Moreton Bay, *F. Mueller.*
N. S. Wales. Clarence river, *Beckler.*

I have seen no specimens answering to the above description taken from Mueller Arg.'s character.

29. **P. grandisepalus,** *F. Muell.; Muell. Arg. in Linnæa* xxxiv. 72, *and in DC. Prod.* xv. ii. 369. Apparently shrubby, glabrous but glaucous, the branches somewhat angular. Leaves distichous, very shortly petiolate, oblong or obovate-oblong, obtuse or mucronulate. Stipules minute. Flowers monœcious, the males 2 or 3 together on exceedingly short

pedicels, the females solitary, the pedicels rather longer but shorter than the fruiting perianth. Male perianth-segments ovate, petal-like, ¾ line, or the inner ones nearly 1 line long. Stamens 3, the filaments free or very shortly united at the base, recurved at the end; anther-cells parallel. Female perianth-segments enlarging under the fruit to about 2 lines. Styles divided to about the middle into 2 branches. Capsule depressed-globular, smooth, scarcely furrowed, about 2½ lines diameter. Seeds longitudinally striate, but sometimes obscurely so.

N. Australia. Fitzmaurice river, *F. Mueller.*—This and several of the following species are difficult to characterize, although they appear to be really distinct. The enlargement of the fruiting perianth is sometimes very variable.

30. **P. Carpentariæ,** *Muell. Arg. in Linnæa* xxxiv. 72, *and in DC. Prod.* xv. ii. 70. An apparently erect shrub, with rigid but slender virgate branches, hoary when young as well as the foliage with a minute tomentum. Leaves oblong-elliptical or obovate-oblong, scarcely petiolate, under ½ in. long. Flowers monœcious, the males 2 or 3 together, sessile or nearly so, the females solitary on very short pedicels, not above 1 line long under the fruit. Male perianth-segments narrow, pubescent outside, nearly 1 line long. Glands large. Stamens 3, the filaments erect and rather thick but free; anther-cells parallel. Female perianth-segments larger than the males and but little enlarged after flowering. Styles 3, erect or scarcely spreading, very shortly 2-lobed. Capsule globular, somewhat depressed, glabrous and smooth, about 2 lines diameter. Seeds smooth.

N. Australia. Tableland, Arnheim's Land and Roper river, Gulf of Carpentaria, *F. Mueller.*

31. **P. Fuernrohrii,** *F. Muell. in Trans. Phil. Soc. Vict.* i. 15, *and in Hook. Kew Journ.* viii. 332. Stems erect, branching, 1 to 1½ ft. high, hard and almost woody at the base but perhaps annual, hoary as well as the foliage with a minute tomentum. Leaves not distichous, very shortly petiolate or almost sessile, obovate-oblong, rarely above ½ in. long. Flowers monœcious, the females solitary on pedicels of about 1 line, alone or with 1 or 2 males on shorter pedicels. Male perianth-segments pubescent, about ¾ line long. Glands rather large. Stamens 3, the filaments free; anther-cells parallel. Female perianth-segments at first about 1 line, but enlarging to from 1½ to 2 lines especially in the Queensland specimens. Disk entire or broadly lobed. Ovary pubescent. Styles 3, divided to about the middle. Capsule pubescent, depressed-globular. Seeds smooth.—Sond. in Linnæa xxviii. 566; Muell. Arg. in DC. Prod. xv. ii. 373.

Queensland. Cleveland Bay, *A. Cunningham;* Dawson river, *F. Mueller;* Rockhampton and Rockingham Bay, *Dallachy;* Rockhampton, *O'Shanesy;* Herbert's Creek and Saunders' Creek, Expedition range, *Bowman;* Barcoo, *Schneider.*

N. S. Wales. Sandy gravelly banks of the Murray river, *F. Mueller* (with the fruiting perianth less enlarged).

S. Australia. Between Lake Eyre and the river Finke, *E. Giles.*

W. Australia? Sharks Bay, a var. with orbicular obovate leaves according to Mueller Arg., but perhaps a different species. I have not seen the specimens.

32. **P. hebecarpus,** *Benth.* An apparently erect rigid shrub, with virgate terete branches, more or less hoary tomentose or villous as well as the foliage. Leaves scarcely distichous, very shortly petiolate or almost sessile, mostly erect, oblong or elliptical, from under ½ in. to nearly 1 in. long. Flowers apparently diœcious, solitary, on very short pedicels, lengthening out to 1 or 2 lines under the fruit. Male perianth-segmenths narrow, petal-like but hirsute outside, 1½ lines long. Glands large. Stamens 3, the filaments erect and closely contiguous but free. Female perianth larger than in the males, the segments broad, herbaceous, tomentose outside and attaining 2 to 3 lines when in fruit. Styles 3, spreading, divided to the middle into 2 branches. Capsule globular, scarcely 3 lines diameter, smooth but more or less sprinkled with hairs. Seeds smooth.

N. Australia. Gulf of Carpentaria, *F. Mueller.*
Queensland. Burdekin river, *F. Mueller* (a rather more villous specimen).

33. **P. lacunarius,** *F. Muell. in Trans. Phil. Soc. Vict.* i. 14, *and in Hook. Kew Journ.* viii. 332. A small diffuse or much-branched annual, our specimens mostly under 6 in., glabrous but often very glaucous, the branches flattened or angular. Leaves very shortly petiolate, oblong or linear-cuneate, contracted at the base, under ½ in. long. Stipules minute. Flowers very small, monœcious, usually 1 female with 2 or 3 males on very short pedicels, lengthening out under the fruit to nearly 1 line. Male perianth of 6 ovate often coloured white or reddish segments of about ¼ line. Glands small. Stamens 3, the filaments free; anther-cells parallel. Female perianth rather larger, the segments about ½ line long or slightly larger under the fruit. Styles short, spreading, 2-lobed. Capsule glabrous, depressed-globular, 3-furrowed. Seeds finely striate longitudinally.—Sond. in Linnæa xxviii. 566; Muell. Arg. in DC. Prod. xv. ii. 370.

N. S. Wales. Junction of the Murray and Darling rivers, *F. Mueller;* Darling desert, *Godwin and Dallachy.*

34. **P. trachyspermus,** *F. Muell. in Trans. Phil. Soc. Vict.* i. 14, *and in Hook. Kew Journ.* viii. 210. A glabrous glaucous annual, with ascending branching stems about 6 in. high in the specimens seen. Leaves almost sessile, broadly oblong, obtuse, ¼ to ½ in. long. Stipules spreading as in *P. rhytidospermus,* but very minute. Flowers monœcious (*Muell Arg.*), but no males on our specimens. Females solitary in the axils, on very short pedicels. Perianth-segments ovate, spreading, about ¾ line long. Styles 3, very short, spreading, broad and emarginate at the end. Capsule not seen perfect, but from the remains it appears to have been ovoid-globular, about 3 lines long.—Sond. in Linnæa xxviii. 566; Muell. Arg. in DC. Prod. xv. ii. 327.

N. S. Wales. At the Junction of the Murray and Darling rivers, *F. Mueller.*—I have only seen two imperfect specimens, one in the Muellerian the other in the Hookerian herbarium.

35. **P. australis,** *Hook. f. in Hook. Lond. Journ.* vi. 284, *and in Fl. Tasm.* i. 341. A low glabrous undershrub with a woody base and

numerous ascending wiry stems of 3 to 6 in. Leaves not distichous, ovate or obovate, obtuse or acute, coriaceous, flat, $1\frac{1}{2}$ to 3 lines long. Flowers diœcious, the males in clusters of 2 or 3, the females solitary, on pedicels of $\frac{1}{2}$ to 1 line. Male perianth-segments red, ovate-oblong, under $\frac{3}{4}$ in. long. Glands large. Stamens 3, the filaments free; anther-cells parallel. Female perianth-segments nearly 1 line long, narrower and more acute than the males. Disk broadly lobed. Ovary glabrous. Styles 3, deeply divided into 2 branches. Capsule depressed, about $1\frac{1}{2}$ lines diameter. Seeds smooth.—Muell. Arg. in DC. Prod. xv. ii. 373.

N. S. Wales.? Some specimens from Lachlan river, *A. Cunningham*, appear to belong to this species, but the flowers are not in a state for examination.

Tasmania. Roadsides, probably common, but overlooked, *J. D. Hooker*.

36. **P. thymoides,** *Sieb. Pl. Exs.; Sond. in Linnæa* xxviii. 566. A shrub of 1 to 2 ft., more or less pubescent or hirsute, at least the young branches. Leaves not distichous, nearly sessile, from broadly obovate to narrow-cuneate, obtuse mucronate truncate or emarginate, the margins recurved or revolute, coriaceous, rarely above $\frac{1}{4}$ in. and sometimes only 1 to $1\frac{1}{2}$ lines long, the midrib very prominent underneath and sometimes the whole leaf complicate. Stipules small and black. Flowers diœcious, the males 2 or 3 together, the females solitary, on pedicels of $\frac{1}{2}$ to 1 line. Male perianth-segments 6, rarely 5, $\frac{3}{4}$ to nearly 1 line long, obtuse or the inner ones mucronulate. Glands large. Stamens 3, the filaments free; anther-cells parallel. Female perianth larger than the male, with ovate herbaceous segments. Ovary more or less hirsute. Styles 3, deeply divided into 2 branches. Capsule depressed-globular, 2 to $2\frac{1}{2}$ lines diameter. Seeds smooth or minutely tuberculate.—Muell. Arg. in DC. Prod. xv. ii. 372; *P. hirtellus*, F. Muell. Herb.; Muell. Arg. in Linnæa xxxii. 22; *P. ledifolius*, A. Cunn. Herb.

N. S. Wales. Port Jackson to the Blue Mountains, *Sieber, n.* 264, *and Fl. Mixt. n.* 475, *A. Cunningham*, and others; New England, *C. Stuart;* Mudgee, *N. Taylor*; Twofold Bay, *F. Mueller*.

Victoria. Wilson's Promontory, Wombaya, Macalister's and Genoa rivers, *F. Mueller;* Grampians, *Wilhelmi;* Wimmera, *Dallachy*.

S. Australia. Tattiara country, *Woods* (these and the Wimmera specimens much less hirsute, sometimes nearly glabrous).

Var. *glabrata.* Almost glabrous except the young shoots.—Twofold Bay, *Mossman, F. Mueller.*

37. **P. scaber,** *Klotzsch in Pl. Preiss.* i. 179. A bushy shrub attaining 2 to 4 ft., the branches angular, glabrous but often glandular-scabrous. Leaves not distichous, obovate oblong or rarely almost linear, obtuse or mucronate, rounded or tapering at the base, rather coriaceous, mostly under $\frac{1}{4}$ in. long. Stipules small, brown or black. Flowers diœcious, the males clustered, the females solitary or 2 together, all on pedicels shorter than the perianth. Male perianth-segments ovate or oblong, $\frac{1}{2}$ to $\frac{3}{4}$ line long. Glands conspicuous. Stamens 3, the filaments free; anther-cells parallel. Female perianth-

segments narrower and more herbaceous than the males, rather above 1 line long when in fruit. Styles. 3, recurved, undivided. Capsule depressed, slightly 3-furrowed, about 2 lines diameter. Seeds smooth.—Muell. Arg. in DC. Prod. xv. ii. 372.

W. Australia. Cape Riche, *Preiss, n.* 1200, *Drummond, 5th coll. n.* 223; Bald island, *Oldfield;* Fitzgerald river, *Maxwell* (with small narrow leaves); between Esperance Bay and Russel range, *Dempster* (with rather large stipules).

38. **P. indigoferoides,** *Benth.* A shrub with rather slender rigid virgate branches, and numerous almost filiform leafy branchlets of 1 to 2 in. resembling pinnate leaves, the whole hoary with a minute papillose pubescence. Leaves distichous, petiolate, ovate, acute, flat, coriaceous rarely above 2 lines long. Flowers monœcious, mostly solitary, on pedicels of about $\frac{1}{2}$ line. Male perianth-segments pubescent outside, about $\frac{3}{4}$ line long, the 3 inner ones larger and more petal-like than the outer. Glands large. Stamens 3, the filaments free; anther-cells parallel. Female perianth rather larger than the male. Disk lobed. Ovary glabrous, conspicuously 6-furrowed. Styles 3, erect, shortly bifid. Fruit not seen.

N. Australia. York Sound, N.W. Coast, *A. Cunningham.*

39. **P. aridus,** *Benth.* An erect much-branched rigid shrub, the smaller branchlets almost filiform, quite glabrous. Leaves distichous, shortly petiolate, oblong, coriaceous, mostly about $\frac{1}{4}$ in. long. Flowers apparently monœcious but both sexes solitary in different axils, on exceedingly short pedicels scarcely lengthened under the fruit. Male perianth-segments (perhaps not yet fully developed) ovate, petal-like, $\frac{1}{2}$ line long. Glands small. Stamens 3, the filaments free; anther-cells parallel. Female perianth slightly enlarged under the fruit, glabrous, almost coriaceous. Capsule depressed-globular, glabrous, 2 lines diameter, 3-celled. Styles not seen.

N. Australia. Barren shores of Brunswick Bay and Port Warrender, Vansittart Bay, N.W. Coast, *A. Cunningham.*

40. **P. Gunnii,** *Hook. f. in Hook. Lond. Journ.* vi. 284, *and Fl. Tasm.* i. 341. A tall shrub, sometimes almost arborescent, quite glabrous, with slender somewhat angular branches. Leaves distichous, obovate or orbicular, rarely obovate-oblong, very obtuse or retuse, mostly about $\frac{1}{2}$ in. but sometimes $\frac{3}{4}$ in. long. Flowers monœcious, in axillary clusters of 3 or more males and 1 female, on slender pedicels of 2 to 3 lines. Male perianth-segments 6, ovate, about $\frac{1}{2}$ line long. Glands prominent. Stamens 3, the filaments free, variable in length; anther-cells parallel. Female perianth scarcely larger than the male. Ovary 3-celled, glabrous. Styles free, bifid. Capsule 2 lines diameter, not furrowed.—Muell. Arg. in Linnæa xxxii. 20, and in DC. Prod. xv. ii. 368.

N. S. Wales. Twofold Bay, *F. Mueller.*

Victoria. Between Yowaka and Sealer's Cove, Mount Hunter, Broadribb river, *F. Mueller.*

Tasmania. Dense forests at George Town, Circular Head and Rocky Cape, *Gunn;* King s Island, *Herb. F. Mueller.*

The specimens with obovate leaves can scarcely be distinguished from *P. Gasstrœmii*, except by the free stamens.

Var. *saxosus*, F. Muell. More rigid, with fewer flowers.—*P. saxosus*, F. Muell. in Linnæa xxv. 441.

Victoria. Wimmera, *Dallachy*.

S. Australia. Flinders Range, Cudnaka, towards Lake Torrens, *F. Mueller*.

SECT. 5. EUPHYLLANTHUS, *Muell. Arg.*—Herbs shrubs or trees of varied habit. Leaves usually distichous. Stamens 3, the filaments free or more or less united; anthers free, the cells short, more or less diverging or opening transversely, and often separated by a broad connective. Glands present. Ovary 3-celled. Styles free. Capsule dry.

This is rather an artificial section than a natural group, some species bearing a close resemblance to corresponding ones in *Paraphyllanthus*, and only distinguished by their anthers; the flowers are, however, usually much smaller than in that section.

41. **P. simplex,** *Retz; Muell. Arg. in DC. Prod.* xv. ii. 391. A glabrous annual or perennial, with decumbent ascending or erect stems, rarely above 1 ft. high, flattened when young. Leaves distichous, almost sessile, lanceolate or almost linear and acute or the lower ones oblong and more obtuse, rarely above ½ in. long. Stipules very small, brown or white. Flowers monœcious or almost diœcious, the males in clusters of 3 to 6, on filiform pedicels of ½ to ¾ line, with occasionally a single female on a filiform pedicel of 3 to 4 lines, the females when without males often 2 or even 3 from the same axil, all turned to one side. Male perianth-segments 6, spreading, coloured, not ¼ line long. Glands prominent. Stamens 3, the filaments free; anther-cells globular, opening obliquely or transversely. Female perianth-segments longer and narrower than in the males, attaining ½ line under the fruit. Styles 3, more or less deeply 2-branched. Capsules depressed, glabrous, smooth, scarcely 1½ lines diameter. Seeds usualy punctate or tuberculate when quite ripe, but sometimes paler coloured and smooth, although apparently full grown.

Queensland. Endeavour river, *A. Cunningham;* Rockingham Bay, *Dallachy;* Rockhampton, *Bowman, O'Shanesy*.

Var. *leiospermus*. Stems 1 to 1½ ft. high, and evidently annual. Pedicels shorter than usual. Seeds almost or quite smooth.—Narran river, *Mitchell*.

P. Beckleri, Muell. Arg. in Linnæa xxxiv. 74, and in DC. Prod. xv. ii. 390, from **N. S. Wales.** Clarence river, *Beckler*, which I have not seen, is also said to differ from *P. simplex* only in the smooth seeds, which here no more than in *P. calycinus* appear to be available as a specific distinction.

P. conterminus, Muell. Arg. in Linnæa xxxii. 31, and in DC. Prod xv. ii. 389, from New Holland, *Hodgson*, *n.* 215, which I have not seen, is said clearly to resemble *P. simplex*, but besides the smooth seeds of *P. Beckleri* to differ in its diœcious flowers. Mueller Arg. however appears to have seen the female only, and many specimens of *P. simplex* are to be seen without any male flowers, which are always very deciduous, and often probably absent from the first.

42. **P. filicaulis,** *Benth.* oA small glabrous plant, with a perennial woody base and numerous filiform stems, from 1 to 6 in. long, the habit approaching that of *P. australis* but more slender. Leaves obovate or the lower ones orbicular, under ¼ in. long. Stipules minute.

Flowers monœcious, minute, on filiform pedicels of ½ to 1 line, lengthening to near 2 lines under the fruit. Male perianth-segments ovate, coloured, under ¼ line long. Glands small. Stamens 3, the filaments free; anther-cells globular, divergent. Female perianth-segments narrower and rather longer than in the males. Disk truncate and lobed. Styles 3, bifid. Capsule glabrous and smooth, under 1 line diameter.

N. S. Wales. New England, *C. Stuart.*

43. **P. minutiflorus,** *F. Muell. Herb.; Muell. Arg. in Linnæa* xxxiv. 75 *and in DC. Prod.* xv. ii. 398. A glabrous annual, with filiform procumbent or ascending stems from a few inches to above 1 ft. long. Lower leaves broadly ovate or almost orbicular, 2 to 3 lines long, the upper ones oblong-lanceolate or almost linear, ¼ to ½ in. long. Flowers exceedingly minute, the females on filiform pedicels of ½ to 1½ lines, the males on shorter pedicels, the female perianth not ¼ line long and the male still smaller. Glands apparently distinct in both sexes. Stamens 3, the filaments free; anthers not seen perfect. Styles short, deeply 2-lobed. Capsules depressed, under 1 line diameter. Seeds smooth.

N. Australia. Upper Victoria river, *F. Mueller*.—Apparently the same species, small young plants of 1 to 2 in., with orbicular leaves, Port Darwin, *Schultz, n.* 326, and elongated specimens with longer pedicels and rather broad small leaves, York Sound, N. W. Coast, *A. Cunningham.*

Var.? *gracillimus.* Filiform branches very slender, 1 ft. long. Leaves all narrow, 2 to 4 lines long.—*P. gracillimus*, F. Muell. in Herb. Hook.

Queensland. Moreton Bay, *F. Mueller.*

The above are probably all forms of one species, but with the minuteness of the flowers it is difficult to establish definite characters from the imperfect specimens in our herbaria.

SECT. 6. REIDIA.—Trees or shrubs. Leaves distichous. Male perianth of 4 segments. Stamens 2, the filaments united in a central column, anthers with separate cells having the appearance of 4 1-celled anthers verticillate round the top of the column. Female perianth of 4 to 6 segments. Styles 3.—*Reidia.* Wight; *Eriococcus*, Hassk.

44. **P. Armstrongii,** *Benth.* A glabrous shrub or tree, the branchlets slender, 4 to 8 in. long, with thin distichous leaves resembling pinnate leaves crowded at the ends of the branches. Leaves very shortly petiolate or almost sessile, ovate or ovate-lanceolate, membranous, mostly about 1 in. long. Stipules minute. Flowers monœcious, the males several together, on capillary pedicels of 2 to 3 lines, the females solitary on pedicels scarcely longer but thickened towards the end. Male perianth rotate, spreading to rather more than 1 line diameter, consisting of 4 broadly rhomboidal crenate segments opposite in pairs. Glands large. Anther-cells radiating from a short central column which protrudes slightly beyond them, each cell opening transversely in 2 small valves. Female perianth of 6 ovate dentate segments ½ line long. Disk broadly cup-shaped, half as long as the perianth. Ovary glabrous, with 3 exceedingly short bifid styles. Fruit not seen.

N. Australia. Port Essington, *Armstrong.*

18. BREYNIA, Forst.

(Melanthesa, *Blume;* Melanthesopsis, *Muell. Arg.*)

Flowers monœcious, axillary, solitary or few together. Male fl.: Perianth turbinate, flat-topped, the small orifice in the centre almost closed by 6 short lobes. Stamens 3, united in a central column, without any rudimentary ovary; anthers 2-celled, adnate to the column, the cells parallel, opening longitudinally in 2 valves. Female fl.: Perianth turbinate or campanulate, with 6 very short lobes or teeth, sometimes minute or obsolete. Ovary sessile or shortly stipitate, 3-celled, with 2 ovules in each cell, thick and fleshy above the cells. Styles 3, very short, erect or slightly spreading and entire in the Australian species. No glands or disk in either sex. Fruit a globular or depressed indehiscent berry. Seeds triangular, with a straight inner angle and a curved back, the hilum small, the lower end with a large nearly closed ventral cavity between the inner and outer coating. Albumen not very copious. Embryo curved, the cotyledons broad, parallel to the back of the seed.—Shrubs or small trees, the smaller branches slender, the foliage usually but not always drying black. Leaves alternate, petiolate, usually broad, entire. Flowers small, on short pedicels. Fruits usually red.

The genus is generally spread over tropical Asia and the Pacific islands. Of the four Australian species, one and perhaps two are also in the Indian Archipelago, the two others appear to be quite endemic. The genus is allied to the section *Glochidion* of *Phyllanthus*, but readily distinguished by the peculiar male perianth, and by the more baccate fruit. The section *Melanthesopsis*, with longer spreading divided styles, has not as yet been detected in Australia. The seeds appear to me to be the same in both sections without anything that can be properly called an arillus. The hilum at the upper end is very small, the large cavity at the lower end appears to lie between the inner and outer coating of the seed, both of them crustaceous except at the lower end of the cavity where the outer one is membranous and wears away leaving a small opening.

Fruiting perianth spreading flat to a diameter of about 3 lines . 1. *B. cernua.*
Fruiting perianth broadly turbinate or concave, enlarging to about 2 lines diameter. Styles very short erect or spreading on the obtuse fruit . . . . 2. *B. oblongifolia.*
Fruiting perianth scarcely enlarged.
  Ovary and capsule more or less contracted into a stipes at the base and crowned with three protuberances surrounding the styles . . . . 3. *B. stipitata.*
  Ovary and capsule sessile, the ovary tapering at the top, the capsule suddenly contracted into a beak . . . . 4. *B. rhynchocarpa.*

1. **B. cernua,** *Muell. Arg. in DC. Prod.* xv. ii. 439. A glabrous shrub, with the broad almost orbicular leaves and the flowers of *B. stipitata*, but the female perianth rather larger at the time of flowering, very shortly broadly and retusely 6-lobed, and when in fruit spreading out quite flat to the diameter of fully 3 lines. Capsule sessile, globular, without appendages, crowned by the very short styles.—*Melanthesa cernua*, Dcne. Herb. Tim. Descr. 155.

N. Australia. N.W. Coast, *A. Cunningham.*—The specimens have female flowers only, with unripe fruits, or fruiting perianths from which the capsule has already fallen off, but as far as they go, they agree perfectly with the typical Timor specimens of *B. cernua*. A specimen from Point Pearce, *F. Mueller*, may also be the same species, with the fruiting perianth not so much developed, but the fruit is not yet quite ripe.

Queensland. Cape York, *Daemel.*

2. **B. oblongifolia,** *Muell. Arg. in DC. Prod.* xv. ii. 440. A glabrous shrub, attaining 10 to 15 ft., with slender branches. Leaves petiolate, ovate or broadly oblong, obtuse, ½ to 1 in. long. Stipules small, rather rigid, acute. Flowers monœcious, the females solitary, with or without 1 or 2 males, the males often in clusters of 2, 3 or more, and the cluster sometimes growing out into a short raceme, with a rhachis of ½ to 1 line, covered with imbricate stipule-like bracts. Pedicels usually about 1 line long. Male perianth nearly 1 line long, broadly turbinate, flat-topped with the orifice closed, the stamens quite included, the anthers covering the greater portion of the central column. Female perianth spreading and shortly and broadly 6-lobed, about 1 line diameter when in flower, enlarging to 2 lines under the fruit but remaining concave, not spreading flat as in *B. cernua*. Styles short, entire, erect or spreading. Capsule sessile, globular, about 3 lines diameter, obtuse, without any protuberances round the styles.—*B. cinerascens*, Baill. Adans. vi. 344.

Queensland. Broad Sound, *R. Brown;* Brisbane river, Moreton Bay, *F. Mueller;* Percy island, *A. Cunningham* (with broader leaves); Rockhampton, *O'Shanesy, Thozet;* Cape York, *Daemel.*

N. S. Wales. Port Jackson to the Blue Mountains, *R. Brown, Sieber, n.* 566, *A. Cunningham,* and others; Hunter's river, *Oldfield;* New England, *C. Stuart.*

3. **B. stipitata,** *Muell. Arg. in DC. Prod.* xv. ii. 442. A tall glabrous shrub. Leaves petiolate, ovate or almost orbicular, very obtuse, ¾ to 1½ in. long. Flowers both male and female usually solitary. Male perianth of *B. oblongifolia*. Female perianth broadly turbinate, very shortly and obtusely sinuate-lobed, scarcely enlarged although more open under the fruit. Capsule globular, contracted at the base into a stipes sometimes very short sometimes half as long as the capsule, crowned by 3 more or less confluent protuberances forming a fleshy ring round the short styles and about their length, both the stipes and the terminal protuberances already apparent on the ovary at the time of flowering.

N. Australia. Islands of the Gulf of Carpentaria, *R. Brown, Henne;* Port Darwin, *Schultz, n.* 546, 581; Prince of Wales islands, *R. Brown.*

Queensland. Rockingham Bay, *Dallachy;* Cleveland Bay, *Bowman* (with narrower leaves).

B. Muelleriana, Baill. Adans. vi. 344, from Rockingham Bay, *Dallachy*, from the very imperfect specimen in Herb. F. Mueller, appears to be a slight variety of *B. stipitata*, with much larger leaves.

4. **B. rhynchocarpa,** *Benth.* Apparently a shrub with the habit of *B. oblongifolia*, but the specimens not drying so black, and assuming a glaucous hue. Leaves broadly ovate or orbicular, very obtuse, mostly about 1 in. long. Male flowers only seen very young, but apparently normal. Female flowers solitary, on very short pedicels. Perianth narrow-turbinate, under 1 line long at the time of flowering, truncate

and entire or with 3 or 4 minute and distant teeth. Ovary nearly sessile, tapering at the top into 3 short distinct concave erect styles, stigmatic on the inner surface. Capsule sessile or nearly so, globular, terminating abruptly in a distinct narrow-conical beak, the perianth scarcely enlarged although opened out rather broader.

N. Australia. King's Sound, N.W. Coast, *Hughan.*

19. SECURINEGA, Juss.

(Fluggea, *Willd.*)

Flowers diœcious, in axillary clusters. Male fl.: Perianth divided to the base into 5 petal-like segments. Stamens 5 or sometimes 4, exserted, alternating with as many glands, and surrounding a central 2-fid or 3-fid pistil without any ovary, but often as long as the stamens; anthers with 2 parallel cells opening longitudinally in 2 valves. Female fl.: Perianth of the males. Disk flat, with a free dentate margin. Ovary 3-celled, with 2 ovules in each cell. Styles 3, recurved and bifid. Fruit dry or scarcely succulent, the pericarp thin, irregularly separating into cocci. Seeds triangular, with the inner edge straight, the back semicircular; testa crustaceous, with a ventral cavity between the inner and outer coating. Albumen rather scanty, curved round the cavity of the seed. Embryo also curved, the cotyledons broad, parallel to the back of the seed.—Shrubs. Leaves alternate, petiolate, distichous, entire. Flowers very small. Fruits red.

The genus is spread over the warmer regions of Asia and Africa, one species reaching as far north as Spain. Both the Australian species are common in E. India, and one has the wide range of the genus. The seeds, in the section *Fluggea* at least, to which the Australian species belong, have the peculiar structure of those of *Breynia*. The genus is, however, readily known by the perianth; the exserted stamens, the rudimentary pistil, the dry fruit, &c.

Branches unarmed. Leaves above 1 in. long . . . . 1. *S. obovata.*
Branches often spinescent. Leaves ½ to ¾ in. long, often emarginate or very obtuse . . . . 2. *S. Leucopyrus.*

1. **S. obovata,** *Muell. Arg. in DC. Prod.* xv. ii. 449. A tall unarmed shrub, quite glabrous but sometimes glaucous, the smaller branches often angular when young. Leaves ovate, usually broad and sometimes almost orbicular, rarely broadly oblong, very obtuse, 1 to 2 in. long or when very luxuriant nearly 3 in., prominently penniveined and the numerous reticulate veinlets often also prominent underneath. Flowers minute, the males usually very numerous in the cluster, the females fewer, both on filiform pedicels of 1 to 2 lines. Perianth in both sexes about ½ line long, the 3 inner segments rather larger than the outer. Stamens longer than the perianth, the anthers opening outwards. Styles rather broad. Capsule red, depressed-globular, not exceeding 2 lines in diameter.—*Xylophylla obovata,* Willd. Enum. Hort. Berol. 329; *Leptonema melanthesoides,* F. Muell. in Hook. Kew Journ. ix. 17; *Fluggea melanthesoides,* F. Muell in Trans. Bot. Soc. Edinb. vii. 490, and very numerous other synonyms given by Muell. Arg. l.c.

N. Australia. Regent's river and Cygnet Bay, N. W. coast, *A. Cunningham;* King's Sound and Collier Bay, *Chapman*, *Hughan;* Hierson island, *Gregory's Expedition;* Victoria and Fitzmaurice rivers, *F. Mueller;* Islands of the gulf of Carpentaria, *R. Brown;* Sweers island and Albert river, *Henne;* Port Darling, *Schultz, n.* 578.

Queensland. Cape York, *M'Gillivray;* Gilbert river and Howick's group, *F. Mueller;* Port Denison, *Fitzalan;* Rockingham and Edgecombe Bays, *Dallachy;* Broad Sound and Bowen river, *Bowman;* Kennedy district, *Daintree;* Flinders river, *Sutherland;* Port Mackay, *Nernst.*

The species is common in tropical Asia and Africa.

2. **S. Leucopyrus,** *Muell. Arg. in DC. Prod.* xv. ii. 451. A large straggling shrub, quite glabrous, with numerous small rigid branchlets occasionally terminating in a spine. Leaves ovate obovate or almost orbicular, very obtuse or emarginate, smaller and more membranous than in *S. obovata*, and usually not above ½ in. long. Flowers and fruit entirely those of *P. obovata.*—*S. virosa,* Baill. Adans. vi. 334, and several synonyms given by Muell. Arg. l.c.

Queensland. Gilbert river, *F. Mueller;* Rockhampton, *Dallachy*, *O'Shanesy;* Bowen river, *Bowman.*—Common in many parts of East India.

20. NEORŒPERA, Muell. Arg.

Flowers monœcious, in axillary clusters. Male fl.: Perianth divided to the base into 5 or 6 petal-like segments. Stamens 5 or 6, exserted, surrounding a broad central irregularly-lobed disk or abortive ovary; anthers with 2 parallel cells, opening longitudinally in 2 valves. Female fl.: Perianth deeply divided into 6 lobes, narrower than in the male. Disk shortly lobed. Ovary 3-celled, with 2 ovules in each cell. Styles 2, clavate or broad, undivided. Capsule globular, separating into 3 2-valved coriaceous cocci. Seeds ovate-oblong, slightly compressed, without any carunculus. Testa smooth and shining, without any internal cavity. Albumen rather copious; embryo nearly straight, with broad cotyledons.—Shrubs. Leaves alternate, shortly petiolate, entire, coriaceous. Flowers not so small as in *Securinega*. Capsule much larger.

The genus is limited to Australia. Baillon reduces it to a section of *Securinega*, but the structure of the fruit and seed appears to differ far too much to sanction the union.

Leaves elliptical-oblong, mostly about 1 in. long. Styles elongated clavate . . . . 1. *N. buxifolia.*
Leaves cuneate-oblong or almost obovate, ½ to ¾ in. long. Styles short broad and thick . . . . 2. *N. Banksii.*

1. **N. buxifolia,** *Muell. Arg. in DC. Prod.* xv. ii. 489. A glabrous shrub. Leaves elliptical-oblong, obtuse, entire, coriaceous, shining, not exceeding 1 in., very much like those of some varieties of *Buxus sempervirens.* Male flowers numerous in the clusters, on pedicels of 3 to 4 lines. Perianth about 1 line long, of 5 or 6 segments, the inner ones larger than the outer, very concave or almost cucullate; stamens exserted. Female flowers solitary in the male clusters, on rather longer and stouter pedicels. Perianth-segments rather longer

and narrower, shortly united at the base. Styles 3, rather long, clavate at the end but not divided. Capsule globular, about 3 lines diameter.—*Rœpera buxifolia*, F. Muell. Herb.; *Securinega Muelleriana*, Baill. Adans. vi. 333.

Queensland. Princhester Creek, *Bowman;* Lizard island, *Walter.*—Some specimens, also without flowers, from Endeavour river, *A. Cunningham*, and referred by him to *Sersalisia obovata*, appear to belong to the species.

2. **N. Banksii,** *Benth.* A twiggy glabrous shrub of several feet. Leaves cuneate-oblong, very obtuse or emarginate, rarely mucronate, contracted at the base and very shortly petiolate or almost sessile, coriaceous, slightly veined, $\frac{1}{2}$ to $\frac{3}{4}$ in. long. Flowers few in the clusters, the males rather smaller than in *N. buxifolia*, and the stamens not exserted in our specimens, in which, however, the flowers are not yet full blown. Female flowers on pedicels of about $\frac{1}{2}$ in., but only seen in fruit, which is the same as in *N. buxifolia*, except that the styles are short and very broad. Seeds not seen quite ripe.—*Phyllanthus Banksii*, A. Cunn. Herb.

Queensland. Sandy ridges, north shore, Endeavour river, *A. Cunningham*,

21. HEMICYCLIA, Wight et Arn.

Flowers diœcious, axillary. Male fl.: Perianth of 4 or 5 much imbricate segments, the inner ones usually more petal-like concave and larger than the outer. Stamens indefinite (4 to 23) inserted round a broad central concave entire or undulate-lobed disk; filaments free; anthers with 2 parallel cells, opening longitudinally in 2 valves. Female fl.: Perianth of the males, or rather larger. Disk flat, with a free margin. Ovary obliquely 1-celled, with 2 ovules. Style or stigma single, broadly reniform or semi-orbicular, flat or recurved, entire or emarginate. Fruit an indehiscent drupe, with a succulent mesocarp and a bony endocarp. Seed usually solitary, oblong, furrowed down one side; testa rather thin; albumen copious. Embryo straight or nearly so, with broad flat cotyledons and a short narrow radicle.—Trees or shrubs. Leaves alternate, petiolate, entire, coriaceous when full grown. Flowers solitary or few together, small, pedicellate, the male clusters sometimes apparently forming a short raceme from the abortion of the leaves on the very short flowering branches.

The genus contains but few species, dispersed over the East Indian Peninsula, Ceylon, and the Eastern Archipelago. Of the three Australian species, one appears to be the same as the commonest of the Indian ones, the two others are endemic. All three are, however, very closely allied to each other.

Filaments exserted, much longer than the small ovoid anthers . 1. *H. sepiaria.*
Filaments very short; anthers twice as long, oblong, not exceeding the perianth.
  Ovary glabrous. 2. *H. australasica.*
  Ovary densely villous . . . . 3. *H. lasiogyna.*

1. **H. sepiaria,** *W. & Arn.; Muell. Arg. in DC. Prod.* xv. ii. 487, var.? *oblongifolia.* A shrub of 6 to 9 ft., the young shoots minutely

pubescent, the adult foliage glabrous. Leaves petiolate, ovate-oblong or oblong-lanceolate, obtuse, coriaceous and shining when full-grown, but most of those on the flowering specimens still young and membranous, finely veined underneath, 1½ to 2½ in. long. Male flowers several together in axillary clusters sometimes growing out into short racemes, the filiform pedicels about 2 lines long. Perianth-segments 4, broad, about 1 line long. Stamens 6 to 8 in the flowers examined, inserted round a hollow disk, with the margin undulate as in *H. australasica*. No female specimens seen of the Australian variety.

N. Australia. On the beach, Port Darwin, *Schultz*, *n.* 746.—The species is common in Ceylon from the sea-coast to an elevation of 1500 ft., and appears also to have an extended range in the Peninsula. The Australian specimens differ slightly in the narrower leaves as well as in the disk, but are probably a variety only; the stamens are entirely those of *H sepiaria*, and not of the two following species.

2. **H. australasica,** *Muell. Arg. in DC. Prod.* xv. ii. 487. A spreading tree attaining 40 ft., rarely reduced to a shrub, the young shoots slightly pubescent, but soon becoming glabrous, sometimes rather glaucous. Leaves petiolate, from broadly ovate to ovate-oblong, obtuse, coriaceous, often shining above, finely veined underneath, 1½ to 3 in. long. Flowers solitary or few together in axillary clusters, or the males sometimes forming short racemes with a rhachis from under 1 line to 3 or 4 lines, the pedicels 1 to 2 lines long in the males, rather longer in the females. Male perianth-segments broad and concave, especially the inner ones, 1¼ to 1½ lines long. Stamens varying from 5 to 10 on the same specimen; anthers oblong, longer than the very short filaments, and not exceeding the perianth. Margin of the disk usually undulate. Female perianth rather larger than the male. Fruit ovoid-globular, 5 to 6 lines long, very smooth, red and succulent, with a bony endocarp.—*H. sepiaria* var. *australasica*, F. Muell. Fragm. iv. 119.

Queensland. Islands of Torres' Straits, *Henne;* Edgecombe and Rockingham Bays, *Dallachy;* Burdekin river, *F. Mueller;* Cleveland Bay, *Bowman;* Kennedy district, *Daintree;* Rockhampton, *O'Shanesy;* Cape Sidmouth, *Brasier*.

N. S. Wales. Clarence river, *Beckler*.

Var.? with longer narrower leaves, a tree of 100 ft Lord Howe's island, *C. Moore*.—There are also among the Queensland plants specimens with narrow leaves, and others with larger or smaller fruits, which in the absence of male flowers I am unable to refer with certainty to this species.

F. Mueller, and after him Baillon, refer the whole species to the East Indian *H. sepiaria*, Wight and Arn.; but notwithstanding much general resemblance, it appears to differ essentially in the stamens and in some minor particulars.

3. **H. lasiogyna,** *F. Muell. Fragm.* iv. 119. A tree with the habit of *H. australasica*, but the leaves usually larger, 2 to 3 in. long, membranous at the time of flowering, but becoming coriaceous when in fruit. Flowers rather larger than in *H. australasica*, the perianth-segments mostly fringed or ciliate. Anthers large, oblong, on very short filaments as in that species, but often bearing a few hairs, and the ovary always densely villous.

N. Australia. Port Essington, *Leichhardt, Armstrong;* Port Darwin, *Schultz*, *n.* 700, 742 (females) *and* 692 (males).

22. BRIEDELIA, Willd.

Flowers monœcious, in axillary clusters or solitary. Male fl.: Calyx deeply divided into 5 segments, valvate in the bud, spreading when in flower. Petals 5, scale-like, stipitate or spathulate, the small lamina usually broad and dentate. Disk broad, with a free entire or slightly lobed margin. Stamens 5, inserted on a central column arising from the disk, the filaments spreading horizontally under an abortive or lobed style which terminates the column without any ovary. Anther-cells parallel, opening longitudinally. Female fl.: Calyx of the males or with longer narrow segments. Disk the same but with the addition of an inner erect margin or cup closely surrounding the ovary. Ovary 2-celled or very rarely and exceptionally 3-celled, with 2 ovules in each cell. Styles distinct or connate at the base, more or less 2-lobed or nearly entire. Fruit a small berry or drupe, with a succulent indehiscent epicarp, the endocarp rather hard or crustaceous, separating into 2 indehiscent cocci or pyrenes. Seeds usually solitary in each pyrene, with a longitudinal furrow on the inner face; albumen copious; embryo nearly straight, with broad flat cotyledons and a short narrow radicle.—Shrubs or trees. Leaves alternate, petiolate, entire, with fine parallel diverging primary veins and transverse veinlets prominent on both sides. Stipules small. Flowers small, sessile or very shortly pedicellate, and surrounded by small scale-like bracts, the males and females in the same or separate clusters. Berries or drupes small, red or black.

The genus extends over the warmer regions of Asia and Africa. Of the four Australian species, two are also Asiatic, the two others appear to be endemic.

Whole plant glabrous.
  Flowers few together. Male calyx-segments 1 line long. Staminal column not half so long as the filaments. Styles very short . . 1. *B. exaltata.*
  Flowers in dense clusters. Male calyx-segments ½ line long. Staminal column nearly as long as the filaments. Styles rather long . . . . 2. *B. ovata.*
Young shoots and underside of the leaves tomentose-pubescent.
  Flowers in dense clusters. Female calyx about ¾ line long . . . 3. *B. tomentosa.*
  Flowers solitary or 2 or 3 together. Female calyx at least 1 line long . . . . 4. *B. faginea.*

1. **B. exaltata,** *F. Muell. Fragm.* iii. 32. A tree of 60 to 70 ft. perfectly glabrous. Leaves shortly petiolate, ovate-lanceolate, acute or rather obtuse, with much more numerous primary veins than in *B. tomentosa,* 2 to 4 in. long, somewhat glaucous underneath, often drying black. Flowers few together and almost sessile, the floral leaves often deciduous or abortive. Male calyx-segments above 1 line long, thin and spreading. Disk large. Staminal column very short, the filaments at least twice as long. Berries black, globular, much larger than in *B. ovata.* Styles exceedingly short, but perhaps not perfect in the specimens seen.—*B. ovata* var. *exaltata,* Muell. Arg. in DC. Prod. xv. ii. 495; *Amanoa ovata,* Baill. Adans. vi. 336.

N. S. Wales. Clarence river, *Beckler*; Richmond river, *Herb. F. Mueller*; Tweed river, *Guilfoyle.*

2. **B. ovata,** *Dcne.*; *Muell. Arg. in DC. Prod.* xv. ii. 495. A tall shrub, with slender branches, quite glabrous. Leaves on very short petioles, broadly ovate-elliptical, very obtuse or with a very short obtuse acumen, 2 to 4 in. long, the primary veins usually more distant than in *B. tomentosa.* Flowers sessile or nearly so, numerous in the clusters and precisely similar to those of *B. tomentosa,* the staminal column about as long as the free part of the filaments. Berry globular, somewhat didymous, rather larger than in *B. tomentosa.* Styles rather long.

N. Australia. Sims island, North coast, *A. Cunningham.*—The species is also in Timor and Malacca. A. Cunningham's specimens entirely agree with those described by Decaisne from Timor.

3. **B. tomentosa,** *Blume*; *Muell. Arg. in DC. Prod.* xv. ii. 501. A tree with rather slender branches, minutely tomentose-pubescent when young as well as the underside of the leaves, but often becoming soon glabrous. Leaves from elliptical-oblong to ovate-elliptical, membranous, with 7 to 15 primary divergent veins on each side of the midrib, and the transverse veinlets also conspicuous, 1½ to 3 in. long. Male flowers densely clustered, sessile or nearly so. Calyx-segments lanceolate, acute, horizontally spreading, rather above ½ line long. Petals about half as long, stipitate or spathulate, broad and lobed. Disk broad, with a rather thick flat free and entire margin. Staminal column slender, the free part of the filaments about as long, radiating from the top of the column round the central abortive 3- to 5-lobed pistil. Female flowers on very short thick pedicels. Calyx rather larger than in the males and the petals more entire. Disk with a double margin, the outer one flat and entire or nearly so, the inner one forming a short cup immediately round the ovary and usually 5-lobed. Ovary 2-celled. Styles shortly and thickly 2-lobed, connivent or spreading. Fruit nearly globular, ripening usually only 1 seed in each cell, and the cocci or rather pyrenes scarcely separating.—*Amanoa tomentosa,* Baill. Adans. vi. 336.

N. Australia. Victoria river, *F. Mueller*; Port Essington, *Armstrong*; Port Darwin, *Schultz, n.* 46, 101, 111, 166.

Queensland. Rockhampton, *O'Shanesy.*

Var. *ovoidea.* Fruits ovoid. Wood island, *Gulliver.*

The species is also in East India and the Archipelago, extending northward to South China.

4. **B. faginea,** *F. Muell. Herb.* A tall shrub or small tree, the slender branches and underside of the leaves tomentose-pubescent or at length glabrous. Leaves ovate or elliptical, very obtuse or almost acute, 1 to 2 in. long, firmer than in *B. tomentosa,* and on some branches under 1 in. long and obovate or almost orbicular. Stipules lanceolate. Bracts small. Flowers closely sessile, solitary or 2 together, mostly females in our specimens, but the males appear to be also solitary. Male calyx-segments rather broad, obtuse, scarcely above ½ line long.

Petals entire or nearly so. Stamens of *B. tomentosa*. Female calyx-segments narrow and at least 1 line long, the inner disk large. Fruit red, globular, fully 2 lines diameter, hard but not thick, separating into 2 cocci or pyrenes, and ripening only one seed in each.—*Amanoa faginea*, Baill. Adans. vi. 336.

Queensland. Rockhampton, *Dallachy*, *Bowman*; Keppel Bay, *Dallachy*; Port Denison, *Fitzalan*.

B. Leichhardtii, Baill. Etud. Euph. 584; Muell. Arg. in DC. Prod. xv. ii. 499 (*Amanoa Leichhardtii*, Baill. Adans. vi. 336), from Mount Cameroons, Moreton Bay, *Leichhardt*, of which the male flowers only are known, is undistinguishable by any character given from *B. faginea*. If it prove really to be the same, the name *B. Leichhardtii* has the priority, but ought perhaps to be rejected on account of the insufficiency of the character.

23. CLEISTANTHUS, Hook. f.

Flowers monœcious, in sessile clusters, axillary or in leafless spikes. Male fl.: Calyx deeply divided into 5 lobes or segments, valvate in the bud, spreading when in flower. Petals 5, stipitate or spathulate, with a small lamina usually broad and dentate. Disk broad with an entire or slightly lobed free margin. Stamens 5, inserted on a central column arising from the disk, the filaments diverging or spreading horizontally under an abortive lobed style terminating the column without any ovary; anther-cells parallel, opening longitudinally in 2 valves. Female fl.: Calyx less deeply divided than in the males, the base forming a cup under the ovary. Disk adnate with an entire sometimes scarcely prominent free margin and within it an erect cup or margin close under the ovary. Ovary 3-celled, with 2 ovules in each cell. Styles 3, distinct, bifid. Capsule globular or depressed, 3-furrowed, separating into 3 2-valved deciduous cocci, leaving, besides the central axis, a broad persistent base. Seeds with a rather scanty albumen; cotyledons broad, rather thin, often more or less folded.—Trees or shrubs. Leaves alternate, coriaceous, entire, the primary pinnate veins not so prominent as in *Briedelia*, arcuate and anastomosing far within the margin, the veinlets reticulate, not transverse.

The genus extends, like *Briedelia*, over the warmer regions of Africa, Asia, and the South Pacific islands, but the Australian species appear to be all endemic. It has been united by Baillon with *Amanoa*, but differs considerably in the fruit as well as in the venation of the leaves The abortion of the floral leaves in many species gives them a peculiar inflorescence, different from that of any other Australian *Phyllantheæ*, and approaching that of *Amanoa*, from which *Cleistanthus* like *Briedelia* differs essentially in the valvate perianth, the structure of the seed, and other characters.

Flower-clusters all axillary, small. Leaves mostly under 2 in. Stamens very shortly united.
- Capsule stipitate. Leaves obtuse . . . . 1. *C. Cunninghamii.*
- Capsule sessile. Leaves mostly acuminate . . . . 2. *C. apodus.*

Flower-clusters mostly in leafless interrupted ferruginous spikes. Leaves above 2 in. long. Staminal column as long as the filaments.
- Calyx segments 1½ lines long. Leaves green on both sides. Capsule glabrous . . . . 3. *C. Dallachyanus.*
- Calyx-segments scarcely 1 line long. Leaves pale or glaucous underneath. Young capsule villous . . . . 4. *C. semiopacus.*

1. **C. Cunninghamii,** *Muell. Arg. in DC. Prod.* xv. ii. 506. A tall shrub, quite glabrous or the young shoots slightly pubescent. Leaves petiolate, ovate-oblong or elliptical, rather thinly coriaceous, obtuse or rarely obtusely and obscurely acuminate, mostly 1½ to 2 in. long, often glaucous underneath. Flowers in dense axillary clusters, sessile or nearly so. Male calyx glabrous. Petals small, obovate-cuneate. Stamens very shortly united at the base. Female calyx with a broadly turbinate base, the lobes spreading, under 1 line long. Disk lining the turbinate base and produced into a short cup round the ovary. Ovary very villous, the tapering base half included in the base of the calyx. Styles rather long. Fruit depressed-globular, tridymous, at first very villous, but becoming nearly glabrous when ripe, borne on a stipes usually exceeding the calyx but variable in length.—*Lebediera Cunninghamii*, Muell. Arg. in Linnæa xxxii. 80; *Amanoa Cunninghamii*, Baill. Adans. vi. 335.

N. Australia? An imperfect specimen from Victoria river, *F. Mueller*, may be this species.

Queensland. Brisbane river, Moreton Bay, *A. Cunningham*, *Fraser*, *F. Mueller*; Burnett river, *F. Mueller*; Rockhampton, *Dallachy*.

N. S. Wales. Hastings river, *Fraser*, *Beckler*; Richmond, Tweed, and Clarence rivers, *C. Moore*.

This plant is allied to *C. stipitatus*, Muell. Arg., from New Caledonia, but the leaves are much less coriaceous, and it is readily distinguished by the densely villous ovary.

2. **C. apodus,** *Benth.* A straggling shrub or small bushy tree, usually quite glabrous, nearly resembling *C. Cunninghamii*, but the leaves more ovate, and usually acute or acuminate, the male flowers rather smaller, the calyx-segments scarcely above 1 line long, the female calyx-lobes broader, 1 line long. Ovary hirsute with a few long hairs. Styles united at the base in a short column, the bifid branches spreading. Capsule closely sessile within the persistent calyx, quite glabrous, 3 to 4 lines diameter.

Queensland. Cape York, *M'Gillivray*, *Daemel*; Rockhampton Bay, *Dallachy*.

3. **C. Dallachyanus,** *Baill. in Herb. F. Muell.* A handsome tree, the inflorescence and sometimes the young shoots ferruginous-pubescent, the adult foliage glabrous. Leaves ovate, obtuse or more frequently obtusely acuminate, rounded at the base, rather thickly coriaceous, shining above, scarcely glaucous underneath, mostly 2 to 4 in. long. Flowers sessile in sessile clusters, which are sometimes distant in the axils of floral leaves of ½ to 1 in., more frequently nearer together and from the abortion of the floral leaves forming more or less interrupted spikes of 1 to 3 in. Bracts small but very broad, obtuse and concave, enveloping the buds, ferruginous-villous as well as the rhachis. Male calyx glabrous, the segments rather thin, 1½ lines long. Petals not half so long, very broad and more or less stipitate. Filaments united in a column to more than half their length. Female flowers "on distinct branches of the same tree" (*R. Brown*), only seen very imperfect, being much injured in our specimens by insects. Ovary

glabrous (*Baillon*). Capsule closely sessile.—*Amanoa Dallachyana*, Baill. Adans. vi. 335.

Queensland. Northumberland and Cumberland islands, *R. Brown;* Rockhampton, *Dallachy, Thozet.*

4. **C. semiopacus,** *F. Muell. Herb.* A tree? with the branches more slender than in *C. Dallachyanus*, but the inflorescence and young shoots ferruginous-pubescent as in that species, the adult foliage glabrous but glaucous underneath. Leaves ovate, acuminate, rather thickly coriaceous, 2 to 4 in. long. Flowers smaller than in *C. Dallachyanus*, the clusters closely sessile in interrupted leafless spikes of 1 to 2 in., and sometimes paniculate, or the lower clusters in the axils of floral leaves. Calyx-segments pubescent outside, nearly 1 line long. Petals small, broad, entire or toothed. Staminal column as long as the free part of the filaments. Ovary and young capsule sessile, densely pubescent. Outer margin of the disk scarcely prominent under the very short cup. Fruit depressed-globular, tridymous, nearly 3 lines diameter, ferruginous-villous or at length nearly glabrous.

Queensland. Rockingham Bay, *Dallachy.*

TRIBE 5. CROTONEÆ.—Flowers distinct, both sexes with a perianth, or the males rarely without any. Ovules 1 in each cell. Embryo with broad cotyledons and a narrow radicle. Trees, shrubs, or herbs. Leaves usually much larger than in the *Stenolobeæ*, and often toothed. Flowers, at least the males, in spikes racemes or panicles, very rarely reduced to clusters. Stamens usually indefinite, few or many.

24. CROTON, Linn.

Flowers monœcious or rarely diœcious in terminal racemes. Male fl.: Calyx of 5, rarely 4 or 6 segments, imbricate or almost or quite valvate in the bud. Petals as many and usually as long as the calyx-segments. Glands small, alternating with the petals. Stamens indefinite (5 to above 30 in the Australian species) inserted on a rather broad usually hairy receptacle or disk; filaments free, inflected in the bud below the anther, erect and usually exceeding the expanded flower; anthers adnate to a connective continuous with the filaments, with 2 parallel cells opening longitudinally. Female fl.: Calyx-segments usually narrower than in the males and sometimes enlarged. Ovary 3-celled (very rarely 2- or 4-celled), with one ovule in each cell. Styles divided into 2 or 4 branches. Capsule separating into 3 deciduous 2-valved cocci. Seeds smooth, with a small caruncular; testa crustaceous or hard. Albumen copious; cotyledons broad.—Trees, shrubs, or rarely herbs, usually clothed or sprinkled with stellate hairs or scales, rarely quite glabrous. Leaves alternate or very rarely opposite, sometimes almost verticillate immediately under the inflorescence, petiolate, entire, variously toothed or very rarely lobed, penniveined or rarely 3- or more-nerved at the base, with 2 or more small sessile or stipitate glands at the top of the petiole or base of the lamina. Stipules usually minute. Flowers usually clustered along the rhachis of a terminal raceme, the bracts very small.

The genus is a very large one, extending over the tropical regions of both the New and the Old World. Of the seven Australian species one is represented by a distinct variety in New Caledonia and another in the Fiji and Philippine islands, the five others appear to be quite endemic.

Stamens not more than 12. Leaves penniveined, rarely irregularly 3-nerved at the base.
  Leaves densely clothed underneath with a stellate scaly or silvery tomentum.
    Female calyx 3 lines long, male only 1 line. Styles with 2 elongated branches . . . . 1. *C. Schultzii.*
    Female calyx under 1 line as well as the male
      Styles with 2 rather broad branches . . . . 2. *C. insularis.*
      Styles with 4 elongated branches.
        Leaves all alternate . . . . 3. *C. phebalioides.*
        Leaves all or nearly all opposite . . . . 4. *C. opponens.*
  Leaves hoary on both sides with a close stellate tomentum . 5. *C. tomentellus.*
  Leaves quite glabrous or very sparingly sprinkled when young with stellate hairs or scales.
    Leaves rather thin. Stamens 10 to 12. Capsule globular, scarcely furrowed . . . . 6. *C. Verreauxii.*
    Leaves coriaceous. Stamens 5 to 8. Capsule longer than broad, scarcely furrowed . . . . 7. *C. acronychioides.*
    Leaves coriaceous. Stamens about 10. Capsule deeply 3-furrowed, 3-lobed at the top . . . . 8. *C. triacros.*
Stamens 20 to 30 or more. Leaves broad, 5- or 7-nerved at the base, densely stellate-tomentose . . . . 9. *C. arnhemicus.*

1. **C. Schultzii,** *Benth.* A shrub of 8 to 12 ft., the young branches inflorescence and underside of the leaves silvery-white or reddish with a close scaly tomentum. Leaves broadly ovate, shortly and obtusely acuminate, rounded or slightly peltate at the base, penniveined or imperfectly 3- or 5-nerved, 2 to 4 in. long, on a petiole of ½ to 1 in. in our specimen, the upper surface hoary when young, sprinkled when full grown with a small scaly pubescence. Racemes short and dense, with a thick rhachis, the upper part male, with a few female flowers in the lower part, the pedicels exceedingly short. Male calyx-segments very broad, obtuse, imbricate, about 1 line long. Petals rather broad. Stamens about 11, on a hairy receptacle. Female calyx-segments narrow, 3 lines long. Styles 3, shortly united at the base, deeply divided into 2 rather long entire branches.

N. Australia. Port Darwin, *Schultz*, *n.* 609 (a single specimen). In the large female calyx this species resembles the Archipelago *C. argyratum* and the E. Indian *C. reticulatum* and its allies, but differs in the form of the leaves, the short racemes, sessile female flowers, &c.

2. **C. insularis,** *Baill. Adans.* ii. 217. A tall straggling shrub or small tree, the branches inflorescence and underside of the leaves silvery-white or slightly reddish with a close scaly tomentum. Leaves ovate to lanceolate, obtuse, entire or scarcely sinuate, rounded or tapering at the base, finely and often obscurely penniveined, the upper surface green, but sprinkled with a few small scales, mostly 2 to 3 in.

long, on petioles of $\frac{1}{4}$ to $\frac{1}{2}$ in. Racemes 3 to 4 in. long, the upper portion male, the female flowers occupying the lower clusters, and often 1 or 2 with the males higher up, all on pedicels of 1 to 2 lines, lengthening under the fruit to 4 to 6 lines. Male calyx-segments $\frac{3}{4}$ line long, very slightly imbricate or almost valvate. Petals about as long. Stamens about 11; anther-connective rather broad. Female calyx-segments thick, rather obtuse, $\frac{3}{4}$ line long. No petals. Styles 3, diverging from the base, deeply divided into 2 erect broad lobes with recurved margins. Capsule about 3 lines diameter.—Muell. Arg. in DC. Prod. xv. ii. 527; *C. phebalioides*, A. Cunn. Herb. not of F. Muell.

Queensland. Broad Sound, *R. Brown;* Brisbane river, Moreton Bay, *A. Cunningham*, *Fraser*, *Leichhardt*, *F. Mueller;* Rockingham Bay, *Dallachy;* Burdekin river, *F. Mueller;* Rockhampton, *Thozet*, *Dallachy*, and others; Queensland Woods, *London Exhibition*, 1862, *n.* 82.

N. S. Wales. Blue Mountains, *Miss Atkinson;* Breakfast Creek, *Leichhardt.*

3. **C. phebalioides,** *F. Muell.; Muell. Arg. in Flora* 1864 (*Oct.*) 485, *and in DC. Prod.* xv. ii. 581. A tree attaining 40 to 50 ft., with slender weak often pendulous branches, silvery-white as well as the inflorescence and underside of the leaves with a close scaly tomentum. Leaves petiolate, lanceolate, or the larger ones ovate-lanceolate, and the smaller ones narrow oblong, obtuse or almost acute, entire or with very small distant teeth, the upper surface green, mostly $1\frac{1}{2}$ to 3 in. long. Racemes 1 to 3 in. long, the flowers usually numerous, mostly males, with here and there a female in the same cluster; in some smaller-leaved specimens the flowers few in shorter racemes, the male pedicels scarcely above 1 line, the female 2 to 3 lines long. Male calyx-segments almost or quite valvate, about 1 line long. Petals as long or rather longer. Stamens 10 or 11. Female calyx-segments lengthening to $1\frac{1}{2}$ lines. Styles divided to the base, or nearly so, into 4 narrow rather long branches. Capsule 3 to 4 lines diameter, hirsute with stellate hairs, and often shortly muricate, slightly tridymous.—*C. stigmatosus*, F. Muell. Fragm. iv. 140 (Nov. 1864); Muell. Arg. in Linnæa xxxiv. 107, and in DC. Prod. xv. ii. 580.

Queensland. Burdekin river, *F. Mueller;* Port Denison, *Fitzalan;* Edgecombe Bay, *Dallachy;* Rockhampton, *Bowman.*

N. S. Wales. Clarence and Richmond rivers, *Beckler.*

The species is also in New Caledonia. The original specimens of *C. phebalioides*, F. Mueller, from the Burdekin, have narrow mostly obtuse entire leaves, the primary veins inconspicuous; the Edgecombe Bay specimens have lanceolate acuminate leaves, entire and almost nerveless; the Clarence river ones have larger ovate-lanceolate usually toothed leaves, with the veins more prominent underneath; but in some specimens the foliage is so variable as to prevent the separating the above forms even as marked varieties.

4. **C. opponens,** *F. Muell. Herb.* A single specimen in herb. F. Mueller has the foliage and indumentum of *C. phebalioides*, but the leaves are all opposite, or nearly so, and the capsules on axillary peduncles of 2 or 3 lines are much larger, ovoid, nearly $\frac{1}{2}$ in. long, densely villous with short stellate hairs, scarcely furrowed, and mostly 4-celled.

Calyx persisting under the capsule, of 5 broad imbricate sepals. Styles evidently 4-lobed, but none of them perfect on the specimen.

Queensland. The collector's name and precise station not given. The species requires further elucidation from more perfect specimens. The opposite leaves are exceptional in the genus and may be accidental in the specimen, or the plant may prove to belong to some other genus notwithstanding its close general resemblance to *C. phebalioides.*

5. **C. tomentellus,** *F. Muell. Fragm.* iv. 141. A shrub, the smaller branches foliage and inflorescence hoary with a short close stellate tomentum, nearly the same on both sides of the leaves: Leaves ovate, broadly elliptical or ovate-oblong, obtuse or shortly acuminate, entire, 2 to 4 in. long, on petioles of $\frac{3}{4}$ to above 1 in. long, the glands at the base of the lamina usually shortly stipitate. Racemes 2 to 6 in. long, with one or more females in most of the lower and middle clusters, the upper clusters entirely males, which however are all fallen off in our specimens. Stamens about 10, with glabrous filaments (*Muell. Arg.*). Female flowers on pedicels of 1 to 2 lines. Calyx segments scarcely 1 line long and not enlarged under the fruit. Styles (not very perfect in our specimens) deeply divided into 2 bifid branches. Capsule rather depressed, tridymous, 3 to 4 lines broad, tomentose or shortly hirsute with stellate hairs. Caruncului of the seeds small.—Muell. Arg. in Linnæa xxxiv. 108, and in DC. Prod. xv. ii. 591.

N. Australia. Victoria river, *F. Mueller*, and probably the same species but the specimens in leaf only, Careening Bay, N.W. Coast, *A. Cunningham.*

6. **C. Verreauxii,** *Baill. Etud. Euph.* 357. A small tree, either quite glabrous or the smaller branches and foliage sprinkled with a few scattered stellate hairs or scales. Leaves from almost ovate to oblong-elliptical or lanceolate, obtuse or acuminate, entire or dentate, rounded or tapering at the base, green on both sides, 2 to 4 in. long in most specimens, but occasionally the larger ones twice that size, the petioles also very variable in length. Racemes slender, rarely above 2 in. long. Flowers few in the clusters, the lower ones chiefly female, the upper chiefly or entirely male, on pedicels of 1 to 2 or rarely 3 lines long, and sometimes the racemes wholly males or chiefly females. Calyx-segments acute, valvate in the bud, $\frac{1}{2}$ to $\frac{3}{4}$ line long in the males, rather longer in the females. Petals fringed-ciliate with long woolly hairs. Stamens 10 to 12. Styles rather thick, divided to about the middle into 2 undivided branches. Ovary tomentose or hirsute with stellate hairs. Capsule nearly globular, variable in size, sprinkled with stellate hairs or at length glabrous.—Muell. Arg. in Linnæa xxxiv. 117, and in DC. Prod. xv. ii. 620; F. Muell. Fragm. iv. 141.

N. Australia. Islands of the Gulf of Carpentaria, *R. Brown;* Port Essington, *Armstrong;* Port Darwin, *Schultz, n.* 620 *and* 680. In these specimens the young parts have more stellate hairs or scales, and the flowers are rather smaller and more numerous than in the N. S. Wales and Queensland ones. The racemes are in some specimens shorter in others longer and looser than usual.

Queensland. Brisbane river, Moreton bay, *A. Cunningham, Fraser, F. Mueller, C. Stuart,*

N. S. Wales. Paterson's, Hunter, and Williams rivers, *R. Brown;* Hastings,

Clarence, M'Leay, and Richmond rivers, *Beckler*, *C. Moore*, and others; Tweed river, *C. Moore;* Blue Mountains, *A. Cunningham*, *Woolls;* Illawarra, *A. Cunningham*, *Macarthur*, *Harvey*.

The species is perhaps endemic in Australia, for the Philippine island specimens referred to it by Mueller Arg. appear to me to differ more from it than the *C. lævifolius*, Blume, from the Archipelago, which Mueller Arg. retains as distinct. Possibly, however, they may all be varieties of one species.

7. **C. acronychioides,** *F. Muell. Fragm.* iv. 142. A shrub or tree, the young shoots and inflorescence more or less sprinkled with a scaly tomentum, the adult foliage usually quite glabrous. Leaves shortly petiolate, elliptical oblong or almost ovate, entire or slightly sinuate-crenate, coriaceous, shining above, penniveined with fine much-anastomosing veins, 2 to 4 in. long, the basal glands sessile. Racemes 1 to 2 in. long, terminal as in the rest of the genus, but with the flowering branches often so short as to appear axillary, the female flowers few, usually only in the lowest cluster close to the floral leaves, the other clusters all male, the pedicels very short. Calyx-segments nearly 1 line long, rather broad, obtuse, slightly imbricate in the bud, somewhat ciliolate. Petals narrow, ciliate. Receptacle hairy. Stamens 5 to 8. Female calyx-segments above 1 line long and narrow. Ovary densely hirsute. Styles broad, divided at least to the middle into 2 mostly bifid branches. Capsule 5 lines long and about 4 lines broad, more or less scaly. Seeds with a small caruncuius.—Baill. Adans. iv. 300.

Queensland. Rockhampton and surrounding districts, *Dallachy*, *Bowman*, *Thozet*.

8. **C. triacros,** *F. Muell. Fragm.* vi. 185. A tree or shrub, quite glabrous or the young shoots and inflorescence very sparingly scaly tomentose. Leaves ovate or elliptical, obtuse or shortly acuminate, entire or obscurely sinuate-crenate, rather coriaceous, smooth, penniveined, with fine and distant primary veins and obscure reticulations, 2 to 6 in. long or even more. Racemes sometimes very short, but some fruiting ones 3 or 4 in. long, often several together at the ends of the branches, some entirely or nearly entirely male, others entirely or nearly entirely female. Pedicels under 1 line long. Calyx-segments nearly 1 line long, imbricate in the bud. Stamens about 10. Styles rather deeply divided into 2 entire branches. Capsule sprinkled with stellate scales, tridymous, obtusely 3-lobed at the top with a deep central depression. Caruncuius of the seeds very small.

Queensland. Rockingham Bay, *Dallachy*.

9. **C. arnhemicus,** *Muell. Arg. in Linnæa* xxxiv. 112, *and in DC. Prod.* xv. ii. 599. A rather slender shrub of 5 or 6 ft., or a small straggling tree, clothed with a stellate tomentum, dense and soft on the young branches inflorescence and underside of the leaves, more scattered on the upper surface. Leaves orbicular-cordate or broadly ovate in the typical form, obtuse or rarely with a short point, crenate, prominently 5- or 7-nerved at the base, with pinnate primary and transverse secondary veins, 3 to 5 in. long and nearly as broad, or

smaller on the side branches. Racemes 3 to 6 in. long, the flowers usually numerous in the clusters, the lower ones chiefly females, the upper ones chiefly or entirely males. Pedicels varying from 1 to 3 lines. Calyx-segments broad, obtuse, rather above 1 line long, imbricate in the bud. Petals scarcely longer, ciliate-hairy. Stamens 20 to 30 or even more, on a hairy receptacle, the filaments glabrous. Styles divided to the base into 2 long slender entire or very shortly 2-lobed branches. Capsule globular, not furrowed, hirsute with stellate hairs, fully 3 lines diameter.

N. Australia. Islands of the Gulf of Carpentaria, *R. Brown;* Victoria and Fitzmaurice rivers and Sea-range, *F. Mueller;* Port Darwin, *Schultz, n.* 48, 563, 684.

Var. *urenæfolius*, Baill. Adans. vi. 300. Leaves more ovate and often acuminate, usually 5-nerved. Flowers rather smaller.

N. Australia. Port Darwin, *Schultz, n.* 186.

Queensland. Cape York, *M'Gillivray;* Port Denison, *Fitzalan;* Edgecombe Bay, *Dallachy;* Gilbert river, *Daintree.*

25. ALEURITES, Forst.

Flowers monœcious, in a terminal panicle. Male fl.: Calyx entire and closed in the bud, splitting into 3 or rarely 2 segments. Petals 5, contorted in the bud. Disk with 5 or 10 slightly prominent lobes or glands. Stamens indefinite, on a central receptacle or disk without any rudimentary ovary; anthers erect in the bud, the cells parallel and adnate to a connective continuous with the filament, opening longitudinally in 2 valves. Female fl.: Calyx and petals of the males. Disk of 5 minute glands. Ovary 2- to 5-celled, with 1 ovule in each cell. Styles 2 to 5, deeply divided into 2 branches. Fruit large, somewhat fleshy, the endocarp hard, tardily separating into cocci. Seeds nearly globular, with a distinct outer somewhat cartilaginous coating, the inner coating bony; albumen oleaginous; cotyledons broad, flat, with a short narrow radicle.—Trees with a stellate tomentum. Leaves alternate, petiolate, large, entire or lobed. Flowers in terminal panicles, the females usually few, terminating the main branches.

A genus of few species, natives of tropical Asia and the Pacific islands. The only Australian species is also widely spread over the Archipelago.

1. **A. moluccana,** *Willd.; Muell. Arg. in DC. Prod.* xv. ii. 723. A tree attaining sometimes a considerable size, the young foliage densely ferruginous-tomentose, becoming nearly glabrous when full-grown. Leaves crowded at the ends of the branches immediately under the panicle, broadly ovate-rhomboidal or ovate-lanceolate, obtuse or acuminate, rarely narrow-lanceolate, entire undulate-crenate or 3-, 5- or 7-lobed, sometimes nearly 1 ft. long and broad but in herbaria usually 4 to 6 in., 3-, 5- or 7-nerved, the primary veins pinnate with transverse secondary veins as in *Croton arnhemicus.* Flowers numerous, in broad terminal much-branched panicles, the pedicels short. Calyx tomentose, opening usually in 3 segments, about 1½ lines long. Petals obovate, about 3 lines long. Stamens 15 to 20, on a convex hairy receptacle,

the filaments also hairy, scarcely longer or sometimes shorter than the anthers. Female flowers (not seen in the Australian specimens) nearly similar to the males as to calyx and corolla. Styles deeply divided into 2 branches. Fruit fully 2 inches diameter.—*Jatropha moluccana*, Linn.; *Aleurites triloba*, Forst., and several other synonyms, as given by Muell. Arg. l.c.

Queensland. Rockingham Bay, *Dallachy*.—Widely spread over the Eastern Archipelago and the islands of the South Pacific, and sent also from various tropical regions, where however it is generally planted.

26. CLAOXYLON, A. Juss.

Flowers diœcious or rarely monœcious, in axillary racemes. Male fl.: Calyx at first globular, opening to the base into 3 rarely 4 valvate segments. Petals none. Stamens indefinite, on a central receptacle or disk, intermixed with glands or lobes of the disk, without any central rudimentary ovary; filaments not inflexed; anther-cells distinct, erect, opening longitudinally from the apex downwards. Female fl.: Calyx less deeply divided than in the males into 3 or 4 valvate lobes. Disk with a free entire or lobed margin. Ovary 3-celled or rarely 2-celled, with 1 ovule in each cell. Styles short, entire, free or united at the base. Capsule separating into 2-valved cocci. Seeds without any caruncülus, the outer coating loose and membranous, the inner testa crustaceous. Albumen copious. Cotyledons broad, with a narrow radicle.—Trees or shrubs, glabrous or sparingly pubescent with short appressed hairs, the foliage often taking a reddish tint when dry. Leaves alternate, petiolate, usually large, entire or toothed, penniveined. Racemes solitary or 2 together, shorter than the leaves. Flowers small, the males few together in clusters, the female solitary. Bracts minute.

The genus is spread over tropical Asia and Africa. The Australian species are all probably endemic, but the characters are difficult to ascertain without good specimens of both sexes.

Leaves long and narrow, on petioles under $\frac{1}{4}$ in. Flowers monœcious, on pedicels of $\frac{1}{4}$ to $\frac{1}{2}$ in. Ovary glabrous	1. *C. angustifolium.*
Leaves various, on petioles of $\frac{1}{2}$ to 1 in. Flowers diœcious or nearly so, on pedicels under 2 lines. Interstaminal glands glabrous.	
Racemes in the axils of young leaves. Ovary glabrous. Styles united at the base	2. *C. tenerifolium.*
Racemes in the axils of full-grown leaves. Ovary pubescent. Styles free	3. *C. australe.*
Leaves large on petioles of 1 to 2 in. Flowers diœcious, on very short pedicels. Interstaminal gland and ovary pubescent .	4. *C. Hillii.*

1. **C. angustifolium,** *Muell. Arg. in Linnæa* xxxiv. 165, *and in DC. Prod.* xv. ii. 786. A shrub of 5 or 6 ft., glabrous except a minute pubescence on the inflorescence. Leaves narrow-lanceolate, acuminate, irregularly toothed, tapering at the base, rather firm and smooth, 4 to 8 in. long and rarely above 1 in. broad, on a petiole of only 1 to 3 lines, the basal glands very small. Flowers monœcious, in racemes of about

1 in., the rhachis slender almost filiform, the males in clusters of 3 to 6 on pedicels of 2 to 3 lines, with occasionally a female in the same cluster on a pedicel twice as long. Male perianth of 3 segments, about ½ line long. Stamens 10 to 12, intermixed with small narrow glands; filaments thick, nearly as long as the perianth; anther-cells small, erect. Ovary glabrous or minutely pubescent; styles exceedingly short, tooth-like. Young capsule tridymous, glaucous and quite glabrous, on a pedicel of above ½ in.—*Mercurialis angustifolia*, Baill. Adans. vi. 322.

Queensland. Cumberland isles and Port Denison, *Fitzalan.*—Single specimens in Herb. F. Mueller.

2. **C. tenerifolium,** *F. Muell. in Baill. Adans.* vi. 323. A tree of 20 to 30 ft., the young shoots and inflorescence sparingly and minutely pubescent, otherwise glabrous. Leaves mostly ovate and acuminate, rarely more elliptical, dentate, tapering at the base, 3 to 4 in. long on a petiole of 1 in. or more when full grown, but in the flowering specimens smaller and still young and membranous, penniveined with more or less of a reddish-purple hue underneath at least when dry. Flowers diœcious, the males in racemes of 1½ to 2 in. on slender pedicels varying from ½ line to above 1 line, the females in shorter racemes on thicker pedicels. Male perianth-segments 3, membranous, reflexed, about 1 line long. Stamens 12 to 20, about as long as the perianth, intermixed with globular glands. Female perianth divided to the middle into 3 valvate lobes. Disk with a broad thin almost petal-like broadly lobed margin. Ovary glabrous. Styles short, united at the base in a deciduous cone, spreading in the upper half. Capsule tridymous, nearly 3 lines diameter.—*Mercurialis tenerifolia*, Baill. Adans. vi. 323.

Queensland. Rockhampton, *Dallachy*, *O'Shanesy;* Broad Sound and Cleveland Bay, *Bowman.*

3. **C. australe,** *Baill. Etud. Euph.* 493. A tall shrub or straggling tree of 25 to 30 ft., the young shoots sparingly the inflorescence more copiously pubescent, or rarely quite glabrous. Leaves oblong, broad or narrow, more rarely ovate, obtuse or rarely shortly acuminate, dentate, tapering at the base, 3 to 6 in. long, on a petiole of from ½ to 1 in., rather firm when full grown, green on both sides or rarely reddish-purple underneath, the basal glands very variable. Flowers diœcious, the male racemes 2 to 3 in. long, the females much shorter, the pedicels very short. Male perianth-segments 3, about 1 line long. Stamens usually above 20, about as long as the perianth, intermixed with short ovate glabrous glands. Female perianth rather thicker and less deeply divided. Disk with a broad free ciliate margin. Ovary more or less pubescent. Styles distinct, at first very short and thick, at length spreading and about ½ line long. Capsule nearly 3 lines diameter, usually pubescent.—Muell. Arg. in DC. Prod. xv. ii. 788; F. Muell. Fragm. iv. 142; *Mercurialis australis*, Baill. Adans. vi. 322.

Queensland. Brisbane river, Moreton Bay, *A. Cunningham*, *Fraser*, *F. Mueller;* Port Bowen, *A. Cunningham;* Wide Bay, *Leichhardt ?* (specimens very bad).

N. S. Wales. Port Jackson to the Blue Mountains, *Woolls*, *Miss Atkinson;*

Sydney woods, *Macarthur*, *Paris Exhibition*, 1855, *n.* 22; Hastings and Clarence rivers, *Beckler* and others; New England, *C. Stuart;* southward to Illawarra, *A. Cunningham*, *Harvey*, *Ralston*.

Var. *latifolia.* Leaves large and broad.—Rockhampton, *O'Shanesy*, males only.

Var. *laxiflora.* Leaves long and narrow. Racemes longer and looser with longer pedicels than usual.—Tweed river, *C. Moore.*

Var *dentata.* Leaves coarsely and deeply toothed.—Macleay river, *Beckler.*

4. **C. Hillii,** *Benth.* A tree of 20 to 30 ft., the young shoots and inflorescence pubescent, the adult foliage glabrous, and often assuming a purplish hue. Leaves ovate, shortly acuminate, dentate, 6 to 8 in. long and 3 to 4 in. broad on petioles of 1 to 2 in. in Hill's and M'Gillivray's specimens, but little more than half that size in Dallachy's, and still very young in Daemel's specimens. Flowers diœcious. Racemes in the upper axils of the previous year's wood or at the base of young shoots, the males 2 to 4 in. long, the flowers 3 to 5 together in distinct clusters, on very short pedicels. Perianth villous, of 3 valvate segments. Stamens 12 to 20, the intermediate glands ciliate-hairy. Female racemes under 2 in. long, the flowers not numerous, solitary within the bracts, on pedicels of about 1 line. Perianth of 3 short broad segments. Disk of 3 broad distinct segments (or petals ?) alternating with the perianth-lobes and shorter than them. Ovary densely pubescent. Styles distinct, spreading, acute, less than ½ line long. Capsule tridymous, at length nearly glabrous, scarcely 3 lines diameter. Seeds globular, reticulate-rugose, about 1 line diameter, without any caruncles.

Queensland. Cape York, *M'Gillivray*, *Daemel;* Albany island, *W. Hill* (all males); Rockingham Bay, *Dallachy* (females).—I am not certain of having correctly referred Dallachy's female specimens to the same species as the Cape York males, or whether they may not belong to some variety of *C. tenerifolium* or *C. australe*, with more pubescent flowers and inflorescence. In the male flowers the glands appear to be always glabrous in *C. tenerifolium*, ciliate-hirsute in *C. Hillii.*

27. ACALYPHA, Linn.

Flowers monœcious or rarely diœcious. Male fl.: clustered in axillary spikes, with a small bract under each cluster. Perianth of 4 valvate segments. No petals or glands. Stamens 8 or rarely 8 to 16, inserted on a raised central receptacle, without any rudimentary ovary; filaments free; anther-cells distinct, linear, wavy or tortuous, attached by one end. Female fl.: 1 to 4 together within a leafy bract, the bracts solitary or spicate. Perianth of 3, rarely 4 imbricate segments. Ovary 3-celled, with 1 ovule in each cell. Styles distinct, finely branched.—Shrubs or trees or in species not Australian herbs. Leaves alternate, usually dentate. Flowers very small, the males and females in separate spikes or the females solitary in separate axils, or one or more at or near the base of the male spikes.

A large genus dispersed over the tropical and subtropical regions of both the New and the Old World. The three Australian species appear to be endemic.

Villous shrub. Leaves 1 to 3 in. long, ovate or broadly lanceolate.
Female flowering bracts 1 or more at the base of the male spikes or in separate axils . . . . 1. *A. nemorum.*
Glabrous or pubescent slender shrubs or trees. Leaves small, oblong.
Female flowering bracts sessile at the base of the males or in separate axils, or if pedicellate with abnormal deeply divided muricate fruits. Male clusters approximate . . . . 2. *A. eremorum.*
Female flowering bracts on filiform peduncles with normal capsules. Male clusters distant in filiform spikes . . . . . . 3. *A. capillipes.*

1. **A. nemorum,** *F. Muell.; Muell. Arg. in Linnæa* xxxiv. 38, *and in DC. Prod.* xv. ii. 858. A shrub of from 3 or 4 ft. to twice that height, the young shoots more or less softly villous, the adult foliage sparingly so. Leaves ovate-lanceolate or oblong, obtuse or scarcely acuminate, crenate, 3- or 5-nerved at the base, penniveined with transverse veinlets, those of the principal branches 2 to 3 in. long, on petioles from under ½ in. to 1 in. long, those of the lateral branches smaller, on short petioles. Male spikes slender, pedunculate, 1 to 2 in. long, the flowers in clusters of 10 to 15 or more, on exceedingly short pedicels, with a minute deeply fringed bract under each cluster. Perianth-segments 4, about ¼ line long. Stamens usually 8, not longer than the perianth. Female flowers 1 to 3 at the base of some of the male spikes or in separate axils, and sometimes several crowded in a short spike, each within an orbicular crenate bract, attaining sometimes ½ in. diameter. Perianth-segments 3 or 4, very small. Ovary hirsute. Styles long, fringed with capillary lobes.—*A. Cunninghamii,* Muell. Arg. in Linnæa xxxiv. 35, and in DC. Prod. xv. ii. 861.

Queensland. Brisbane river, Moreton Bay, *A. Cunningham, Leichhardt, F. Mueller, C. Stuart;* Burnett river, *F. Mueller;* Wide Bay, *Bidwill.*

N. S. Wales. Hastings and Clarence rivers, *Beckler* and others; Richmond river, *C. Moore;* New England, *Leichhardt, C. Stuart.*

The species is exceedingly variable in the indumentum, sometimes very dense and soft, sometimes scarcely any, in the breadth of the leaves and length of the petiole, in the female bracts few or many, crowded and clustered, or distant in a short interrupted spike, the teeth of the bracts also variable in number and breadth.

2. **A. eremorum,** *Muell. Arg. in Flora* 1864, 440, *and in DC. Prod.* xv. ii. 863. A rigid shrub, with virgate or divaricate rather slender branches occasionally spinescent at the end, glabrous or the young shoots and foliage pubescent. Leaves small and distant, very shortly petiolate, oblong, crenate, contracted at the base, ½ to 1 in. long, or on some of the lateral branches much smaller and almost obovate. Male racemes slender, pubescent, from under ½ in. to nearly 1 in. long, the flowers minute, in dense usually approximate clusters, the bracts very minute. Female flowers solitary, dimorphous, mostly nearly sessile, alone or near the base of the male spikes, within a bract of 2 to 2½ lines diameter deeply toothed or lobed. Perianth-segments 3, minute but broad. Ovary pubescent. Styles fringed in the upper half. Capsule tridymous, nearly glabrous, about 1 line diameter. Some female flowers are however borne on a filiform pedicel of 4 to 5 lines, the bract then

very small and the capsule deeply divided into 3 muricate obovoid cocci.

Queensland. Brisbane river, *Fraser;* scrub on the Burdekin, *F. Mueller;* Rockhampton, *Dallachy.*

3. **A. capillipes,** *F. Muell.; Muell. Arg. in Linnæa* xxxiv. 40, *and in DC. Prod.* xv. ii. 823. A tall shrub or small tree, with divaricate slender branches, the smaller ones often acicular and spinescent, resembling the spinescent specimens of *A. eremorum*, with which Baillon unites it as a variety. It is more glabrous, the leaves rather broader and thinner, the male spikes filiform, with still smaller flowers in distant clusters, the females apparently all solitary on filiform peduncles of $\frac{3}{4}$ to 1 in., with a normal orbicular bract of nearly 2 lines diameter, the capsule glabrous and tridymous but the cocci not deeply separate as in the pedunculate fruits of *A. eremorum*, the styles with much more numerous capillary branches than in that species.

N. S. Wales. Clarence river, *Beckler.*

28. ADRIANA, Gaudich.

(Trachycaryon, *Kl.*)

Flowers diœcious, in terminal spikes. Male fl.: Perianth globular and closed in the bud, opening in 4 or 5 valvate segments. No petals or glands. Stamens very numerous, crowded on a slightly raised central receptacle, without any rudimentary ovary; filaments very short; anthers linear, erect, the cells adnate, parallel, opening longitudinally in 2 valves, the connective produced beyond them into a papillose point or linear appendage. Female perianth of 6 or 8 segments, imbricate in about 2 rows. Ovary 3-celled, with 1 ovule in each cell. Styles 3, distinct or very shortly connate at the base, bifid, densely covered or fringed with much raised or linear papillæ. Capsule separating into 2-valved cocci. Seeds ovoid, with a small carunculus. Testa crustaceous. Albumen copious. Cotyledons flat, much broader than the radicle.—Erect shrubs, glabrous or stellate-tomentose. Leaves alternate or opposite, 3- or 5-nerved, coarsely toothed and often 3-lobed. Male spikes usually rather long and interrupted, the flowers sessile in clusters of 3 to 6 in the axil of an ovate or lanceolate bract. Female spikes usually very short and dense, sessile or very shortly pedunculate within the last leaves.

The genus is endemic in Australia. The species are all very closely allied to each other, and might easily be reduced to varieties of two species. F. Mueller thinks that even these might be united in a single one, for which he proposes the name of *A. Dampieri.*

Leaves all or nearly all alternate on rather long petioles. (Each species glabrous or tomentose.)
  Leaves or their middle lobe ovate-lanceolate or broad, often acuminate. Styles free, slender, longer than the capsule. Eastern or tropical species . . . . 1. *A. acerifolia.*

Leaves deeply 3-lobed, the middle lobe oblong or ovate-oblong, obtuse, not acuminate. Styles free rarely exceeding the capsule. North-western and Western species . . . . 2. *A. tomentosa.*
Leaves oblong or lanceolate, obtuse, rarely 3-lobed. Styles shortly united at the base. Desert species . . . . . . 3. *A. Hookeri.*
Leaves all opposite, sessile or on very short petioles.
Leaves quite glabrous on both sides . . . . 4. *A. quadripartita.*
Leaves white-tomentose underneath . . . . 5. *A. Klotzschii.*

1. **A. acerifolia,** *Hook. in Mitch. Trop. Austr.* 371. A rather coarse shrub of 3 or 4 ft., usually hoary or white with a stellate tomentum, sometimes dense and mixed with longer stellate hairs even on the upper surface of the leaves, sometimes very close or almost mealy, usually more sparing or almost wanting on the upper side of the leaves, or very rarely the whole plant glabrous and reddish. Leaves alternate, on rather long petioles, 3- or 5-nerved at the base, very variable in shape, usually either ovate-lanceolate and coarsely toothed or deeply 3-lobed with ovate-lanceolate coarsely toothed lobes, the middle lobe the longest, the larger 3-lobed leaves often 4 to 6 in. long, the upper ones often much smaller. Male spikes 2 to 3 in. long; perianth-segments spreading, membranous, about 1½ lines long; anthers about 1 line long on very short glabrous filaments. Female spikes very short and dense, sometimes contracted into a head, sessile within the floral leaves or petiolate bracts. Perianth-segments usually 6 but varying from 5 to 8, herbaceous, 2 to 3 lines long. Ovary and back of the styles more or less stellate-tomentose. Styles 4 to 6 lines long, divided to about the middle into 2 branches. Capsule 4 to 5 lines diameter, very obtuse. Seeds smooth.—Muell. Arg. in DC. Prod. xv. ii. 890; Baill. Adans. vi. 312; *A. heterophylla,* Hook. in Mitch. Trop. Austr. 124; *Trachycaryon Cunninghamii,* F. Muell. in Trans. Phil. Soc. Vict. i. 15, and in Hook. Kew Journ. viii. 209.

N. Australia. Victoria and Fizmaurice rivers, *F. Mueller.*
Queensland. Maranoa and Balonne rivers, *Mitchell;* Rockhampton, *Dallachy, Bowman;* North Kennedy district, *Daintree;* Moreton Bay, *C. Stuart.*
N. S. Wales. Hunter, Paterson, and Williams rivers, *R. Brown;* Port Jackson to the Blue Mountains, *Woolls, Miss Atkinson;* Lachlan river, *A. Cunningham;* Darling river to Cooper's Creek, *Nielson;* New England, *C. Stuart;* Mount Lindsay, Peel and Namoi rivers, *C. Moore;* Hunter river, *Backhouse;* Hastings river, *Beckler.*
Victoria. Snowy and Buchan rivers, *F. Mueller.*

Var. *glabrata.* Glabrous or nearly so in all its parts.—*A. glabrata,* Gaudich. in Ann. Sc. Nat. ser. 1, v. 223, and in Freyc. Voy. Bot. 487.—A few specimens from Queensland, N.S. Wales, and Victoria.

2. **A. tomentosa,** *Gaud. in Ann. Sc. Nat. ser.* 1, v. 223, *and in Freyc. Voy. Bot.* 487, t. 116. A shrub of 2 to 4 ft. very nearly allied to *A. acerifolia,* and united with it by F. Mueller. It is also similarly variable in indumentum, but usually more glabrous than that species. Leaves almost always deeply 3-lobed, with narrower and more obtuse lobes, the whole leaf rarely above 2 in. long. Flowers as in *A. acerifolia* or rather longer and the styles shorter and thicker. Capsules larger and the seeds as far as known, with a pitted testa.—Muell. Arg. in DC. Prod.

xv. ii. 891; *A. acerifolia γ puberula,* Muell. Arg. l.c.; *A. Gaudichaudi,* Baill. Adans. vi. 312 (partly).

N. Australia. N.W. coast: Carew river, *A. Cunningham;* Depuech island, *Bynoe;* Nichol Bay and De Grey river, *Ridley's Expedition;* Point Larrey, *Hughan.*
W. Australia. Sharks Bay, *Milne, Denham;* Port Gregory and Murchison river, *Oldfield.*

3. **A. Hookeri,** *Muell. Arg. in DC. Prod.* xv. ii. 891. A glabrous or minutely tomentose shrub, more slender than *A. acerifolia.* Leaves alternate or here and there opposite, petiolate, oblong or oblong-lanceolate, obtuse, coarsely toothed and 1 to 1½ in. long, or a few of the larger ones 3-lobed with the central lobe like the undivided leaves and the lateral ones short and broad. Flowers fewer than in *A. acerifolia,* on shorter spikes, the females often solitary, the perianth and capsule smaller than in that species. Styles shortly united at the base, more slender than in *A. tomentosa,* not so long as in *A. acerifolia.*—*Trachycaryon Hookeri,* F. Muell. in Trans. Phil. Soc. Vict. i. 16, and in Hook. Kew Journ. viii. 210.

Victoria. Murray river, *F. Mueller;* North-west districts, *L. Morton;* Wimmera, *Dallachy.*

4. **A. quadripartita,** *Gaudich. in Freyc. Voy. Bot.* 489. A shrub of 2 to 6 ft., quite glabrous in all the specimens seen. Leaves all opposite, sessile or very shortly petiolate, ovate ovate-lanceolate or oblong, acute or obtuse, coarsely toothed, 3-nerved at the base, mostly 1½ to 2 in., rarely 3 in. long. Spikes short and few flowered as in *A. Hookeri,* and the styles similarly united at the base. Capsule glabrous, minutely stellate-hairy or almost muricate. Seeds smooth.—Muell. Arg. in DC. Prod. xv. ii. 892, *Croton quadripartitus,* Labill. Pl. Nov. Holl. ii. 73, t. 223; *Trachycaryon Billardieri,* Klotzsch in Pl. Preiss. i. 175; *Adriana Billardieri,* Baill. Etud. Euph. Atl. 6, t. 2, f. 19 to 20.

Victoria. Port Phillip, *R. Brown, Gunn, Harvey;* Cape Otway, *Herb. Hook.*
Tasmania. *Labillardière.*
W. Australia. Point Henry near the sea, *Oldfield;* towards Cape Riche, *Drummond, 5th coll. n.* 224 *and* 225, *and in Herb. F. Muell.* 239; Esperance Bay, *Maxwell;* Swan river, *Preiss, n.* 1206, *Oldfield;* Port Gregory, *Oldfield.*

5. **A. Klotzschii,** *Muell. Arg. in DC. Prod.* xv. ii. 892. A shrub of 3 to 4 ft., with the opposite almost sessile leaves of *A. quadripartita,* of which it may be a variety distinguished by the leaves, white-tomentose underneath, the female perianth-segments shorter and more obtuse, and the styles longer and free from the base as in *A. acerifolia.*—*Trachycaryon Klotzschii,* F. Muell. in Trans. Phil. Soc. Vict. i. 15, and in Hook. Kew Journ. viii. 209.

Victoria. Wilson's Promontory, *F. Mueller.*
S. Australia. Memory Cove and Port Lincoln, *R. Brown;* Rivoli Bay, *Robertson;* Encounter Bay, *Whittaker;* near Adelaide, *F. Mueller, Blandowski;* Port Lincoln, *Wilhelmi;* Venus Bay and Kangaroo island, *Waterhouse.*

29. ALCHORNEA, Swartz.

(Cladodes, *Lour.*; Cœlebogyne, *J. Sm.*)

Flowers diœcious or rarely monœcious, in terminal or axillary racemes or spikes. Male fl.: Perianth globular and closed in the bud, opening in 4, rarely 3 or 2, valvate segments. No petals or glands. Stamens 8 or more, rarely 4, in the centre of the flower, without any rudimentary ovary; filaments free or very shortly united; anthers versatile, the cells parallel, opening longitudinally in 2 valves. Female fl.: Perianth of 4, rarely 3, 5 or 6 segments or lobes, imbricate in the bud. No disk (except in one Madagascar species). Ovary 2- or 3-celled, with 1 ovule in each cell. Styles entire or 2-branched, free or shortly connate at the base. Capsule separating into 2-valved cocci. Seeds without any caruncului. Testa crustaceous. Albumen copious. Cotyledons flat, much broader than the radicle.—Trees or shrubs. Leaves alternate, petiolate, undivided, toothed or in one species almost lobed, with 2 or more glands on the under side at the base of the lamina. Stipules small, usually subulate, or none. Male spikes slender, interrupted, often paniculate, the flowers very small, clustered along the rhachis, with a small bract under each cluster. Female spikes or racemes usually single, the flowers solitary within each bract.

The genus is spread over the tropical regions of both the New and the Old World. A considerable number of its species have been proposed by various botanists as distinct genera enumerated by Mueller Arg. as synonyms. Amongst these, in uniting them, he has selected Swartz's name as the oldest. Baillon has for reasons which he does not give, substituted Loureiro's name *Cladodes*, which, however, is two years more recent. The two Australian species are endemic.

Leaves coriaceous, with broad short prickly-pointed lobes. Styles short, broad, spreading flat on the top of the ovary . . . . . 1. *A. ilicifolia.*
Leaves thin, with obtuse or shortly pointed teeth. Styles erect, narrow, connate at the base . . . . 2. *A. Thozetiana.*

1. **A. ilicifolia,** *Muell. Arg. in Linnæa* xxxiv. 170, *and in DC. Prod.* xv. ii. 906. A glabrous straggling evergreen shrub, attaining 12 to 15 ft. Leaves ovate or rhomboidal, broadly sinuate-toothed or shortly lobed, the teeth or lobes terminating in prickly points, coriaceous, penniveined and reticulate, resembling those of a holly, 1½ to 3 in. long, tapering into a short petiole. Flowers diœcious. Racemes axillary or lateral, the males slender and often several together on a short common rhachis or leafless branch, the females solitary and under 1 in. long. Male perianth-segments 4, about ¾ line long. Stamens 8, not exceeding the perianth. Female flowers on thick pedicels of ½ to 1 line, the perianth-segments rather smaller than in the males. Ovary 3-celled. Styles broader than long, flat, closely spreading over the top of the ovary. Capsule depressed-globular, 3- or 2-celled, 3 to 4 lines diameter.—*Cœlebogyne ilicifolia,* J. Sm. in Trans. Linn. Soc. xviii. 512. t. 36; *Cladodes ilicifolia,* Baill. Adans. vi. 321.

Queensland. Brisbane river, Moreton Bay, *A. Cunningham, F. Mueller;* Rockhampton, *Dallachy, O'Shanesy, Bowman.*

N. S. Wales. Cabramatta, *Woolls;* northward to Clarence, Hastings, and Macleay rivers, *Beckler;* southward to Illawarra, *Harvey;* Sydney woods, *Paris Exhibition*, 1854, *M'Arthur*, *n.* 56.

The shrub, celebrated for its parthenogenetic properties, having reproduced itself from seed in European gardens through several generations from female plants alone without the intervention of any male flowers, has been the subject of numerous papers by Caspary, Karsten, A. Braun, Baillon, and others. They have, however, added very little to the facts detailed by J. Smith in the above quoted memoir. Al. Braun of Berlin has observed that the seeds have occasionally two embryos united at the base, which I also found to be the case in some seeds from a Queensland specimen which I examined.

2. **A. Thozetiana,** *Baill. in Herb. F. Muell.* A glabrous shrub of 2 to 4 ft. Leaves, in the typical form, ovate obovate or rhomboidal, sharply toothed, but the teeth more numerous less deep and not so pungent as in *A. ilicifolia*, and the leaf thinner, 1½ to 3 in. long, penniveined and reticulate, tapering at the base and shortly petiolate, the basal glands often obscure and sometimes deficient. Stipules, also in the typical form, very small and subulate as in *A. ilicifolia.* Male flowers unknown. Female racemes terminal, 2 to 4 in. long, the flowers distant. Perianth unequally 4- or 5-lobed, the lobes acute, ½ to ¾ line long. Ovary usually 3-celled. Styles short, erect, narrow but flat, shortly united at the base. Capsule tridymous, about 4 lines diameter.—*Cladodes Thozetiana*, Baill. Adans. vi. 321.

Queensland. Rockhampton, *Thozet.*

Var. *longifolia.* Leaves 3 to 4 in. long, acuminate, the teeth obtuse or with very small points. Stipules setiform, the upper ones sometimes 2 lines long. Female racemes and fruits as in the typical form.—Rockingham Bay, *Dallachy.*

30. TRAGIA, Linn.

Flowers monœcious, in terminal or lateral racemes. Male fl.: Perianth globular in the bud, of 5, 4 or 3 valvate segments. Disk none or with a slightly prominent margin. Petals none. Stamens numerous or few, the filaments free or connate, sometimes very short. Anthers dorsally attached, the cells parallel, opening longitudinally in 2 valves. Rudimentary ovary none or small and obscure. Female fl.: Perianth of 6 or fewer, rarely 7 or 8, imbricate segments, entire or pinnately divided. Ovary 3-celled or rarely 4- or 5-celled, with 1 ovule in each cell. Styles erect and connate at the base, free and entire at the end. Capsule separating into 2-valved cocci. Seeds globose, without any caruncuIus. Testa crustaceous. Albumen copious. Cotyledons flat, much broader than the radicle.—Twining or climbing perennials or undershrubs, rarely erect annuals, usually hispid with stinging hairs. Leaves alternate, petiolate, toothed, often cordate, 3- or 5-nerved. Flower small, the racemes normally terminal, but often leaf-opposed from the elongation of the lateral shoot, or apparently axillary from the shortness of the flowering branch, the males in the upper, the females in the lower, part of the raceme, all usually solitary in the axil of a small bract.

The genus is spread over the tropical and subtropical regions of both the New and the Old World. The only Australian species appears to be endemic.

1. **T. Novæ-hollandiæ,** *Muell. Arg. in Linnæa* xxxiv. 180, *and in DC. Prod.* xv. ii. 929. A twining herb, attaining several feet, more or less hispid with simple rigid appressed or spreading stinging hairs. Leaves petiolate, ovate or ovate-lanceolate, acute, coarsely toothed, 3- or 5-nerved and broadly or deeply cordate at the base, penniveined, 1½ to 3 in. long. Stipules small. Racemes slender, 1 to 2 in. long. Flowers solitary in the axils of small narrow bracts, the lower ones female and distant, the upper ones male, all on very short pedicels. Male perianth-segments 5, rather thick, acute, ¾ line long. Stamens varying from 3 to 5, inserted within the margin of a broad disk, the filaments exceedingly short, the anther-cells almost stipitate. Female perianth-segments usually 6, more acutely acuminate than the males, but imbricate in the bud, at least 1 line long. Styles 3, erect, and connate to above the middle, recurved at the end and entire. Capsule tridymous, densely setose or nearly glabrous, about 4 lines diameter.

Queensland. Broad Sound, *R. Brown*, *Bowman*; Brisbane river, Moreton Bay, *A. Cunningham*, *Leichhardt*, *F. Mueller*; Logan river, *Fraser*; Rockhampton, *Dallachy*, *O'Shanesy*, *Bowman*, *Thozet*; Rockingham Bay, *Dallachy*.

Like most species of the genus, this plant is noted by O'Shanesy as "stinging like the common Nettle." On Dallachy's labels, however, I find the memorandum "does not sting like a Nettle." Whether there be a mild variety, or whether it loses its stinging properties on some occasions, or whether there has been some error on the part of Dallachy remains to be ascertained.

31. MALLOTUS, Lour.

(Rottlera, *Roxb.*; Echinus, *Lour.*)

Flowers diœcious or rarely monœcious, in terminal panicles or axillary racemes or spikes. Male fl.: Perianth globular and closed in the bud, opening in 3 or 4 valvate segments. No petals or glands. Stamens indefinite, usually numerous, on a central receptacle, without any rudimentary ovary; filaments free or very shortly united; anther-cells distinct, parallel, separated by a broad connective or attached in the centre to a small connective, opening longitudinally in 2 valves. Female fl.: Perianth more or less deeply 3- to 5-lobed or minutely toothed and at length spathaceous. Ovary 2- or 3-celled with 1 ovule in each cell. Styles free or very shortly united at the base, spreading, undivided, the upper or inner stigmatic surface fringed with raised papillæ or processes. Capsule separating into 2-valved cocci. Seeds usually globular, not carunculate; albumen copious; cotyledons flat, much broader than the radicle.—Shrubs or trees. Leaves alternate or rarely opposite, petiolate, usually broad and often large, entire toothed or lobed, sometimes peltate, usually 3- or more-nerved at the insertion of the petiole with 2 or sometimes more flat almost immersed glands on the upper surface near the base, sometimes however very obscure or obsolete. Male flowers very shortly pedicellate, clustered along the rhachis of the spikes, with a small bract under each cluster; females usually solitary within each bract and more pedicellate.

The genus is generally spread over tropical Asia and Africa. Of the nine Australian species four are also more or less generally spread over tropical Asia, extending northwards to South China, the other five are endemic. Loureiro's generic names are both of the same date; Mueller Arg. preferred that of *Mallotus*, because *Echinus* is clearly inapplicable to many of the species, and is moreover a well known Latin name for the hedgehog, and for the sea urchin. Baillon prefers the other name, but does not say for what reason he proposes to substitute it for *Mallotus*, previously adopted by Mueller.

Anther-cells (small) separated by a broad connective. Capsules echinate with long soft processes.
- Leaves mostly alternate. Inflorescence terminal. Capsule processes long and soft.
  - Tomentum of the plant soft and loose. Capsules very densely echinate with long crowded processes . . . . . 1. *M. ricinoides.*
  - Tomentum close and white. Capsules echinate with fewer scattered processes . . . . 2. *M. paniculatus.*
- Leaves opposite. Racemes axillary. Capsule processes setiform 3. *M. claoxyloides.*

Anther cells centrally attached to a small connective. Capsules tomentose, without processes.
- Panicles terminal.
  - Leaves oblong or ovate-lanceolate, 3 to 6 in. long. Capsules mostly 3-celled.
    - Leaves mostly alternate, minutely tomentose underneath. Capsules with a red tomentum . . . . 4. *M. philippinensis.*
    - Leaves alternate, green and glabrous on both sides, without small glands. Capsule tomentose . . . . 5. *M. angustifolius.*
    - Leaves often opposite, coriaceous, shining, glabrous except the small glands underneath. Capsules glandular, not red . . . . 6. *M. polyadenus.*
  - Leaves broadly ovate-rhomboidal. Capsules mostly 2-celled, the tomentum not red . . . . 7. *M. repandus.*
- Racemes or interrupted spikes simple, axillary at the base of the young shoots.
  - Leaves nearly orbicular, the transverse veinlets prominent underneath . . . . 8. *M. nesophilus.*
  - Leaves ovate and acute or ovate-lanceolate, white underneath, the veins fine . . . . 9. *M. discolor.*

1. **M. ricinoides,** *Muell. Arg. in Linnæa* xxxiv. 187, *and in DC. Prod.* xv. ii. 963. A tall shrub or spreading tree, more or less clothed with a stellate often floccose tomentum, soon wearing off from the upper surface of the leaves, often very dense on the underside as well as on the branches and inflorescence. Leaves broadly ovate or orbicular, acuminate, entire or slightly sinuate, either peltately attached near the base, or the petiole quite basal, 3- 5- or 7-nerved, with 2 glands near the base, varying in our specimens from 4 to 10 in. diameter. Spikes terminal, more or less paniculate, the central one sometimes 6 in. long or more, the lateral ones shorter. Flowers nearly sessile, the males clustered with occasionally a female in the same cluster, the females solitary within their bract, and alone or with 1 or 2 males. Male perianth-segments about 1 line long. Stamens about as long and very numerous, the anthers small, very much shorter than the filaments, the cells separated by a broad connective. Female perianth usually 5-cleft, but very soon concealed under the woolly processes of the ovary. Styles 3, tomentose outside, densely fringed and glabrous on the inner or upper surface. Capsules very densely covered with long

soft tomentose-villous processes, forming a dense moss-like mass of ½ in. or more in diameter.—*Croton ricinoides*, Pers. Syn. ii. 586; *C. mollissimus*, Geisel. Crot. Monogr. 73; *Echinus mollissimus*, Baill. Adans. vi. 316; *Mallotus pycnostachys*, F. Muell. Fragm. iv. 138, and *M. Zippellii*, F. Muell. l.c. 139, and numerous other synonyms quoted by Muell. Arg. l.c.

Queensland. Rockingham Bay, *Dallachy;* Mount Elliott, *Fitzalan.*—Extends also over the Eastern Archipelago to the Philippines and South China.

Persoon's and Geiseler's specific names both bear the same date, 1807. Mueller Arg. adopted the former under the impression that it was a year older, but 1806 is the date of Persoon's first volume only. Baillon gives no reason, however, for substituting Geiseler's name.

2. **M. paniculatus,** *Muell. Arg. in Linnæa* xxxiv. 189, *and in DC. Prod.* xv. ii. 965. A tall spreading shrub or small tree, the branches inflorescence and underside of the leaves white or ferruginous with a short close stellate sometimes almost scaly tomentum. Leaves on long petioles, ovate-rhomboid or almost orbicular, acutely acuminate, entire or obscurely sinuate or rarely lobed, not peltate, green and glabrous on the upper surface except when very young, 3- or 5-nerved at the base with 2 glands, 3 to 6 in. long. Flowers monœcious or almost diœcious, in terminal broadly pyramidal panicles of 6 in. to 1 ft., the males clustered along the branches, the females solitary within each bract, all on very short pedicels. Male perianth-segments about 1 line long. Anthers small, with a broad connective as in *M. ricinoides.* Female perianth rather longer than the male, usually 5-lobed with acute or acuminate lobes. Styles much shorter than in *M. ricinoides.* Capsule 3-celled, 3 to 4 lines diameter, tomentose and muricate with soft closely-tomentose processes, which are few and distant from each other, not densely covering the whole capsule as in *M. ricinoides.*—*Croton paniculatus*, Lam. Dict. ii. 207; *Mallotus chinensis*, Lour., and other synonyms quoted by Muell. Arg. l.c.

Queensland. Rockingham Bay, *Dallachy.*—Extends over the Eastern Archipelago to the Philippines and S. China.

3. **M. claoxyloides,** *Muell. Arg. in Linnæa* xxxiv. 192, *and in DC. Prod.* xv. ii. 972. A tall straggling shrub or small tree, or sometimes a handsome tree (*Dallachy*), the branches and foliage seabrous with scattered stellate hairs. Leaves opposite, but those of each pair often unequal, petiolate, mostly ovate or elliptical and acuminate, but varying from broadly-oblong to almost orbicular, 2 to 6 in. long, obtuse or acute, penniveined and often more or less distinctly 3-nerved at the base, green on both sides, with 2 or more glands near the base sometimes almost obsolete. Stipules rigidly setiform, short. Flowers diœcious, the males sessile or shortly pedunculate in 2 or 3 dense clusters collected in a head or in a short dense or scarcely interrupted axillary spike, the females 3 to 6 together in an umbel-like cluster on a common peduncle of ½ to 1 in., the pedicels at first short and thick, but attaining under the fruit the length of the peduncle. Male perianth-segments about 1 line long. Stamens numerous, the anthers

small with a broad connective as in *M. ricinoides*. Female perianth-segments lanceolate, attaining 2 lines under the fruit. Styles rather short, densely fringed on the inner face. Capsule 3-dymous, nearly ½ in. diameter, muricate with rather rigid setiform processes.—*Echinocroton claoxyloides*, F. Muell. Fragm. i. 32; *Echinus claoxyloides*, Baill. Adans. vi. 315; *Plagianthera? affinis*, Baill. Etud. Euph. 424.

Queensland. Brisbane river, Moreton Bay, *Fraser*, *W. Hill and F. Mueller*, *C. Stuart;* Rockhampton, *Dallachy;* Wide Bay, *Bidwill;* Lizard island, *A. Cunningham*.
N. S. Wales. Archers Creek, *Leichhardt;* Richmond river, *Herb. F. Mueller*.

Var. *ficifolia*, Baill. Leaves broader, often orbicular and 3- or 5 nerved, entire or coarsely-toothed. Male flowers large.—Rockhampton, *Dallachy*, *Bowman*.

Var. *macrophylla*. Leaves broadly ovate, acuminate, 4 to 8 in. long, usually 3-nerved at the base. Stipules longer.—Rockingham Bay, *Dallachy*.

4. **M. philippinensis,** *Muell. Arg. in Linnæa* xxxiv. 196, *and in DC. Prod.* xv. ii. 980. A tree often acquiring a considerable size, the branches and inflorescence more or less ferruginous-tomentose. Leaves on long petioles, oblong ovate-lanceolate or almost ovate, acuminate or obtuse, entire, contracted or rounded and 3-nerved at the base, 3 to 6 in. long, more coriaceous than in the preceding species, the upperside glabrous, with obscure glands near the base, the under surface pale or ferruginous with a minute tomentum, the principal veins ferruginous tomentose. Flowers diœcious, the racemes terminal or in the upper axils, the males more branched than the females, all much shorter than the leaves. Male perianth-segments membranous, about 1 line long. Filaments short, anthers rather large, the cells attached in the centre to a short connective often tipped with a red gland. Female perianth ovoid-tubular, 4- or 5-toothed, enclosing the ovary. Styles short, oblong, densely fringed on the inner face. Capsule tridymous, 3 to 4 lines diameter, covered with a red stellate tomentum without any processes. Seeds nearly globular.—*Croton philippinensis*, Lam. Dict. ii. 206; *Echinus philippinensis*, Baill. Adans. vi. 314; *Rottlera tinctoria*, Roxb. Pl. Corom. ii. 36, t. 168, and other synonyms quoted by Muell. Arg. l.c.

Queensland. Brisbane river, Moreton Bay, *A. Cunningham*, *F. Mueller;* Wide Bay, *Bidwill;* Pine river, *Fitzalan;* Rockhampton, *Thozet;* Rockingham and Edgecombe Bays, *Dallachy*.
N. S. Wales. Hastings and Clarence rivers, *Beckler*, *C. Moore*, and others; Northern woods, N. S. Wales, *London Exhibition*, 1862, *n.* 52.

The species is widely spread over tropical Asia, extending northwards to South China.

5. **M. angustifolius,** *Benth.* A small tree, glabrous, except the inflorescence and sometimes the young shoots pubescent with short spreading hairs. Leaves alternate, but often crowded at the ends of the branches so as to appear opposite or verticillate, on petioles varying in the same cluster from under ½ in. to above 1 in. long, oblong-elliptical or almost lanceolate, acuminate, slightly and irregularly dentate, 4 to 8 in. long, and rarely 2 in. broad, rather rigid, penniveined, green on both sides without the small glands of the under surface of most *Malloti*. Flowers diœcious, in short sessile panicles at the ends of the branches. Bracts small, lanceolate or ovate, entire.

Male perianth dividing into 4 lobes. Stamens numerous; anther-cells attached dorsally by a connective shorter than themselves. Female perianth of 5 imbricate acute segments. Ovary villous, 3-celled. Styles united at the base, spreading and bifid, but not long. Capsule tridymous, tomentose, not muricate, above 4 lines diameter.

Queensland. Rockingham Bay, *Dallachy.*

6. **M. polyadenus,** *F. Muell. Fragm.* vi. 184. A tall straggling shrub or small tree, glabrous except a minute scaly tomentum on the inflorescence. Leaves either opposite and unequal in the pair or alternate, shortly petiolate, oblong elliptical or ovate-lanceolate, obtuse or acuminate, coriaceous and sometimes shining above, covered underneath with the minute glands of *M. nesophilus* and of *M. discolor*, 3 to 6 in. long, penniveined, and sometimes 3- or 5-nerved at the base, the 2 to 4 glands of the upper surface very uncertain or obsolete. Flowers sometimes monœcious, but the two sexes in different spikes or racemes, and usually on different specimens, the racemes 1 to 4 in. long in the forks or upper axils, or forming a terminal panicle. Male flowers clustered, the pedicels rarely 1 line long. Filaments short. Anther-cells attached to a small connective. Female flowers solitary within the bracts, on pedicels at first short, but lengthening to 2 or 3 lines. Perianth divided nearly to the base into very small segments not enlarged under the fruit. Styles recurved and closely appressed to the ovary, the inner or upper surface very shortly fringed-papillose. Capsule tridymous, about 4 lines diameter, glabrous except a few of the scale-like glands of the rest of the plant.

Queensland. Cape York, *W. Hill;* Rockingham Bay, *Dallachy.*

7. **M. repandus,** *Muell. Arg. in Linnæa* xxxiv. 197, *and in DC. Prod.* xv. ii. 981. A large tree, the young branches inflorescence and underside of the leaves softly stellate-tomentose, with longer hairs often intermixed. Leaves on rather long petioles, broadly ovate rhomboidal or almost orbicular, acuminate, entire or obscurely sinuate-toothed, 2 to 4 in. long, slightly peltate, 3- or rarely 5-nerved at the insertion of the petiole, the flat glands of the upper surface obscure or obsolete, the minute glands of the underside almost concealed by the tomentum, darker and more glabrous above. Flowers diœcious, the racemes in terminal panicles, the central one in the males often 6 in. long. Male flowers clustered, the pedicels 1 to 2 lines long. Stamens numerous; anther-cells attached to a small connective. Female flowers (in Indian specimens) solitary within each bract, the perianth turbinate with 3 or 4 lanceolate lobes. Styles about 1 line long, the inner surface very densely fringed. Capsule 2-celled, about 4 lines broad, coriaceous, closely tomentose, without processes.

Queensland. Rockhampton and Rockingham Bay, *Dallachy.*—The species is also widely spread in tropical Asia. The Australian specimens are all males; they agree better with the tomentose ones from the East Indian Peninsula, than with the more glabrous ones from Timor and New Caledonia.

8. **M. nesophilus,** *F. Muell.; Muell. Arg. in Linnæa* xxxiv. 196, *and in DC. Prod.* xv. ii. 981. A tall shrub, the branches inflorescence and underside of the leaves whitish or slightly ferruginous with a close stellate tomentum. Leaves alternate, very broadly ovate or orbicular, mostly obtuse and entire, rarely sinuate and very shortly acuminate, the larger ones sometimes 4 in. long on petioles as long, but usually about half that size, and on lateral shoots much smaller on short petioles, 3-nerved or obscurely 5-nerved at the base, the upper surface very sparingly tomentose or at length glabrous, with 2 flat glands, the under surface minutely glandular, the primary veins and transverse reticulations prominent. Flowers diœcious, in single racemes in the lower axils of young shoots, the males 2 to 3 in. long, with clustered flowers on very short pedicels, the female racemes shorter, the flowers solitary within each bract. Stamens numerous, the anther-cells attached to a small connective. Female perianth narrow-turbinate, divided to about the middle into 3 or 4 lanceolate lobes. Styles short and broad, very densely fringed on the inner face. Capsule covered like that of *M. philippinensis* with an orange-red tomentum, but smaller and most frequently didymous and 2-celled, rarely 3-celled.—*Echinus nesophilus*, Baill. Adans. vi. 314.

N. Australia. Islands of the North coast, *Henne*, *Flood;* Port Darwin, *Schultz*, *n.* 881; Port Essington, *Armstrong;* Caledon Bay, *Gulliver.*

Queensland. Albany island, *W. Hill*, and Cape York, *Daemel* (with longer male pedicels); Cape Flinders, *A. Cunningham.*

9. **M. discolor,** *F. Muell.* A tall tree, the branches inflorescence and underside of the leaves white with a short close tomentum, with more or less of longer hairs on the principal veins underneath. Leaves on rather long petioles, ovate or ovate-lanceolate, acute or acuminate, thinner and smoother than in *M. repandus*, 2 to 3 in. long, becoming quite glabrous on the upper side, rounded or acute and 3-nerved at the base, not at all peltate, the flat glands rather prominent, the small glands of the under surface very numerous as in the three preceding species. Male racemes (only seen on O'Shanesy's specimens) slender simple, 3 to 4 in. long, in the lower axils of the young shoots as in *M. repandus.* Flowers smaller than in that species, very shortly pedicellate. Stamens numerous, the anther-cells attached to a small connective. Female flowers and fruits unknown.—*Rottlera discolor*, F. Muell. in Coll. Northern Woods N.S. Wales Lond. Exhib. n. 82; *Macaranga mallotoides* var. F. Muell. Fragm. iv. 140.

Queensland. Rockhampton (rare), *O'Shanesy;* Moreton Bay, Queensland woods, *London Exhibition*, 1862, *n.* 83, *W. Hill.*

N. S. Wales. Clarence river, Mountain brush forests, *London Exhibition*, 1862, *n.* 82.

This plant is reduced by F. Mueller, Fragm. vi. 185, to a variety of *M. repandus*, from which, however, as far as the specimens go, it appears to be constantly distinct. It must, however, remain doubtful until the females are known. Baillon, Adans. vi. 317, refers it to *Macaranga involucrata*, but he had probably only seen the Exhibition specimens, which had no flowers.

32. MACARANGA, Thou.

(Mappa, *A. Juss.*)

Flowers diœcious or rarely monœcious in axillary or rarely terminal spikes racemes or panicles. Male fl.: Perianth globular and closed in the bud, opening in 3 or 4 valvate segments. No petals or glands. Stamens indefinite, usually numerous, on a central receptacle without any rudimentary ovary; filaments free or very shortly united; anthers terminal, 4-lobed, 2-celled, opening in 4 valves, or loculicidally in 2 valves. Female fl.: Perianth ovoid oblong or almost globular, truncate or shortly toothed, opening obliquely into a broad cup or laterally into a spathe. Ovary 1-, 2- or 3-celled, with 1 ovule in each cell. Styles free or shortly united at the base, undivided, minutely papillose or fringed with processes on the inner surface. Capsule separating into 2-valved cocci. Seeds usually globular, not carunculate; albumen copious; cotyledons flat, much broader than the radicle.—Shrubs or trees. Leaves alternate, petiolate, usually broad and often large, entire toothed or lobed, often peltate or 3- or more nerved at the insertion of the petiole, but sometimes penniveined without lateral nerves. Male flowers clustered along the rhachis or branches of the inflorescence, sessile or very shortly pedicellate, with an entire toothed or fringed bract under each cluster. Female flowers in fewer clusters and few in the cluster or solitary, the bract usually longer than in the male, and often fringed or ciliate.

The genus is generally spread over tropical Asia and Africa. Of the five Australian species two are also in the Eastern Archipelago, one of them extending over the greater part of East India and S. China, the other three are endemic. The genus is closely allied to *Mallotus*, although most species differ in habit, and some have a different female perianth, or fewer stamens, or the ovary reduced to a single cell and ovule, but the only constant character is that of the anthers.

Leaves oblong elliptical or lanceolate, penniveined or scarcely 3-nerved at the base. Bracts very small and entire.
- Inflorescences including the peduncle under 1 in. Ovary not muricate, 2-celled. Styles short . . . . 1. *M. Dallachyi.*
- Inflorescences 2 to 4 in. long. Ovary muricate, usually 2-celled. Styles long and filiform.
  - Leaves mostly toothed, pale underneath with numerous minute scales. Female perianth lobes lanceolate, free . . 2. *M. subdentata.*
  - Leaves mostly entire, green underneath with few scattered minute scales. Female perianth usually spathaceous . . 3. *M. inamœna.*

Leaves very broad, palminerved Bracts as long as or longer than the flowers and toothed or fringed.
- Leaves slightly peltate. Stipules subulate. Ovary 1- or 2-celled . . . . 4. *M. involucrata.*
- Leaves deeply peltate. Stipules broad. Ovary usually 3-celled . . . . 5. *M. Tanarius.*

1. **M. Dallachyi,** *F. Muell.* A shrub variously described as small and straggling or tall, glabrous except a small scaly tomentum on the inflorescence and young branches. Leaves elliptical or oblong, acuminate, penniveined, usually with a few depressed glands on the upper

surface near the base, 3 to 5 in. long, on a petiole of ½ to 1 in Male spikes only seen in one specimen intermixed with the females, ½ to 1 in. long. Bracts and flowers very small, but the anthers entirely those of *Macaranga*. Perianth-segments usually 3, about ½ line long, glabrous or sprinkled with a very few scales. Female flowers 2 to 4 together almost clustered on axillary peduncles varying from ¼ to 1 in. Bracts small, entire. Perianth short and cup-shaped, opening laterally or rarely in 2 lobes. Ovary 2-celled, scaly-tomentose, without processes. Styles linear, thickened at the base, not fringed, rather acute, about 1 line long.—*Mallotus Dallachyi*, F. Muell. Fragm. vi. 184; *Echinus Dallachyanus*, Baill. Adans. vi. 314.

Queensland. Rockingham bay, *Dallachy.*—The specimens examined by Baillon had female flowers only, but with them were loose male flowers and remains of spikes which belong probably to *Mallotus polyadenus*, thence Baillon's mistake in referring *M. Dallachyi* to *Mallotus*.

2. **M. subdentata,** *Benth.* A shrub or small tree, glabrous except a minute scaly tomentum on the inflorescence and young shoots. Leaves oblong narrow-elliptical or lanceolate, acuminate, irregularly sinuate-toothed or rarely almost entire, penniveined, green on both sides but paler underneath from the minute scales they are more or less covered with, 3 to 9 in. long, on petioles varying from ½ to 1 in. Male flowers not seen. Female peduncles axillary, elongated, but much shorter than the leaves, bearing 2 to 5 flowers, of which 1 or 2 sessile or nearly so, and 2 or 3 on long pedicels. Perianth of 4 or 5 narrow-lanceolate segments scarcely 1 line long and quite distinct from the base. Ovary and capsule usually 2-celled, scaly-tomentose and shortly echinate. Styles filiform, scarcely thickened at the base, often above ½ in. long.

Queensland. Rockingham bay, *Dallachy.*—Very near *M. inamœna*, with which I had probably confounded it in the specimens of F. Mueller's collection first examined. After having returned them I received the specimens now described, which appear to me to be too distinct in the female perianth as well as in the foliage to be left as a mere variety. The two forms, however, require further comparison in both sexes.

3. **M. inamœna,** *F. Muell.* A tall shrub or small tree, the branches foliage and inflorescence scabrous-pubescent, the short hairs scarcely or irregularly stellate. Leaves oblong or narrow-elliptical, acuminate, entire, penniveined, green on both sides, the small glands underneath few and scattered, 3 to 5 in. long on a petiole of ½ to 1 in. Male spikes in the upper axils or several together at the ends of the branches, 2 to 4 in. long, the flowers clustered within small broad densely tomentose bracts, the clusters at length distant along the rhachis. Perianth-segments about ¾ line long. Stamens 20 to 30. Female peduncles axillary or lateral, elongated, bearing sometimes 3 pedicellate flowers at the end with another occasionally lower down, sometimes dichotomous with 1 to 3 flowers at the end of each branch. Pedicels at first very short, lengthening to from ½ to 1 in. Perianth of 4 or 5 narrow linear segments 1 to 2 lines long, united in a spathaceous calyx splitting open on one side (or rarely in the uppermost flower separate?). Ovary

and capsule usually 2-celled, rarely 3-celled, echinate. Styles long and slender, thickened at the base, attaining sometimes 5 lines.—*Mallotus inamœnus*, F. Muell. Herb.

Queensland. Rockingham Bay, *Dallachy.*

4. **M. involucrata,** *Baill. Etud. Euph.* 432. A tree attaining sometimes 50 to 60 ft., but often small and straggling, the branches inflorescence and underside of the leaves softly pubescent. Leaves broadly ovate-rhomboidal, acuminate, entire or slightly sinuate and denticulate, usually slightly peltate, 3- 5- or 7-nerved, with 2 to 4 flat glands on the upper side, 3 to 6 in. long and sometimes as broad, on a petiole of 2 to 3 in. Stipules subulate. Male spikes in axillary panicles not much branched and shorter than the leaves, the flowers almost sessile, in dense clusters within an oblong or ovate toothed bract of about 2 lines. Perianth-segments about ¾ line long, villous outside. Stamens 10 to 15. Female inflorescence simple or nearly so, the peduncle axillary or lateral, shorter than the leaves, with a single terminal or several distant clusters. Bracts broadly ovate or cordate, acute, toothed, often ½ in long. Perianth sessile, short, broadly and obliquely cup-shaped. Ovary 1- or 2-celled, more or less muricate with soft processes. Styles 1 or 2, long, glabrous, fringed on the upper side. Fruit (as yet unripe in our specimens) globular, 1-celled and 1-seeded, losing the processes of the ovary and sometimes quite glabrous and smooth.—Muell. Arg. in DC. Prod. xv. ii. 1011; *Urtica involucrata*, Roxb. Fl. Ind. iii. 592; *Macaranga mallotoides*, F. Muell. Fragm. iv. 139; *M. asterolasia*, F. Muell. l.c. 140, Baill. Adans. vi. 317.

Queensland. Endeavour river, *A. Cunningham;* Port Molle, *M'Gillivray;* Rockingham Bay and Mount Elliott, *Dallachy;* Cape York, *Daemel.*—The species is also in the Eastern Archipelago, but not in Bengal, except as cultivated in the Calcutta Garden.

5. **M. Tanarius,** *Muell. Arg. in DC. Prod.* xv. ii. 997. A tall erect shrub, glabrous or the young parts minutely pubescent in the typical form, the branches and petioles often very glaucous. Leaves peltate, very broadly ovate or orbicular, acuminate or rarely obtuse, entire or obscurely sinuate, from 3 to 4 in. to nearly 1 ft. diameter, pale or almost canescent underneath, with about 9 nerves radiating from the top of the long petiole, penniveined from the larger nerves with transverse veinlets. Stipules ovate-lanceolate, acute, ½ in. long, with broad membranous margins. Male panicles often much branched but shorter than the leaves, the flowers pedicellate in the clusters. Bracts ovate-lanceolate, concave, 2 to 3 lines long, fringed with long cilia. Perianth-segments scarcely ½ line long. Female peduncles simple, bearing few clusters. Perianth obliquely cup-shaped. Bracts ovate-cordate, often 4 to 6 lines long, deeply fringed. Ovary muricate with glabrous entire or bifid processes. Styles rather long, papillose or shortly fringed. Capsule 3-celled, coriaceous, shortly and sparingly muricate, about 4 lines diameter.—*Ricinus Tanarius*, Linn.; *Mappa tanaria*, Spreng. Syst. iii. 878.

N. Australia. Port Essington, *Armstrong, Leichhardt.*

Queensland. Northumberland islands, *R. Brown ;* Howick's Group, *F. Mueller;* Port Denison and Rockingham Bay, *Dallachy, Fitzalan;* Broad Sound and head of Isaacs river, *Bowman;* Rockhampton, *Thozet;* Moreton Bay, *W. Hill, F. Mueller;* Liverpool river, *Gulliver.*

N. S. Wales. Tweed river, *Guilfoyle.*

Var. *tomentosa,* Muell. Arg. Leaves softly tomentose. Capsules rather larger.—Rockingham Bay, *Dallachy.*

The species extends over East India and the Archipelago northward to South China.

33. CODIÆUM, Rumph.

Flowers usually monœcious in axillary or terminal racemes. Male fl.: Calyx of 5 or 6 membranous segments much imbricate in the bud. Petals 5 or 6, very short. Glands as many as petals, alternating with them. Stamens indefinite, inserted on a slightly raised receptacle, without any rudimentary ovary; anthers with the cells divaricate or placed back to back and confluent at the top. Female fl.: Calyx-segments thicker or larger than in the males. Disk obscurely lobed. Ovary 3-celled, with 1 ovule in each cell. Styles free, spreading or recurved, undivided. Capsule globular, separating into 2-valved cocci.—Shrubs or trees, quite glabrous. Leaves alternate, petiolate, entire. Flowers small, the males usually clustered but few together, the females solitary within each bract in separate racemes, or, in a species not Australian, at the base of the male raceme.

The genus comprises a few species from East India and the Archipelago, including the only Australian one which, if correctly identified, has a wide range over the whole area, but in some places perhaps cultivated only.

1. **C. variegatum,** *Blume; var.* moluccanum, *Muell. Arg. in DC. Prod.* xv. ii. 1119. A tall shrub or small tree, quite glabrous. Leaves from obovate-oblong to narrow elliptical or oblong-spathulate, 4 to 8 in. long, on petioles of ½ to 1 in., penniveined, green on both sides or especially in the typical form blotched or variegated with white. Flowers in long axillary or lateral racemes, and usually a male and female raceme in the same axil, the former longer than the latter. Male flowers in clusters of 2 to 6, on pedicels of 1 to 2 lines. Calyx-segments nearly orbicular and nearly 1 line diameter. Petals usually not above ⅓ as long as broad, or sometimes rather longer than broad, slightly dentate. Glands about the same length, broad, thick and truncate. Stamens about 20. Female flowers on thick pedicels of 1 to 4 lines. Calyx-segments shorter and thicker than in the males. Disk obscurely lobed. Styles short thick and recurved. Capsule glabrous and smooth, 3 to 4 lines diameter.—*C. obovatum,* Zoll.; Baill. Adans. vi. 303; *C. moluccanum,* Dcne. Herb. Tim. Descr. 157.

Queensland. Mount Elliott and Seaview Range, Rockingham Bay, *Dallachy.*—The same variety also in Timor and Java, and the species widely spread over East India and the Archipelago, but often cultivated only. I do not feel, however, certain that Mueller Arg. is right in referring this broad-leaved Australian form to the real *C. variegatum* (*C. chrysosticton,* Spreng Syst. iii. 865, *C. pictum,* Hook. Bot. Mag. t. 3051, and other synonyms given by Muell. Arg. l. c.). For although there is not always much difference in the breadth of the leaf, the general shape is not the same, and the styles are much shorter and thicker in this broad-leaved form than in the common *C. variegatum.*

34. BALOGHIA, Endl.

Flowers diœcious or monœcious, in unisexual short terminal racemes. Male fl.: Calyx 4- or 5-lobed, the lobes imbricate in the bud or very short. Petals as many as calyx-lobes. Disk with a thick raised undulate or irregularly lobed border. Stamens indefinite, the filaments shortly united or inserted on a raised or conical central receptacle or column without any rudimentary ovary; anthers dorsally attached, with 2 distinct parallel cells opening outwards and longitudinally in 2 valves. Female fl.: Calyx petals and disk as in the males. Ovary 3-celled, with 1 ovule in each cell. Styles 3, distinct or shortly united at the base, spreading, deeply divided into 2 branches (at least in the Australian species). Fruit globular, the exocarp often fleshy or succulent, the endocarp rather hard, separating into 3 2-valved cocci. Seeds oblong, with a small caruncle; albumen copious, cotyledons flat, longer than and at least twice as broad as the radicle.—Trees or shrubs, glabrous except sometimes the flowers. Leaves opposite or alternate, coriaceous, finely veined. Flowers few, not small, the racemes sometimes almost reduced to umbels; bracts very deciduous, with 1 flower within each.

The genus contains but few species, chiefly from New Caledonia, including the two found also in Australia. Mueller Arg reduces it to a section of *Codiæum*, but the habit, inflorescence, perianth and stamens, and perhaps the styles, are quite different. Baillon thinks it scarcely distinct from *Ricinocarpus*, but besides the habit, the embryo is that of the *Crotoneæ*, not of the *Beyerieæ.*

Leaves opposite. Petals glabrous . . . . 1. *B. lucida.*
Leaves alternate or scattered. Petals densely woolly tomentose inside . . . . 2. *B. Pancheri.*

1. **B. lucida,** *Endl. Prod. Fl. Norf.* 84, *and Iconogr.* t. 122, 123. A tall shrub or small tree, perfectly glabrous. Leaves opposite, very shortly petiolate, oblong obovate-oblong or elliptical, obtuse or obtusely acuminate, rigidly coriaceous and shining, the primary veins numerous, fine but prominent, transverse and anastomosing. Flowers few together in short loose sessile terminal racemes, the males and females usually on separate branches, but sometimes the lower 1 or 2 pair female and the upper 2 or 3 pair males; the pedicels opposite, $\frac{1}{4}$ to $\frac{1}{2}$ in. long, solitary in the axils of very small bracts. Calyx deeply divided into 5 lanceolate lobes, varying to 4 only in the males, or rarely to 6 in the females, 2 to $2\frac{1}{2}$ lines long. Petals oblong or lanceolate, nearly twice as long. Disk in both sexes with a thick irregularly lobed undulate margin. Stamens numerous, the filaments very shortly united in a conical or oblong column or receptacle. Styles divided almost to the base into 2 branches. Capsule hard, globular, $\frac{1}{2}$ to $\frac{3}{4}$ in. diameter, somewhat tridymous, with a furrow bordered by 2 narrow ridges on the back of each coccus.—*Codiæum lucidum,* Muell. Arg. in DC. Prod. xv. ii. 1116.

Queensland. Rockingham Bay, *Dallachy;* Rockhampton, *O'Shanesy;* Moreton Bay, *F. Mueller.*

N. S. Wales. Hastings and Macleay rivers, *Beckler;* Clarence and Richmond rivers, *C. Moore, London Exhibition* 1862, *n.* 46; Illawarra, *A. Cunningham, McArthur;* Sydney woods, *Paris Exhibition* 1867, *n.* 185; Lord Howe's island, *Milne.*

The species is also in Norfolk Island and in New Caledonia.

2. **B. Pancheri,** *Baill. Adans.* ii. 214. A slender tree of 50 to 60 ft., glabrous except the flowers. Leaves alternate or here and there opposite, crowded at the end of the branches, obovate or obovate-oblong, obtuse, coriaceous, of a shining green, prominently veined as in *B. lucida,* but the veins much more oblique, 1½ to 3 in. long. Male flowers in very short terminal sessile racemes almost reduced to an umbel. Pedicels slender, glabrous, 3 to 4 lines long. Calyx broadly cup-shaped, very shortly sinuate-lobed or almost truncate and sometimes irregularly splitting, tomentose on the margin. Petals nearly 3 lines long, glabrous outside, but the inner or upper surface very densely covered with a soft white loose tomentum or wool. Stamens indefinite but rather fewer than in *B. lucida,* the central column more prominent and tomentose-villous, the free part of the filaments glabrous or nearly so. Female flowers and fruits unknown to me, but described by Baillon as having the generic character.—*Codiæum Pancheri,* Muell. Arg. in DC. Prod. xv. ii. 1117.

Queensland. Scrubs near Kilcoy, *Herb. F. Mueller,* the collector not named, a single specimen agreeing precisely with a specimen in Herb. Hook. from New Caledonia, where the specimens were gathered on which the species was founded.

35. CARUMBIUM, Reinw.

(Omalanthus, *A. Juss.;* Wartmannia, *Muell. Arg.*)

Flowers monœcious, in terminal racemes. Male fl.: Perianth small, at first irregularly truncate or shortly lobed, often dividing into 2 broad lobes. No glands or petals. Stamens few, inserted on a central receptacle, without any rudimentary ovary; filaments free; anther-cells distinct, divaricate or placed back to back, opening longitudinally in 2 valves. Female perianth nearly similar to the male, usually deciduous. Ovary 2-celled, with 1 ovule in each cell. Styles 2, linear, divergent, undivided, papillose on the inner surface. Capsule compressed, didymous, somewhat fleshy, indehiscent or tardily opening in 2 valves along the back of the cocci. Seeds with a fleshy arillus or caruncula.—Glabrous trees or shrubs. Leaves alternate, petiolate, broad, entire. Stipules membranous, often large but very deciduous. Flowers very small, the males in small clusters occupying the greater part or the whole of the raceme, the females solitary within each bract, one or few at the base of some of the male racemes, or alone.

The genus has but few species, limited to the Indian Archipelago and the islands of the South Pacific. Of the two Australian species, one ranges generally over the area of the genus, the other is endemic.

Capsule quite smooth. Seeds half-enveloped in a fleshy arillus. Bracts with 2 large glands . . . . 1. *C. populifolium.*
Capsule bearing usually 2 to 6 short conical processes or tubercles. Seeds with a thick fleshy caruncula. Bracts with villous glands . . . . 2 *C. stillingiæfolium.*

1. **C. populifolium,** *Reinw.; Miq. Fl. Ind. Bat.* i. part ii. 414. A tall shrub or small tree, quite glabrous. Leaves broadly ovate-triangular or rhomboidal, acuminate, glaucous, prominently penniveined and often turning red underneath, 2 to 4 in. or on luxuriant shoots 6 in. long, and often as broad, on petioles usually about the same length. Stipules lanceolate, ½ to 1 in. long, but so deciduous as to be rarely seen except on the very young shoots, which being usually at the base of the inflorescence, these stipules have been described as barren bracts. Racemes 1 to 4 in. long. Bracts small, entire or denticulate, with 2 large glands at their base. Male flowers 3 to 6 together, on pedicels of 1 or rarely 2 lines. Perianth when young broadly cup-shaped and entire, expanding horizontally to a diameter of 1 to 1½ lines, nearly flat and often splitting on one side or into 2 unequal lobes, and when pressed laterally in drying appearing often 2-auriculate at the base. Stamens 6, or fewer in the lateral flowers. Female flowers few at the base of the raceme, on pedicels varying from ¼ to 1 in. Perianth like that of the males, but very deciduous. Capsule glaucous, didymous, 4 to 5 lines broad, opening very tardily along the margins or back of the cocci. Seeds more or less enveloped in a fleshy arillus or carunculus.—*C. populneum,* Muell. Arg. in DC. Prod. xv. ii. 1144, with the synonyms adduced; *C. Sieberi,* Muell. Arg. in Linnæa xxxii. 85, and in DC. Prod. xv. ii. 1145, Baill. Adans. vi. 326; *C. platyneuron,* Muell. Arg. in DC. l.c.; *C. pallidum,* Muell. Arg. in Linnæa xxxii. 85; *Omalanthus populifolius,* Grah. in Bot. Mag. t. 2780, F. Muell. Fragm. i. 32.

Queensland. Shoalwater Bay, *R. Brown;* Moreton Bay, *F. Mueller;* Crocodile Creek, *Bowman;* Rockhampton and Rockingham Bay, *Dallachy.*

N. S. Wales. Port Jackson to the Blue Mountains, *R. Brown, Sieber, n.* 640, and others; northward to Hastings, Clarence, and Richmond rivers, *Beckler, Henderson,* and others; southward to Illawarra, where the reddish tint of the young shoots gives a peculiar character to the aspect of some of the valleys, *A. Cunningham, Lownes;* Twofold Bay, *F. Mueller.*

Victoria. Eastern extremity of Gipps' Land, *F. Mueller.*

The species is spread over the Eastern Archipelago, and some of the Pacific islands. I am unable to distinguish even as varieties the three forms described as species by Mueller Arg in the Prodromus. The foliage is exceedingly variable in size and consistence according to age and luxuriance.

2. **C. stillingiæfolium,** *Baill. Adans.* vi. 325. A glabrous shrub of 4 or 5 ft. rarely growing out into a small tree, the branches and foliage much more slender than in *C. populifolium.* Leaves broadly ovate-triangular or almost rhomboidal, usually acute but scarcely acuminate, whitish or glaucous underneath, 1 or 2 in. long, on a petiole sometimes shorter sometimes longer than the lamina. Stipules of *C. populifolium,* but, like the leaves, smaller. Racemes very slender, 1 to 2 in. long. Bracts ovate and acute or lanceolate and acuminate, without any or with very small glands, and sometimes borne with the cluster on a short peduncle. Male pedicels filiform, ½ to above 1 line long, the flowers very small. Female flowers few at the base of the raceme, on pedicels of ½ in. or more. Capsule about 3 lines broad, usually but not always muricate with a few very short conical processes, rather more

readily dehiscent than that of *C. populifolium*. Seeds with a short fleshy carunculus.—*Omalanthus stillingiæfolius*, F. Muell. Fragm. i. 32; *Wartmannia stillingiæfolia*, Muell. Arg. in DC. Prod. xv. ii. 1147.

Queensland. Brisbane river, Moreton Bay, *A. Cunningham*, *Fraser*, *F. Mueller*.
N. S. Wales. New England, *C. Stuart;* Manning river, *C. Moore.*

This species, so closely resembling *C. populifolium* in general habit and characters, has been generically distinguished by Mueller Arg. chiefly on account of the appendage of the seed, supposed to be in one a carunculus, in the other a true arillus. Although so different in size these appendages are shown by Baillon to be in both species of the same nature and origin.

36. SEBASTIANIA, Spreng.

(Gymnanthes, *Sw.;* Microstachys, *A. Juss.;* Elachocroton, *F. Muell.*)

Flowers monœcious, in terminal or leaf-opposed racemes or spikes. Male fl.: Perianth small, variously divided into 2 or 3 lobes or segments imbricate in the bud or open. No petals or glands. Stamens 2 or 3, inserted on a central receptacle without any rudimentary ovary; filaments free; anther-cells distinct, divaricate or placed back to back, opening longitudinally in 2 valves. Female perianth of 3 segments. Ovary 3-celled, with 1 ovule in each cell. Styles 3, linear, undivided, free or very shortly connate at the base. Capsule separating in 2-valved cocci, leaving a central persistent axis. Seeds ovoid or oblong, caruncuIate.—Shrubs trees or (in the Australian species) annual or suffrutescent herbs. Leaves alternate, often minutely serrulate. Male flowers 2 or 3 together in clusters occupying the greater part or the whole of the raceme, females usually solitary or few at the base of the spike.

The genus is rather a large one in America, with a single species spread over tropical Asia and Africa, which is also the only Australian one. Baillon unites the genus with *Excœcaria*, to which it is certainly nearly allied, but the presence of the carunculus on the seed is accompanied by some differences in the habit and flowers, which appear to justify the separation.

1. **S. chamelæa,** *Muell. Arg. in DC. Prod.* xv. ii. 1175. An annual or perennial, with a hard woody base and erect or ascending branching virgate stems of 1 to 1½ ft., glabrous and often glaucous. Leaves oblong-linear narrow-elliptical or lanceolate, mostly obtuse, minutely serrulate, 1 to 2 in. long, tapering into a very short petiole. Male spikes slender, 1 to 1½ in. long, mostly leaf-opposed. Bracts very small, acute, with 2 large more or less stipitate glands sometimes as long as the point of the bract. Flowers 1 or 2 within each bract. Perianth-segments about ¼ line long. Stamens 3, almost exserted. Female flowers usually solitary at the base of the male spike or lateral on the branch without any males, rarely 2 or 3 together within a separate bract, the bracts and perianths rather larger than in the males. Styles undivided, not very long. Capsule ovoid-truncate, tridymous, about 3 lines long, sometimes quite smooth but more frequently with 2 lines of prominent glands or conical processes on the back of each coccus.—

Tragia chamelæa, Linn.; *Excæcaria chamelæa*, Baill. Adans. vi. 323; *Elachocroton asperococcus*, F. Muell. in Hook. Kew Journ. ix. 17.

N. Australia. Islands of the Gulf of Carpentaria, *R. Brown*, *Henne*; Victoria river, *F. Mueller*; Goulburn islands, *A. Cunningham*; Port Darwin, *Schultz*, *n.* 539; Beagle Bay, N.W. Coast, *Hughan*.

Queensland. Endeavour river, *A. Cunningham*; Baines creek, *F. Mueller*; Connor's and Bowen rivers, *Bowman*; Rockingham Bay, *Dallachy*; Cape York, *M'Gillivray*.

This, the only Old World species, is widely spread over tropical Asia and Africa, but does not extend to America.

37. EXCÆCARIA, Linn.

Flowers monœcious or diœcious, in terminal or axillary racemes or spikes. Male fl.: Perianth of 3 or 2 very small segments. No petals or glands. Stamens 3 or 2, forming almost the whole flower, without any rudimentary ovary; filaments free or shortly united at the base; anther-cells distinct, placed back to back, opening longitudinally in 2 valves. Female fl.: Perianth usually more distinctly 3-lobed or 3-partite than the male. Ovary 3- or 2-celled, with 1 ovule in each cell. Styles linear, undivided, free or shortly united at the base. Capsule dividing into 2-valved cocci, or (in species not Australian) somewhat fleshy and almost indehiscent. Seeds globular or ovoid, without any caruncula. Albumen copious; cotyledons flat, much broader than the radicle.—Trees or shrubs, with a very acrid milky juice. Leaves alternate, entire or crenulate. Flowers very small, the males clustered 2 or 3 together along the rhachis of the raceme or spike, or sometimes solitary within each bract. Stamens exserted. The females in separate shorter racemes, or in species not Australian at the base of the male racemes or spikes.

The genus, if taken to include *Sapium*, is generally spread over the tropical regions of both the New and the Old World. Of the three Australian species or varieties, one is common on the sea-coasts of tropical Asia, the other two are endemic.

Leaves obovate or broadly elliptical, obtuse, entire or crenate, 2 to 3 in. long . . . . 1. *E. Agallocha.*
Leaves ovate-lanceolate or ovate, obtusely acuminate, crenate, 1 to 3 in. long . . . . 2. *E. Dallachyana.*
Leaves narrow-oblong, very obtuse, entire, $\frac{1}{2}$ to 1 in. long . . . 3. *E. parvifolia.*

1. **E. Agallocha,** *Linn.*; *Muell. Arg. in DC. Prod.* xv. ii. 1220. A small tree, quite glabrous. Leaves mostly obovate or broadly elliptical, obtuse or shortly and obtusely acuminate, rounded or contracted at the base, entire or somewhat crenate, thick and shining when old, 2 to 3 in. long on a petiole of $\frac{1}{4}$ to $\frac{1}{2}$ in. Spikes or racemes usually in the axils of the previous year's leaves or at the old nodes, solitary or 2 or 3 together, 1 to $1\frac{1}{2}$ in. long. Male flowers rather crowded, the bracts rather thick, with two more or less distinct glands inside at the base. Within the bract the 2 or 3 stamens are supported on a short stipes, with a small lanceolate scale on each side at the base, and two or three still smaller alternating with the filaments, which are from $\frac{3}{4}$ to 1 line long. Female flowers in short racemes on separate specimens,

the pedicels 1 to 2 lines long. Perianth of 3 small acute rather thick lobes. Capsule tridymous, about 3 lines diameter.

N. Australia. Islands of the Gulf of Carpentaria, *R. Brown*, *Sweers;* salt-water banks of Victoria river, *F. Mueller;* Goulburn island, *A. Cunningham;* Port Darwin, *Schultz*, *n.* 597, 677.

Queensland. Common along the coast from Rockhampton and Broad Sound to Cape York, *Dallachy*, *Bowman*, *M'Gillivray*, and others.

This species appears to be a common maritime tree in tropical Asia.

2. **E. Dallachyana,** *Baill. Adans.* vi. 324, as a var. of *E. Agallocha.* Nearly allied to *E. Agallocha*, and perhaps really only an inland variety. Leaves ovate-lanceolate or ovate, obtusely acuminate, crenate, 1 to 3 in. long, less coriaceous and the veins more prominent and reticulate underneath. Flowers both male and female apparently the same as in *E. Agallocha.*

Queensland. Burnett river, *F. Mueller;* Rockhampton common, always in the scrub, *Dallachy*, *Bowman*, *Thozet.*

3. **E. parvifolia,** *Muell. Arg. in Flora* 1864, 433, *and in DC. Prod.* xvii. 1221. Very nearly allied to *E. Agallocha*, and reduced by Baillon like the last to a variety of that species, with narrow oblong very obtuse entire leaves of ½ to 1 in., tapering into a short petiole. The male racemes are also smaller, ½ to 1 in. long. Female flowers and fruit unknown.

N. Australia. Common round the Gulf of Carpentaria, *R. Brown*, *F. Mueller*, *Landsborough.*—" Gutta-percha tree" of the latter.

ORDER CIX. URTICEÆ.

Flowers unisexual or very rarely polygamous. Perianth simple and calyx-like, of 3 to 5 segments (rarely reduced to 1 or 2) imbricate or induplicate-valvate in the bud. Stamens in the males as many as perianth-segments, and opposed to them, very rarely fewer or more; filaments short and erect or longer and inflexed in the bud; anthers usually with 2 parallel cells opening longitudinally. Ovary in the females free or rarely more or less adnate to the perianth, 1-celled. Style simple or more or less deeply divided into 2 branches or 2 distinct styles, stigmatic in the upper portion, or sometimes the style reduced to a sessile fringed or tufted stigma. Ovule 1, erect and orthotropous, or laterally attached and amphitropous, or pendulous and anatropous, the micropyle always superior. Fruit (of each separate flower) a small berry drupe nut or indehiscent utricle, and sometimes the fruits of a whole inflorescence united in a succulent syncarp, surrounding or subtended by or enclosed in a fleshy receptacle. Seed with a membranous testa, with or without albumen. Embryo straight curved or spirally involute, the cotyledons flat or folded, the radicle superior.—Trees shrubs or herbs very varied in habit and foliage. Leaves alternate or opposite, entire toothed or rarely divided, penniveined and often 3-nerved. Stipules present, but usually very deciduous. Flowers small, in cymes clusters or heads, rarely solitary, the clusters or heads often racemose or paniculate, the receptacle of the heads very variously shaped and often bordered by an involucre of small bracts.

A very large Order, spread over the New and the Old World, chiefly tropical, but a few species extending into temperate regions, both in the northern and the southern hemisphere, a very few only to be met with in cold climates. Of the seventeen Australian genera, eight have the general distribution of the Order, seven are generally spread over the Indian Archipelago, most of them extending more or less over tropical Asia and Africa and the Pacific islands, one, *Pseudomorus*, appears only to be found out of Australia in New Caledonia and Norfolk island, and one, *Australina*, only in New Zealand and in tropical Africa. Not one is endemic in Australia.

Some important groups of this Order have not yet been worked up for the Prodromus, the subjoined tribes and their characters have therefore reference chiefly to the few Australian genera; the Urticeæ proper have, however, been very carefully monographed by Weddell. He gives the name of *cystoliths* to certain calcareous concretions under the epidermis of the leaves, which, when linear, assume in the dried specimens the aspect of appressed superficial hairs, although really within the substance of the leaf. The form of these cystoliths has in some Urticeæ been made use of as a specific character, but they are, I believe, dot-like in all the Australian species.

TRIBE 1. **Celtideæ.**—*Flowers often polygamous, in axillary or lateral cymes. Filaments short, erect or slightly incurved in the bud. Styles or style-branches 2, equal. Ovule pendulous. Embryo curved, the cotyledons often folded over the incumbent radicle. Trees or shrubs.*

Flowers polygamous, the fertile ones frequently hermaphrodite. Perianth-segments imbricate in the bud. Style-branches (or styles) linear-oblong or dilated, truncate or 2-lobed . . . . . 1. CELTIS.
Flowers polygamous, the fertile ones frequently hermaphrodite. Male perianth-segments induplicate-valvate. Style-branches (or styles) short, involute and persistent on the small fruit . . . 2. TREMA.
Flowers unisexual (monœcious). Perianth-segments imbricate in the bud. Style-branches (or styles) subulate . . . . 3. APHANANTHE.

TRIBE 2. **Artocarpeæ.**—*Flowers unisexual in dense spikes or heads, or crowded on or inclosed in a fleshy receptacle. Stamens erect or slightly incurved in the bud. Styles undivided or 2-branched. Ovule pendulous or laterally attached. Embryo curved or straight. Trees or shrubs, with a milky juice.*

Flowers enclosed in a globular ovoid or pear-shaped receptacle closed at the small orifice by small bracts . . . . 4. FICUS.
Flowers in globular heads on a fleshy receptacle. Fruit a syncarp formed of the enlarged fleshy perianths and receptacle . . . 5. CUDRANIA.
Male flowers densely crowded on a broad receptacle, females solitary. Fruit an ovoid mass consisting of the consolidated involucre and pericarp, the tips of the bracts alone free . . . . 6. ANTIARIS.

TRIBE 3. **Moreæ.**—*Flowers unisexual in dense spikes or heads. Stamens inflected in the bud. Styles usually 2-branched. Ovule pendulous or laterally attached. Embryo incurved or involute. Trees or shrubs, very rarely herbs.*

Male flowers in dense spikes. Females in globular heads, their perianths urceolate with a small orifice. Style-branches elongated 7. MALAISIA.
Male flowers in dense spikes. Females few in very small spikes or heads, their perianth of 4 segments. Style-branches elongated 8. PSEUDOMORUS.
Flowers in globular androgynous heads. Style elongated with a small branch or tooth at the base. Stem herbaceous . . . . 9. FATOUA.

TRIBE 4. **Euurticeæ.**—*Flowers unisexual, in cymes clusters or rarely in heads. Stamens inflected in the bud. Styles undivided. Ovule erect. Embryo straight or nearly so. Trees shrubs or herbs.*

SUBTRIBE 1. **Procrideæ.**—*Plants not stinging. Female perianth deeply lobed.*

Flowers densely crowded on a flat or concave receptacle with an involucre of several bracts. Stigma tufted. Leaves opposite or rarely alternate . . . . 10. ELATOSTEMMA.

SUBTRIBE 2. **Boehmerieæ.**—*Plants not stinging. Female perianth either tubular and minutely toothed, enclosing or adnate to the ovary, or rarely minute or none.*

Shrubs or trees. Flower-clusters in axillary spikes or rarely solitary and sessile.
 Stigma linear, persistent . . . . 11. BOEHMERIA.
 Stigma linear, deciduous . . . . 12. PIPTURUS.
Herbs. Flowers in axillary sessile solitary cymes or clusters.
 Stigma linear, deciduous. Male perianth-lobes or segments concave or abruptly inflected at the top. Leaves entire. Bracts very small . . . . 13. POUZOLSIA.
 Stigma tufted. Male perianth-segments nearly flat. Leaves entire. Bracts united at the base into an involucre . . . 14. PARIETARIA.
 Stigma linear. Male perianth with 1 large outer lobe. Stamen 1. Leaves toothed. No bracts . . . . 15. AUSTRALINA.

SUBTRIBE 3. **Urereæ.**—*Plants more or less armed with stinging hairs. Female perianth 4-lobed, 2 lobes usually larger than the 2 others.*

Herbs with opposite leaves. Stigma tufted . . . . 16. URTICA.
Trees or shrubs with alternate leaves. Stigma linear . . . . 17. LAPORTEA.

TRIBE I. CELTIDEÆ.—Flowers in axillary or lateral cymes often polygamous, the hermaphrodite or female flowers sometimes solitary. Filaments short, erect or slightly incurved in the bud. Styles always deeply divided into 2 equal branches. Ovule pendulous or laterally attached near the top. Embryo curved, the cotyledons often folded over the incumbent radicle. Trees or shrubs, not milky-juiced.

1. CELTIS, Linn.

(Solenostigma, *Endl.*)

Flowers polygamous, in axillary or lateral cymes. Perianth in both sexes of 4 or 5 segments imbricate in the bud. Stamens in the male and hermaphrodite flowers as many as perianth-segments, not exceeding the perianth, but slightly incurved in the bud. Disk hairy, with a rudimentary pistil in the males. Ovary in the female and hermaphrodite flowers inserted on a hairy disk. Style-branches (or distinct styles) 2, equal, linear oblong or dilated, recurved and papillose on the upper or inner surface, truncate or 2-lobed in the Australian species. Ovule pendulous or laterally attached near the top. Drupe ovoid or globose. Embryo curved, the cotyledons broad, conduplicate or rarely flat, incumbent on or embracing the ascending radicle.—Trees or shrubs. Leaves alternate, more or less 3-nerved, entire in the Australian species, dentate in some others. Stipules small and deciduous.

The genus is dispersed over the temperate and subtropical regions of both the New and the Old World, with a few tropical species, chiefly in mountainous districts. Both the Australian species extend to the Indian Archipelago.

Leaves broad, strongly 3-nerved, scarcely acuminate. Cotyledons flat or nearly so . . . . 1. *C. philippinensis*.
Leaves ovate-lanceolate, the lateral nerves scarcely prominent. Cotyledons conduplicate . . . . 2. *C. paniculata*.

1. **C. philippinensis,** *Blanco, Fl. Filip.* 197. A tall shrub or stunted tree, or according to some collectors a fine tree, quite glabrous or with a minute pubescence on the young shoots. Leaves on petioles of 1 to 4 lines, ovate or broadly elliptical, shortly acuminate, entire, rounded and sometimes rather oblique at the base, varying from 1½ to 3 in in the more rigid broader leaved specimens, to 3 to 5 in. in luxuriant ones, coriaceous, green on both sides, very prominently and strongly 3-nerved, with very fine reticulations. Cymes loose, ½ to ¾ in. diameter, those with all male flowers more crowded than the polygamous ones. Perianths pedicellate, the segments broad, ¾ to nearly 1 line long. Stigmatic branches of the style (or styles) broadly cuneate truncate or 2-lobed, sometimes broader than long, sometimes much longer than broad and often the two of the same flower dissimilar. Drupe ovoid, 3 or 4 lines long, the endocarp bony. Embryo curved lengthwise but the cotyledons transversely flat in the seeds examined, not conduplicate as in *C. paniculata*.—Planch. in Ann. Sc. Nat. ser. 3, x. 306; *C. strychnoides*, Planch. l.c.

N. Australia. Clermont, Vansittart, and Careening bays, N.W. Coast, *A. Cunningham;* Victoria river, *F. Mueller;* King's Sound, *Hughan;* islands of the Gulf of Carpentaria and opposite mainland, *R. Brown*, and others; Port Essington, *A. Cunningham.*

Queensland. Sunday island, *A. Cunningham;* Port Denison, *Fitzalan, Dallachy.*

The species extends over the Archipelago to South China. The smaller more rigid-leaved specimens upon which the *C. strychnoides* was founded appear to have grown in open dry situations. Cunningham's larger leaved specimens are stated by him to have been from moist shady situations. Cuming's Philippine island specimens have still larger, less coriaceous leaves. The Hongkong plant I have referred to the same species appears to have smaller fruits, but perhaps not yet full grown.

2. **C. paniculata,** *Planch. in Ann. Sc. Nat. ser.* 3, x. 305. A large tree, quite glabrous. Leaves from ovate-lanceolate to elliptical-oblong, acuminate, more or less cuneate at the base and often oblique or slightly falcate, entire, coriaceous, smooth, penniveined and 3-nerved at the base but the midrib alone prominent, the lateral veins or nerves short and fine or almost evanescent. Cymes sometimes dense and few-flowered, sometimes loose and 1 in. broad. Drupes smaller than in *C. philippinensis*. Style-branches truncate or 2-lobed, often unequally so. Cotyledons very broad and folded over the ascending radicle.—*Solenostigma paniculatum*, Endl. Prod. Fl. Norf. 42; *S. brevinerve*, Blume Mus. Bot. ii. 67; *C. sp. nova, n.* 32, Planch. in Ann. Sc. Nat. ser. 3, x. 305; *C. ingens*, F. Muell. Fragm. iv. 88.

N. Australia. Islands of the Gulf of Carpentaria, *R. Brown, Henne.*

Queensland. Broad Sound, *R. Brown;* Endeavour river, *A. Cunningham;* Moreton bay, *F. Mueller;* Rockhampton, *Thozet, O'Shanesy;* Edgecombe bay and Port Denison, *Dallachy;* Curtis island, *Henne;* Keppel bay, *Thozet.*

N. S. Wales. Hastings, Clarence, and Richmond rivers, *Beckler*, *C. Moore* (London Exhibition 1862, *n.* 93, *C. opaca*); Kiama, *Harvey.*

In C. Moore's collection from Lord Howe's Island, specimens marked *Elatostemma sesquipedalis*, *n.* 34, appear to be a variety of *Celtis paniculata*, with the leaves rather thicker and all very obtuse, the cymes are very small and dense, and I only see male flowers, but they are mostly still in young bud.

2. TREMA, Lour.

(Sponia, *Commers.*)

Flowers polygamous, in small axillary cymes. Male perianth of 5, rarely 4 segments, induplicate-valvate in the bud. Stamens as many as perianth-segments, erect in the bud; filaments very short. Ovary rudimentary or more or less developed. Female perianth-segments nearly flat and slightly imbricate when entirely without stamens, more or less concave and induplicate in the hermaphrodite flowers. Style deeply divided to the base into 2 linear rather thick branches (or styles), hirsute along the inner stigmatic surface, and not exceeding the length of the ovary. Ovule pendulous, laterally attached near the top. Fruit a very small drupe, usually crowned by the persistent involute styles, with a slightly succulent exocarp, and a bony endocarp often pitted outside. Seed pendulous, with a thin testa and fleshy albumen. Embryo linear, curved, with a superior radicle.—Trees or tall shrubs. Leaves alternate, more or less distichous and often oblique, denticulate, 3-nerved and penniveined, the principal primary veins starting usually from both sides of the midrib and from the outer side only of the lateral veins. Flowers and fruits small, the cymes or even the whole specimens often unisexual.

The genus is widely spread over the tropical and subtropical regions of the New and the Old World. Numerous species have been published, but the greater number of them differ only in the indumentum which is often very variable in the same species, and their number will have to be very much reduced. F. Mueller regards the three following, which I have admitted amongst the Australian forms, as varieties of one species. If that be the case, the whole would have to be reduced to the *T. orientalis*. One of the three appears to me to agree so nearly with some Asiatic varieties of that species as to be inseparable except as a variety. Another also cannot be distinguished from a common Asiatic form; the third is generally admitted to be endemic in Australia, but it is very difficult to point out any definite character by which it differs from two or three Archipelago and Indian species. In the delineation of these presumed species I regret much not having been able to wait for the publication of the 17th vol. of the Prodromus containing Planchon's revision of the genus (under Commerson's name of *Sponia*). Dr. Planchon has, however, kindly transmitted to me copies of those articles which relate to the species I have here admitted. It appears that the materials he had at his disposal were very scanty as to Australian stations, and although he has much reduced the species he had originally proposed, he has still felt himself obliged to admit many upon very slight and vague distinctions. The genus appears indeed to be almost as susceptible of extension or reduction as the European *Rubi.*

It is with much regret also that I have found myself obliged to depart from the Prodromus in the nomenclature of the genus, but I cannot but agree with Blume that Loureiro's character is quite as definite as the original one given to Commerson's

Sponia at a later date. There is indeed as much reason for adopting Loureiro's name in this case as in those of *Mallotus*, *Dichroa*, *Centipeda*, &c., now so generally admitted.

Leaves green and scabrous on both sides, sprinkled with scattered hairs or nearly glabrous . . . . 1. *T. aspera.*
Leaves green and glabrous above, and smooth or scabrous, white or hoary tomentose underneath . . . . 2. *T. orientalis.*
Leaves softly pubescent above, densely velvety-pubescent or hirsute underneath . . . . 3. *T. amboinensis.*

1. **T. aspera,** *Blume, Mus. Bot.* ii. 58. A slender tree of 15 to 25 ft. or a shrub of 8 to 10 ft., the branches more or less pubescent with short rigid appressed or scarcely spreading hairs. Leaves shortly petiolate, obliquely ovate ovate-oblong or ovate-lanceolate, acuminate, regularly serrate-crenate, rounded or slightly cordate at the base, 3-nerved and obliquely penniveined, membranous, sometimes rather rigid, green on both sides or pale underneath, scabrous, usually more or less hirsute on the principal veins underneath and often sprinkled on both sides with short scattered hairs. Flowers small, in short cymes sessile or shortly pedunculate in the axils, sometimes all males but more frequently a few or several female or hermaphrodite ones in the same cymes. Male perianth scarcely 1 line diameter when open and the female smaller. Styles varying from half the length to the length of the ovary. Drupes ovoid, scarcely compressed, obtuse or rather acute, varying from under $1\frac{1}{2}$ lines to nearly 2 lines in length.—*Celtis aspera,* Brongn. in Duperr. Voy. Coq. 213, t. 48; *Sponia aspera,* Planch. in Ann Sc. Nat. ser. 3, x. 318.

N. Australia. Victoria river, *F. Mueller*, with small very strongly veined leaves and compact cymes.

Queensland. Brisbane river, Moreton Bay, *W. Hill*, *F. Mueller*, and others, and thence to Rockhampton, Rockingham and Wide bays, and Burdekin river, *F. Mueller*, *Bidwill*, *Thozet*, and others; Northumberland islands, *R. Brown;* sent by *Maitland* as a poison plant from Rockhampton.

N. S. Wales. Port Jackson to the Blue Mountains, *R. Brown*, *Woolls*, and others; northward to Hastings and Macleay rivers, *Beckler;* New England, *C. Stuart;* southward to Illawarra, *A. Cunningham*, *Harvey*, and others, and Twofold Bay, *F. Mueller;* Sydney Woods, London Exhibition 1862, *M'Arthur*, *n.* 75.—Some of the Macleay river specimens have the small compact cymes and small leaves of those from Victoria river.

Var. *viridis.* Leaves larger and less hairy, but no other difference. *Sponia viridis,* Planch. in Ann. Sc. Nat. ser. 3, x. 319; *Trema viridis*, Blume Mus. Bot. ii. 58.—Port Essington, *Armstrong;* Port Darwin, *Schultz*, *n.* 1, 8, 183, 788; Rockingham Bay, *Dallachy*. This variety is referred by Planchon (DC. Prod. xvii. ined.) to the widely-spread *T. virgata*, Blume (*Celtis virgata*, Roxb., *Sponia virgata*, Planch.), from which it is certainly not easily distinguished. Brongniart's name has, however, the right of priority. The *S. timorensis*, Dcne. from Timor, belongs to the same group as a variety or closely allied species.

2. **T. orientalis,** *Blume Mus. Bot.* ii. 62. A tree sometimes attaining 40 ft., the branches pubescent. Leaves ovate-lanceolate or lanceolate, acuminate, toothed, cordate at the base, green and scabrous or almost smooth on the upper surface and usually glabrous, white or hoary underneath with a close almost silvery tomentum or short soft pubescence,

varying from 2 to 6 in. long. Male cymes usually broad and many flowered, with the perianths larger than in *T. aspera*, the cymes as well as the flowers smaller when several or all of them are fertile.—*Celtis orientalis*, Linn.; *Sponia orientalis*, Planch. in Ann. Sc. Nat. ser. 3, x. 323.

Queensland. Albany island, *W. Hill;* Port Molle, *M'Gillivray;* Port Denison, *Fitzalan, Dallachy;* Rockhampton, *Thozet*, and others.
N. S. Wales. Paramatta, *Woolls.*

The species is widely spread over East India and the Archipelago. The Australian specimens belong mostly to a form with long leaves almost smooth on the upper surface and with numerous flowers larger than usual, and the cymes themselves larger and broader. They appear not to have been seen by Planchon, who would probably have considered the variety as specifically distinct. But, amongst the numerous specimens I have had before me, I met with several which, in one or more of the above characters, were entirely conformable to Asiatic or Archipelago specimens of the typical *T. orientalis.*

3. **T. amboinensis,** *Blume Mus. Bot.* ii. 61. A fine tree of 40 ft., the branches densely pubescent or villous. Leaves shortly petiolate, ovate or ovate-lanceolate, acuminate with long points, more equally rounded or cordate at the base than in *T. aspera*, 2 to 4 in. long, rather thick, softly pubescent above, densely velvety-pubescent or villous underneath. Cymes compact, the bracts usually more acuminate than in *T. aspera.*—*Sponia amboinensis*, Planch. in DC. Prod. xvii. ined.; *S. velutina*, Planch. in Ann. Sc. Nat. ser. 3, x. 327; *Trema velutina*, Blume Mus. Bot. ii. 62.

N. Australia. North coast, *A. Brown.*
Queensland. Rockingham bay, *Dallachy.*

The species is widely spread over East India and the Archipelago, extending northward to South China. It is on the authority of Planchon that I refer this very common Archipelago species to the original *Celtis amboinensis*, Willd. He believes also that this may be the typical *Trema cannabina*, Lour.

3. APHANANTHE, Planch.

Flowers monœcious, the males in axillary cymes, the females solitary or 2 together. Perianth in both sexes of 4 or 5 segments, imbricate in the bud. Stamens in the males 4 or 5, the filaments short, slightly incurved in the bud. Pistil rudimentary. Styles in the females deeply divided into linear acute entire branches papillose-hirsute inside. Ovule pendulous or laterally attached near the top. Drupe ovoid, slightly compressed, the endocarp crustaceous. Seed nearly globular; testa membranous; albumen little or none. Embryo curved or involute, the outer larger cotyledon enclosing the smaller one.—Tree or shrub. Leaves alternate, penniveined. Stipules very small or none. Male cymes in the axils of the old leaves, female flowers sessile or shortly pedicellate in the lower axils of the young shoots.

The genus is limited to the single Australian species, which is also in the Philippines, and probably in some of the islands of the Archipelago. It is closely allied to *Gironniera*, which, has, however (as far as known to me) a different habit, diœcious flowers, and more subulate style-branches.

1. **A. philippinensis,** *Planch. in Ann. Sc. Nat. ser.* 3, x. 337. A tree or tall shrub, glabrous or scabrous-pubescent. Leaves shortly petiolate broadly ovate to elliptical, acute or almost obtuse, rigidly membranous or coriaceous, scabrous, the primary veins very prominent underneath and although anastomosing near the margin generally produced into small rigid mucronate teeth, the whole leaf usually 1 to 2 in. long, but on some barren specimens the leaves larger, ovate-lanceolate, truncate or almost cordate at the base, the marginal teeth more prominent, on other specimens the leaves smaller broader and deeply divided into pungent-pointed lobes. Male cymes almost sessile but loose. Perianth-segments broad, concave, ciliolate. Anthers half exerted when fully out. Female perianth-segments narrower. Fruit ovoid, acuminate, about 3 lines long.—*Taxotrophis rectinervis*, F. Muell. Fragm. vi. 192; *Sponia ilicifolia*, S. Kurz in Flora, 1872, 448.

Queensland. Brisbane river, Moreton bay, *F. Mueller;* Queensland woods, London Exhibition 1862, *W. Hill, n.* 86; Rockhampton, *O'Shanesy;* Rockingham bay, *Dallachy.*

N. S. Wales. Clarence river, *Wilcox, Beckler;* Clarence and Richmond brushes, Northern woods, London Exhibition 1862, *C. Moore, n.* 33.

TRIBE II. ARTOCARPEÆ.—Flowers unisexual, in dense unisexual or androgynous spikes or heads, or crowded on or inclosed in a fleshy receptacle. Stamens erect or slightly incurved in the bud. Styles undivided or unequally rarely equally 2-branched. Ovule pendulous or laterally attached. Embryo curved or straight. Trees or shrubs, with a milky juice.

4. FICUS, Linn.

(Urostigma *and* Covellia, *Gasp.*)

Flowers unisexual, minute, enclosed in a hollow globular ovoid or pear-shaped receptacle called a *fig* or *synœcium*; the minute orifice closed by bracts turned inwards, or the first rows erect outwards. Male flowers usually near the mouth of the receptacle, very rarely in separate receptacles, and often very few. Perianth of 3 to 6 lobes or segments, imbricate in the bud, rarely reduced to a single one. Stamens 1, 2, or rarely more, opposite the perianth-segments; anthers 2-celled or the cells confluent at the apex. Female perianth usually with narrower segments than the male and sometimes very much reduced or almost none. Styles usually lateral, at least after the growth of the ovary, filiform with a terminal peltate oblique or elongated and unilateral stigma, sometimes unequally 2-branched in species not Australian. Ovule pendulous or laterally attached near the top. Fruiting receptacle usually enlarged, but remaining closed, the small seed-like nuts surrounded by the membranous or succulent persistent perianth. Embryo curved, in a fleshy albumen usually rather scanty.—Trees or shrubs with the juice usually milky. Leaves alternate or opposite, entire or lobed, penniveined and usually more or less distinctly 3-nerved at the base. Stipules usually very deciduous, convolute on the young buds. Receptacles usually in pairs, or solitary by the abortion of one of each pair, either axillary or on the old wood, and then often forming clusters or racemes on short leafless branchlets. Bracts usually 3, often small and scale-like either at the base of the receptacle or along the pedicel below it. Bracts within the receptacle subtending the flowers usually very numerous, varying with the perianth in consistence and colour, those near the orifice of the receptacle usually rather larger, without flowers, and closing the orifice, the outermost rows sometimes exserted and erect, but usually horizontal or inflexed, those subtending the flowers sometimes very minute or replaced by hairs or setæ or obsolete. Male flowers usually fewer than the females, and in the upper part of the receptacle, sometimes numerous and intermixed with the females or in separate receptacles.

A very large genus, spread over the tropical and subtropical regions of the New and the Old World, but most abundant in the Indian Archipelago. Of the thirty-four Australian species at least eight extend into the Archipelago, and most of these also into East India, and two more may possibly be varieties only of a common Asiatic species, the remaining twenty-four are all endemic as far as I have been able to ascertain; but it is possible that on the general elaboration of this difficult genus, now in the hands of M. Bureau, some further identifications of Australian and Archipelago species may be effected.

SECT. 1. **Urostigma.**—*Male perianth 3-merous, rarely 5–6-merous. Stamen* 1; *anther-cells distinct or confluent. Female perianth 4 6-merous. Stigma (in the Australian species) elongated, acute. Leaves alternate, entire, usually coriaceous. Receptacles usually axillary.*

* *Receptacle setose inside between the flowers. Male perianth 5–6-merous. Stamen exserted.*

Leaves ovate-cordate, densely pubescent underneath . . . . 1. *F. colossea.*

** *Receptacle bracteate inside between the flowers. Male perianth 3-merous, longer than the stamens.*

Leaves with rather distant principal primary veins and numerous transverse reticulations, with a few smaller fine primary veins between the principal ones.
  Receptacles oblong, sessile. Stipules and young shoots usually hairy . . . . 2. *F. pilosa.*
  Receptacles globular or turbinate.
    Petioles ½ to 1 in. long.
      Receptacles sessile or on a peduncle of 1 line, not exceeding 5 lines diameter.
        Leaves obtuse or shortly and obtusely acuminate (N. coast species) . . . . 3. *F. nesophila.*
        Leaves abruptly and shortly acuminate (Queensland species) . . . . 4. *F. Cunninghamii.*
      Receptacles on peduncles of 2 lines, ½ to ¾ in. diameter . 5. *F. Henneana.*
    Petioles under ¼ in. long. Peduncles very short . . . . 6. *F. validinervis.*

(*F. macrophylla* and occasionally some other species of the following group, approach those of the present group in venation.)

Leaves with numerous parallel primary transverse veins all equal or every third or fourth more prominent.
  Leaves thinly coriaceous, mostly under 3 in.
    Receptacles sessile or on very short peduncles.
      Petioles under 3 lines. Leaves usually broad and very obtuse. Receptacles ½ in diameter . . . . . . 7. *F. retusa.*

Petioles ¼ to ½ in. Leaves oblong-lanceolate or elliptical, scarcely acuminate. Receptacles ¼ in. diameter . . 8. *F. eugenioides.*
Petioles ¼ to ½ in. Leaves ovate or broadly elliptical, acuminate. Receptacle under ½ in diameter . . . 9. *F. benjaminea.*
Receptacles pedunculate, ½ in. diameter. Leaves of *F. benjaminea* . . . . 10. *F. Muelleri.*
Leaves more coriaceous, obtuse or shortly and obtusely acuminate, rarely under 3 in. long.
Receptacles pedunculate.
Leaves softly pubescent underneath. Receptacles villous, scarcely umbonate . . . . 11. *F. leucotricha.*
Leaves glabrous. Receptacles glabrous, prominently umbonate . . . . 13. *F. puberula.*
Receptacles sessile or on very short thick peduncles.
Leaves mostly 3 to 4 in. long.
Leaves ferruginous underneath when young . . . 12. *F. rubiginosa.*
Leaves glabrous or minutely pubescent and pale . . 14. *F. platypoda.*
Leaves glabrous, with very short petioles and very prominent veins (doubtful species) . . . . 15. *F. dictyophleba.*
Leaves mostly 6 to 10 in. long, glabrous . . . . . 16. *F. macrophylla.*

SECT. 2. **Eusyce.**—*Male perianth of* 5 *or* 6 *lobes or segments, rarely reduced to* 1. *Stamens* 1, 2 *or more; anther-cells distinct. Female perianth* 4–6-*merous. Stigma* (*in the Australian species*) *undivided, peltate, oblique or oblong. Leaves alternate or opposite, entire, toothed or lobed, often deciduous. Receptacles axillary or on the old wood.*

Leaves smooth, at least on the upper side, or scarcely scabrous.
Leaves large (½ to 1½ ft.), glabrous. Stigma not peltate.
Leaves obovate-oblong or elliptical. Stipules narrow, rigid. Stigma obtuse, scarcely thickened. . . . . 17. *F. magnifolia.*
Leaves cordate-ovate. Stipules membranous. Stigma oblong, thick . . . . 18. *F. ehretioides.*
Leaves under 6 in. long.
Stipules and young shoots silky-hairy or hoary. Stigma oblique, lanceolate.
Creeping or climbing shrub. Leaves mostly cordate-ovate, strongly reticulate underneath . . . . 19. *F. pumila.*
Small tree with pendulous branches. Leaves long, lanceolate, acuminate . . . . 20. *F. coronulata.*
Quite glabrous. Leaves elliptical or oblong ovate.
Petioles slender, 2 to 3 lines long. Leaves prominently reticulate underneath. Stigma oblique . . . . . 21. *F. leptoclada.*
Petioles ½ in. long. Leaves with scarcely conspicuous veinlets. Style short. Stigma peltate . . . . . 22. *F. depressa.*
Petioles thick, under 3 lines. Stigma peltate. Male perianth of 1 narrow segment . . . . 23. *F. philippinensis.*
Underside of the leaves and receptacles softly villous. Stigma peltate . . . . 24. *F. mollior.*
Leaves very scabrous.
Receptacles unisexual, the males oblong-cylindrical, the females ovoid or globose. Leaves opposite or alternate . . 25. *F. stenocarpa.*
Receptacles androgynous, ovoid or globose.
Leaves all alternate, pubescent or villous underneath. Receptacles villous, the orifice broad with exserted bracts . . 26. *F. aspera.*
Leaves frequently opposite. Receptacles globular, glabrous or rarely pubescent.
Leaves mostly glaucous, rigid, ovate or orbicular, the margins aculeate.
Branches glabrous . . . . 27. *F. orbicularis.*
Branches hirsute or pubescent . . . . 28. *F. aculeata.*

Leaves entire or sinuate-crenulate, not aculeate.
Leaves of the flowering branches ovate, 2 to 8 in. long, petioles ½ to 1 in. . . . . 29. *F. opposita.*
Leaves of the flowering branches obovate, 2 to 3 in. long, petioles very short . . . . 30. *F. scobina.*

SECT. 3. **Covellia.**—*Male perianth of 3 or 4 broad segments enveloping each other, enclosing 1 large anther with distinct cells. Female perianth very small or more rarely exceeding the stipes of the ovary. Style glabrous, short, with a peltate or oblique stigma. Leaves usually large. Stipular scar prominent. Receptacles chiefly on the old wood.*

Leaves all opposite. Receptacles ¾ to 1 in. diameter, not ribbed . . . . 31. *F. hispida.*
Leaves all or mostly alternate. Receptacles about ½ in. diameter, 6-ribbed.
Leaves scabrous, 4 to 10 in. long. Young shoots pubescent or hispid . . . . 32. *F. fasciculata.*
Leaves 3 to 4 in., glabrous and smooth as well as the branches 33. *F. casearia*
Leaves all alternate, glabrous and smooth. Receptacles 1 to 1½ in. diameter, not ribbed. Stigma oblique . . . . . . 34. *F. glomerata.*

Miquel, in the Journ. Bot. Neerl. 1861, 234, mentions his *U. stipulosum*, Miq., as from Hastings river, *Beckler*. I can find nothing in Beckler's collections which I am able to refer to the Philippine island plant originally described as *U. stipulosum*, and entered as *Ficus stipulosa* in the Ann Mus. Lugd. Bat. iii. 287. Miquel has also in the Journ. Bot. Neerl. 1861, 240, described an *Urostigma? subglaucinum* from Rockhampton, of which he had seen leaves only, and doubts its belonging to the genus. It is therefore omitted from the enumeration in the Ann. Mus. Lugd. Bat. There is, however, a *Ficus Fitzalani*, Miq in Journ. Bot. Neerl. 1861, 242, from Cape Cleveland, *Fitzalan*, which he has included in the Annales although described also from leaves only. I find nothing amongst Fitzalan's plants in F Mueller's collection answering to the description nearer than some of the forms of *F. platypoda*, but they have more numerous veins than are mentioned by Miquel, and no *Ficus* can be satisfactorily identified without the fructification.

SECT. I. UROSTIGMA —Male perianth of 3 lobes or segments (except in *F. colossea*). Stamen 1; anther-cells distinct or more frequently confluent at the apex. Female perianth of 4 or 5 rarely 6 lobes or segments. Stigma undivided, elongated, acute, filiform, or slightly broader towards the base. Leaves alternate, entire, usually coriaceous, glabrous or softly pubescent or villous, not at all or scarcely scabrous. Receptacles usually axillary.

1. **F. colossea,** *F. Muell. Herb.* A tree "attaining more than 100 ft. with immense abutments and a spreading head, and therefore named *Abbey-tree* by the colonists," the young shoots, petioles, and underside of the leaves densely and softly pubescent or villous. Leaves alternate, ovate cordate, acuminate, entire, mostly 4 to 5 in. long and 3 to 4 in. broad, minutely pubescent, and at length nearly glabrous on the upper surface, the primary veins rather distant, prominent underneath as well as some of the transverse veinlets, the basal pair not very prominent. Petioles 1 to 1½ in. long. Receptacles, only seen loose and their attachment not noted, turbinate-globose, 6 to 8 lines diameter, shortly pubescent, on peduncles of 1 to 2 lines. Flowers within the receptacle intermixed with hairs or setæ about as

long as the perianths, which are brown as well as the bracts. Male flowers very numerous. Perianth stipitate, with 5 or 6 ovate or oblong obtuse lobes. Stamen 1, the anther exserted, with 2 distinct parallel cells, the connective often slightly produced beyond them. Female flowers nearly sessile. Perianth-segments 5 (or 6?) nearly equal. Style very slender, with a filiform stigma, slightly dilated towards the base.

Queensland. Herbert river, Rockingham bay, *Dallachy.*

2. **F. pilosa,** *Reinw. in Blume Bijdr.* 446. A tall tree, the adult foliage and fruits and even the young leaves usually glabrous, the stipules alone on the terminal bud covered with ferruginous hairs sufficient to justify the name, very rarely the petioles and principal veins of the leaves also hairy. Leaves alternate, on petioles of $\frac{1}{2}$ to 1 in., elliptical-oblong, shortly and abruptly acuminate, rounded or rarely slightly cordate at the base, the larger ones 6 in. to nearly 1 ft. long and 3 to 5 in. broad, coriaceous, the principal primary veins rather distant and very prominent, as well as the fine smaller veins and transverse reticulations, with 1 or 2 pairs of basal oblique veins. Receptacles usually in pairs, closely sessile in the axils or below the leaves, oblong, about $\frac{3}{4}$ in. long, the small broad external bracts almost concealed under the base of the receptacle. Male flowers stipitate, intermixed with the females, and with numerous bracts without setæ. Perianth of 3 unequal segments hairy inside at the base. Stamen 1, the anther not exceeding the perianth, with 2 distinct parallel cells. Female flowers more sessile. Stigma subulate, acute, slightly dilated towards the base.—Miq. in Ann. Mus. Lugd. Bat. iii. 285; *Urostigma pilosum,* Miq. Fl. Ind. Bat. i. part ii. 351; *F. ellipsoidea,* F. Muell. Herb.

Queensland. Albany island, Cape York, *F. Mueller;* Rockingham bay, *Dallachy.* The species is also in Timor and Java, and probably in other islands of the Archipelago. It is readily distinguished from all other Australian species by the shape of the receptacle.

3. **F. nesophila,** *Miq. in Ann. Mus. Lugd. Bat.* iii. 286. A shrub or small tree, quite glabrous or with a very minute pubescence on the stipules and young buds. Leaves on petioles of $\frac{3}{4}$ to $1\frac{1}{2}$ in. or sometimes even 2 in., ovate or oblong-elliptical, very shortly and obtusely acuminate, rounded or almost truncate at the base, 3 to 6 in. long, $1\frac{1}{2}$ to 3 in. broad, coriaceous and shining above, with rather distant primary veins prominent underneath, the lowest pair starting from the base and more oblique, the smaller veins and reticulations conspicuous but fine. Receptacles solitary or in pairs, in the lower axils or at the leafless nodes of the previous year's shoots, nearly sessile or on peduncles of 1 line, globular, 4 to 5 lines diameter when ripe but much smaller on most specimens, the outer subtending bracts short and broad. Male flowers few in the receptacles opened. Perianth stipitate, with 3 segments, brown as well as the bracts. Stamen 1, shorter than the perianth; anther-cells parallel. Female flowers very numerous.

Stigma subulate, acute, slightly dilated towards the base.—*Urostigma nesophilum*, Miq. in Journ. Bot. Neerl. 1861, 237.

N. Australia. Cambridge gulf and Enderby's island, N.W. coast, *A. Cunningham;* Nichol bay, *Gregory's Expedition;* King's Sound and Collier bay, *Chapman;* islands of the gulf of Carpentaria, *R. Brown, Gulliver;* Port Darwin, *Schultz, n.* 552, 882, 887.

Queensland. Quail island, *Flood;* Rockingham bay, *Dallachy.*

This may prove to be a variety only of *F. Cunninghamii*, with more coriaceous leaves, the points much less prominent and more obtuse, and both species are perhaps too closely allied to the wide-spread Indian *F. infectoria.*

4. **F. Cunninghamii,** *Miq. in Ann. Mus. Lugd. Bat.* iii. 286. A large robust tree of 80 ft. quite glabrous. Leaves on petioles of ½ to 1 in., from ovate to oblong-elliptical, shortly and abruptly acuminate, rounded truncate or scarcely cordate at the base, 4 to 5 in. long, 2 to 2½ in. broad, coriaceous, shining above but often less so than in *F. nesophila*, the primary distant veins prominent underneath, the smaller veins fine, anastomosing and often scarcely conspicuous. Stipules narrow, very deciduous, ½ to ¾ in. long. Receptacles solitary or 2 together in the lower axils, closely sessile concealing the small outer bracts, globular and smooth, 4 to 5 lines diameter. Male flowers few under the bracts near the orifice; perianth stipitate, of 3 brown segments, the single stamen rather shorter than the perianth. Female flowers numerous, sessile or nearly so; stigma filiform, acute, slightly dilated towards the base.—*Urostigma Cunninghamii*, Miq. in Hook. Lond. Journ. vi. 560; *U. Fraseri*, Miq. in Hook. Lond. Journ. vi. 561; *Ficus Fraseri*, F. Muell. Fragm. vi. 195; *Urostigma psychotriæfolium*, Miq. in Hook. Lond. vi. 561; *Ficus psychotriæfolia*, Miq. in Ann. Mus. Lugd. Bat. iii. 286.

Queensland. Brisbane river and Percy island, *A. Cunningham;* Brisbane and Bremer rivers, *Fraser;* Keppel bay, *R. Brown;* Rockhampton, *Bowman, O'Shanesy, Dallachy;* Rockingham bay, *Dallachy.*

The species is perhaps too closely allied to the East Indian *F. infectoria*, differing slightly in the form and especially in the acumination of the leaf. *F. psychotriæfolia* was described by Miquel from a single specimen with one damaged receptacle in Herb. Hooker from Brisbane river, *Fraser.* His *F. Fraseri* was founded on a specimen in leaf only from nearly the same locality in Herb. Hooker, which he afterwards in Ann. Mus. Lugd. Bat. iii. 287, refers to the Philippine island *F caulobotrya*, Miq., which however, besides an inflorescence unknown in any Australian species, is also readily distinguished by the venation of the leaves. I can see no difference at all as far as the specimens go between *F. Fraseri, F. psychotriæfolia*, and some acknowledged forms of *F. Cunninghamii.*

5. **F. Henneana,** *Miq. in Ann. Mus. Lugd. Bat.* iii. 216. A shrub or slender tree quite glabrous. Leaves on petioles of 1 to 1½ in., oval or oblong-elliptical, obtuse or very shortly and obtusely acuminate, entire, rounded or cordate at the base, 3 to 5 in. long, 1½ to 2½ broad, rather thinly coriaceous, the primary veins distant and prominent, the basal pair very oblique, the others spreading, the veinlets conspicuous but scarcely prominent. Stipules lanceolate, acuminate, glabrous. Receptacles (in pairs?) on peduncles of about 2 lines, globular, ½ to ¾ in. diameter, smooth but mottled with white, the subtending bracts very

deciduous, leaving a truncate margin under the ripe fig. Male flowers few, shortly stipitate, the perianth trimerous, with 1 large anther on a very short filament, and the subulate stigma of the females entirely those of the section.

N. Australia. Maria island and Caledon bay, *Gulliver.*
Queensland. Booby island, Torres Straits, *A. Cunningham, Henne;* also perhaps the same species, Rockingham bay, *Dallachy.*

The species differs from *F. nesophila* chiefly in the receptacles twice as large on longer peduncles.

6. **F. validinervis,** *F. Muell. Herb.* A small tree, quite glabrous. Leaves elliptical or oblong, abruptly acuminate, entire, rounded or cuneate at the base, 4 to 8 in. long, 1½ to 2½ broad, coriaceous, the primary rather distant veins and the smaller intervening ones as well as the transverse reticulations very prominent underneath, the basal pair not very different from the others, the petiole thick, not above 2 or 3 lines long. Receptacles in the lower axils on peduncles of about 1 line, globular, 4 to 5 lines diameter on our specimens but not yet ripe, the internal structure apparently the same as in *F. Cunninghamii,* and the styles certainly those of *Urostigma.* I could find however no male flowers in the only receptacle I could open, not a perfect one.

Queensland. Rockingham bay, *Dallachy.*—This has the short petioles of *F. philippinensis,* but is evidently an *Urostigma,* and is remarkable for the very prominent venation of its very coriaceous leaves.

7. **F. retusa,** *Linn. Mant.* 129. A small or large tree, quite glabrous. Leaves on rather broad petioles of 2 to 3 lines, varying from broadly obovate or almost orbicular and very obtuse or retuse to oval-elliptical or almost oblong, rounded or very shortly contracted at the base, 2 or 3 in. long and 1 to 2 in. broad, more coriaceous than in *F. benjaminea,* the transverse primary veins as in that species fine and parallel but not so numerous and more anastomosing. Receptacles sessile, in pairs, globular, attaining about ½ in. diameter when ripe, the external bracts nearly orbicular, about 1 line diameter. Perianths and stigma entirely those of *Urostigma.* Anther not exceeding the perianth (the cells distinct and parallel?).—*F. nitida,* Thunb.; Wight Ic. t. 642; *Urostigma pisiferum, U. ovoideum, U. nitidum* and *U. retusum,* Miq. in Hook. Lond. Journ. vi. 580, 581, 582.

N. Australia. Islands of the Gulf of Carpentaria, *R. Brown.*
Queensland. Rockingham bay, *Dallachy.*

This species extends over East India, the Archipelago, and New Caledonia, reaching northward to the Philippines and South China.

8. **F. eugenioides,** *F. Muell.; Miq. in Ann. Mus. Lugd. Bat.* iii. 286. A small tree, quite glabrous. Leaves on petioles of ¼ to ½ in., oblong lanceolate or elliptical-oblong, obtuse or scarcely acuminate, tapering at the base, 1½ to 2½ in. long, ½ to 1 in. broad, entire, coriaceous, with numerous fine parallel primary veins diverging from the midrib, which is alone prominent with sometimes a pair of very oblique basal veins. Stipules narrow. Receptacles sessile, mostly in pairs in the lower axils

or at the nodes below the leaves, globular, about 3 lines diameter, the external bracts very short and broad. Male flowers intermixed with the females, the perianth stipitate and obtusely 3-lobed or rarely 4-lobed, filament very short adnate to one of the lobes, anther-cells confluent at the apex, and when open the anther becoming reniform or almost transverse as in *F. rubiginosa*. Stigma subulate, acute, more or less dilated at the base.—*Urostigma eugenioides*, Miq. in Journ. Bot. Neerl. 1861, 238.

Queensland. Northumberland islands, *R. Brown;* Albany island, *F. Mueller, W. Hill;* Rockhampton, *Dallachy, Thozet;* Crocodile Creek and Berseker Range, *Bowman.*

N. S. Wales. Tweed river, *Guilfoyle.*

Var. *puberula.* Young shoots slightly pubescent, but as far as the specimens show, not otherwise differing from *F. eugenioides. F. brachypoda*, Miq. in Ann. Mus. Ludg. Bot. iii. 287; *Urostigma brachypodum*, Miq. in Hook. Lond. Journ. vi. 562.

N. Australia. York Sound, N.W. coast, *A. Cunningham.*

9. **F. benjaminea,** *Linn.; Miq. in Ann. Mus. Lugd. Bat.* iii. 288. A large elegant tree with slender pendulous branches "weeping like the weeping willow" (*Dallachy*), quite glabrous. Leaves on petioles of ¼ to ½ in., ovate or ovate-oblong, acuminate, rounded or cuneate at the base, 2 to near 4 in. long, entire, thinly coriaceous, with numerous fine transverse and parallel primary veins, and slightly reticulate between them. Stipules narrow, under ½ in. long. Receptacles sessile, solitary or in pairs in the lower axils, globular, about ½ in. diameter when ripe or rather smaller, the subtending bracts orbicular, concave, short and rather rigid. Male flowers not numerous, intermixed with the females. Perianth trimerous; anther not exceeding the perianth, the cells distinct and parallel. Stigma subulate.—*Urostigma benjamineum*, Miq. in Hook. Lond. Journ. vi. 583; *Ficus neglecta*, Dcne. Herb. Tim. Descr. 166.

Queensland. Rockingham bay, *Dallachy.*

10. **F. Muelleri,** *Miq. in Ann. Mus. Lugd. Bat.* iii. 287. A glabrous tree. Leaves on rather slender petioles of ½ to ¾ in., ovate or elliptical-oblong, shortly acuminate, rounded at the base, 2 to 3 in. long, entire, thinly coriaceous, with rather numerous fine parallel primary veins less equal and more reticulate than in *F. benjaminea.* Stipules small, narrow, acuminate. Receptacles in pairs closely sessile or on exceedingly short peduncles, globular, attaining above ½ in. diameter, the internal structure as in *F. benjaminea.*—*Urostigma Muelleri*, Miq. in Journ. Bot. Neerl. 1861, 235.

N. S. Wales. Hastings river, *Beckler.*—This may prove to be a variety of *F. benjaminea*, but, as far as the very few specimens show, it appears to have narrower leaves on longer petioles, with a slightly different venation and larger receptacles.

11. **F. leucotricha,** *Miq. in Ann. Mus. Lugd. Bat.* iii. 285. A small tree, the flowering branches and petioles hirsute with spreading white hairs. Leaves shortly petiolate, ovate broadly oblong or elliptical, obtuse or very obtusely and obscurely acuminate, rounded or scarcely cordate at the base, entire, 3 to 5 in. long, 1½ to 2½ broad,

rigidly coriaceous, pubescent, but the down almost disappearing on the upper side, remaining soft and dense underneath, the primary veins rather numerous, fine, and almost transverse. Stipules long and narrow. Receptacles axillary, usually in pairs, on peduncles of 2 to 4 lines, globular, somewhat rugose, very villous, attaining ½ in. diameter, the subtending bracts ovate, about ½ in. long, but already fallen away from almost all the specimens seen. Male flowers intermixed with the females towards the orifice; perianth stipitate 3-merous, with one large anther with parallel cells. Female flowers nearly sessile; perianth 4-merous. Stigma linear and acute, but rather short. Bracts and perianths as in most species of the section dark brown when dry —*Urostigma leucotrichum*, Miq. in Journ. Bot. Neerl. 1861, 234; *Ficus lanata*, F. Muell. Herb.

N. Australia. Islands of the Gulf of Carpentaria, *R. Brown;* Sea Range, *F. Mueller.*

F. Mueller distinguished two varieties, *microcarpa*, in which the receptacles are about 4 lines, and *macrocarpa*, in which they are fully 6 lines diameter; but on examination the former appear to be not yet full grown, with the flowers in bud or only just expanded, leaving a central cavity; whilst in the larger form the fruits are ripe, completely filling the receptacle. The bracts subtending the receptacle appear to be larger in this than in any other Australian species.

12. **F. rubiginosa,** *Desf.; Vent. Jard. Malm.* t. 114. A tree of considerable size, with spreading branches, throwing out woody roots, which descend to the ground, forming pillars as in the Indian Banyan tree (*F. indica*), the young shoots and petioles more or less ferruginous-pubescent. Leaves on petioles of ½ to 1 in., oval or elliptical, obtuse or very shortly and obtusely acuminate, entire, rounded or very slightly cordate at the base, 3 to 4 in. long, and 2 to 2½ broad when full grown, coriaceous, glabrous above, more or less ferruginous-pubescent underneath, with numerous parallel very divergent primary veins, of which 10 to 12 on each side of the midrib rather more prominent than the others, and the basal pair more oblique. Stipules narrow-acuminate. Receptacles axillary, mostly in pairs, on thick broadly turbinate peduncles of 1 line or rather more, globular, about 4 or 5 lines diameter, usually marked with prominent warts. Subtending bracts broad, membranous, about 2 lines diameter, very deciduous. Male flowers intermixed with the females. Bracts acuminate, brown as well as the perianths. Anther-cells confluent at the apex into a single reniform cell, and at length very divergent so as to appear to open transversely. Stigma linear and acute, not very long.—Bot. Mag. t. 2939; *F. australis*, Willd. Sp. Pl. iv. 1138; *Urostigma rubiginosum*, Gaspar Nov. Gen. Fic. 7, quoted in his Ricerch. Caprif. 82, t. 7, f. 6 to 13.

N. S. Wales. Port Jackson and Blue Mountains according to several herbaria, but no collector's name given; Hunter's river, *R. Brown;* New England, *C. Stuart;* Hastings and Clarence rivers, *Beckler;* Lord Howe's island, *C. Moore.*

It is by some mistake that Miquel has quoted the plates of Ventenat and of the Botanical Magazine as *F. ferruginea*; they are both correctly named *F. rubiginosa. F. ferruginea*, Desf., was published as a distinct species, which Miquel believes to have been of American origin; but Bureau refers it to the true *F. rubiginosa*. The specimens from Lord Howe's island were received under the ms. name of *F. columnaris*, accompanied by a sketch of the habit of the tree with its grove of columnar adventive roots; but I can find nothing to distinguish them from the N. S. Wales *F. rubiginosa*.

13. **F. puberula,** *A. Cunn.; Miq. in Ann. Mus. Lugd. Bat.* iii. 287. A tree with the habit of *F. platypoda*, and apparently almost as variable in the leaves, rather large and broad or smaller and narrower, always obtuse or shortly and obtusely acuminate, coriaceous, glabrous or very slightly pubescent, and not ferruginous, with the venation of *F. platypoda*, the young shoots and stipules most frequently pubescent. Receptacles globular and smooth, about 4 to 5 lines diameter, like those of *F. platypoda* but on peduncles of 3 to 4 lines, and usually distinctly umbonate.—*Urostigma puberulum*, Miq. in Hook. Lond. Journ. vi. 562, t. 23; *U. vitellinum*, Miq. in Journ. Bot. Neerl. 1861, 237; *Ficus vitellina*, Miq. in Ann. Mus. Lugd. Bat. iii. 288.

N. Australia. York Sound, N.W. coast, *A. Cunningham;* Port Walcot, *C. Harper;* Fitzmaurice river, *F. Mueller.*

14. **F. platypoda,** *A. Cunn.; Miq. in Ann. Mus. Lugd. Bat.* iii. 287. A small tree of robust growth, perfectly glabrous in all its parts in the typical form, more or less pubescent in several varieties, but not ferruginous. Leaves in the typical form on broad petioles of about ½ in., ovate, obtuse, entire, rounded or slightly cuneate at the base, or the lower ones almost cordate, 2½ to 4 in. long, and 2 to 2½ broad, thickly coriaceous with numerous transverse parallel primary veins, the principal ones not distant, and the basal pair not very conspicuous. Receptacles axillary, mostly in pairs, sessile or on peduncles not exceeding 1 line, globular, not warted, without any umbonate prominence, 4 to 5 lines diameter. Male flowers few, intermixed with the females towards the orifice. Perianths all stipitate. Anther-cells contiguous at the apex, but scarcely confluent in the flower examined. Stigma linear-subulate and acute, or sometimes in the same receptacle shorter and more obtuse.—*Urostigma platypodum*, Miq. in Hook. Lond. Journ. vi. 561.

N. Australia. York Sound and Vansittart's Bay, *A. Cunningham.*

The following forms may some of them, when better known, prove to be sufficiently distinct to be received as species :—

Var. *lachnocaulon.* Closely resembling the typical form except that the ends of the branches and petioles are pubescent, and the under surface of the leaves also slightly so; the petioles particularly short.—*Urostigma lachnocaulon*, Miq. in Journ. Bot. Neerl. 1861, 238. *Ficus lachnocaula*, Miq. in Ann. Mus. Lugd. Bat. iii. 287.—Australia, *Baudin s Expedition*, probably from the N.W. Coast; Port Darwin, *Schultz*, *n.* 411.

Var. ? *minor*, Miq. Glabrous. Leaves elliptical-oblong, 2 to 3 in. long, and 1 to 1½ in. broad, the petioles rather longer than in the typical form and the receptacles on very short peduncles.

N. Australia. N.W. Coast, *Bynoe;* Nicol Bay, *Gregory's* and *Ridley's Expeditions.*

Var. ? *petiolaris.* Glabrous. Leaves usually larger than in the typical form, on petioles of 1½ to 2½ in. Stipules very long. Receptacles rather small, on short peduncles.

Queensland. "Brisbane and Hastings rivers" (probably Brisbane river), *Fraser;* Narra river, *Leichhardt;* Rockhampton and Cape river, *Bowman;* Cape Cleveland, *Burdekin Expedition;* Port Denison, *Fitzalan;* and nearly the same form but with smaller leaves, Maranoa river, *Mitchell.*

Var.? *mollis.* Leaves of the last variety but velvety-pubescent on both sides as well as the young branches.

Queensland. Rockingham bay, *Dallachy.*

Var.? *subacuminata.* Leaves large on long petioles as in the last two forms, pubescent, but not so much so as in the var. *mollis,* and tapering above the middle so as to be sometimes almost ovate-lanceolate.

Queensland. Whitsunday island, *Henne.*

The two last varieties are referred in Herb. F. Mueller to the *F. Leichhardtii,* Miq. in Ann. Mus. Lugd. Bat. iii. 287 (*Urostigma Leichhardtii,* Miq. in Journ. Bot. Neerl. 1861, 235), of which, however, the typical specimens from Cleveland Bay are not in F. Mueller's collection unless they may be the glabrous ones quoted under the var. *petiolaris.* If the last three varieties with long petioles be admitted as a species distinct from the North-Western ones with short petioles, it should receive Miquel's name of *F. Leichhardtii.*

15. ? **F. dictyophleba,** *F. Muell.; Miq. in Ann. Mus. Lugd. Bat.* iii. 218. This supposed species, described from leaves only, very much resembles the broad-leaved glabrous forms of *F. platypoda,* differing in the shorter petiole, usually from ¼ to nearly ½ in. long, the more coriaceous shining leaves, with numerous veins much more prominent.

Queensland. Islands off Cape Flattery, *F. Mueller.*

16. **F. macrophylla,** *Desf.; Pers. Syn. Pl.* ii. 609. A large tree with a broad head, quite glabrous, closely allied to *F. rubiginosa,* and especially to the var. *petiolaris* of *F. platypoda,* but with much larger leaves. These are oval-elliptical or broadly oblong, obtuse or obtusely acuminate, entire, 4 to 10 in. long, and 3 to 4 in. broad, coriaceous, with numerous transverse parallel primary veins, but with the principal ones at some distance from each other more prominent than the intermediate fine ones. Stipules often above 2 in. long. Receptacles nearly globular or somewhat pear-shaped, ¾ to 1 in. diameter, on peduncles of 3 to 4, the internal structure entirely as in *F. platypoda.*—*Urostigma macrophyllum,* Miq. in Hook. Lond. Journ. vi. 560.

Queensland. Pine river, *Leichhardt;* Moreton Bay, *Hort. Kew.;* Mount Dryander, *Fitzalan.*

N. S. Wales. Hunter's and Paterson's rivers, *A. Brown;* Macleay and Bellinger rivers, *C. Moore.*

Miquel in Ann. Mus. Lugd. Bat. iii. 287, refers to this species the *F. Huegelii,* Kunth and Bouché, or *Urostigma Huegelii,* Miq. in Hort. Lond. Journ. vi. 586, and the *U. squamellosum,* Miq. in Journ. Bot. Neerl. 1861, 239.

Some specimens labelled "Brush forests along the coast," and exhibited as *F. macrophylla* in the collection of Northern woods, London Exhibition 1862, *n.* 85, *C. Moore,* are evidently the *F. elastica,* Linn., which has leaves of the same size but with a more abrupt acumination, and the numerous parallel veins much more equal and approximate. These specimens may have been taken from a botanical garden to illustrate the specimens of wood of the true *F. macrophylla,* collected on the coast, for we have no corroborative evidence of *F. elastica* being an Australian species.

Sect. 2. Eusyce.—Male perianth of 5 or 6 lobes or segments, rarely reduced to a single one. Stamens 1, 2 or more; anther-cells distinct and parallel, not confluent at the apex, and not exceeding the perianth. Female perianth of 4 to 6 lobes or segments, as long as or longer than the ovary, Stigma in the Australian species undivided, either peltate or oblique or oblong and obtuse, in species not Australian 2-lobed. Leaves alternate or opposite, entire toothed or lobed, often deciduous. Receptacles axillary or on the old wood.

17. **F. magnifolia,** *F. Muell. Fragm.* iv. 50, *partly*. A handsome tree, the foliage slightly scabrous, but otherwise glabrous. Leaves alternate, obovate-oblong or elliptical, shortly acuminate, entire, contracted towards the base, but rounded or cordate at the base itself, mostly 1 to 1½ ft. long, and 6 to 8 in. broad, but smaller on some of the lateral branches, the distant primary diverging veins and transverse anastomosing veinlets prominent underneath, the basal pair very oblique. Stipules narrow, rigid and often persistent. Receptacles on peduncles of about ½ in., mostly in pairs on the previous year's or older woods, depressed globular, ½ to ¾ in. diameter. Inner bracts and perianths white almost hyaline, the perianths of 4 or 5 very unequal segments. Male flowers few, with 2 or 3 stamens, and in one case I found a stamen and an ovary within the same perianth. Style glabrous, with a small obtuse shortly oblong stigma.

Queensland. Mount Elliott and Rockingham bay, *Dallachy*, *Fitzalan.*

Fitzalan originally gathered this species without fructification, and named it *F. magnifolia.* Dallachy found fruits on trees of *F. hispida*, which he took to be the same, and sent them as the fruits of *F. magnifolia* to F. Mueller, who thus described the species as a *Covellia.* Dallachy, however, afterwards found out his mistake, and sent further specimens of *F. magnifolia*, with its own receptacles, which are entirely those of *Eusyce.*

18. **F. ehretioides,** *F. Muell. Herb.* A tree of 40 to 60 ft., quite glabrous or with a very slight pubescence on the young branches. Leaves alternate, on petioles of 1 to 5 in., cordate-ovate, acuminate, entire or sinuate-toothed, 6 to 10 in. long, and 4 to 6 in. broad, membranous, with few distant primary diverging veins, and a few fine transversely anastomosing veinlets prominent underneath, the one or two lowest pair of veins starting from the base more oblique but not more prominent. Receptacles on the old wood, apparently obovoid or turbinate-globular, and about 1 in. diameter, but only seen broken up and the flowers much injured. I could not find the males, and most of the females were far advanced in fruit and much eaten; the styles appear to be glabrous, not very long, with a thick oblong obtuse oblique glabrous stigma.

Queensland. Rockingham bay, *Dallachy.*

19. **F. pumila,** *Linn. Spec. Pl.* 1515. A prostrate or climbing shrub, often closely clinging to rocks, trees, or buildings, and then the branches frequently flattened, the young shoots and stipules more or less silky-pubescent, the adult foliage glabrous, or nearly so. Leaves alternate, distichous, on very short petioles, ovate, obtuse, entire or slightly sinuate, rounded or cordate, and often unequal at the base, rigid when full-grown, nearly smooth above, the primary distant veins

and numerous reticulated veinlets very prominent underneath, with 1 or 2 pairs of oblique basal veins, the leaves mostly 2 to 3 in. long on the principal branches, under 1 in. and broader in proportion on slender creeping barren shoots. Stipules about ½ in. long, deciduous. Receptacles (on Chinese specimens) on thick peduncles of 1 to 6 lines, globular or slightly turbinate, 1½ to 2 in. diameter. Male flowers not seen. Style glabrous, with an oblique terminal shortly lanceolate stigma.—*F. stipulata*, Thunb. Miq. in Hook. Lond. Journ. vii. 439.

Queensland? · A specimen in Herb. F. Mueller, without station, is labelled "The Wide Bay creeper."

N. S. Wales. Hunter's river, *C. Moore*.

The species, as limited by Miquel, extends over the Indian Archipelago to South China, but fruiting specimens are very rare in herbaria, and in several localities it is an introduced plant. Whether the Australian specimens are really indigenous may be in some measure uncertain. The closely allied *F. erecta*, Thunb., sometimes regarded as a variety only, extends over tropical Asia generally.

20. **F. coronulata,** *F. Muell.; Miq. in Journ. Bot. Neerl.* 1861, 242. A small tree with pendulous branches, the young shoots slightly hoary pubescent, otherwise glabrous. Leaves alternate, lanceolate, acuminate, entire, contracted towards the base, 4 to 6 in. long, ½ to 1 in. broad, membranous, not scabrous, with rather numerous transverse primary veins, but without any oblique basal pair, the petiole rather broad, 3 to 4 lines long. Receptacles in the specimens seen solitary at the lower nodes below the leaves, on pedicels of ¼ to ½ in., ovoid, contracted into a short neck formed as in *F. aspera* by the erect bracts of the broad orifice, becoming at length nearly globular and nearly ½ in. diameter. Bracts within the receptacle and perianths white-hyaline. Male flowers not seen. Style glabrous, with a terminal oblique slightly dilated stigma.—*F. salicina*, F. Muell. Fragm. iv. 49.

N. Australia. Victoria and Fitzmaurice rivers, *F. Mueller*.

21. **F. leptoclada,** *Benth.* A "beautiful tree of about 40 ft." with slender branches, quite glabrous. Leaves elliptical-oblong, acuminate, entire, cuneate at the base, 2 to 4 in. long, ¾ to 1½ in. broad, not scabrous, with rather distant primary veins and transverse reticulations prominent underneath, the lowest pair of veins oblique from a little above the base, the petioles 2 to 3 lines long. Stipules narrow, about as long as the petioles, membranous and deciduous, or here and there rigid and persistent. Receptacles shortly pedunculate in the axils, usually solitary, globular, 3 to 4 lines diameter, the bracts of the orifice forming a short broad neck, the external bracts small and scale-like alternating on the peduncle. Perianth-segments and inner bracts narrow and often brown as in the section *Urostigma*. Style glabrous, dilated at the apex into a short oblique or almost peltate stigma. I could find no male flowers in the two receptacles I opened.

Queensland. Rockingham bay, *Dallachy*.

22. **F. depressa,** *Benth.* A tree, quite glabrous. Leaves alternate, on flattened petioles of ½ in. or more, obovate-oblong, about 3 in. long

and 1½ in. broad in the imperfect specimen seen, thinly coriaceous, smooth, the primary veins rather distant, the smaller veins and transverse veinlets fewer and less conspicuous than in any other Australian species, the basal pair of veins prominent and very oblique. Receptacles axillary, in pairs, on slender pedicels of 2 to 4 lines, depressed-globular, flat-topped, 4 to 5 lines diameter, smooth or sparingly verrucose, very hollow inside. Empty bracts near the orifice orbicular and ciliate, those under the flowers small especially in the lower part of the fig. Flowers all closely sessile, the males few near the orifice. Perianth-segments 3, very broad and enveloping each other as in the section *Covellia*. Stamens 1 or 2, filaments very short and dilated into a cuneate connective bordered by 2 distinct cells. Female perianth of 2, 3 or 4 quite distinct segments about as long as the ovary, which is sessile on a broad base, the style short, lateral, with a peltate stigma.

Queensland. Mount Elliott, *Fitzalan.*—The internal structure of the receptacle is in many respects the same as in *F. mollior*, approaching that of *Covellia;* but in *F. depressa* there are no setæ between the flowers, and the foliage is different from that of *F. mollis*, at least as far as it can be ascertained from the single specimen received by post from F. Mueller.

23. **F. philippinensis,** *Miq. in Hook. Lond. Journ.* vii. 435. A small tree, our specimens quite glabrous. Leaves on very short rather thick petioles, elliptical or oblong, acuminate, entire, contracted at the base, 3 to 4 in. long and about 1½ in. broad, thinly coriaceous, smooth, the principal primary veins rather distant, almost transverse, prominent underneath as well as the fine smaller veins and transverse reticulations, the basal pair very oblique but not very conspicuous. Stipules 3 to 4 lines long, acute. Receptacles on exceedingly short peduncles, solitary or in pairs, globular, not exceeding 3 lines diameter, quite glabrous, the bracts of the orifice not prominent. Perianths and internal bracts white-hyaline. Male flowers few, consisting of a single stamen in the axil of and shortly adnate to the single lanceolate or oblanceolate perianth-segment. Female perianth of 2 to 4 short narrow segments. Style short, glabrous, with a terminal concave peltate stigma.

Queensland. Family island, Rockingham bay, *Dallachy.*—A single specimen in Herb. F. Mueller, but quite similar to Archipelago specimens. The species extends northward to the Philippine islands, and is also in New Caledonia.

24. **F. mollior,** *F. Muell. Herb.* A tall tree, the young branches petioles and underside of the leaves softly pubescent or villous. Leaves alternate, on petioles of about ½ in.; oblong elliptical or almost ovate, acuminate, entire, contracted towards the base but usually cordate at the base itself, 4 to 6 in. long, 2 to 3 in. wide, glabrous and rather smooth above, the rather distant primary veins and transverse reticulations prominent underneath, with 1 or sometimes 2 pairs of basal veins, oblique but usually finer than some of the primary veins higher up. Receptacles on peduncles of 1 to 2 lines, axillary, solitary or in pairs, globular or somewhat obovoid, 4 to 5 lines long, shortly villous, the external subtending bracts small and scale-like; those round the orifice sometimes protruding almost as much as in *F. aspera.* Perianths and

inner bracts dark brown as in *Urostigma*. Male flowers very few amongst the larger bracts near the orifice, the 3 perianth-lobes broad and enveloping each other, enclosing a single stamen as in *Covellia*. Female flowers intermixed with the long rigid hairs or setæ of the receptacle as in *F. colossea*, the perianth-segments narrow and unequal, scarcely exceeding the ovary. Style short, glabrous, with a terminal concave peltate stigma.

Queensland. Rockingham bay, *Dallachy*.

25. **F. stenocarpa,** *F. Muell. Herb.* A tree, apparently nearly glabrous, but all parts especially the leaves very scabrous, with a minute scattered rigid pubescence. Leaves opposite or more frequently alternate, on rather slender petioles of $\frac{1}{4}$ to $\frac{1}{2}$ in., ovate obovate or elliptical, obtuse or acuminate, entire, rounded or contracted at the base and sometimes oblique, rigidly membranous, green on both sides, 3 to 5 in. long, distantly penniveined, with a basal pair of oblique veins. Stipules small and narrow. Receptacles on peduncles of 2 to 4 lines, solitary or in pairs in the lower axils, and unisexual, the males cylindrical and 4 or 5 lines long, the females ovoid or nearly globose and fully as large, all minutely scabrous like the rest of the plant. Outer bracts scale-like on the peduncle below the receptacle. Perianths in both sexes white-hyaline as well as the bracts, the segments oblong, the bracts rather broad or fringed. Stamens usually 2. Style ending in a linear hirsute stigma obtuse and coloured towards the end.

Queensland. Rockingham bay, *Dallachy;* Fitzroy island, *Walter*.

26. **F. aspera,** *Forst. Prod.* 76. A tree, sometimes described as small, sometimes as attaining 80 to 100 ft., the young branches petioles and inflorescences hispid with short hairs. Leaves on short petioles, oblong-elliptical, shortly acuminate, often irregularly toothed above the middle, rounded often oblique and sometimes emarginate at the base. 3 to 6 in. long and $1\frac{1}{2}$ to $2\frac{1}{2}$ broad, very scabrous above, pubescent or hispid underneath, the primary divergent veins distant and prominent underneath as well as the transverse reticulations, the basal pair of veins more oblique than the others. Receptacles axillary, on peduncles of 1 to 2 lines, solitary or in pairs, ovoid globular or urceolate, usually densely hispid, about 4 to 6 lines diameter, the orifice rather broad, with very numerous lanceolate bracts of $\frac{1}{2}$ line, the outer rows erect forming a kind of neck to the receptacle, the succeeding rows closing the orifice and the inner ones reflexed as in other figs. Outer bracts subtending the receptacle near the top of the peduncle small, hispid, with green tips, very deciduous. Male flowers numerous intermixed with the females. Perianth stipitate, with 5 or 6 narrow concave segments. Stamens 2 to 4; anther-cells parallel, one often abortive in 1 or 2 of the stamens. Female perianths on a shorter stipes. Style glabrous with a terminal truncate or slightly peltate stigma.—Miq. in Hook. Lond. Journ. vii. 425.

Queensland. Brisbane river, Moreton bay, *A. Cunningham, Leichhardt, F. Mueller*, and others.

N. S. Wales. Port Jackson to the Blue Mountains, *Woolls*, *Miss Atkinson;* Hunter's river, *R. Brown;* Maitland, *Backhouse;* Clarence and Hastings rivers, *Beckler;* New England, *C. Stuart;* southward to Illawarra, *A. Cunningham* and others, and Twofold bay, *F. Mueller*, the most southern point reached by any *Ficus.*

Var. *subglabra.* Indumentum of the receptacle so short as to appear at first sight glabrous. Brisbane river, *F. Mueller:* Hunter's river, *R. Brown.*

27. **F. orbicularis,** *A. Cunn.; Miq. in Hook. Lond. Journ.* vii. 426. A shrub usually of 4 or 5 ft., growing into a small tree of about 10 ft., glabrous or sparingly pubescent on the young shoots. Leaves alternate or rarely opposite, on petioles of 1 to 3 lines, very broadly ovate or orbicular, obtuse, more or less bordered by minute rigid teeth or callosities, mostly 1½ to 2 in. long and 1 to 1½ broad, but larger on luxuriant barren shoots, rigid and very scabrous above, nearly smooth underneath, with few distant primary veins, and 3-nerved from the prominence of the basal pair, the reticulate veinlets fine and little conspicuous. Stipules small and narrow. Receptacles axillary, on pedicels very short or nearly as long as the petioles, solitary or rarely in pairs, globular, 3 to 4 lines diameter, usually scabrous and sometimes sparingly pubescent. Subtending bracts small and scale-like, 1 or 2 usually on the pedicel below the receptacle. Inner bracts and perianths white-hyaline. Perianth-segments usually 5, narrow, three inner ones especially in the females much longer than the others. Male flowers few and only 1 stamen with a large oblong anther in those I examined. Style short, glabrous, with a terminal peltate stigma.—*F. indecora*, Miq. in Hook. Lond. Journ. vii. 426 (with leaves not quite so broad), and *F. Beckleri*, Miq. in Journ. Bot. Neerl. 1861, 241 (as to the N.-Western species).

N. Australia. Careening bay, Enderby island and Dampier's Archipelago, N.W. Coast, *A. Cunningham;* King's Sound, *Hughan;* Nichol bay, *Gregory's Expedition;* Victoria and Fitzmaurice rivers, *F. Mueller;* Port Darwin, *Schultz, n.* 407.

28. **F. aculeata,** *A. Cunn.; Miq. in Hook. Lond. Journ.* vii. 426. A tall shrub, very closely allied to *F. orbicularis*, and perhaps a variety, differing chiefly in the branches pubescent or hirsute. Leaves usually larger, more cordate at the base, still more scabrous above, and often tomentose, the margins more aculeate. Receptacles rather larger and more densely pubescent or hirsute, on exceedingly short peduncles.

N. Australia. South Goulburn island, *A. Cunningham;* Nichol bay, *Ridley's Expedition;* King's Sound and Collier bay, *Chapman.*

Var. *micracantha.* Leaves all or nearly all opposite. *F. micracantha*, Miq. in Ann. Mus. Lugd. Bat. iii. 221.—Islands of the Gulf of Carpentaria, *R. Brown, Henne, Gulliver;* Cairncross island, *F. Mueller.*

29. **F. opposita,** *Miq. in Hook. Lond. Journ.* vii. 426. A tall shrub or small tree, the young branches and underside of the leaves softly and densely pubescent. Leaves mostly opposite, exceedingly variable in size and shape, in the typical specimens broadly cordate-ovate and about 2 in. long on petioles not exceeding ½ in., in others ovate, ovate-oblong, or ovate-lanceolate, 6 to 8 in. long, on petioles of ½ to 1 in.,

all obtuse or acuminate, entire or very slightly undulate-crenulate, very scabrous above, distantly penniveined with the lowest pair starting from very near the base, the transverse veinlets and reticulations prominent underneath; on some barren branches the leaves are hastately 3-lobed with 1 long lanceolate central and 2 short lateral lobes. Stipules about 2 lines long. Receptacles axillary, solitary or in pairs, at first somewhat pear-shaped, at length nearly globular and about ½ in. diameter. Peduncles varying from 1 to 3 lines, the scale-like bracts usually at some distance from the fig, but sometimes close to it. Flowers entirely those of *F. orbiculata*.—*F. indecora*, Miq. in Journ. Bot. Neerl. 1861, 242, as to the specimens from Clarence river.

Queensland. Keppel and Shoalwater bays and Broad Sound, *R. Brown*; Bremer river, *Fraser*; Rodd's bay, *A. Cunningham*; estuary of the Burdekin, *F. Mueller*; Port Denison, *Fitzalan*; Rockingham bay, *Dallachy*; Rockhampton, *Bowman*.

N. S. Wales. New England, *C. Stuart*; Clarence river, *Beckler*.

30. **F. scobina,** *Benth.* A shrub or small tree of 8 to 20 ft., remarkable for the extreme asperity of both sides of the leaves as well as of the petioles and young branches, otherwise glabrous. Leaves alternate or rarely opposite, on very short petioles, mostly obovate or obovate-oblong, and very obtuse or shortly acuminate, rounded or contracted at the base, and 2 to 3 in. long, and ¾ to 1½ in. broad, but passing into oblong-elliptical, and on luxuriant barren branches sometimes oblanceolate, 4 to 5 in. long, and ½ to ¾ in. broad, often slightly and irregularly sinuate-toothed at the end, the distant primary veins and transverse reticulate veinlets prominent underneath, without any distinct basal pair of veins. Receptacles mostly solitary, globose, the largest on our specimens 4 lines diameter, scabrous like the rest of the plant, but without hairs, the external bracts small and scale-like, either close under the receptacle or along the short peduncle. Male flowers few near the orifice. Perianth-segments of both sexes narrow and unequal, and as well as the bracts white-hyaline as in *F. orbiculata*. Stamens 1 with a large 2-celled anther, and sometimes a second smaller one. Style glabrous, with a terminal truncate or peltate stigma.

N. Australia. Lizard island, *A. Cunningham*; Port Essington, *Leichhardt*; Port Darwin, *Schultz*, *n.* 6, 410, 499.—Although allied in some respects to the three preceding species, this differs so much in aspect, in the shape of the leaves and in their short petioles, that it can scarcely be considered as a variety only, nor can I identify it with any of the Indian scabrous species to which it bears some resemblance.

SECT. 3. COVELLIA.—Male perianth of 3 or 4 broad segments enveloping each other, and enclosing 1 large anther with distinct parallel cells. Female perianth very small or none, or growing out nearly to the length of the ovary, and then very thin and transparent, truncate or with short very obtuse lobes. Style glabrous or with very few hairs, short, with a peltate or oblique stigma. Leaves usually large, the stipular scar prominent. Receptacles chiefly on the old wood.

31. **F. hispida,** *Linn. f. Suppl.* 442. A small tree, remarkable for the young branches, when luxuriant, very hollow and contracted at the

nodes, the foliage and branches more or less sprinkled or covered with short stiff hairs. Leaves all opposite, broadly oblong-elliptical or almost ovate, usually acuminate, rounded or cordate at the base, mostly 6 to 10 in. long and 4 to 5 broad, but very variable in size and shape, entire or sinuate-toothed, the indumentum scabrous above, soft underneath, the primary veins distant, prominent underneath as well as the transverse veinlets, the basal pair very oblique. Receptacles either in pairs in the lower axils or more frequently in leafless clusters or racemes on the older wood, globose or somewhat turbinate, ¾ to 1 in. diameter, "white," more or less hirsute. Male flowers amongst the larger bracts near the orifice, the segments brown-hyaline, very broad, each one completely enveloping the next in the bud. Female perianth at length nearly as long as the ovary, with very obtuse or truncate lobes, but so thin and closely appressed as to be easily overlooked. Ovary stipitate. Style with a few hairs; stigma peltate.—*F. oppositifolia*, Willd. Spec. Pl. iv. 1151; Roxb. Corom. Pl. t. 124; Wight Ic. t. 638; *Covellia oppositifolia*, Gasp., and many other synonyms quoted by Miq. in Ann. Mus. Lugd. Bat. iii. 296.

N. Australia. Brunswick Bay, N.W. coast, *A. Cunningham*.

Queensland. Northumberland islands and Broad Sound, *R. Brown* (without figs, but apparently this species); Rockingham and Edgecombe bays, *Dallachy*.

32. **F. fasciculata,** *Muell. Herb.* A shrub of 10 to 15 ft., the young shoots hispid with short stiff hairs. Leaves usually alternate, ovate or broadly elliptical, shortly acuminate, irregularly sinuate-crenate, rounded contracted or slightly cordate at the base, 8 to 10 in. long and 4 to 5 broad in some specimens, much smaller in others, the distant primary veins and transverse reticulations prominent underneath, slightly scabrous above, more so underneath, and the veins often pubescent or hispid. Receptacles in pairs, sometimes axillary, but most frequently several together on short leafless branches, forming oblong clusters or short racemes on the old wood, nearly globular, about ½ in. diameter, tubercular-scabrous, marked with about 6 prominent longitudinal ribs, and often depressed at the orifice. Bracts within the receptacle minute excepting near the orifice. Male flowers few under the innermost developed bracts, entirely like those of *F. hispida*. Female perianth scarcely any, or at length shorter than the stipes of the ovary and truncate. Style short, glabrous, with a large peltate concave stigma.

Queensland. Rockhampton, Fitzroy river, *Bowman, Thozet, Dallachy*.

Var.? *subopposita*. Leaves here and there opposite and branches hollow as in *F. hispida*, but with the small 6-ribbed receptacles of *F. fasciculata*.—Rockingham bay and Mackay river, *Dallachy*.

33. **F. casearia,** *F. Muell. Herb.* A tree with a broad spreading head, quite glabrous. Leaves alternate, on petioles of ¼ to ½ in., ovate or elliptical, acuminate, entire, rounded or cuneate at the base, 3 to 5 in. long, 1½ to 2½ in. broad, membranous, not scabrous, the principal primary veins distant, slightly prominent underneath, the basal oblique

pair very small or obsolete, the smaller veins and transverse reticulations very fine or inconspicuous. Stipules rather long, acuminate, deciduous. Receptacles usually below the leaves, nearly sessile or on peduncles of 2 to 4 lines, apparently obovoid when young, depressed-globular when full grown, and attaining about ½ in. diameter, quite glabrous, with 6 longitudinal raised ribs outside, very hollow inside, the flowers very small, but their structure entirely as in *F. fasciculata*.

Queensland. Rockingham Bay, *Dallachy;* Fitzroy island and Endeavour river, *Herb. Delessert.*—F. Mueller distinguishes two forms, *F. casearia*, with obovoid pedunculate receptacles, and *F. glochidioides*, with almost sessile depressed-globular ones; but some specimens are intermediate as to the peduncles, and the shape of the receptacles depends perhaps upon age; the foliage and internal structure of the figs is precisely the same in both. The species is very near *F. fasciculata* in the ribbed receptacles, but the foliage is very different.

34. **F. glomerata,** *Willd. Spec. Pl.* iv. 1148. A large tree, glabrous or the young shoots slightly pubescent. Leaves alternate, on petioles of 1 to 2 in., from ovate or ovate-lanceolate to oblong-elliptical, shortly acuminate, entire, rounded at the base, 3 to 5 in. long, 1 to 2½ in. broad, rigidly membranous, the primary veins distant and prominent underneath, the lowest pair rather more oblique, but starting from above the base, the transverse reticulations very fine and often inconspicuous. Stipules lanceolate, often ½ in. long, membranous, crowded on the young shoots, and often persistent even when the leaves are full grown. Receptacles "in thick bunches or spikes on the principal stems all over the tree," globular or somewhat turbinate, 1 to 1½ in. diameter, slightly mealy or downy when young, "crimson when ripe and edible." Subtending bracts small. Male perianths of broadly orbicular hyaline segments closely enveloping each other, and enclosing 1 or 2 equal or unequal stamens, the anther-cells distinct and parallel. Female perianth very short, but more developed than in the last two species. Style glabrous, with an oblique terminal stigma, sometimes very short, but varying to lanceolate.—Roxb. Corom. Pl. ii. t. 123; Wight Ic. t. 667; *Covellia glomerata*, Miq. in Hook. Lond. Journ. vii. 465; *Ficus vesca*, F. Muell.; Miq. in Journ. Bot. Neerl. 1861, 243.

N. Australia. Fitzmaurice river, *F. Mueller.*

Queensland. Northumberland islands, *R. Brown;* Port Denison, *Fitzalan;* Rockingham Bay, *Dallachy;* Rockhampton, *Bowman, Thozet, Dallachy.*

The species is common in moist rich soil in East India, and extends probably over the Indian Archipelago under some other name, if I am really correct in referring the Australian plant to the Indian species. I have not succeeded in detecting any tangible difference.

5. CUDRANIA, Trec.

Flowers diœcious, in globular heads intermixed with small bracts, the receptacles more or less fleshy. Male perianth of 4 or 5 narrow segments, dilated and concave at the apex. Stamens 4 or 5, filaments erect and flattened. Rudimentary pistil subulate. Female perianth of 4 segments imbricate in the bud. Style simple, with a filiform stigma. Fruit a syncarp formed of the enlarged somewhat fleshy perianths and receptacle. Nuts free but enclosed within the perianths and partially immersed in the receptacle, the pericarp crustaceous. Seeds nearly globular, the testa membranous; albumen very scanty or none. Cotyledons broad and thick, folded over the long incurved radicle.—More or less climbing shrub, often armed with axillary spines (abortive peduncles). Leaves alternate. Flower-heads axillary.

The genus is probably limited to a single species, extending from Eastern Africa over East India and the Archipelago and northward to Japan.

1. **C. javanensis,** *Trecul in Ann. Sc. Nat. ser.* 3, viii. 123. A shrub usually climbing, but sometimes said by Dallachy to be a straggling shrub or small tree, the Australian specimens all quite glabrous except the inflorescence. Leaves petiolate, oblong or elliptical, usually acute or acuminate but sometimes obtuse, quite entire, 1½ to 3 in. long, penniveined and reticulate, but the veins impressed not prominent. Spines straight or recurved, attaining about 1½ in. Flower-heads solitary or 2 together, on peduncles of 1 to 3 lines, the males 2 to 3 lines diameter, the females smaller at first but attaining 6 to 8 lines when in fruit, and usually slightly tomentose.—Bureau in Ann. Sc. Nat. ser. 5, xi. 378, with the several synonyms adduced; *Maclura javanica*, Miq.; Blume Mus. Bot. ii. 83, t. 31; *Morus calcar-galli*, A. Cunn. Herb.

Queensland. Brisbane river, Moreton Bay, *A. Cunningham, F. Mueller;* Rockhampton, *O'Shanesy;* Stewart's Creek, *Bowman;* Rockingham Bay, *Dallachy;* Mackay river, *Sutherland.*

N. S. Wales. Cabramatta and Hunter's river, *Woolls;* Sydney woods, Paris Exhibition 1855, *McArthur, n.* 76; Kiama, *Harvey.*

6. ANTIARIS, Leschen.

Flowers monœcious, the males densely crowded on a broad receptacle bordered by an involucre of small bracts, the females solitary. Male perianth of 4 rarely 3 segments, imbricate in the bud. Stamens 4 rarely 3, the filaments very short. Female flowers consisting of an involucre of several bracts, in irregular rows, adnate to the ovary and closely combined in an ovoid mass, the tips alone free, without any intervening perianth. Style divided nearly to the base into 2 long filiform stigmatic branches. Ovule pendulous. Fruit consisting of the enlarged more or less succulent involucre and pericarp. Seed pendulous, with a crustaceous testa and no albumen. Embryo straight, with thick plano-convex cotyledons, and a short superior radicle.—Trees or shrubs with a milky juice. Leaves alternate, distichous. Stipules small, deciduous. Inflorescence axillary.

The genus consists of few species natives of tropical Asia and Africa and of the islands of the South Pacific, and includes the celebrated Upas tree of the Archipelago. The Australian species, in as far as known, appears to be endemic.

1. **A. macrophylla,** *R. Br. in Flind. Voy.* ii. 602, t. 5. A tall shrub or very small much branched tree, quite glabrous. Leaves broadly oblong, shortly acuminate, entire, slightly and unequally cordate at the base, coriaceous, 4 to 8 in. long, 2 to 4 in. wide, penniveined with pro-

minent very spreading primary veins, the petioles ¼ to ½ in. long. Flower-heads several on a very short common peduncle, the lowest female, the others male, on short pedicels. Male receptacle flat or concave, about 5 lines diameter, the marginal bracts ciliate, at first inflexed, at length reflexed. Flowers sessile, the perianth shorter than the marginal bracts, with spathulate connivent segments. Female involucre glabrous. Fruit the size of a small plum, dark red, the flesh full of a white milky juice.

N. Australia. Shores of Company's Island opposite Arnhem's Land, *R. Brown*. —I have found no specimen of this plant in any of the herbaria I have had access to, and have therefore no means of comparing it with the closely allied Archipelago species. The above description is taken from Brown's elaborate account and Bauer's plate in Flinders' Voyage.

TRIBE 3. MOREÆ.—Flowers unisexual, in dense spikes or heads. Stamens inflected in the bud. Styles usually 2-branched. Ovules pendulous or laterally attached. Embryo incurved or involute. Trees or shrubs, very rarely herbs.

7. MALAISIA, Blanco.

(Cephalotropis, *Blume;* Dumartroya, *Gaudich.*)

Flowers diœcious, the males in oblong or cylindrical spikes, the females in globular heads on a small receptacle. Male perianth deeply divided into 3 or 4 lobes or segments, valvate in the bud. Stamens 3 or 4, the filaments elongated, inflected in the bud. Female perianth urceolate, with a small orifice, enclosing the ovary. Style with 2 elongated stigmatic branches. Ovule pendulous. Fruiting-head not much enlarged, the nuts enclosed in the slightly succulent perianths. Seed with a thin testa and very scanty albumen. Cotyledons very unequal, curved over the ascending radicle, the larger one embracing the smaller one in its concave surface.—A straggling tree or woody climber, with a milky juice. Leaves alternate, usually entire. Stipules small, deciduous. Inflorescence axillary.

The genus appears to be limited to the single Australian species, extending over the Indian Archipelago and islands of the South Pacific to the Philippines.

1. **M. tortuosa,** *Blanco, Fl. Felip.* 789. A small straggling tree with its upper branches twining according to some collectors, or a tall climbing shrub, glabrous or the young shoots and inflorescences slightly pubescent. Leaves shortly petiolate, oblong-elliptical or almost ovate, very obtuse or acuminate, coriaceous, prominently penniveined, 1½ to 3 in. long. Male spikes solitary or 2 together, sessile or shortly pedunculate, dense, often curved, 2 to 6 lines long. Female heads 1½ to 2 lines diameter or rather larger when in fruit, tomentose, solitary on short peduncles or forming little axillary racemes (short leafless flowering branches) always much shorter than the leaves. Bracts numerous, concave, the prominent dorsal pubescent gibbosities densely imbricate. Flowers mostly rudimentary, only 2 or 3 in the head perfect, concealed

under the bracts except the long filiform exserted style-branches.—Bureau in Ann. Sc. Nat. ser. 5, xi. 369, with the following Australian besides several other synonyms; *M. Cunninghamii*, Planch. in Ann. Sc. Nat. ser. 4, iii. 293, F. Muell. Fragm. vi. 193; *M. scandens*, *M. viridescens* and *M. acuminata*, Planch. l.c. 293, 294; *Dumartroya fagifolia*, Gaudich. in Voy. Bonite, t. 97; *Cephalotropis javanica*, Blume, Mus. Bot. ii. 76.

N. Australia. Port Darwin, *Schultz*, *n.* 396, 745.

Queensland. Brisbane river, Moreton Bay, *A. Cunningham*, *F. Mueller*, and others; thence to Rockhampton, Rockingham and Edgecombe Bays, and the Burdekin, *F. Mueller*, *Thozet*, *Dallachy*, *Fitzalan*, and others; Wide Bay, *Bidwill*.

N. S. Wales. Camden Harbour, *C. Moore;* Richmond river, *Henderson;* Illawarra, *Backhouse;* "Crow-ash" of the Colonists, *F. Mueller;* Clarence river, *Beckler* (with acuminate leaves sometimes toothed near the apex and 3 to 4 in. long); Lord Howe's Island, *C. Moore* (with large ovate acuminate or ovate-lanceolate leaves and remarkably large male spikes, the females not seen).

8. PSEUDOMORUS, Bureau.

Flowers monœcious (or sometimes diœcious?), the males in dense cylindrical spikes, the females few in very short spikes almost reduced to heads. Male perianth of 4 segments, imbricate in the bud. Stamens 4, the filaments elongated, inflected in the bud. Pistil rudimentary. Female perianth of 4 segments, not enlarged after flowering. Style divided to the base or nearly so into 2 linear stigmatic branches. Ovule pendulous. Fruit a small drupe, surrounded by the persistent perianth and crowned by the style branches; the epicarp succulent, the endocarp crustaceous or almost woody. Seed pendulous with a membranous testa; albumen very scanty or none; embryo curved; cotyledons semi-globular; radicle incumbent, ascending.—Tree or shrub with a milky juice. Leaves alternate. Stipules small, deciduous. Inflorescences axillary.

The genus is limited to the single Australian species, which extends into New Caledonia and Norfolk Island.

1. **P. Brunoniana,** *Bureau in Ann. Sc. Nat. ser.* 5, xi. 372. A tall shrub or small tree, attaining sometimes 30 or 40 ft., glabrous or nearly so. Leaves very shortly petiolate, elliptical ovate-lanceolate or lanceolate, mostly acuminate, denticulate, 1½ to 4 in. long, penniveined, often slightly pubescent on the underside especially on the nerves and sometimes scabrous above. Spikes solitary in the axils, sessile or shortly pedunculate, the males not above 1 in. long in the majority of the Australian specimens, but in a few of these and especially in the Norfolk Island variety lengthening to 2 or even 3 in. Perianth-segments about ½ line long. Female spikes very small, usually containing only 3 or 4 flowers, each within a minute bract. Berries globular, the size of a currant or sometimes rather larger.—*Morus Brunoniana*, Endl. Atakta, t. 32; *M. pendulina*, F. Bauer, in Endl. Prod. Fl. Norf. 40; *Streblus Brunoniana* and *S. pendulina*, F. Muell. Fragm. vi. 192.

Queensland. Brisbane river, Moreton Bay, *A. Cunningham*, *F. Mueller* and others; Rockhampton, *O'Shanesy, Bowman;* Rockingham Bay, *Dallachy;* also in *Leichhardt's* collection and Queensland Woods, London Exhibition 1862, *W. Hill, n.* 85.

N. S. Wales. Cabramatta, *Woolls;* northward to Macleay and Clarence rivers, *Beckler;* New England, *C. Stuart;* southward to Illawarra, *A. Cunningham, Harvey, Ralston;* Sydney woods, Paris Exhibition 1855, *M'Arthur, n.* 43, *C. Moore, n.* 53; Clarence and Richmond brushes, London Expedition 1862, *C. Moore, n.* 38.—"Used by the natives for boomerangs," *F. Mueller.*

9. FATOUA, Gaudich.

Flowers monœcious, in loose globular heads, the males and females intermixed, the rhachis not dilated into a receptacle but very shortly branched, the outer bracts forming an irregular involucre. Male perianth campanulate, deeply 4-lobed. Stamens 4, the filaments inflexed; anther-cells parallel. Rudimentary pistil very small. Female perianth with narrower lobes than the male. Style lateral, tapering into a long filiform papillose stigma, with a tooth or abortive branch at the base. Ovule pendulous. Fruit surrounded by the persistent perianth, obliquely globular, slightly compressed, the pericarp thinly crustaceous. Seed pendulous; testa membranous; albumen none; embryo curved, with broad flat equal cotyledons, and a long incumbent radicle.—Stems herbaceous. Leaves alternate, toothed. Flower-heads axillary.

The genus consists of the single Australian species dispersed over the Indian Archipelago and the South Sea Islands, extending northward to Japan.

1. **F. pilosa,** *Gaudich. in Freycin. Voy. Bot.* 509; *Voy. Bonite,* t. 84. Stems herbaceous, sometimes hard or almost woody at the base (1 to 2 ft. high?). Leaves petiolate, ovate ovate-lanceolate or lancolate, acuminate, crenate-toothed, rounded or cordate at the base, 1 to 2½ in. long, more or less pubescent and scabrous on both sides. Stipules ovate or lanceolate, often 2 to 3 lines long and remaining long persistent, but sometimes small or very deciduous. Flower-heads shortly pedunculate and usually solitary in the axils, 2 to 3 lines diameter. Bracts small, membranous. Male flowers pedicellate, the females almost sessile.—Bureau in Ann. Sc. Nat. ser. 5, xi. 375 with the numerous synonyms adduced.

N. Australia. Port Essington, *Armstrong.*

Tribe 4. Euurticeæ.—Flowers unisexual, in cymes clusters or rarely in heads. Stamens inflected in the bud. Styles undivided. Ovule erect, orthotropous. Embryo straight or nearly so. Albumen usually very scanty or none. Trees shrubs or herbs.

The Order Urticeæ is often limited to this tribe, but none of the characters given are constant, and the habit is not always distinct. Thus *Fatoua* in that respect nearly resembles several *Euurticeæ, Elatostemma* has the receptacle of *Artocarpeæ, Pseudomorus* and *Pipturus* are very nearly allied to each other, &c.

Subtribe 1. Procrideæ.—Plants not stinging. Female perianth deeply lobed.

10. ELATOSTEMMA, Forst.

Flowers monœcious or diœcious, in dense unisexual heads, the receptacle usually flat or concave, surrounded by an involucre of several bracts. Male perianth of 4 or 5 distinct segments imbricate in the bud, each with a dorsal point at or near the top. Stamens 4 or 5, the filaments more or less adnate to the segments, the central rudimentary pistil usually very small; anthers with parallel cells placed back to back. Female perianth minute, of 3, rarely 4 segments. Stigma sessile, tufted. Nut ovate, slightly compressed.—Herbs rarely shrubby at the base, without stinging hairs. Leaves usually distichous, alternate or if opposite one of each pair much smaller than the other, sessile or very shortly petiolate, oblique or falcate, and unequal at the base or broadly semicordate. Flower-heads solitary in the axils, the outer bracts usually 4 or 5, the flowers very small and numerous, shortly pedicellate and intermixed with numerous narrow bracts or bracteoles.

The genus is spread over the tropical and subtropical regions of the Old World. The two Australian species appear to be endemic, although closely allied to one or two common Asiatic and South Sea Island species. The distinctive characters in the genus are very uncertain and difficult.

Coarse plant. Leaves 3 to 6 in. long. Male heads ½ to 1 in. diameter. Females on short peduncles or nearly sessile . . 1. *E. reticulatum.*
Slender plant. Leaves 1 to 2 in. long. Male heads 2 to 3 lines diameter. Females on slender peduncles . . . . 2. *E. stipitatum.*

1. **E. reticulatum,** *Wedd. in Ann. Sc. Nat. ser.* 4, i. 188, *Monogr. Urt.* 302, *and in DC. Prod.* xvi. i. 176. A coarse straggling herb, the stems often rooting at the base, attaining 1 to 2 ft. and sometimes branched, the whole plant in the typical form either quite glabrous or with a few rigid hairs on the midrib of the leaves underneath. Leaves usually 3 to 6 in. long, nearly sessile, broadly lanceolate oblong-lanceolate or obliquely ovate-elliptical, slightly falcate or curved, more or less acuminate, coarsely toothed, narrow towards the base especially the inner side, the outer side decurrent nearly to the base of the petiole, and often semicordate, penniveined, with 6 to 9 primary veins on each side, the lower ones not at all or scarcely more oblique than the succeeding, and always shorter. Flower-heads variable in size, the males usually ½ to near 1 in. diameter when fully out, both sexes nearly sessile or the males on peduncles varying from ½ to 3 in. in the typical form, the involucre of 4 or 5 acute or acuminate bracts nearly as long as the flowers. Perianth about 1 line long, of 5 broad segments. Female heads more globular, rarely above 3 or 4 lines diameter and the involucre much shorter. Nuts about ½ line long, contracted at the base almost into a stipes as long as the minute perianth.

Queensland. Brisbane river, Moreton Bay, *F. Mueller.*
N. S. Wales. Hunter's river, *Leichhardt;* Richmond river, *Henderson;* Clarence river and Mount Lindsay, *Beckler;* New England, *C. Stuart;* Illawarra, *A. Cunningham, Backhouse.*

Var. *pubescenti-hirta,* from the brush of Kiri, *Leichhardt.*

Var. *sessile.* Male flower-heads sessile or nearly so, but with the leaf-venation and glabrous surface of *E. reticulatum.*—Macleay river, *Beckler.*

Var. *grande.* Leaves 6 to 8 in. long and broader than in the other varieties. Flower-heads on long peduncles.—*E. sessile* var. *grande,* Wedd. in DC. Prod. xvi. i. 173.—Lord Howe's island, *Milne, M'Gillivray, C. Moore.*

I have little doubt but that all the above belong to one species, closely allied to the *E. sessile,* Forst., of the South Sea islands, but without the marked almost triplinerved venation of that species. The only approach to it is in some of the leaves of the Lord Howe's island plants, but even there the lowest primary vein is shorter than the following one, and the peduncles are peculiarly long. The Macleay river specimens have the sessile heads, but not the leaf-venation of *E. sessile;* its male flowers are 5-merous, not 4-merous.

2. **E. stipitatum,** *Wedd. in Ann. Sc. Nat. ser.* 4. i. 190; *Monogr. Urt.* 322, t. 9 D, f. 11-16, *and in DC. Prod.* xvi. i. 186. A small slender diffuse or prostrate herb, the stems simple or branched, rooting at the base, a few inches or rarely nearly a foot long. Leaves quite sessile, oblong oblong-lanceolate or almost ovate, very shortly acuminate, rather deeply toothed, from under 1 in. to nearly 2 in. long, the lower margin expanded into a rounded almost decurrent auricle, the principal veins as well as the stems more or less hispid. Flower-heads small, all pedunculate, the males 2 to 3 lines or very rarely nearly 4 lines diameter, the bracts much narrower than in *E. reticulatum.* Flowers nearly as in that species. Female heads still smaller.

N. S. Wales. Hastings and Clarence rivers, *Beckler;* Mount Lindsay, *W. Hill.*

SUBTRIBE 2. BOEHMERIÆ.—Plants not stinging. Female perianth either tubular and minutely toothed, enclosing or adnate to the ovary, or rarely minute or none.

11. BOEHMERIA, Jacq.

Flowers monœcious or diœcious, in globular usually unisexual clusters, with small scarious bracts. Male perianth of 4, rarely 3 or 5, segments or lobes, valvate in the bud. Stamens as many as perianth-segments, with a central clavate or globose rudimentary pistil. Female perianth tubular, compressed, more or less dilated below, usually contracted at the orifice, with a 2- or 4-toothed border. Nut dry, included in the persistent perianth, free or more or less adnate to it, sessile or stipitate, with a long linear-filiform papillose-hirsute style or stigma, protruding from the perianth and persistent on the ripe nut. Seed more or less albuminous.—Shrubs or small trees. Leaves alternate or opposite, petiolate, 3-nerved, toothed. Stipules usually deciduous. Flower-clusters axillary, either sessile and solitary or in simple interrupted spikes or short branched panicles.

The genus is spread over the tropical and subtropical regions of both the New and the Old World. No species has as yet been detected on the main land of Australia; the subjoined one is endemic in Lord Howe's island.

1. **B. calophleba,** *F. Muell. Fragm.* viii. 11. A bushy shrub of 6 to 8 ft., the branches minutely pubescent with appressed hairs. Leaves alternate, ovate-lanceolate, somewhat acuminate, obtusely

serrate, rounded or slightly cordate and 3-nerved at the base, 2 to 3 in. long, glabrous except a few hairs along the principal veins, green above with the dot-like cystoliths very conspicuous, very white underneath except the impressed nerves and reticulate veinlets. Flowers in globular axillary sessile clusters, all females in our specimens, smaller than in *B. australis*. Fruiting perianth much compressed, the short neck minutely 4-toothed at the orifice. Nut sessile and completely enclosed in the perianth but free from it, the linear persistent hirsute style protruding to about the length of the perianth itself.

N. S. Wales. Lord Howe's island near the base of Mount Lingbird and sides of Mount Gower, *C. Moore*.—The whiteness of the underside of the leaves recalls that of *B. nivea*, but in their shape as well as in the inflorescence the species is much more nearly allied to the Norfolk island *B. australis*, Endl.

12. PIPTURUS, Wedd.

Flowers diœcious, in dense globular clusters, with very small bracts. Male perianth 4- or 5-lobed, with ovate acute lobes. Stamens as many as perianth-lobes, with a central club-shaped rudimentary pistil. Female perianth ovoid, contracted and minutely toothed at the orifice, becoming slightly succulent as well as the receptacle when in fruit. Ovary enclosed in and adnate to the perianth, the linear elongated style or stigma hirsute on one side, and deciduous after flowering. Fruit forming a small succulent globular syncarp, the slightly succulent perianths, however, quite distinct at the top. Seeds with a very scanty albumen.—Trees or tall shrubs. Leaves alternate, entire or toothed, often white underneath, 3- or 5-nerved. Stipules axillary, deeply bifid, deciduous. Flower-clusters axillary, solitary, or in interrupted spikes.

The genus comprises but few species, dispersed over the Indian Archipelago, the Pacific islands, and the Mascarene group. The only Australian species has the general area of the genus.

1. **P. argenteus,** *Wedd. in DC. Prod.* xvi. i. 235[19]. A tree usually small but sometimes attaining 50 ft., the young branches and underside of the leaves hoary or white with a close tomentum. Leaves on rather long petioles, ovate, shortly acuminate, slightly toothed or crenulate, rounded or very rarely slightly cordate at the base, 3- or 5-nerved, glabrous above, 3 to 6 in. long. Flower-clusters distant along the rhaches of single axillary spikes, shorter than or not much exceeding the petioles, and sometimes reduced to a single axillary cluster, quite globular, forming when in fruit a white edible syncarp of 2 to 3 lines diameter, although not nearly so succulent as those of *Morus*, and appearing dry in the dried specimens.—*Urticæ argentea*, Forst. Prod. 65; *R. propinquus*, Wedd. Monogr. Urt. 447, t. 15 D.

Queensland. Pine Head, *A. Cunningham;* Dawson river, *F. Mueller;* Fitzroy island, *M'Gillivray*, *C. Moore*, and others; Cape York, *Daemel;* Kennedy district, *Daintree;* Rockingham Bay, *Dallachy;* Rockhampton, *Dallachy*, *Bowman*, *Thozet.*
N. S. Wales. Richmond river, *Fawcett.*
The species is also in the Indian Archipelago and in the Pacific islands.

13. POUZOLSIA, Gaudich.

(Memorialis, *Ham.;* Gonostegia, *Turcz.;* Hyrtanandra, *Miq.*)

Flowers monœcious or rarely diœcious, in sessile usually androgynous clusters, with small scarious bracts. Male perianth of 4 or 5 lobes or segments, valvate in the bud. Stamens 3, 4 or 5, with a club-shaped or obovoid rudimentary pistil. Female perianth usually ovate, contracted at the orifice, with a 2- or 4-toothed border, often enlarged in fruit and sometimes winged but not succulent. Nut enclosed in the persistent perianth, the linear filiform style deciduous.—Herbs undershrubs or shrubs. Leaves alternate or the lower ones or rarely all opposite, entire or very rarely toothed. Stipules free, usually persistent. Flower-clusters axillary, solitary or in interrupted spikes.

The genus is spread over the tropical regions of both the New and the Old World. The Australian species are both of them East Indian and represent the two sections into which the genus has been divided and which have been adopted as genera by Weddell as well as by several other botanists. I have, however, followed F. Mueller in reuniting them, the sole character derived from the shape of the male perianth is one of little importance, and the differences in habit, however marked in the typical species of each section, do not run through the whole of the species.

SECT. 1. **Pouzolsia.**—*Male perianth-segments concave at the top, but rounded and obtuse on the back.*

Stems diffuse. Leaves all alternate . . . . 1. *P. indica.*

SECT. 2. **Memorialis.**—*Male perianth-segments abruptly bent in near the top, with an acute transverse dorsal angle.*

Stems elongated. Lower leaves opposite . . . . 2. *P. quinquenervis.*

SECT. 1. POUZOLSIA.—Male perianth-segments concave at the top but rounded and obtuse on the back, giving the whole perianth a globular form.

1. **P. indica,** *Gaudich.; Wedd. Monogr. Urt.* 398, t. 13, *and in DC. Prod.* xvi. i. 220, *var.* tetraptera *subvar.* pentandra, *Wedd.* A diffuse perennial with the habit of a *Parietaria*, more or less pubescent or hirsute, the stems usually from 6 in. to 1 ft. long. Leaves alternate or the lower ones rarely opposite, shortly petiolate, ovate, acute, entire, sometimes all under ½ in., rarely nearly 1 in. long, 3-nerved, with the lateral nerves undivided. Flowers few in the clusters, the males and females mixed. Perianths hispid, the males of 5 segments in most of the Australian specimens examined. Fruiting female perianth about ½ line long, sometimes ovoid and equally 8- or 10-ribbed, but others in the same axils with 4 of the ribs produced into broad wings. Nuts black and shining.—*P. arnhemica*, F. Muell. Fragm. iv. 87.

N. Australia. Sea Range, *Wilson;* Sunday island, *A. Cunningham.*
Queensland. Rockingham bay, *Dallachy.*

The species is common in East India and the Archipelago, with the male flowers usually 4-merous. The Sea Range specimens correspond to the *P. procumbens*, Wight Ic. t. 2099, n. 35 and to the *P. minor*, Wight Ic. t. 2100, n. 43, the former with 10-ribbed, the latter with 4-winged fruiting perianths, both correctly reduced by Weddell to *P. indica* and although both are usually described as having 4-merous male flowers, Blume, Mus. Bot. ii. 231, expressly says that *M. procumbens* is usually 5-merous, though sometimes 4-merous. The single Rockingham Bay specimen has rather larger leaves, the male flowers are all fallen away and the fruiting perianths appear to be all ribbed without wings. Cunningham's Sunday island specimens, with the foliage of Dallachy's, have some of the male flowers 3-androus or 4-androus, the fruiting perianths variously deformed, with a larger nut, not black although containing a perfect embryo. Should the 5-merous flowers of *P arnhemica* be really established as a specific distinction, it would fall into *P. Rothiana*, Bl., which Weddell suspects might, with *P. diffusa*, Wight, and *P. confinis*, Bl., be only 5-merous varieties of *P. indica.*

Parietaria oppositifolia, F. Muell. Fragm., iv. 88, appears to me to be a young luxuriant specimen of *Pouzolsia indica* var. *alienata*, Wedd., with some of the leaves opposite, broad, obtuse, rounded or broadly cordate at the base. Very similar specimens occur occasionally amongst the Indian ones.

SECT. 2. MEMORIALIS, *Ham.*—Male perianth-segments abruptly bent in near the top, with an acute transverse dorsal angle, giving a perfectly flat or concave top to the perianth.

2. **P. quinquenervis,** *Benn. Pl. Jav. Rar.* 66. A perennial with ascending or erect stems (of 2 ft. or more ?) scarcely branched except at the base, quite glabrous or the angles of the stems and margins and nerves of the leaves very slightly pubescent. Leaves on very short petioles or almost sessile, the lower ones opposite the upper ones alternate, all lanceolate and entire, the lower ones 2 to 2½ in. long and 3- or 5-nerved, the upper ones often smaller and scarcely 3-nerved, but not reduced to small bracts as in some species. Flower-clusters rather loose, the two sexes usually mixed but in different proportions in different specimens. Male perianth nearly 1 line diameter, with a concave almost saucer-like broad apex, usually 5-merous. Female fruiting perianth with 3 or 4 rarely 2 or 5 usually unequal wings, and quite glabrous. Nut black, smooth and shining.—*Memorialis quinquenervis*, Ham.; Wedd. in DC. Prod. xvi. i. 235[8]; *Hyrtanandra, Pouzolsia* or *Memorialis lythroides*, F. Muell. Fragm. v. 194.

Queensland. Lagoons, Rockingham bay, *Dallachy;* table-land of South Alligator river and Upper Lind river, *Leichhardt.*

The species is also in Sikkim, Khasia and Nepal. It is closely allied to the common *P. hirta*, differing in being much more glabrous, the petioles much shorter or scarcely any, and the wings of the fruiting perianth opaque and apparently veinless. It is also very near to *M. pentandra*, Benn., and *M aquatica*, Benn., which differ in the upper leaves being reduced to small bracts. With regard to the sectional (or generic) name, if it be thought that that of *Memorialis*, Ham., was insufficiently published in Wallich's Catalogue, that of *Gonostegia*, Turcz., would take precedence over Miquel's *Hyrtanandra.*

14. PARIETARIA, Linn.

(Freirea, *Gaud.*)

Flowers polygamous, in axillary cymes often reduced to loose clusters, the external bracts more or less united at the base. Perianth of the hermaphrodite and male flowers deeply divided into 4 rarely 3 lobes, with as many stamens, in the females with a more developed tube and smaller lobes without stamens. Ovary in the hermaphrodite and female flowers free within the perianth-tube, in the males rudimentary. Style filiform or very short, with a densely tufted linear-spathulate recurved stigma, deciduous after flowering. Nut enclosed in the variously enlarged perianth, that of the hermaphrodite flowers usually cylindrical, not succulent.—Annual or perennial much-branched herbs, pubescent with more or less prehensile hooked hairs. Leaves alternate, entire, 3-nerved or triplinerved. Stipules none.

The genus is dispersed over the greater part of the globe, the only Australian species being one of the widest range.

1. **P. debilis,** *Forst. Prod.* 73. A diffuse more or less pubescent branching annual of 6 in. to above 1 ft. Leaves on slender petioles, ovate or cordate, obtuse or acuminate, membranous, green on both sides or hoary underneath, 3-nerved from the base, varying from under ½ in. long in some specimens to above 1 in. in others. Flowers 3 to 7 together in axillary cymes almost reduced to sessile clusters, the 3 or 4 external bracts shortly united at the base. Perianth-lobes usually 4 but sometimes 3 only. Fruiting perianth ovoid, that of the central hermaphrodite flower, although enclosing a perfect nut, often remaining unchanged and scarcely above ½ line long, those of the lateral female flowers more or less enlarged, sometimes ¾ to 1 line long. Style scarcely any besides the dense stigmatic oblong tuft of hairs.—Hook. f. Fl. Tasm. i. 344; Wedd. in DC. Prod. xvi. i. 235[45]; *P. squalida,* Hook. f. in Hook. Lond. Journ. vi. 285; *Freirea australis,* Nees in Pl. Preiss. i. 638.

N. Australia. Sea Range, *F. Mueller.*

Queensland. Brisbane river, Darebin creek, and Buchan river, *F. Mueller.*

N. S. Wales. Argyle county, *A. Cunningham;* Hunter's river, *Miss Scott;* Shoalhaven gullies, *C. Moore;* New England, *C. Stuart.*

Victoria. In fissures of rocks, *F. Mueller;* Curling creek, *Robertson;* Little river, *Fullagar;* Murray river, *Dallachy.*

Tasmania. Launceston and various places along the coast, *Gunn;* Flinders island, *Milligan;* King's island, *F. Mueller.*

S. Australia. St. Vincent's gulf, *F. Mueller;* Gawler ranges, *Sullivan;* Lake Gillies, *Burkitt;* Kangaroo island, *F. Mueller.*

W. Australia. Swan river, *Drummond,* 1*st coll. n.* 734, *Oldfield;* Vass and Kalgan rivers, *Oldfield;* Oldfield river, *Maxwell;* Rottenest island, *Preiss, n.* 2399.

The species is spread over the tropical and temperate regions both of the New and the Old World, with the exception of Europe and Northern Asia. It was distinguished by Gaudichaud generically from the common *P. officinalis,* under the name of *Freirea,* on account of its sessile stigma, a character which however does not hold good in the *P. mauritanica,* which is in other respects so closely allied to *P. debilis.*

15. AUSTRALINA, Gaudich.

Flowers monœcious, in the same or different axillary clusters, the males few together (1 to 5) sessile on a very short common peduncle, the females sessile or on very short pedicels. Male perianth irregularly bilabiate, the outer lip inflexed in the bud. Stamen 1, without any rudimentary pistil. Female perianth ovoid-tubular, obscurely 5-toothed. Ovary enclosed in the perianth but free, with a linear style villous especially on one side and persistent. Nut enclosed in the persistent perianth.—Diffuse or prostrate branching perennial herbs, with the habit of *Parietaria*. Leaves petiolate, alternate or rarely opposite, crenate or obtusely toothed, membranous, 3-nerved. Stipules small.

Besides the Australian species, one of which is also in New Zealand, there is a third in tropical Africa.

Stems filiform, prostrate. Leaves orbicular, mostly 2 to 4 lines diameter 1. *A. pusilla.*
Stems ascending. Leaves, except the lowest, ovate, 1 to 2 in. long . 2. *A. Muelleri.*

1. **A. pusilla,** *Gaudich. in Freyc. Voy. Bot.* 505. Stems filiform, creeping along or hanging from rocks, usually under 6 in. long, and much branched. Leaves nearly orbicular, broadly crenate, 2 to 4 lines diameter or rarely ½ in. when very luxuriant. Male flowers solitary or 2 together on a common peduncle about as long as the perianth, which is nearly 1 line long when open, with 1 large stamen within the large outer lip. Female flowers 2 to 5 together in the same or different axils, each on a very short pedicel, without any common peduncle. Bracts very small and few or none. Fruiting perianth scarcely above ½ line long, ovate, compressed.—Hook. f. Fl. Tasm. i. 345; Wedd. in DC. Prod. xvi. i. 235[59]; *Urtica pusilla*, Poir. Dict. Suppl. iv. 224; *Australina Novæ Zelandiæ* and *A. Tasmanica*, Hook. f. Fl. Nov. Zel. i. 226.

Tasmania. Circular Head and banks of the Acheron, *Gunn;* Macquarrie Harbour and Hampshire hills, *Milligan;* South Port, *C. Stuart.*

The species is also in New Zealand preserving the same characters. I find in some axils 2 male peduncles each bearing 2 flowers and only 1 or 2 females, in others 4 or 5 females and a single male 1-flowered peduncle, or sometimes the axils are wholly male or wholly female.

2. **A. Muelleri,** *Wedd. Monogr. Urt.* 545, *and in DC. Prod.* xvi. i. 235[60]. A much stouter plant than *A. pusilla*, creeping at the base, with ascending or erect stems of 6 in. to 1 ft., a few of the lower leaves sometimes small and orbicular, but those of the flowering stems always ovate or ovate-lanceolate, obtuse or almost acute, coarsely crenate-toothed, 1 to 1½ in. long. Male peduncles short, bearing 2 to 4 flowers, all sessile in a whorl so as to resemble a single flower, the perianth greener and more hirsute than in *A. pusilla*. Female flowers in dense sessile clusters, often numerous, in separate axils or on separate plants from the males, or rarely 1 or 2 in the male axils. Fruiting perianth ½ to ¾ line long.—*Urtica Tasmanica*, F. Muell. First Gen. Rep. 18 (as to the Victorian plant).

Victoria. Buffalo and Dandenong Ranges, Mount Disappointment, Delatite river, Apollo Bay, etc., *F. Mueller;* who considers this as a variety only of *A. pusilla*, but I have seen no connecting form.

SUBTRIBE 3. UREREÆ.—Plants more or less armed with stinging hairs. Female perianth 4-lobed, 2 opposite lobes usually larger than the 2 others.

16. URTICA, Linn.

Flowers monœcious or diœcious, clustered but distinct. Male perianth depressed-globular in the bud, deeply divided into 4 concave nearly equal segments. Stamens 4; with a small rudimentary pistil. Anthers oblong-reniform. Female perianth with 2 outer segments usually smaller than the 2 inner, which become enlarged and ovate when in fruit. Stigma sessile or nearly so, tufted or divided into long cilia. Nut small, dry, enclosed in or surrounded by the persistent perianth.—Herbs or rarely, in species not Australian, shrubs, more or less sprinkled or covered with rigid stinging hairs. Leaves opposite, toothed, 5- 7- or rarely 3-nerved. Stipules interpetiolar, free or the 2 on each side united into one. Flower-clusters in axillary simple or branched interrupted spikes or panicles, shorter than the leaves, and often shorter than the petioles.

The genus is widely spread over the extra-tropical regions of both the northern and southern hemispheres in the New as well as the Old World, more rare within the tropics. The only indigenous Australian species is also in New Zealand, and is a close representative of a common northern one.

Perennial. Male and female flowers in distinct inflorescences . . . 1. *U. incisa.*
Annual. Male and female flowers in the same inflorescences . . . . *2. *U. urens.*

1. **U. incisa,** *Poir.; Wedd. in DC. Prod.* xvi. i. 52. A perennial, with stems usually weak or decumbent at the base, sometimes slender and rarely 1 ft. long, sometimes trailing to the length of 10 or 12 ft., glabrous between the stinging hairs. Leaves usually on long petioles, lanceolate or almost linear, acute, deeply toothed, somewhat cordate at the base and 1 to 3 in. long, rarely ovate-lanceolate. Inflorescence usually simple or branching into 2 at the very base, the males mostly as long as or longer than the petiole, the clusters sometimes lengthening into short branches, the females shorter and more dense, but in some specimens the female inflorescences are long, slender, and branched, in others the males are reduced to dense sessile clusters. Male perianth $\frac{3}{4}$ to 1 line diameter, glabrous or slightly scabrous; female perianth smaller when in flower, but the inner segments enlarging to $\frac{3}{4}$ line round the nut, which is ovate, slightly compressed, as long as or rather longer than the persistent perianth.—Hook. f. Fl. Tasm. i. 343; *U. lucifuga,* Hook. f. in Hook. Lond. Journ. vi. 285.

Queensland. Covering an island in Fitzroy river, *O'Shanesy*, a single specimen not in flower.

N. S. Wales. Paramatta, *Woolls;* Macleay and Clarence rivers, *Beckler;* Newcastle, *Leichhardt.*

Victoria. Marshy plains on the Yarra, *F. Mueller*, *Robertson;* Little river, *Fullagar.*

Tasmania. Common in various parts of the island, chiefly in moist shady places, *J. D. Hooker;* Flinders island, *Gunn;* King's island, *Neat.*

S. Australia. Strathalbyn, *Herb. F. Mueller*, collector not named.

This species, which is also in New Zealand, is usually distinguished from the northern *U. dioica*, Linn., by the weaker stems glabrous between the stinging hairs, narrower leaves and larger male flowers; some northern specimens, however, of *U.*

dioica come very close to the Australian plant. The true *U. dioica* is mentioned by F. Mueller as among the introduced plants, but all the Australian specimens so named which I have seen in his or other herbaria appear to me to be referrible rather to the indigenous *U. incisa.*

*2. **U. urens,** *Linn.; Wedd. in DC. Prod.* xvi. i. 40. An annual with erect or ascending branched stems rarely above 1 ft. high, glabrous with the exception of the rigid stinging hairs. Leaves petiolate, ovate or elliptical, deeply and regularly toothed, 1 to 2 in. long. Inflorescences contracted into loose axillary clusters seldom exceeding the petioles, the males and the females intermixed in the same clusters, of the same structure as in *U. incisa*, except that the larger segments of the female perianth are ciliate on the margin and usually bear a single dorsal stinging hair.

A European weed now common near buildings in various parts of **Victoria, Tasmania,** and **S. Australia.**

17. LAPORTEA, Gaudich.

Flowers diœcious or monœcious, clustered but distinct. Male perianth depressed-globular in the bud, deeply divided into 4 rarely 5 segments. Stamens 4, rarely 5, with a rudimentary pistil; anther-cells reniform. Female perianth of 4 lobes or segments, the 2 inner ones usually larger than the outer and dissimilar from each other. Stigma simple, elongated, incurved. Nut more or less compressed and oblique, enclosed in or surrounded by the persistent perianth.—Trees, shrubs, or (in species not Australian) herbs, sprinkled or covered with rigid stinging hairs. Leaves alternate. Stipules entire or bifid, deciduous. Flower-clusters in axillary or, in species not Australian, terminal panicles, usually shorter than the leaves.

The species are distributed over the warmer regions of the New and the Old World, but chiefly in the Indian Archipelago and Pacific Islands, one exceptionally herbaceous species is, however, North American and extratropical. The Australian species all appear to be endemic. They all belong to the section *Sarcopus*, Wedd., comprising trees or shrubs with the majority of the fruiting pedicels thickened, fleshy and incurved, assuming frequently in the specimens the appearance of little grubs, and the lobes of the female perianths very unequal.

Leaves broadly ovate, cordate at the base, glabrous above, pubescent underneath . . . . 1. *L. gigas.*
Leaves ovate, not cordate, nearly glabrous . . . . 2. *L. photiniphylla.*
Leaves broad, more or less peltate at the base, pubescent or villous on both sides . . . . 3. *L. moroides.*

1. **L. gigas,** *Wedd. Monogr. Urt.* 129, t. 3 *and* 4, *and in DC. Prod.* xvi. i. 82. A large tree, exceeding 80 ft. the trunk erect, with a soft juicy fibrous wood and smooth ash-coloured bark, supported at the base by prominent angles or almost wing-like buttresses. Leaves broadly cordate-ovate, obtuse or shortly acuminate, entirely or obscurely or more distinctly sinuate-toothed, often above 1 ft. long and nearly as broad, glabrous above or nearly so, or sprinkled with a few stinging hairs, more or less pubescent or villous underneath. Stipules ovate-

lanceolate, very deciduous. Flowers diœcious, clustered along the branches of rather loose axillary panicles rarely exceeding the petioles. Male perianth scarcely ½ line diameter. Female perianth-lobes acute and hirsute, the 2 outer ones larger than the 2 inner. Nuts much flattened, very oblique, the pedicel and perianth much thickened into a curved fleshy mass.—*Urtica gigas*, A. Cunn. Herb.; *Urera rotundifolia*, Wedd. in Ann. Sc. Nat. ser. 4, i. 177, and *U. excelsa*, Wedd. l.c. 178.

Queensland? Some specimens in leaf only from Brisbane river, *A. Cunningham*, may possibly belong to this species.

N. S. Wales. Abundant in the Illawarra district, *A. Cunningham*, *Lownes*; Clarence river, *Beckler*. Some specimens also in leaf only, from Glendon and Three-mile scrub, *Leichhardt*, may possibly be a variety of this species with the leaves more toothed.

2. **L. photiniphylla,** *Wedd. Monogr. Urt.* 138, *and in DC. Prod.* xvi. i. 83. A fine tree of 60 ft. with a straight soft-wooded stem of 30 to 40 ft. (*Dallachy*). Leaves ovate or almost elliptical, obtuse or more frequently acuminate, entire or sinuate-toothed, not cordate, 3-nerved, nearly glabrous but sometimes sprinkled with a few stinging hairs. Panicles axillary, often exceeding the petioles. Flowers of *L. gigas*, but more glabrous. Nuts rather larger, the pedicels usually but not always enlarged into an incurved fleshy mass.—*Urtica photiniphylla*, A. Cunn. Herb.; *Fleurya photiniphylla*, Kunth. Ind. Sem. Hort. Berol. 1846, 11 (*Weddell*).

Queensland. Brisbane river, Moreton bay, *A. Cunningham*, *F. Mueller*; Rockhampton, *Thozet*, *O'Shanesy*, *Dallachy*; Nurrum-nurrum, *Leichhardt*.

N. S. Wales. Newcastle, *Harvey*; Clarence river, *Beckler*; Northern woods, London Exhibition 1862, *n.* 107.

Cunningham's specimens may possibly include two species, one with perfectly smooth branches, the other (in leaf only) with the branches covered with prickles. Seeman's *L. vitiensis*, from the Fiji islands, referred to this species by Weddell, appears to me to differ specifically in the venation of the leaves and in the pedicels always remaining unenlarged.

3. **L. moroides,** *Wedd. Monogr. Urt.* 142, *and in DC. Prod.* vii. i. 88. A tall shrub or small tree, with most virulent stinging hairs. Leaves broadly ovate, peltate at the base, shortly acuminate, more prominently toothed than the other species and more pubescent or villous on both sides, 6 to 8 in. long and almost as broad. Panicles axillary, rarely exceeding the petioles. Male flowers not seen; females of *L. gigas*. Fruits densely clustered, the pedicels rather less thickened but succulent, the whole fruiting panicle of a rich reddish-purple (*Dallachy*).—*Urtica moroides*, A. Cunn. Herb.

Queensland. *A. Cunningham*; Port Denison, *Fitzalan*; Mount Elliott and Edgecombe and Rockingham bays, *Dallachy*; Cleveland bay, *Herb. F. Mueller*, collector not named.

ORDER CX. CASUARINEÆ.

Flowers unisexual, the males in cylindrical spikes, the females in globular or ovoid spikes or cones, both sexes sessile and solitary in the axils of whorled bracts, the bracts of each whorl united into a toothed sheath enclosing the base of the whorl of flowers, each flower within 2 small persistent bracteoles placed right and left. Male fl.: Perianth of 1 or 2 (anterior and posterior) concave or hood-shaped segments, breaking off at their narrow base as they are forced off by the development of the stamen. Stamen 1, the filament folded in the bud, erect and exserted from the sheath of bracts when fully out. Anther with 2 large distinct cells, placed back to back, and opening in 2 valves. Female fl.: Perianth none. Ovary minute, 1-celled. Style very shortly entire, with 2 long filiform stigmatic branches. Ovules 2, collateral, semi-anatropous, laterally attached above the base of the cell. Fruit a globular ovoid or cylindrical compact cone, formed of the enlarged thickened more or less woody bracts and bracteoles, the bracteoles closed against each other over the unripe nut, often protruding beyond the bracts, and opening as two valves when ripe. Nut much compressed laterally, smooth and shining, produced at the apex into a membranous often transparent wing, traversed longitudinally by an opaque nerve (the base of the style). Seeds solitary, erect, laterally attached by a funicle showing at its apex the second abortive ovule; testa membranous; no albumen; embryo with 2 large flattened cotyledons and a small superior radicle.—Trees or shrubs, with leafless wiry or rigid erect or pendulous branches and usually numerous deciduous verticillate branchlets, often articulate at the nodes. Leaves replaced by very small scales in whorls of 4 to 16 often united at the base, their midribs decurrent to the next node forming so many ribs or angles to the internode; the scales of each node alternating with those of the nodes immediately below and above, so that when the internodes are very short (as in the inflorescences), then the bracts, axillary flowers, &c., are superposed in twice as many series as there are parts of each whorl. Male spikes terminating deciduous branchlets, or almost sessile along the permanent branches, the female cones usually sessile or only shortly pedunculate, but in both sexes there are always 1, 2 or more barren internodes and sheaths below the flowering ones. The number of parts of each whorl, whether on the branches, branchlets, spikes, or cones, are the same in the same plant, or rarely fewer on the weaker branches and branchlets than on the stronger ones. Anthers and style-branches usually purple or red, the latter elegantly fringing the female flower-cones.

The Order consists but of a single genus, chiefly Australian, but extending also over tropical Asia from East Africa to the Indian Archipelago and South Pacific islands. The Australian species appear to be all endemic except *C. equisetifolia*, which ranges over nearly the whole area of the Order, and has become naturalized in many tropical and subtropical regions of the New and the Old World. Except the New Caledonian types there are very few species, and possibly none besides *C. nodiflora* from the South Pacific, which are not Australian.

The Order is a very distinct one, the floral structure may be nearly that of *Urticeæ*, but the remarkable vegetative characters have no nearer parallel than amongst some *Coniferæ*. The perianth-segments, sometimes called inner bracteoles, are generally described in the generic character as two, an anterior and a posterior; but in some individuals or species there is only one, the posterior one. I have found it so, as figured in the Flora Tasmanica, in *C. stricta* (*C. quadrivalvis*), and in some other species I have found both the anterior and posterior one; but these organs are very delicate and difficult to examine in dried specimens. I have not therefore been able to verify it in a sufficient number to avail myself of it in the specific characters. From the careful observations of Bornet mentioned in Decaisne and Lemaout's Traité de Botanique, p. 533, it would appear that the one is almost constant in *C. stricta* (*C. quadrivalvis*, Labill.), the two in *C. torulosa*.

1. CASUARINA, Linn.

Character and distribution the same as those of the Order.

SECT. 1. **Leiopitys.**—*Whorls* 7–16-*merous, rarely* 6-*merous. Cone-valves usually prominent, thickened and keeled or angled on the back or with a short broad and smooth dorsal protuberance* (*rugose in* C. Fraseriana).

Whorls usually 10–12-merous (varying 9 to 16). Cone-valves thickened into a dorsal angle or keel.
Cone-valves very prominent, ovate.
Branches usually prominently ribbed. Cones 1 in. diameter or more.
Male-spikes usually long, the sheath-teeth shortly pointed. Eastern species . . . . 1. *C. stricta.*
Male-spikes not exceeding 1 in., the sheath-teeth with long fine points. Western species . . . . 2. *C. trichodon.*
Branches terete. Male-spikes not exceeding 1 in. Cones about ½ in. diameter, the cones villous on the back . . . . 3. *C. glauca.*
Cones very compact, the valves small and thick, scarcely prominent. Western species . . . . 4. *C. Huegeliana.*
Whorls usually 7-merous (varying 6 to 8). Cone-valves usually with a short transverse dorsal appendage sometimes very short or obsolete.
Trees usually monœcious. Cone-valves very prominent.
Male spikes cylindrical imbricate. Cones nearly globular, the valves pubescent on the back . . . . . 5. *C. equisetifolia.*
Male-spikes moniliform. Cones usually oblong or small, the valves glabrous. Branchlets very slender.
Cones ½ in. diameter or more, usually oblong . . . 6. *C. suberosa.*
Cones not above 4 lines diameter, globular . . . . 7. *C. Cunninghamiana.*
Shrubs (always?) diœcious. Cone-valves scarcely prominent.
Cone-valves with the dorsal protuberance short and smooth. East and west species . . . . 8. *C. distyla.*
Cone-valves with the dorsal protuberance as long as the valve and rugose. Western species . . . . . . 9. *C. Fraseriana.*

SECT. 2. **Trachypitys.**—*Whorls* 4–5-*merous, rarely* 6-*merous. Cone-valves rarely prominent beyond the thick broad obtuse dorsal protuberances, which are very rugose or divided into tubercles* (*nearly smooth in* C. Decaisneana).

Male-spikes and branchlets slender.
Cones quite glabrous.
Male-spikes 3 to 6 lines long. Cone-valves with the dorsal protuberance divided into 1 large and 2 small tubercles . . . . 10. *C. nana.*
Male-spikes 2 to 4 lines. Cone-valves with the dorsal protuberance divided into about 6 nearly equal tubercles . . . . 11. *C. humilis.*
Cones more or less villous.
Sheath-teeth very short. Cone-valves slightly protruding beyond several small dorsal tubercles.
Branchlets nearly terete. Sheath-teeth of the male spikes scarcely acute. Eastern species . . . . 12. *C. torulosa.*

Branchlets very acutely 2- or 4-angled. Sheath-teeth of the male spikes acuminate. Western species . 13. *C. decussata.*
Sheath-teeth 1 to 3 lines long. Cone-valves not exceeding the single large dorsal tubercle . . . . 14. *C. Decaisneana.*
Male spikes short. Branchlets short, rigid, divaricate and intricate.
Branchlets of several short articles. Male spikes 2 to 3 lines long. (Cones unknown) . . . . 15. *C. Drummondiana.*
Branchlets mostly of 1 article. Male spikes 1 to 2 lines long. Cone-valves with several nearly equal dorsal tubercles . 16. *C. microstachya.*

SECT. 3. **Acanthopitys.**—*Cone-valves very prominent, the dorsal protuberance produced into a rigid point.*

Whorls 10–12-merous. Habit of *C. stricta.* Point of the dorsal protuberance as long as the valve . . . . 17. *C. bicuspidata.*
Whorls 4–6-merous. Branchlets short. Point of the dorsal protuberance much longer than the valve.
Branchlets spreading, consisting of several nearly equal internodes . . . . 18. *C. thuyoides.*
Branchlets erect, rigid, consisting of one short basal internode and one long terminal pungent-pointed one . . 19. *C. acuaria.*

SECT. 1. LEIOPITYS.—Whorls (of stem-angles, sheath-teeth and flowers) 7- to 16-merous, rarely 6-merous. Cone-valves usually prominent, thickened on the back into a prominent keel or angle, or into a short broad transverse usually smooth protuberance.

1. **C. stricta,** *Ait. Hort. Kew.* iii. 320, *not of Miq.* A small tree with the branchlets usually if not always pendulous notwithstanding the name, sometimes reduced to a tall dense shrub, the branches more robust than in *C. equisetifolia*, the whorls of 9 to 12 parts, the internodes often ½ in. long or more, the ribs rather prominent, the sheath-teeth acute or shortly acuminate. Flowers diœcious. Male spikes some of them terminating deciduous branchlets of several inches, others almost sessile on the permanent branch, often more than 2 in. long, dense when young, the sheaths scarcely overlapping when full-grown. Fruit-cones globular or ovoid, often 1 in. diameter or even more, the valves much protruded, ovate-triangular or almost ovate-oblong, thickened into a smooth dorsal prominent angle or keel.—Willd. Spec. Pl. iv. 190; *C. quadrivalvis*, Labill. Pl. Nov. Holl. ii. 67, t. 218; Miq. Rev. Cas. 71, t. 9, 10, and in DC. Prod. xvi. ii. 334; Hook. f. Fl. Tasm. i. 347; F. Muell. Fragm. vi. 18; *C. macrocarpa*, A. Cunn. Herb.; *C. cristata*, Miq. Rev. Cas. 70, t. 10 A; *C. Gunnii*, Hook. f., Miq. in Ned. Kruidk. Arch. iv. 100.

N. S. Wales. Lachlan river and Twofold Bay, *A. Cunningham.*

Victoria. From Wendu Vale, *Robertson*, to Wilson's Promontory and Gipps' Land, *F. Mueller.*

Tasmania. Abundant throughout the island, excepting towards the north-west, *J. D. Hooker.*—The *oak* of the colonists.

S. Australia. St. Vincent's Gulf, *Blandowski;* Bugle Range, Port Lincoln, Kangaroo island, *F. Mueller.*

The Lachlan river specimens distinguished by Miquel as two varieties under the names of *macrocarpa* and *cristata* were probably all gathered from one tree; the cones do not appear to me to be larger than are many of the Tasmanian specimens. The *C. Gunnii*, formerly published by Miquel and reduced by Hooker to a var. *spectabilis* of *C. quadrivalvis*, differs in nothing but a more robust habit.

2. **C. trichodon,** *Miq. in Pl. Preiss.* i. 641; *Rev. Cas.* 63, t. 8, *and in DC Prod.* xvi. ii. 336. A tall erect shrub closely resembling *C. stricta*, the whorls 8- to 10-merous, the ribs of the internodes prominent, and the cones the same as in that species, but the sheath-teeth mostly terminate in long fine points, the male spikes, not much above 1 in. long, are closely imbricate, with the long spreading almost hair-like points of the sheath-teeth very conspicuous. Valves of the cones ovate, thickened into a dorsal smooth angle as in *C. stricta.*

W. Australia. King George's Sound to Cape Riche, *Drummond, 4th coll n.* 239, *Preiss. n.* 2001, also perhaps *n.* 2000, but not the loose fruits sent with it, *Harvey, Maxwell, F. Mueller.*

3. **C. glauca,** *Sieb. in Spreng. Syst.* iii. 803. A tree, often attaining a considerable size, the specimens sometimes very difficult to distinguish from those of *C. stricta*, the internodes however much less prominently ribbed, and becoming glaucous in open dry situations. Whorls usually 10- to 12-merous, but the parts varying from 9 to 16, the sheath-teeth short on the deciduous branchlets, more acuminate on the permanent branches. Male spikes rather dense, $\frac{1}{2}$ to 1 in. long, the sheath-teeth sometimes ciliate with points almost as long as in *C. trichodon*, but often much shorter. Cones usually subglobose, flat-topped, rarely much above $\frac{1}{2}$ in. diameter, the valves very prominent, often pubescent on the back, thickened into a smooth dorsal prominent angle or keel. In some forms, including Sieber's typical specimens, the cones are smaller, with very numerous rather smaller valves very regularly arranged.—Miq. Rev. Cas. 76, t. 11, and in DC. Prod. xvi. ii. 334; *C. torulosa*, Miq. Rev. Cas. 75, t. 11 B, but not of Ait.

Queensland. Brisbane river, Moreton bay, and between Mackenzie and Suttor rivers, *F. Mueller;* Rockhampton, *O'Shanesy.*

N. S. Wales. Port Jackson to the Blue Mountains, *R. Brown, Sieber, n.* 425, *Woolls;* Liverpool plains and New England, *Leichhardt;* Lachlan river and Liverpool plains, *A. Cunningham;* from the Lachlan and Darling to the Barrier Range, *Victorian and other Expeditions.*

Victoria. Avoca and Murray rivers, *F. Mueller;* Wimmera, *Wilson.*

S. Australia. Flinders' Range and Murray Desert, *F. Mueller.*

In the specimens from the interior mentioned by Miquel in Ned. Kruidk. Archiv. iv. 100, as *C. pauper*, F. Muell., and in those from Queensland, the cones are more like those of *C. equisetifolia* than those of Sieber's typical specimen.

4. **C. Huegeliana,** *Miq. in Pl. Preiss.* i. 640; *Rev. Cas.* 68, t. 11 A, *and in DC. Prod.* xvi. ii. 335. A shrub or small tree, apparently diœcious, with the terete branches, 8- to 12-merous whorls, short sheath-teeth, and glaucous aspect of *C. glauca*, but the male spikes appear to be longer more slender and not so much imbricate, with shorter sheath-teeth and the cones are mostly cylindrical, $\frac{3}{4}$ to 1 in. long, very closely packed, the small valves much less protruded, and their thick backs slightly rugose-pubescent.

W. Australia. Murchison river, *Oldfield;* King George's Sound or Swan river, *Preiss, n.* 2006, *Drummond*, and others.

The only specimens seen in fruit are those of Oldfield and some loose fruits with specimens, otherwise imperfect, in Herb. F. Mueller. The male specimens of Preiss and the female ones of Drummond differ in their much more slender branches. Drummond's n. 232, with male amenta only, may belong to *C. distyla*, or possibly to *C. bicuspidata*. Some specimens with shorter cones from Murchison river, are referred by Miquel to *C. obesa*, Miq. in Pl. Preiss. i. 640, which he afterwards, in DC. Prod. xvi. ii. 334, reduces to *C. glauca*, and they certainly appear in some measure intermediate between that species and *C. Huegeliana*. The eastern specimens named by Miquel in Herb. Hook. *C. obesa* are certainly a common form of *C. glauca*. The species altogether requires revision from better materials.

5. **C. equisetifolia,** *Forst. Char. Gen.* 103, t. 52. A tree attaining a large size, but so frequently cut that it is generally met with much smaller, the principal branches elongated and spreading or ascending, the smaller ones often pendulous, glabrous or tomentose when young. Whorls usually 7-merous, but the parts varying from 6 to 8. Sheath-teeth short and acute or sometimes fine-pointed on the persistent branches, the internodes rarely exceeding ¼ in. the ribs rather prominent. Flowers diœcious, male spikes about ¾ in. long, terminating slender deciduous branchlets, the sheaths of the spikes closely imbricate. Fruit-cones very shortly pedunculate on the persistent branches, globular, usually about ½ in. diameter, the valves protruding about 1 line, broadly ovate, obtuse, pubescent outside, without any or with a very obscure dorsal protuberance at the base.—Miq. Rev. Cas. 43, t. 5, and in DC. Prod. xvi. ii. 338, with the several (non-Australian) synonyms there given.

N. Australia. North coast, *A. Cunningham;* islands of the Gulf of Carpentaria, *Henne;* Escape Cliffs, *Hulse;* Port Darwin, *Schultz, n.* 9, 225.

Queensland. Northumberland islands, *R. Brown;* Cape Bedford and islands off Cape Flattery, *F. Mueller;* Entrance island, Endeavour Straits, *Leichhardt;* Percy islands, *Walter;* Rockingham and Edgecombe bays, *Dallachy;* Port Denison, *Fitzalan.*

Var. *incana.* Young shoots very tomentose. *C. incana*, A. Cunn. Herb.—Port Macquarrie, *A. Cunningham, Leichhardt;* Moreton island, *C. Moore.* This variety appears also to be found in New Caledonia, from a specimen received from the Paris herbarium.

The species is widely spread over East India and the Archipelago. With *C. distyla* it has long been cultivated in gardens, where varieties have arisen which have been described as new Australian species. Amongst them *C excelsa*, Dehnh.; Miq. Rev. Cas. 23, t. 1 F., is referred by Miquel in DC. Prod. xvi. ii. 344, positively to *C. equisetifolia*, and *C. truncata*, Willd., *C. sparsa*, Tausch., and *C. tortuosa*, Hortul., are supposed to be either this species or *C. distyla*.

6. **C. suberosa,** *Ott. and Dietr.; Miq. Rev. Cas.* 54, t. 6, *and in DC. Prod.* xvii. ii. 337. A tree of 30 to 40 ft., the specimens closely resembling those of *C. equisetifolia*, the whorls similarly 7-merous or the parts varying from 6 to 8, and often monœcious like that species. Branches seldom if ever corky, the branchlets usually slender and quite glabrous. Male spikes much more slender and interrupted, the short sheaths not overlapping those next above. Fruit-cones more frequently tending to become ovoid or oblong, truncate at both ends, the valves more prominent, glabrous or nearly so, with usually a short broad thick but almost scale-like transverse dorsal protuberance at the base rarely extending nearly to the apex of the valve.—Hook. f. Fl. Tasm. i. 348, t. 96; *C. leptoclada*, Miq. Rev. Cas. 41, t. 4, and

in DC. Prod. xvi. ii. 339; *C. mæsta*, F. Muell. in Miq. Stirp. Nov. Holl. Muell. 2.

Queensland. Percy islands, *A. Cunningham*, *Denham;* Brisbane river and Moreton island, *F. Mueller;* Stradbrooke island, *Fraser;* Ipswich, *Nernst;* Rockhampton, *O'Shanesy;* Rockingham bay, *Dallachy;* Cape York, *W. Hill.*

N. S. Wales. Port Jackson, *R. Brown*, *J. D. Hooker*, *Woolls;* Argyle County, *A. Cunningham;* Clarence, Hastings, and Macleay rivers, *Beckler;* Richmond river, *Fawcett;* Severn river, *Leichhardt;* New England, *C. Stuart.*

Victoria. Yarra river and Dandenong mountains, *F. Mueller*, *Robertson*, *Adamson.*

Tasmania. Common in the island on stony hills (the specimens rather more robust than the northern ones), *J. D. Hooker.*—Known in some of the colonies as the Oak or Swamp-Oak.

C. rigida, Miq. Rev. Cas. 61, t. 7 D, and in DC. Prod. xvi. ii. 337, as to Fraser's specimens, may be *C. suberosa;* the more robust Tasmanian specimens belong to *C. distyla.* A species very nearly allied to *C. suberosa* is found in New Caledonia.

7. **C. Cunninghamiana,** *Miq. Rev. Cas.* 56, t. 6, *and in DC. Prod.* xvi. ii. 335. A tree with slender branches, closely resembling *C. equisetifolia* and *C. suberosa* in the aspect of the specimens and number of parts of the whorls, and possibly a variety of the one or the other, but the fruiting-cones are much smaller, scarcely exceeding 4 lines in diameter in any of the specimens seen, globular, very regular, glabrous, with prominent valves. Male spikes (in Herb. R. Brown) slender like those of *C. suberosa* but more dense.

Queensland. Woods of the London Exhibition 1862, *Hill*, *n.* 9; Gilbert river, *Daintree.*

N. S. Wales. Hunters river, *R. Brown;* Sydney woods, Paris Exhibition 1855, *M'Arthur*, *n.* 134; Glendon, *Leichhardt;* Hastings river, *Beckler;* Nepean river, *Woolls;* between the Darling and the Barrier range, *Goodwin and Dallachy;* Twofold bay, *F. Mueller.*

8. **C. distyla,** *Vent. Jard. Cels.* t. 62. A diœcious shrub, usually only 2 or 3 ft. high, with erect or spreading branches, but in favourable situations attaining the size of a small tree, the branches glabrous or more or less pubescent. Whorls usually 7-merous, but the parts varying from 6 to 8, the teeth short. Male spikes on deciduous branchlets of 1 to 3 in. or almost sessile on the persistent branches, 1 to 1½ or rarely 2 in. long, more or less moniliform, the sheaths not overlapping and the teeth not at all or scarcely acuminate. Fruit-cones sessile or nearly so, oblong, from under ¾ to about 1 in. long, the broad very obtuse almost truncate valves slightly prominent though considerably longer than the dorsal protuberance, which is broad entire and smooth.—Miq. Rev. Cas. 57, t. 7; Hook. f. Fl. Tasm. i. 348; *C. stricta*, Miq. in DC. Prod. xvi. ii. 336, not of Aiton; *C. Muelleriana*, Miq. in Ned. Kruidk. Arch. iv. 99; *C. rigida*, Miq. in DC. Prod. xvi. ii. 337, as to the Tasmanian specimens.

N. S. Wales. Port Jackson to the Blue Mountains, *R. Brown*, *Sieber*, *n.* 326, *Fl. Mixt. n.* 605 and others; Lachlan river, *A. Cunningham.*

Victoria. Melbourne, *Adamson;* Wilson's promontory, Cape Howe, Gipps' Land, *F. Mueller.*

Tasmania. Common throughout the island, ascending to 4000 ft., *J. D. Hooker.*

S. Australia. St. Vincent's Gulf, *Blandowski;* Onkaparinga range, Mount Torrens, etc., *F. Mueller;* Kangaroo island, *Waterhouse.*

W. Australia. King George's Sound and adjoining districts (the cones usually very regular, ½ to ¾ in. long in some specimens, above 1 in. in others), *R. Brown, Drummond,* (*4th coll.?*) *n.* 231, 232, *Harvey* and others; Kalgan and Tweed rivers, *Oldfield;* Swan river, *Drummond, 1st coll.*

The above quoted *C. Muelleriana* and *C. Baxteriana,* Miq. Rev. Cas. 37, t. 3 C, referred to *C. suberosa* as varieties by Miq. in DC. Prod. xvi. ii. 338, both from West Australia, appear to me both to belong rather to *C. distyla.* I do not understand why Miquel reduced this to Aiton's *C. stricta,* which by the character given in the Hortus Kewensis "vaginis multifidis" in contradiction to the "vaginis 7-fidis" of *C. equisetifolia,* as well as from the inspection of the original specimen is evidently Labillardière's *C. quadrivalvis,* notwithstanding the inapplicability of Aiton's name.

Var. *paludosa,* a low shrub with the branchlets smaller and more slender than in the typical form, the male spikes ½ to 1 in. long, closer than in most specimens of *C. distyla* and the cones usually shorter, varying from ½ to ¾ in.—*C. paludosa,* Sieb. in Spreng. Syst. iii. 803; Miq. Rev. Cas. 64, t. 8 B, and in DC. Prod. xvi. ii. 338; *C. pumila,* Ott. and Dietr.; Miq. Rev. Cas. 66, t. 8 C.; *C. dumosa,* A. Cunn. Herb.—Port Jackson to the Blue Mountains, *Sieber, n.* 329; Argyle County, *A. Cunningham's* Twofold Bay, and Genoa river, and perhaps Port Phillip, *F. Mueller;* North west interior of Victoria, *L. Moreton.*

9? **C. Fraseriana,** *Miq. Rev. Cas.* 59, t. 6 D, *and in DC. Prod.* xvi. ii. 337 *partly.* A tall erect diœcious shrub or small tree, nearly allied to *C. distyla,* and very difficult to distinguish from it without the cones. These are in the typical form nearly globular, ¾ in. diameter, very woody, the valves shortly prominent, the dorsal protuberances very rugose though not divided regularly into distinct tubercles, in another specimen they are more nearly of the shape of those of *C. distyla,* but 1 in. long or more, and again some detached cones in the Kew Museum are more like the typical ones but larger with more numerous parts to the whorl.—*C. torulosa,* Miq. in Pl. Preiss. i. 639, not of Ait.

W. Australia. King George's Sound, *Preiss, n.* 2000, *F. Mueller, Oldfield.*—The species requires further elucidation from better specimens. It has nearly the fruit of *C. torulosa* with the branches and whorls of *C. distyla.* It ought perhaps to be referred to the following section.

Sect. 2. Trachypitys.—Whorls (of stem-angles, sheath-teeth and flowers) 4- or 5-merous, rarely 6-merous. Cone-valves rarely prominent beyond the thick broad obtuse dorsal protuberances, which are very rugose or divided into tubercles.

10. **C. nana,** *Sieb. in Spreng. Syst. Veg.* iii. 804. A densely branched erect diœcious shrub, the branchlets short slender and terete, the ribs scarcely prominent, the parts of the whorls usually 5, varying from 4 to 6. Male spikes slender, compact, ¼ to ½ in. long, the sheaths short and imbricate. Fruit-cones sessile, from nearly globular and 4 lines diameter to oblong-cylindrical and ¾ in. long, glabrous or slightly hispid, the valves not at all protruding, the dorsal protuberance large and obtuse forming usually one large triangular tubercle and two smaller ones to each valve, giving the surface of the cone an appearance of being regularly divided into numerous triangular compartments.—Miq. Rev. Cas. 29, t. 2 B, and in DC. Prod. xvi. ii. 340.

N. S. Wales. Rocky hills, Blue Mountains, *Sieber*, *n.* 328, *A. Cunningham*, *Woolls.*

11. **C. humilis,** *Ott. and Dietr.*; *Miq. in DC. Prod.* xvi. ii. 340. A diœcious shrub of 2 to 6 ft., erect or the lower branches diffuse, very nearly allied to the eastern *C. nana*, the whorls usually 4-merous, with very short sheath-teeth, but the angles of the internodes are much more prominent although usually obtuse and never so acute as in *C. decussata.* Male-spikes small and dense, with short imbricate sheaths as in *C. nana*, but usually still shorter, varying from 2 to 4 lines long. Cones oblong or cylindrical, mostly about ¾ in. long, but sometimes lengthening to 1 in., glabrous or very slightly ciliate, the valves not produced beyond the dorsal tubercles, but these tubercles are usually 6 to each valve, all small and nearly equal, forming on the cone 3 longitudinal rows between each line of valves, sometimes pressed into 2 rows when the valves are open.—*C. Preissiana*, Miq. in Pl. Preiss. i. 640, and Rev. Cas. 31, t. 2; *C. selaginoides*, Miq. in Pl. Preiss. i. 643, and Rev. Cas. 35, t. 3 B; *C. Lehmanniana*, Miq. in Pl. Preiss. i. 639, and Rev. Cas. 33, t. 3 A, and *C. tephrosperma*, Hort. Hamb. Miq., Rev. Cas. 31, all referred to *C. humilis*, by Miq. in DC. Prod. xvi. ii. 340; *C. nana*, A. Cunn. Herb. not of Sieb.

W. Australia. King George's Sound to Swan River, *Drummond*, 1*st coll. n.* 732, 733, *also n.* 238 *and* 333, *Preiss*, *n.* 2008, and according to Miquel, *n.* 2003, 2005 *and* 2009, *A. Cunningham*, *Baxter*, *F. Mueller* and others; Murchison river, *Oldfield;* Dirk Hartog s Island, *Milne.*

C. ramuliflora, Otto and Dietr. Miq. Rev. Cas. 38, t. 4 A, is believed by Miquel to be an abnormal cultivated form of *C. humilis* with 6-merous whorls and terminal cones. It may however perhaps be rather referrible to *C. suberosa.*

12. **C. torulosa,** *Ait. Hort. Kew.* iii. 320. A small tree, diœcious or sometimes monœcious, with more slender branchlets than any other species except some forms of *C. suberosa*, the ribs scarcely prominent. Whorls 4-merous or very rarely (only in cultivated specimens?) the parts increasing to 5, the sheath-teeth very short. Male spikes very slender, ½ to 1 in. long, terminating deciduous branchlets, compact although the short sheaths scarcely overlap. Cones nearly globular but flat-topped, about ¾ in. diameter, the valves very woody, broad, slightly protruding, villous on the back, the dorsal protuberance divided into numerous small nearly equal tubercles.—Miq. in DC. Prod. xvi. ii. 341, but not of his Rev. Cas.; *C. tenuissima*, Sieb. in Spreng. Syst. iii. 804, Miq. Rev. Cas. 39, t. 4.

Queensland. Brisbane river, Moreton bay, and Burnett river, *F. Mueller;* Stradbrooke island, *Fraser;* Berseker range, *O'Shanesy* (with very corky bark); Rockhampton, *Thozet;* Rockingham bay, *Dallachy;* Mount Elliott, *Fitzalan;* Queensland woods, London Exhibition 1862, *Hill*, *n.* 7.

N. S. Wales. Port Jackson, *R. Brown*, *Sieber*, *n.* 327, *Woolls;* Newcastle, *Leichhardt;* Hastings and Macleay rivers, *Beckler;* Sydney woods, Paris Exhibition 1859, *C. Moore*, *n.* 59, and London Exhibition 1862, *C. Moore*, *n.* 72.

S. Australia. Port Lincoln, *R. Brown.*

13. **C. decussata,** *Benth.* Branches apparently elongated and slender as in *C. torulosa*, to which this species is closely allied, and the

whorls similarly 4-merous, but the angles very acute and two opposite ones more prominent than the two others, giving them the decussate appearance of the Polynesian *C. nodiflora.* Male spikes only seen on one specimen, terminating slender branchlets or here and there sessile, ½ to ¾ in. long, the sheath-teeth acuminate and imbricate. Cones as in *C. torulosa,* nearly globular, flat-topped, ¾ in. diameter, very woody, the broad valves very slightly protruding beyond the rather numerous obtuse small and nearly equal dorsal tubercles, ferruginous-villous between the tubercles.

W. Australia. Towards Cape Riche, *Drummond, 5th coll. n.* 434.

14. **C. Decaisneana,** *F. Muell. Fragm.* i. 61. A tree of 30 to 40 ft. the branches terete, rigid, scarcely ribbed, the internodes above ½ in. long. Whorls 4-merous, the sheath-teeth somewhat paleaceous, lanceolate, fine-pointed, 1 to 2 lines long, or on the young permanent branches often 3 lines, erect and never recurved. Male spikes unknown. Cones ovoid, very shortly pedunculate, 1½ in. long and 1 in. diameter, very woody, tomentose-villous, the thick valves not protruding beyond the broad thick undivided dorsal tubercles.—Miq. in DC. Prod. xvi. ii. 340.

N. Australia. Desert interior, near Mount Mueller, *F. Mueller.*

15? **C. Drummondiana,** *Miq. Rev. Cas.* 26, t. 1 D, *and in DC. Prod.* xvi. 343. A bushy shrub, with the divaricate rigid intricate branchlets of *C. microstachya,* but the whorls are mostly 5-merous, the internodes shorter, less constricted at the nodes, the male spikes 2 to 3 lines long, and mostly supported on branchlets of ¼ to ½ in., although sometimes nearly sessile. Cones unknown.

W. Australia. *Drummond,* probably towards Cape Riche.

16. **C. microstachya,** *Miq. in Pl. Preiss.* i. 642, *Rev. Cas.* 24, t. 1 E, *and in DC. Prod.* xvi. ii. 343. A rigid much branched diœcious shrub, in flower when only 6 in. to 1 ft. high, but said to attain 3 ft., emitting usually at every node short branchlets either of a single internode or again branched, without any of the long simple deciduous branchlets of most species. Whorls 4-merous, the internodes 4-sided, with obtuse often transversely rugose angles, constricted at the nodes, the sheath-teeth small and acute. Male spikes mostly sessile at the nodes, or terminating exceedingly short rigid branchlets, ovoid or globular, 1 or rarely 2 lines long. Cones sessile, globular or ovoid, ¼ to ½ in. long, the valves not protruding beyond the broad dorsal tubercular-rugose protuberances, which are usually divided into 3 scarcely prominent compartments.

W. Australia. King George's Sound and adjoining districts, *Fraser, Preiss, n.* 1997 and 1999, *F. Mueller;* Gordon river, *Maxwell;* Swan river, *Drummond,* 1*st coll., Preiss.*

SECT. 3. ACANTHOPITYS.—Whorls (of stem-angles, sheath-teeth, and flowers) 10–12-merous or 4–5-merous. Cone-valves very prominent, the dorsal protuberance produced into a rigid point as long as or longer than the valve.

17. **C. bicuspidata,** *Benth.* A stout shrub (or tree ?) with numerous whorled erect branchlets resembling those of *C. stricta.* Whorls mostly 10-merous, but probably varying from 9 to 12 parts, the internodes striate, the teeth setaceous. Male amenta terminating long branchlets, but only seen very young, then from ½ to 1 in. long, the sheath-teeth finely setaceous as in *C. trichodon.* Cones larger (1 to 1½ in. long) sessile; ovoid-globular, the valves very prominent, rigidly acuminate, thickened and smooth on the back, the keel-like protuberance of the lower part produced into a free point of the length and shape of the point of the valve itself.

S. Australia. Flinders Island, *R. Brown.*
W. Australia, *Roe,* two cones in the Kew Museum without indication of the precise station.

18. **C. thuyoides,** *Miq. in Pl. Preiss.* i. 641; *Rev. Cas.* 27, t. 2 A, *and in DC. Prod.* xvi. ii. 343. A straggling or spreading monœcious shrub of 3 to 5 ft. in height, with numerous spreading slender branchlets rarely much above 1 in. long, consisting of several short nearly equal terete internodes, the ribs scarcely prominent. Whorls 4-merous or 5-merous. Male spikes usually about 2 lines long, consisting of 3 to 6 sheaths, rather close, with short teeth. Cones sessile or nearly so, globular, about ½ in. diameter, the bracts very short and broad with a minute point, the valves prominent, broad, obtuse, the dorsal protuberance thick at the base but smooth, and produced into a lanceolate-acuminate pungent point much longer than the valve.—*C. echinata,* R. Br. Herb.

W. Australia. Lucky bay, *R. Brown;* Quangen district and Hay river, *Preiss, n.* 2004; King George's Sound and to the eastward towards Cape Riche, *Drummond,* 3*rd coll. n.* 233, *Baxter, Harvey, Maxwell, F. Mueller.*

19. **C. acuaria,** *F. Muell. Fragm.* vi. 16. A shrub? with erect rigid branches, the deciduous branchlets very numorous, erect, slender but rigid, all consisting of only 2 internodes, the lowest only 1 to 1½ lines long, the terminal one 1 in. long or more, ending in a pungent point. Whorls 4-merous. Sheath-scales membranous and very acute. Male spikes (only seen young) clustered at the ends of the branches, already 3 to 4 lines long, but probably longer when in flower, cylindrical, slender, with imbricate sheaths. Cones solitary or clustered 2 to 4 together at the end of more or less developed branches, globular, ¾ to 1 in. diameter. Bracts thickened into a broad triangular apex with a small rigid point. Valves much protruding, obtuse, the dorsal protuberance thick and villous at the very base, produced into an external bract tapering into a long rigid point much exceeding the valve.—*C. oxyclada,* Miq. in DC. Prod. xvi. ii. 344.

W. Australia. *Drummond,* 4*th coll. n.* 240, 241.

Order CXI. PIPERACEÆ.

Flowers hermaphrodite or unisexual, in closely packed spikes or rarely racemes, each with a subtending bract often stipitate. Perianth none. Stamens 2 to 10, inserted on the rhachis or at the base of the ovary. Ovary (except in the tribe *Saurureæ*, which is not Australian) 1-celled, with 1 erect ovule; stigmas 1 to 6, sessile. Fruit a small indehiscent berry. Seed solitary with a farinaceous or fleshy albumen. Embryo minute at the upper end.—Herbs shrubs or climbers, sometimes succulent, often articulate at the nodes. Leaves alternate or rarely opposite or whorled, entire, usually minutely pellucid dotted, with or without stipules. Flowers small, the spikes terminal, leaf-opposed or rarely axillary.

The Order is generally distributed over the tropical and subtropical regions of the New and the Old World, consisting chiefly of the two Australian genera, both of which range over the general area of the Order.

Anthers of 2 distant cells placed back to back, each opening in 2 valves. Stigmas 2 to 4, usually 3. Shrubs trees or climbers, rarely tall herbs . . . . 1. Piper.
Anther cells confluent, the whole anther opening in 2 valves. Stigma tufted. Herbs often small . . . . 2. Peperomia.

The spurious dissepiment which so frequently divides the young anther-cell more or less perfectly into two is usually very prominent in Piperaceæ, and the anther is therefore sometimes described as 4-celled in *Piper*, and 2-celled in *Peperomia*.

1. PIPER, Linn.

(Potomorphe *and* Macropiper, *Miq.*)

Flowers unisexual or hermaphrodite, the subtending bract peltate or adnate. Stamens in the Australian species 2 or 3; anthers of 2 distinct cells placed back to back, each opening outwards in 2 valves. Ovary 1-celled, with a single ovule. Stigmas 3, or in species not Australian 2, 4 or more. Berries sessile or stipitate.—Shrubs trees or woody climbers, rarely tall herbs, the branches usually articulate at the nodes. Leaves alternate. Spikes usually leaf-opposed and solitary, rarely clustered or solitary on short axillary peduncles or branches.

The genus ranges over all tropical lands, slightly extending to the southward in Australasia and S. America. Of the six Australian species one has the wide range of the genus, another is also in New Zealand and the South Pacific islands, the other four are endemic, and there may be one or two more species amongst those of which we have as yet the leaves only.

Spikes solitary or 2 or more together on axillary peduncles or short branches. Bushy erect shrubs or tall herbs.
Flowers hermaphrodite. Leaves 6 to 12 in. broad . . . 1. *P. subpeltatum.*
Flowers unisexual. Leaves 2 to 4 in. broad . . . . . . 2. *P. excelsum.*
Spikes all solitary and leaf-opposed. Stems (in the Australian species) climbing. Spikes unisexual.
Spikes all dense, ½ to 1 in. long. Bracts peltate. Anthers exserted. Berries stipitate. Leaves membranous, glabrous 3. *P. Novæ-Hollandiæ.*
Male spikes unknown. Females 1 to 3 in. long. Berries stipitate. Leaves pubescent underneath . . . . 4. *P. Banksii.*

Male spikes slender, 3 to 5 in. long, the coils of flowers separated by the adnate bracts. Anthers exserted. Females unknown . . . . 5. *P. triandrum.*
Male spikes very dense, 1 to 1½ in. long. Bracts peltate, closely imbricate. Females ovoid-oblong, ¼ in. long. Leaves rather coriaceous . . . . 6. *P. hederaceum.*

1. **P. subpeltatum,** *Willd.; Cas. DC. Prod.* xvi. i. 333. A "large herb or shrub of 3 to 5 ft.," glabrous or with a minute pubescence on the petioles and nerves of the leaves and on the inflorescence. Leaves on long petioles, orbicular-cordate or almost reniform, shortly and acutely acuminate, membranous, prominently many-nerved at the base, with 1 or 2 primary veins on each side of the midrib higher up, 6 to 12 in. broad, the petiole shortly sheathing at the base. Spikes usually 2, sometimes 3 to 7 together, shortly pedicellate on a common axillary peduncle of ⅓ to 1 in., and often 2 common peduncles in the same axil, the spikes very dense, 1 to 3 in. long. Bracts peltate, the terminal laminæ triangular or semilunar, hirsute at the margin. Flowers hermaphrodite. Berries obovate- 3-gonous, glandular-pubescent.—*Potomorphe subpeltata,* Miq., Wight Ic. t. 1925.

Queensland. Rockingham bay, *Dallachy.*—Widely spread over the tropical regions of both the New and the Old World.

2. **P. excelsum,** *Forst.; Cas. DC. Prod.* xvi. i. 334. A small bushy glabrous shrub in Lord Howe's Island, attaining 20 ft. in other stations. Leaves petiolate, orbicular-cordate, shortly acuminate, 7- or 9-nerved from the base, 2 to 4 in. diameter, varying in some stations to ovate. Spikes unisexual, solitary or 2 together, terminating short peduncles or branchlets in the upper axils. Males 2 or 3 in. long, about 1½ lines diameter, not nearly so dense as in *P. subpeltatum,* the flowers small and very numerous. Bracts peltate, the lamina orbicular and sessile. Stamens 2 or occasionally 3. Female spikes shorter and thicker than the males. Stigmas 3. Berries glabrous.—*Macropiper excelsum,* Miq. Syst. Pip. 221.

N. S. Wales. Lord Howe's Island, *M'Gillivray, n.* 726 (the specimen numbered 970, from the same collection, quoted in the Prodromus, as from the same station, is from Raoul island in the Kermadoc group).—The species extends over New Zealand, Norfolk Island, and several of the South Pacific islands.

3. **P. Novæ Hollandiæ,** *Miq. Pip. Nov. Holl.* 8 (*from Meddel. K. Akad. Vetensk. Amsterd.* ser. 2, ii.). A "tall dichotomous plant climbing against trees in dense forests." Leaves on the barren branches usually broadly ovate, equally cordate, acuminate, 7-nerved, 3 to 4 in. long; on flowering branches not so broad, unequal at the base, scarcely cordate, and quintupli- or septupli-nerved. Spikes unisexual, leaf-opposed, the males ½ to nearly 1 in. long, on slender peduncles rather shorter than the spike, very dense, 1 to 1½ lines diameter. Bracts orbicular-peltate, the lamina much undulate when the flowers are fully out. Stamens 2 or 3, shortly exserted. Fruiting-spike not longer than the males. Berries ovoid, red, 2 to 3 lines long, on stipules at least twice that length when full-grown.—Cas. DC. Prod. xvi. i. 343.

Queensland. Brisbane river, Moreton Bay, *F. Mueller, W. Hill, Dallachy;* Rockhampton *Thozet;* Rockingham bay, *Dallachy.*

N. S. Wales? Port Macquarrie, *Backhouse* (leaves only).

P. Paramattense, Cas. DC. Prod. xvi. i. 353, from Paramatta, *Huegel,* may probably be the same species, at least I find no character given to distinguish it. I have seen no specimen for comparison.

4. **P. Banksii,** *Miq. Pip. Nov. Holl.* 9. A tall woody climber, the branches and upper surface of the leaves glabrous or nearly so. Leaves ovate, acuminate, oblique and often slightly cordate at the base, membranous, septuplinerved, 3 to 6 in. long, more or less sprinkled or villous underneath with scattered hairs. Spikes unisexual, the males not seen, the females leaf-opposed, 1 to 3 in. long, on peduncles rarely exceeding ½ in. Berries red, ovoid or almost globular, 2 to 2½ lines long, on stipules of about the same length.—Cas. DC. Prod. xvi. i. 342.

Queensland. Endeavour river, *Banks and Solander;* Mackay and Murray rivers, Rockingham bay, *Dallachy.*

5. **P. triandrum,** *F. Muell. Fragm.* v. 197. A "bushy climber," quite glabrous. Leaves shortly petiolate, ovate, acuminate, unequal at the base but scarcely cordate, membranous, quintupli- or septuplinerved, 3 to 5 in. long. Male spikes leaf-opposed, slender, 3 to 5 in. long. Bracts adnate to the rhachis, with a very small free margin separating the coils or rings of flowers. Stamens 3, the anthers slightly protruding. Female spikes unknown.—Cas. DC. Prod. xvi. i. 365.

Queensland. Mackay river, *Dallachy.*

6. **P. hederaceum,** *A. Cunn.; Cas. DC. Prod.* xvi. i. 366. A "magnificent woody climber ascending to the tops of trees 150 to 180 ft. high," quite glabrous. Leaves shortly petiolate, ovate, shortly acuminate, unequal but not cordate at the base, more coriaceous than in any other Australian species, quintupli- or septupli-nerved, 2 to 3 in. long. Male spikes leaf-opposed, 1 to 1½ in. long, 1 to 1½ lines diameter, very dense. Bracts broadly orbicular, shortly peltate, sessile or nearly so, closely imbricate and concealing the stamens in all our specimens, but perhaps not yet fully out. Stamens 2 (or nearly 3?). Female spikes ovoid-oblong, about ¼ in. long. Bracts like those of the males, but the ovaries shortly protruding, with 3 small stigmas. Very young berries not stipitate.

N. S. Wales. Illawarra, *A. Cunningham, Backhouse;* Sydney woods, Paris Exhibition 1855, *Macarthur, n.* 14.

P. australasicum, Cas. DC. Prod. xvi. i. 353, from Huegel's collection, is probably the same species, remarkable for the firm consistence of its leaves, and very distinct in its floral characters. The supposed persistent character of the anthers, on which account Casimir DC. placed the *P. hederaceum* in his section *Apopiper,* was in this instance founded on the observation of imperfect specimens, as I learn from M. C. De Candolle himself.

2. PEPEROMIA, Ruiz and Pav.

Flowers hermaphrodite, the subtending bract peltate. Stamens 2; anther-cells confluent at the apex, the whole anther opening in 2 valves.

Ovary 1-celled with a single ovule. Stigma tufted or capitate. Berries sessile or scarcely stipitate.—Herbs sometimes very small, rarely tall or shrubby at the base. Leaves alternate opposite or whorled, usually succulent or membranous and pellucid-dotted. Spikes slender, terminal axillary or rarely leaf-opposed. Flowers and berries very small.

The genus is spread over the tropical and subtropical regions of the New and the Old World, but is especially rich in American species. Of the two Australian species one is also in the Pacific islands, the other extends over nearly the whole area of the genus.

Leaves usually opposite, pubescent, thin when dry . . . . . . 1. *P. leptostachya.*
Leaves in whorls of four, glabrous or nearly so, coriaceous when dry 2. *P. reflexa.*

1. **P. leptostachya,** *Hook. and Arn. Bot. Beech.* 96. Stems shortly decumbent and rooting at the base, ascending or erect, from a few inches to about 1 ft. long, more or less pubescent with scattered hairs as well as the foliage. Leaves opposite or rarely appearing whorled from the close approximation of two whorls under the branches, ovate elliptical or obovate, obtuse or obtusely acuminate, under ½ in. long in rocky exposed situations, above 1 in. in shady woods, cuneate or rounded at the base, thinly membranous when dry, 5-nerved but the lateral nerves near the margin and sometimes very faint. Spikes very slender, terminal or in the upper axils, 2 to 3 in. long or even more. Bracts very small, peltate, almost sessile. Berries usually pubescent.—Cas. DC. Prod. xvi. i. 448; Miq. Pip. Nov. Holl. 6; *P. Baueriana* var. *Brisbaniana*, Cas. DC. l.c. 414.

Queensland. Brisbane river, Moreton bay, *F. Mueller;* Rockhampton, common in the scrubs, *O'Shanesy, Bowman, Dallachy;* Rockingham bay, *Dallachy.*

N. S. Wales. Newcastle, *Harvey;* New England, *C. Stuart;* Illawarra, *A. Cunningham.*

The species is also in the Pacific islands, and is very closely allied to an East Indian one.

2. **P. reflexa,** *A. Dietr.; Cas. DC. Prod.* xvi. i. 451. A small erect or diffuse herb, said to be annual, 2–3-chotomously branched, more or less succulent and quite glabrous or slightly pubescent. Leaves in whorls of 4, very shortly petiolate or almost sessile, ovate rhomboidal or almost orbicular, very obtuse, ¼ to ½ in. long, fleshy when fresh, coriaceous when dry, the veins very obscure. Spikes terminal, dense, shortly pedunculate, ½ to 1½ in. long. Bracts almost sessile, peltate. Ovary half-immersed, with a capitate stigma. Berries exserted.—Miq. Pip. Nov. Holl. 7; Wight Ic. t. 1923.

N. S. Wales. Blue Mountains, *Miss Atkinson, Woolls;* Newcastle, *Leichhardt;* New England, *C. Stuart;* Hastings river, *Beckler;* Tweed river, *Guilfoyle;* Illawarra, *A. Cunningham;* Lord Howe's island, *C. Moore.*—Common in most tropical countries.

ORDER CXII. ARISTOLOCHIACEÆ.

Flowers hermaphrodite. Perianth herbaceous, adherent to the ovary at the base, with a superior variously shaped entire or lobed limb, the lobes valvate in the bud. Stamens 5, 6, 8 or more, inserted round the base of the style; anthers 2-celled opening outwards. Ovary inferior, 3- to 6-celled, with several ovules in each cell. Style simple, with an entire or lobed terminal stigma. Fruit a capsule, or rarely succulent. Seeds angular or compressed. Embryo minute, in the apex of a fleshy albumen.—Herbs or rarely shrubs, often climbing. Leaves alternate, without stipules. Flowers usually axillary, solitary or racemose.

A small Order, common to the New and the Old World, chiefly tropical, with a few species dispersed over the temperate regions of the northern hemisphere, and a very few extratropical South American ones. The only Australian genus, the principal one of the Order, has the same general geographical range.

1. ARISTOLOCHIA, Linn.

Perianth constricted over the ovary, then dilated into an utricle enclosing the stamens and pistil, and produced above the utricle into a limb usually unilabiate; or in a few species not Australian unequally bilabiate or 3-lobed. Stamens adnate to the style; anthers 6 or rarely 5, sessile or on very shortly free filaments. Stigma 3-, 5- or 6-lobed. Capsule usually 6-celled and opening in 6 valves from the base upwards, the pedicel itself also splitting.—Climbers or rarely erect herbs or undershrubs. Flowers axillary, solitary or in clusters or racemes. Perianth very large in some species not Australian.

The genus is widely distributed over the warmer and temperate regions of the globe. Of the five Australian species one is a common South Asiatic one, the other four are endemic.

Woody climbers, with coriaceous reticulate leaves.
  Leaves much acuminate, deeply cordate at the base. Perianth-lip broadly triangular . . . . 1. *A. deltantha.*
  Leaves obtuse or scarcely acuminate, scarcely cordate at the base. Flowers unknown . . . . 2. *A. prævenosa.*
Small trailing herbs. Leaves membranous. Flowers solitary. Perianth-lip linear-lanceolate.
  Leaves 1 to 2 in. long, oblong or ovate, cordate . . . . . . 3. *A. pubera.*
  Leaves 2 to 5 in. long, linear or linear-lanceolate . . . . . . 4. *A. Thozetii.*
Climbing herbs. Leaves membranous. Flowers in axillary racemes or clusters . . . . 5. *A. indica.*

1. **A. deltantha,** *F. Muell. Fragm.* vi. 179. A tall climber, apparently woody, the branches at first pubescent, but becoming glabrous. Leaves shortly petiolate, oblong or oblong-lanceolate, acutely acuminate, deeply cordate at the base, 3 to 5 in. long, coriaceous, smooth and shining above, peniveined, 3- or 5-nerved and very strongly reticulate underneath. Pedicels and very young buds very hirsute. Perianth (a single one in Herb. F. Muell.) "very dark in coloured stripes," hirsute, the tube about 4 lines long, the utricle very oblique, scarcely stipitate, the upper portion as long as the utricle, curved and dilated upwards, the lamina or lip forming an almost equilateral obtuse-angled triangle, 5 lines broad. Style hemispherical, with 6 erect stigmatic lobes, without any external transverse ring. Anthers sessile and equidistant. Fruit (only seen in an imperfect state) "yellow, at least

3 in. long and 1 in. diameter, with 6 prominent longitudinal ribs. Seeds and pulp yellow" (*Dallachy*).

Queensland. Rockingham bay, *Dallachy.*

2. **A. prævenosa,** *F. Muell. Fragm.* ii. 166. A tall climbing shrub, the young branches and principal veins of the underside of the leaves ferruginous-pubescent or nearly glabrous. Leaves petiolate, from ovate-oblong and 3 to 4 in. long, to narrow-oblong and 8 to 10 in., obtuse or obtusely acuminate, rounded or slightly cordate at the base, coriaceous, shining above, penniveined and more or less distinctly 3- or 5-nerved, with numerous prominent transverse and reticulate veinlets underneath. Flowers unknown. Capsule (according to F. Mueller) ovoid-globular, 1 in. long. Seeds compressed, tubercular-rugose.—Duchartre in DC. Prod. xv. i. 496.

N. S. Wales. Clarence river, *Beckler.*—The only specimens I have seen are without flowers or fruit.

3. **A. pubera,** *R. Br. Prod.* 349. A small prostrate or trailing herb, with an apparently perennial base, the stems sometimes attaining 1 to 2 ft., sometimes not above 2 or 3 in., more or less pubescent, or sometimes quite glabrous. Leaves on rather long petioles, ovate ovate-lanceolate or oblong-pandurate, obtuse, cordate at the base with broad rounded auricles, 1 to 2 in. long. Pedicels axillary, solitary, 1-flowered, 1 to 2 lines long below the ovary, which at the time of flowering is scarcely distinguishable from it, usually bearing a very small bract near the base. Perianth "dark crimson," pubescent or glabrous outside the tube, 6 to 7 lines long, shortly constricted below the oblique utricle, slender and cylindrical above it, the lip lanceolate, about as long as the tube. Style broadly hemispherical, with 6 short narrow erect stigmatic lobes, surrounded at the base by a ring of gland-like horizontal lobes, immediately over the sessile anthers. Capsule stipitate, obovoid-globular, about $\frac{3}{4}$ in. long when perfect, but in some specimens much smaller. Seeds cymbiform, the outer convex surface elegantly tubercular-punctate.—*A. strictiflora*, Duch. in DC. Prod. xv. i. 484.

Queensland. Brisbane river, *W. Hill;* Rockhampton, *O'Shanesy;* Elliot river and Nerkool creek, *Bowman;* Rockingham bay, *Dallachy;* in herb. *R. Brown* without indication of the station.

N. S. Wales. Clarence river, *Beckler.*

4. **A. Thozetii,** *F. Muell Fragm.* ii. 167. A trailing or prostrate herb, closely allied to and perhaps a variety of *A. pubera*, usually rather stouter. Leaves linear or linear-lanceolate, broader at the base and cordate or hastate, usually 2 to 4 in. long; a few of the lower ones rarely almost ovate-lanceolate. Flowers apparently the same as in *A. pubera*, but not seen perfect. Fruit of *A. pubera*, and equally variable in size.—Duch. in DC. Prod. xv. i. 484.

Queensland. Rockhamptom, *Thozet, O'Shanesy;* Keppel bay, *Thozet;* Herbert's Creek, *Bowman;* Rockingham bay, *Dallachy.*

Var.? *angustissima.* Leaves very narrow and not dilated at the base, pedicels more slender and bracts longer. Flowers only seen very young.

N. Australia. Port Darwin, *Schultz, n.* 547, a single specimen. *A. Baueri,* Duch. in DC. Prod. xv. i. 484, is probably the same species.

5. **A. indica,** *Linn.; Duch. in DC. Prod.* xv. i. 479, *var.?* magna, *F. Muell. Fragm.* vi. 180. A tall but apparently herbaceous glabrous twiner. Leaves in the typical form usually ovate-oblong or almost obovate, obtuse, truncate or cordate at the base, and 2 to 3 in. long; in the Australian variety broader, acutely acuminate, more deeply cordate, 3 to 6 in. long, membranous, 5- or 7-nerved. Flowers in short axillary racemes, sometimes almost contracted into clusters, the pedicels usually longer than the common rhachis, and each with a small bract at the base. Perianth not seen in the Australian variety, in the typical form with a nearly globular utricle of 2 lines diameter, the tube slender and about 3 lines long above the utricle, the lip oblong-linear or narrow-lanceolate, obtuse, ¾ to near 1 in. long, slightly dilated and almost 2-auriculate at the base. Capsule ovoid, 1 to 1½ in. long. Seeds flat, obtusely triangular, slightly tuberculate in the centre, with a broad smooth margin.—R. Br. Prod. 349.

Queensland. Endeavour river, *Banks and Solander (Herb. R. Brown); Rockingham bay, Dallachy.*—The latter specimens are in leaf only with two detached capsules, the identification is therefore doubtful. The species is widely distributed over East India and the Archipelago.

ORDER CXIII. CUPULIFERÆ.

Flowers monœcious. Males in spikes or catkins. Perianth of 1 or several usually unequal scales, segments, or lobes. Stamens 1 or more, with or without a central rudimentary pistil; filaments slender; anthers 2-celled. Female flowers solitary or few together, surrounded by scales or bracts either remaining free or more frequently united in an entire or lobed involucre often enclosing the fruits, and sometimes growing out into setæ or prickles. Perianth-tube adnate to the ovary, the limb usually 6-toothed. Ovary inferior, 1-celled or more or less perfectly 3- or more-celled. Styles as many as cells, simple, stigmatic in the upper portion. Ovules 1 or 2 in each cell, erect or pendulous. Fruit consisting of one or more nuts placed upon, or more or less enclosed in, the usually enlarged persistent involucre. Seeds usually solitary in each nut, without albumen. Embryo various, the radicle usually superior.—Trees or shrubs. Leaves alternate, penniveined, with or without stipules. Male catkins usually falling off entire.

The Order is almost limited to the northern hemisphere in the New as well as the Old World, with the exception of the single Australian genus, which belongs to the temperate and colder regions of both the northern and the southern hemispheres.

1. FAGUS, Linn.

Male flowers in globular pendulous catkins within small scales falling off very early, or rarely solitary. Perianths shortly stalked within each catkin-scale, campanulate, 4- to 6-lobed, containing 8 to 16 stamens, with protruding filaments. Female catkins globular, almost sessile, the scales linear, with numerous closely-packed filiform inner scales, all empty except the innermost and forming an involucre round 2 to 4 sessile flowers in the centre of the catkin. Perianth-limb of 4 or 5 short lobes. Ovary 3-celled, with 2 pendulous ovules in each cell. Styles 3. Nuts 2 to 4, angled or winged, enclosed in a hard prickly involucre, composed of the combined scales of the catkin, and opening in 4 valves.—Trees or rarely shrubs. Leaves alternate, coriaceous, penniveined, frequently plicate and toothed. Stipules usually deciduous. Male catkins usually in the lower axils, the females in the upper ones.

The genus comprises 2 or 3 European or North American species, and several Antarctic ones from South America and New Zealand. The Australian species are all endemic.

Leaves strongly plicate, with the veins very prominent underneath. Stipules persistent, saccate at the base . . . . 1. *F. Gunnii.*
Leaves flat, coriaceous, the veins slightly prominent. Stipules membranous, very deciduous.
Leaves of flowering branches broad, very obtuse, $\frac{1}{4}$ to $\frac{1}{2}$ in. long . . . . 2. *F. Cunninghamii.*
Leaves of flowering branches ovate, shortly acute, $\frac{3}{4}$ to 1 in. long . . . . 3. *F. Moorei.*

1. **F. Gunnii,** *Hook. f. in Hook. Ic. Pl. t.* 881; *Fl. Tasm.* i. 346. A dense bushy shrub, often covering the ground with an impenetrable scrub 5 to 8 feet high, the young branches minutely pubescent and the foliage usually sprinkled with a few hairs on the ribs. Leaves very shortly petiolate, broadly ovate or almost orbicular, regularly crenate, strongly plicate, the primary veins very prominent underneath, mostly about $\frac{1}{2}$ in., rarely $\frac{3}{4}$ in. long. Stipules from a gibbous almost saccate base lanceolate-subulate and more persistent than in any other species. Flowers unknown. Fruiting involucres solitary and sessile in the upper axils, ovoid, about 4 lines long, the valves lanceolate, obtuse, coriaceous, each with 4 to 6 recurved obtuse dorsal scales (or tips of the adnate scales). Nuts usually 3, broadly 3-winged, or the central one 2-winged.—A. DC. Prod. xvi. ii. 120.

Tasmania. Summit of Mount Olympus, at an elevation of 4600 to 5000 ft., *Gunn;* Mount Lapeyrouse, *C. Stuart;* Mount Sorrell, Macquarrie harbour, *Milligan.*

2. **F. Cunninghamii,** *Hook. Journ. Bot.* ii. 152, t. 7. A tree attaining a height of 200 ft. and a girth of 40 ft. with a very dense foliage, reduced on high bleak mountains to a dense shrub of a few feet, glabrous or the branches minutely pubescent. Leaves evergreen, very shortly petiolate, broadly ovate deltoid rhomboid or orbicular, flat or slightly convex, coriaceous, the midrib and primary veins scarcely prominent underneath, $\frac{1}{4}$ to $\frac{1}{2}$ in. long, or on barren shoots twice that size and less coriaceous. Stipules membranous, very deciduous. Male flowers solitary and pedicellate in the lower axils of small axillary branchlets. Perianth 6-lobed, about $1\frac{1}{2}$ lines long.

Stamens about 8, the anthers oblong, longer than the filaments. Female involucre sessile in the upper axils, containing each 3 flowers. Perianth-limb of 3 small teeth on the angles of the ovary. Stigmas capitate. Fruiting involucre about three lines long, the dorsal scales or recurved tips of the catkin-scales narrow and terminating in a gland. Nuts usually 2 with 3 wings and a central flat one with 2 wings.—Hook. f. Fl. Tasm. i. 346; A. DC. Prod. xvi. ii. 122.

Victoria. Upper Yarra, Mounts Useful and Bawbaw, Cape Otway, Wilson's Promontory, *F. Mueller*.

Tasmania. Derwent river, *R. Brown;* common especially in mountainous and western humid districts, forming a large proportion of the forest and ascending to 4000 ft. as a shrub.—"Myrtle tree" of the colonists, *J. D. Hooker*.

3. **F. Moorei,** *F. Muell. Fragm.* v. 109. A "beautiful tree 150 ft. high, the trunk of some of them 70 feet to the branches," closely allied to *F. Cunninghamii*, but at once known by the leaves which are 1 to 2 in. long on the barren shoots, $\frac{3}{4}$ to 1 in. on the flowering branches, ovate or ovate-lanceolate, acute or a few only of the lowest obtuse, flat and coriaceous as in *F. Cunninghamii*, but with more numerous and rather more prominent primary veins. Flowers unknown. Fruits of *F. Cunninghamii*, but larger, the involucres mostly about 5 lines long.

N. S. Wales. On high mountain slopes forming dense forests at the head of Bellinger river and Bealsdown Creek, a few at the head of Macleay river, *C. Moore* (in herb. F. Muell.).—Received also from Mr. Moore himself under the name of *F. Caronii*, Moore.

ORDER CXIV. SANTALACEÆ.

Flowers hermaphrodite or more or less diœcious. Perianth-tube adnate to the ovary, either entirely or at the base only, or the adnate part reduced to the broad base of the ovary; simple or showing a slightly prominent ring outside the limb above the ovary, the limb of 3 to 5 rarely 6 segments, valvate in the bud. Stamens as many as the lobes and opposite to them, inserted at their base or within the free part of the tube; anthers 2-celled, the cells opening longitudinally, sometimes confluent and apparently opening in 4 valves. Ovary inferior or if superior attached by a broad base, usually 1-celled, with 2 to 5 ovules suspended from a free placenta, but the structure often obscure or apparently homogeneous before fecundation; stigma terminal, entire or lobed, sessile on the ovary or borne on a short style. Fruit an indehiscent nut drupe or berry, with a single erect seed. Albumen fleshy; embryo straight, usually very small near the top of the albumen, with a superior radicle and small cotyledons.—Herbs or shrubs rarely trees, usually glabrous or nearly so. Leaves alternate or rarely opposite, entire, without stipules, often reduced to minute scales or very deciduous. Flowers usually small sometimes minute, in terminal or lateral heads cymes or spikes, or rarely solitary.

The Order is widely dispersed over the temperate parts of the globe with a few tropical species. Of the seven Australian genera one has a general distribution over the

P 2

extra-tropical regions of the Old World, two extend into New Zealand and tropical Asia, and four are limited to Australia. It is nearly related to *Olacineæ*, above described in the first volume, and to *Loranthaceæ*, in the third volume; the latter Order might indeed have been better placed in the present group.

Perianth-tube adnate at the base, with the upper campanulate ovoid or cylindrical portion superior.
  No scales or glands between the stamens. Fruit a small nut. Herbs or small shrubs, with alternate linear leaves . . . . 1. Thesium.
  Scales or glands alternating with the stamens at the base of the perianth-limb. Fruit a drupe. Trees or shrubs, with flat leaves mostly opposite . . . . 2. Santalum.
Perianth-tube adnate, the lobes divided to the ovary or to a broad epigynous disk.
  Anthers with 2 parallel cells opening longitudinally. Fruit a drupe. Trees or shrubs, with opposite flat leaves . . . . 3. Fusanus.
  Anthers terminal, with cells confluent, opening in 2 or 4 lobes. Leaves alternate, minute and scale-like, or deciduous or none. Flowers minute.
    Flowers solitary or clustered, surrounded by 2 to 4 scale-like bracts . . . . 4. Choretrum.
    Flowers in little spikes or clusters or rarely solitary, each subtended by a single scale-like bract often very deciduous 5. Leptomeria.
  Anthers with 2 parallel cells opening longitudinally. Habit and inflorescence of *Leptomeria* . . . . 6. Omphacomeria.
Perianth inferior, the lobes divided to the broad base or dilated summit of the peduncle.
  Flowers diœcious, usually 3- or 4-merous, the females solitary, the males in clusters. Leaves alternate, linear-terete or minute and scale-like . . . . 7. Anthobolus.
  Flowers polygamous, inflorescence and foliage of *Leptomeria*, or the leaves developed and flat . . . . 8. Exocarpus.

(The flowers in the last five genera very much smaller than in the first three.)

1. THESIUM, Linn.

Flowers hermaphrodite. Perianth-tube adnate at the base, the free portion campanulate or tubular, with 5 or rarely 4 persistent lobes, with a tuft of hairs inside at the base of or behind the stamens. Stamens inserted near the base of the lobes; filaments short; anthers with 2 parallel cells opening longitudinally. Ovary inferior; placenta filiform often flexuose, with 3 ovules suspended from near the apex; style more or less elongated with a terminal obtuse or capitate stigma. Fruit a small nut, usually ribbed outside and crowned by the persistent free portion of the perianth.—Herbs or small shrubs, with slender wiry stems. Leaves alternate, usually linear. Flowers small, of a greenish yellow, solitary or in small cymes, pedunculate in the axils, but the peduncle usually adnate at the base to the subtending leaf, with 2 bracts on the short free portion.

The genus is widely dispersed over the temperate and warmer regions of the Old World, the species particularly numerous in South Africa. The only Australian species appears to be the same as an East Asiatic one.

1. **T. australe,** *R. Br. Prod.* 353. A glabrous perennial, with ascending or erect wiry branches, rarely above 1 ft. high. Leaves

linear, often above 1 in. long, but the upper ones much shorter and more slender, and a few of the lowest short and broad. Peduncles 1-flowered, very short, adnate at the base to the subtending leaf. Perianth scarcely above 1 line when in flower, cylindrical; lobes 5, about as long as the tube. Style reaching to the level of the anthers, with a capitate stigma. Nut ovoid or nearly globular, 1 to $1\frac{1}{4}$ lines long, marked when dry with 8 to 10 longitudinal ribs more or less branched into intermediate reticulations, and crowned by the small persistent upper portion of the perianth.—A. DC. Prod. xiv. 653; Hook. f. Fl. Tasm. i. 337.

Queensland. Moreton bay, Dawson and Burnett rivers, *F. Mueller.*

N. S. Wales. Nepean river and Cow pastures, *R. Brown;* Hunter's river, *American Exploring Expedition;* Bokhara Creek, *Leichhardt;* Arne river, *Beckler;* New England, *C. Stuart.*

Victoria. Port Phillip, *R. Brown;* Delatite and Ovens rivers, Lake Omeo, *F. Mueller;* Wendu Vale, *Robertson.*

Tasmania. Derwent river, *R. Brown;* also *Laurence.*

I am unable to distinguish from this species the *T. chinese,* Turcz.; A. DC Prod. xiv. 649, from N. China, or the *T. decurrens,* Bl., A. DC. l.c. 652, from Japan, Formosa, and Loo-choo. Taking the flowers at the same age, I find the same shape and proportions of the perianth-tube and lobes in all three.

2. SANTALUM, Linn.

Flowers hermaphrodite. Perianth-tube adnate at the base, the free portion campanulate or ovoid, lined with the disk which is entirely adnate but produced between each two stamens into a spathulate or ovate triangular scale, the lobes 4 or rarely 5, with a tuft of hairs inside behind each stamen. Stamens inserted at the base of the lobes, the filaments usually longer than the intervening scales; anthers 2-celled, the cells parallel turned inwards and opening in longitudinal slits. Ovary semi-inferior, with an erect placenta with 2 or 3 adnate ovules free only at the lower extremity, the placenta produced above them into a long point. Style elongated, with a small 2- or 3-lobed stigma. Fruit a globular drupe, the epicarp fleshy but not thick, marked above the middle or on the summit with a circular scar left by the deciduous perianth-lobes, the endocarp hard and usually rugose.—Trees or shrubs. Leaves opposite or rarely alternate, petiolate, entire, coriaceous or somewhat fleshy, the midrib only or when old a few lateral pinnate veins conspicuous. Flowers rather larger than in the other genera, in small axillary or terminal trichotomous panicles usually shorter than the leaves and sometimes almost reduced to simple racemes. Bracts very small and scale-like, falling off long before the flowers open.

Besides the Australian species which appear to be endemic, there are a few nearly allied ones in East India, the Eastern Archipelago, and the N. and S. Pacific islands. The species require a careful revision from complete specimens; many of those in herbaria, especially the Australian ones, are very doubtful, from the absence of flowers or of fruits, or of both.

Flowers several in the panicle. Perianth 3 lines long or more.
Fruit $\frac{1}{2}$ to $\frac{3}{4}$ in. diam., marked with the scar of the perianth-lobes considerably below the summit . . . . 1. *S. lanceolatum.*

Flowers few (rarely above 3) on the peduncles. Perianth rarely 2 lines long. Fruit 3 to 4 lines diameter, the terminal scar enclosing a small area.

Leaves ovate, usually broad and rather acute . . . . . . 2. *S. ovatum.*
Leaves oblong or lanceolate, obtuse . . . . 3. *S. obtusifolium.*

1. **S. lanceolatum,** *R. Br. Prod.* 356. An erect shrub from 2 or 3 to 15 ft. high, or sometimes a small tree, with pendulous or spreading branches. Leaves mostly oblong or lanceolate, rather acute and shortly tapering into a petiole of 2 or 3 lines, the limb usually 1½ to 2½ in. long but occasionally varying much in size and breadth, rarely obtuse, the lateral veins often conspicuous on old leaves. Flowers rather large, in trichotomous panicles in the upper axils or at the ends of the branches, rarely exceeding the leaves. Perianth 3 to 3½ lines long, the adnate turbinate base very short, the lobes about as long as the campanulate free portion. Anthers oblong, rather large, on short filaments alternating with broad thick obtuse or spathulate scales or glands. Drupe obovoid-globular, ½ to ¾ in. diameter, the circular scar of the limb considerably below the summit.—A. DC. Prod. xiv. 682; F. Muell. Fragm. i. 85; *S. oblongatum,* R. Br. Prod. 355, A. DC. l.c. 683.

N. Australia. Islands of the Gulf of Carpentaria, *R. Brown;* Upper Victoria river, *F. Mueller;* Port Darwin, *Schultz, n.* 517.
Queensland. Endeavour river, *Banks and Solander, A. Cunningham;* Port Denison, *Fitzalan, Dallachy;* Rockhampton, *Thozet,* and others; Balandool river, *Looker;* Nerkool Creek, *Bowman;* Armadillo, *Barton.*
N. S. Wales. Gwydir river, *Leichhardt.*

S. venosum, R. Br. Prod. 355; A. DC. Prod. xiv. 683, from Arnhem bay, *R. Brown,* does not appear to be specifically distinct from the broad-leaved forms of *S. angustifolium.*

Var *angustifolium.* Leaves narrow, often with a fine incurved point.
N. S. Wales. From the Darling river to Cooper's Creek, *Dallachy and Goodwin, Neilson;* New England, *C. Stuart*

2. **S. ovatum,** *R. Br. Prod.* 355. An erect shrub of 7 to 10 ft. Leaves ovate, mostly acute, rather thin, 1 to 2 or rarely 2½ in. long, on rather long petioles. Flowers few together in loose cymes scarcely longer than the petioles, either in the upper axils or at the ends of the branches, and then 3 cymes together a central and 2 lateral ones, the pedicels very short, and often only 3 flowers on each peduncle. Perianth dark purple, scarcely 2 lines long, the adnate base turbinate, the free part of the tube campanulate; lobes 5 or 4, scarcely as long as the free part of the tube. Scales or glands alternating with the stamens, large, ovate, dark purple. Stigma distinctly 3-lobed. Fruit globular, 3 to 4 lines diameter, the scar of the apex enclosing a small area.—A. DC. Prod. xiv. 683.

N. Australia. Arnhem N. bay, *R. Brown;* Port Darwin, *Schultz, n.* 678, 714, 771.—I have only seen fruiting specimens in Herb. Brown and Banks, the description of the flowers and some other particulars are taken from Brown's MS. notes, and from Schultz's specimens which are not in fruit. The species appears to have the fruit of *S. obtusifolium,* with the foliage near that of some forms of *S. lanceolatum,* of which it may possibly prove to be a very broad-leaved small-fruited variety.

3. **S. obtusifolium,** *R. Br. Prod.* 356. A tall slender shrub of livid aspect (*A. Cunn.*). Leaves opposite or the uppermost rarely alternate, linear-oblong lanceolate or broadly oblong, obtuse, rather thick, the margins often revolute in drying, 1 to 2 in. long. Flowers few, in small, shortly pedunculate axillary racemes or cymes, the short pedicels or lateral branches rarely bearing 2 or 3 flowers. Perianth not 2 lines long, the tube campanulate, shortly adnate at the base, the free part much longer; lobes shorter than the tube, triangular, concave. Scales or glands alternating with the stamens, ovate or triangular. Ovary semisuperior, with an elongated style, the placenta acuminate as in *S. lanceolatum.* Fruit globular, 3 or 4 lines diameter, the scar of the apex enclosing a small area.—A. DC. Prod. xiv. 682.

N. S. Wales. Hawkesbury river, *R. Brown;* Paramatta, rare, *Woolls;* Illawarra, *A. Cunningham, Shepherd.*

3. FUSANUS, Linn.

(Eucarya, *Mitch.*)

Flowers hermaphrodite. Perianth-tube adnate, turbinate, shortly produced beyond the ovary into a broad open free portion, lined by the sinuately 4-lobed disk, the margin of which is continuously free inside the stamens, the perianth-lobes 4, with a tuft of hairs behind each stamen. Filaments short, inflected over the notches of the disk; anthers short, with 2 parallel cells opening longitudinally. Ovary inferior, with an erect rather thick placenta, scarcely acuminate at the top, the 2 or 3 adnate ovules distinct only at the base and the whole difficult to separate from the fleshy ovary before fecundation. Style very short and conical or scarcely any, with 2 or rarely 3 distinct terminal stigmas. Fruit a globular drupe, crowned by the persistent perianth-lobes or rarely by the scar only of the fallen lobes, the epicarp more or less fleshy or succulent, the endocarp hard and usually rugose or pitted.—Trees or shrubs with the habit foliage and inflorescence of *Santalum,* but with smaller flowers. Bracts small and very deciduous so as to be rarely seen.

The genus is limited to Australasia, there being besides the Australian species only one known from New Zealand. It is united by De Candolle with *Santalum,* but the perianth, the disk, the ovary, and the style appear to me to be sufficiently distinct to maintain the Linnæan genus adopted by Brown, and to show an approach almost as much to *Choretrum* as to *Santalum.*

Leaves mostly acuminate. Panicles terminal. Perianth-lobes persistent till the fruit is nearly ripe . . . . 1. *F. acuminatus.*
Leaves mostly acute. Panicles axillary. Perianths pedicellate, the lobes falling off immediately after flowering . . . . . 2. *F. persicarius.*
Leaves mostly obtuse. Panicles axillary, dense. Perianths nearly sessile, the lobes persistent till the fruit is nearly ripe . . . 3. *F. spicatus.*
Leaves mostly obtuse. Fruits nearly sessile, the lobes falling off immediately after flowering . . . . 4. *F. crassifolius.*

1. **F. acuminatus,** *R. Br. Prod.* 355. A tall shrub or a tree of 20 to 30 ft. Leaves opposite, lanceolate, acute or sometimes when young

with a short hooked point, mostly 2 or 3 in. long and tapering into a petiole of 2 or 3 lines, but very variable in size and breadth, coriaceous, with the lateral veins often prominent when old. Flowers rather numerous, in a terminal pyramidal panicle scarcely longer than the leaves, but in some of the western specimens much reduced. Perianth spreading to about 2½ lines diameter, the lobes somewhat concave even when open. Free margin of the disk very prominent, broadly rounded between the stamens which curve over the notches. Anthers very short. Style exceedingly short and conical or scarcely any, with a deeply 2- or 3-lobed stigma. Fruit globular, ½ to ¾ in. diameter, with a succulent epicarp, and a hard bony much pitted endocarp, the perianth-lobes persisting on the top till the fruit is nearly or quite ripe.—*Santalum acuminatum*, A. DC. Prod. xiv. 684; *S. Preissianum*, Miq. in Pl. Preiss. i. 615, A. DC. l.c., F. Muell. Fragm. i. 85; *S. cognatum*, Miq. l.c. 616, A. DC. l.c.; also *Fusanus acuminatus*, Miq. l.c. 617, according to F. Muell.

N. S. Wales. Murray and Darling rivers, *Victorian and other Expeditions*; Mudgee, *N. Taylor*.

Victoria. N.W. district, *L. Morton*; Mount Korong, *Herb. F. Mueller*; also in *Mitchell's* collection.

S. Australia. Memory Cove and Port Lincoln, *R. Brown*; Murray river near Morunda, *F. Mueller*.

W. Australia. Fowler's bay, *R. Brown*; from the Kalgan to Swan and Murchison rivers, *Oldfield*, *Maxwell*, *Drummond*, *n.* 227, 152, *Preiss*, *n.* 2098, 2102.

Var.? *angustifolia*. Leaves narrow. Flowers rather larger and fewer. Fruit not seen. – *Santalum angustifolium*, A. DC. Prod. xiv. 685.—W. Australia, *Drummond*, *n.* 430, and perhaps also *n.* 218.

2. **F. persicarius**, *F. Muell.* A tall shrub or small tree, with opposite or scattered lanceolate or linear leaves, often very much like those of *F. acuminatus*, but the lateral veins less conspicuous or quite evanescent, and more frequently terminating in a hooked point. Panicles mostly or all in the upper axils and shorter than the leaves, the primary branches bearing usually a small cyme of few flowers. Perianth almost rotate, opening to about 2 lines diameter, the margin of the disk less prominent than in *F. acuminatus*, but the structure of the flowers otherwise the same. Fruit globular, differing according to F. Mueller in the epicarp not succulent and the endocarp minutely not coarsely pitted, a character rarely to be verified in dried specimens, which however show that the perianth-lobes fall away immediately after flowering, leaving only a scarcely prominent circular scar enclosing a larger terminal area than the persistent lobes of *F. acuminatus*.—*Santalum persicarium*, F. Muell. in Trans. Vict. Inst. 1855, 41; Fragm. i. 86.

Victoria. Murray desert, *F. Mueller*; Wimmera, *Dallachy*; Mount Dispersion, *Mitchell*; Wendu vale, *Robertson*.

S. Australia. Near Mount Baker, *F. Mueller*.

W. Australia. Gordon river, *Maxwell*.

F. diversifolius, Miq. in Pl. Preiss. i. 617 (*Santalum diversifolium*, A. DC. Prod. xiv. 684) from W. Australia, Preiss, n. 2111, will probably prove to be the same species, and the name would have the right of priority, but the several species of *Fusanus* are so imperfectly represented in our herbaria, that it is as yet impossible to establish correctly their respective limits.

Eucarya Murrayana, Mitch. Three Exped. ii. 100, with a woodcut, is most probably this species.

3. **F. spicatus,** *R. Br. Prod.* 355. A tree attaining 30 ft., with spreading but not pendulous branches. Leaves opposite, from oblong-linear to broadly oblong or almost lanceolate, obtuse or rarely acute, contracted into a short petiole, thick, with the midrib usually very prominent underneath, the lateral veins rarely conspicuous, mostly 1 to 2 in. long. Flowers in axillary more or less branching spike-like panicles rarely as long as the leaves, rather crowded and mostly sessile, 3 to 5 together on the short secondary branches. Perianth-tube turbinate, about 1 line long, the broad epigynous disk with a prominent free margin; lobes triangular, about as long as the tube. Style very short, with 2 rarely 3 stigmas or stigmatic lobes. Fruit globular, $\frac{3}{4}$ to nearly 1 in. diameter, crowned till nearly ripe by the persistent perianth-lobes enclosing a rather broad area, the putamen nearly smooth.—*Santalum spicatum*, A. DC. Prod. xiv. 685; *S. cygnorum*, Miq. in Pl. Preiss. i. 615; A. DC. l.c.

S. Australia. Spencer's Gulf, *R. Brown;* Marble Ranges, *Wilhelmi?* (See *F. crassifolius.*)

W. Australia. Salt and Gardner rivers, *Maxwell;* N. of Stirling range, *F. Mueller;* and thence to Swan and Murchison rivers, *Oldfield, Harvey, Preiss, n.* 2103, *Gardner, n.* 226; Shark's bay and Dirk Hartog's island, *Milne.*

4. **F. crassifolius,** *R. Br. Prod.* 355. An erect shrub of 2 or 3 ft., the penultimate branches compressed or 4-angled. Leaves opposite, on very short petioles, linear, obtuse or scarcely acute, entire, fleshy, the margins recurved in the dried specimens, 1 to 2 in. long. Peduncles axillary, few-flowered at the top, the pedicels very short. Drupes 3 to 4 lines diameter in the specimens but apparently unripe, ovoid-globular, crowned by an entire rim enclosing an area smaller than in *F. persicaria.*—*Santalum crassifolium*, A. DC. Prod. xiv. 685.

N. S. Wales. Paramatta and Hunter's river, *R. Brown*, from his Herbarium and MS. notes, the station (M) in the Prodromus being evidently a mistake. The specimens are in fruit only, the fruits solitary, almost sessile, on axillary peduncles of 3 or 4 lines, with the scars of other fallen flowers.

S. Australia? The specimen from Marble Ranges, *Wilhelmi*, quoted above under *F. spicatus*, may possibly belong to *F. crassifolius.*

4. CHORETRUM, R. Br.

Perianth-tube adnate, turbinate, slightly produced above the ovary, lined by the truncate or sinuately 4- or 5-lobed disk, and the border of the tube more or less prominent outside the base of the limb; segments of the limb 4 or 5, of a somewhat different texture from the tube, inflected or thickened at the end. Stamens inserted near the base of the segments; filaments short; anthers terminal, with confluent cells opening out in 4 valves. Ovary inferior, with an epigynous disk lining the free part of the perianth-tube. Style very short, with an entire or slightly 2-lobed stigma. Fruit a globular or ovoid drupe, crowned by the persistent perianth-lobes, the epicarp succulent, the endocarp hard and rather thick.—Shrubs with numerous slender or rigid apparently leafless branches, the leaves all reduced to minute scales usually deciduous. Flowers minute, solitary or in little clusters along the branches, surrounded by an involucre of 2 to 4 or more minute scale-like bracts.

The genus is limited to Australia.

Flowers 2 to 5 together in shortly pedunculate or almost sessile lateral distant clusters . . . . 1. *C. glomeratum.*
Flowers solitary within each involucre.
Branches rigid terete. Flowers approximate in spikes . . . 2. *C. spicatum.*
Branches slender terete. Outer rim of the perianth-tube scarcely prominent . . . . 3. *C. lateriflorum.*
Branches slender acutely angular. Outer rim of the perianth-tube prominent . . . . 4. *C. Candollei.*

C. oxycladum, F. Muell. Fragm. i. 21, from Port Lincoln, *Wilhelmi*, is scarcely referrible to the genus. It is a rigid leafless spinescent shrub, with the aspect of *Leptomeria aphylla*, but the flowers are sessile and solitary or in pairs within very short broad connate bracts. The perianths are all closed and ovoid, a line long or more, and when opened they show 5 valvate lobes without stamens or hairs inside, but in the centre of the flower are very numerous minute stamens without any rudimentary ovary. They may all possibly be in a monstrous state. If not, the plant must belong to some very different Order.

1. **C. glomeratum,** *R. Br. Prod.* 354. An erect shrub, sometimes scarcely 1 ft. high, sometimes almost arborescent, with numerous erect slender wiry angular branches. Leaves reduced to minute subulate deciduous scales. Flowers smaller than in *C. lateriflora*, 2 to 5 together sessile and clustered on a common peduncle of ½ to 1 line, each cluster surrounded by 3 or 4 minute almost orbicular bracts. Perianth about ¾ line long, the broadly turbinate adnate tube not above half the length of the lobes, the external margin very slightly prominent. Drupe when dry 2 to 4 lines long, globular or slightly ovoid.—A. DC. Prod. xiv. 676; Miq. in Pl. Preiss. i. 608; Endl. Iconogr. t. 45.

N. S. Wales. Croker's Range and Wellington valley, *A. Cunningham;* Macquarrie river, *Bowman;* Castlereagh river, *C. Moore;* Mudgee, *N. Taylor.*
Victoria. Light river, Port Elliott, *F. Mueller.*
S. Australia. Memory Cove, Port Lincoln, *R. Brown;* Encounter bay, *Whittaker;* Victoria Lake, *F. Mueller;* Kangaroo island, *Waterhouse.*
W. Australia. *Drummond, 3rd coll. n.* 199, 200; Cape Riche, *Harvey;* Point Henry and Blackwood river, *Oldfield;* base of Stirling Range, *F. Mueller;* Swan river, *Preiss, n.* 2091.

Var. *chrysanthum.* Flowers rather larger and more yellow. I am quite unable to detect any other difference.—*C. chrysanthum*, F. Muell. in Hook. Kew Journ. viii. 205, and Pl. Vict. t. 81, ined.; A. DC. Prod. xiv. 676.—Murray and Avoca desert, *F. Mueller, Behr;* near Wheal Barton Mines, *F. Mueller.*

2. **C. spicatum,** *F. Muell. Fragm.* i. 21. An erect shrub, the branches terete and when in flower thicker and more rigid than in the other species, the young ones slender with numerous closely appressed linear-lanceolate scale-like leaves of about 1 line, which soon fall away. Flowers on the previous year's branches, sessile, solitary at each node

but usually numerous so as to form a rather close spike, each flower surrounded by about 4 broad somewhat jagged bracts, half as long as the perianth, with some smaller ones outside. Perianth about 1 line long, the external rim of the adnate tube prominent. Fruit rather small, ovoid-globular.

Victoria. Murray desert, *F. Mueller;* Wimmera, *Dallachy;* near Reedy Creek, *Irvine;* Wendu Vale, *Robertson.*
S. Australia. Kangaroo island, *Bannier.*

3. **C. lateriflorum,** *R. Br. Prod.* 354. A shrub, with erect slender broom-like branches, terete and slightly striate, the raised decurrent lines not nearly so prominent as in *C. Candollei* and more continuous. Scale-like leaves very small, spreading and more persistent than in that species. Flowers very shortly pedunculate along the ends of the branches, solitary within each subtending bract but surrounded on the peduncle by 4 nearly equal bracts and some smaller ones outside, all shorter than the perianth-tube. Perianth scarcely 1 line long, the lobes about as long as the adnate tube, the marginal rim of which is prominent round the base of the lobes.—*C. pauciflorum*, A. DC. Prod. xiv. 676; *Leptomeria aphylla*, Sieb. Pl. Exs., not of R. Br.

N. S. Wales. Upper Clarence river, *Beckler;* New England, *C. Stuart;* Port Jackson, Berrima and Mittagong, *Woolls;* Lachlan river, *A. Cunningham.*
Victoria. Mitta Mitta, Latrobe and Murray rivers, *F. Mueller.*
W. Australia. King George's Sound, *R. Brown*, the specimens closely resembling F. Mueller's Latrobe river ones.

4. **C. Candollei,** *F. Muell. Herb.* An erect glabrous shrub, attaining several feet, with numerous very slender wiry broom-like branches, acutely angled by short prominently raised lines decurrent from the subulate minute scale-like leaves, which are very deciduous, the branches appearing leafless at the time of flowering. Inflorescence bracts and perianths of *C. lateriflorum*, except that the rim or margin of the adnate perianth-tube is scarcely prominent outside the lobes.—*C. lateriflorum*, A. DC. Prod. xiv. 675, not of R. Br.; *Leptomeria Billardieri*, Sieb. Pl. Exs., not of R. Br.

Queensland. Wide bay, *Bidwill;* Darling Downs, *Law;* Moreton bay, *F. Mueller.*
N. S. Wales. Port Jackson to the Blue Mountains, *A. Cunningham, Sieber, n.* 133, *and Fl. Mixt. n.* 525; Liverpool plains, *Leichhardt;* New England, *C. Stuart;* Castlereagh river, *C. Moore;* southward to Illawarra, *A. Cunningham, Shepherd;* Mudgee, *N. Taylor.*

5. LEPTOMERIA, R. Br.

Perianth-tube adnate, usually narrow, the border not at all or scarcely prominent outside; segments of the limb 5 or rarely 4, inflated or somewhat thickened at the end. Stamens inserted near the base of the segments; filaments short; anthers terminal, with confluent cells opening out into 4 valves or lobes or into an almost entire disk, the centre often very prominent and angled or lobed. Ovary inferior, with a more or less distinct epigynous disk. Style very short, with an entire or slightly lobed stigma. Fruit a small globular or ovoid fleshy or dry indehiscent drupe, crowned by the persistent perianth-lobes.—Shrubs with numerous slender or rigid apparently leafless branches, the leaves all reduced to minute alternate scales usually deciduous, or rarely longer linear terete and persistent. Flowers minute, in little terminal or lateral spikes racemes or clusters, each one sessile or shortly pedicellate in the axil of a single minute scale-like deciduous or persistent bract, or rarely the bracts developed into leaves and the raceme into a leafy branch.

The genus is limited to Australia.

SECT. 1. **Xeromeria.**—*Bracts subtending the flowers falling off long before the flower expands.*

* *Western species.*

Branches spinescent. Spikes dense, few-flowered, the rhachis about 1 line. Fruit dry. Perianth-lobes whitish . . . 1. *L. spinosa.*
Branches rigid, with raised decurrent lines. Spikes loose, the rhachis 2 to 6 lines. Fruit dry. Perianth-lobes whitish . 2. *L. Preissiana.*
Branches slender, terete. Spikes dense, the rhachis 1 to 2 lines. Fruit succulent. Perianth brown yellow . . . . 3. *L. pauciflora.*
Branches angular. Spikes dense, the flowers closely sessile on the notched or excavated rhachis. Fruit dry . . . 4. *L. scrobiculata.*

** *Eastern species.*

Racemes loose, often ½ to ¾ in. long. Perianth-lobes fulvous when dry. Disk not lobed . . . . 5. *L. acida.*
Racemes loose, ½ to ¾ in. long. Perianth-lobes white when dry. Disk lobed . . . . 6. *L. Billardieri.*
Racemes 3 to 4 lines long, rather dense. Perianth-lobes dark when dry. Disk slightly lobed. Branches rigid, often spinescent . . . . 7. *L. aphylla.*
Racemes reduced to dense almost sessile clusters of few flowers 8. *L. glomerata.*

SECT. 2. **Oxymeria.**—*Bracts persistent under the flowers.*

Bracts very spreading and mostly recurved . . . . 9. *L. squarrulosa.*
Bracts lanceolate erect or slightly spreading, not recurved. Stem-leaves persistent . . . . 10. *L. Cunninghamii.*
Bracts concave, contracted at the base. Stem-leaves deciduous 11. *L. empetriformis.*

Species insufficiently known. Stem-leaves persistent, 1½ to 3 lines long; flowers solitary in the upper axils.

Leaves and bracts semiterete, not rigid . . . . 12. *L. axillaris.*
Leaves semiterete, at length pungent. Bracts smaller, elliptical-lanceolate . . . . 13. *L. laxa.*
Leaves and bracts obovate or obovate-oblong . . . . 14. *L. obovata.*

SECT. 1. XEROMERIA, *Endl.*—Bracts subtending the individual flowers falling off long before the flowers expand, and rarely to be seen except in the nascent spike.

1. **L. spinosa,** *A. DC. Prod.* xiv. 678. A much-branched rigid shrub, leafless at the time of flowering, the minute linear-subulate scale-like leaves soon falling off from the young shoots, the adult branches terete, slightly striate with raised lines, the shorter ones often

spinescent at the end. Flowers very small and few in little lateral spikes or clusters, the rigid rhachis rarely 2 lines, and usually scarcely 1 line long, recurved after the flowers are fallen, and then resembling the leaves of *L. squarrulosa.* Perianth scarcely above ½ line long, the narrow turbinate tube sessile or nearly so, separated from the lobes by a distinct line, the lobes apparently white. Anthers nearly of *L. Cunninghamii.* Fruit small, globular, crowned by the persistent perianth-lobes.—*Choretrum spinosum,* Lehm. in Pl. Preiss. i. 609.

W. Australia. *Drummond, n.* (*2nd coll.?*) 226 *and* 230; Hay district, *Preiss, n.* 2105; Gordon and Swan rivers and Cape Naturaliste, *Oldfield; *Fitzgerald Range, *Maxwell;* Stirling Range, *F. Mueller.*—This species as to the perianth connects *Leptomeria* with *Choretrum,* but the inflorescence and bracts are those of *Leptomeria.*

Var.? *leptoclada.* Branches much more slender, rarely spinescent.—Dirk Hartog's island, *A. Cunningham.*

2. **L. Preissiana,** *A. DC. Prod.* xiv. 678. Branches erect, rather rigid, terete, slightly striate with raised lines, leafless at the time of flowering, the minute scale-like linear-subulate leaves on the young slender shoots falling off very early. Flowers in little lateral rather loose racemes, mostly on the previous year's branches, the rhachis 2 to 6 lines. Bracts minute, ovate-lanceolate, slightly denticulate, falling off so early that they are never seen on specimens in full flower. Perianth apparently white, scarcely 1 line long, the narrow turbinate adnate portion tapering into a very short pedicel, the lobes quite continuous without any external rim.—*Choretrum Preissianum,* Miq. in Pl. Preiss. i. 608.

W. Australia. Swan river, *Oldfield, Preiss, n.* 2101.

3. **L. pauciflora,** *R. Br. Prod.* 354. An erect shrub, attaining 3 to 5 ft., with erect or spreading terete branches, quite leafless at the time of flowering, the minute narrow scale-like leaves falling off very early from the young shoots. Flowers in little lateral spikes or racemes, the rhachis 1 to 3 lines long. Bracts minute, broadly ovate or obovate, concave, falling off very early so as to be rarely seen on flowering specimens. Perianth greenish-white, about 1 line long, the lobes as long as the narrow tube. Fruit succulent.—A. DC. Prod. xiv. 680; *L. aphylla,* A. DC. l.c. 677, partly but not of R. Br.; *L. Lehmanni,* Miq. in Pl. Preiss. i. 614, A. DC. l.c. 678.

W. Australia. King George's Sound and adjoining districts, *R. Brown, A. Cunningham, Oldfield, F. Mueller, Drummond, 2nd. coll. n.* 229, *3rd coll. n.* 197; *Preiss, n.* 2107, 2121; Vasse river, *Oldfield.*

4. **L. scrobiculata,** *R. Br. Prod.* 354. A shrub with numerous slender slightly angular branches, leafless at the time of flowering except on the young shoots, where the minute linear-lanceolate acute scale-like leaves persist rather longer than in the preceding species. Flowers rather numerous, in dense spikes, the rhachis varying from 2 to 4 lines in length, each flower sessile and more or less immersed in a cavity of the rhachis, the subtending bracts ovate acute and very deciduous, leaving sometimes a tooth-like scar. Perianth-tube shortly

turbinate, the lobes rather longer. Epigynous disk very prominent, but entire, without the distinct glandlike lobes of *L. Billardieri*. Stigma shortly 5-lobed.—A. DC. Prod. xiv. 680; *L. ericoides*, Miq. in Pl. Preiss. i. 611, A. DC. l.c. 679; *L. chrysadena*, Miq. l.c. 612, A. DC. l.c.

W. Australia. King George's Sound and adjoining districts, *R. Brown*, *Preiss*, *n.* 2117, 2124, *Oldfield*.

Miquel describes the bracts as persistent in his *L. ericoides*, which I do not find to be the case in the specimens I have seen, except at the top of the young spike.

5. **L. acida,** *R. Br. Prod.* 353. An erect broom-like shrub, attaining 6 to 8 feet, the virgate branches much more angular than in *L. Billardieri*, leafless at the time of flowering, the minute linear-lanceolate scale-like leaves falling off very early from the young shoots. Spikes more slender than in *L. Billardieri*, often ½ to ¾ in. long and the flowers much smaller. Perianth-tube narrow-turbinate, tapering at the base, but sessile, the limb of the same fulvous colour as the tube, barely ¾ line diameter when open, the lobes much hooded, with reflexed tips. Epigynous disk somewhat angular but not lobed, stigma minutely lobed.—A. DC. Prod. xiv. 677; Endl. Iconogr. t. 74.

Queensland. Moreton island, *F. Mueller*.

N. S. Wales. Port Jackson to the Blue Mountains, *R. Brown*, *Sieber*, *n.* 132, and many others; northward to Hastings river, *Beckler;* southward to Twofold bay, *F. Mueller*.

Victoria. Genoa Peak, *F. Mueller*.

6. **L. Billardieri,** *R. Br. Prod.* 354. An erect broom-like shrub, attaining sometimes 6 or 7 ft. though sometimes under 2 ft., the branches rather slender and angular, but not so much so as in *L. acida*, leafless at the time of flowering, the minute linear-lanceolate scale-like leaves falling off early from the young shoots. Spikes or racemes lateral, loose, ¼ to ½ in. long. Bracts ovate-lanceolate, acute, concave, falling off long before the flowering. Perianth-tube tapering into a distinct pedicel, the limb whitish when dry, spreading to about 1 line diameter. Epigynous disk distinctly lobed. Stigma minutely 5-lobed. Drupe small, with a fleshy scarcely succulent epicarp.—A. DC. Prod. xiv. 677; Hook. f. Fl. Tasm. i. 357; *Thesium drupaceum*, Labill. Pl. Nov. Holl. i. 68, t. 93.

N. S. Wales. Blue Mountains, *Fraser;* Tweed river, *C. Moore*.—The flowers are smaller than in the Tasmanian specimens, but distinctly stipitate, the perianth-limb white and the disk lobed as in the typical Tasmanian form.

Tasmania. Port Dalrymple, *R. Brown;* in poor moist soil, especially sandy places near the N. coast, *J. D. Hooker*.

7. **L. aphylla,** *R. Br. Prod.* 354. An erect shrub of 3 to 4 ft. with rigid spreading branches often spinescent at the end, quite terete without prominent ridges, leafless at the time of flowering, and I have not succeeded in finding any young shoots with scale-like leaves still persistent. Flowers rather numerous, in lateral racemes of 3 or 4 lines, the rhachis rather thick. Perianth-tube narrow-turbinate, the lobes dark-coloured when dry, opening to nearly 1 line in diameter. Epi-

gynous disk prominent, obtusely angled or almost lobed. Stigma minutely 5-lobed. Fruit ovoid, the epicarp succulent.—A. DC. Prod. xiv. 677 as to Brown's synonym; *L. pungens*, F. Muell. in Trans. Vict. Inst. 1855, 41.

Victoria. Grampians, *Wilhelmi;* Murray river and Mount Korong, *Herb. F. Mueller*

S. Australia. Memory Cove, *R. Brown;* Serra Range, Guichen bay, *F. Mueller;* Bethanie, *Behr.*—In Brown's specimen the flowers are smaller than in the others, but as yet in bud only.

8. **L. glomerata,** *F. Muell.; Hook. f. Fl. Tasm.* ii. 370. A much lower and more rigid shrub than *L. Billardieri*, rarely exceeding 1 ft., the branches thicker, with obtuse slightly prominent angles, the minute scale-like leaves falling off early from the young shoots. Spikes or racemes exceedingly short or reduced to almost sessile clusters of 3 or 4 flowers, the rhachis rarely 2 to 3 lines long. Perianth-tube tapering into a very short pedicel inserted in a slight notch of the rhachis, the lobes white or red when dry, spreading to nearly 1 line in diameter. Epigynous disk obscurely lobed. Stigma slightly 5-lobed.

Tasmania. South Port, *C. Stuart;* south of Huon river, *Milligan.*

L. Billardieri, var. *humilis*, Hook. Fl. Tasm. i. 337, from Lake St. Clair, *Gunn*, appears to be rather this species, with the rhachis of the spike slightly elongated but never exceeding 3 lines.

SECT. 2. OXYMERIA, *Endl.*—Bracts subtending each flower in the spike usually persisting at least until the flowers expand.

9. **L. squarrulosa,** *R. Br. Prod.* 354. A shrub of 1 to 3 ft. with numerous divaricate branchlets, the branches terete, more or less striate with slightly prominent lines. Leaves linear-terete or triquetrous, persistent, more or less recurved at the end, 1 to 1½ lines long. Flowers fulvous, forming little loose lateral spikes of 4 to 8 lines, the bracts or floral leaves subtending each flower persistent at the time of flowering, very spreading or recurved, ovate-lanceolate acute and concave, about ½ line long. Perianth scarcely above ½ line long, the lobes longer than the distinctly pedicellate adnate portion. Fruit globular, smooth, not ribbed, about 1 line diameter, crowned by the persistent perianth lobes.—A. DC. Prod. xiv. 679; *L. Brownii*, Miq. in Pl. Preiss. i. 612.

W. Australia. King George's Sound and adjoining districts, *R. Brown, A. Cunningham, Drummond, 3rd coll. n.* 198, *Preiss, n.* 2109, *F. Mueller*, and others.

10. **L. Cunninghamii,** *Miq. in Pl. Preiss.* i. 611. An erect shrub, attaining 2 or 3 ft., with slender virgate slightly angular branches. Leaves persistent, linear-terete, 2 or 3 or rarely 4 to 5 lines long on the main stems, much smaller and flatter on the flowering branches. Flowers forming little loose leafy spikes, usually ¼ to ½ in. long or rather longer but sometimes growing out into leafy branches, the subtending bracts or floral leaves persistent, scarcely 1 line long, lanceolate, acute, concave, sometimes slightly contracted at the base. Perianth

not 1 line long, the lobes as long as the narrow adnate part which tapers into a short pedicel. Anthers in this and several allied species, although opening in 4 lobes as in all *Leptomeriæ*, have the lobes less distinct than in some species, and the central connective very prominent. Fruit gobular, crowned by the persistent perianth-lobes, small and ribbed when immature, but not seen ripe.—A. DC. Prod. xiv. 679.

W. Australia. Swan river, *Preiss, n.* 2096; Swan and Vasse rivers, *Oldfield;* also *Drummond* (*2nd coll.?*) *n.* 228; King George's Sound, *Muir.*

11. **L. empetriformis,** *Miq. in Pl. Preiss.* i. 610. An erect much branched shrub, of 1 to 2 ft., the branches terete, often more or less sprinkled or covered with minute glandular papillæ but not really hirsute and the papillæ varying much even on different parts of the same specimen. Leaves terete, rather fleshy, contracted at the base, 1½ to 2½ lines long, deciduous on the main branches. Flowers in loose leafy spikes, the subtending bracts or floral-leaves smaller and more persistent than the stem-leaves, rather broad, flattened or concave, acute, much contracted at the base. Perianth about ¾ line long, the lobes at least as long as the tube, which tapers into a very short pedicel.—A. DC. Prod. xiv. 680; *L. hirtella*, Miq. in Pl. Preiss. i. 610; A. DC. l.c. 679.

W. Australia. Swan river, *Preiss, n.* 2094, 2113.

12. **L. axillaris,** *R. Br. Prod.* 354. A divaricately branched shrub of 1½ ft. with terete branches. Leaves linear-terete, persistent, rather thick, 1½ to 3 lines long. Flowers very shortly pedicellate, solitary in the upper axils and very much shorter than the leaves, the raceme forming a leafy branch. Perianth opening to ⅔ line diameter, 5-lobed. Anthers 2-celled. Nut small, globular, crowned by the persistent perianth-lobes.—DC. Prod. xiv. 680.

W. Australia. Rocky hill, Lucky bay, *R. Brown.* I have taken the description of the flowers from Mr. Brown's notes, his specimens have unexpanded flowers, and I only see one nut, from which, however, the perianth-lobes are fallen away. A specimen with a single nut from near Port Enolo, *J. Forest*, in Herb. F. Mueller, may possibly belong to the same species.

13. **L. laxa,** *Miq. in Pl. Preiss.* i. 612. A small shrub with several erect slender simple stems of about ½ ft., leafless at the base. Leaves in the upper part erect, scattered, semiterete, at first soft, at length hardening and pungent, about 4 or 5 lines long, those of the lateral branches shorter. Spikes spreading, the flowers distant, with minute elliptic-lanceolate bracts. Drupe dry, ovoid, a little more than 1 line long, striate, crowned by the remains of the 5 perianth-lobes.—A. DC. Prod. xiv. 678.

W. Australia. *Preiss, n.* 2120. I have not seen any specimen answering to the above, and it was described by Miquel from a single one past flowering and bearing a single fruit.

14. **L. obovata,** *Miq. in Pl. Preiss.* i. 613. A shrub with grey angular branches. Leaves scattered, alternate, obovate or obovate-

elliptical, flat or concave, fleshy, 2½ to 3 lines long. Flowers solitary in the axils of bracts similar to the leaves but smaller, forming leafy spikes of ½ to 1 in. Flowers with their pedicel ½ line long. Lobes of the epigynous disk very prominent.—A. DC. Prod. xiv. 680.

W. Australia. *Preiss, n.* 2108, *Drummond, 4th coll. n.* 254, neither of which specimens have I seen. The above character is taken from Miquel's and De Candolle's.

6. OMPHACOMERIA, A. DC.

Flowers unisexual by abortion. Perianth-tube short, adnate, or none in the males, segments of the limb 4 or 5 scarcely inflexed at the end. Stamens inserted near the base of the segments; filaments short; anthers with 2 distinct parallel cells opening longitudinally, apparently empty or abortive in the females. Ovary inferior, abortive in the males, the epigynous disk more or less conspicuous. Style very short, with a distinctly 2-lobed stigma. Fruit a globular fleshy indehiscent drupe, crowned by the persistent perianth-lobes.—Apparently leafless shrubs with the habit of *Leptomeria* and *Choretrum*, the specimens not showing even the small scales of those genera. Flowers minute, lateral, the females solitary, the males in little clusters, usually on separate specimens, but sometimes a few male clusters on a female specimen, both sexes sessile in a concave disk without distinct bracts.

The genus is limited to Australia.

Branches rigid but rather slender, terete . . . . 1. *O. acerba.*
Branches more rigid and shorter, prominently striate or angled . . 2. *O. psilotoides.*

1. **O. acerba,** *A. DC. Prod.* xiv. 681. An erect broom-like leafless shrub of 2 to 4 ft., with rigid but elongated and rather slender branches, terete and only slightly striate, with gland-like nodes, but I have been unable to discover any leaves on any of our specimens. Female flowers solitary in a concave slightly prominent disk. Ovary thick, scarcely ½ line long, the perianth-lobes scarcely longer and as broad as long, the anthers apparently empty and smaller than in the males. Male flowers in almost sessile clusters of 3 to 5, the perianth without any distinct tube or ovary, the lobes or segments as in the females and the anthers perfect; the central disk flat, with a slightly prominent entire rudimentary style. Drupes ovoid, 3 to 4 lines long, with a succulent epicarp.—*Leptomeria acerba,* R. Br. Prod. 354.

N. S. Wales. Port Jackson to the Blue Mountains, *R. Brown, A. Cunningham,* and others.

Victoria. Genoa Peak and mountains on the Mitta Mitta, *F. Mueller,* specimens both male and female, and on one of the latter a few male clusters.

2. **O. psilotoides,** *A. DC. Prod.* xiv. 681 (*partly*). Very closely allied to *O. acerba,* and perhaps a variety only, with shorter more rigid branches very prominently striate or angled.

N. S. Wales. Blue Mountains? *Sieber, n.* 134; the specimens seem male only, with precisely the clustered flowers of *O. acerba.*

I have not seen any specimens of A. Cunningham's answering to this species. The Tasmanian plant of Gunn's referred to it by A. DC. is the *Leptomeria glomerata,* F.

Muell., which at first sight closely resembles *O. psilotoides*, but the flowers are hermaphrodite, with the perianth-tube, stamens and style of *Leptomeria*, whilst in *Omphacomeria* the anthers are much nearer to those of *Exocarpos*.

7. ANTHOBOLUS, R. Br.

Flowers diœcious. Perianth free, divided to the broad base into 3, 4 or rarely 5 segments, more or less inflected or concave at the end. Male fl.: Stamens inserted near the base of the segments; filaments very short; anthers with 2 distinct cells opening longitudinally, turned inwards in the bud, but opening out back to back. Ovary free, thick and fleshy; stigma sessile, pulvinate, obscurely lobed (or furrowed by the pressure of the margins of the perianth-segments). Drupe ovoid or oblong, sessile on the thickened pedicel, the exocarp succulent but not thick; endocarp crustaceous or rather hard.—Glabrous shrubs, with rigid or slender branches. Leaves either linear-terete and persistent or minute scale-like and deciduous. Flowers very small, pedicellate on a common axillary peduncle, the males usually 3 to 5 together, the females solitary or 2 together. Bracts at the base of the pedicels minute and very caducous.

The genus is limited to Australia.

Leaves linear terete or filiform, persistent.
  Leaves slender. Perianth usually 3-merous. Endocarp not pitted. Tropical species.
    Branches terete or nearly so . . . . 1. *A. filifolius.*
    Branches angular . . . . 2. *A. triqueter.*
  Leaves rather thick. Perianth usually 4-merous. Endocarp pitted. Western species . . . . 3. *A. foveolatus.*
Branches leafless, the minute scale-like leaves falling off from the very young shoots . . . . 4. *A. leptomerioides.*

1. **A. filifolius,** *R. Br. Prod.* 357. A tall shrub, with slender nearly terete branches. Leaves linear-filiform, ¾ to 2 in. long. Male flowers (which I have not myself seen) 3 or 4 together on a common peduncle of 3 lines, the pedicels about 1 line long, the bracts very small and deciduous. Female flowers solitary or 2 together, the common peduncle and pedicel each about 1 line long or often twice as long when in fruit. Perianth 3-merous, scarcely ¾ line long. Ovary thick, with a pulvinate stigma. Fruit ovoid, rather smaller than in *A. foveolatus*, the endocarp smooth, not pitted.—A. DC. Prod. xiv. 687.

N. Australia. Islands of the Gulf of Carpentaria, *R. Brown;* Fitzmaurice river, *F. Mueller.*

2. **A. triqueter,** *R. Br. Prod.* 357. Very closely allied to *A. filifolius*, and probably a variety only, the stems and leaves rather thicker and the young branches slightly angular, the female flowers and fruits rather larger.—A. DC. Prod. xiv. 687.

Queensland. Endeavour river, *Banks and Solander*, a single specimen in Herb. R. Brown.

3. **A. foveolatus,** *F. Muell. Fragm.* i. 212. An erect shrub of 6 to 8 ft., with virgate spreading or sometimes pendulous branches, terete

or slightly angular when young. Leaves all linear-terete, acute, $\frac{3}{4}$ to $1\frac{1}{2}$ in. long. Male flowers 2 or 3 together on axillary peduncles of 3 to 4 lines, the perianth about $\frac{3}{4}$ line long, 4-merous, the disk slightly 4-lobed. Females only seen in fruit, then solitary on a lengthened and much thickened peduncle. Fruit ovoid. 3 or 4 lines long, the endocarp marked with very small scattered pits, otherwise as in *A. leptomerioides*.

W. Australia. Murchison river, *Oldfield, Drummond, 6th coll. n.* 216.

4. **A. leptomerioides,** *F. Muell. Fragm.* i. 21. A shrub of several feet, with rigid broom-like terete branches, often pungent at the extremity, and leafless at the time of flowering, the minute linear scale-like leaves falling off from the very young shoots. Male flowers in sessile clusters of about 4 or 5, each one on a pedicel of $\frac{1}{2}$ to $\frac{3}{4}$ line, the perianth about as long, 4-merous, the anthers rather large. Female flowers solitary (or 2 together?), the perianth broadly cylindrical, very shortly 4-lobed, without any stamens. Ovary thick and fleshy, with a thick pulvinate stigma quite enclosed in the perianth but free. Drupe oblong, 3 to 4 lines long, the exocarp not thick, the endocarp hard and smooth. Embryo straight, linear-terete, more than $\frac{3}{4}$ the length of the albumen, the cotyledons at least as long as the radicle.

Queensland. Burdekin, Suttor and Burnett rivers, *F. Mueller.*

8. EXOCARPUS, Labill.

Flowers hermaphrodite or males by the abortion of the ovary. Perianth free, divided to the broad base into 5 rarely 4 segments, slightly concave at the end. Stamens inserted near their base; anther-cells distinct, adnate to a very short broad filament and either nearly parallel and turned inwards, or divergent and marginal opening longitudinally. Ovary free, thick fleshy and somewhat conical, reduced in the male flowers to a flat disk. Stigma sessile, rather small, entire or obscurely lobed. Drupe or nut ovoid or nearly globular, resting on the enlarged usually succulent pedicel, the epicarp thin and not readily detached from the crustaceous or hard endocarp, the perianth-lobes either persistent round the base of the fruit or deciduous leaving the enlarged apex of the pedicel truncate. Seed erect, with a very thin testa; albumen copious; embryo minute near the apex, slightly divided at the lower end into 2 minute cotyledons.—Trees or shrubs. Leaves alternate or rarely opposite, often reduced to minute scales or very deciduous, rarely enlarged and persistent. Flowers minute, in small axillary spikes sometimes reduced to sessile clusters, each flower sessile or nearly so, in a notch of the rhachis or in the axil of a minute scale-like bract, one only or rarely 2 or 3 in the spike fertile with the pedicel rapidly enlarged, the others falling off without any enlargement of the semi-abortive ovary.

Of the eight Australian species one extends over the Eastern Archipelago, the others are endemic. The genus, has also one species from New Zealand, one from Norfolk Island, one from the Sandwich Islands, and apparently one from Madagascar. Some of the leafless species closely resemble some species of *Leptomeria*, but are at once distinguished by the free ovary. Some species with the fruiting pedicel very succulent are known to the colonists by the name of "native Cherry."

Spikes cylindrical, mostly shortly pedunculate.
Leaves ovate, flat, 1 to 2 in. long . . . . 1. *E. latifolia.*
Leaves numerous, linear, 8 to 10 lines long . . . . . . 2. *E. odorata.*
Leaves reduced to minute tooth-like spreading persistent scales . . . . 3. *E. cupressiformis.*
Leaves linear-subulate, 1 to 2 lines long and deciduous, or rarely rather longer and persistent . . . . 4. *E. spartea.*
Spikes very short and scarcely pedunculate, the rhachis pubescent. Branches stout, often spinescent. Leaves reduced to minute ovate deciduous scales . . . . 5. *E. aphylla.*
Spikes reduced to sessile clusters of 2 or few flowers.
Tall erect shrub, with flattened leafless branches . . . . 6. *E. homaloclada.*
Tall erect shrub, with slender angular branches. Leaves minute, subulate, very deciduous . . . . 7. *E. stricta.*
Procumbent much branched shrub, with terete rigid branches. Leaves reduced to minute alternate tooth-like persistent scales . . . . 8. *E. humifusa.*
Prostrate much branched dwarf shrub. Leaves reduced to minute tooth-like scales mostly opposite . . . . . 9. *E. nana.*

1. **E. latifolia,** *R. Br. Prod.* 356. A small tree, the young parts slightly hoary with a minute stellate or almost scaly pubescence. Leaves alternate, petiolate, from broadly ovate to oval-oblong, very obtuse, coriaceous, with several more or less distinct nerves diverging from the base, 1 to 2 in. long. Spikes rather slender, mostly about ½ in. long, shortly pedunculate, solitary or several in a short raceme in the upper axils. Flowers 5-merous or rarely 4-merous, not closely packed. Fruit ovoid, 3 to 4 lines long, on a thickly turbinate truncate pedicel of above 2 lines.—A. DC. Prod. xiv. 688; *E. miniata*, Zipp. and *E. luzoniensis*, Presl; A. DC. l.c.; *E. ovata*, Schnitzl. Iconogr. ii. t. 108***

N. Australia. Islands of the Gulf of Carpentaria, *R. Brown*, *Henne;* islands and mainland, N. Coast, *A. Cunningham;* Point Pearce and Upper Victoria river, *F. Mueller;* Port Darwin, *Schultz, n.* 358.

Queensland. Keppel bay and Shoal bay, *R. Brown;* Wide bay, *Bidwill;* Burdekin river, *F. Mueller;* Port Denison, *Fitzalan;* Rockingham bay and Rockhampton, *Dallachy;* Bowen river, *Bowman;* Kennedy district, *Daintree* (with leaves 2½ in. long and broad).

N. S. Wales. Tweed river, *Guilfoyle.*

The species is generally dispersed over the Eastern Archipelago to the Philippine islands.

2. **E. odorata,** *A. DC. Prod.* xiv. 689. An erect densely branched shrub. Leaves crowded, linear or linear-lanceolate, acute or obtuse, mostly about ½ in. long on the flowering branches, sometimes twice as long or more on barren ones. Spikes axillary, 2 to 3 lines long in our specimens, said to be twice as long in Preiss's. Bracts minute and tooth-like. Perianth-segments more frequently 4 than 5, triangular, about ¼ line long. Anther-cells nearly globular, parallel. Stigma nearly sessile, rather broad, scarcely lobed. Fruit nearly globular, about 1 line diameter, resting in the slightly enlarged broadly cup-shaped perianth, the pedicel only slightly thickened.—*Leptomeria odorata*, Miq. in Pl. Preiss. i. 613.

W. Australia. Sussex district, *Preiss*, *n.* 2093 (whose specimens I have not seen); near Busselton, *A. and E. Pries.*

3. **E. cupressiformis,** *Labill. Voy.* i. 155, t. 14. Usually a tree of about 20 ft., the very numerous green wiry rigid or filiform apparently leafless branches sometimes collected in a dense conical head, sometimes loose and pendulous at the extremities, all terete but more or less furrowed. Leaves reduced to minute alternate scales. Flowers minute, in little terminal or lateral very shortly pedunculate spikes of 1½ to 3 lines, each one sessile in a notch of the rhachis or in the axil of a minute tooth-like bract. Perianth-segments 5, about ¼ line long. Anthers divergent, adnate to the margin of a broad almost triangular filament. Ovary immersed in and continuous with the broad disk; stigma sessile, 2-lobed. The great majority of the flowers, although with apparently perfect stigmas remain sessile and soon fall off, a few only (usually no more than one in each spike) after fecundation are raised on an obconical pedicel, which, under the small ovoid globular fruit, enlarges to 2 or 3 lines, becoming thick red and succulent.—R. Br. Prod. 356; A. DC. Prod. xiv. 689; Hook. f. Fl. Tasm. i. 336; *Leptomeria acerba*, Sieb. Pl. Exs. not of R. Br.

Queensland. Sandy Cape, *R. Brown;* Moreton bay, *F. Mueller*, *C. Stuart;* Rockhampton, *O'Shanesy*.

N. S. Wales. Port Jackson, *R. Brown*, *Sieber*, *n.* 136; "Cherry tree," Woods N. S. Wales, London Exhibition 1862, *n.* 161; Hastings river, *Beckler;* New England, *C. Stuart*.

Victoria. Port Phillip, *R. Brown*, *Gunn;* Melbourne, *Adamson;* Yarra river and Dandenong, *F. Mueller;* Ballarook forest, *Whan;* Seven Hill, *Hinteracker;* Ararat, *Green*.

Tasmania. Port Dalrymple, *R. Brown;* common in most parts of the island, but rare in the N. West, *J. D. Hooker*.

S. Australia. Memory Cove, *R. Brown;* near Adelaide, *Blandowski;* Mount Torrens and Mount Flinders, *F. Mueller;* Kangaroo island, *Waterhouse*.

W. Australia? A specimen from Wilson s Inlet, *Oldfield*, seems to belong to this species, but it is in flower only, and the spikes are much longer and more slender than usual.

4. **E. spartea,** *R. Br. Prod.* 356. An erect shrub of 6 to 8 ft. or small tree of 15 to 20 ft., the branches usually rather slender erect or horizontal and pendulous at the ends, scarcely furrowed but often somewhat angular. Leaves alternate, distant, linear-subulate, usually 1 to 2 lines long, acute and recurved at the end, sometimes a few of them smaller and deciduous, sometimes rather thicker and 4 to 6 lines long, in a few N.W. specimens short and thick, in all often falling off before the fruit ripens. Flower-spikes 2 to 4 lines long, usually rather slender, often more than one in the same axil and generally flowering from near the base. Flowers mostly 4-merous. Fruit ovoid or oblong, red, the thick succulent pedicel usually shorter than the fruit itself.—A. DC. Prod. iv. 690; F. Muell. Pl. Vict. t. 88 (ined.); *E. glandulacea*, Miq. in Pl. Preiss. i. 619; A. DC. Prod. xiv. 689; *E. spicata*, DC. l.c. (from the character given); *E. pendula*, F. Muell. in Trans. Vict. Inst. 1855, 42.

Queensland. Head of Flinders river, *Bowman*.

N. S. Wales. Murray and Darling desert, *Herb. F. Mueller*.

Victoria. Scrub in the N.W. and along the Murray, *F. Mueller, L. Morton.*
S. Australia. Near Enfield, *F. Mueller;* beyond Salt Creek, *Behr;* Port Lincoln, *Wilhelmi;* York Peninsula, *Miss Salmon.*
W. Australia. King George's Sound, *R. Brown, A. Cunningham,* and many others, and thence to Swan and Murchison rivers *Oldfield, Preiss, n.* 2125 *and* 2106 (the latter incautiously referred by Miq. in Pl. Preiss. i. 614 to his *Leptomeria Lehmanni*), *Drummond,* 1*st coll. n.* 728.

5. **E. aphylla,** *R. Br. Prod.* 357. An erect much-branched shrub of 4 to 6 ft., growing out sometimes into a small tree, with stout rigid terete finely-furrowed branchlets, sometimes spinescent at the end. Leaves reduced to minute ovate appressed scales, distant and very deciduous. Flower-spikes ovoid or oblong, sessile or very shortly pedunculate, very dense, 1 to 2 lines long, the rhachis usually slightly pubescent. Flowers 5-merous. Fruit rather small, ovoid-globular, the short broad thickened peduncle more or less succulent and at length truncate by the fall of the perianth-lobes.—A. DC. Prod. xiv. 690; *E. leptomerioides,* F. Muell.; Miq. Stirp. Nov. Holl. 7; A. DC. l.c.

Queensland. Peak Downs, *Herb. F. Mueller;* Armadillo, *Barton.*
N. S. Wales. Field's and Liverpool plains, *A. Cunningham, Fraser;* Lachlan and Darling rivers, *Victorian and other Expeditions.*
Victoria. Avoca, *F. Mueller.*
S. Australia. Islands off the S. coast and Memory Cove, *R. Brown;* scrub on the Murray, *Behr,* and thence to St. Vincent's Gulf, *F. Mueller;* York Peninsula, *Miss Salmon.*
W. Australia, *Drummond,* 3*rd coll. n.* 101.

E. dasystachys, Schlecht. in Linnæa, xx. 580, from the author's description, must be this species; but the specimens so named in Herb F. Mueller appear to me to belong some to *E. stricta,* and others to *E. spartea.*

6. **E. homaloclada,** *Moore and Muell. in F. Muell. Fragm.* viii. 9. An erect glabrous shrub of 10 to 15 ft., the flowering branches very flat, nearly 1 line broad, finely striate, notched at the nodes but without any leaves on the flowering specimens. F. Mueller thinks however that a barren specimen with oblong-linear or lanceolate obtuse leaves of 1 to 1½ in. may belong to the same species. Flowers in small clusters of 2 or 3 at the upper notches and often 2 clusters at the same notch on exceedingly short thick peduncles, the flowers sessile, 4-merous or 5-merous. Fruit not seen.

N. S. Wales. Lord Howe's island, *C. Moore.*—The habit of the flowering specimens is precisely that of a *Psilotus;* the species has no resemblance or affinity (excepting as to generic characters) with the Norfolk island *E. phyllanthoides,* but is very closely allied to *E. stricta.*

7. **E. stricta,** *R. Br. Prod.* 357. An erect glabrous shrub of several feet, the branches slender and striate, but usually with 2 or 3 very prominent angles, leafless at the time of flowering. Leaves on the very young shoots only, subulate, ½ to nearly 2 lines long, leaving as they fall off a minute triangular tooth-like base. Flowers in small sessile axillary clusters, often only 2 or 3 together, and sometimes 2 clusters of 2 or 3 flowers each in the same axil, mostly 4-merous, but sometimes 4-merous and 5-merous in the same clusters. Fruit nearly

globular, smaller than in *E. cupressiformis*, rarely above 2 lines diameter and the thickened pedicel much smaller and white (or red ?).—A. DC. Prod. xiv. 690; Hook. f. Fl. Tasm. i. 336.

N. S. Wales. Port Jackson, *R. Brown;* Hastings and Macleay rivers, *Beckler;* New England, *C. Stuart;* Blue Mountains, *A. Cunningham, Fraser;* Berrima, *Woolls;* Illawarra, *Shepherd;* Darling river, *Victorian Expedition, Mrs. Ford;* Twofold bay and near Cape Howe, *F. Mueller.*

Victoria. Buffalo range, Delatite river, Wilson's Promontory, *F. Mueller;* Wendu vale, *Robertson.*

Tasmania. Derwent river and Port Dalrymple, *R. Brown;* common in poor land, *J. D. Hooker.*

S. Australia. Memory Cave, *R. Brown.*

Var. *syrticola*, F. Muell. Branches stouter, less angular, approaching those of *E. aphylla*, but the inflorescence as in *E. stricta*, although the clusters may not be so closely sessile, and the pedicel appears to be red.—Sandy shores of Lake Alexandrina and Rivoli bay, *F. Mueller;* Cape Nelson and Spencer's Gulf, *Herb. F. Mueller.*

8. **E. humifusa,** *R. Br. Prod.* 356. A rigid shrub, prostrate or spreading over rocks nearly level with the ground, the branches shortly ascending, rather thick, furrowed, terete or slightly compressed. Leaves reduced to small triangular acute toothlike scales, persistent and all alternate. Flowers 2 or 3 together in sessile axillary clusters, usually 4-merous, but sometimes 5-merous. Fruit ovoid-oblong, about 2 lines long, the succulent pedicel apparently red.—A. DC. Prod. xiv. 691.

Tasmania. Summit of Table Mountain (Mount Wellington), *R. Brown;* Western Mountains, *Archer;* Mount Lapeyrouse, *C. Stuart.*

9. **E. nana,** *Hook. f. in Hook. Lond. Journ.* vi. 281. A dwarf prostrate shrub, with numerous short deeply-furrowed somewhat compressed branches. Leaves reduced to small tooth-like scales, persistent and mostly opposite or nearly so. Flowers apparently unisexual, the males 2 together sessile in the axils, 5-merous and scarcely above ½ line in diameter, the anther-cells almost parallel, the disk broad and 5-angled. Females only seen in fruit and then solitary, the drupe or nut ovoid, smooth, scarcely 2 lines long, the thickened succulent pedicel about as long, the perianth-lobes persistent under the fruit.—A. DC. Prod. xiv. 691; *E. humifusa*, Hook. f. Fl. Tasm. i. 336, not of R. Br.

Victoria. Summit of Cobberas mountains, at an elevation of 6000 ft., *F. Mueller.*

Tasmania. St. Patrick's river, *Gunn.*

Order CXV. BALANOPHOREÆ.

Flowers unisexual. Male flowers: Perianth 3-cleft, the lobes or segments valvate in the bud, or rarely no perianth. Stamens as many as perianth-segments and opposite to them, the filaments united, or where there is no perianth stamens free and 2 only or solitary; anthers 2- or several-celled or rarely 1-celled. Ovary none. Female flowers minute, without any apparent perianth, or the ovary produced at the top into a small lobed limb. Ovary 1-celled, with a simple terminal stigma. Fruit a minute utricle nut or drupe, enclosing an adherent seed.—Stout succulent leafless root-parasites. Stem reduced to a tuberous often lobed rhizome, with short thick erect scapes. Leaves replaced by concave scales. Flowers in dense thick terminal spikes, usually very numerous and closely packed.

A small Order, chiefly tropical, both in the New and the Old World. One monotypic genus is found as far north as the Mediterranean, and another is in New Zealand in the south. The Australian genus extends over E. India and the Eastern Archipelago.

1. BALANOPHORA, Forst.

Male perianth regular, usually of 3 or 4 but varying from 2 to 8 segments. Stamens all united; anthers 4-celled. Female flowers consisting of a naked ovary, terminating in a single style. Scapes bearing alternate or imbricate scales and a diœcious or monœcious spike; when monœcious, the males occupying the lower, and the females the upper portion.

The genus comprises eight species, natives of E. India, the Archipelago, and the S. Pacific islands; the only Australian one is found also in the New Hebrides.

1. **B. fungosa,** *Forst. Char. Gen.* 99, t. 50. Rhizome short, thick, irregularly lobed, with a minutely granular surface. Scapes thick, 2 to 4 in. high. Scale-like leaves ovate, obtuse, concave, membranous, $\frac{1}{2}$ to $\frac{3}{4}$ in. long. Female flowers exceedingly numerous and minute, very densely packed in a globular terminal white head of about 1 in. diameter, the males occupying a loose ring under it, 3 or 4 lines broad, each one on a pedicel of $1\frac{1}{2}$ to 2 lines. Perianth-lobes 3 or 4, spreading or reflexed, oblong, concave, about 1 line long. Anthers forming a globular mass, on a short stipes consisting of the united filaments.—Hook. f. in Trans. Linn. Soc. xxii. 46, t. 8; *Cynomorium balanophora*, Willd. Spec. Pl. iv. 177.

Queensland. Rockingham bay, "on the roots of trees," *Dallachy;* Gould island in the same bay, *M'Gillivray.* Also in Tanna island, New Hebrides.—I have not myself been able to verify the structure of the flowers, but have taken the characters chiefly from Hooker's above-quoted Memoir.

ORDER CXVI. CONIFERÆ.

Flowers monœcious or diœcious, the males in deciduous catkins, the females in cones or solitary, all without any perianth. Male catkins consisting of several usually numerous scale-like stamens, opposite and decussate, or in alternating whorls, or in dense spires imbricated round a common axis, each stamen consisting of a connective more or less contracted into a stipes at the base and dilated at the apex; anther-cells 2 or more, adnate to the stipes or pendulous from under the scale-like apex, opening longitudinally in 2 valves. Female cones consisting either of opposite verticillate or spirally arranged imbricate scales, with 12 or more erect or inflexed naked orthotropous or anatropous ovules (erect or recurved pistils according to some theorists) within each scale; or of a fleshy cup or receptacle with 1 or 2 exserted ovules (or pistils). Fruit (or syncarp according to some) the more or less enlarged and hardened or succulent cone. Seeds (or fruits) often winged; testa (or pericarp) hard crustaceous or membranous; albumen fleshy; embryo in the axis, straight, with two or more cotyledons; radicle terete, often attached by a folded thread.—Trees or shrubs often resinous, the wood without medullary rays or vascular tissue proper; wood-cells studded with disks. Leaves, sometimes reduced to small scales, opposite whorled or spirally arranged or in genera not Australian alternate, and sometimes clustered 2 to 5 together in membranous sheaths. Male catkins solitary or clustered, terminal or rarely axillary, female cones usually lateral on short peduncles or terminating reduced branchlets.

An extensive Order spread over nearly the whole globe, especially in the northern hemisphere, but within the tropics chiefly confined to mountainous regions. The eleven Australian genera are all limited to the southern hemisphere except *Podocarpus*, which extends northwards to Japan and to the West Indies, and that genus and *Araucaria* are in South America as well as in the Old World. Three other genera, *Dammara*, *Dacrydium*, and *Phyllocladus* reach from New Zealand to the Archipelago. *Frenela* extends only to New Caledonia, and the five remaining ones are endemic, mostly monotypic or nearly so.

In describing the female organs of this Order, I have made use of the terminology corresponding to Brown's view of their homology, without however intending to decide the question which is still the subject of keen controversy. The nucleus in this and other gymnospermous Orders lies within and is often partially adnate to an utricle which has at the apex an orifice open at the time of flowering, so as to admit of the pollen-tubes penetrating to the nucleus, but closed over it after fecundation, without any real style or stigma or any perianth. Whether, on the one hand, the nucleus alone represents the ovule and seed, and the utricle is formed of two carpellary leaves, or whether, on the other, the whole forms the ovule or seed, of which the ovule is the integument, is a problem which does not appear to admit of a decisive solution. Strong arguments have been adduced on each side by eminent observers, none of them absolutely convincing, for whichever view be adopted there still remains a broad distinction between these organs in Gymnosperms and any such supposed homologues in Angiosperms. As long, therefore, as the theoretical explanation remains unsettled, it seems best, in describing the generic modifications, to make use of the terminology which is most conformable to actual appearance, and to consider these organs as naked ovules and seeds, not as pistils and nuts. The genera are also here arranged solely with a view to the practical distinction of the few represented in Australia, without reference to the tribes founded on the theoretical significance of the coverings of the seed, whether discoid, carpellary, or integumental; for, besides that the Australian genera are insufficient to give any clear explanation of these differences, the position of one or two of the monotypes is as yet uncertain.

Ovules (or carpels) in the axils or on the inner surface of the scales.
- Leaves (often reduced to scales) verticillate or opposite. Cones with only the innermost or 2 innermost pairs or whorls of scales bearing ovules.
  - Leaves or scales in whorls of 3 or 4. Ovules numerous within each scale . . . . 1. FRENELA.
  - Leaves or scales in whorls of 3. Ovules 1 or 2 within each scale . . . . 2. ACTINOSTROBUS.
  - Leaves (small, thick, and imbricate) opposite and decussate. Ovules 1 or 2 within each scale . . . . 3. DISELMA.
- Leaves (small, thick, and imbricate) opposite and decussate. Cones with several series of scales bearing 1 ovule each . 4. MICROCACHRYS.

Leaves (rarely reduced to scales) spiral or scattered. Cone-scales flat, hardened at the apex. Seeds compressed, the outer integument appressed, often winged.
Male amenta small. Stamens with a slender stipes and 2 anther-cells. Cones small. Ovules few. Seeds 2-winged . . . 5. ARTHROTAXIS.
Male amenta dense, cylindrical. Anther-cells more than 5. Cones large. Ovules 1 to each scale.
Seeds obovate-oblong, adnate to the scale at the base, not winged . . . . 6. ARAUCARIA.
Seeds oblong, free from the scale, winged on one side . 7. DAMMARA.
Leaves (sometimes reduced to scales) spiral or scattered. Cones small, scales thickened and concave, with 1 ovule each. Seeds small, the outer integument membranous, contracted into a neck.
Leaves very small, thick, closely imbricate.
Seeds seated in a short membranous cup . . . . . 8. DACRYDIUM.
Seeds without any external membranous cup . . . . 9. PHEROSPHÆRA.
Leaves reduced to small scattered or almost verticillate scales, with axillary leaf-like rhomboidal flat branchlets or phyllodia . . . . 10. PHYLLOCLADUS.
Ovules 1 or 2, exserted from an oblong fleshy receptacle. Leaves alternate or opposite, usually distichous or flat, with a prominent midrib . . . . 11. PODOCARPUS.

1. FRENELA, Mirb. (partly).

(Callitris, *Vent.* (partly); Leichhardtia, *Steph.*; Octoclinis, *F. Muell.*)

Flowers monœcious. Male amenta cylindrical oblong or ovoid, the stamens in whorls of 3 or rarely 4, imbricate in twice as many vertical rows, the scale-like apex ovate orbicular or slightly peltate; anther-cells 2 to 4. Female amenta of 6 rarely 8 scales, more or less distinctly arranged at the time of flowering in 2 whorls without any enlarged outer empty scales. Ovules (or carpels) several within each scale, in 3 vertical series, sessile and erect. Fruiting-cone globular ovoid or pyramidal, the 6 rarely 8 scales enlarged and hardened, shortly united at the base, apparently arranged in a single whorl, and opening in as many valves, either all equal and strictly valvate, or three alternate ones smaller and sometimes overlapping the others on the margin. Fertile seeds usually few only in each cone, compressed, with a hardened integument, the margins produced into 2 unequal wings or rarely only 1 wing developed, or very rarely a third wing also prominent on one face, the abortive seeds mostly enlarged and very flat with winged margins; in some species there is a more or less prominent central columella usually 3-angular or 3-lobed, and sometimes apparently formed of abortive ovules. Cotyledons 2, rarely 3.—Trees or shrubs, with slender terete or 3- or rarely 4-angled branches. Leaves in whorls of 3 rarely 4, those of the young plants sometimes acicular though short, but generally reduced to minute acute scales, the decurrent midribs forming the angles of internodes as in *Casuarina*. Male amenta usually small, solitary or clustered at the ends of the branches, and rarely a few lateral ones. Female cones on short thick peduncles or branchlets, solitary or clustered, ripening usually the second year, and persisting many years after the seeds have fallen.

Besides the Australian species, which are all endemic, there are one or two from New Caledonia.

Fruit-cones angular, the junction of the valves prominent.
  Cones pyramidal, acuminate, about 1 in. long, 6-valved. Eastern species . . . . 1. *F. Parlatorei.*
  Cones globular or scarcely pyramidal, ¾ in. diameter, mostly 8-valved. Leaves often acicular. Eastern species . . . 2. *F. Macleayana.*
  Cones globular, ½ to ¾ in. diameter, 6-valved. Western species 3. *F. Roei.*
Fruit-cones globular, strictly valvate, the junction of the valves neither prominent nor furrowed.
  Internodes angular. Cone-valves nearly equal. Western species . . . . 4. *F. Drummondii.*
  Internodes terete, slender, the ribs broad and very obtuse. Cone-valves alternately smaller. East and West species . 5. *F. robusta.*
  Internodes angular. Cone-valves alternately smaller. Eastern species . . . . 6. *F. Muelleri.*
Fruit-cones globular or oblong, more or less furrowed at the junction of the valves, the three smaller valves often slightly overlapping the others.
  Cones globular, the inner larger valves dilated at the apex, with the dorsal point near the centre . . . . 7. *F. rhomboidea.*
  Cones ovoid or oblong, the inner larger valves not much dilated, with the small dorsal point near the end.
    Male amenta usually solitary. Cones rarely above ½ in. diameter . . . . 8. *F. Endlicheri.*
    Male amenta usually in threes. Cones usually ¾ to nearly 1 in. long . . . . 9. *F. australis.*

1. **F. Parlatorei,** *F. Muell. Fragm.* v. 186. A tree attaining 60 ft., with the angular internodes of *F. Endlicheri*, but the teeth or scales more prominent and acute. Male amenta unknown. Fruit-cones on short stout peduncles, ovoid-pyramidal, acuminate, about 1 in. long, the valves 6, nearly equal, very thick, smooth outside or nearly so, forming prominent angles at their junction. Fertile seeds with one large wing and one small one sometimes obsolete.—Parlat. in DC. Prod. xvi. ii. 447; *Callitris Parlatorei*, F. Muell. l.c.

Queensland. Darlington Range, *W. Hill.*

2. **F. Macleayana,** *Parlat. in DC. Prod.* xvi. ii. 446. A tall pyramidal tree with spreading branches. Leaves in whorls of 4 or sometimes 3, developed on the lower or sometimes on nearly all the branches into rigid linear-triquetrous almost pungent-pointed spreading laminæ of 2 to 4 lines, reduced in some of the upper branches to the minute scales or teeth of the other species, the angles of the internodes very prominent. Male amenta 2 to 4 lines long. Fruit-cones sessile, nearly globular or slightly pyramidal, about ¾ in. diameter, the valves 8 or sometimes 6, nearly equal, thick, with a small dorsal point near the end, their junction forming prominent angles before the cone opens. Fertile seeds with one of the wings usually large, the other small or obsolete.—*Octoclinis Macleayana*, F. Muell. in Trans. Phil. Inst. Vict. ii.

22, with a plate; *Leichhardtia Macleayana*, Sheph. Cat. Pl. Cult. Sydn. 1851, 15, as quoted by F. Mueller.

N. S. Wales. Port Macquarie, *Macleay;* Hastings river, *Thozet.*

The original specimens have none but the acicular leaves. Thozet's specimens (without fructification) have similar acicular leaves in whorls of 4 on the lower branches, whilst the upper ones resemble those of *F. Endlicheri*, except in their number in the whorls. W. Hill's Brisbane specimens (without fruit) appear to me to belong to *F. Endlicheri*, which has also frequently acicular leaves on the lower branches.

3. **F. Roei,** *Endl. Syn. Conif.* 36. A shrub or tree with flexuous branchlets, stouter than in any other species, the internodes very angular. Male amenta unknown. Fruit-cones shortly pedunculate or terminating short branchlets, nearly globular, truncate or intruded at the base, ½ to ¾ in. diameter, the 6 valves very thick and strictly valvate, forming at their junction prominent obtuse angles on the cone before it opens, as in the preceding species, but unequal as in several of the following ones, nearly smooth outside, the larger valves with a prominent dorsal conical point below the apex as in *F. rhomboidea.* Fertile seeds usually with 1 large and 1 small wing.—Parlat. in DC. Prod. xvi. ii. 448; *F. subcordata*, Parlat. in Enum. Sem. Hort. Flor. 1862, 24, and in DC. Prod. xvi. ii. 446.

W. Australia. King George's Sound or to the eastward, *Baxter*, *Drummond*, *4th coll. n.* 235; foot of the Stirling Range, *F. Mueller.*—I have not seen Roe's specimens, originally described by Endlicher, and quote the name on Parlatore's authority, our specimens agreeing with the description except as to the wings of the seeds, which Endlicher describes as narrow, but they are often variable in many species. Roe gathered also the *F. robusta*, to the specimens of which Parlatore inadvertently gave the name of *F. Roei* in Herb. Hook. Baxter's plant is certainly the same as *Drummond's;* the species is readily known by its stout branchlets.

4. **F. Drummondii,** *Parlat. in DC. Prod.* xvi. ii. 448. A shrub or tree with the angular internodes of *F. Endlicheri.* Male amenta unknown. Fruit-cones on stout short peduncles, mostly solitary, strictly globular, without prominent angles, about ½ in. diameter or rather more, the valves thick, nearly equal or alternately rather shorter and more acute, strictly valvate, smooth or slightly rugose on the back, with a minute dorsal point below the apex.

W. Australia. Towards Cape Riche, *Drummond, 5th coll., n.* 433; Salt, Gardner, and Fitzgerald rivers and Esperance bay, *Maxwell.*

5. **F. robusta,** *A. Cunn.; Mirb. in Mem. Mus. Par.* xiii. 74. A tree of considerable size, often exceeding 90 ft. (*Fraser*), sometimes reduced to a tall shrub, the crowded branchlets short and erect, often slender and glaucous, the internodes terete or with very obtuse angles, never so prominent as in the other species, the scales or teeth small and acute. Male amenta solitary or in threes, 2 to 4 lines long, more slender and looser than in *F. rhomboidea* and *F. Endlicheri.* Fruit-cones solitary or few together, nearly globular and usually about 1 in. diameter, neither angled nor furrowed, the valves 6, alternately about ¼ shorter, strictly valvate, smooth or more or less verrucose on the back, without any dorsal point. Seeds usually 2-winged, the central columella often

somewhat prominent.—Parlat. in DC. Prod. xvi. ii. 450; *Callitris robusta*, R. Br. Herb.; *F. propinqua*, A. Cunn. (*Callitris propinqua*, R. Br.), Mirb. in Mem. Mus. xiii. 74; *F. glauca* (*Callitris glauca*, R. Br.), Mirb. l.c.; *F. crassivalvis*, Miq. Stirp. Nov. Holl. Muell. 1; *Callitris Preissii*, Miq. in Pl. Preiss. i. 643; *F. canescens*, Parlat. in DC. Prod. xvi. ii. 448; *F. Gulielmi*, Parlat. l.c. 449?

N. Australia. York Sound, Regent's river, and Brunswick bay, N.W. coast, *A. Cunningham;* Mackenzie river, *F. Mueller;* Port Darwin, *Schultz*, *n.* 438.

Queensland. Rockingham bay, *Dallachy;* Gilbert river, *Daintree;* Nogoa river, *Bowman.*

N. S. Wales. Blue Mountains, *A. Cunningham;* and on all the barren lands of the interior from thence to the Darling and Murray rivers and to the Barrier range, *A. Cunningham*, *Fraser*, *Mitchell*, *Victorian and other Expeditions;* New England, *C. Stuart;* Mount Lindsay, *Fraser.*

Victoria. Mount Brown, Spencer's Gulf, Kangaroo island, *R. Brown;* Port Phillip and Murray river, *F. Mueller.*

S. Australia. Enfield, St. Vincent's and Spencer's Gulfs, *F. Mueller.*

W. Australia. Middle island, Goose island bay, *R. Brown;* King George's Sound and adjoining districts, *Baxter*, *Oldfield*, *F. Mueller;* Swan river, *Preiss*, *n.* 1310, *Drummond*, *1st coll.*, also (*3rd coll.?*) *n.* 186; Rottenest island, *A. Cunningham;* also in Roe's and other collections.

Var. *microcarpa.* Cones small, but varying from under ½ in. to nearly ¾ in., the valves usually more unequal, and a central columella more developed, triangular and varying from scarcely one line to nearly the length of the valves. *F. microcarpa*, A. Cunn. Herb.; *F. intratropica*, F. Muell. Herb.; *F.* or *Callitris columellaris*, F. Muell. Fragm. v. 198; Parlat. in DC. Prod. xvi. ii. 451; *F. Moorei*, Parlat. in DC. Prod. xvi. ii. 449. To this belong most of the northern specimens, also Richmond and Clarence rivers, *Beckler*, *Henderson*, *C. Moore;* Moreton island, Northern woods, Paris Exhibition 1855, *C. Moore*, *n.* 62.

Var. *verrucosa.* Cones large, with large warts on the backs of the valves. *F. verrucosa*), A. Cunn. (*Callitris verrucosa*, R. Br.), Mirb. in Mem. Mus. Par. xiii. 74; Parlat. in DC. Prod. xvi. ii. 448; *F. tuberculata* (*Callitris tuberculata*, R. Br.), Mirb. l.c. Found with the smooth-valved form in the interior of N. S. Wales, in S. Australia and in W. Australia.

6. **F. Muelleri,** *Parlat. in DC. Prod.* xvi. ii. 450. A tree attaining 20 to 30 ft., with the angular internodes and clustered male amenta of *F. australis*, but the branches stouter and the fruit-cones more like those of *F. robusta*, globular, ¾ to 1 in. diameter, neither angled nor furrowed, the valves 6, very thick, strictly valvate, rugose outside, with a minute dorsal point below the summit, the smaller valves about half the breadth of the larger ones though not very much shorter.—*F. fruticosa*, A. Cunn. Herb., but probably not *Callitris fruticosa*, R. Br.

N. S. Wales. Port Jackson, *F. Mueller*, *Thozet;* Blue Mountains, *A. Cunningham*, *Frazer*, *Miss Atkinson.*—The species requires further investigation.

7. **F. rhomboidea,** *Endl. Syn. Conif.* 36. A tree described sometimes as 20 to 25 ft. high, sometimes as double that height, the branches rather slender, often drooping, angular when young, the small scales or teeth much more acute than in *F. australis*. Male amenta solitary or 3 together, small or loose. Fruit-cones often clustered on short branches, globular, not exceeding ½ in. diameter in the typical forms; valves 6, alternately smaller, the larger ones dilated into a broadly rhomboidal apex with a short conical protuberance about the centre

and usually rugose, the alternate ones much shorter, with a broad base and slightly overlapping the others on the margin, at least when young, the unopen cone furrowed at the junctions. Seeds 2-winged, the breadth of the wings exceedingly variable.—Parlat. in DC. Prod. xvi. ii. 447; Hook. f. Fl. Tasm. i. 352; *Callitris rhomboidea*, R. Br. in Rich. Conif. 47, t. 18; *Frenela Ventenatii*, Mirb. in Mem. Mus. Par. xiii. 74; *Thuya australis*, Poir. Dict. Suppl. v. 302; *Cupressus australis*, Desf. Cat. Hort. Par. ed. 3, 355, not of Persoon; *Callitris cupressiformis*, Vent. Nov. Gen. Dec. 10; *Frenela australis*, Endl. Syn. Conif. 37, not of Brown; *Callitris arenosa*, Sweet, Hort. Brit. 473; *Frenela arenosa*, A. Cunn.; Endl. Syn. Conif. 38; Parlat. in DC. Prod. xvi ii. 451; *F. triquetra*, Spach, Suit. Buff. xi. 345; Endl. Syn. Conif. 36; *F. attenuata*, A. Cunn. Herb.

Queensland. Moreton island, *F. Mueller;* Stradbrooke island (*Fraser?*).

N. S. Wales. Port Jackson, *R. Brown, J. D. Hooker;* Sydney woods, Paris Exhibition 1855, *Macarthur, n.* 151; New England, *C. Stuart;* and southward to Illawarra, *Shepherd;* N. S. Wales woods, Southern district, London Exhibition 1862, *n.* 99; Twofold bay, *F. Mueller.*

Victoria. Grampians, *Fisher.*

S. Australia. Sturt's river, Onkaparinga, *Blandowski.*

Var. *Tasmanica.* Cones ½ to ¾ in., thick and rugose at the back, the dorsal point or prominence less distinct.—Oyster Bay Pine.

Tasmania. Oyster bay, *Gunn*, and others.

Var. *mucronata.* Dorsal conical point of the larger cone-valves very prominent.—Grampians, *F. Mueller;* Mount Sturgeon, *Robertson.*

8. **F. Endlicheri,** *Parlat. in DC. Prod.* xvi. ii. 449. A tree of 60 to 100 ft., closely resembling *F. australis* and *F. rhomboidea* as to its angular branchlets and small scales or teeth, except in the young plant which has sometimes acicular leaves like those of *F. Macleayana.* Male amenta usually solitary, short and compact. Fruit-cones usually clustered on short branches and of the size of those of *F. rhomboidea*, but nearer in shape to those of *F. australis*, about ½ in. diameter, the 3 larger valves but little or not at all dilated upwards, the dorsal point very near the end, smooth or scarcely rugose, the 3 smaller ones often slightly overlapping, and the cone furrowed at the junctions before it opens. Seeds varying in the breadth of the wings.—*F. fruticosa*, Endl. Syn. Conif. 36 (*Parlatore*); *F. pyramidalis*, A. Cunn.; Sweet Hort. Brit. 473; *F. calcarata*, A. Cunn. (*Callitris calcarata*, R. Br.), Mirb. in Mem. Mus. Par. xiii. 74.

Queensland. Wide bay, *Bidwill*, also probably *Octoclinis Backhousii*, from Moreton Bay, Queensland woods, London Exhibition 1862, *W. Hill, n.* 4 (without flower or fruit).

N. S. Wales. Lachlan river and Liverpool plains, *A. Cunningham, Leichhardt;* Berrima, *Woolls;* Darling river, *H. Law.*

Victoria. Futter's Range, *F. Mueller.*

Var. *mucronata.* Cone-valves produced into a thick almost terminal point.—*F. Gunnii*, var. *mucronata*, Parlat. in DC. Prod. xvi. ii. 450.—Mount Mitchell, *Beckler.*

9. **F. australis,** *R. Br.; Mirb. in Mem. Mus. Par.* xiii. 74, *not of Endl.* A bush or small tree of 20 to 25 ft. with erect dense branches,

the internodes prominently triangular, with small teeth or scales. Male amenta very small, usually 3 together. Fruit-cones ovoid or oblong, $\frac{3}{4}$ to 1 in. or very rarely only $\frac{1}{2}$ in. long, the 3 larger valves scarcely dilated upwards, smooth on the back with the dorsal point nearly terminal, the three smaller ones slightly overlapping on the margins, the cone furrowed at the junctions before opening. Seeds broad, the margins equally or unequally winged.—Hook. f. Fl. Tasm. i. 352, t. 97; *Callitris oblonga*, Rich. Conif. 49, t. 18, f. 2; *C. Gunnii*, Hook. f. in Hook. Lond. Journ. iv. 147; *Frenela Gunnii*, Endl. Syn. Conif. 38; Parlat. in DC. Prod. xvi. ii. 450, also according to Parlatore, *F. variabilis*, Carr. and *F. macrostachya*, Gord.

Tasmania. Port Dalrymple, *R. Brown;* abundant on the gravelly banks of the South Esk river near Launceston, &c., *J. D. Hooker*. It is probably a Tasmanian specimen of this species that R. Brown had originally designated under the name of *Callitris fruticosa*, which does not occur in his herbarium.

2. ACTINOSTROBUS, Miq.

Flowers monœcious. Male amenta oblong, the stamens in whorls of 3, imbricate in 6 vertical rows; anther-cells 2 to 4. Female amenta globular or acuminate, the scales imbricate in whorls of 3, all closely appressed at the time of flowering, those of the 2 innermost whorls alone bearing each 1 or 2 erect ovules at the base. Fruit-cones ovoid-globular or acuminate, the 6 inner much enlarged scales becoming almost valvate in a single whorl, with 6 or 12 of the outer barren scales more or less enlarged and closely appressed or adnate to their base. Fertile seeds usually only one to each scale, 3-winged, the abortive ones also more or less enlarged but only 2-winged, the central columella more or less developed or obsolete.—Densely branched shrubs. Leaves in whorls of 3, very short, thick, rigid and acute; or on the smaller branches appressed obtuse and 3-gonous. Amenta and cones on very short peduncles, or almost sessile, in the axils of the leaves.

The genus is endemic in Western Australia. F. Mueller proposes to reunite it with *Callitris* and *Frenela*, but the habit, the numerous imbricate scales of the female amenta, and the reduced number of ovules seem to justify the retaining it as distinct.

Scale-like apex of the stamens very obtuse. Fruit-cones globular or obtusely acuminate . . . . 1. *A. pyramidalis.*
Scale like apex of the stamens acutely acuminate. Fruit-cones contracted at the top into a neck with short spreading terminal points . . . . 2. *A. acuminatus.*

1. **A. pyramidalis,** *Miq. in Pl. Preiss.* i. 644. A densely branched pyramidal glabrous shrub. Lower leaves sometimes acicular and 3 or 4 lines long, those of the main branches acute and spreading, but only 1 or 2 lines long or even shorter, on the smaller branchlets often still shorter appressed and obtuse. Male amenta 1 to 2 lines long, the scale-like apex of the stamens orbicular, very obtuse and not keeled. Female amenta when as yet only 2 or 3 lines diameter consisting of 4 to 6 whorls of 3 scales each, all imbricate in alternate series, but as

the cone enlarges, those of the 2 inner ovule-bearing whorls become strictly valvate and either remain very obtuse or become shortly acuminate; the cone attains ½ in. diameter or more and each fertile valve has a broad sterile one enlarged to 2 or 3 lines diameter so closely appressed to its base as to appear adnate, and sometimes a second outer one enlarged to nearly half its size.—Endl. Conif. 40; Parlat. in DC. Prod. xvi. ii. 444; *Callitris actinostrobus,* F. Muell. Rep. Burdek. Exp. 19.

W. Australia. King George's Sound, *Baxter;* and thence to Swan river, *Preiss, n.* 1311, *Drummond, 1st coll. and 3rd coll. n.* 234, *Oldfield;* Murchison river, *Oldfield.*

2. **A. acuminatus,** *Parlat. Enum. Sem. Hort. Flor.* 1862, 25, *and in DC. Prod.* xvi. ii. 445. A small erect densely branched shrub, our specimens with the root not exceeding 1 ft., with the leaves of *A. pyramidalis.* Male amenta all terminating developed branchlets, the scale-like apex of the stamens larger than in *A. pyramidalis,* keeled, and terminating in a fine acute point. Fruit-cones of the size and structure of those of *A. pyramidalis,* but contracted at the top into a distinct neck, each valve terminating in a short spreading point.

W. Australia. Between Moore and Murchison rivers, *Drummond, 6th coll. n.* 225.

3. DISELMA. Hook. f.

Flowers diœcious, the amenta terminal. Male amenta ovoid or oblong consisting of 3 or 4 pairs of opposite stamens, the stipes very short, the scale-like imbricate apex triangular and coriaceous; anther-cells 2. Female amenta of 2 pair of opposite scales, with 2 erect ovules at the base of each of the inner ones. Fruit-cones small, globular. Seeds 3-winged.—Erect shrubs with small opposite closely appressed leaves.

The genus is limited to a single species, endemic in Tasmania.

1. **D. Archeri,** *Hook. f. Fl. Tasm.* i. 353, t. 98. An erect densely branched shrub of 5 to 15 ft. Leaves closely imbricate but strictly opposite and decussate, very obtuse, thick and keeled, about ½ line long. Male amenta erect, 1 to 2 lines long, scarcely thicker than the branchlets with their leaves. Young female cones purplish in the dried state, about 1 line long and broad, consisting of 2 pairs of very obtuse or truncate scales, the 2 outer empty ones rather smaller than the 2 inner fertile ones. Seeds not yet ripe but already enlarged to the length of the scales.—Parlat. in DC. Prod. xvi. ii. 462.

Tasmania. Western Mountains, Lake St. Clair, Falls of the Meander, &c., *Gunn.* With the foliage of *Microcachrys* this is more nearly allied to *Frenela* in fructification. The mistake, owing to which the female plant was described as prostrate, although cleared up by Archer, has been omitted to be corrected in the Prodromus.

4. MICROCACHRYS, Hook. f.

Flowers diœcious, the amenta terminal. Male amenta ovoid, consisting of several pairs of opposite stamens, the stipes very short, the scale-like imbricate apex ovate almost acute. Female amenta consisting of several pairs of loosely imbricate small scales with one recurved ovule within each scale. Fruit-cones small, ovoid, the scales succulent. Seeds nearly erect, not winged, the outer integuments more or less fleshy.—Prostrate shrub, with small opposite closely appressed leaves.

The genus is limited to a single species, endemic in Tasmania.

1. **M. tetragona,** *Hook. f. Fl. Tasm.* i. 358, t. 100. A prostrate densely branched shrub. Leaves closely imbricate, but strictly opposite and decussate, very obtuse, thick and keeled, and ½ line long on the branchlets, more acute and ¾ line long on some of the older branches. Male amenta small and recurved. Fruit-cones recurved or almost erect, nearly 3 lines long, the scales loosely imbricate, concave, thick succulent and scarlet when ripe. Seeds becoming almost erect, the outer integument fleshy at least at the base.—Bot. Mag. t. 5576; *Arthrotaxis? tetragona*, Hook. Ic. Pl. t. 560; *Dacrydium tetragonum*, Parlat. in DC. Prod. xvi. ii. 496.

Tasmania. Summits of the Western Mountains, *Gunn, Archer, F. Mueller.*

The foliage of this plant is so entirely that of the *Diselma* which grows on the same mountains, that it was at first taken for the female of the same species, although the one is always prostrate and he other erect; the cones are, however, totally different, and much nearer to those of *Pherosphæra* and *Dacrydium*, differing in the fleshy outer integuments of the seed. In the younger stages, this outer integument is shortly cup-shaped, when ripe it probably has enlarged, although I find only two integuments altogether to the seed; its development requires further investigation. For an explanation of the confusion which had been made between *Microcachrys*, *Diselma*, and *Pherosphæra*, see Hook. f. Fl. Tasm. i. 355, under *Pherosphæra*.

5. ARTHROTAXIS, Don.

Flowers monœcious on different branches, the amenta terminal and small. Male amenta consisting of numerous spirally imbricated stamens, the stipes slender, the scale-like apex oblong-sagittate and peltate; anther-cells 2. Female amenta of spirally-imbricate scales, with 3 to 6 pendulous ovules within each. Fruit-cones small, globular, the scales woody, contracted at the base, thickened upwards, incurved and acuminate, or with a dorsal point at the apex. Seeds few under each scale, ovate, compressed, with a transverse hilum and two longitudinal wings, the integument crustaceous. Cotyledons 2.—Densely branched trees. Leaves small, in close spires, either very short obtuse and appressed or lanceolate and looser. Fruit-cones sessile.

The genus is limited to the three Tasmanian species, but is so nearly allied to the Chinese *Cunninghamia*, that Zuccarini proposed the union of the two.

Leaves closely appressed, very obtuse, 1 to 1½ lines long. Fruit-scales orbicular at the apex . . . . 1. *A. cupressoides.*
Leaves looser, acute, 1½ to 2 lines long. Fruit-scales shortly acuminate . . . . 2. *A. laxifolia.*
Leaves loose, incurved, acute, 3 to 4 lines long. Fruit-scales lanceolate at the apex . . . . 3. *A. selaginoides.*

1. **A. cupressoides,** *Don in Trans. Linn. Soc.* xviii. 173, t. 13, f. 2. A tree of 20 to 40 ft. in height, with sometimes a girth of 15 ft., the branches ascending, with more or less distichous branchlets, sometimes apparently opposite. Leaves closely appressed and densely covering the branches, broad, very obtuse, thick and keeled, 1 to 1½ lines long and broad. Fruit-cones rarely ½ in. diameter when open, the dilated apex of the scales nearly orbicular, with a short dorsal point. Seeds usually about 3 to each scale.—Hook. Ic. Pl. t. 559; Hook. f. Fl. Tasm. i. 354; Parlat. in DC. Prod. xvi. ii. 433; *Cunninghamia cupressoides*, Zucc. in Sieb. Fl. Jap. ii. 9; *Arthrotaxis imbricata*, Maule (Parlatore).

Tasmania. Lake St Clair, Western Mountains, Pine river, *Gunn*, and others; Lake Teuton, *F. Mueller*.

2. **A. laxifolia,** *Hook. Ic. Pl.* t. 573. A tree of 25 to 30 ft. closely allied to *A. cupressoides*, from which it differs in the leaves less closely appressed although imbricate, acute, and mostly about 2 lines long, and the cones rather larger with the scales more acuminate, thus forming an approach as it were to the *A. selaginoides*.—Hook. f. Fl. Tasm. i. 354; Parlat. in DC. Prod. xvi. ii. 434; *A. Doniana*, Park. (Parlatore).

Tasmania. Near the summits of the Western Mountains, at an elevation of 3000 to 4000 ft., *Gunn*, *Archer*, *F. Mueller*.

The leaves of young plants sent by Gunn are not elongated as in *A. selaginoides*.

3. **A. selaginoides,** *Don in Trans. Linn. Soc.* xviii. 172, t. 14. A stouter tree than the two other species, attaining 45 ft. Leaves loosely imbricate, lanceolate, acute, keeled, incurved, 3 to 4 lines long, those of the young seedlings more linear, spreading, ½ in. long. Fruit-cones ½ to ¾ in. diameter, the scales terminating in a lanceolate point. Seeds usually 4 to 6 under each scale.—Hook. Ic. Pl. t. 574; Hook. f. Fl. Tasm. i. 354; Parlat. in DC. Prod. xvi. ii. 434; *Cunninghamia selaginoides*, Zucc. in Sieb. Fl. Jap. ii. 9; *Arthrotaxis alpina*, Van Houtte (Parlatore).

Tasmania. Western Mountains, at an elevation of 3000 to 4000 ft., *Gunn*, *Archer*, *F. Mueller*.

6. ARAUCARIA, Juss.

Flowers diœcious or rarely monœcious, the amenta terminal. Male amenta cylindrical; stamens numerous, spirally imbricated, contracted at the base, with an ovate or lanceolate incurved scale-like apex; anther-cells 6 to 20, in 2 rows. Females with a single reflexed ovule within each scale. Fruit-cones large, ovoid or globular, the scales very numerous, closely imbricate, the margins usually attenuated into wings at the base, the apex thickened and woody, with a raised transverse line often produced into a lanceolate or pungent point. Seeds flattened, obovoid-oblong, not winged, adnate to the scale at the base, free at the apex. Embryo with 2 cotyledons, sometimes deeply divided so as to appear to be 4.—Trees often very lofty, the branches almost verticillate.

Leaves in close spires, flat or on sterile branches vertically compressed, short and rigidly acicular or lanceolate and longer, pungent-pointed or rarely obtuse, with a prominent midrib. Fruit-cones in some species attaining a very large size.

The genus ranges over extratropical and subtropical South America, New Zealand, and some of the South Pacific Islands. The Australian species are both endemic.

Leaves rigidly acicular, 2 to 6 lines long. Fruit-cones about 3 in. long . . . . 1. *A. Cunninghamii.*
Leaves lanceolate, ¾ to 1½ in. long. Fruit-cones about 9 in. long 2. *A. Bidwilli.*

1. **A. Cunninghamii,** *Ait. in Sweet, Hort. Brit.* 475. A tree with a pyramidal or somewhat flattened head, attaining in some situations 150 to 200 ft., in others remaining much smaller. Leaves crowded in dense spires, rigidly acicular and very acute, those of the barren branches often spreading, straight, vertically compressed, with the dorsal rib decurrent and ¼ to ½ in. long, those of the flowering branches from a broad adnate base triquetrous or lanceolate, incurved and rather shorter. Male amenta sessile, cylindrical, very dense, 2 to 3 in. long and 3 to 4 lines diameter, the scale-like apex of the stamens ovate-rhomboidal and acute. Fruit-cones ovoid, about 3 in. long and 2 in. diameter, the scales (including their marginal wings) broadly cuneate, the broad hard apex terminating in a lanceolate spreading or recurved rigid point.—Parlat. in DC. Prod. xvi. ii. 372.

Queensland. Port Bowen, *R. Brown;* Brisbane river, Moreton bay, extending 80 miles inland, and northward to lat. 14°, *A. Cunningham, Leichhardt;* Rockhampton, *Herb. F. Mueller;* Burdekin river, *Fitzalan.* Known as the "Moreton bay Pine."
N. S. Wales. Hastings and Clarence rivers, *Beckler.*

In general aspect and in foliage the tree much resembles the Norfolk Island Pine (*A. excelsa*), but the cones are very different.

2. **A. Bidwilli,** *Hook. Lond. Journ. Bot.* ii. 503, t. 18. A tree, attaining from 100 to 150 ft. in height, with a remarkably stout trunk and smooth bark, the branches usually in whorls of about 16, crowded at the top of the tree. Leaves in crowded spires, lanceolate and about ¾ to 1½ in. long on some barren branches, ovate-lanceolate and ½ in. long on the flowering branches, smooth and shining, of a pale colour when dry, with a broad midrib prominent underneath. Male amenta very dense, appearing sessile in some of the upper axils from the shortness of the flowering branchlets, 2 to 3 in. long and 4 to 5 lines diameter, the imbricate scale-like apices of the stamens triangular, acute, about 1 line broad. Fruit-cones erect on the topmost branches, ovoid-globose, about 9 in. long and 7 in. diameter, the scales loosely imbricate, about 4 in. long and 3 broad, tapering towards their winged base, the terminal points recurved and spinescent. Seeds obovate, 2 to 2½ in. long and ¾ in. broad.—Parlat. in DC. Prod. xvi. ii. 371.

Queensland. Brisbane range, N.W. of Moreton bay, *Bidwill;* between Cleveland and Rockingham bays, *W. Hill;* Condamine, Dawson, and Burnett rivers, *Leichhardt.*

7. DAMMARA, Rumph.

(Agathis, *Salisb.*)

Flowers diœcious, the amenta sessile or nearly so. Male amenta axillary or lateral, cylindrical, surrounded by a few imbricate scales at the base; stamens numerous, in close spires, the imbricate scale-like apices thick, clavate or orbicular and slightly incurved. Anther-cells 5 to 16, cylindrical, pendulous, in 1 or 2 transverse rows. Female amenta lateral or terminal, the scales numerous, with 1 reversed ovule within each. Fruit-cones large, ovoid-globular, the scales closely imbricate, deciduous, flattened, very broadly cuneate, the margins more or less attenuated into wings, the apex slightly thickened, coriaceous or scarcely woody. Seeds oblong or cuneate, free, flattened, truncate or emarginate at the end, one margin produced into a horizontal erect or decurved wing.—Trees with spirally arranged flat leaves.

Besides the Australian species which is endemic, there are one from East India and the Archipelago, one from New Zealand, and two or perhaps three from New Caledonia.

1. **D. robusta,** *C. Moore; F. Muell. in Trans. Pharm. Soc. Vict.* ii. 174. A tree, attaining a height of nearly 150 ft., the branches nearly verticillate. Leaves ovate-lanceolate or oblong-lanceolate, shortly acuminate or almost obtuse, rounded or tapering at the base and contracted into a very short petiole, 2 to 3 or rarely 4 in. long, rigidly coriaceous, very finely striate. Male amenta about 1½ in. long, sessile within a few broad orbicular or reniform bracts of about 2 lines diameter. Fruit-cones ovoid-globular, about 4 in. long and 3 in. diameter, the scales as broad as long, the lateral wings more or less indented on each side at the base so as to leave marginal deflexed auricles. Seeds nearly ½ in. long, truncate or emarginate at the apex, one angle sometimes produced into a short broad point, the other into an oblong erect wing as long as the seed itself.—Parlat. in DC. Prod. xvi. ii. 375.

Queensland. Scattered through the dense forest country near Wide bay, *Bidwill*, and others.

It is probably through a slip of the pen that Parlatore (in DC. Prod. xvi. ii. 376) includes New Holland in the area of the Kauri Pine, *D. australis*, Lam., for Mercury bay, the special station given, is in New Zealand.

8. DACRYDIUM, Soland.

Flowers diœcious or rarely monœcious. Male amenta ovoid or cylindrical. Stamens several, spirally arranged, imbricate, very shortly contracted at the base, the apex incurved. Anther-cells 2, opening outwards in 2 valves. Female amenta of a very few small fleshy scales in a short spike or reduced to a single one, each with a single reversed ovule. Fruit-cones small. Seeds erect or nearly so, seated in a short membranous cup or disk, ovoid-oblong, the outer integument membranous, the inner crustaceous.—Trees or shrubs. Leaves small and closely imbricate, or on the young plants longer and linear. Amenta small and terminal.

The genus is dispersed over the Indian Archipelago, New Caledonia, and New Zealand; the Tasmanian species is endemic.

1. **D. Franklinii,** *Hook. f. in Hook. Lond. Journ.* iv. 152, t. 6, *and Fl. Tasm.* i. 357, t. 100. A tree attaining sometimes 100 ft. though generally 60 to 80 ft. high, with a dense spreading head and pendulous branchlets. Leaves small and scale-like, acute and spreading on the young plant, closely appressed obtuse with a very prominent keel and only ½ line long on the branchlets of the tree, all spirally arranged but often quadrifarious. Male amenta (only seen in an imperfect state) erect, cylindrical, consisting of about 12 to 15 stamens. Fruit-cones very short, decurved, consisting of 4 to 8 loosely imbricate persistent scales scarcely longer than the leaves. Seeds globular, about 1 line diameter, seated in a very shallow membranous cup or disk, the outer integument membranous with a minute orifice, the inner one crustaceous.—Parlat. in DC. Prod. xvi. ii. 495; *D. Huronense*, A. Cunn. Herb.

Tasmania. Southern and western coasts of the island, *A. Cunningham*, and others. "Huron Pine" of the colonists.

9. PHEROSPHÆRA, Archer.

Flowers diœcious. Male amenta ovoid-globular. Stamens several, spirally arranged, very shortly contracted at the base, the incurved apex not so broad as the anther. Anther-cells 2, parallel, opening outwards in 2 valves. Female amenta ovate, with several spirally arranged scales, and a single erect ovule within each. Fruit-cones ovoid, the scales thickened at the base, concave. Seeds (as yet unripe) erect, ovoid-oblong, the outer integument green, loose, contracted into a neck open and crenulate at the orifice and sometimes longitudinally winged outside, the inner integument of a firmer consistence.—A shrub with small imbricate leaves. Amenta small, terminal.

The genus is limited to a single species, endemic in Tasmania.

1. **P. Hookeriana,** *Archer in Hook. Kew Journ.* ii. 52 (*partly*). A densely branched erect shrub. Leaves closely imbricate, often falling into 4 or 5 rows, but not opposite and strictly decussate as in *Diselma* and *Microcachrys*, thick, very obtuse, keeled, about ½ line long and broad on the flowering branches, often ¾ line long on the older ones. Male amenta erect, 1 to 1½ lines long, with about 10 to 15 stamens. Female cones decurved, scarcely 1½ lines long, with about 4 to 8 scales, thickened at the base into an obtuse external protuberance, acuminate at the apex but obtuse, about ½ line long. Seed small.—Hook. f. Fl. Tasm. i. 355, t. 99; Parlat. in DC. Prod. xvi. ii. 497.

Tasmania. Mountains near Lake St. Clair, *Gunn;* high alpine flats, Mount Field East, *F. Mueller.*

10. PHYLLOCLADUS, Rich.

(Thalamia, *Spreng.*)

Flowers monœcious (or sometimes diœcious?). Male amenta cylindrical, surrounded by bracts at the base. Stamens imbricated, contracted into a very short stipes, the scale-like apex small; anther-cells 2, adnate. Female amenta of a very few scales in a short spike or reduced to a single one, each with a single erect ovule or the upper scales empty. Fruiting scales thick and fleshy, enclosing the base of the seed. Seed erect within a cup-shaped disk, ovoid, the outer integument membranous, not winged; inner one crustaceous. Cotyledons . . .—Trees with expanded leaf-like flat rigid toothed or lobed branchlets or *cladodia*, the real leaves reduced to small appressed scales. Amenta small, terminal, the males often clustered, the females very small.

Besides the Australian species which is endemic, there is one from New Zealand and another from Borneo.

1. **P. rhomboidalis,** *Rich. Conif.* 130, t. 3. A slender tree, attaining 60 ft. but reduced to a shrub on the summits of mountains, the persistent branches more or less verticillate, the cladodia or deciduous leaf-like branchlets cuneate or rhomboidal, obtuse, obtusely toothed or lobed, $\frac{3}{4}$ to 2 in. long, the real leaves or scales very small and subulate or fine-pointed. Male amenta usually 2 or 3 together, 3 to 4 lines long. Females globular and about 2 lines long or sometimes lengthening out in fruit to 3 or 4 lines, with 1, 2, or 3 fertile scales surmounted by 1 or 2 barren ones. Seeds scarcely exceeding the scales.—Hook. f. Fl. Tasm. i. 358; Parlat. in DC. Prod. xvi. ii. 499; *P. Billardieri*, Mirb. in Mem. Mus. Par. xiii. 76; *P. asplenifolia*, Hook. f. in Hook. Lond. Journ. iv. 151; *Podocarpus asplenifolia*, Labill. Pl. Nov. Holl. ii. 71, t. 221; *Thalamia asplenifolia*, Spreng. Syst. iii. 890.

Tasmania. Derwent river, *R. Brown;* common in dense forests in the mountainous and southern parts of the island, *J. D. Hooker*.

11. PODOCARPUS, L'Her.

Flowers diœcious or rarely monœcious. Male amenta cylindrical. Stamens numerous, slightly contracted at the base, the scale-like apices closely imbricate; anther-cells 2. Female amenta of 2 to 4 bracts or scales more or less succulent and united with the rhachis in an oblong receptacle, unequally 2- or 4-toothed at the apex. Ovules 1 or 2, exserted, reversed and adnate to an erect stipe from within the larger teeth or bracts of the receptacle. Seeds drupaceous, the nucleus enclosed in a double integument, the outer one succulent, the inner one long. Embryo with 2 short cotyledons and an inferior radicle.—Trees or shrubs. Leaves alternate or rarely opposite, usually distichous and flat, with a prominent midrib. Buds scaly. Amenta axillary or terminal, solitary or several together, sessile or shortly racemose.

The genus is dispersed over the tropical and subtropical regions of the Old World, from South Africa and New Zealand to Japan, and over the whole of South America. The Australian species are all endemic.

Leaves broadly oblong linear or lanceolate, $1\frac{1}{2}$ to 10 in. long. Male amenta 1 to $1\frac{1}{2}$ in. long . . . . 1. *P. elata.*

Leaves linear, acute, 1½ to 3 in. long. Male amenta 2 to 4 lines long.
Leaves pungent-pointed. Male amenta clustered in the axils. Eastern species . . . . 2. *P. spinulosa.*
Leaves very acute but scarcely pungent. Male amenta usually solitary. Western species . . . . 3. *P. Drouyniana.*
Leaves linear, not exceeding ½ in. Male amenta small and solitary 4. *P. alpina.*

1. **P. elata,** *R. Br.; Mirb. in Mem. Mus. Par.* xiii. 75. A tree of 50 to 100 ft. Leaves oblong-linear or broadly linear-lanceolate, very variable in size, on some specimens with young flowers 1½ to 2 in. long and ¼ in. wide and quite straight, in the ordinary forms 3 to 6 in. long and 4 to 6 lines broad, straight or slightly falcate, on some barren specimens 8 to 10 in. long ¾ in. wide and much falcate, acute or rather obtuse, the midrib prominent, the petiole very short. Male amenta clustered 2 or 3 together, sessile, 1 to 1½ in. long, surrounded by several short scales or bracts. Female peduncles 2 to 3 lines long, solitary in the axils of the lower leaves or more frequently of small bracts at the base of the year's branches. Fruiting receptacle oblong, 4 to 6 lines long, with usually only one seed, ovoid or globular, 4 to 6 lines diameter.—Parlat. in DC. Prod. xvi. ii. 517; *P. ensifolia,* R. Br.; Mirb. l.c.; Parlat. l.c.; *P. falcata,* A. Cunn. Herb.

Queensland. Cape Grafton, *A. Cunningham;* Brisbane river, Moreton bay, *F. Mueller:* Rockingham bay, *Dallachy.*

N. S. Wales. Hunter's and Paterson rivers, *R. Brown;* Macquarrie river, *Fraser;* Hunter's river, *Leichhardt;* Hastings and Clarence rivers, *Beckler;* Richmond river, *Henderson;* Illawarra, *A. Cunningham, Macarthur.*

2. **P. spinulosa,** *R. Br.; Mirb. in Mem. Mus. Par.* xiii. 75. A much-branched erect shrub. Leaves linear, rigid, more or less pungent-pointed, 1½ to 2 in. long. Male amenta numerous in sessile axillary clusters and not above 2 to 3 lines long. Female peduncles about 2 lines long, in the axils either of the lower leaves or of scale-like bracts below the leaves of the young branches, with 2 small opposite bracteoles immediately under the oblong 2-lobed receptacle. Seeds larger than in *P. elata.*—Parlat. in DC. Prod. xvi. ii. 513; *Taxus spinulosa,* Sm. in Rees Cycl. xxxv.; *P. pungens,* Caley; Don in Lamb. Pin. ed. 2, 123 (Parlatore).

N. S. Wales. Port Jackson, *Caley, Woolls,* and others. "Native Plum or Damson."

3. **P. Drouyniana,** *F. Muell. Fragm.* iv. 86, t. 31. A shrub or tree, with virgate branches. Leaves crowded, linear, 2 to 3 in. long, sharp-pointed but not so rigid as those of *P. spinulosa.* Male amenta oblong-cylindrical, 2 to 4 lines long, usually solitary in the axils, terminating scaly peduncles or branchlets of ¼ to ½ in. Female peduncles also solitary, about ½ in. long, the receptacles when in fruit about ½ in. long, either equally 2-lobed with 2 seeds, or unequally so with 1 seed exserted from the larger lobe, the seeds larger even than in *P. elata.*—Parlat. in DC. Prod. xvi. ii. 514.

W. Australia, *Drummond,* 2nd coll. n. 153, 154, 3rd coll. n. 199, 200; Vasse river, *Oldfield;* swampy plains, Tone river, *Maxwell.*

4. **P. alpina,** *R. Br.; Mirb. in Mem. Mus. Par.* xiii. 75. A straggling densely-branched shrub, usually low, but sometimes attaining 10 to 12 ft. Leaves crowded, linear, straight or falcate, rigid, varying from $\frac{1}{4}$ in. long and obtuse to $\frac{1}{2}$ in. and acute, especially on luxuriant barren branches. Male amenta 2 to 3 lines long, usually solitary and sessile or nearly so in the axils. Fruits much smaller than in any other species, the fleshy receptacle about $1\frac{1}{2}$ lines long, sessile in the axil, the ovoid seed not much longer.—Hook. f. Fl. Tasm. i. 356; Parlat. in DC. Prod. xvi. ii. 520; *P. Lawrencii*, Hook. f. in Hook. Lond. Journ. iv. 151.

Victoria. Mount Butler, Hardinge's range, Cobberas mountains, at an elevation of 3000 to 6000 ft., *F. Mueller.*

Tasmania. Mount Wellington (Table mountain), *R. Brown;* Mountain localities at an elevation of 3000 to 4000 ft., *J. D. Hooker.*

The *P. Lawrencii* can scarcely be considered as a distinct variety, for it appears to be the form assumed by the luxuriant barren branches of young plants.

ORDER CXVII. CYCADEÆ.

Flowers unisexual, without any perianth. Male flowers forming catkins or cones consisting of numerous spirally arranged imbricated scales (or stamens), more or less cuneate, bearing on the concealed portion of their under surface numerous sessile or rarely stipitate anther-cells, each opening in 2 valves; the upper imbricate and exposed part of the scales hardened and often much thickened, the apex truncate or more or less produced into an incurved or recurved point or lanceolate appendix. Female cones consisting of numerous scales, imbricate at least when young, either with one pendulous ovule (or carpel) on each side of the thickened and hardened apex, or with 3 or more erect ovules (or carpels) in marginal notches below the flattened acuminate and usually dentate or pinnatifid apex. Fruiting-cone enlarged and either remaining imbricate with 2 pendulous seeds to each scale, or the scales with marginal seeds spreading as the central shoot is developed within the cone. Seeds naked (or nuts) with a thick or hard outer coating or integument and a fleshy albumen, in a central cavity of which the straight embryo is suspended by a long folded cord. Cotyledons 2, undivided.—Palm-like plants, with a thick globose and underground or erect and cylindrical woody stem, simple or rarely slightly branched, marked with the scars or bases of the old leaves. Leaves forming a crown at the apex of the stem, once or twice pinnate. Cones sessile or very shortly pedunculate within the crown of leaves.

The Order extends over tropical America, subtropical and southern Africa, and tropical Asia. Of the three Australian genera one is also in Asia and Africa, the other two are endemic. The theoretical significance of the outer coating of the ovules and seeds, whether carpellary or seminal, is, as in the *Coniferæ*, still the subject of contention.

Leaves simply pinnate. Pinnæ linear, with a prominent midrib. Female scales elongated, woolly, with 2 or more erect ovules on each side in marginal notches . . . . 1. CYCAS.
Leaves simply pinnate. Pinnæ linear, with several longitudinal scarcely prominent nerves. Female scales with 1 pendulous ovule on each side under the thickened acuminate apex . . . . 2. MACROZAMIA.
Leaves doubly pinnate. Pinnules obliquely ovate or broadly falcate, with scarcely prominent veins. Cones of *Macrozamia*, but the apex of the scales truncate . . . . 3. BOWENIA.

1. CYCAS, Linn.

Male cones oblong-ovoid or globular, the scales cuneate, hard, the thickened apex more or less produced into a straight or incurved point. Female cones at first globular, but opening out by the growth of the central shoot, the scales elongated, tomentose or woolly, flat, bearing on each margin 2 to 5 ovules, erect in distant notches, the apex of the scale dilated acuminate and toothed or pinnatifid, the scales at first loosely imbricate, at length spreading or recurved. Seeds large, erect. Leaves simply pinnate, the pinnæ numerous, linear, with a prominent midrib, circinnate in vernation.—Leaves long, simply pinnate, the pinnæ numerous, linear, with the midrib prominent underneath.

The genus extends over the Indo-Australian region, reaching Madagascar and the east coast of Africa to the westward and Japan to the northward. The Australian one or more species are believed to be endemic, but the distinctive characters are very insufficiently known.

1. **C. media,** *R. Br. Prod.* 348. Trunk sometimes attaining 8 to 10 ft. sometimes twice that height, rarely branched at the top. Leaves 2 to 4 ft. long or even more, the pinnæ very numerous, straight or falcate, obtuse or pungent-pointed, flat or slightly concave above when young, prominently keeled underneath, the margins often at length recurved, mostly slightly decurrent on the rhachis, glabrous or slightly pubescent when young, the longer ones varying from 3 to 8 inches, the lower ones shorter and more contracted at the base, the lowest passing into small prickles which are sometimes very few or scarcely any, sometimes continued almost to the base of the petiole. Cones variable in size, but apparently smaller than in *C. circinnalis*, which the species otherwise resembles. Seeds 1 to 1½ in. long, glabrous—A. DC. Prod. xvi. ii. 527.

N. Australia. North-west and north coasts, *A. Cunningham;* Port Essington, *Armstrong;* Escape cliffs, *Hulls*.

Queensland. Burnett and Dawson rivers, *F. Mueller;* Cape Upstart, *Burdekin Expedition;* Rockingham bay and Mount Elliott, *Dallachy;* Rockhampton, *Thozet;* Castlereagh bay (*W. Hill?*).

Three Australian species of *Cycas* have been described, the above *C media*, *C. angulata*, Br. Prod. 348, A. DC. Prod xvi. ii. 527, and *C. gracilis*, Miq. in Versl. K. Akad. Wet. Amsterdam, xv. 366, A. DC. l.c. 528; but whether these three are really distinct, and by what characters they are to be separated from one or the other of the common Asiatic species, we have unfortunately no materials for determining. These allied Asiatic ones, *C. circinnalis*, Linn. (*C. sphærica*, Roxb.) and *C. Rumphii*, Miq. (*C. circinnalis*, Roxb.) are distinguished most readily by the scales of their male cones, which are unknown in the great majority of the Australian specimens. I have only seen two, both in F. Mueller's collection; in both the scales are much smaller than in the Indian species. In one, belonging to Hull's Escape Cliff specimens, they are rather narrow, 1½ in. long, the anther-cells commence almost at the base and occupy fully ¾ of the scale; in the other, collected by the Burdekin Expedition, the scales are still smaller, 1 to 1¼ in. long, and the anther-cells cover rather less of the under surface, not reaching to the base, and ceasing rather lower down. These were described by Miquel as his *C. gracilis*, but there are no leaves with them. In the *C. gracilis* of our gardens, the fronds are small with the rhachis slightly furfuraceous. In Castlereagh bay specimens similar small leaves are slightly woolly-pubescent; in almost all other specimens the full-grown leaves are very glabrous. We have female cone-scales of several specimens varying much in the number of ovules or seeds they bear, 2 to 5 on each margin; but I am quite unable to connect these differences with any characters derivable from the leaves. It is much to be urged on resident botanists in tropical Australia carefully to collect and to match with accuracy male and female cones with the leaves of all these species or varieties of *Cycas*, in order to determine their systemic value.

2. MACROZAMIA, Miq.

Cones of both sexes ovoid oblong or cylindrical, or the females rarely nearly globular, the scales hard, more or less thickened at the apex, with an erect spreading or rarely recurved point, either broad and short or elongated and narrow. Scales of the females with one pendulous ovule and seed on each side.—Trunk and leaves of *Cycas*, except that the pinnæ have no midrib, but are more or less distinctly striate, especially on the under side, with several parallel equal veins, the whole leaf occasionally slightly twisted in some species, but not constantly so in any one.

The genus is limited to Australia, and there represents the South African *Encephalartos*, with which F. Mueller proposes to reunite it. The latter genus has, however, a much more rigid habit and very obtuse or truncate cone-scales, which, together with the geographical distribution, seems to warrant the following Miquel and A. De Candolle, in maintaining the two genera as distinct. Still less does it seem advisable again to reduce these Old World forms to the American genus *Zamia*, characterized by the articulate attachment of the pinnæ as well as by the cones. It is, however, a much more difficult matter to characterize the species of *Macrozamia*. With regard to two of them, *M Perowskiana* and *M. Paulo-Gulielmi*, there can be no doubt, but the remainder are very puzzling; for although we have at least three apparently distinct forms of fructification, and at least twice as many marked forms of foliage, they are very rarely matched with certainty in our wild specimens, and very few have produced cones in our garden collections. Thus, after having spent much time over the genus, it remains with me a matter of great doubt whether, besides the two above mentioned, we have really one variable species, or what number from two to six more or less constantly distinct. Most if not all the *Macrozamiæ* when very young have their pinnæ frequently denticulate at the end, as represented by Miquel, Monogr. Cycad. t. 6, under the name of *Encephalartos tridentatus*.

Pinnæ very narrow, often nearly terete. Cones small, rarely above 4 in. Fruit very woolly . . . . 1. *M. Paulo-Gulielmi.*
Pinnæ flat, inserted on the margins of the rhachis, contracted at the base, the larger ones usually above 3 lines broad. Cones 4 to 10 in., glabrous. Trunk glabrous or rarely loosely woolly.
  Rhachis of the leaves usually raised longitudinally between the pinnæ. Cone-scales much flattened.
    Eastern species, the insertion of the pinnæ mostly longitudinal. Points of the scales usually short . . . . 2. *M. spiralis.*

Western species, the insertion of the pinnæ very oblique or almost transverse. Cones large, with long points to the upper or to all the scales . . . . 3. *M. Fraseri.*
Rhachis of the leaves very flat between the pinnæ and often broad. Cone-scales very thick. Eastern subtropical species . . . . 4. *M. Miquelii.*
Pinnæ inserted by their broad base along the centre of the upper surface of the rhachis, scarcely separated by a very narrow line. Cones large, pubescent, the scale points broad and often recurved . . . . 5. *M. Perowskiana.*

See also doubtful forms under *M. spiralis* and *M. Miquelii.*

1. **M. Paulo-Gulielmi,** *F. Muell. Fragm.* i. 86. Trunk scarcely raised from the ground, covered with the woolly imbricate base of the old petioles. Leaves otherwise glabrous, 1 to 3 ft. long, the rhachis narrow but often flat on the top; pinnæ numerous, very narrow and often almost terete, contracted and sometimes callous at the base, the longer ones 6 to 8 in. long and 1 to 1½ lines broad, thick and obscurely veined. Cones on woolly peduncles of 1 to 3 in., the males oblong-cylindrical, scarcely above 3 in. long, the scales about 4 lines broad, somewhat thickened at the apex, with a short point. Fruiting cones about 4 in. long and fully 2 in. thick, the larger scales about 1 in. broad and rather thick, those of the lower part of the cone narrower and thicker, the apex almost rhomboidal with a very short point.—A. DC. Prod. xvi. ii. 536; *Encephalartos Paulo-Gulielmi*, F. Muell. in Trans. Pharm. Soc. Vict. ii. 91; Miq. in Versl. K. Akad. Wet. Amsterdam, xv. 374.

Queensland. Moreton bay, Maranoa, *W. Hill;* Wide bay, *Leichhardt;* Mackenzie river, *Mrs. Cobham.*
N. S. Wales. New England, *C. Stuart.*

Some specimens in the herbarium marked "*Macrozamia tenuifolia,*" Sydney, *Macleay,* appear to belong to the same species, so also possibly the *M. tenuifolia* of our gardens.

2. **M. spiralis,** *Miq. Monogr. Cycad.* 36, t. 4, 5. Trunk short. Leaves glabrous, 2 to 4 ft. long, the rhachis usually more or less raised longitudinally on the upper surface between the two rows of pinnæ. Pinnæ numerous, flat, straight or slightly falcate, the larger ones 8 to 10 in. long and 3 to 4 lines broad, marked on the underside with longitudinal parallel veins more prominent than in *M. Miquelii,* slightly contracted and callous at the base, inserted longitudinally and the lower margin slightly decurrent, the lower pinnæ much smaller and sometimes passing into a few small teeth, a gland-like callosity at the base of the pinnæ on the upper side sometimes prominent, sometimes very obscure. Male cones 6 to 10 in. long, 1½ to 2 in. thick, the scales much flattened, the apex however thicker than in *M. Preissii,* about ½ in. broad, tapering into an incurved point very short on the lower scales, ¾ in. long or more on the upper ones. Fruiting-cones varying much in size but usually shorter and thicker than the males, the apex of the larger scales 1 to 1½ in. broad and usually 3 or 4 lines thick, with an incurved point usually short.—A. DC. Prod. xvi. ii. 535; *Zamia spiralis,* R. Br. Prod. 348 (partly); *Encephalartos spiralis,* Lehm. Pugill. vii. 13.

Queensland. Brisbane river, Moreton bay, *F. Mueller.*
N. S. Wales. Port Jackson, *R. Brown, Woolls*, and others; Taylor's Range, *Leichhardt;* Richmond river, *C. Moore;* Springsure, *Wuth.*

This is probably the only species which in the Eastern districts descends to Port Jackson; but in the imperfect state of our specimens, whether in herbaria, museums or gardens, it is not always easy to distinguish them from forms or varieties of the *M. Preissii* or *M. Miquelii*, nor to determine whether the following are varieties or distinct species.

Var.? *corallipes.* The callous base of the pinnæ of a bright red. Cones of *M. spiralis.*—*M. corallipes*, Hook. f. in Bot. Mag. t. 5943, from garden specimens. The rhachis is figured as very flat, with marginal pinnæ, almost as in *M. Miquelii*, but narrower. In a plant in Kew Gardens the rhachis is narrower, and the bases of the pinnæ are losing their red colour. *M. Macleayi*, Miq. in Nieuw. Bijdr. Cycad. 53, Nouv. Mater. Cycad. 58, also described from garden specimens, must be the same variety. He observes that the red base of the pinnæ is not constant.

Var.? *secunda.* Leaves small, the pinnæ very narrow and rigid, all erect from the upper surface of the narrow rhachis, so as to be parallel instead of spreading and distichous, and said to be so in the living state.—Dry situations, Reedy Creek, N. S. Wales, *C. Moore.* A form possibly due to the dry station, but a variety in cultivation has a similar aspect. The cones are unknown.

Var.? *cylindracea.* *M. cylindrica* is a garden name for a small form with the narrow foliage nearly of *M. Paulo-Gulielmi*, but with a glabrous trunk and more terete rhachis. Cones unknown.

Var.? *diplomera*, F. Muell. Leaf-pinnæ narrow, almost as in *M. Paulo-Gulielmi*, but the trunk glabrous, and many of the pinnæ divided to below the middle into 2 branches. Cones unknown.—Castlereagh river, *C. Moore;* Maitland, *Rucker.*

3. **M. Fraseri**, *Miq. Monogr. Cycad.* 37. A western species very nearly allied to the *M. spiralis*, and from leaves alone often very difficult to distinguish from it. It is usually much taller, the trunk often attaining 10 to 12 ft., the leaves are usually more rigid, the rhachis raised between the rows of pinnæ as in *M. spiralis*, and the pinnæ more or less contracted and callous at the base, but their insertion, especially in the lower ones, is more oblique and sometimes almost transverse, they are also frequently shorter with a more rigid acute point and their longitudinal veins are finer and less conspicuous. The cones are larger and longer, the scales much flattened, the upper ones of the males tapering into a narrow erect point almost subulate of 1½ to 2 in. in the females into a lanceolate point of 2 to 3 in. —*Zamia spiralis*, R. Br. Prod. 348 (partly); *M. spiralis*, Miq. Monogr. Cycad. 36 (as to the Western plant only); *Cycas Riedlei*, Gaudich. in Freyc. Voy. Bot. 434; *Encephalartos Fraseri*, Miq. Versl. K. Akad. Wet. Amst. xv. 368; *M. Preissii*, Lehm. Pugill. viii. 31; Heinzel in Nov. Act. Nat. Cur. xxi. 203, t. 10 to 13; Miq. in Linnæa xix. 415, t. 2, 3; A. DC. Prod. xvi. ii. 535; *Encephalartos Preissii*, F. Muell. in Journ. Pharm. Soc. Vict. ii. 90.

W. Australia, *Preiss;* Swan river, *Oldfield;* King George's Sound, *R. Brown;* Stokes bay, *Maxwell.*

I am unable to perceive any grounds for distinguishing two or more Western species. Oldfield's specimens (on which was founded the *M. Oldfieldii*, Miq. Nieuw. Bijdr. Cycad. 53, or Nouv. Mater 58; A. DC. Prodr. xvi. ii. 535; *Encephalartos Oldfieldii*, Miq. in Versl. K. Akad. Wet. Amst. xv. 370), consist of leaves only, which are smaller and stiffer, with more rigid and shorter pinnæ than in some other specimens. Some cones that I have seen of Preiss's are nearly 1 ft. long.

M. Macdonelli, F. Muell. Fragm. ii. 179, v. 49; or *Encephalartos Macdonelli*, F. Muell. in Miq. Versl. K. Akad. Wet. Amst. xv. 376, from Neale's river in Central Australia, *M'Douall Stuart*, is described only from half a dozen pinnæ with a fragment of their rhachis, and these show nothing to distinguish them from those of *M. Fraseri.*

4. **M. Miquelii,** *F. Muell.; A. DC. Prod.* xvi. ii. 535. Fronds 2 to 4 ft. long, the base of the petiole in the typical form densely covered with a loose floccose wool readily rubbed off, the upper surface of the rhachis very flat between the rows of pinnæ and often in the lower part ½ in. broad; pinnæ usually longer than in *M. spiralis*, straight or falcate, the longitudinal veins finer and less prominent, contracted and more or less callous at the base and their insertion at the rhachis quite marginal, the lowest often reduced to small teeth. Male cones cylindrical, 6 to 8 in. long, 2 to 2½ in. thick, the scales thickened into a woody rhomboidal apex almost as thick as broad, with the ascending points in the centre very short, almost obsolete on the lower scales, ¼ to ½ in. long on a few of the upper ones. Female cones about as long and thicker, the scales fewer, their apex at least ¾ in. broad and ½ in. thick, and mostly with the transverse appendage tapering into an erect linear-lanceolate point of 1 to 2 in.—*Encephalartos Miquelii*, F. Muell. Fragm. iii. 38.

Queensland. Moreton bay, *W. Hill;* Rockhampton, *Thozet.*

The above typical form seems to be characterized by the woolly base of the petioles, the broad flat rhachis, and the thick scaled cones; but there are other specimens where these characters do not appear to be conjoined. Some, from Queensland and perhaps also from N. S. Wales, have the cones and in most respects the foliage of *M. Miquelii*, but with no trace of the wool on the base of the petioles. These have found their way into some of our gardens and museums under the names of *M. Macleayi* and *M. Mackenzii*, and have also in our gardens been sometimes misnamed *M. Fraseri.* Most of them, however, are either without cones, or with no certainty that the cones have not been mismatched.

5. **M. Perowskiana,** *Miq. Cycad. N. Holl.* This is the largest and most distinct of Australian *Macrozamiæ.* Trunk 18 to 20 ft. high and at least 1 ft. thick. Leaves 7 to 12 ft. long, the petioles angular, glabrous or pubescent at the base; pinnæ 1 to 2 ft. long in the larger leaves, ½ in. broad below the middle, very obscurely and finely marked with parallel veins, only slightly contracted at the base and inserted longitudinally along the centre of the upper surface of the rhachis, without any or only a very narrow line separating the two rows, the upper ones gradually shorter. Male cones ovoid, 4 to 6 in. long, 3 to 4 in. diameter, the apex of the scales 1 to 1½ in. broad, very thick and produced into a short triangular or lanceolate almost obtuse point. Female cones 8 to 16 in. long and very thick, the scales shorter and broader than in the males, the apex tomentose-pubescent, often 2 in. broad, tapering into a short and very obtuse or rather longer and lanceolate recurved point. Seeds very oblique, about 2 in. long and 1 in. broad.—*Lepidozamia Perowskiana*, Regel in Bull. Soc. Imp. Nat. Mosc. 1857, i. 184, t. 4; *Macrozamia Denisonii*, F. Muell. Fragm. i. 41, 243; A. DC. Prod. xvi. ii. 536; *Encephalartos Denisonii*, F. Muell. in Journ. Pharm. Soc. Vict. ii. 90; Miq. in Versl. K. Akad. Wet. Amst. xv. 371.

Queensland. Between Cleveland and Rockingham bays, *W. Hill;* Rockingham bay, *Dallachy;* Expedition range, *A. C. Gregory.*

Specimens occur in herbaria and in our gardens of two varieties, differing in the greater or less distinctiveness of the veins of the pinnulæ, and bearing often the names of *Catakidozamia Macleayi* and *C. Hopei*, Hill.

3. BOWENIA, Hook.

Male cones oblong-cylindrical; females globular, the scales broad at the apex, rather thick, truncate, the females with one pendulous ovule and seed on each side.—Trunk of *Macrozamia.* Leaves bipinnate, with long petioles and rhachis, the pinnules petiolulate, broad, oblique, without any midrib.

The genus is limited to the single Australian species, differing from *Macrozamia* only in foliage and in the absence of the point to the cone-scales.

1. **B. spectabilis,** *Hook. Bot. Mag.* t. 5398. Trunk thick, scarcely raised above the ground, marked with the scars of the old leaves, the whole plant glabrous. Leaves attaining in outline a length of 3 or 4 ft. and spreading to at least half that breadth, loosely bipinnate, the primary pinnæ clustered 3 to 5 together a little below the middle of the common petiole or rhachis, with 1 or 2 distant ones on each side higher up, the rhachis nearly terete, each pinna often a foot long or more, bearing 9 to 20 segments ovate or ovate-lanceolate, oblique or falcate acuminate, tapering into a short petiole, marked with numerous parallel scarcely prominent veins, 2 to 4 in. long. Cones very shortly pedunculate, the males $1\frac{1}{2}$ to 2 in. long, $\frac{3}{4}$ to 1 in. diameter; fruiting females nearly globular, 3 to 4 in. diameter in the specimens seen, the scales with a narrow base between the seeds expanded into a broad thick truncate apex which appears somewhat fleshy in the unopened cone, but when ripe and dry is hard, not so thick, and fully 1 in. broad.—F. Muell. Fragm. v. 171; A. DC. Prod. xvi. ii. 534.

Queensland. Endeavour river, *A. Cunningham;* Rockingham bay, *W. Hill, Dallachy.*

Class II. MONOCOTYLEDONS.

Stem not distinguishable into pith, wood and bark, but, when perennial, consisting of bundles of fibres irregularly imbedded in cellular tissue, with a firmly adherent rind outside. Seeds with one cotyledon, the embryo undivided, the young stem developed from a sheath-like cavity on one side.—Herbs or if arborescent the stem usually undivided or sparingly branched and crowned by a tuft of large leaves. Leaves usually alternate or radical, entire with simple parallel veins, the base usually encircling or sheathing the stem or the base of the next leaf, pinnate or otherwise divided in some *Palms* and *Aroideæ*, occasionally lobed in a very few species, and net-veined in a few genera of several Orders. Parts of the flower most frequently in threes, or in a few Orders the perianth wanting, or the parts reduced in number when irregular, or in twos or fours in *Naiadeæ*.

Order CXVIII. HYDROCHARIDEÆ.

Flowers mostly unisexual. Perianth of 3 or 6 segments, either all petal-like or the three outer ones herbaceous and usually smaller, with a tube adherent to the ovary in the females, without any tube in the males. Stamens 3 to 12 or rarely more. Anthers 2-celled. Ovary inferior, either 1-celled with 3 parietal placentæ, or more or less perfectly divided into 3, 6, or 9 cells. Styles 3, 6, or 9, with entire or 2-cleft stigmas. Ovules numerous, ascending or pendulous, orthotropous or anatropous, attached to placentas lining the walls or dissepiments of the ovary. Fruit indehiscent, membranous or fleshy, ripening under water. Seeds several or many, without albumen. Embryo straight, the plumule more or less lateral, the radicle next the hilum.—Aquatic herbs, entirely submerged or the lamina of the leaves floating. Leaves undivided. Flowers enclosed when young in a spathe, either of 1 to 3 leaves or tubular and 2- or 3-lobed, the males 1, 2 or more in the spathe, the females solitary.

The Order has a wide range over the tropical and temperate regions both of the New and the Old World, and of the five Australian genera three have the general area of the Order, the other two are limited to the Old World.

Tufts of radical leaves and scapes floating. Leaves with a broad lamina. Male spathes 2- or 3 flowered. Ovary and fruit 6-celled . . . 1. Hydrocharis.
Tufts of radical leaves and scapes fixed to the bottom of the water.
  Leaves with a broad lamina. Flowers hermaphrodite, solitary. Ovary partially 6-celled . . . . 2. Ottelia.
  Leaves elongated without any lamina. Flowers unisexual.
    Male flowers several in the spathe, exserted. Perianth of 6 segments . . . . 3. Blyxa.
    Male flowers numerous, crowded in a head shorter than the spathe. Perianth of 3 segments . . . . 4. Vallisneria.
Leaves small, verticillate along the floating stems. Spathes sessile, both males and females 1-flowered . . . . 5. Hydrilla.

1. HYDROCHARIS, Linn.

Flowers diœcious. Males 2 or 3 together in a spathe of 2 bracts. Outer segments of the perianth green, inner ones larger and petal-like. Stamens usually 9 with anthers, and 3 barren filaments, the filaments united in pairs. Female flowers solitary and pedicellate within the spathe. Ovary and fruit 6-celled. Styles 6, each with 2 stigmatic branches.—Stems floating, with tufts of radical leaves and peduncles. Leaves bearing a cordate lamina.

The genus is limited to a single species spread over Europe and the temperate regions of Asia, the Australian specimens presenting no apparent difference.

1. **H. morsus-ranæ,** *Linn. Spec. Pl.* 1466. Stems floating, resembling the runners of creeping plants, with floating tufts of radical leaves peduncles and submerged roots. Leaves on long petioles expanded into a sheath at the base, orbicular, entire, cordate at the base, but less deeply so in the Australian than in the European specimens, rather thick, 1 to 2 in. diameter. Male spathes ½ to 1 in. long, on peduncles of about the same length, the flowers shortly exserted. Inner perianth-segments white, 4 or 5 lines long, outer ones about half as long and green, all very broad. Stamens united in six pairs, the pairs opposite the outer segments with both filaments bearing anthers, those opposite the inner segments with the inner filament barren. Anther-cells bordering a broad connective. Female spathe sessile, the flower on a rather long pedicel enlarged at the top into an inferior ovary, the perianth nearly the same as in the males. Fruit ovoid, somewhat fleshy, under ½ in. long.—L. C. Rich. in Mem. Inst. Fr. 1811, t. 9; Reichb. Ic. Fl. Germ. t. 62.

Queensland. Wide bay, *Bidwill.* F. Mueller states that he has also received it from two different localities in Queensland. I have not seen his specimens, nor do I know whether there may be any reason to suppose that it is an introduced plant.

2. OTTELIA, Pers.

(Damasonium, *Schreb., not of Juss.*)

Flowers hermaphrodite, solitary and sessile within a tubular 2-lobed spathe. Outer perianth-segments green, inner ones larger and petal-like. Stamens 6 or more; anthers linear. Styles or stigmas 6, 2-lobed. Ovary and fruit more or less completely 6-celled, with numerous ovules and seeds.—Submerged herbs, the radical leaves and peduncles in tufts at the bottom of the water. Leaves mostly or all bearing an oblong ovate or broad-cordate lamina.

The genus is spread over tropical Asia and Africa with one American species. Of the three Australian species, one is the common Indian one, the two others apparently endemic.

Leaf-lamina broadly cordate. Spathe winged . . . . 1. *O. alismoides.*
Leaf-lamina ovate or oblong. Spathe not winged.
  Spathe firm, the outer perianth-segments protruding and persistent on the fruit . . . . 2. *O. ovalifolia.*

Spathe thinly membranous, deeply 2 lobed, the lobes much longer than the fruit . . . . 3. *O.? tenera.*

1. **O. alismoides,** *Pers. Syn. Pl.* i. 400. Leaves on long petioles dilated and tufted at the bottom of the water, the lamina submerged or floating, orbicular-cordate and about 6 in. diameter in the Australian specimens, varying in Indian ones to broadly ovate. Peduncles usually long. Spathe ovoid-oblong, nearly 1½ in. long, shortly 2-lobed at the top, and bearing 5 or 6 longitudinal herbaceous wings, 2 or 3 of them 1 to 2 lines broad, the others usually narrower. Ovary and fruit about the length of the spathe. Outer perianth-segments green, oblong-lanceolate, 4 to 5 lines long; inner ones white, veined, under 1 in. diameter. Stamens 6 to 9.—L. C. Rich. in Mem. Inst. Fr. 1811, t. 7; *Stratiotes alismoides*, Linn. Spec. 754; *Damasonium indicum*, Willd. Sp. Pl. ii. 276; Roxb. Corom. Pl. t. 185; Bot. Mag. t. 1201.

N. Australia. Roper river, *F. Mueller;* Creeks in the neighbourhood of Rockhampton, *Bowman, O'Shanesy, Watson;* Kennedy district, *Daintree.*—The species is widely dispersed over East India.

2. **O. ovalifolia,** *L.C. Rich. in Mem. Inst. Fr.* 1811, 78. Habit of *O. alismoides*, but the leaf-lamina ovate or oblong, 2 to 4 in. long when perfect, obtuse, rounded at the base and not at all or scarcely cordate. Peduncles sometimes very short sometimes above 1 ft. long, varying probably according to the depth of the water. Spathe almost coriaceous, about 1½ in. long, either quite smooth or with 2 or 3 slightly prominent, longitudinal nerves, but not winged in any of the specimens seen Outer perianth-segments green, ¾ to 1 in. long; inner ones pale yellow 1½ to 2 in. diameter in the only specimens in which they are perfect. Stamens 9 to 15.—*Damasonium ovalifolium*, R. Br. Prod. 344; *D. cygnorum*, Planch. in Ann. Sc. Nat. ser. 3, xi. 82.

N. Australia. Albert river, *Henne.*
Queensland. Rockhampton, *Bowman, Thozet;* Mount Elliott, *Fitzalan.*
N. S. Wales. Port Jackson, *R. Brown;* Penrith, *Backhouse;* north of Lachlan river, *A. Cunningham;* Clarence river, *Wilcox.*
Victoria. Glenelg or Wendu river, *Robertson;* Yarra river, *Adamson, F. Mueller;* Lake Alexandria, *F. Mueller.*
W. Australia. Swan river, *Drummond,* 1st coll.

3. **O.? tenera,** *Benth.* Leaf-lamina ovate or oblong as in *O. ovalifolia*, but smaller, and the whole plant much more slender. Spathe thinly membranous, rarely above 1 in. long, deeply divided into 2 lobes which project much beyond the capsule, forming an oblique double lanceolate point. Capsule thin and narrow, the seeds rather large, oblong, the placentas scarcely protruding into the cavity, the fruit bearing at the end some withered remains of the perianth, but I have been unable to find flowers in any of our specimens; the generic identity is therefore in some measure doubtful.

Victoria. Wendu river, *Robertson.*
W. Australia, *Drummond,* 4th coll, n. 322.

3. BLYXA, Thou.

Flowers usually diœcious. Males several, protruding when open from a tubular 2-toothed spathe. Perianth-segments all linear, the 3 outer ones green, the 3 inner longer and petal-like. Stamens 8 or 9; anthers linear. Female flowers solitary in the sheath, the ovary sessile; perianth-tube above the ovary long and filiform, the segments as in the males or narrower. Style exserted, with 3 linear stigmas. Ovary narrow, with parietal placentas. Fruit narrow, enclosed in the spathe. Seeds many.—Submerged herbs, the leaves long and grass-like without laminæ, acute and entire, tufted with the peduncles at the bottom of the water.

Besides the Australian species, which is spread over tropical Asia, there is another from the Mascarene islands.

1. **B. Roxburghii,** *Rich. in Mem. Inst. Fr.* 1811, 77, t. 5. Leaves entirely submerged, long and narrow like those of *Vallisneria spiralis*, but more acute and not serrulate. Spathes both male and female on long slender peduncles, the males about 1½ to 2 in. long, containing several flowers, but usually only 1 to 3 protrude at the same time, on pedicels several lines longer than the spathe. Outer perianth-segments about 2½ lines long, the inner twice as long. Stamens about 8, the filaments short, the linear anthers not exceeding the outer perianth-segments. Female spathe usually longer and more slender than the male, the ovary much shorter, but the perianth-tube projecting above ½ in. from the spathe, the segments narrower than in the males. Fruit narrow, entirely enclosed in the spathe, varying from 1 to 2 in., the seeds not very numerous.—*Vallisneria octandra*, Roxb. Pl. Corom. ii. 34, t. 165; *Blyxa octandra*, Planch. Mss.; Thw. Enum. Pl. Ceyl. 332.

N. Australia. Robinson river, Gulf of Carpentaria, *F. Mueller;* Port Darwin, *Schultz, n.* 423.

Queensland. Water-holes, Rockingham bay, and Burdekin river, *Dallachy.*

S. Australia. Waters near Lake Torrens, *F. Mueller.*

Our specimens are not sufficiently perfect to show the form of the female perianth, nor whether it has the three stamens mentioned by Decaisne and by Thwaites, but which do not appear to be always present in the Indian plant.

4. VALLISNERIA, Linn.

Flowers diœcious. Males minute and very numerous in an ovoid-globular 3-lobed spathe. Perianth-segments 3. Stamens 1 to 3; anthers with 2 globular cells. Female flowers solitary and sessile in a narrow tubular 3-toothed spathe. Perianth-tube not produced above the ovary; segments 3. Staminodia (or inner perianth-segments?) 3, small, bifid, alternating with the perianth-segments. Stigmas 3, broad, 2-dentate or bifid. Ovary narrow, with 3 parietal placentas. Fruit narrow-cylindrical, enclosed in the spathe. Seeds numerous, cylindrical.—Submerged herbs, the leaves and peduncles tufted at the bottom of the water, the leaves very long, without any lamina.

The genus is generally distributed over the tropical and temperate regions of the New as well as the Old World, the Australian species being the one most common over the whole area.

1. **V. spiralis,** *Linn. Sp. Pl.* 1441. Leaves entirely submerged, very long and narrow when the water is deep, short in shallow water, obtuse or acute and more or less serrulate at the end with minute teeth, or sometimes perhaps quite entire. Male spathes about 3 lines long, on a peduncle usually short, but said to break off and enable the flower to float to the surface and fecundate the female. Flowers minute, apparently forming an ovoid or globular head not quite so long as the spathe, but the pedicels really 3 or 4 times as long as the minute perianth. Female spathe usually about ½ in. long, very narrow, on a spirally coiled filiform peduncle, which unfolds so as to carry the flower to the surface till after fecundation, when it contracts and brings the ovary down to the bottom to mature. Perianth very small but larger than in the males. Fruiting spathe only slightly enlarged.—Rich. Mem. Inst. Fr. 1811, t. 3; Reichb. Ic. Fl. Germ. t. 60; Hook. f. Fl. Tasm. ii. 37; *V. spiralis* and *V. nana*, R. Br. Prod. 345.

N. Australia. Albert and Roper rivers, *F. Mueller;* Arnhem's Land, *R. Brown.*

Queensland. Burnett river, *F. Mueller;* between Cleveland and Rockingham bays, *W. Hill;* Fitzroy river, *O'Shanesy*, several feet long.

N. S. Wales. New England, *C. Stuart;* lagoons near Richmond, *Wilhelmi.*

Victoria. Wendu river, *Robertson;* Avon, Tambo, Mitta-Mitta, and Murray rivers, *F. Mueller.*

Tasmania. Very common in the South Esk river, *Gunn.*

S. Australia. Torrens river, *F. Mueller.*

The Australian specimens I have seen prove to be all females; I have described the males from European ones. Brown's *V. nana* was founded on a dwarf narrow-leaved form, such as I have seen from some other countries. The leaves of his specimens are as described by him mostly acute and all quite entire, but one or two are obtuse, and in the larger forms from New South Wales, Victoria, and Tasmania the shorter submerged leaves are sometimes acute, and occasionally, whether obtuse or acute, without any or scarcely any of the minute serratures which usually characterize them.

5. HYDRILLA, Rich.

Flowers diœcious, both sexes solitary in a short tubular spathe. Male perianth shortly pedunculate, the outer segments ovate and green, three inner ones oblong-linear and petal-like. Stamens 3, with reniform anthers. Female perianth with a long filiform tube above the ovary, the segments all petal-like and less unequal than in the males. Style elongated, with 3 filiform stigmas. Ovary 1-celled, with 3 parietal placentas. Fruit cylindrical, linear, with few seeds.—Submerged herbs with branching stems and short verticillate leaves. Spathes sessile in the axils.

The genus is restricted by Caspary to the single species common in still and slowly running waters of the tropical and temperate regions of the Old World, the Australian form being the typical one originally described from India.

1. **H. verticillata,** *Casp. in Monatsber. Akad. Berl.* 1857. Stems leafy throughout, much branched and floating under water in large

masses. Leaves all in whorls of 4 to 8, except a single small sheathing one at the base of each branch and a pair only next above it, all oblong-lanceolate or broadly linear, 2 to 4 lines long and serrulate in the Australian as in the typical Indian form, very narrow-linear in some Asiatic varieties. Male spathes not seen; the above character taken from Roxburgh and others. Female spathes sessile and solitary, shorter than the leaves, the perianth-tube $\frac{1}{2}$ to $\frac{3}{4}$ in. long in the Australian specimens, the outer perianth-segments nearly $1\frac{1}{2}$ lines long, the inner ones shorter and narrower. Fruit linear, with few seeds.—*Serpicula verticillata*, Linn. f. Suppl. 416; Roxb. Corom. Pl. t. 164; *Hydrilla ovalifolia*, Rich. in Mem. Inst. Fr. 1811, 76, t. 2; *Udora australis*, F. Muell. Second Gen. Rep. 16.

N. Australia. Gilbert and Roper rivers, *F. Mueller*.
Queensland. Water-holes near Rockhampton, *Bowman*, *O'Shanesy*; near Herbert river, *Dallachy*; Mount Elliott, *Fitzalan*.
N. S. Wales. Richmond river, *Fawcett*; Balfour's station, *Leichhardt*.
Victoria. Murray river, *F. Mueller*.

Order CXIX. SCITAMINEÆ.

Flowers hermaphrodite or unisexual, irregular. Perianth superior, normally of 6 parts in two series, the 3 outer ones united in a 3-toothed or 3-lobed tube or upper lip, or in genera not Australian free, the 3 inner ones variously combined with the outer perianth or more frequently with the staminodes or stamen, or the lower one free. Stamens normally 3 or 6, but in most genera only 1, in *Musa* only 5, bearing anthers, the others either wanting or converted into barren petal-like staminodes often called an inner corolla, or short and linear; in most genera one of them is usually larger, broadly petal-like and on the side opposed to the perfect stamen and is then called the *labellum*. Anthers 2-celled or in genera not Australian 1-celled. Ovary inferior, 3-celled, with 1 or more ovules in each cell or rarely 1-celled. Style simple with a terminal entire or lobed stigma. Fruit a berry or a fleshy or dry capsule. Seeds albuminous.—Herbs, usually with a perennial rhizome. Stem short or rarely elongated unless formed of the convolute leaf-sheaths, and then often attaining a considerable height. Leaves entire, with long sheathing petioles, the limb often very large, with very numerous parallel veins diverging from the midrib. Flowers often very showy, in spikes racemes or panicles, on a radical or terminal scape or peduncle.

A considerable tropical or subtropical Order, common to the New and the Old World. The Australian genera are all Asiatic also, and two of them extend to America. Of the nine species, however, here enumerated, I have seen specimens of three only, the specific descriptions of the other six are taken from a sheet of Mueller's Fragmenta, received at the moment of placing this portion of the copy in the printer's hands, whilst the generic characters have necessarily been extracted from other works and checked by Asiatic specimens, and may therefore not always agree precisely with the plants here referred to them.

TRIBE 1. **Musaceæ.**—*Inner and outer perianth more or less combined, or each 3-parted to the base. Perfect stamens usually 5.*

Perianth of 2 segments, the upper outer one 5-lobed at the end, the lower inner one much smaller	1. MUSA.

TRIBE 2. **Zingiberaceæ.**—*Outer perianth or calyx 3-toothed or spathaceous; inner perianth or corolla free from it, combined at the base with the filament into a tube. One perfect upper stamen and one large petal-like lower staminode or labellum opposite it.*

Flowers in dense spikes with imbricate bracts, on a short scape separate from the leafy stems or tufts.	
Flowers 3 to 5 within each bract. Corolla with 2 inner lobes or petal-like staminodes	2. CURCUMA.
Flowers solitary within each bract. Corolla without inner lobes.	
Connective produced into a 3-lobed appendage beyond the anther. Corolla-lobes as long as the tube	3. AMOMUM.
Connective not produced beyond the anther. Corolla-tube slender, much longer than the lobes	4. ELETTARIA.
Flowers in a loose raceme or thyrsus terminating a leafy stem.	
Labellum flat, spreading	5. ALPINIA.
Flowers in a thick dense spike with imbricate bracts, terminating a leafy stem. Labellum convolute erect.	
Ovary 3-celled. Labellum longer than the corolla	6. COSTUS.
Ovary 2-celled. Labellum shorter than the corolla	7. TAPEINOCHEILOS.

TRIBE 1. MUSACEÆ.—Inner and outer perianth more or less combined, or each 3-parted to the base. Perfect stamens usually five.

1. MUSA, Linn.

Flowers usually unisexual. Perianth of 2 segments, the outer one formed of the 3 outer and 2 of the inner parts, tubular in the bud but open from the base on the lower side, petal-like, with 5 teeth or short lobes of which 2 inner ones usually smaller than the three outer; the inner perianth-segment (or third inner part) much shorter and usually recurved. Perfect stamens 5, with linear anthers, the sixth either wanting or forming a filiform staminode adnate to the inner perianth-segment. Ovary 3-celled, with numerous ovules. Style elongated, clavate, with a concave stigma. Fruit oblong or cylindrical, often curved, more or less succulent and indehiscent. Seeds numerous, with a coriaceous testa and a broad concave hilum; albumen copious, mealy or almost granular.—Tall often almost tree-like herbs, the convolute sheaths of their very long large leaves forming a stem of considerable height. Scapes protruding from the centre of the leaf-sheath. Flowers clustered in the axils of large coloured bracts, forming a long terminal spike or raceme, either drooping, or, in species not Australian, erect.

The genus is limited to the tropical regions of the Old World. The only Australian species is believed to be endemic.

1. **M. Banksii,** *F. Muell. Fragm.* iv. 132. "A species of moderate height, very readily stoloniferous, with the habit of *M. paradisiaca*," to which it is evidently very closely allied if really distinct. Leaves

"5 or 6 ft. long, on a petiole of 1½ to 2 ft.," not including probably the sheathing base. Raceme apparently recurved. Bracts oblong, rather obtuse, 3 to 4 in. long, a few of the lower ones much longer and acuminate. Flowers from 10 to 20 within each bract, on short pedicels. Male perianth (from a wild specimen) conspicuously striate with numerous parallel longitudinal veins, the outer convolute segment about 1¼ in. long and 5 lines broad if spread open, shortly 5-lobed, the 3 outer lobes lanceolate, about 2 lines long, the 2 inner ones oblong and rather shorter; the lower inner segment 4 or 5 lines long and at least 3 lines broad, striate like the outer one, but with a prominent midrib (or adnate staminode?) produced into a rather long point beyond the segment. Stamens nearly as long as the outer segment. Ovary rudimentary, with a style about as long as the stamens, slightly clavate at the end. Female flowers (from a specimen cultivated in the Melbourne Botanic Garden) with an ovary of nearly ¾ in. Outer perianth-segment under 1 in. long, the outer lobes narrower and more acute than in the males, the inner lobes linear; lower inner segment narrow-lanceolate, above ½ in. long. Anthers linear but smaller than in the males. Style rather shorter than the outer perianth-segment, thickly clavate towards the end, with a large concave oblique unequally lobed stigma. Fruits (from the wild specimens) in bunches of 12 to 20, each one on a stipes of 1 to 2 in., cylindrical, straight or slightly incurved, 4 to 5 in. long, ½ to ¾ in. diameter when dry. Seeds about 2 lines diameter, irregularly angular from pressure; testa coriaceous, marked with a broad concave hilum; albumen very white mealy almost granular.

Queensland. Mount Elliott, *Fitzalan;* Rockingham bay, *Dallachy.*—Gærtner's figure, which F. Mueller thinks may have been taken from this species, appears to me to represent the true *M. paradisiaca.* I find no record of Banks and Solander having seen any *Musa* in Australia. If they had brought fruits home they would surely have been mentioned by R. Brown, either in the Prodromus or in his notes on the geographical distribution of the genus in his Observations on the Botany of Congo.

TRIBE 2. ZINGIBERACEÆ.—Outer perianth or calyx 3-toothed or 3-lobed or spathaceous; inner perianth or corolla free from it, combined with the filament into a tube, the limb 3-lobed with sometimes 2 inner additional lobes or staminodia. One perfect upper stamen, with a 2-celled anther embracing the style, and one large petal-like labellum or lower staminode opposite it and sometimes 2 short linear staminodes (stylodes of Horaninow) at the base of the style.

2. CURCUMA, Linn.

Flowers hermaphrodite. Calyx tubular, 3-toothed; corolla-tube longer than the calyx, the limb of 3 outer lobes of which the upper one is broader than the 2 others, and 2 inner lobes or staminodes resembling the 2 outer lower lobes and adnate at the base on their inner margins to the short broad petal-like filament of the perfect stamen. Labellum broad and petal-like; two short linear staminodes at the base of the style. Anther oblong-linear, the 2 cells folded round the summit of the style and produced at the base into auricles or spurs. Ovary 3-celled. Style filiform, with a capitate stigma. Capsule 3-celled, loculicidally 3-valved. Seeds several, arillate.—Herbs with a perennial rhizome and clustered fleshy roots. Leaves with convolute sheathing bases. Scapes simple, with a thick erect spike. Bracts concave or saccate and imbricate at the base, with broad spreading often coloured ends. Flowers yellow, 3 to 5 within each bract.

The genus is generally distributed over tropical Asia, the only Australian species is, as far as hitherto known, endemic.

1. **C. australasica,** *Hook. f. Bot. Mag.* t. 5620. Rhizome with a cluster of white cylindrical tuberous roots. Leaves 1 to 1½ ft. long, lanceolate or narrow-elliptical, acute, tapering into a long sheathing petiole. Scape lengthening to about 6 in. below the broad spike, which attains from 5 to 7 in., the upper bracts 1 to 1½ in. long, with broad spreading rose-coloured ends, the lower bracts green, closely erect and broadly saccate at the base, with short broad spreading ends. Flowers of a pale yellow, about as long as the lower bracts. Corolla-tube twice as long as the calyx, dilated upwards, the upper lobe concave and broad, the lateral outer ones and upper inner one or staminodes broadly oblong; labellum broadly orbicular almost reniform, notched and undulate on the margin. Anther-auricles narrow and acute.

Queensland. Cape York, *Daemel*, *Gulliver;* Cape Sidmouth, *C. Walter*.—The dried specimens too much crushed to admit of a careful examination of the structure of the flower; the above description taken chiefly from that given by Dr. Hooker in the Botanical Magazine.

3. AMOMUM, Linn.

Calyx tubular, spathaceous or 3-lobed at the top. Corolla-limb as long as the tube, 3-lobed, the dorsal lobe broader than the lateral ones, without inner lobes. Labellum large, flat, entire or lobed. Filaments flat, the connective produced beyond the anther-cells into a 3-lobed appendage, the lateral lobes divaricate, the middle one erect, entire or notched; anther-cells embracing the style. Two small linear staminodes at the base of the style. Ovary 3-celled. Style filiform, clavate at the end, with a concave stigma. Fruit succulent or opening in 3 valves. Seeds arillate.—Herbs with creeping rhizomes. Leaves on barren stems often several feet high. Flowering scapes short, with sheathing scales. Spike short, with broad imbricate bracts. Flowers usually large, one within each bract.

The genus is widely spread over the tropical regions of both the New and the Old World. The only Australian species, which I have not seen, is believed to be endemic.

1. **A. Dallachyi,** *F. Muell. Fragm.* viii. 25. Leaves lanceolate, narrow-pointed, often above 1 ft. long and nearly 2 in. broad. Scapes very short. Bracts shorter than the calyx, a few larger ones at the base of the spike, the largest 1½ in. long. Calyx about 1 in long, membranous, striate-veined, shortly 3-toothed. Corolla yellow, the lobes

about an inch long and rather longer than the tube, the dorsal one broadly ovate, the lateral ones oblong. Labellum as long as the corolla, $\frac{3}{4}$ in. broad, orbicular, contracted at the base, shortly and obtusely 3-lobed, thickened along the centre. Middle lobe of the connective-appendage shorter than the lateral ones. Capsule green, nearly globular, unequally muricate, tardily opening in 3 valves. Seeds rather numerous.

Queensland. Rockingham bay, *Dallachy.*—The above description extracted from that given by F. Mueller.

4. ELETTARIA, White.

Calyx tubular, 2- or 3-toothed. Corolla-limb 3-lobed, shorter than the slender tube, without inner lobes. Labellum large, flat, entire or lobed. Filament flat, but the connective not produced beyond the anther-cells; anther-cells embracing the style. Two small linear staminodes at the base of the style. Ovary 3-celled. Style filiform, clavate at the end, with a concave stigma. Fruit succulent or opening in 3 valves. Seeds arillate.—Herbs with the habit of *Amomum*, but usually more slender, with smaller flowers, fewer in a looser spike.

The genus extends over tropical Asia, the Australian species, which I have not seen, is believed to be endemic.

1. **E. Scottiana,** *F. Muell. Fragm.* viii. 24. Rhizome woody, the leafy stems attaining 12 ft. or more, including the long lanceolate leaves, which are $1\frac{1}{2}$ to 2 ft. long and 3 to 4 in. broad in the middle. Scapes about 2 in. long, with imbricate sheathing scales, passing into involucral bracts of about 2 in., silky-pubescent outside. Bracts under each flower membranous, pellucid, 2 to 3 in. long. Calyx the length of the bracts. Corolla-tube 3 to 4 in. long, very slender, the lobes red, nearly equal, oblong, at least $\frac{1}{2}$ in. long. Labellum rather longer than the corolla-lobes, ovate, entire. Connective not extending beyond the apex of the anther-cells. Capsule ovate or ellipsoid, opening tardily in 3 valves. Seeds numerous.

Queensland. Rockingham bay, *Dallachy.*—The above description extracted from that given by F. Mueller.

5. ALPINIA, Linn.

(Hellenia, *Willd.*)

Flowers hermaphrodite. Calyx tubular, 3-toothed, often spathaceous. Corolla-tube longer than the calyx, the limb of 3 usually unequal lobes. Labellum broad, flat, usually exceeding the corolla-lobes. Filament narrow, often filiform, the connective not at all or only very shortly produced beyond the anther-cells. Two short linear staminodes at the base of the style. Ovary 3-celled, with many ovules. Style filiform. Stigma terminal, concave. Fruit globular or obovoid, succulent or with a crustaceous or dry pericarp, indehiscent or rarely obscurely 3-valved. Seeds not very numerous.—Erect herbs with a tuberous rhizome and leafy stems. Flowers in a terminal raceme thyrsus or raceme-like panicle. Bracts usually deciduous.

A tropical genus limited to the Old World. The three Australian species, of which I have only seen one, are believed to be endemic. The genus *Hellenia*, united with *Alpinia* by Roscoe, was again separated by Brown, chiefly with reference to the *H. cœrulea*, Br., which differs from most other species, even from Willdenow's *H. chinensis*, in the short terminal appendage to the anther and the crustaceous not baccate fruit. But *A. Allughas* has precisely the same fruit, and the projection of the connective is so short as to be of very little importance. I had therefore in the Flora Hongkongensis proposed reuniting the genera, and F. Mueller in his MSS. notes also suggests the consolidation.

Raceme simple. Anther-connective not produced beyond the cells.
 Capsule tardily 3-valved . . . . 1. *A. racemigera.*
Thyrsus or panicle narrow. Anther-connective produced beyond the cells.
 Flowers under 1 in. long. Anther-connective very shortly produced. Fruit crustaceous indehiscent . . . . 2. *A. cœrulea.*
 Flowers above 1 in. long. Anther-appendage obovate 2 lines long. Capsule ellipsoid, 3-valved . . . . 3. *A. arctiflora.*

1. **A. racemigera,** *F. Muell. Fragm.* viii. 27. A low species. Leaves long-lanceolate, acuminate, about 1 ft. long and $2\frac{1}{2}$ in. broad, on a minutely biauriculate sheath without any intervening petiole. Flowers pale yellow, numerous in a single raceme of $\frac{1}{2}$ ft. or shorter, the rhachis and pedicels pubescent. Bracts lanceolate, shorter than the pedicels which are 2 to 3 in. long. Calyx unequally divided to the middle into 3 lobes, about 4 lines long. Corolla glabrous, about 8 lines long, the lobes about as long as the tube, nearly equal but the upper one more concave. Labellum orbicular-rhomboidal, 3 to 4 lines diameter. Filament broadly linear, $\frac{3}{4}$ line long; anther nearly 1 line, the connective not produced beyond the cells. Style capillary, with a minute stigma. Capsule ovoid, red, 4 to 7 lines long, opening tardily in 3 valves. Seeds shining, 2 to 7 in each cell.

Queensland. Rockingham bay, *Dallachy*.—I have seen no specimen. The above description is an abridgment of the one given by F. Mueller.

2. **A. cœrulea,** *Benth.* Leafy stems attaining 4 or 5 ft. Leaves oblong-lanceolate, often above 1 ft. long and 2 in. broad, acutely acuminate, shortly petiolate above the sheath, which ends in a broad obtuse erect auricle of 3 or 4 lines. Thyrsus terminal, 4 to 8 in. long, appearing almost as a simple raceme when first flowering, but most of the peduncles, though short, developing 2 to 6 flowers, on pedicels of $\frac{1}{2}$ to $\frac{3}{4}$ in., the whole peduncle usually much shorter. A convolute bract of $\frac{1}{2}$ in. or less under each pedicel or branch, and 2 or 3 long lanceolate sheathing bracts at the base of the panicle, often 2 in. long. Calyx narrow, 4 to 5 lines long, usually split on the lower side. Corolla-tube slender, 6 to 7 lines long, lobes oblong-linear, about 4 lines long, the dorsal one scarcely broader than the others. Labellum longer than the lobes, broadly orbicular or almost reniform, about $\frac{1}{2}$ in. diameter. Connective produced beyond the anther in a rounded or truncate appendage, not 1 line long. Fruit globular, indehiscent, about $\frac{1}{2}$ in.

diameter, with a brittle crustaceous pericarp. Seeds few in each cell, closely packed, with a small arillus.—*Hellenia cærulea*, R. Br. Prod. 308.

Queensland. Shoalwater bay and Northumberland islands, *R. Brown;* Endeavour river, *A. Cunningham;* Moreton island, *F. Mueller;* Fitzroy island, *C. Walter;* very common about Rockingham bay, *Dallachy*, and Rockhampton, *O'Shanesy.*

N. S. Wales. Hunter's river, *R. Brown;* Tweed river, very common, *Guilfoyle;* Hastings and Clarence rivers, *Beckler;* Richmond river, *Henderson, Fawcett.*

3. **A. arctiflora,** *F. Muell. Fragm.* viii. 25. Stems attaining 12 ft. Leaves long-lanceolate about 1½ ft. long and 2½ to 4 in. broad, contracted into a very short petiole above the long sheathing base. Panicle narrow and dense, pubescent as well as the under side of the leaves. Outer bracts few, rather above 1 in. long, those subtending the peduncles 1½ in. long or shorter. Calyx about 1 in. long, narrow, dilated upwards and acutely 3-lobed. Corolla white, pubescent outside, the lobes about 4 lines long, the tube longer. Labellum longer than the corolla, 2-lobed, at least ½ in. long and broad. Connective produced beyond the anther-cells into a cuneate-obovate appendage of about 2 lines. Style glabrous. Capsule ellipsoid, 3-valved, many-seeded.—*Hellenia arctiflora*, F. Muell. l.c.

Queensland. Rockingham bay, *Dallachy.*—I have seen no specimen, the above character is taken from that given by F. Mueller.

6. COSTUS, Linn.

Flowers hermaphrodite. Calyx tubular, shortly 3-lobed. Corolla-tube short, the limb with 3 erect lobes. Labellum convolute, erect, usually large. Filament broad and petal-like, continuous with the connective and produced laterally and beyond the cells into a broad appendage. No staminodes. Ovary 3-celled, with numerous ovules. Style filiform, the stigma dilated, flatly 2-lobed, with 2 dorsal appendages. Capsule 3-celled, opening loculicidally in 3 valves. Seeds arillate— Herbs with an erect leafy stem and a dense terminal spike. Bracts broad, imbricate, with 1 flower in the axil of each.

The genus is spread over the tropical regions of both the New and the Old World. The Australian species, which I have not seen, is believed to be endemic.

1. **C. Potieræ,** *F. Muell. Fragm.* iv. 164. Stems about 10 ft. high. Leaves on very short petioles, oblong-lanceolate, acutely acuminate, 4 to 9 in. long, 1½ to 2½ in. broad, sprinkled underneath with appressed hairs, the sheath shortly produced above the petiole, truncate and ciliate. Spike dense, globose or ovoid-globose, 2 to 3 in. long. Bracts scarlet, shorter than the calyx. Calyx about 1 in. long, 3-toothed. Corolla yellow, the lobes 8 to 10 lines long and nearly equal. Labellum 1 in. long or rather larger, orbicular, undivided, striate along the centre. Filament and connective produced into a narrow border and a long terminal appendage beyond the anther-cells. Capsule crowned by the persistent calyx, about ½ in. long, opening loculicidally in 3 slits.

Queensland. Rockingham bay, *Dallachy.*—The above description extracted from that given by F. Mueller.

7. TAPEINOCHEILOS, Miq.

Flowers hermaphrodite. Calyx tubular, shortly 3-lobed. Corolla-tube short, the limb 3-lobed, the dorsal lobe rather broader and shorter than the lateral ones. Labellum erect, concave, shorter than the corolla, obscurely 3-lobed. Filament broadly petal-like, continuous with the connective and shortly produced beyond the anther-cells. No staminodia. Ovary 2-celled, with numerous ovules. Style filiform, the stigma dilated, flatly 2-lobed, without appendages. Fruit (dry ?) apparently indehiscent, crowned by the persistent coriaceous calyx, 2-celled. Seeds angular, with a small arillus.—Herb with the habit of *Costus*.

The genus is limited to a single species, native of the island of Ceram in the Indian Archipelago, the Australian plant, which I have not seen, is believed to be identical as to species.

1. **T. pungens,** *Miq. in Ann. Mus. Lugd. Bat.* iv. 101, t. 4. Stem about 2 ft. high, covered with sheathing scales, some of which emit from their axils leafy branches. Leaves ovate-oblong, terminating abruptly in a narrow rigid point. Spike terminating the main stem very dense, 6 in. long and 3 in. broad in our specimen, larger as figured by Miquel. Bracts rigid, striate, imbricate, the broad ends recurved, of a rich crimson. Flowers yellow, scarcely exceeding the bracts. Calyx about 10 lines long. Corolla 1¾ in. long. Labellum much shorter. Fruiting calyx rigidly coriaceous, compressed but thick, above 1 in. long, the lobes recurved and rigid, the free tube as long as and quite continuous with the adnate base.—*Costus pungens*, Teysm. and Binnend.

Queensland. Rockingham bay, *Kennedy, Dallachy* (*F. Mueller*).—Not having seen the Australian specimens, I have taken the above character from Miquel's elaborate description and figure and from a fruiting specimen received from the Botanic Garden, Calcutta.

ORDER CXX. ORCHIDEÆ.

Flowers hermaphrodite. Perianth superior, irregular or rarely regular, of 6 petal-like or green segments, all free or variously united; 3 outer ones called *sepals* all similar and erect or spreading, or the *dorsal* one (next to the main axis unless the flower is reversed by a twist of the ovary) more concave or otherwise different from the 2 *lateral* ones (lower ones unless the flower is reversed), which are always similar to each other; 2 inner segments called *petals*, similar to each other, one on each side of the dorsal sepal, and sometimes connivent or connate with it into a *galea*, sometimes similar to the lateral sepals, or different from all the sepals; the sixth segment or third petal called the *labellum*, different from all the others (except in *Thelymitra* and *Apostasia*), inserted between the lateral sepals at the base of the column, and exceedingly varied in size, shape, lobes, calli, fringe, or other appendages. In the centre of the flower or somewhat under the dorsal sepal is the *column*, consisting of the combined andrœcium and pistil; at the apex attached to the dorsal margin is usually one anther, erect incumbent on or adnate to the apex of the column or to the back of the stigma, with 2 cells on its inner face or almost marginal. In the genus *Cypripedium* which is not Australian, and in *Apostasia*, there are two lateral perfect anthers. Pollen either waxy granular or mealy, usually more or less distinctly collected into 1, 2, or 4 pairs of pollen-masses, either oblong or tapering to the upper end, free or attached by their narrow end, either directly or by a linear or filiform *caudicle*, to a gland on the rostellum or apex of the stigma, or sessile on that apex without a gland. In front of the column either at or near its apex or lower down is a concave or rarely convex viscid stigma, the upper margin often produced into an erect appendage called the *rostellum* sometimes very short, sometimes as long as or longer than the anther; each side of the column towards the front shows a longitudinal angle, often expanded in the whole or part of its length into a wing, sometimes continued behind the anther into a hood over it, or expanded into two auricles or appendages (sometimes described as staminodia) one on each side between the stigma and the anther, sometimes continued into a basal projection of the column or even continuous with raised lines on the labellum, or in *Thelymitra* the two wings shortly joined in front of the column, or in *Apostasia*, the style normally cylindrical with a terminal stigma. Ovary inferior, 1-celled or with 3 parietal placentæ, or in *Apostasia* perfectly 3-celled with innumerable ovules. Fruit capsular, opening in 3 valves or longitudinal slits, or very rarely succulent and indehiscent. Seeds minute, fusiform or rarely winged, resembling fine sawdust. Embryo a solid, apparently homogeneous body.—Herbs usually perennial, either *terrestrial* with underground rhizomes creeping or producing annually renewed tubers or thick clustered fibres, or *epiphytical* with creeping rhizomes and (often fleshy) fibrous roots adhering to the surface of rocks or trunks or branches of trees. Leaves either alternate and sheathing at the base and sometimes distichous, on flowering stems which when epiphytical are sometimes thickened into *pseudo-bulbs*, or in radical tufts at the base of the flowering stems or in tufts or on pseudo-bulbs distinct from the flowering ones, entire or very rarely lobed (see *Acianthus*). Flowering stems scapes or peduncles annually renewed, either proceeding directly from the rhizomes or axillary on perennial leafy stems or pseudo-bulbs, bearing usually one or more scarious or membranous sheathing scales, either without any leaves, or when leafy 1 or 2 sheathing scales below the leaves and often 1 or more above them. Flowers either solitary and terminal or 2 or more in a terminal raceme or spike, either simple or branching into a panicle, each flower sessile or more frequently pedicellate within a bract, but without bracteoles on the pedicel.

The Order is one of the most natural and sharply defined, as well as the most numerous amongst Monocotyledons after Gramineæ, and abundantly distributed over the whole globe, rare only in some high Alpine or extreme Arctic and Antarctic regions. The 48 Australian genera may be geographically divided into two groups. 28 genera, comprising one-third of the total number of species, including the whole of the tribes Malaxideæ, Vandeæ, Bletideæ, Arethuseæ, the first group of Neottideæ, and the Ophrydeæ belong to the tropical Asiatic Flora, represented in Australia by endemic or frequently by identical species. These are all tropical or eastern, some extending down to Tasmania, but none found in West Australia; five of these genera are also in New Zealand The remaining 20 genera, comprising two-thirds of the species, are essentially Australian, belonging to three Australian groups of Neottideæ; four of these genera are however represented by single or very few species in the Indian Archipelago, and eleven have New Zealand congeners, sometimes identical in species.

In the elaboration of the genera and other groups endemic in or represented in Australia, I have had little to do but to follow in the footsteps of Brown and Lindley, to the whole of whose typical specimens I have had free access, and in the identification of species I have been materially assisted by the verbal communications as well as by the published labours (especially the Beiträge above quoted) of the younger Reichenbach, as well as by the admirable illustrations of Hooker's Flora Tasmanica. In this case, however, as in the rest of the present work, I have made it a rule to work out the descriptions of genera as well as of species, in the first instance, from the specimens themselves wherever they admitted of examination, and afterwards to check them by those of the great authorities on the Order. I have thus had no alterations of any note to propose in the circumscription of generic or subgeneric groups, which in Australia are very fairly defined, but I have had sometimes to choose between the different views of Lindley and of Reichenbach as to the grade to be assigned to these groups, always so much a matter of individual opinion. As to species, I have had for examination, especially in the very rich herbarium placed at my disposal by F. Mueller, a far greater mass of excellent materials than any of my predecessors, showing for instance how great is the range of variation exhibited in the precise form, markings, and processes of the labellum in many of the commoner terrestrial Australian Orchids, as in the Mediterranean Ophrydes. I have therefore felt obliged very much to reduce the number of published species of *Thelymitra*, *Diuris*, *Prasophyllum*, *Microtis*, and *Caladenia*. I must however admit that the circumscription of several of these species is still far from satisfactory; the forms can often scarcely be ascertained accurately from dried specimens, and colours very rarely. It is to be hoped, therefore, that the revision of these and some other genera will be taken up by resident botanists who have an opportunity of studying them in a fresh state.

TRIBE. 1 **Malaxideæ.**—*Anther lid-like, incumbent, usually deciduous. Pollen-masses waxy,* 2, 4, *or rarely* 8, *without caudicles or gland. Epiphytes or rarely terrestrial with a creeping rhizome.*

Anther-cells longitudinal. Lateral sepals not dilated at the base. Labellum embracing or adnate to the column at the base.
- Column elongated. Leaves at or near the base of the stem. Flowers very small . . . . 1. LIPARIS.
- Column very short. Leaves distichous. Flowers minute . . 2. OBERONIA.

Anther-cells longitudinal. Lateral sepals dilated at the base and forming with the basal projection of the column a pouch or spur.
- Pollen-masses 2, or 4 in pairs.
  - Labellum with a broad erect base usually expanded into lateral lobes. Stems or pseudobulbs bearing both leaves and peduncles (except the sect. *Rhizobium*) . . . . 3. DENDROBIUM.
  - Labellum distinctly unguiculate, not lobed. Leaves (on pseudobulbs) and peduncles from distinct nodes of the rhizome . . . . 4. BOLBOPHYLLUM.
- Pollen-masses 8, flowers minute . . . . 5. PHREATIA.

Anther-cells nearly transverse. Leaf long. Raceme on a long recurved peduncle . . . . 6. PHOLIDOTA.

(The pollen-masses are almost waxy without caudicles in one species of *Galeola*.)

TRIBE 2. **Vandeæ.**—*Anther lid-like, incumbent, usually deciduous. Pollen-masses waxy,* 4 *in pairs, on a single or double caudicle attached to a gland. Epiphytes or terrestrial with creeping rhizomes.*

Epiphytes. Caudicle single.
- Sepals and petals united to the middle. Very small plants with minute flowers . . . . 7. TÆNIOPHYLLUM.

Sepals and petals free.
Labellum with a fleshy protuberance underneath between or beyond the lateral lobes . . . . 8. SARCOCHILUS.
Labellum with a hollow spur or pouch near the base.
Labellum 3-lobed, the spur with a reflexed or horizontal scale inside . . . . 9. CLEISOSTOMA.
Labellum 3-lobed, the spur without any internal scale. Flowers large . . . . 10. VANDA.
Labellum undivided, the spur without any internal scale. Lateral sepals dilated at the base . . . . 11. SACCOLABIUM.
Terrestrial with short creeping rhizomes. Caudicle single.
Sepals and petals erect. Labellum scarcely saccate at the base, the disk with longitudinal raised lines . . . . . 12. GEODORUM.
Sepals and petals spreading. Labellum with a short pouch or spur at the base, marked with cristate or bearded veins . . 13. EULOPHIA.
Terrestrial with short creeping rhizomes. Caudicle bipartite.
Labellum gibbous and adnate to the column at its base . . 14. DIPODIUM.

TRIBE 3. **Bletideæ.**—*Anther lid-like, incumbent, usually deciduous. Pollen-masses waxy, 4 or 8, tapering at the base, separately attached and sessile on short caudicles, or on a short dichotomous caudicle. Terrestrial with creeping or rarely tuberous rhizomes or rarely epiphytes. Sepals and petals nearly equal, free and spreading. Flowers often large.*

Pollen-masses 4 in pairs, sessile on a gland. Usually epiphytes . 15. CYMBIDIUM.
Pollen-masses 8, on very short separate caudicles without any gland. Terrestrial. Leaves long, plicate . . . . 16. SPATHOGLOTTIS.
Pollen-masses 8, on a short dichotomous caudicle without any gland. Terrestrial. Leaves large, plicate. Labellum spurred, convolute round the column . . . . 17. PHAIUS.
Pollen-masses 8, separately attached to a divisible gland. Terrestrial. Leaves large, plicate. Labellum usually spurred, connate with the column at the base . . . . 18. CALANTHE.

TRIBE 4. **Arethuseæ.**—*Anther lid-like, incumbent, usually deciduous. Pollen granular or mealy. Terrestrial or rarely epiphytes. Stems in the Australian genera or sections leafless at the time of flowering.*

Large epiphytes with paniculate flowers. Labellum curved round the column. Anther with a broad dorsal appendage . . . 19. GALEOLA.
Terrestrial. Flowers nodding, in a simple raceme.
Column very short. Sepals and petals free. Anther with a thick fleshy terminal appendage . . . . 20. EPIPOGUM.
Column elongated (longer than the anther).
Sepals and petals united in a 5-lobed cup or tube . . . . 21. GASTRODIA.
Sepals and petals free . . . . 22. POGONIA.

TRIBE 5. **Neottieæ.**—*Anther erect or bent forward, persistent but free from the rostellum. Pollen granular or mealy. Terrestrial herbs with simple stems (except* Corymbis) *bearing 1 or more leaves or rarely leafless, and a single spike raceme or single flower.*

Column very long and persistent. Sepals and petals very long narrow and deciduous. Leaves large and strongly ribbed. Panicles axillary. . . . . 23. CORYMBIS.
Column very short, with wings either very broad or produced between the anther and rostellum into lateral lobes or appendages.
Rhizome creeping. Leaves petiolate. Spike slender.
Flowers reversed, the lateral sepals forming a hood over the labellum . . . . 24. RAMPHIDIA.
Flowers erect, the dorsal sepal and petals forming a hood over the column. Stigma in a deep pouch . . . . . . 25. GOODYERA.

Rhizome with ovoid or oblong tubers or rarely thick clustered fibres. Leaves sessile, few or only one.
Flowers small, sessile in a dense spiral spike. Sepals and petals erect or spreading at the tips only . . . . . . 26. SPIRANTHES.
Flowers racemose. Sepals broad and petal-like.
Dorsal sepal concave. Petals much smaller. Labellum densely fringed on the surface . . . . 27. CALOCHILUS.
Sepals petals and labellum all alike and spreading. Column-wings connected at the base in front. . . . 28. THELYMITRA.
Sepals and petals alike and spreading. Labellum with 2 lobes on the claw and a tuft of linear processes at the base of the lamina . . . . 29. EPIBLEMA.
Flowers racemose. Lateral sepals narrow-linear and long. Labellum 3-lobed at or near the base.
Dorsal sepal embracing the column at the base, erect and open at the end . . . . 30. DIURIS.
Dorsal sepal hood-shaped and incurved over the column . 31. ORTHOCERAS.
Flowers racemose, reversed. Sepals usually narrow. Labellum undivided.
Labellum dilated and enclosing the column at the base. Leaves flat. Flowers large. No caudicle . . . . 32. CRYPTOSTYLIS.
Labellum contracted or clawed rarely gibbous at the base. Leaves terete. Flowers often small. Caudicle linear . 33. PRASOPHYLLUM.
Flowers spicate, small and green, not reversed. Labellum entire or 2-lobed. Lateral lobes of the column very small. Leaves terete. No caudicle . . . . 34. MICROTIS.
Column short. Labellum tubular, erect under the hood-shaped dorsal sepal, with or without a reflexed lamina, lateral sepals and petals very narrow or minute. Small plants with 1 broad leaf and 1 large flower . . . . 35. CORYSANTHES.
Column elongated (longer than the anther) semiterete or longitudinally winged. Stems simple, with 1 rarely 2 or 3 or no leaves. Rhizome with annually renewed tubers.
Labellum unguiculate, the lamina peltate or produced beyond its insertion into a basal appendage.
Dorsal sepal hood-shaped. Lateral sepals united at the base. Column semiterete with an oblong stigma about the middle and 2 hatchet-shaped wings above it. Labellum with a basal appendage . . . . 36. PTEROSTYLIS.
Sepals and petals all linear.
Labellum rather broadly peltate. Column with a broad petal-like wing its whole length . . . . 37. CALEANA.
Labellum hammer-shaped. Column with 1 or 2 pairs of narrow auricles . . . . 38. DRAKÆA.
Labellum-lamina sessile or not produced beyond its insertion on the claw.
One ovate-cordate leaf. Lateral sepals narrow. Petals short. Labellum undivided without fringes or erect calli, but two adnate ones at the base . . . . 39. ACIANTHUS.
One ovate or lanceolate leaf. Lateral sepals oblong, stipitate. Labellum-lamina or middle lobe very convex and villous . . . . 40. ERIOCHILUS.
One to three ovate or lanceolate leaves. Dorsal sepal broad. Labellum thickened along the centre, glabrous or papillose, without calli. Column not winged . . . . 41. LYPERANTHUS.
No leaves on the flower-stem. Sepals and petals nearly equal, connivent. Labellum with 2 longitudinal raised lines without calli. Column winged . . . . 42. BURNETTIA.
One broad leaf. Lateral sepals and petals very narrow and spreading. Labellum with two adnate calli at the base continued in raised lines. Column winged . . . . . . 43. CYRTOSTYLIS.

One linear or oblong leaf. Lateral sepals as broad as or broader than the petals. Labellum with raised calli or fringes variously arranged. Column winged . . . . . 44. CALADENIA.
Two oblong leaves. Lateral sepals narrower than the petals. Labellum with raised calli or fringes. Column winged . 45. CHILOGLOTTIS.
One oblong or lanceolate leaf. Sepals and petals nearly equal, petal-like and spreading. Labellum without calli or fringes except 1 or 2 linear-clavate processes erect against the column . . . . 46. GLOSSODIA.

TRIBE 6. **Ophrydeæ.**—*Anther adnate to the top of the column over the stigma, the cells usually forming 2 lobes. Pollen-masses 2, granular, attached by caudicles to one or two glands or pouches over the stigma. Terrestrial herbs, rhizomes with annually renewed tubers. Stems simple leafy. Flowers spicate.*

Labellum spurred, with 2 linear processes of the column incumbent on its base . . . . 47. HABENARIA.

TRIBE 7. **Apostasieæ.**—*Anthers 2, lateral near the base of the style, with a dorsal rudimentary or rarely perfect anther. Stigma terminal. Pollen granular. Terrestrial herbs.*

Stem leafy. Panicles (as in *Corymbis*) axillary. Perianth of 6 equal segments . . . . 48. APOSTASIA.

TRIBE 1. MALAXIDEÆ.—Anther lid-like, incumbent on the apex of the column which has usually a membranous margin, usually deciduous. Pollen-masses waxy, 2, 4, or rarely 8, without caudicles or gland. Epiphytes or rarely terrestrial, with a creeping rhizome.

1. LIPARIS, Rich.

(Sturmia, *Endl.*)

Sepals and petals all free and spreading, equal and similar or the petals and dorsal sepal narrower. Labellum shortly embracing or united with the column at the base, erect or ascending, entire. Column elongated, incurved, the apex winged. Anther terminal, lid-like. Pollen-masses 4, waxy, obovoid, equal in pairs in the two cells, which are sometimes not closely contiguous.—Terrestrial or epiphytical plants, the stems sometimes thickened at the base into small pseudo-bulbs. Leaves at or near the base of the stem. Flowers greenish-yellow white or faintly tinged with red, in a terminal pedunculate raceme.

The genus is widely spread over the tropical and subtropical regions of the Old World, with one northern species found in both hemispheres. The Australian species are, as far as known, all endemic.

Sepals all narrow and nearly equal.
  Sepals and petals 3½ to 5 lines long. Labellum broadly oblong.
    Flowers white . . . . 1. *L. reflexa.*
    Flowers yellow . . . . 2. *L. cuneilabris.*
  Sepals and petals 3 lines long. Labellum broadly obovate-cuneate . . . . 3. *L. cœlogynoides.*
Lateral sepals broadly oblong, falcate. Dorsal sepals and petals longer, narrow-linear. Labellum broadly oblong . . . . . 4. *L. habenaria.*

1. **L. reflexa,** *Lindl. Bot. Reg. under n.* 882. Stems from a shortly creeping rhizome, thickened at the base, the whole plant including the

raceme varying from 3 or 4 in. to nearly 1 ft. high when luxuriant. Leaves almost distichous, the 2 or 3 lower ones reduced to acute sheathing scales, 1 to 3 upper ones oblong-lanceolate, 3 to 5 in. long and about ½ in. broad, with a short sheathing base, those of the barren pseudobulbs longer and contracted into a long sheathing base. Flowers yellowish white in loose racemes. Bracts lanceolate-subulate or subulate. Pedicels with the ovary 4 to 5 lines long. Buds falcate. Sepals and petals 4 to nearly 5 lines long, the sepals narrow-lanceolate, the petals linear. Labellum about as long as the sepals, the erect base embracing the base of the column but free from it, the lamina broadly oblong, concave, the margins slightly undulate and dilated at the base into 2 auricles or short lateral lobes embracing the apex of the column, the apex obtuse or retuse and sometimes denticulate-ciliate, the disk with 2 more or less marked longitudinal raised lines. Column narrow, incurved, about ½ as long as the sepals, the margins slightly dilated at the top into membranous wings.—Reichb. f. Beitr. 46; *Cymbidium reflexum*, R. Br. Prod. 331; *Sturmia reflexa*, F. Muell. Fragm. ii. 72, iii. 165.

N. S. Wales. Port Jackson, *R. Brown*, *Woolls;* Clarence river, *Beckler;* New England, *C. Stuart*; Tweed river, *Fitzgerald.*

2. **L. cuneilabris,** *F. Muell. Fragm.* iv. 164. Habit stature and loose racemes of rather large flowers of *L. reflexa*, but the flowers said to be yellow. Leaves more acute and tapering than in that species. Bracts shorter. Buds rather longer and narrower. Labellum shaped as in *L. reflexa*, of which this is probably a variety.—*Sturmia cuneilabris*, F. Muell. l.c.

Queensland. Rockingham bay, *Dallachy.*

3. **L. cœlogynoides,** *F. Muell. Fragm.* ii. 71. Habit nearly of *L. reflexa*, but the leaves more rigid and acute and those of the flowering stems more contracted at the base. Racemes much more slender than in *L. reflexa*, and the flowers smaller, the peduncle and rhachis flattened and almost winged in the specimens (possibly from pressure in drying?). Bracts lanceolate, acuminate. Ovary and pedicels about 3 lines long. Flowers (white?), very delicate. Sepals and petals linear, about 3 lines long, the petals about half as broad as the sepals. Labellum as long as the petals, the short base embracing the column, the lamina broadly obovate-cuneate, the broad end slightly denticulate. Column half as long as the sepals, narrow, incurved, 2-winged at the top.—*Sturmia cœlogynoides*, F. Muell. l.c.

N. S. Wales. Clarence river, *Beckler.*

4. **L. habenarina,** *F. Muell. Fragm.* iv. 131. Habit and foliage nearly of *L. reflexa*, but taller, often above 1 ft. high. Leaves several, tapering at both ends, mostly acute, ½ to 1 in. broad, very few of them reduced to sheathing scales. Racemes long and rigid. Pedicels short, erect or spreading, the flowers much smaller than in *L. reflexa.* Bracts short. Lateral sepals about 2 lines long, broadly oblong-falcate, obtuse;

dorsal sepal and petals linear, about 3 lines long, the petals still narrower than the sepal. Labellum as long as the lateral sepals, recurved from the middle, broadly oblong, shortly embracing and adnate to the column at the base, obtuse or retuse, the disk with 2 small prominent callosities. Anther-cells at some distance from each other in the anther-case.—*Sturmia habenarina*, F. Muell. l.c.

Queensland. Rockingham bay, *Dallachy*.—Very closely allied to the *L. ferruginea*, Lindl. in Gard. Chron. 1848, 55, from Borneo and Malacca, and perhaps a variety; but that species has a dark brown labellum showing its colour in the dried state, and from a note in Herb. Lindley, has no calli on the disk, whilst the *L. habenarina* has the labellum of the same pale yellow as the rest of the flower.

2. OBERONIA, Lindl.

Sepals free, nearly equal and erect, or the dorsal one smaller and reflexed. Petals narrower or shorter than the sepals. Labellum sessile, concave, entire or variously divided, often cushion-like or keeled at the base and usually embracing the column. Column very short, terete, contracted at the base, the apex with angular margins. Anther terminal, lid-like. Pollen masses 4, waxy, closely contiguous in pairs and often falling away in one mass, sometimes oblique and unequal.—Epiphytical plants with very short or, in species not Australian, elongated stems, not usually thickened into pseudobulbs. Leaves distichous, equitant. Flowers very small, in terminal pedunculate dense racemes, the pedicels short, the bracts small.

The genus is spread over tropical Asia and the Indian Archipelago, extending westward to the Mascarene islands, and eastward to the South Pacific. Of the two Australian species one has a wide range over East India and the Archipelago, the others appear to be endemic. The genus has been reunited by Reichenbach with the European *Malaxis*, from which it differs but very little in the structure of the flowers, but the vegetative characters and geographical distribution appear to be sufficient to maintain it as distinct.

Bracts ovate, fringed-ciliate. . . . . 1. *O. iridifolia.*
Bracts lanceolate, fine-pointed, entire . . . . 2. *O. palmicola.*

1. **O. iridifolia,** *Lindl. Gen. and Sp. Orch.* 15, *and Fol. Orchid.* Leafy stems very short and thick, rarely lengthening to 1 in. Leaves 3 to 7, sometimes none of them exceeding 3 in., in other specimens above 6 in. long and rather broad. Racemes as long as or longer than the leaves, rather dense but very slender when in flower, the minute flowers more or less distinctly collected in closely approximate whorls. Bracts ovate, scarious, mostly denticulate. Pedicels at length $\frac{3}{4}$ line long. Labellum broad, more or less fringed at least at the base and often 2-lobed at the end, about $\frac{1}{2}$ line long. Sepals smaller. Capsule $1\frac{1}{2}$ to 2 lines long, prominently angled.—*Malaxis iridifolia*, Reichb. f. in Walp. Ann. vi. 208.

Queensland. Brisbane river, *F. Mueller;* Rockhampton, especially Crocodile Creek, *Bowman, Thozet, O'Shanesy, Dallachy;* also in East India and the Archipelago.

2. **O. palmicola,** *F. Muell. Fragm.* ii. 24. A small delicate almost stemless species with the habit of some of the smaller E. Indian ones.

Leaves 5 to 7, lanceolate, 1 to nearly 2 in. long. Racemes very slender, 2 to 3 in. long, the minute flowers very numerous, clustered in distinct whorls. Bracts as long as the flowers, lanceolate with fine points and often ciliate. Sepals and petals about $\frac{1}{4}$ line long, lanceolate, acute, the petals narrower than the sepals. Labellum about as long as the sepals, with 2 broad (entire ?) lateral lobes, the middle lobe rhomboidal, rather broader than long. Fruiting pedicels $\frac{1}{2}$ line long. Capsule nearly 1 line.—*Malaxis palmicola*, F. Muell. Fragm. vii. 30.

Queensland. Brisbane river, *Kellemay;* Rockhampton? *Dallachy.*
N. S. Wales. On the trunks of Palms, Hastings river, *Beckler;* Clarence river, *C. Moore;* Bellinger range, *Fitzgerald.*

3. DENDROBIUM, Swartz.

Sepals nearly equal in length, the lateral ones very obliquely dilated at the base and connate with a projection from the base of the column into a pouch or spur. Petals usually nearly the length of the upper sepal or rather longer. Labellum articulate at the end of or (in species not Australian) shortly connate with the basal projection of the column, concave at the base, with the margins gradually expanded into 2 lateral lobes usually embracing the top of the column, and a central terminal lobe usually spreading or recurved, or the lateral and terminal lobes confluent in an entire concave or spreading lamina, the disk usually bearing longitudinal raised plaits. Column not very long, winged or toothed at the top. Anther terminal, lid-like. Pollen masses 4, in collateral pairs, usually equal and free or slightly coherent.—Rhizome tufted or creeping on trees or rocks; stems elongated and branching or simple and thick, sometimes reduced to short pseudobulbs, and usually bearing both leaves and racemes or 1-flowered peduncles. Flowers often rather large and showy, rarely very small.

A large genus ranging over the warmer regions of both the New and the Old World, one species found as far south as New Zealand. Of the 24 Australian species one only has been identified with certainty with an exotic species, the *D. hispidum* of Vanikoro, in the South Pacific, although another from the same island, of which the flower is unknown, has been conjectured to be the same as the Australian *D. striolatum;* the remaining 22 appear to be strictly endemic.

SECT. 1. **Dendrocoryne.**—*Stems simple, elongated or short, bearing* 2 *or more flat or channelled leaves at or towards the end. Racemes* 1 *to* 3, *apparently terminal or nearly so, or only in the uppermost axils.* (*Stems more leafy with short axillary racemes in* D. agrostophyllum *and* D. Smilliæ.)

Petals obovate, broader than the sepals. Racemes few-flowered, on long peduncles. Flowers pink or lilac.
  Basal pouch of the flower with a prominent spur underneath, forming a double spur.
    Petals $\frac{3}{4}$ in. broad. Labellum middle-lobe very broad and obtuse . . . . 1. *D. bigibbum.*
    Petals $\frac{1}{4}$ in. broad. Labellum middle-lobe oblong-lanceolate, acute or mucronate . . . . 2. *D. dicuphum.*
  Basal pouch scarcely gibbous on the lower side . . . . 3. *D. Sumneri.*

Petals narrower or not broader than the sepals. Flowers white, yellow, brown-red, or spotted with red.
Leaves flat. Large species. Racemes above 6 in. and often above 1 ft. long, with numerous rather large flowers.
Petals and sepals obtuse, very much undulate and spreading.
Leaves broad. Bracts $\frac{1}{4}$ to $\frac{1}{2}$ in. long . . . . 4. *D. undulatum.*
Leaves narrow. Bracts minute . . . . 5. *D. Johannis.*
Petals and sepals lanceolate, acute, not undulate and almost connivent . . . . 6. *D. speciosum.*
Leaves flat. Racemes under 6 in., with a slender rhachis.
Stems usually attenuate towards the base. Sepals and petals with long slender points, 3 or 4 times as long as the labellum.
Stems prominently 4-angled . . . . 7. *D. tetragonum.*
Stems terete, many-angled . . . . 8. *D. æmulum.*
Stems not much or not at all attenuate at the base. Sepals and petals lanceolate not above $\frac{1}{3}$ longer than the labellum.
Central lobe of the labellum broad, almost reniform.
Flowers reddish purple. Spur conical, straight . . 9. *D. Kingianum.*
Flowers yellow. Spur short, broad and incurved . 10. *D. gracilicaule.*
Central lobe of the labellum linear-lanceolate. Spur rather long nearly straight . . . . 11. *D. Moorei.*
Stems rather slender, leafy, with short axillary racemes.
Sepals and petals broad, about as long as the labellum 12. *D. agrostophyllum.*
Leaves flat. Stems long and thick. Racemes several, short and dense. Spur very obtuse, longer than the lanceolate sepals and petals . . . . 13. *D. Smilliæ.*
Leaves linear, channelled. Stems short. Racemes long pedunculate. Sepals and petals linear-oblong or spathulate . . . . 14. *D. canaliculatum.*

SECT. 2. **Monophyllæa.**—*Stems tufted, short and thick, terminating in a single leaf with* 1 *to* 3 *flowers at its base.*

Leaves flat, oblong or lanceolate.
Ovary and fruit smooth . . . . 15. *D. monophyllum.*
Ovary and fruit muricate . . . . 16. *D. hispidum.*
Leaves very thick, ovoid-oblong, with tuberculate ribs . . . 17. *D. cucumerinum.*

SECT 3. **Rhizobium.**—*Rhizomes creeping, with scarious sheathing scales. Leaves solitary in the axils of the scales, either sessile on a broad disk or on a very short turbinate protuberance. Racemes few-flowered, also from the rhizome.*

Leaves flat, coriaceous, very acute. Sepals lanceolate, about 5 lines long without the spur . . . . 18. *D. pugioniforme.*
Leaves very thick and fleshy, almost acute. Sepals oblong-lanceolate, about 4 lines without the pouch . . . . 19. *D. rigidum.*
Leaves very thick and fleshy, obtuse. Sepals linear or narrow linear-lanceolate, 7 to 10 lines long . . . . 20. *D. linguiforme.*

SECT. 4. **Strongyle.**—*Stems branched, usually rather slender. Leaves distant, terete. Racemes (few-flowered) or peduncles lateral.*

Sepals and petals narrow, 1 in. long. Spur from $\frac{1}{4}$ to near $\frac{1}{3}$ as long. Labellum much shorter than the sepals, the middle lobe acuminate . . . . 21. *D. teretifolium.*
Sepals and petals lanceolate, under $\frac{3}{4}$ in. Spur short, very obtuse. Labellum nearly as long as the sepals, the middle lobe ovate-oblong . . . . 22. *D. striolatum.*

Sepals and petals narrow, about $\frac{1}{2}$ in. Spur about half as long. Labellum as long as the sepals, the middle lobe acuminate . . . . 23. *D. Mortii.*
Sepals and petals lanceolate, about $\frac{1}{4}$ in. Spur about as long. Labellum nearly as long as the sepals, the middle lobe short and broadly reniform . . . . 24. *D. Bowmanii.*

D. complanatum, A. Cunn. in Lindl. Bot. Reg. 1839, Misc. 34, from the neighbourhood of the Brisbane river, is described as having the flattened stems and distichous leaves of the section *Aporum*, Lindl., but as the flowers were never seen, and no specimen of the foliage has been preserved for comparison, it will be impossible to identify the plant. The mention of the yellowish tinge of the foliage might lead one to suppose that it may have been the *Oberonia iridifolia.*

D. Fellowsii, F. Muell. Fragm. vii. 63, is described from very insufficient materials—the summit of a stem, with the rhachis of a few racemes from which the flowers are all fallen away with 2 loose leaves, which would all indicate the *D. Smilliæ.* Dallachy saw no plants in flower on the trees, but picked up a few yellow flowers from the ground which I cannot distinguish from those of *D. gracilicaulis.* It is very doubtful therefore whether the *D. Fellowsii* may not be a compound of these two species.

D. minutissimum, F. Muell. Fragm. v. 95, from a single locality near Botany Bay, is mentioned by name only, and I have seen no specimens.

SECT. 1. DENDROCORYNE, *Lindl.*—Stems simple, elongated or short and sometimes thickened into oblong pseudobulbs, bearing 2 or more flat or channelled leaves at or near the end, the lower part with scarious thin sheathing scales which usually soon wear away, leaving annular scars. Racemes 1 to 3, apparently terminal or nearly so (owing to the arrest of the terminal shoot), or only in the upper axils.

1. **D. bigibbum,** *Lindl. in Paxt. Fl. Gard.* iii. 25, *f.* 245. Stems in the cultivated plant 6 to 8 in. long, slightly contracted towards the base, bearing in the upper part 3 or 4 lanceolate leaves of 3 or 4 in., and in the older plants swollen at the base in a short pseudobulb. Raceme apparently terminal on a peduncle of 6 to 8 in., with 3 to 10 large flowers of a deep lilac on pedicels of $\frac{3}{4}$ to 1 in. Sepals ovate or ovate-lanceolate, acute, 9 to 10 lines long and about 5 lines broad in our specimens, the lateral ones produced with the basal projection of the column into an obtuse pouch with a conical straight or curved obtuse spur on the lower side forming a double spur as in *D. bicuphum.* Petals broadly obovate almost orbicular, fully $\frac{3}{4}$ in. broad. Labellum rather shorter than the sepals, with the large lateral lobes forming the very broad base of *D. bicuphum,* but the middle lobe at least as broad as long, very obtuse or retuse and not so long as the broad base, the disk with 3 to 5 raised longitudinal lines fringed or crested from the middle upwards.—Bot. Mag. t. 4898.

Queensland. Mount Adolphus, Torres Straits, *Thomson.*—Only known from specimens cultivated by Loddiges.

2. **D. dicuphum,** *F. Muell. Fragm.* viii. 28. Stems strongly ridged and furrowed, sometimes 3 to 4 in long, rather thick, equal or scarcely contracted towards the base, sometimes shortened into a conical pseudobulb. Leaves few on the upper part of the stem, lanceolate or

linear-lanceolate, 3 to 6 in. long. Raceme erect, on an apparently terminal peduncle of 6 in. to 1 ft., with 3 or 4 (probably pink or purple) flowers, on pedicels of about ½ in. Sepals lanceolate, acutely acuminate, 7 to 8 lines long, the lateral ones forming at the base, with the basal projection of the column, a pouch as in other *Dendrobia*, but also emitting from the under side an obtuse hollow spur 1 to 1½ lines long, forming a distinctly double spur to the flower. Petals obovate, acute, rather longer and broader than the sepals, but not above 3 lines broad in our specimens. Labellum nearly as long as the sepals, attached at the junction of the 2 lobes of the spur, scarcely clawed but mobile, the broad lateral lobes forming a truncate base nearly 5 lines broad, the middle lobe oblong-lanceolate, acute or mucronate and not 2 lines broad; the disk with 3 raised longitudinal lines or plates more or less fringed or crested and extending some way along the middle lobe, and occasionally 1 or 2 shorter additional fringed lines.

N. Australia. Liverpool river, *Gulliver;* and probably the same, but our specimen without flowers, Port Darwin, *Schultz*, *n.* 412.

3. **D. Sumneri,** *F. Muell. Fragm.* vi. 94. Stems elongated (22 in. in the imperfect specimen), not deeply furrowed. Leaves (detached) lanceolate, 3 to 4 in. long. Raceme with its long peduncle about 8 in. long, the rhachis flexuose, with few distant pink flowers, on pedicels of 3 to 4 lines. Sepals ovate-lanceolate, acute, nearly ½ in. long, the spur not half so long, broadly conical, slightly gibbous near the base on the upper side, but not forming the double spur of *D. bigibbum.* Petals as long as the sepals, broadly obovate. Labellum shorter than the sepals, the claw very much dilated from the base and expanded into broadly obovate lateral lobes, the middle lobe rather smaller, at least as broad as long, the disk with slightly raised lines fringed with raised processes between the lateral lobes and extending very shortly on the middle lobe.

Queensland. Near Cape York, *Jardine;* a very imperfect specimen in Herb. F. Mueller.

4. **D. undulatum,** *R. Br. Prod.* 332. A stout species growing in large tufts and attaining with the racemes several ft., the stems often swollen in the middle. Leaves bifarious, ovate or elliptical, obtuse or emarginate, 2 to 4 in. long, flat but thick and somewhat undulate or the margins recurved. Racemes from the upper part of the stem often above 1 ft. long, the flowers numerous, rather large, on pedicels often exceeding 1 in. Bracts lanceolate or linear-lanceolate. Sepals and petals nearly similar, speading, linear-oblong, obtuse, very much undulate, of a dingy brown usually bordered with yellow, about 1 in. long, the short broad basal pouch ending in a short curved or straight obtuse spur. Labellum shorter than the sepals, the lateral lobes large, erect, nearly flat, the middle lobe small, broadly lanceolate or oblong, recurved and undulate; the disk with 5 raised lines or plates of a light violet colour, of which 2 more prominent especially near the base and sometimes 7 immediately below the middle lobe. Column short.—

Lindl. Gen. and Sp. Orch. 87; Reichb. f. in Walp. Ann. vi. 298, Beitr. 47; F. Muell. Fragm. i. 87; *D. discolor*, Lindl. Bot. Reg. 1841, t. 52.

Queensland. Endeavour river, and many other places along the coast, *Banks and Solander*, *R. Brown*, *A. Cunningham;* Port Curtis, *M'Gillivray;* Curtis island, *Thozet;* Rockingham bay, *Dallachy;* Port Denison, *Fitzalan* (racemes 1½ ft. long, with above 30 flowers); Cape York, *Daemel;* Albany island, *A. C. Gregory.*

The species is said to be a native also of Java, but only on garden authority. It is not included in any enumeration of Archipelago Orchids, nor have we any extra-Australian specimen.

5. **D. Johannis,** *Reichb. f. in Gard. Chron.* 1865, 890; *Xen. Orch.* ii. 165. Very near *D. undulatum*, and perhaps a variety, differing in the much narrower acuminate leaves, in the minute bracts, and the flowers rather smaller, the undulate petals and sepals narrow, about ¾ in. long, of a uniform dull brown, the labellum yellow.—Bot. Mag. t. 5540.

Queensland? Cultivated by Veitch from Northern Australia with *D. canaliculatum* (*D. Tattonianum*). Of this I have only seen the plate quoted, and two loose flowers in Reichenbach's herbarium.

6. **D. speciosum,** *Sm. Exot. Bot.* i. 17, t. 10 (*the flowers grossly misrepresented*). Stems very thick and fleshy, 6 in. to 1 ft. high. Leaves few (2 to 5) distichous towards the apex of the stem, ovate or oblong, thick, flat or slightly undulate, 3 to 6 in. long. Racemes apparently terminal, often above 1 ft. long, the sheathing scales at the base of the peduncle 1 to 1½ in., the bracts very small, ovate or lanceolate. Flowers numerous, rather large, pale yellow, on pedicels of 1 to 2 in. Sepals and petals nearly equal, erect or slightly open, usually incurved, ¾ to 1 in. long, lanceolate but varying in breadth, the lateral ones incurved, forming with the basal projection of the column a short broad pouch. Labellum considerably shorter than the sepals, nearly white spotted with purple, the lateral lobes short and broad, the middle lobe broader than long, very obtuse or retuse. Column white, often spotted with purple.—R. Br. Prod. 332; Lindl. Gen. and Sp. Orch. 87; Reichb. f. Beitr. 48; Bot. Mag. t. 3074; Bot. Reg. t. 1610.

Queensland. Port Bowen, *R. Brown;* Brisbane river, *W. Hill;* Rockhampton, *Bowman*, *O'Shanesy;* Rockingham bay, *Dallachy.*
N. S. Wales. Port Jackson, *Banks and Solander*, *Caley*, *Woolls*, *Vicary;* northward to Hastings river, *Tozer;* southward to Cape Howe, *Walter.*
Victoria. Nangatta mountains and Genoa river, *F. Mueller.*

D. Hillii, Hook. Bot. Mag. t. 5261, as suggested by F. Mueller, Fragm. iv. 175, appears to be a slight variety of *D. speciosum*, with the sepals and petals narrower and longer than those figured t. 3074, but shorter and rather narrower than those of several other specimens.

7. **D. tetragonum,** *A. Cunn. in Bot. Reg.* 1839, *Misc.* 33. Stems from a creeping or tufted rhizome numerous, forming small pseudobulbs at the base, from 1 or 2 in. to above 1 ft. long, very prominently 4-angled, rather slender in the lower portion, thickened above the middle or near the apex. Leaves 2 or 3 near the summit of the stem, oblong or broadly lanceolate, acute, 2 to 4 in. long. Racemes above the leaves 1 or 2, short and loose, bearing each only 1 to 3 yellowish

green flowers bordered with brownish red on pedicels of ½ to 1 in. Bracts small and narrow. Sepals from a broad triangular base of ¼ in. suddenly contracted into a linear almost filiform point of 1 to nearly 1½ in. the basal spur ascending, thick and very obtuse, 2½ lines long. Petals much shorter than the sepals, linear-filiform with a slightly dilated lanceolate base. Labellum nearly ½ in. long, pale yellow streaked with narrow bands of crimson, the lateral lobes broad and prominent, the middle one larger, almost rhomboidal, shortly and acutely acuminate. Disk with 3 raised lines or narrow plates scarcely undulate, the central one more raised and alone continued on the base of the middle lobe.—Lindl. Bot. Reg. 1841, Misc. 2; Bot. Mag. t. 5956; F. Muell. Fragm. i. 87.

Queensland. Dry shaded woods, Moreton bay, *A. Cunningham;* Islands of Moreton bay, *F. Mueller;* Rockhampton, *Bowman, O'Shanesy;* Rockingham bay, *Dallachy.*

N. S. Wales. Hastings, Macleay, and Clarence rivers, *Beckler;* Tweed river, *Guilfoyle.*

8. **D. æmulum,** *R. Br. Prod.* 333. Stems terete, rather thick, sometimes tapering into a long thin base with a small pseudobulb as in *D, tetragonum.* Leaves 2 or 3 near the summit, ovate or oblong. Racemes 1 to 3 at the end between the leaves, 2 or 3 in. long, the rhachis slender, the bracts small and lanceolate. Flowers rarely above 6 in the raceme, on slender pedicels of ½ in. or less. Sepals narrow lanceolate, almost linear, often nearly 1 in. long, striate, the basal pouch or spur short and broad, turned upwards. Petals narrow-linear, as long as the sepals. Labellum scarcely above ¼ in. long, contracted into a claw at the base, the lateral lobes broad short and acute, the middle lobe recurved, ovate, much undulate, with a small acute point; the disk with 3 raised lines or plates between the lateral lobes merging into a single broad much undulate one extending to the end of the middle lobe.—Lindl. Gen. and Sp. Orch. 87; Reichb. f. Beitr. 49; F. Muell. Fragm. i. 213; Bot. Mag. t. 2906.

Queensland. Brisbane river, *Bailey.*

N. S. Wales. Port Jackson, *R. Brown;* Blue Mountains, *Miss Atkinson, Vicary;* New England, *C. Stuart;* Hastings river, *Beckler;* Richmond river, *Henderson;* Tweed river, *Guilfoyle.*

The figure above quoted from the Botanical Magazine represents the stems thicker and shorter, and the sepals and petals shorter and broader than they are in the wild specimens.

9. **D. Kingianum,** *Bidw. in Lindl. Bot. Reg.* 1844, *Misc.* 11, 1845, t. 61. Stems usually 3 to 6 in. high, striate with prominent angles, thickened at the base. Leaves at the summit of the stem 3 to 5, lanceolate or oblong-lanceolate, acute, 3 to 4 in. long. Racemes within or above the leaves 1 to 3, longer than the leaves. Flowers of a reddish purple, on pedicels of ¼ to ½ in. Bracts very small. Sepals broadly lanceolate, the lower ones much falcate, 4 lines long in some specimens, fully 5 in others. Petals about as long, but narrower. Spur conical, slightly incurved, about 3 lines long. Labellum not much shorter than the sepals, not undulate, the lateral lobes very prominent, almost oblong, obtuse, the

middle lobe scarcely longer, but very broad, almost reniform, the disk with 3 raised lines or plates extending to the base of the middle lobe, but not beyond.—Bot. Mag. t. 4527; F. Muell. Fragm. iii. 60.

Queensland. Moreton bay? *Bidwill.*
N. S. Wales. On rocks and trees, Biron, *Leichhardt;* New England, *C. Stuart.*

10. **D. gracilicaule,** *F. Muell. Fragm.* i. 179. Closely resembles *D. Kingianum* in habit, foliage and all essential characters. Leaves rather more rigid and broader than in that species, racemes shorter, and the flowers (sometimes but not always rather smaller) of a dingy yellow spotted with red, the spur or pouch shorter, broader and more curved.—*D. elongatum,* A. Cunn. in Bot. Reg. 1839, Misc. 33; Lindl. l.c. 1841, 21, but not of Lindl. Gen. et Sp. Orch. 77; *D. brisbanense,* Reichb. f. in Walp. Ann. vi. 299.

Queensland. Moreton bay, *F. Mueller.*
N. S. Wales. Macleay and Clarence rivers, *Beckler;* Lord Howe's island, *C. Moore,* and others.

11. **D. Moorei,** *F. Muell. Fragm.* vii. 29. Closely resembles *D. Kingianum* and *D. gracilicaule,* and scarcely to be distinguished from them without the flowers, which are very different. Stems 3 to 8 in. high, strongly marked with prominent angles and furrows, sometimes equally thick throughout, sometimes attenuate or thickened at the base. Leaves 2 to 4 in. long. Racemes usually exceeding the leaves, 1 or 2 at the end of the stem. Bracts lanceolate or subulate, 1 to 4 lines long. Pedicels $\frac{1}{4}$ to $\frac{1}{2}$ in. Sepals linear-lanceolate, varying in different specimens from 5 to 7 lines, besides the spur which is straight or slightly curved and 3 to 4 lines long. Petals as long as the sepals but narrower at the base. Labellum attached to the end of the basal projection of the column, with a narrow claw as long as the spur, expanded below the apex of the very short column into 2 short lateral lobes, the middle lobe linear-lanceolete, complicate, not undulate, at least $\frac{2}{3}$ as long as the sepals; disk with slightly raised lines only between the short lateral lobes.

N. S. Wales. Trees and rocks, Lord Howe's island, *C. Moore, Eclipse Expedition.*

12. **D. agrostophyllum,** *F. Muell. Fragm.* viii. 28. Stems about 1 ft. high, rather slender, leafy from the middle upwards. Leaves lanceolate, rather thin, 2 to 3 in. long. Racemes axillary and distant, the rhachis not 1 in. long. Flowers yellow, on pedicels of about $\frac{1}{2}$ in. Sepals broadly lanceolate or the dorsal one oblong, about 4 lines long. Petals obovate-oblong, as long as the sepals. Labellum sessile, nearly 5 lines long and broad and almost square, the lateral lobes obliquely rhomboidal, separated by a narrow sinus from the terminal one, which is more than twice as broad as long; the disk with a raised plate or callus along the centre below the middle.

Queensland. Rockingham bay, *Dallachy.*

13. **D. Smilliæ,** *F. Muell. Fragm.* vi. 94. Stems 1 to 2 ft. long, thick, very prominently angled and furrowed, some of the specimens marked only with the annular scars of the scarious sheathing scales, others with the short persistent sheaths of fallen leaves. Leaves oblong or lanceolate, 2 to 4 in. long. Racemes from the uppermost nodes 2 to 3 in. long, the flowers "crimson tipped with green," crowded almost from the base on pedicels of 3 to 6 lines. Bracts very small. Sepals ovate-lanceolate, not 3 lines long without the spur, which is 4 or 5 lines long, straight, somewhat dilated towards the end and very obtuse. Petals rather smaller than the sepals. Labellum with a long broad claw, expanded at the apex into a concave complicate almost hood-shaped lamina, shorter than the sepals, broader than long, entire or broadly and shortly 2-lobed. Pollen-masses of the genus, closely cohering in pairs.

Queensland. Rockingham bay, *Dallachy.*—The species is evidently closely allied to the *D. viridiroseum*, Reichb. f., described from garden specimens said to have been brought from Java, which, however, has more acute sepals and a somewhat differently shaped labellum.

14. **D. canaliculatum,** *R. Br. Prod.* 333. Stems or pseudobulbs usually rather thick and not above 1 to 2 in. high. Leaves at the summit 2 to 6, linear, thick, almost semicylindrical but grooved on the upper side, 4 to 8 in. long, often 3 lines broad at the base but tapering to the end. Racemes in the upper axils often a foot long including the peduncle, but flowering only in the upper portion, the rhachis slender, the sheathing scales of the peduncle small, the bracts still smaller. Pedicels slender, $\frac{1}{2}$ to $\frac{3}{4}$ in. long. Sepals and petals linear, white tipped with yellow, the sepals 5 or 6 lines long, the lateral ones slightly falcate and produced at the base with the basal projection of the column into a conical spur. Petals rather longer than the sepals and contracted at the base. Labellum about $\frac{2}{3}$ the length of the sepals, with 3 lobes of a rich mauve colour, the lateral ones prominent and obtuse, the middle one nearly orbicular or rather broader than long, with a short point, the disk with 3 prominently raised lines or plates, much undulate between the lateral lobes and ending on the middle lobe in richly coloured nearly orbicular laminæ.—Lindl. Gen. and Sp. Orch. 91; Reichb. F. Beitr. 49; F. Muell. Fragm. iii. 126; *D. Tattonianum*, Batem. in Gard. Chron. 1865, 890; Bot. Mag. t. 5537.

Queensland. Endeavour river, *Banks and Solander;* Rockingham bay, *Dallachy.* "Flowers fragrant."—There are no specimens preserved of Banks and Solander's plant, but from the drawing in the British Museum there is no doubt of its identity with *D. Tattonianum,* although the colour of the flowers is rather duller.

SECT. 2. MONOPHYLLÆA.—Stems usually numerous, short and thick, terminating in a single leaf, with 1 to 3 flowers also at the apex of the stem at the base of the leaf.

15. **D. monophyllum,** *F. Muell. Fragm.* i. 189. Stems from a creeping rhizome numerous, erect, thick, narrow-conical, 1 to 2 in. long, with very prominent ribs and furrows, at least when dry. Leaf

apparently terminal, oblong or lanceolate, flat, 2 to 3 in. long. Raceme also solitary and apparently terminal, about as long as the leaf, with 2 or 3 "yellow" flowers on pedicels of 3 or 4 lines. Bracts minute. Sepals broadly lanceolate, acute, about 3 lines long, besides the broad obtuse spur about 2 lines long and slightly curved upwards. Petals as long as the sepals but narrower especially at the base. Labellum "deep yellow," nearly as long as the sepals, the lateral lobes small, the middle one broadly triangular or almost rhomboidal and obtuse; the disk without raised lines below the lobes, but 1 or 3 raised calli between the lateral lobes, sometimes produced into short undulating raised lines or plates on the middle lobe.—*D. tortile*, A. Cunn. in Lindl. Bot. Reg. 1839, Misc. 33, name only, not *D. tortile* Lindl. from Moulmein.

Queensland. On the upper branches of lofty trees, Brisbane river, Moreton bay, *A. Cunningham*, *W. Hill;* Glasshouse Mountains, *Beyerley*.

N. S. Wales. Richmond and Clarence-rivers, *Beckler*.

16. **D. hispidum,** *A. Rich. Sert. Astrol.* 13, t. 5. A dwarf plant forming dense tufts. Stems usually from under 1 in. to nearly 2 in. high, not thickened at the base, formed of only 2 or 3 internodes with membranous scarious sheaths. Leaf apparently terminal, oblong or lanceolate, prominently keeled underneath, 1 to 2 in. long. Pedicels at the base of the leaf 1 or 2, slender, 1 to 2 lines long. Ovary densely muricate with bristly processes. Sepals ovate, obtuse, about $1\frac{1}{2}$ lines long besides the spur which is about as long as the free part, the dorsal sepal rather narrower. Petals linear, as long as the sepals. Labellum as long as the sepals, the lateral lobes rather broad, the middle lobe at least as broad as long, with raised wrinkles on its surface but no raised lines on the disk between the lateral lobes. Capsule echinate, ovoid or globular, about 3 lines diameter.—F. Muell. Fragm. vii. 30.

Queensland. On trees, Rockingham bay, *Dallachy*.—Our specimens are in fruit only, but a somewhat withered flower in Herb. F. Mueller shows precisely the structure figured by A. Richard, from the original specimens gathered in the island of Vanikoro.

17. **D. cucumerinum,** *Lindl. Bot. Reg.* 1842, *Misc.* 58; 1843, t. 37. A dwarf species like the two preceding, the tufted stems not exceeding 1 in., with prominent ribs and furrows and annular scars of the sheathing scales. Leaf terminal, ovoid-oblong, fleshy and about as thick as the stem, above 1 in. long, marked with longitudinal raised tuberculate ribs and assuming the aspect of a little cucumber. Pedicels 1 to 3 at the base of the leaf and rather shorter. Sepals and petals yellowish-white, streaked with a reddish-yellow, about $\frac{1}{2}$ in. long. Spur short and conical. Labellum shorter than the sepals, the lateral lobes prominent, almost acute, the middle lobe ovate, shortly acuminate, recurved, with undulate-crisped margins; the disk with 3 or 5 longitudinal raised lines or plates, even between the lateral lobes, much undulate on the middle lobe.—Bot. Mag. t. 4619; F. Muell. Fragm. iii. 59.

N. S. Wales. On *Casuarina* trees near Brownlow hill, Camden, *Woolls*, according to a memorandum received from F. Mueller.—I have seen no specimen; the above character is taken from Lindley's figure and descriptions.

SECT. 3. RHIZOBIUM, *Lindl.*—Stems or rhizomes creeping, with scarious sheathing scales. Leaves solitary in the axils of the scales, either sessile on a broad disk or on a very short turbinate protuberance (a rudimentary stem or pseudobulb). Racemes few-flowered, also from the rhizome.

18. **D. pugioniforme,** *A. Cunn. in Lindl. Bot. Reg.* 1839, *Misc.* 33. Stems or rhizomes rather slender, much-branched, creeping and rooting at the nodes and covered when young by the scarious sheathing scales. Leaves articulate on a very short turbinate protuberance in the axils of the scales, ovate or ovate-lanceolate, tapering into a rigid point, flat but thick and rigid when dry, with the veins scarcely visible, $\frac{3}{4}$ to nearly 2 in. long. Racemes probably short and few-flowered from the remains of the rhachis, for the flowers only seen detached in Cunningham's specimens. Sepals lanceolate, about 5 lines long without the spur, which is straight, obtuse, about 3 lines long. Petals linear, as long as the sepals. Labellum nearly as long as the sepals, rather narrow to above the middle, then expanded into a broadly ovate-triangular acute recurved lamina, very much undulate but scarcely 3-lobed; the disk with 3 raised lines or plates, even on the claw, much undulate on the lamina.—*D. pungentifolium*, F. Muell. Fragm. i. 189.

Queensland. Brisbane river, Moreton bay, *F. Mueller;* Mount Lindsay, *W. Hill.*
N. S. Wales. Blue Mountains, *Woolls, Miss Atkinson;* Hastings river, *Beckler;* New England, *C. Stuart;* southward to Illawarra, *A. Cunningham, Shepherd.*

19. **D. rigidum,** *R. Br. Prod.* 333. Stems or rhizomes shortly creeping, the young shoots covered by the membranous scarious sheathing scales. Leaves apparently closely sessile on a broad base as in *D. linguiforme*, and similarly very thick fleshy and nerveless, mostly 1 to $1\frac{1}{2}$ in. long, oblong and almost acute but not nearly so much so as in *D. pugioniforme.* Raceme about as long as the leaves, the one figured bearing 3 flowers about 5 lines long including the broad obtuse pouch. Sepals "greenish white tinged with red," the dorsal one broadly lanceolate, the lateral ones very broad below the middle; petals as long as the sepals apparently narrow-oblong, "buff-coloured edged with red." Labellum as long as the sepals, the precise form not known.—Lindl. Gen. and Sp. Orch. 85.

Queensland. Endeavour river, *Banks and Solander* —Only known from a drawing in the British Museum, the original sketch, evidently taken at the time, accompanied by pencil notes in Solander's handwriting.

20. **D. linguiforme,** *Swartz in K. Akad. Stockh. N. Handl.* 1800, 247. Stems or rhizomes closely creeping and rooting at the nodes, rather thick and fleshy, the membranous scarious sheathing scales completely covering the young shoots. Leaves articulate on broad circular scars or scarcely raised disks at the axils of the sheaths, ovate or oblong, mostly under 1 in. long, but very thick and fleshy. Racemes from the rhizomes 2 or 3 in. long, with a slender rhachis, bearing from 6 or even fewer to above 20 flowers on filiform pedicels of 3 to 5 lines. Bracts

minute and scale-like. Sepals narrow linear-lanceolate, 7 to 10 lines long, the pouch or spur short broad and turned upwards. Petals narrow-linear, about the length of the sepals. Labellum under 3 lines long, much contracted at the base, the lateral lobes rather broad prominent and obtuse, the middle lobe longer, narrow-ovate, obtuse, the disk with 3 very prominent raised lines or plates, even between the lateral lobes, all three produced and much undulate on the middle lobe.—Sm. Exot. Bot. i. 19 t. 11; R. Br. Prod. 333; Lindl. Gen. and Sp. Orch. 85; Reichb. f. Beitr. 51.

Queensland. Mountains near Brisbane, *Dallachy;* Moreton bay, *W. Hill.*

N. S. Wales. Port Jackson to the Blue Mountains, *Caley* and many others; Hastings river, *Beckler;* Clarence river, *Wilcox;* Durval, *Leichhardt;* New England, *C. Stuart;* Port Macquarrie, *Tozer.*

SECT. 4. STRONGYLE, *Lindl.*—Stems elongated, branched, usually rather slender. Leaves distant, terete. Racemes few-flowered or 1-flowered, peduncles lateral.

21. **D. teretifolium,** *R. Br. Prod.* 333. Stems clustered on a creeping rhizome, elongated, terete, divaricately branched. Leaves few, terminating the branches or clasping the stem, from 3 in. to 1 ft. long in the typical form and 1 to 2 lines thick, straight or curved. Racemes lateral, often branched, very loose, with few white flowers on almost filiform peduncles of ½ to 1 in. Sepals linear-subulate, above 1 in. long, the lateral ones dilated at the base into a conical obtuse spur of 2 to 3 lines. Petals linear-filiform, as long as or slightly longer than the sepals. Labellum about half as long as the sepals, lanceolate, canaliculate, acuminate and recurved, the lateral lobes very small, the disk dotted with red and bearing 3 undulate raised lines or plates. Column dotted with red.—Lindl. Gen. and Sp. Orch. 91; F. Muell. Fragm. i. 89; Reichb. f. Beitr. 51; Endl. Iconogr. t. 99; Bot. Mag. t. 4711; *D. calamiforme,* Lodd. in Lindl. Bot. Reg. 1841, Misc. 9.

Queensland. Rockhampton, *O'Shanesy, Bowman.*

N. S. Wales. Port Jackson, *R. Brown;* Hastings and Clarence rivers, *Beckler, Wilcox;* Port Macquarrie, *Backhouse*; Richmond river, *Henderson;* New England, *C. Stuart.*

D. Fairfaxii, F. Muell. in Sydney Mail, Sept. 21, 1872, 360, with a woodcut, from Mount Tomah, appears to be a slight variety of *D. teretifolium* with leaves 1 to 2 ft. long and some slight differences in the spotting and undulations of the labellum. The flowers are also described as more numerous but not so figured.

22. **D. striolatum,** *Reichb. f. in Hamb. Gartenz.* 1857, 313, *and Xen. Orchid.* ii. 24, t. 109. Stems from a creeping rhizome often elongated as in *D. teretifolium,* but the branches not so divaricate. Leaves terminal or distant, terete, straight or more frequently curved, fleshy, 1 to 3 or even 4 in. long. Peduncles usually solitary and 1-flowered, rarely bearing 2 flowers on pedicels of ½ in. Sepals and petals white with 3 to 5 dark-coloured striæ towards the base; the sepals lanceolate, 8 to 9 lines long including the short basal pouch or spur of the lateral ones, the dorsal sepal rather narrower and the petals still narrower. Label-

lum rather shorter than the sepals, dilated in the middle into 2 broad lobes, the middle lobe ovate-oblong, recurved, the margins undulate-crisped, the disk with 3 undulate longitudinal raised lines or plates. Column white, the margin of the apex more or less 2- or 3-lobed.—*D. teretifolium*, Lindl. Bot. Reg. 1839, Misc. 32, not of R. Br.; *D. Milligani*, F. Muell. Fragm. i. 88, t. 6; Hook. f. Fl. Tasm. ii. 373.

N. S. Wales. Blue Mountains, *Vicary*, *Miss Atkinson;* Cape Howe, *Walter.*
Victoria. Rocks along the Genoa river, *F. Mueller.*
Tasmania. On rocks Flinders island, Streletzky's Peak, &c., *Milligan*, *Ricketts*, *Story.*

Some specimens from Hastings river and from Maryborough are referred to this species in Herb. F. Mueller, but having no flowers it cannot be determined whether they belong to this or to the following species. *D. schœninum*, Lindl. in Paxt. Fl. Gard. i. 134 (name only) is represented in Herb. Lindl. by a single flower which appears to be a slight variety of *D. striolatum*, with the sepals somewhat acuminate.

23. **D. Mortii,** *F. Muell. Fragm.* i. 214 *and* ii. 93 (partly). Stems elongated and branched as in *D teretifolium* but much more slender. Leaves also more slender and somewhat 4-angled when dry, 2 to 6 in. long. Peduncles mostly 1-flowered. Sepals narrow-lanceolate, acuminate, about ½ in. long besides the spur which is 2 to 3 lines long. Petals much narrower. Labellum the length of the sepals, the lateral lobes broad, the middle lobe acuminate. Column short.

N. S. Wales. Hastings river, *Beckler;* Macleay river, *Fitzgerald;* New England, *C. Stuart.*

D. Beckleri, F. Muell. Fragm. v. 95, and vii. 59, seems to me to be precisely the same as the plant originally described by him as *D. Mortii*, although the flowers described in Fragm. ii. 93 must have been from one of Bowman's Bersaker Range specimens, which appear to me to be specifically distinct.

24. **D. Bowmanii,** *Benth.* Stems elongated and branched, more slender than in *D. teretifolium* but not so much so as in *D. Mortii.* Leaves terete, 2 to 4 in. long. Flowers apparently white, generally 2 together on a very short common peduncle, the slender pedicels not above 3 or 4 lines long. Sepals lanceolate as in *D. striolatum*, but the spur nearly as long as the remainder of the sepal, each about 3 lines long. Petals rather narrower. Labellum nearly as long as the sepals, the lateral lobes broad, the middle lobe broadly reniform, undulate crisped and slightly notched, the disk with longitudinal undulate raised lines or plates.—*D. Mortii*, F. Muell. Fragm. ii. 93, at least as to the flowers, but not the original *D. Mortii*, F. Muell. Fragm. i. 214.

Queensland. On trees and rocks, Rodd's bay, *A. Cunningham;* Bersaker Range and Port Cooper, *Bowman.*

4. BOLBOPHYLLUM, Thou.

Sepals erect, free, acuminate, nearly equal, the lateral ones obliquely dilated at the base and connate with the basal projection of the column into a pouch or short spur. Petals usually much smaller than the sepals. Labellum articulate at the end of the basal projection of the column, usually entire and contracted into a claw. Column very short, produced below its insertion, the apex with 2 teeth or horns in front. Anther terminal, lid-like. Pollen masses 4, connate or cohering in pairs, without any gland or caudicle.—Herbs with a creeping rhizome usually covered with thin scarious sheathing scales. Leaves solitary or 2 together on small pseudobulbs. Racemes on 1-flowered peduncles issuing, like the pseudobulbs, from the axils of the sheathing scales of the rhizome.

The genus is spread over the tropical and subtropical regions of the Old World, one species found as far south as New Zealand. The Australian species appear to be all endemic.

Pseudobulbs reduced to a small scarcely prominent disk. Leaves small, ovoid-globular, succulent. Peduncles 1-flowered . .	1. *B. lichenastrum.*
Pseudobulbs oblong. Leaves linear or lanceolate, succulent, without any midrib. Peduncles 1-flowered.	
Leaves linear-lanceolate, 3 to 5 in. long. Peduncle 1 to $1\frac{1}{2}$ in. long	2. *B. nematopodum.*
Leaves linear, very thick and fleshy, 1 to 2 in. long. Peduncle 2 to 4 lines long, with distant bracts	3. *B. Shepherdi.*
Leaves oblong or lanceolate, $1\frac{1}{2}$ to 3 in. long. Peduncle scarcely 2 lines long, with loosely overlapping bracts . .	4. *B. aurantiacum.*
Pseudobulbs ovoid, very deeply wrinkled. Leaves oblong or lanceolate, with a prominent midrib. Peduncles with a raceme of several flowers.	
Peduncles filiform, 1 to 2 in. long. Sepals 2 to $2\frac{1}{2}$ lines long, all equal	5. *B. exiguum.*
Peduncles 3 to 6 in. long. Lateral sepals $\frac{3}{4}$ in., the dorsal one much shorter	6. *B. Elisæ.*

1. **B. lichenastrum,** *F. Muell. Fragm.* vii. 60. Creeping rhizomes forming very dense patches. Pseudobulbs reduced to a small scarcely prominent circular disk, surrounded by the long fringed remains of the scarious sheath. Leaves thick, fleshy, ovoid or almost globular, 2 to 3 lines diameter, irregularly rugose when dry. Flowers very small, "yellow," solitary on peduncles or scapes 2 to 3 lines long, with 1 or 2 scarious sheaths at their base and apparently articulate below the flower. Sepals ovate-lanceolate, about 1 line long, besides the spur, which is broad, obtuse, $1\frac{1}{2}$ lines long. Petals lanceolate, very much narrower, but not much shorter than the sepals. Labellum from the end of the basal projection of the column, with a narrow channelled erect claw, the lamina rather thick, obovate-oblong, obtuse, recurved, nearly as long as the sepals.

Queensland. On rocks and stones, Seaview Range, Rockingham bay, *Dallachy.*

2. **B. nematopodum,** *F. Muell. Fragm.* viii. 30. Creeping rhizomes apparently short and dense. Pseudobulbs closely imbricate, oblong-conical, about $\frac{1}{2}$ in. long. Leaves solitary, linear-lanceolate, flat but thick and succulent, 3 to 5 in. long, 3 to 5 lines broad, attenuate towards the base, without any prominent midrib. Flowers sent by the collector detached, on filiform peduncles of 1 to $1\frac{1}{2}$ in. slightly thickened at the base, very narrow and acute. Flowers yellow, the sepals about 4 lines long, the dorsal one rather shorter than the lateral ones; petals linear-subulate, about 1 line long. Labellum scarcely $1\frac{1}{2}$ lines long,

abruptly contracted into a short claw, the lamina nearly hastate. Column very short, with two narrow erect teeth.

Queensland. Rockingham bay, *Dallachy.*

3. **B. Shepherdi,** *F. Muell. Fragm.* iii. 40. Creeping rhizomes not very intricate, but extending to a considerable breadth, the scarious sheathing scales very conspicuous. Pseudobulbs narrow, 2 to 3 lines long. Leaves solitary, linear, very thick and fleshy, channelled above, convex underneath but not keeled, 1 to 2 in. long. Peduncles 1-flowered, filiform, 2 to 4 lines long, usually with 1 or 2 sheathing scales at the base and 1 similar bract distant from them under the flowers. Sepals about 2 lines long, the lateral ones with a broadly ovate base adnate to the projection of the column and abruptly contracted upwards into a narrow point; dorsal sepal broadly lanceolate at the base and more gradually tapering into the point. Petals ovate-triangular, scarcely above ½ line long. Labellum shorter than the sepals, lanceolate, recurved, channelled above, contracted into a slender claw. Column with 2 subulate teeth.—*Dendrobium Shepherdi,* F. Muell. Fragm. i. 190; *B. Schillerianum,* Reichb. f. in Otto, Hamb. Gartenz. 1860, 423 referred by him to *B. Shepherdi,* in Xen. Orchid. ii. 166 and Beitr. 52.

N. S. Wales. Blue Mountains, *Miss Atkinson, Woolls;* northward to Hastings, Macleay, and Clarence rivers, *Beckler.* Hunter's river, *Leichhardt;* southward to Illawarra, *Shepherd;* and probably the same species from Grose river, *R. Brown;* the specimen has no flower, but thick oblong fleshy leaves of about 1 in., and a single capsule.

4. **B. aurantiacum,** *F. Muell. Fragm.* iii. 39. Nearly allied to *B. Shepherdi,* but a stouter plant. Pseudobulbs small and ovoid. Leaves oblong or oblong-linear, thick but flat, contracted at the base, 1½ to 3 in. long and often ½ in. broad. Peduncles 1-flowered, rarely 2 lines long, covered by the loose scarious sheathing bracts which, although only 2 or 3, overlap each other. Flowers smaller even than in *B. Shepherdi,* but similar in structure, showing however a more decided angle or spur at the end of the basal projection of the column.—*Dendrobium aurantiacum,* F. Muell. Fragm. vii. 98.

Queensland. Moreton bay, *W. Hill;* on trees and rocks about Rockhampton, *Bowman, O'Shanesy.*

N. S. Wales. Narvoo falls, Macleay river, *Fitzgerald.*

The flowers are in some specimens rather crowded on the rhizomes, a fragment of which without leaves led to the mistake of describing the inflorescence as a thyrsoid spike.

5. **B. exiguum,** *F. Muell. Fragm.* ii. 72. Creeping rhizomes "forming a carpet covering large masses of rock." Pseudobulbs ovoid or nearly globular, fleshy angular and furrowed when fresh, very deeply rugose when dry, 2 to 3 lines diameter. Leaves solitary on the pseudobulbs, oblong-linear or lanceolate, contracted at the base, ½ to 1½ in. long, the margins recurved, the midrib prominent underneath. Peduncles filiform, 1 to 2 in. long, bearing 2 to 4 flowers on short filiform pedicels. Sepals lanceolate, 2½ lines long, the lateral ones dilated at the base

into a short broad pouch. Petals scarcely half as long as the sepals. Labellum nearly as long as the sepals, linear, thick and channelled, tapering and slightly recurved towards the end.—*Dendrobium exiguum*, F. Muell. Fragm. v. 95.

N. S. Wales. Blue Mountains, *Caley*, *Woolls;* northward to Hastings river, *Beckler;* Richmond and Tweed rivers, *C. Moore;* Bellinger ranges, *Fitzgerald;* southward to Illawarra, *A. Cunningham.*

Var.? *Dallachyi.* Pseudobulbs rather larger and more ovoid. Flowers white, the sepals and labellum rather broader than in the N. S. Wales specimens and the petals rather larger, but the specimens imperfect.

Queensland. Rockingham bay, *Dallachy.*

The species is very closely allied to the New Zealand *B. pygmæus*, but that has depressed globular pseudobulbs and apparently differently shaped flowers, with a more prominent spur or pouch.

6. **B. Elisæ,** *F. Muell. Fragm.* vi. 120. Rhizome shortly creeping. Pseudobulbs ovoid, very deeply wrinkled and furrowed when dry, usually about ½ in. long. Leaves solitary, narrow-oblong, contracted at the base, mostly 1 to 2 in. long, the midrib prominent underneath as in *P. exiguum.* Racemes including the peduncle 6 to 8 in. high. Flowers much larger than in the preceding species, numerous, white tinged with pink, all turned to one side, on pedicels of 1½ to 3 lines. Bracts small and narrow. Lateral sepals linear-lanceolate, ½ to ¾ in. long, the oblique base adnate to the basal projection of the column forming a short pouch; dorsal sepal not half so long as the lateral ones; petals still shorter, ovate-lanceolate. Labellum purple, about 2 lines long, with a very short broad concave claw, the lamina erect or spreading, oblong, thick and fleshy, grooved on the upper surface. Column short, with 2 prominent teeth.—*Cirrhopetalum Elisæ*, F. Muell. Fragm. vi. 120, t. 57.

N. S. Wales. Blue Mountains, *C. Moore;* Vale of Clwyd, *Vicary;* Clarence river, *Beckler;* New England, *C. Stuart.*—Although technically approaching *Cirrhopetalum* in the shortness of the dorsal sepal, this has not the peculiar inflorescence and habit of that genus, nor yet all its essential characters.

5. PHREATIA, Lindl.

(Plexaure, *Endl.*)

Sepals nearly equal, erect or connivent, the lateral ones dilated at the base and adnate to the basal projection of the column, forming a short pouch. Petals usually smaller than the sepals. Labellum articulate on the basal projection of the column, contracted and concave at the base, the lamina spreading and entire. Column very short, shortly produced at the base, the membranous margin of the apex entire. Anther lid-like, 2-celled. Pollen-masses 8, waxy, slightly cohering by a viscid substance.—Epiphytical herbs, with short leafy stems sometimes thickened into pseudobulbs. Leaves flat or canaliculate, distichous, their persistent bases loosely imbricate. Flowers usually minute, on exceedingly short pedicels, in axillary racemes.

The genus is spread over the Indian Archipelago and the South Sea Islands. The only Australian species is also found in Norfolk island.

1. **P. limenophylax,** *Reichb. f. in Bonplandia* 1857, 54 (*partly*). A dwarf plant, the very short stem covered by the persistent bases of the leaves. Leaves linear, thick and semiterete, channelled on the upper side, 1 to 2 in. long, the dilated base 3 or 4 lines broad. Flowers very minute, yellow, rather crowded and almost sessile, in axillary racemes about as long as the leaves. Sepals about $\frac{1}{4}$ line long, rather broad, acute, the petals rather smaller and narrow. Labellum nearly as long as the sepals, very concave at the base, the lamina spreading, ovate-rhomboidal, entire, the disk with a longitudinal raised line not extending on the lamina. Capsule almost sessile or very shortly pedicellate, ovoid-oblong, about 1 line long.—*Plexaure limenophylax*, Endl. Prod. Fl. Norf. 30; *Oberonia crassiuscula*, F. Muell. Herb.; *Eria limenophylax*, Reichb. f. Xen. Orch. ii. 97, t. 130.

Queensland. On the barks of trees, Rockingham bay, *Dallachy*.—This species is also in Norfolk island. I have not seen the pollen-masses in the Australian specimens the flowers being all too far advanced, but the whole plant, the flowers in every particular, including the anther-cases, agree so perfectly with the detailed analytical drawing of Bauer, of which there is a tracing in Herb. Lindl., that I have no hesitation in referring it to the same species. Reichenbach reduces *Phreatia* to *Eria*, but Lindley does not assent to the union. Lindley appears also to have confounded this species with a South Sea island one which has flat leaves; and Reichenbach by some mistake places the *P. limenophylax* in his group headed "foliis papyraceis;" the leaves are, however, figured by Bauer, and specially described by Endlicher as "carnosula, connato-plicata, intus sulco longitudinali notata," precisely as in the Australian specimens.

6. PHOLIDOTA, Lindl.

Flowers subglobose. Sepals nearly equal, free. Petals smaller. Labellum sessile at the base of the column, concave or almost saccate at the base, entire or 3-lobed, the lateral lobes erect, the middle lobe recurved. Column erect, somewhat hood-shaped at the top and winged in front. Anther terminal, lid-like, 2-celled, the valves almost transverse. Pollen-masses 4, waxy, globular, without any caudicle.—Epiphytical herbs, the rhizome usually shortly creeping, bearing short flowering stems or pseudobulbs, with a single terminal leaf. Flowers rather small, in terminal pedunculate racemes usually recurved. Bracts often rather broad and imbricate in the young raceme.

The genus is spread over East India and the Archipelago, the only Australian species ranging over the greater part of the area of the genus.

1. **P. imbricata,** *Lindl. in Hook. Exot. Fl.* ii. t. 138; *Gen. and Sp. Orch.* 36. Stems short, with a few sheathing scales and a single leaf, the older stems thickened into pseudobulbs. Leaf broadly lanceolate or oblong, acuminate, contracted and convolute at the base, prominently ribbed, often above 1 ft. long. Peduncle long and slender from within the convolute base of the leaf, the flowering part at length recurved and 6 in. long or more. Bracts broadly ovate, obtuse or almost acute, 4 to 5 lines long, complicate and imbricate at first, spreading from the

flexuose rhachis when the flowers are out. Pedicels about 3 lines long. Sepals ovate-lanceolate, about 3 lines; petals rather smaller. Labellum about as long as the sepals, the concave almost globular part erect, bordered by the short broad lateral lobes, the middle lobe broader than long, the margin undulate and more or less distinctly 3-lobed. Column about 1½ lines long, the margins winged upwards. Capsule obovate, about ½ in. long.—Bot. Reg. t. 1213; Wight Ic. t. 907; F. Muell. Fragm. iv. 163.

Queensland. Rockingham bay, *Dallachy.*—This species is common in E. India and the Archipelago, the Australian specimens differing slightly in the rather larger bracts. Dallachy describes the flowers as yellow.

TRIBE 2. VANDEÆ.—Anther lid-like, incumbent, usually deciduous. Pollen-masses waxy, 4 in pairs, on a single or double caudicle attached to a gland. Epiphytes, or terrestrial with creeping rhizomes.

7. TÆNIOPHYLLUM, Blume.

Sepals and petals nearly equal, erect or connivent, connate at the base. Labellum adnate to the column at the base and produced into a short spur or pouch, entire (or shortly 2-lobed ?). Column very short, erect, with 2 teeth in front. Anther terminal, lid-like. Pollen-masses 4 in 2 pairs, with a short slender caudicle. Gland minute.—Epiphytical herbs, almost stemless and leafless or with a tuft of linear leaves. Flowers minute, in small slender racemes.

The genus has been found also in Java and in Ceylon; the Australian species is probably endemic.

1. **T. Muelleri,** *Lindl. Herb.* Stems leafless, scarcely above ½ line long, emitting long linear wavy roots, and 2 or 3 filiform scapes of about ½ in. Bracts minute. Flowers 2 or 3 on exceedingly short pedicels, and the whole flower under 1 line long. Sepals and petals united to about the middle, the petals rather narrower than the sepals. Labellum linear, as long as the sepals, and in the only flower I could examine appeared to be shortly 2-lobed at the end, with a minute tooth between the lobes, the basal pouch or spur obtuse, about ¼ line long.—*Sarcochilus Baileyi*, F. Muell. Herb.

Queensland. On trees near Brisbane, *W. Hill, C. Prentice, Bailey.*

8. SARCOCHILUS, Br.

(Thrixspermum, *Lour.;* Gunnia, *Lindl.*)

Sepals and petals nearly equal, free, spreading, the lateral sepals often more or less dilated at the base and adnate to the basal projection of the column. Labellum articulate at the end of the basal projection of the column without any spur at its base, 3-lobed, the lateral lobes rather large, the terminal one (in the Australian species) very short and tooth-like or cushion-like, with a solid fleshy dorsal protuberance at its base sometimes elongated oblong or conical, sometimes very short; the disk between the lateral lobes with prominent callosities. Column short, erect, produced at the base. Anther terminal, lid-like. Pollen-masses 4 in pairs on a somewhat flattened caudicle.—Epiphytical herbs. Stems short, either covered with the prominent persistent truncate bases of the leaves or leafless. Leaves flat and often falcate, or narrow-linear, or none. Racemes axillary. Bracts small. Capsules usually linear or narrow-oblong.

The genus is spread over East India and the Archipelago, with one New Zealand species, but the Australian ones appear to be all endemic. Reichenbach having identified Loureiro's *Thrixspermum* as a species of *Sarcochilus* has adopted his generic name as the oldest, and as being sufficiently characterized. But the composition of the word is against all rules, and the character given (the fleshy middle lobe of the labellum longer than the lateral ones) will certainly not apply to the Australian species. We cannot therefore admit that this barbarous name of *Thrixspermum* should now be substituted for the universally received one of *Sarcochilus*, nor can we correct it according to the rules of etymology without interfering with *Trichospermum*, Blume, in *Tiliaceæ*.

Leaves oblong, lanceolate or falcate. Middle lobe of the labellum short and toothlike, glabrous.
  Lateral sepals adnate to the base only of the projection of the column which represents a claw to the labellum.
    Sepals and petals narrow-linear, subulate-acuminate, 1 to $1\frac{1}{4}$ in. long . . . . 1. *S. divitiflorus.*
    Sepals and petals oblong, 6 to 7 lines long . . . . 2. *S. falcatus.*
  Lateral sepals adnate to the whole of the projection of the column.
    Sepals (5 to 6 lines) twice as long as the labellum.
      Sepals and petals oval-oblong . . . . 3. *S. Fitzgeraldi.*
      Sepals and petals linear-oblong . . . . 4. *S. olivaceus.*
    Sepals (about 5 lines) but little longer than the labellum 5. *S. parviflorus.*
Leaves narrow-linear or none. Middle lobe of the labellum broad or cushionlike and densely covered with a white pubescence (*Chiloschista*).
  Leaves linear. Lateral sepals adnate to the whole of the projection of the column.
    Sepals nearly 3 lines long. Lateral lobes of the labellum much longer than the middle lobe and the dorsal protuberance . . . . 6. *S. Ceciliæ.*
    Sepals about $1\frac{1}{2}$ lines. Lateral lobes of the labellum shorter than the broad middle lobe and the spurlike dorsal protuberance . . . . 7. *S. Hillii.*
  No leaves. Lateral sepals (about 2 lines) adnate only to the base of the projection of the column which represents a claw to the labellum . . . . 8. *S. phyllorhizus.*

1. **S. divitiflorus,** *F. Muell. Herb.* Stems unknown. Leaves oblong, 3 to 4 in. long and nearly 1 in. broad, flat with prominent nerves. Scape or peduncle about 6 in. long, flowering from below the middle, with a few empty sheathing scales or bracts below the inflorescence. Flowers much longer than in any other species, not very crowded, on short pedicels, the subtending bracts ovate, 1 to $1\frac{1}{2}$ lines long. Sepals and petals very narrow, tapering into a filiform point, 1 to $1\frac{1}{4}$ in. long, apparently pale yellow or white with dark blotches near the base, the lateral sepals narrow-lanceolate towards the base, but

without any prominent spur or point. Labellum with a narrow claw of about 1 line, the lamina with an erect central saccate lobe of about 1 line, the lateral lobes twice as long, oblong, obtuse, clasping the very short column. Pollen-masses 2 on a short caudicle, the pollen somewhat mealy.

N. S. Wales. Macleay river, *Fitzgerald.*—I have only seen racemes and loose leaves of this remarkably distinct species. The flowers at first sight resemble those of *Dendrobium teretifolium*, but the structure is totally different.

2. **S. falcatus,** *R. Br. Prod.* 332. Stems rarely above 2 or 3 in. high, rather stout, covered by the rigid loosely imbricate sheathing bases of the leaves. Leaves oblong, often falcate, 2 to 4 in. long and $\frac{1}{4}$ to $\frac{1}{2}$ in. broad. Peduncles in the lower axils scarcely exceeding the leaves and sometimes shorter. Flowers usually 3 or 4, distant, white. Bracts ovate, about 2 lines long. Pedicels and ovary about $\frac{1}{2}$ in. Sepals and petals nearly equal, oblong, obtuse, 6 to 7 lines long, the lateral sepals adnate to the base only of the basal projection of the column which forms a canaliculate claw to the labellum of about $1\frac{1}{2}$ lines. Labellum ascending from the end of it, the lateral lobes large, ovate, the middle lobe very short broad and almost scale-like or scarcely prominent, with a thick fleshy dorsal protuberance or solid spur; the disk with a transverse 2-lobed scale or callus between the lateral lobes. Column short, with 2 very prominent acuminate angles. Capsules linear, sometimes 3 in. long.—Lindl. Gen. and Sp. Orch. 142; Bot. Reg. t. 1832; F. Muell. Fragm. vii. 97; *Thrixspermum falcatum*, Reichb. f. Beitr. 46.

N. S. Wales. Hunter's, Paterson's, and Williams' rivers, *R. Brown; Macleay* river, *Fitzgerald;* Hastings river, *C. Moore;* Woolongong, *Backhouse;* Illawarra, *A. Cunningham.*

3. **S. Fitzgeraldi,** *F. Muell. Fragm.* vii. 115. Stem foliage and general aspect of *S. falcatus*, the leaves from 3 to 6 in. long. Racemes with the peduncle 6 in. to 1 ft. long, the flowers of the size of those of *S. falcatus*, "snowy white spotted with rich lake or maroon." Bracts small, pedicels $\frac{1}{2}$ in. long or rather more. Sepals and petals nearly equal, contracted at the base, 5 to 6 lines long, the lateral sepals adnate to the whole of the basal projection of the column as in *S. olivaceus*, but the projection shorter and broader than in that species. Labellum not half the length of the sepals, the lateral lobes ovate, falcate, scarcely $1\frac{1}{2}$ lines long, the middle lobe scarcely prominent, the solid dorsal protuberance short and obtuse; disk with a large very prominent callus between the lateral lobes and a smaller one just within the small middle lobe.

N. S. Wales. Narvoo falls, Bellinger river, *Fitzgerald;* Mount Warning, *Guilfoyle.*

4. **S. olivaceus,** *Lindl. Bot. Reg.* 1839, *Misc.* 32. Stems covered with the prominent bases of the leaves as in *S. falcatus*, but generally shorter, under 1 in. long. Leaves oblong, often falcate, 2 to 3 in. long, apparently thinner than in *S. falcatus*. Racemes loose, of 2 or 3 flowers, the rhachis flexuose, not exceeding the leaves. Bracts very small. Sepals

and petals of a dull pale purple or yellowish brown, 5 to 6 lines long linear-oblong, much contracted below the middle, the lateral sepals dilated at the base and adnate to the whole of the basal projection of the column. Labellum white streaked with red, almost sessile, about half as long as the sepals, the lateral lobes oblong-falcate, the middle lobe very short and orbicular, the dorsal solid protuberance ovoid-conical, obtuse; the disk with several very prominent irregular calli between the lateral lobes. Column short, with a long basal projection. Capsule narrow.—F. Muell. Fragm. vii. 97; *S. dilatatus*, F. Muell. Fragm. i. 191; *Thrixspermum olivaceum* and *T. dilatatum*, Reichb. f. Xen. Orchid. ii. 122.

Queensland. Rockingham bay, *Dallachy;* Moreton bay, *W. Hill.*
N. S. Wales. Hastings and Clarence rivers, *Beckler;* Macleay river, *Fitzgerald;* Illawarra, *Shepherd.*

Gunnia picta, Lindl. Bot. Reg. 1838, Misc. 45 (*Sarcochilus pictus*, Reichb. f. in Walp. Ann. vi. 501; *Thrixspermum pictum*, Reichb. f. Xen. Orch. ii. 122), from Sydney, *Hort. Loddiges*, or from Brisbane (*Reichb. f.*), does not appear to me to differ from *S. olivaceus*. The calli of the labellum in this as in *S. parviflorus* vary from specimen to specimen.

5. **S. parviflorus,** *Lindl. Bot. Reg.* 1838, *Misc.* 34. Habit entirely that of the smaller specimens of *L. olivaceus*, the short stems covered with the prominent bases of the fallen leaves. Leaves rather thin, narrow-oblong or falcate, 2 to 3 or rarely 4 in. long. Scapes as long as or rather longer than the leaves, bearing 3 to 6 pale yellowish-green flowers, on pedicels of 2 to 3 lines. Sepals narrow-oblong, 4 to 5 lines long, besides the narrow base of the lateral ones adnate to the projection of the column. Petals rather shorter than the sepals. Labellum nearly sessile at the end of the basal projection of the column, white, more or less tinted with yellow, and spotted or streaked with red, the lateral lobes ovate-oblong often nearly as long as the sepals, the middle lobe very small, the dorsal protuberance or solid spur thickly conical or obovoid, at least half as long as the lateral lobes; the disk with several irregular very prominent calli between the lobes. Capsule linear.—*Gunnia australis*, Lindl. Bot. Reg. under n. 1699; Hook. f. Fl. Tasm. ii. 33, t. 128; *Sarcochilus Barklyanus*, F. Muell. Fragm. i. 89; *S. Gunnii*, F. Muell. l. c. 90; *S. australis*, Reichb. f. in Walp. Ann. vi. 501; *Thrixspermum parviflorum* and *T. australe*, Reichb. f. Xen. Orch. ii. 122.

N. S. Wales. Twofold bay, *F. Mueller* (no flowers seen but apparently this species).
Victoria. Apollo bay, *F. Mueller;* Dandenong range, *Taylor.*
Tasmania. On bushes and small trees in deep gullies and dense forests, Emu bay, Black river, Circular Head, Great Swan Port, &c., *Gunn*, *Milligan*, and others.

6. **S. Ceciliæ,** *F. Muell. Fragm.* v. 42, t. 42. Stems sometimes very short, sometimes elongated to 2 or 3 in. Leaves linear or narrowly linear-lanceolate, thick, 2 to 3 in. long. Racemes longer than the leaves and sometimes attaining 6 to 8 in., rather rigid, erect, bearing above the middle a number of small shortly pedicellate pink flowers. Lateral sepals almost ovate, nearly 3 lines long, adnate to the rather

long basal projection of the column; dorsal sepal of the same length but narrower; petals still narrower. Labellum much shorter than the sepals, the lateral lobes oblong-falcate, the middle lobe very much shorter, thick and woolly-villous on the surface; the dorsal protuberance or solid spur broad, obtuse, half as long as the lateral lobes; the disk with several calli, more or less adnate to the lateral lobes.—*Thrixspermum Ceciliæ*, Reichb. f. Beitr. 71.

Queensland. Rockingham bay, *Dallachy* (with linear leaves); Cleveland bay, *Bowman* (with more lanceolate leaves).

7. **S. Hillii,** *F. Muell. Fragm.* ii. 94, vii. 98. Stems very short. Leaves few, narrow-linear, rather thick, 1 to 3 in. long. Racemes very slender, shorter or scarcely longer than the leaves, with a number of very small white flowers very shortly pedicellate. Sepals ovate-oblong, scarcely 1½ lines long; petals the same length, but narrower. Labellum about 1 line long, sessile on the very short basal projection of the column, the lateral lobes short, almost acute, the middle lobe rather longer, broader than long, retuse, thickly covered on the surface with white wool; the dorsal protuberance or solid spur narrow-conical, longer than the lateral lobes; the disk with several prominent calli.—*Dendrobium Hillii*, F. Muell. Fragm. i. 88, ii. 94; *Thrixspermum Hillii*, Reichb. f. Beitr. 71.

Queensland. Brisbane river, Moreton bay, *W. Hill;* Rockhampton, *Thozet, O'Shanesy.*

N. S. Wales. Paramatta, Camden and Nepean rivers, *Woolls;* Hastings and Clarence rivers, *Beckler.*

8. **S. phyllorhizus,** *F. Muell. Fragm.* v. 201. Apparently stemless and leafless, the irregularly flattened creeping roots spreading from the very short stock and sometimes assuming almost the aspect of leaves. Scapes slender, erect, almost filiform, 3 to 6 in. high, bearing several small flowers on very short filiform pedicels. Sepals and petals obovate or obovate-oblong, about 2 lines long, the lateral sepals as in *S. falcatus* adnate to the base only of the columnar projection, which forms a linear claw to the labellum almost half as long as the sepals. Labellum sessile at the end of this projection or claw, the lateral lobes small, narrow-oblong, clavate, purple; the middle lobe very short and obtuse, almost globular and densely white-woolly on the inner surface as in *S. Ceciliæ* and *S. Hillii;* the dorsal protuberance very short.—*Thrixspermum phyllorhizum*, Reichb. f. Beitr. 71.

Queensland. Cape York, *M'Gillivray, Daemel;* Fitzroy island, *Herb. F. Mueller* (collector not named).

9. CLEISOSTOMA, Blume.

Sepals and petals nearly equal, free, spreading, the lateral sepals sometimes adnate to a basal projection of the column. Labellum inserted at the base of the column or of its basal projection but free from it, with a pouch or spur at its base, undivided inside but with a reflexed or horizontal scale or appendage inside at the orifice, the lamina 3-lobed, the middle lobe usually short and broad, the lateral ones falcate or narrow. Column short, with 2 teeth or lobes at the apex in front. Anther lid-like; pollen-masses 4 in pairs, attached to a somewhat flattened caudicle.—Herbs with the habit of *Sarcochilus* or of *Saccolabium*, the stems short or elongated. Leaves more or less distichous, leaving short sheathing persistent bases. Flowers small, in axillary racemes, the spur of the labellum rather long in the Australian species.

The genus extends over East India and the Archipelago; the Australian species, as far as known, are all endemic.

Column very shortly produced at the base. Spur of the labellum with the inner appendage deflexed and ciliate on the upper or lamina side of the cavity . . . . 1. *C. tridentatum*.
Column with a rather long basal projection. Spur with the inner appendage horizontal and glabrous on the lower or column side of the orifice . . . . 2. *C. Beckleri*.
Column not produced at the base. Spur of the labellum with the inner appendage horizontal and glabrous on the lower or column side of the orifice . . . . 3. *C. Macphersoni*.

1. **C. tridentatum,** *Lindl. Bot. Reg.* 1838, *Misc.* 33. Stems often elongated, rather slender, the persistent bases of the leaves much less prominent than in the *Sarcochili*. Leaves mostly 2 to 3 in. long, linear-oblong or falcate. Racemes slender, flexuose, shorter than the leaves. Bracts very small. Flowers very small, shortly pedicellate. Sepals and petals oblong-lanceolate, about 2 lines long, the dorsal sepal rather broader, the lateral ones and the petals slightly falcate, the lateral sepals adnate to the very short basal projection of the column. Labellum nearly as long as the sepals; spur rather long, deflexed, with a deflexed ciliate membrane inside, on the side next the lamina of the labellum; lateral lobes spreading, falcate and acute; middle lobe very short, obtuse, fleshy, concave. Column exceedingly short, the margin deeply membranous, with 2 anterior narrow teeth. Capsule narrow, 1 to $1\frac{1}{2}$ in. long.—*Saccolabium calcuratum*, F. Muell. Fragm. i. 192; *Sarcochilus calcaratus*, F. Muell. Fragm. ii. 181, vii. 98; *Sarcochilus tridentatus*, Reichb. f. in Walp. Ann. vi. 500.

Queensland. Brisbane river, Moreton bay, *W. Hill, Bailey;* Wide bay, *Leichhardt*.

N. S. Wales. Camden, Bent's Basin, Nepean river, *Woolls;* Hastings and Clarence rivers, *Beckler;* New England, *C. Stuart;* southward to Illawarra, *Ralston*.

2. **C. Beckleri,** *F. Muell. Herb.* Stem and leaves not seen, but said to be an epiphyte with a short rigid stem hanging from trees. Racemes 2 or 3 in. long. Sepals and petals scarcely $1\frac{1}{2}$ lines long, the lateral sepals adnate to a basal projection of the column of about $1\frac{1}{2}$ lines. Labellum at the end of the basal projection; spur narrow-conical, above 1 line long, the orifice half closed by a transverse plate on the basal side; the lamina short and broad, the lateral lobes erect and narrow, almost linear, shorter than the spur, the middle lobe shortly and broadly semiorbicular.

N. S. Wales. Clarence river, *Beckler*.

3. **C. Macphersoni,** *F. Muell. Herb.* Stems short, covered with the very prominent bases of the leaves. Leaves 4 to 6 in. long and at least 1 in. broad, the veins not prominent except the midrib, which forms an acute keel underneath. Spikes rigid, not longer than the leaves, the flowers rather numerous, sessile, "red." Sepals and petals rather thick, about 2½ lines long, all nearly equal. Column not produced at the base. Labellum sessile, the spur oblong, obtuse, rather dilated beyond the middle, 1½ lines long, closed at the orifice by a large ovate plate close under the column; lamina short and broad, the middle lobe orbicular, about 1¼ lines diameter, the lateral lobes shorter, falcate and narrow. Capsule oblong, strongly ribbed.—*Saccolabium Macphersonii*, F. Muell. Fragm. vii. 96.

Queensland. Rockingham bay, *Dallachy.*—This has much the aspect of a small-flowered *Sarcanthus*, but the spur is not divided inside.

10. VANDA, R. Br.

Sepals and petals nearly equal, free, spreading, contracted at the base. Labellum inserted at the base of the column, produced at the base into a pouch or conical spur, the lamina spreading, 3-lobed, the disk smooth or in species not Australian with callosities above the pouch. Column short, thick, erect, with an obtuse or retuse rostellum. Anther terminal, lid-like, 2-celled. Pollen-masses waxy, 4 in pairs or 2 deeply 2-lobed, attached to a linear or cuneate caudicle on a large gland.—Epiphytical herbs, with distichous often thick and coriaceous or fleshy leaves. Racemes lateral. Flowers usually large and showy.

The genus extends over E. India and the Archipelago as far as S. China; the only Australian species is apparently the same as an Archipelago one.

1. **V. Hindsii,** *Lindl. in Hook. Lond. Journ.* ii. 237, *and in Paxt. Mag.* ii. 21. Stems of moderate length, with linear canaliculate leaves of 1 ft. or more. Racemes 6 in. to above 1 ft. in length, with 3 to 10 large flowers, the spreading pedicels often 2 to 3 in. long including the ovary. Sepals and petals nearly 1 in. long, broadly obovate with sinuate margins, contracted into a broad claw, of a pale yellowish-white outside, white inside with purple spots, slightly shaded with yellow at the base and with pink towards the margins. Labellum at least as long as the sepals, convex, rather thick and fleshy, generally purple but with more or less of white towards the base and darker streaks on the disk, the lateral lobes short and broad, the middle lobe much longer, obovate-oblong, emarginate or shortly 2-lobed, without callosities on the disk. Column white.—*V. tricolor*, Lindl. in Bot. Reg. 1847, under t. 59, and in Paxt. Fl. Gard. ii. 20, t. 42; Bot. Mag. t. 4432; *V. suavis*, F. Muell. Fragm. vii. 135, but scarcely of Lindl.

N. Australia. Arnhem's Land, *F. Mueller;* and also in New Guinea and Java, if the determination and synonymy are correct.—Of the Australian plant I have only seen a single flower and leaf in Herb. F. Muell., in which the size and shape of the sepals, petals, and labellum agree precisely with those of *V. Hindsii* and *V. tricolor* in Herb. Lindl. as far as can be judged from dried specimens. Lindley distinguishes

V. Hindsii from *V. tricolor* chiefly by the long raceme with 10 instead of 3 or 4 flowers. In Herb. Hook. the specimen of *V. tricolor* figured in the Magazine has had at least 10 flowers. It is true that Lindley, in Folia Orchidacea, refers this to *V. suavis*, but Reichenbach thinks it more correctly placed under *V. tricolor*. F. Mueller had at first referred his Australian plant to *V. tricolor*, but afterwards determined it as *V. suavis*, on account of the colour of the flowers. His description, however, agrees well with the colours of *V. tricolor* as depicted in a sketch in Herb. Lindl., whilst the true *V. suavis* has not only the petals and sepals with a pure white ground, but their shape as well as that of the labellum is somewhat different. Possibly, however, *V. Hindsii*, *V. tricolor*, and *V. suavis* may all be forms of one species, now frequent in our collections of living Orchideæ under the names of *V. tricolor* or *V. suavis*, and evidently very variable both in the colour and the shape of the perianth.

11. SACCOLABIUM, Lindl.

Sepals and petals nearly equal, free, spreading, the lateral sepals often more or less dilated at the base and adnate to a basal projection of the column. Labellum articulate at the base of the column or at the end of its basal projection, with a hollow spur or pouch at the base, neither internally divided nor with any internal appendage, the lamina usually undivided or without any prominent middle lobe. Column short, erect, often produced at the base. Anther lid-like. Pollen-masses 4 in pairs (or 2 deeply 2-lobed), attached to a caudicle.—Epiphytical herbs. Stems marked with or covered by the truncate persistent bases of the leaves. Leaves flat. Racemes axillary, simple or in species not Australian branched. Bracts small.

The genus is generally distributed over East India and the Archipelago. The only Australian species appears to be endemic.

1. **S. Hillii,** *F. Muell. Fragm.* i. 192. Stems rigid, flexuose, several inches long, covered with the prominent deeply striate bases of the leaves. Leaves distichous, rigid, with prominent nerves, mostly 3 to 5 in. long and ¾ to 1 in. broad. Racemes usually about the length of the leaf, the flowers numerous and small on very short pedicels. Sepals and petals, oblong-linear, not quite 2 lines long, the lateral ones adnate to the short basal projection of the column, falcate as well as the petals, the dorsal one rather longer and incurved. Labellum nearly as long as the sepals, the basal pouch short and broad without any internal appendage, the lamina concave or embracing the column, much broader than long, truncate without any middle lobe, the angles (or lateral lobes) shortly acuminate and somewhat incurved, the disk with an erect conical tooth or callosity immediately above the pouch but not reflexed into it.

Queensland. Brisbane river, Moreton bay, *F. Mueller*, *W. Hill*, *Bailey*.
N. S. Wales. Clarence river, *Beckler*; Tweed river, *Guilfoyle*.

12. GEODORUM, Jacks.

Sepals and petals nearly equal, free, erect. Labellum erect, sessile at the base of the column but free from it, broad, concave and slightly saccate at the base, entire or scarcely lobed. Column short, erect, semiterete. Anther terminal, lid-like, very concave. Pollen masses 2, 2-lobed, waxy, attached to a very short caudicle on a transverse gland. —Terrestrial herbs with a short creeping rhizome. Leafy stems short, sometimes pseudo-bulbous at the base. Leaves rather large, plicate and strongly ribbed, the lower ones reduced to membranous sheathing scales. Scapes from the base of the leafy stem, leafless except the sheathing scales, terminating in a rather dense usually recurved raceme.

The genus extends over East India and the Archipelago. The Australian species is generally supposed to be endemic, but the differences between some of the species are very slight, and require further investigation.

1. **G. pictum,** *Lindl. Gen. and Sp. Orch.* 175. Leafy stems a few inches high, terminating in 2 or 3 ovate-lanceolate leaves of 4 to 8 in., tapering at both ends. Scapes from the axil of a membranous scale close to the base of the leafy stem, shorter or perhaps sometimes longer than the leaves, bearing membranous sheathing scales, several rather large at the base of the scape, distant higher up. Flowers pink, rather numerous in a terminal raceme, reflexed only after the flowers have begun to expand (from F. Mueller's notes). Pedicels short. Bracts linear, white. Sepals and petals oblong, 4 to 5 lines long. Labellum broadly ovate, darkly veined, obtuse and emarginate or very shortly 2-lobed at the end, the margin somewhat undulate, the disk saccate at the base with 2 double raised lines or plates more or less marked at the base often evanescent upwards or confluent into 2 single ones and terminating in a toothed or entire transverse callus below the end of the labellum. Column short, the margin winged. Pollen-masses ovoid-globular, waxy, 2-lobed.—F. Muell. Fragm. iii. 24; Reichb. Beitr. 46; *Cymbidium pictum,* R. Br. Prod. 331.

N. Australia. North coast, *R. Brown;* Port Darwin, *Schultz, n.* 728; Escape Cliffs, *Hulse.*

Queensland. Moreton bay, *Bernays;* Rockhampton, *O'Shanesy, Thozet;* Cleveland bay, *Bowman;* Wide bay, *Bidwill;* Rockingham bay, *Dallachy,* Port Denison, *Fitzalan.*

F. Mueller, Fragm. viii. 31, refers Hulse's specimens to the East Indian *G. dilatatum,* Br., founded on Roxburgh's *Limodorum recurvum,* and in both Hulse's and Schultz's specimens I find the callus near the end of the labellum entire, not toothed or divided as in the majority of the Queensland specimens. I have been unable, however, to find two specimens with their calli and markings the same, and it appears to me most probable that there is but one species in Australia, and that perhaps not really distinct from the Indian *G. dilatatum.*

13. EULOPHIA, R. Br.

Sepals and petals nearly equal, spreading, free or the lateral sepals adnate to the short basal projection of the column. Labellum inserted at the base of the column or its projection but free from it, produced at the base into a short pouch or spur, the lamina 3-lobed or rarely undivided, the disk usually marked with cristate or bearded veins. Column semiterete, with the front angles acute or winged. Pollen-masses 4 in pairs, or 2 and bifid, waxy, attached to a short linear caudicle on a transverse gland.—Terrestrial herbs, with short stems. Leaves distichous, plicate, or sometimes those of the flowering stems reduced to sheathing scales. Racemes terminal or on radical scapes.

The genus is spread over tropical and subtropical Asia and Africa. The Australian species are both endemic.

Labellum strongly and darkly veined, the middle lobe much broader than long . . . . 1. *E. venosa.*
Labellum finely veined, the middle lobe as long as broad . . . . 2. *E. Fitzalani.*

1. **E. venosa,** *Reichb. f. in Herb. Lindl.* An erect leafless herb, with the habit of *Dipodium punctatum*, the sheathing scales imbricate at the base of the stem, the upper ones distant, passing into narrow bracts, often as long as the pedicel and ovary. Flowers several in a terminal raceme, whitish with deep red veins. Sepals broadly lanceolate, 6 to 8 lines long, marked with longitudinal somewhat anastomosing veins, the lateral ones attached to the short basal projection of the column. Petals rather shorter and broader, almost obovate-oblong. Labellum rather longer than the sepals, the spur short, the lateral lobes ovate, strongly veined, the middle lobe twice as long and very much broader than long, almost reniform, elegantly veined, the disk with two longitudinal glabrous raised lines or plates between the lobes, shortly prolonged on the middle lobe which has besides 3 or more short undulate raised lines or plates. Column half as long as the sepals, the dorsal lobe bearing the anther rather long and ovate.—*Dipodium venosum*, F. Muell. Fragm. i. 61.

N. Australia. Providence Hill and Macadam Range, *F. Mueller.*
Queensland. Rockingham bay, *Dallachy;* Port Mackay, *Nernst.*

2. **E. Fitzalani,** *F. Muell. Fragm.* viii. 30. Habit apparently that of *E. venosa*, the single specimen leafless, nearly 1 ft. high, with a sheathing bract near the base about 1½ in. long, the bracts subtending the pedicels reaching to the top of the ovary. The sepals narrow-lanceolate, striate, acuminate-acute, about 7 lines long, the lateral ones attached at the base to the projection of the column. Petals scarcely broader but rather shorter and more obtuse, the veins slightly anastomosing. Labellum as long as the petals, the basal spur short but longer than in *E. venosa*, the veins branching but not so dark and strong as in *E. venosa*, lateral lobes not halfway up and not very prominent; middle lobe large but scarcely broader than long, the margins much undulate and very obtuse; the disk with 4 slightly raised lines quite entire between the lateral lobes, crisped or fringed and extending to about half the length of the middle lobe. Column not half so long as the sepals. Pollen-masses 2, depressed globular.

Queensland. Mount Dryander, *Fitzalan.*

14. DIPODIUM, R. Br.

(Leopardanthus, *Blume;* Wailesia, *Lindl.*)

Sepals and petals nearly equal, free, spreading. Labellum sessile, erect, adnate to the column at its base and then gibbous or produced into a very short pouch, the lamina 3-lobed, the lateral lobes narrow, the middle lobe longer, oblong-ovate or rhomboidal, with a hairy or pubescent patch near the end. Column erect, semicylindrical, the membranous margin variously sinuate or toothed. Anther lid-like. Pollen-masses 2, deeply 2-lobed (or 4 in pairs), lateral, attached to separate caudicles proceeding from a rather large gland.—Terrestrial herbs, the leafy stems when present simple with distichous leaves. Racemes on long leafless scapes or long erect axillary peduncles, with sheathing scales imbricate at the base of the scape or peduncle, the upper distant ones passing into small bracts. Flowers rather large, often spotted.

Besides the two Australian species which appear to be endemic, there are a few from New Caledonia, the Eastern Archipelago, and East India.

No leaves. Scales not numerous, loosely imbricate at the base of the scapes . . . . 1. *D. punctatum.*
Stems with linear-lanceolate leaves. Peduncles axillary . . . 2. *D. ensifolium.*

1. **D. punctatum,** *R. Br. Prod.* 331. A leafless plant with thick fibrous roots and erect stem attaining with the racemes 1 to 2 ft., the sheathing scales few and loosely imbricate and obtuse at the base, distant higher up. Flowers rather large, more or less red and usually but not always spotted with purple, in a terminal raceme sometimes very short sometimes occupying a third of the stem. Sepals and petals oblong-lanceolate, 6 to 8 lines long. Labellum as long or rather longer, the basal pouch or gibbosity very short; lateral lobes below the middle narrow and erect; middle lobe twice as long, obovate-oblong; disk with 2 raised lines very prominent and glabrous at the base, pubescent upwards and ending sometimes in tufts or pubescent scales, the middle lobe with one broad pubescent line or patch. Column half as long as the sepals, the inner face pubescent. Pollen-masses 2, deeply 2-lobed, laterally attached below the subulate ends of the caudicles.—Hook. f. Fl. Tasm. ii. 32, t. 127; Bot. Reg. t. 1980; Reichb. f. Beitr. 45; *Dendrobium punctatum*, Sm. Exot. Bot. i. 21, t. 12.

N. Australia. Port Darwin, *Schultz*, *n.* 623 (with narrow pale coloured sepals and petals, perhaps not spotted).

Queensland. Brisbane river, Moreton bay, *F. Mueller;* Condamine river, *Leichhardt;* Rockhampton, *O'Shanesy;* Armidale, *Perrott;* Burdekin river and Mount Elliott, *Fitzalan.*

N. S. Wales. Hastings and Clarence rivers, *Beckler;* Macleay river, *Fitzgerald;* New England, *C. Stuart.*

Victoria. Upper Yarra and Dandenong Range, *F. Mueller;* Glenelg river, *Robertson.*

Tasmania. Circular Head, *Gunn;* Port Sorell and Cheshunt, in stony and moist places, generally growing near Eucalypti, *Archer.*

S. Australia. Ranges near Mount Lofty, *F. Mueller.*

The pubescence of the labellum appears to be very variable in shape and extent. The closely allied New Caledonian *D. squamatum* differs chiefly in the more closely imbricate appressed and acute scales at the base of the stem.

2. **D. ensifolium,** *F. Muell. Fragm.* v. 42. Stems leafy, from a few inches to above 1 ft. high without the racemes. Leaves distichous, complicate or canaliculate, linear-lanceolate, acute, strongly keeled and

usually prominently ribbed on each side, 3 to 6 in. long, the persistent truncate base usually rather long. Racemes with the peduncle often above 1 ft. long, sometimes appearing at first terminal, and usually only 1 or 2 on the same stem, but really always axillary. Sheathing scales small, distant, with a few imbricate ones at the base of the peduncle. Pedicels with the ovary about ½ in. long. Sepals and petals "pink and spotted," 6 to 8 lines long, the sepals oblong-lanceolate, the petals rather broader and more contracted at the base. Labellum about the length of the sepals, scarcely gibbous at the base but shortly connate with the column as in other species, the lateral lobes placed much below the middle, linear or linear-spathulate, incurved, the middle lobe about twice as long, broadly rhomboidal; the disk with 2 pubescent lines between the lateral lobes confluent into 1 at the base of the middle lobe, and a dense patch of scaly hairs at the end of it. Column not half the length of the sepals, pubescent in front. Pollen-masses 2-lobed, the 2 caudicles long and slender.

Queensland. Rockingham bay, *Dallachy*.

TRIBE 3. BLETIDEÆ.—Anther lid-like, incumbent, usually deciduous. Pollen-masses waxy, 4 or 8, tapering at the base, separately attached and sessile or on short caudicles, or on a short dichotomous caudicle. Terrestrial herbs, with creeping or rarely tuberose rhizomes or rarely epiphytes. Sepals and petals nearly equal, free and spreading. Flowers often large.

15. CYMBIDIUM, Swartz.

Sepals and petals nearly equal, free, spreading. Labellum sessile, free, articulate on the base of the column, or very shortly adnate to it, concave, entire or 3-lobed. Column erect or slightly incurved, semiterete, sometimes narrowly winged. Anther lid-like, very concave, more or less 2-celled. Pollen-masses 2, usually 2-lobed (4 united in pairs), sessile on a somewhat triangular gland.—Plants usually epiphytical. Stems often short and slightly swollen into pseudobulbs. Leaves elongated, keeled, striate. Flowers not small, in loose racemes pedunculate in the lower axils, the peduncle often long with sheathing rigid scales at the base. Bracts usually small.

The genus, as at present understood, comprises tropical and subtropical species, both of the New and of the Old World, but chiefly from the latter. It has not, however, been subject to any recent revision. The Australian species appear to be all endemic.

Labellum 3-lobed with 2 longitudinal raised pubescent or fringed plates on the disk . . . . 1. *C. canaliculatum*.
Labellum 3-lobed without longitudinal plates. Leaves very long and mostly 1 in. broad . . . . 2. *C. albuciflorum*.
Labellum undivided without longitudinal plates . . . . . . 3. *C. suave*.

1. **C. canaliculatum,** *R. Br. Prod.* 331. Leaf-stems or pseudobulbs usually 2 to 4 in. long. Leaves elongated, narrow, keeled, channelled above, striate, the upper ones often 6 in. to 1 ft. long or even more, the lower ones short. Racemes from the lower axils often 1 ft.

long including the peduncle, the sheathing scales at the base rather rigid, the bracts small and spreading. Pedicels ½ to 1 in. long. Sepals and petals oblong or lanceolate, 5 to 7 lines long. Labellum rather shorter than the sepals, distinctly 3-lobed, the lateral lobes decurrent along the claw, the middle lobe broadly ovate or almost rhomboidal, as long as the lower part, papillose on the upper surface; the disk between the lateral lobes with 2 longitudinal raised lines or plates slightly pubescent or shortly fringed. Column about as long as the lateral lobes, slightly incurved, with 2 narrow longitudinal wings.—Lindl. Gen. and Sp. Orch. 164; Bot. Mag. t. 5851; Reichb. f. Beitr. 45.

N. Australia. Fitzmaurice river, *F. Mueller.*

Queensland. Broad Sound, *R. Brown;* Cape York (*Botanical Magazine*); Herbert's Creek, *Bowman;* Cape river, *Fitzalan;* Burnett, *Haly.*

N. S. Wales. Hunter's river, *A. Cunningham;* Richmond river, *Fawcett;* also in *Mitchell's* and *Leichhardt's* collections.

S. Australia. Cooper's Creek (*F. Mueller*), the specimen not seen.

The flowers are brown with green margins according to the Bot. Mag.; yellow, blotched with red according to others; the labellum dull white spotted with red.

2. **C. albuciflorum,** *F. Muell. Fragm.* i. 188. Stems or pseudobulbs often 1 ft. long. Leaves attaining 2 ft. or more and often 1 in. broad, keeled underneath, channelled above, and striate. Racemes including the peduncle 1 to 2 ft. long, axillary, with sheathing scales at their base. Bracts small, at length spreading or reflexed. Pedicels rather rigid, ½ to ¾ in. long. Sepals and petals greenish yellow, about 5 lines long, rather brown outside, more obtuse than in *C. canaliculatum,* the sepals broadly oblong, the petals rather narrower. Labellum nearly as long as the sepals, red at the base, yellowish above, 3-lobed as in *C. canaliculatum* but without the longitudinal plates of that species, of a rather thicker consistence and not quite so broad. Column with a prominent angle in front, the apex truncate.

Queensland. Moreton bay, *W. Hill;* Rockingham bay, *Dallachy;* Mount Dryander, *Fitzalan.*

3. **C. suave,** *R. Br. Prod.* 331. Stems usually short, more densely covered with the imbricate strongly striate bases of the leaves than the two preceding species, and these bases often split up into fibres. Leaves narrow, often above 1 ft. long, keeled and strongly striate. Racemes rather more dense than in *C. canaliculatum,* the sheathing scales at the base of the peduncle more rigid and leaf-like, the flowers rather smaller, green blotched with red. Sepals and petals scarcely 5 lines long, rather acute. Labellum narrower than in *C. canaliculatum,* especially towards the base, undivided or obscurely sinuate 3-lobed, the disk without longitudinal plates but thickened along the centre. Column with 2 narrow wings. Capsule ovoid-globular, scarcely 1 in. long.—Lindl. Gen. and Sp. Orch. 164; F. Muell. Fragm. i. 187; Reichb. f. Beitr. 46.

Queensland. Moreton bay, *C. Stuart.*—Some far advanced specimens from Rockhampton, *O'Shanesy, Dallachy,* with smaller flowers may belong to the same species.

N. S. Wales. Hunter's river, *R. Brown;* Paramatta, *Woolls;* northward to Hastings river, *Beckler;* southward to Illawarra, *Shepherd.*

16. SPATHOGLOTTIS, Blume.

Sepals and petals nearly equal, free, spreading. Labellum articulate at the base of the column, concave or saccate at the base, deeply 3-lobed, the middle lobe contracted at the base and bearing prominent tubercles or calli. Column erect, free, more or less dilated or 2-winged upwards. Anthers terminal, lid-like, 2-celled. Pollen-masses 8, of which 4 usually smaller, waxy, with very short separate caudicles without any common gland.—Terrestrial herbs with subterranean tuberous rhizomes. Leaves usually long, plicate and strongly ribbed. Racemes on erect scapes, leafless except sheathing scales. Bracts usually rather large.

The genus is dispersed over tropical Asia; the only Australian species perhaps endemic, but closely allied to one ranging over the Archipelago.

1. **S. Paulinæ,** *F. Muell. Fragm.* vi. 95. Tubers small. Leaves lanceolate, acuminate, 2 to 3 ft. long, tapering into a long petiole, plicate and strongly ribbed. Scapes attaining 3 or 4 ft. bearing a short raceme of "purple" flowers. Bracts lanceolate, about $\frac{1}{2}$ in. long; pedicels about 1 in. Sepals and petals about 5 lines long. Labellum about as long, very short and concave below the lobes, the lateral lobes linear-oblong, slightly spathulate and incurved, the middle lobe scarcely longer, obovate, obtuse or emarginate, contracted much below the middle, with 3 large prominent calli immediately above the lateral lobes, hairy around and immediately above the calli. Column incurved, not much shorter than the sepals, slightly dilated upwards.—*Bletia Paulinæ*, F. Muell. l.c.

Queensland. Rockingham bay, *Dallachy*.—Very near the *S. plicata*, Blume (*S. lilacina*, Griff.), but the flowers are rather smaller, and the shape of the middle lobe of the labellum different. The few flowers seen were, however, not in a good state.

17. PHAIUS. Lour.

Sepals and petals nearly equal, free, spreading. Labellum broad, produced into a spur at the base, erect and convolute round the column, entire or 3-lobed and more or less spreading at the top. Column semicylindrical, elongated. Anther lid-like. Pollen-masses 8, nearly equal or 4 shorter, waxy, attached to the branches of a dichotomous caudicle, but no gland.—Terrestrial herbs, the leafy stems short and thickened into pseudobulbs or almost stemless. Leaves large. Scapes radical, tall, erect, leafless except sheathing scales imbricate at the base, distant on the stem and passing into the bracts. Flowers large and showy.

The genus is spread over tropical and subtropical Asia. Of the two Australian species or varieties, one is the same as an Archipelago one, the other may be endemic, but is not sufficiently known.

Sepals and petals brown inside . . . . 1. *P. grandifolius.*
Sepals and petals yellow inside . . . . 2. *P. Bernaysii.*

1. **P. grandifolius,** *Lour.; Lindl. Gen. and Sp. Orch.* 126. Stems tufted, usually thickened into short pseudobulbs at the base, bearing 2

or 3 oblong or ovate-lanceolate leaves often above 1 ft. long, narrowed into a long petiole. Scapes radical, 2 to 4 ft. high, bearing a loose raceme of large showy flowers. Sepals and petals broadly lanceolate, 1¾ to near 2 in. long, white outside, cinnamon-brown inside. Labellum nearly as long as the sepals, very broadly obovate, broadly and very obtusely 3-lobed, or notched or shortly acute in the centre, the margins undulate-crisped, white and shaded or streaked with crimson, loosely encircling the column at its base, the spur short narrow and usually curved. Column nearly ¾ in. long.—*Bletia Tankervilliæ*, R. Br. in Bot. Mag. t. 1924; *Phaius australis*, F. Muell. Fragm. i. 42; *P. leucophæus*, F. Muell. Fragm. iv. 163; *P. Carroni*, F. Muell. Pl. Burdek. Exped. 19.

Queensland. Moreton bay and island, *A. Cunningham, M'Gillivray;* Rockingham bay, *A. Cunningham, Dallachy;* Lady Elliott's island, *Burdekin Expedition.*
N. S. Wales. Macleay river, *Fitzgerald;* Tweed river, *Herb. F. Mueller.*

2. **P. Bernaysii,** *Rowl.; Reichb. f. in Gard. Chron.* 1873, 361. Habit stature foliage and inflorescence of *P. grandifolius*, from which it is only to be distinguished by the colour of the flower of a pale yellow inside, the labellum also yellow edged with white. The spur of the labellum appears in the two flowers accompanying the wild specimen to be rather straighter than in the common species, but curved in the cultivated plant. Reichenbach refers it to *P. Blumei*, distinguished by the labellum acute not notched in the centre, but this appears to vary much from specimen to specimen.—*P. Blumei* var. *Bernaysii*, Reichb. f. in Bot. Mag. t. 6032.

Queensland. Moreton bay, *Bernays.*

18. CALANTHE, R. Br.

Sepals and petals nearly equal, free, spreading, the lateral sepals sometimes shortly adnate to the labellum at the base. Labellum connate at the base with the column in a sort of cup, usually produced into a spur at the base, the lamina spreading, lobed or undivided, the disk with several tubercles or callosities opposite the anther. Column erect, the margins connate with the labellum, the rostellum usually rostrate. Anther lid-like. Pollen-masses 8, tapering to the base and there affixed to a divisible gland.—Terrestrial herbs, stemless or nearly so. Leaves large, plicate, usually in tufts of 2 or 3. Scapes in the axils of the outer leaves tall, erect, and many-flowered. Flowers often showy, white or lilac.

The genus is dispersed over tropical Asia and the islands of the Pacific, with one Mexican species; the only Australian species extends over the Archipelago and the East Indian Peninsula.

1. **C. veratrifolia,** *R. Br. in Bot. Reg. under n.* 573. Rhizome shortly creeping, with tufts of 2 or 3 leaves, sometimes forming a very short stem or pseudobulb at the base. Leaves 1 to 2 ft. long, ovate-lanceolate, plicate undulate and strongly ribbed, tapering into a petiole which is again dilated at the base. Scapes usually in the axil of the

outer leaf, 2 to 4 ft. high, the flowers rather crowded near the summit. Pedicels spreading, ½ to 1 in. long, recurved after flowering. Sepals and petals white, obovate-oblong, nearly 5 lines long, the petals usually broader and more contracted at the base than the sepals. Labellum much longer, the spur slender, ¾ in. long and usually pubescent, the lamina 3-lobed with the middle lobe deeply bifid, the 4 lobes oblong and sometimes nearly equal, but variable in breadth as well as in the relative depth to which they are divided, the callosities of the disk yellow. Capsule obovoid-oblong, about 1½ in. long.—Bot. Reg. t. 720; Bot. Mag. t. 2615.

Queensland. Rockingham bay, *Dallachy;* Brisbane river, Moreton bay, *F. Mueller.*

N. S. Wales. Hastings river, *Beckler;* Richmond river, *Henderson;* Tweed river, *Guilfoyle;* Illawarra, *A. Cunningham.*

The Australian specimens, constituting the var. *australis,* Lindl. Fol. Orchid. Calanthe, 8, appear generally to have the lobes of the labellum rather broader than the Indian ones.

TRIBE 4. ARETHUSEÆ.—Anther lid-like, incumbent, usually deciduous. Pollen granular or mealy. Terrestrial or rarely epiphytes. Stems in the Australian genera or sections leafless at the time of flowering.

19. GALEOLA, Lour.

(Erythrorchis, *Blume;* Ledgeria, *F. Muell.*)

Sepals and petals nearly equal in length, connivent or open, the dorsal sepal incurved, the petals narrower. Labellum sessile, broad, incurved round the column, the lateral lobes very short and erect or obsolete, the middle lobe short and broad, undulate-crisped, the disk with 2 raised longitudinal lines, the intervening space pubescent or glabrous. Column elongated, erect, not at all or scarcely winged. Anther lid-like, incumbent, with a broad flat or convex dorsal appendage, 2-celled. Pollen granular-farinaceous or almost waxy, in 2 deeply 2-lobed distinct masses, without any caudicle or gland.—Leafless epiphytes, sometimes climbing to a great extent, the branches flexuose. Flowers in terminal usually pendulous panicles. Bracts at the base of the branches and panicles small or large, but always concave and half-stem-clasping.

Besides the Australian species, one of which is closely allied to a Javanese one, there are two or three others from East India, the Archipelago, and perhaps from New Caledonia. The genus appears to have been quite correctly referred by Reichenbach f. to the *Galeola* of Loureiro, and I should also concur in the retention of *Cyrtosia,* Bl., as distinct.

Bracts scarcely ¼ in. long. Labellum pubescent inside between the raised lines which end in a transverse callus. Pollen almost waxy . . . . 1. *G. cassythoides.*

Bracts 1 to 2 in. long. Labellum glabrous between the raised lines which converge into one, the lamina on each side marked with diverging lines fringed with small linear hairs . . . . 2. *G. foliata.*

1. **G. cassythoides,** *Reichb. f. Xen. Orchid.* ii. 77. Stems leafless, of a chocolate-brown colour, climbing to a great length and closely adhering to the stems of trees, throwing out adventitious rootlets at the nodes opposite the bracts, terminating in long pendulous panicles. Flowers of a brownish or golden yellow, in short racemes or branches of the panicle, quite glabrous and smooth. Bracts at the base of the pedicels and branches ovate-lanceolate, acute, 1 to 2 lines long and those of the flowerless stems scarcely 3 lines. Pedicels and ovary 3 to 4 lines. Sepals 5 to 6 lines long, oblong-lanceolate, the dorsal one incurved, the lateral ones slightly falcate; petals as long as the sepals but linear. Labellum white with transverse coloured bands, scarcely so long as the sepals, sessile, very broad, erect, concave, almost convolute, obscurely 3-lobed, the lateral lobes or obtuse angles short, erect and entire, the middle lobe very short and broad, spreading, undulate-crenate; disk of the erect part with 2 raised longitudinal lines separated by a broad pubescent centre and ending in a transverse callus, the lamina or middle lobe pubescent on the surface at the base and bearing sometimes irregular undulate calli. Anther with a large broad convex or almost hood-like dorsal appendage, 2-celled in front; pollen-masses 2, without caudicles, deeply 2-lobed, but the lobes closely approximate and the consistence almost as waxy as in *Dendrobium.*—*Dendrobium cassythoides,* A. Cunn. in Lindl. Bot. Reg. under n. 1828; *Ledgeria aphylla,* F. Muell. Fragm. i. 239, ii. 167; *Erythrorchis aphylla,* F. Muell. Fragm. ii. 167.

Queensland. Moreton island, *F. Mueller.*
N. S. Wales. Hastings river, *C. Moore, Beckler;* New England, *C. Stuart.*

The species is very closely allied to the Javanese *G. altissima,* which, judging from Blume's figure and description (under *Erythrorchis*), has the same pollen; but there appear on a comparison of specimens, to be sufficient differences in the flowers to keep the two distinct as well-marked varieties, if not as species. I have only seen the pollen-masses in one flower, where they were certainly very much like those of a *Dendrobium.* They are distinctly described as waxy by F. Mueller, and by Blume in his *E. altissima* as "solidiuscula."

2. **G. foliata,** *F. Muell. Fragm.* viii. 31. Climbing habit apparently the same as that of *G. cassythoides,* but not specially described by the collectors, the panicle larger broader and more branched, and the bracts subtending the branches often 1 to 2 in. long, but retaining the ovate-lanceolate shape, the broad stem-clasping base, and apparently the colour and consistence of bracts rather than of true leaves. Sepals and petals lanceolate, $\frac{3}{4}$ in. long, the petals much narrower than the sepals. Labellum broadly obovate, more contracted at the base than in *G. cassythoides,* the erect part broadly cuneate, with two raised lines along the centre but glabrous between them, the two lines converging into a single one on the lamina, this lamina or upper spreading portion of the labellum very broad, the margins undulate-crisped, the surface of the whole labellum on each side of the smooth centre fringed with several lines of small linear calli rather than hairs, the lines transverse on the claw, longitudinal or diverging on the lamina. Anthers with the broad dorsal appendage of *G. cassythoides* but flatter, and the pollen distinctly

granular, in two masses deeply divided into somewhat distant oblong lobes, giving the mass somewhat of a horseshoe shape. Capsule 7 to 8 lines long. Seed winged.—*Ledgeria foliata*, or *Erythrorchis foliata*, F. Muell. Fragm. ii. 167.

Queensland. Pine river, *Fitzalan;* Rockingham bay, *Dallachy.*
N. S. Wales. Clarence river, *C. Moore* (in fruit only).

This fine species has precisely the pollen figured by J. D. Hooker in the *Cyrtosia* (*Erythrorchis*) *Lindleyana*, Ill. Himal. Pl. t. 22, and very different from that of *G. cassythoides;* but notwithstanding this apparently important difference in the two species, which would technically place them in different tribes of the order, it is difficult not to regard them as congeners, especially as both appear to have the exceptionally winged seeds, and probably the same remarkable habit.

20. EPIPOGUM, Gmel.

Sepals and petals free, nearly equal, narrow, erect or spreading. Labellum sessile, large, ovate, concave, with a short obtuse spur at the base. Column very short, the margin membranous. Anther lid-like, with a large thick terminal appendage. Pollen-masses 2, granular, attached to the gland by long caudicles.—Leafless terrestrial herbs, with a thick and fleshy or branching and coral-like rhizome. Scapes simple, ascending or erect, with a few scarious scales, not green. Flowers white (or sometimes pink ?) in a terminal raceme usually nodding or pendulous.

The genus has very few species scattered in few individuals over a great part of the Old World. The only Australian one is also in tropical Asia and Africa.

1. **E. nutans,** *Lindl. in Journ. Linn. Soc.* i. 177. Stem ascending from a thick rhizome, 6 to 9 in. high, with 2 or 3 empty scarious bracts besides those which subtend the pedicels, all ovate-lanceolate, acute, 3 or 4 lines long. Flowers white, on short pedicels in a raceme occupying the greater part of the plant. Dorsal sepal and petals lanceolate, nearly 4 lines long, very thin, connivent, lateral sepals narrower. Labellum sessile on a broad base enclosing the column, as long as the sepals, broadly ovate and very concave, entire, the spur about 1 line long, the disk with 2 obscure rows of papillæ along the centre. Appendage at the end of the anther as large as the anther itself.—*Galera nutans*, Blume, Bijdr. 415, Coll. Orchid. t. 52, 54 E; *Podanthera pallida*, Wight Ic. t. 1759 (represented much larger in all its parts than the Australian specimen); *E. Guilfoylii*, F. Muell. Fragm. viii. 30.

N. S. Wales. Tweed river, *Guilfoyle*, the specimens agreeing precisely with some of those from East India, where the species appears to be widely scattered, as it has also been found in tropical Africa.

21. GASTRODIA, R. Br.

Sepals and petals united in a 5-lobed tube or cup, gibbous at the base under the labellum. Labellum shorter than the perianth, shortly adnate to it at the base along the centre, entire or with 2 obtuse auricles near the base, oblong, the margins undulate, the disk with 2 longitudinal raised lines or plates confluent upwards into a single one. Column elongated, the apex concave, with a membranous margin. Anthers lid-like, incumbent, very shortly stipitate, deciduous, the cells contiguous. Pollen granular. Stigma on a short protuberance at the base of the column.—Herbs parasitical on roots, leafless and not green. Scapes simple, erect, with short loosely sheathing scales. Flowers white, in a terminal raceme.

Besides the Australian species which is endemic, there is one in New Zealand, and a few from the Indian Archipelago and East India have been recently associated with it.

1. **G. sesamoides,** *R. Br. Prod.* 330. Stems 1 to 1½ ft. high, the sheathing scales loose and very obtuse or shortly acute, 2 to 3 lines long, approximate at the base of the stem, distant higher up. Raceme erect, usually 1 to 4 in. long, but sometimes much longer. Bracts scarious, very broad and obtuse, shorter than the pedicels. Flowers white or brownish outside, on pedicels of 2 or 3 lines. Perianth varying from 6 to 8 lines long, the lobes short and broad. Labellum scarcely shorter than the perianth, broadly oblong, very obtuse, much undulate. Column nearly as long as the labellum, angular, the basal stigmatic protuberance very prominent. Capsule obovoid-turbinate.—Endl. Iconogr. t. 5; Lindl. Gen. and Sp. Orch. 384; Hook. f. Fl. Tasm. ii. 31, t. 126 (the stigmatic protuberance overlooked by the artist); Reichb. f. Beitr. 44.

Queensland. Moreton bay, *W. Hill.*

N. S. Wales. Port Jackson to the Blue Mountains, *R. Brown*, *A. Cunningham*, and others. "Dry rocky situations and sandy forest grounds," *A. Cunningham.*

Victoria. Mountains of Upper Barwan, Apollo bay, M'Alister river, top of Mount William, *F. Mueller.*

Tasmania. Not uncommon in dense humid forests, *J. D. Hooker.*

22. POGONIA, Juss.

Petals and sepals free, erect connivent or somewhat open, equal or the petals smaller. Labellum erect, concave, undivided or lobed, the disk crested papillose or bearded. Column elongated, semiterete, usually angular or winged upwards. Anther lid-like, incumbent, sessile or shortly stipitate. Pollen granular, cohering in 2 entire or 2-lobed masses, free or attached to the rostellum by an elastic web.—Terrestrial herbs, forming a spherical tuber under ground. Flowering stems or scapes in the Australian section leafless besides scarious sheathing scales, erect, bearing a raceme (usually one-sided) of pendulous pedicellate flowers, sometimes reduced to a single flower, usually red (sometimes blue?). Leaves developed later than the flowers, solitary on separate stems, those of the Australian species not as yet observed.

The genus is a widely-spread one, having been originally founded on North American species. The section *Nervilia*, to which the Australian species belong, characterized by the separate development of the solitary leaf and of the flowering stems, is spread over tropical Asia and the Mascarene islands as well as those of the Archipelago; the typical form or section *Eupogonia*, with leafy flowering stems, is North American, with one Japanese species.

Flowers solitary on the scape, 6 to 7 lines long. Labellum 3-lobed, the middle lobe contracted at the base, the disk papillose . . 1. *P. uniflora.*
Flowers 2 to 6 in the raceme, 7 to 8 lines long. Labellum shortly 3-lobed, the middle lobe broad at the base, the disk bearded . 2. *P. holochila.*
Flowers 2 or 3 in the raceme, nearly 1 in. long. Labellum quite entire, broadly obovate, the disk smooth . . . . 3. *P. Dallachyana.*

1. **P. uniflora,** *F. Muell. Fragm.* v. 201. Flowering stem very slender, under 6 in. long, with few rather long loose sheathing scales and terminating in a single flower, blue according to Dallachy, pink according to F. Mueller. Pedicel very short, the junction with the stem marked by a small bract, wanting on some specimens. Sepals and petals 6 to 7 lines long, linear, rather acute, the lateral sepals somewhat broader or linear-lanceolate. Labellum nearly as long as the sepals, 3-lobed, the lateral lobes broadly triangular and obtuse, the middle lobe at least as long, ovate-oblong, obtuse, much contracted at the base; disk with a papillose line, narrow and double at the base, expanded upwards and extending partially on to the middle lobe. Column very slender at the base, broadly 2-winged upwards.

Queensland. Rockingham bay, *Dallachy.*

2. **P. holochila,** *F. Muell. Fragm.* v. 200. Flowering stems slender, 6 to 10 in. high, with a few rather long sheathing scarious bracts, the upper ones sometimes almost leafy at the end. Flowers 2 to 6, on slender pedicels of 1 to 3 lines, apparently pink, but like those of *P. uniflora* said by Dallachy to be blue. Bracts linear, rather long. Sepals and petals 7 to 8 lines long, the lateral sepals oblong-linear, the dorsal sepal and petals narrow-linear, the petals more contracted at the base. Labellum nearly as long as the sepals, broadly ovate, shortly sinuate-3-lobed, the lateral lobes broadly rounded, the middle one smaller, rather broader than long; disk with a bearded line extending to halfway along the middle lobe. Column slender, very shortly winged at the apex.

Queensland. Rockingham bay, *Dallachy.*—With a very few specimens are some of the young leaf-stems, with the leaf as yet insufficiently developed to judge of its form, but apparently cordate.

3. **P. Dallachyana,** *F. Muell. Herb.* Stems about 6 in. high, with 2 or 3 long loose sheathing scales. Flowers 2 or 3, on short pedicels crowded at the end of the stem. Bracts linear. Sepals said to be reddish, nearly 1 in. long, narrow-lanceolate, acuminate, the petals rather shorter and narrower. Labellum about $\frac{3}{4}$ in. long, broadly obovate, quite entire, contracted and embracing the column at its base; the disk without any raised papillose or bearded lines. Column 3 to 4 lines long, dilated and winged at the apex only.

Queensland. Rockingham bay, *Dallachy.*

TRIBE 5. NEOTTIEÆ.—Anther erect or bent forward, persistent but free from the rostellum, sessile or stipitate. Pollen granular or mealy, in 1 or 2 masses in each cell, with or without a caudicle. Terrestrial herbs with simple stems (except *Corymbis*) bearing 1 or more leaves or rarely leafless, and a simple spike raceme or single flower.

In the characters given of the following genera I have rarely mentioned the number of masses in which the pollen-grains cohere, for the cohesion is usually so slight that, in drying the specimens it is so far destroyed by pressure, that there are but very few cases where I have been able to observe the unbroken masses.

23. CORYMBIS, Thou.

(Corymborchis, *Thou.*)

Sepals and petals nearly equal, linear and dilated above the middle, all spreading or the dorsal one more erect. Labellum about as long, narrow, channelled, dilated at the end into a short recurved lamina, the disk with 2 longitudinal raised lines. Column elongated, terete, clavate at the end, with 2 erect lateral lobes, the stigma and rostellum acuminate, as long as the anther. Anther erect behind the stigma, acuminate, 2-celled; pollen granular, in 2 masses attached by a caudicle to a peltate gland.—Tall terrestrial herbs, with a fibrous rhizome. Leaves large, strongly ribbed. Flowers in short axillary somewhat corymbose panicles, the column persistent on the capsule.

The genus extends over tropical Asia and Africa, and appears also to be represented in Brazil. The Australian species extends over the Indian Archipelago and perhaps into East India.

The persistent column in this genus is said to elongate on the fruit after flowering. Of this the dried specimens show no sign. I have never seen the column on the capsule longer than the petals of the same species or variety. Possibly there may have crept in some error derived from the comparison of fruiting specimens of the very long-flowered Mauritius species with the shorter flowers of other species.

1. **C. veratrifolia,** *Reichb. f. in Flora* 1865, 184. Stems erect or somewhat climbing, attaining 5 to 10 ft. Leaves distichous, oblong-elliptical, acuminate, 9 to 18 in. long, very strongly ribbed, tapering into a short sheathing petiole. Flowers white, in axillary panicles not half so long as the leaves, erect and almost sessile along its spreading branches. Bracts linear-lanceolate, rarely $\frac{1}{2}$ in. long. Sepals and petals nearly $\frac{3}{4}$ in. long, the dorsal one rather broader at the base than the others, more erect and concave. Labellum rather longer than the sepals, broadly lanceolate, concave, about 2 lines broad, with entire margins, slightly dilated at the end into a short broad acute crisped lamina, not very perfect in the Australian specimens seen; the disk with 2 long raised lines, curving outwards on the lamina and ending in small callosities. Column at the time of flowering as long as the sepals, and not elongated afterwards, persisting on the oblong capsule after the sepals and petals have fallen.—*Corymborchis veratrifolia*, Blume, Orchid. 125, t. 42, 43.

Queensland. Rockingham bay, *Dallachy.*—Certainly the same species as the one gathered by Cuming in the Philippine islands, which appears to extend over the Indian Archipelago, and even to be identical with one gathered by Fritz Mueller at St. Catherine's in Brazil, but well distinguished by Blume from Thouars' original Mascarene and African species, which has the flowers above 2 in. long.

24. RAMPHIDIA, Lindl.

Flowers reversed. Lateral sepals erect, broader at the base, forming a hood over the labellum; dorsal sepal and petals reflexed. Labellum erect, sessile, undivided, concave. Column very short, with a narrow appendage on each side connecting the base of the anther and stigma into a cup. Anther erect. Pollen-masses granular. Rostellum bifid, as long as the anther.—Terrestrial herbs, with a creeping rhizome and ascending leafy stem. Flowers small, in a slender terminal point.

The genus is spread over East India and the Archipelago, the only Australian species having nearly the range of the genus.

1. **R. tenuis,** *Lindl. in Journ. Linn. Soc.* i. 182. Stems ascending to from 6 in. to 1 ft. Leaves on slender petioles, dilated at the base into a broad loose scarious sheath of $\frac{1}{4}$ to $\frac{1}{2}$ in., the lamina elliptic-oblong to ovate-lanceolate, $1\frac{1}{2}$ to 3 in. long, and usually 2 or 3 narrow empty sheathing-bracts above the leaves. Flowers very small, in a slender spike of 3 or 4 in., the bracts almost subulate, the rhachis as well as the ovaries and perianths loosely pubescent with short hairs, the rest of the plant glabrous. Lateral sepals rather broad, about $1\frac{1}{4}$ lines long. Labellum scarcely if at all connate with the column, nearly as long as the lateral sepals, broadly oblong, very concave, very obtuse, the margins and end inflexed, entire or slightly crisped. Anther and rostellum about equal in length and half as long as the labellum.

Queensland. Mackay river and Rockingham bay, *Dallachy.*—The specimens agree precisely with the one originally described by Lindley from the Philippine islands.

25. GOODYERA, R. Br.

Dorsal sepal and petals erect connivent and often connate; lateral sepals as long, spreading. Labellum sessile, embracing the column and sometimes adnate to it at the base, concave or almost saccate, undivided, often fringed or hairy inside at the base but without appendages. Column short, the stigma very concave or pouch-like, and connected with the anthers by lateral membranes, forming a second pouch between the rostellum and the anther, the rostellum entire and lanceolate or divided into two lobes. Anthers erect. Pollen-masses granular, attached to long caudicles.—Terrestrial herbs, with a creeping rhizome and weak ascending leafy stems. Flowers in terminal spikes, loose in the Australian species, dense and one-sided in the typical northern ones.

The genus is generally distributed over the tropical and temperate regions of the Old World. Of the two Australian species one appears identical with an Asiatic one, the other, as far as known, is endemic.

Rostellum divided into 2 long linear lobes without appendages at the base. Sepals and petals 3 to 4 lines long . . . . . . 1. *G. viridiflora.*
Rostellum lanceolate, undivided, with a gland-like appendage on each side at the base. Sepals and petals $1\frac{1}{2}$ lines long . . . 2. *G. polygonoides.*

1. **G. viridiflora,** *Blume, Orchid.* 41, 9 C. Stems ascending from 6 in. to 1 ft. Leaves on rather long petioles, dilated at the base into a loose scarious sheath, the lamina ovate-lanceolate or ovate-oblong, acute or shortly acuminate, sometimes slightly cordate at the base, 1½ to 3 in. long; and usually 2 or 3 empty scarious bracts above the leaves. Flowers rather distant, in a spike of 2 to 4 in., the bracts broad, membranous, subulate-acuminate, often as long as the ovary. Dorsal sepal and petals 3 to 4 lines long, the petals broader and more obtuse than the sepal, and usually cohering to it; lateral sepals obliquely lanceolate. Labellum as long as the sepals, broadly ovate, very concave, rather copiously fringed inside with short cilia near the base. Rostellum divided into linear-lanceolate erect lobes as long as the long acuminate anther, stigma pouch-like, almost truncate.—*Neottia viridiflora,* Blume Bijdr. 408; Lindl. Gen. and Sp. Orch. 494; *Georchis cordata,* Lindl. Gen. and Sp. Orch. 496; *Georchis viridiflora,* F. Muell. Fragm. viii. 29.

Queensland. Rockingham bay, *Dallachy.*—This species certainly belongs to the genus *Georchis* of Lindley, which Blume appears to be right in reducing to *Goodyera,* unless the latter be restricted to the original *G. repens* and *G. pubescens.* Reichenbach identified the specimens with the *G. cordata,* Lindl., from Sikkim, which only differs in the leaves being rather more cordate at the base, and it also appears to be identical with Blume's plant from Java, of which, however, we have no specimen.

2. **G. polygonoides,** *F. Muell. Fragm.* viii. 29. Stems from a creeping rhizome ascending, about 8 in. long including the raceme. Leaves petiolate above the loose scarious sheath, oblong-lanceolate, 1½ to 3 in. long. Spike loose, the flowers small on a long ovary attenuate upwards. Bracts lanceolate, about as long as the ovary. Sepals and petals about 1½ lines long, ovate or ovate-lanceolate, the sepals acute, the petals thin and adnate to the dorsal sepal. Labellum not very broad, acutely acuminate, rather longer than the sepals, very concave, without hairs or appendages inside. Column short, the stigma membranous, large concave and undulate, but not pouch-like. Rostellum lanceolate, undivided, with an almost gland-like appendage on each side at the base on the margin of the membrane connecting the stigma and anther.

Queensland. Rockingham bay, *Dallachy*.—A single specimen in Herb. F. Mueller. The aspect of the plant is that of *Rhomboda,* Lindl., the gland-like or crest like appendages at the base of the rostellum those of some species of *Hetæria* (*Etæria* or *Ætheria*), Blume; but the labellum has not the internal appendages of those two genera. Neither of the above species agree precisely in these technical characters with the typical *Goodyeræ,* but probably a general revision upon good materials would require the extension of the genus even beyond the limits proposed by Blume.

26. SPIRANTHES, Rich.

Dorsal sepal and petals erect, connivent or slightly coherent in an upper lip or galea, or the ends alone spreading. Lateral sepals free and more spreading, all nearly equal. Labellum sessile or nearly so, embracing the column by its broad base, undivided, often dilated at the end, the disk with 2 tubercles at the base. Column short, with a small erect appendage or lobe on each side. Anther erect; pollen-masses 4, granular or mealy, sessile on a gland. Rostellum short, bifid.—Terrestrial herbs, with oblong underground tubers or thick clustered fibres. Flowers small, spirally arranged in a terminal spike. Stems leafy, or sometimes at the time of flowering with sheathing scales only.

A considerable genus generally diffused over the temperate and tropical regions of the globe; the only Australian species has a wide range over Asia and a part of Europe.

1. **S. australis,** *Lindl. Gen. and Sp. Orch.* 464. Rhizome short, with a cluster of thick fibres or oblong tubers. Stem glabrous below the inflorescence, 6 in. to above 1 ft. high. Lower leaves linear or narrow-lanceolate, varying in length from 1½ to 4 in. as well as in breadth, the upper ones reduced to sheathing scales. Spike spiral, very dense or rather loose. Flowers sessile, generally described as pink with a white labellum; bracts usually about as long as the ovary, the rhachis ovary and sometimes the perianths pubescent. Sepals and petals varying from 1½ to 2 lines in length, the lips spreading, the lateral sepals obscurely dilated near the base but not saccate. Labellum as long as the sepals, the broad base quite sessile or sometimes appearing raised on a very short claw with a tubercle on each side, concave, often slightly contracted above the column, then expanded into a short broad undulate crisped or almost fringed lamina. Anther scarcely acuminate but much longer than the rostellum.—Wight Ic. t. 1724; Hook. f. Fl. Tasm. ii. 15; Reichb. Ic. Fl. Germ. t. 476; *Neottia australis,* R. Br. Prod. 319.

Queensland. Brisbane river, Moreton bay, *F. Mueller;* Armidale, *Riley.*
N. S. Wales. Port Jackson, *R. Brown, A. Cunningham, Woolls;* Blue Mountains, *Miss Atkinson;* New England, *C. Stuart;* Macleay river, *Fitzgerald*; Clarence river, *Beckler;* Richmond river, *Fawcett,* but in most places said to be very rare.
Victoria. Mitta-Mitta, Broadribb and Snowy rivers and Lake Omeo, *F. Mueller;* Portland, *Crouch.*
Tasmania. Circular Head, *Gunn;* Cheshunt, *Archer;* Swanport, *Story.*

The species is also in New Zealand and in a great part of tropical and temperate Asia, extending to some parts of Europe.

27. CALOCHILUS, R. Br.

Dorsal sepal erect, rather broad, concave; lateral sepals about as long, broadly lanceolate, spreading; petals much shorter, broadly falcate. Labellum as long or longer, undivided, contracted at the base, the margin and whole surface densely fringed except a narrow terminal point or ligula. Column short, with a rather broad wing more or less produced behind the anther but not beyond it. Anther bent forward at the base, usually recurved towards the end, obtuse or obtusely acuminate or rostrate. Pollen-masses granular. Rostellum much shorter than the anther.—Terrestrial glabrous herbs, with ovoid underground tubers. Leaf usually solitary, long and narrow, but usually 2 or 3 erect almost leaf-like sheathing bracts on the stem. Flowers few in a terminal raceme, green or yellowish, with more or less of purple especially the labellum.

The genus is limited to Australia. In the column it is nearly allied to *Thelymitra*, but differs widely in the perianth.

Anther rostrate. Column-wing quite open in front, with a gland on each side within its anterior angles or lobes . . . . . . 1. *C. campestris.*
Anther shortly rostrate. Column-wing usually with a gland on each side as in *C. campestris*, and open in front, but connected by a transverse raised line across the base of the labellum . . 2. *C. Robertsoni.*
Anther very obtuse. Column-wing without any gland, open in front. Labellum with 2 short longitudinal intramarginal erect plates or auricles near the base . . . . 3. *C. paludosus.*

F. Mueller, Fragm. v. 96, proposes to unite all the species of this genus into one under the name of *C. australianus.*

1. **C. campestris,** *R. Br. Prod.* 320. Stem usually rather stout, from under 1 ft. to above 1½ ft. high, with a rather long leaf and 3 to 5 flowers, but sometimes with the habit of *C. paludosus.* Sepals in the typical form 4 to 5 lines long, the petals much shorter. Labellum ½ in. long or more, obovate or obovate-oblong, the margins and surface covered with long purple fringes except near the base where (in the same typical form) there is a raised plate or thickened surface quite smooth and extending more or less along the centre of the narrow part of the labellum, and the end is produced into a linear or lanceolate smooth often flexuose point varying much in length. Column-wings rather broad and dilated in front into a variously-shaped lobe or angle with a dark-coloured gland inside, the two wings not at all connected in front. Anther bent forward at the base, recurved and acuminate or rostrate upwards.—Lindl. Gen. and Sp. Orch. 459; Bot. Mag. t. 3187; Hook. f. Fl. Tasm. ii. 15, t. 106 A; Reichb. f. Beitr. 21; *C. herbaceus*, Lindl. l.c.

Queensland. Shoalwater bay, *R. Brown* (belonging probably to the var. *grandiflora*).
N. S. Wales. Port Jackson, *R. Brown*, *Woolls.*
Tasmania. Rocky Cape, Woolnorth, *Gunn;* Port Sorell, *Archer;* Huon river, *Oldfield;* Oyster Cove, *Milligan;* Southport, *C. Stuart.*

Var. *grandiflora.* Flowers 1 to 3, larger than in the common form, and altogether resembling *C. paludosus*, the labellum covered with fringes or linear calli from the base, but without the intramarginal appendages of that species, and the glands of the column-wings as in *C. campestris.*—Moreton island, *M'Gillivray;* also some specimens from Port Jackson, *Woolls;* and probably Macleay river, *C. Moore.* Brown's original Port Jackson specimens belong to the above-described typical forms, and correspond with several of Woolls', as also with the majority at least of Tasmanian ones, with Lindley's *C. herbaceus*, and with the figure in the Botanical Magazine from a drawing made in Tasmania.

2. **C. Robertsoni,** *Benth.* A stout species, with the habit of the larger specimens of *C. campestris*, but the leaf usually broader. Sepals acuminate, fully ½ in. long in the specimens seen; petals also acuminate, more than half as long. Labellum fringed all over, the terminal smooth point short. Column-wings with a more or less distinct gland on each side in front as in *C. campestris*, but the two wings connected at the

base by a transverse raised plate across the base of the labellum, of which I see no trace in the two other species. Anther shortly and obtusely rostrate.

Victoria. Heaths on Glenelg river, *Robertson;* Mount M'Ivor, *Herb. F. Mueller;* Bendigo, *Oldfield;* and probably a specimen from Dandenong, *F. Mueller*, with the flower too far advanced for examination.

3. **C. paludosus,** *R. Br. Prod.* 320. Usually more slender than *C. campestris*, with a long leaf and only 2 or 3 rarely 4 flowers, often but not always larger than in that species. Sepals usually 7 to 8 lines long; petals not half so long, strongly veined. Labellum covered with the long fringes or cilia, shorter and much crowded towards the base, with 2 longitudinal but short and much raised plates near the base resembling auricles but intramarginal, not strictly marginal as figured (C 3) in Endlicher's plate, the terminal smooth point of the labellum usually long flexuose and linear. Column-wing produced behind the anther to about its length, broadly rounded in front on each side, without the glands of *C. campestris.* Anther as broad as long, very obtuse, neither acuminate not rostrate.—Lindl. Gen. and Sp. Orch. 459; Endl. Iconogr. t. 14; Reichb. f. Beitr. 22.

N. S. Wales. Port Jackson, *R. Brown*, *Woolls;* Hunter's river, *R. Brown;* Blue Mountains, *Miss Atkinson.*

28. THELYMITRA, Forst.

Sepals and petals all nearly equal and spreading. Labellum similar to the sepals and petals and spreading with them. Column erect, rather short, very broadly winged, the wings either reaching to the base of the anther with an erect usually thick entire lateral lobe or appendage on each side of it, or dilated at the end and united behind the anther, sometimes extended into a broad lobed hood over it, the lateral lobes often penicillate or crested, the wings below the anther embracing the column but open in front except at the base where they are united and sometimes are produced into a short tooth between the labellum and column. Anther erect or bent forward between the lateral lobes or under the hood; the cells distinct, the connective produced into a broad appendage sometimes elongated and entire or shortly bifid; pollen-masses granular, without any or with a very small caudicle.—Terrestrial herbs, glabrous or very rarely pubescent on the leaf-sheaths, with ovoid underground tubers. Leaf solitary, usually with a rather long sheath, the lamina linear, lanceolate or rarely almost ovate, often rather thick, but not terete; empty bracts 1 or 2 along the stem. Flowers usually several in a terminal raceme, sometimes reduced to 1 or 2, blue purple red or yellow, occasionally with white varieties.

One of the Australian species extends over New Zealand, New Caledonia, and the Indian Archipelago, and there are three or perhaps four species peculiar to New Zealand; the genus is otherwise endemic in Australia. It is remarkable for the labellum perfectly resembling and taking its place as one of the petals, and quite detached from the column, from which it is separated by the annular base of the wing.

SECT. 1. **Cucullaria.**—*Column-wing produced behind and beyond the anther into a broad hood over it, variously lobed or fringed at the top.*

Hood with the 2 extreme lateral lobes penicillate (bearing a tuft of cilia). Flowers blue, purple or white.
Hood with 3 short denticulate or fringed lobes between the penicillate ones and shorter than them.
Middle lobe of the hood crested on the back . . . . . . 1. *T. ixioides.*
Middle lobe of the hood smooth on the back . . . . . . 2. *T. canaliculata.*
Hood with 1 entire or bifid lobe between the penicillate ones, usually longer than them, broad and concave.
Hood densely crested on the back. Leaf very broadly lanceolate . . . . 3. *T. crinita.*
Hood smooth on the back. Leaf linear or linear-lanceolate.
Tall robust plant. Leaves usually rather broad. Middle lobe of the hood usually 2-fid and papillose-denticulate . 4. *T. aristata.*
Plant usually slender, with narrow leaves. Middle lobe of the hood scarcely notched and entire . . . . 5. *T. longifolia.*
Hood with the 2 extreme lateral lobes cristate, but without tufts of cilia. Flowers yellow, often spotted with purple.
Leaf villous, broad. Middle lobe of the hood undivided, cristate at the end, and a transverse crest inside at the base . . 6. *T. villosa.*
Leaf glabrous, narrow. Middle lobe of the hood 3-fid without the internal crest . . . . 7. *T. tigrina.*
Hood deeply fringed with linear lobes, with a club-shaped appendage on the back. Flowers yellow or brown.
Dorsal appendage of the hood crested at the end. Perianth-segments narrow-lanceolate . . . . 8. *T. stellata.*
Dorsal appendage of the hood smooth tubercular or notched at the end. Perianth-segments broadly lanceolate . . . . . 9. *T. fusco-lutea.*

SECT. 2. **Macdonaldia.**—*Column-wing broadly produced behind the anther, but much shorter than it, and not hood-shaped. Slender flexuose herbs, with the habit of* T. antennifera.

Two extreme lateral lobes of the truncate column-wing prominent and denticulate or fringed. Flowers pink . . . . 10. *T. carnea.*
Column-wing broadly truncate, slightly sinuate, but the lobes scarcely prominent. Flowers yellow . . . . 11. *T. flexuosa.*

SECT. 3. **Biaurella.**—*Column-wing not produced behind the anther, but with 2 prominent erect lateral lobes as long as or longer than the anther, and often connected by a short crest behind it.*

Habit of *T. carnea.* Flowers yellow. Lateral lobes of the column long, erect, and spathulate . . . . 12. *T. antennifera.*
Habit nearly of *T. carnea.* Flowers deep-coloured (purple or red?) Lateral lobes of the column oblong, erect, curved, denticulate, not spathulate.
Leaf-sheaths glabrous . . . . 14. *T. Macmillani.*
Leaf-sheaths pubescent . . . . 14. *T. variegata.*
Habit of *T. ixioides.* Flowers blue.
Lateral lobes of the column longer than the anther, involute . 15. *T. venosa.*
Lateral lobes of the column not exceeding the anther, oblong or lanceolate, convolute or thickened . . . . 16. *T. cyanea.*

SECT. 1. CUCULLARIA.—Column-wing produced behind and beyond the anther into a broad hood over it, variously lobed or fringed at the top.

1. **T. ixioides,** *Sw. in K. Akad. Stockh. Handl.* 1800, 228, t. 3 L, *and in Schrad. Neu. Journ.* 58, t. 1 L. Stem usually above 1 ft. high, with 1

long linear or linear-lanceolate flat or channelled leaf, and 1 or 2 shorter ones, and several flowers pedicellate in a raceme of 4 to 6 in., but like several other species very variable in stature and number of flowers, sometimes slender with 1 or 2 flowers like the var. *pauciflora* of *T. longifolia.* Sepals petals and labellum elliptical-oblong, 9 to 10 lines long in the common Port Jackson variety, smaller in others especially the western ones. Column about 2 lines long, the broad wing extending behind and beyond the anther, shortly adnate to it at the base, and forming a broad hood over it with 2 lateral lobes at the angles shortly linear erect and bearing a dense tuft of white cilia, and 3 rather shorter lobes between them, all truncate and denticulate, and the central one with a crest of several rows of dorsal calli. Anther produced into an incurved point much longer than the rostellum.—Sm. Exot. Bot. i. 55, t. 29; R. Br. Prod. 314; Lindl. Gen. and Sp. Orch. 522; Hook. f. Fl. Tasm. ii. 6, t. 103 B; Reichb. f. Beitr. 7; *T. iridioides*, Sieb., and *T. juncifolia*, Lindl. l.c.; *T. lilacina*, F. Muell., referred by Lindl. in Linnæa xxvi. 242 to *T. canaliculata.*

Queensland. Archer's station, Moreton bay, *Leichhardt.*
N. S. Wales. Port Jackson, *R. Brown*, *Sieber*, *n.* 168, *Woolls*, and others; New England, *C. Stuart;* southward to Twofold bay, *F. Mueller.*
Victoria. E. Gipps Land, *Walter;* near Brighton, rare, *F. Mueller.*
Tasmania. Abundant throughout the Colony, *J. D. Hooker*, and others.
W. Australia. Swan river, *Drummond, Maxwell;* Upper Kalgan river, Monjerup and Peronjerup, *F. Mueller;* Lake Maxwell and Peronjerup ranges, *Oldfield.*

These Western specimens have usually smaller flowers and rather narrower, more channelled leaves than most but not all of the Eastern ones. They all appear to me to have the middle lobe of the column hood-crested on the back as in the typical *T. ixioides;* but possibly on the living plant characters might be found to connect them rather with *T. canaliculata* than with *T. ixioides.* They certainly include *T. campanulata*, Lindl. Swan Riv. App. 49, Gen. and Sp. Orch. 521.

2. **T. canaliculata,** *R. Br. Prod.* 314. Habit and few rather small flowers of the slender narrow-leaved forms of *T. ixioides*, and the floral characters the same except that the central lobe of the column-hood is broader, more external, and though much denticulate has no dorsal crest.—Lindl. Gen. and Sp. Orch. 522; Reichb. f. Beitr. 7.

W. Australia. King George's Sound, *R. Brown;* Albany, *F. Mueller.*—The station is given in the Prodromus as (T) by some mistake, probably typographical. The specimens in Herb. Brown and Bauer's drawing all give King George's Sound. *T. cornicina*, Reichb. Beitr. 54, also from King George's Sound, appears to me to be the same species, which further observation of the fresh plant may prove to be a variety only of *T. ixioides.*

T. media, R. Br. Prod. 314, Reichb. f. Beitr. 7, from Port Jackson, *R. Brown*, with the same aspect as some varieties of *T. ixioides*, or of *T. longifolia*, is like *T. canaliculata* distinguished from the former by the want of any crest on the back of the central lobe of the column-hood, and this crest is not figured in the plate of *T. ixioides* in Fl. Tasm. I have not been able to examine Brown's specimens of *T. media*, but in all those resembling it both from Port Jackson and from Tasmania I have uniformly found the dorsal crest.

T. epipactoides, F. Muell. Fragm. v. 174, from Port Philip, which I have not seen, would appear from the character given to be a broad-leaved variety of *T. ixioides*, or of *T. canaliculata.*

T. campanulata, Endl. in Pl. Preiss. ii. 14, which I have not seen, is referred by Reichb. f. Beitr. 55, to *T. canaliculata.*

3. **T. crinita,** *Lindl. Swan Riv. App.* 49, *Gen. and Sp. Orch.* 521. Usually above 1 ft. high, the leaf ovate-lanceolate or very broadly lanceolate, 1½ to 3 in. long. Raceme loose, of several rather large blue flowers. Sepals and petals varying from ½ to ¾ in., usually obtuse. Column-wing produced behind and beyond the anther into a broad hood, the 2 extreme lateral lobes as in *T. longifolia*, bent forward and penicillate but shorter than the centre, which is broadly 2-lobed as in that species but very densely cristate on the back with shortly linear papillæ or calli.—Endl. in Pl. Preiss. ii. 14; *T. ovata*, F. Muell. Fragm. vi. 84.

W. Australia. Swan river, *Drummond, 1st coll.*, *Preiss, n.* 2194; Vasse, Gordon, and Kalgan rivers, *Oldfield; King George's Sound, F. Mueller; Lake Muir, Muir.*

F. Mueller appears to have distinguished his *T. ovata* from the colour of the tuft of ciliæ on the lateral lobes of the column-hood, violet instead of white, which, however, appears to be a very inconstant character.

4. **T. aristata,** *Lindl. Gen. and Sp. Orch.* 521. Usually tall and leafy, the lower leaf linear-lanceolate, rather broad and sometimes very long, and two of the empty sheathing bracts long and loose with leafy points. Flowers several, like those of the stouter forms of *T. ixioides* and *T. longifolia.* Column-wings produced behind and beyond the anther into a broad hood over it, the 2 extreme lateral lobes penicillate, bent forward, and shorter than the centre as in *T. longifolia*, but these 2 central lobes more deeply divided than in that species and denticulate or crested on the edge, and usually pale-coloured in the dried state.—*T. grandis*, F. Muell.; *T. angustifolia*, Hook. f. Fl. Tasm. ii. 5, not of R. Br.

Victoria. Wendu vale, *Robertson;* Melbourne, *Adamson;* Darebin Creek, *F. Mueller.*

Tasmania. Circular Head, &c., *Gunn;* South Huon river, *Oldfield;* Southport, *C. Stuart;* Rocky Cape, M'Quarrie harbour, *Milligan.*

S. Australia. Mount Gambier, Rivoli bay, Bugle ranges, *F. Mueller.*

W. Australia. Swan river, *Drummond, 1st coll.*; Kalgan river, *Oldfield;* between Esperance bay and Russell Range, *Dempster.*—These Western specimens which represent the *T. macrophylla*, Lindl Swan Riv. App. 49 Gen. and Sp. Orch. 519, seem to connect the species with the typical *T. longifolia*, according to Reichb. f. Beitr. 56. Preiss's n. 2187, referred by Endl. in Pl. Preiss. ii. 14, to *T. macrophylla*, is not to be distinguished from *T. crinita.*

5. **T. longifolia,** *Forst. Char. Gen.* 98, t. 49. Stature and foliage varying more even than that of *T. ixioides.* What we may consider as the typical form is rather tall, with a long narrow leaf and a raceme of several rather large flowers nearly represented by *T. nuda* and *T. angustifolia* of Brown, whilst his *T. pauciflora* is a smaller slender form with very few (often only one) smaller flowers, and some northern specimens are tall and vigorous with broader leaves coming very near to those of *T. aristata*, but none of these in the dried state can well be sorted into distinct varieties, and none of them show any difference in the perianth from that of *T. ixioides*, the colour variously described as blue, lilac or pink. Column-wing produced behind and over the anther into a broad

hood, usually conspicuous for its dark colour, the 2 extreme lateral lobes penicillate as in *T. ixioides*, but bent forward and shorter than the broad centre, which is entire emarginate or shortly 2-lobed with the margin entire and smooth.—*T. Forsteri*, Sw. in K. Akad. Stockh. Handl. 1800, 228, and in Schrad. N. Journ. i. 57; Lindl. Gen. and Sp. Orch. 520; *T. nuda*, R. Br. Prod. 314, Lindl. l.c.; Hook. f. Fl. Tasm. ii. v. t. 103 A; Reichb. f. Beitr. 8; *T. pauciflora*, R. Br. l.c. Reichb. f. l.c.; *T. arenaria* and *T. versicolor*, Lindl. l.c. 519, 520; *T. graminea*, Lindl. Swan Riv. App. 49, Gen. and Sp. Orch. 521.

Queensland. Endeavour river, *Banks and Solander;* Port Bowen, *R. Brown;* Archers station, Moreton bay, *Leichhardt;* Rockingham bay, *Dallachy;* Armidale *Perrot;* also from the *Burdekin Expedition.*

N. S. Wales. Port Jackson, *R. Brown*, *Woolls;* New England, *C. Stuart.*

Victoria. Wendu Vale, *Robertson;* and numerous stations from Melbourne to the Grampians, Wimmera, and the Murrumbidgee, *F. Mueller*, and others.

Tasmania. Port Dalrymple, *R. Brown;* abundant throughout the Colony, *J. D. Hooker*, and others.

S. Australia. Mountains around St. Vincent's Gulf, *F. Mueller*, *Behr;* Spencer's Gulf, *F. Mueller.*

W. Australia. Swan river, *Drummond*, 1st *coll.*; Hampden, *Clarke.*

There is some confusion about Brown's *T. angustifolia* and *T. canaliculata*, owing to his having in the first instance given the former name to the Western plant he afterwards published as *T. canaliculata*, and in his Prodromus *T. angustifolia* is marked (J) for Port Jackson, and *T. canaliculata* (T) for tropical Australia; whilst his herbarium shows that it is *T. angustifolia* that ought to have been marked (T) and *T. canaliculata* from King George's Sound should have been marked (M). The Port Bowen specimens of *T. angustifolia* have the central lobe of the hood sometimes slightly denticulate, showing an approach in this respect to *T. aristata*, but not at all in the foliage; on the other hand, some of the New England and Queensland specimens of *T. longifolia* have the broad leaves nearly of *T. aristata*, but not at all the column-hood. A few specimens from near King George's Sound, *F. Mueller*, and a very few from other localities, have the central lobe of the column-hood more distinctly 2-lobed, but are not otherwise different.

6. **T. villosa,** *Lindl. Swan. Riv. App.* 49, t. 8 C, *Gen. and Sp. Orch.* 521. A rather stout species, usually above 1 ft. high, remarkable for its ovate obovate or ovate-oblong leaf, villous on both sides especially underneath, as well as the sheath, the rest of the plant including the empty sheathing bracts glabrous. Flowers rather large, yellow dotted with purple, in a loose raceme. Sepals and petals acute, usually nearly $\frac{3}{4}$ in. long. Column-wings produced behind and over the anther into a broad hood, the extreme lateral lobes densely cristate or fringed and curved forward but not white-penicillate, the centre prominent, broad and entire as in *T. longifolia*, but densely cristate-dentate on the back at the end, and also with a transverse crest of small calli inside at its base.—*T. pardalina*, F. Muell. Fragm. v. 94.

W. Australia. Swan river, *Drummond*, 1st *coll.*; Stirling Range, *F. Mueller;* Perongerup, *Mr. Knight;* eastward to Esperance bay and Cape Le Grand, *Maxwell.*

7. **T. tigrina,** *R. Br. Prod.* 315. Stems slender, attaining about 1 ft., the leaf very narrow and channelled. Flowers small, 2 to 4 in the raceme "yellow and spotted." Sepals and petals about 4 lines long. Column-wing broad, produced behind and beyond the anther, the ex-

treme lateral lobes oblong, densely papillose-bearded but not with the white cilia of the penicillate species, the three middle lobes broader, shorter, fringed with similar papillæ or calli.—Lindl. Gen. and Sp. Orch. 521.

W. Australia. King George's Sound, *R. Brown;* Swan river, *Drummond, 3rd coll. n.* 308.

8. **T. stellata,** *Lindl. Swan Riv. App.* 49, *Gen. and Sp. Orch* 519. A stout species of 1 to 1½ ft. with an ovate-lanceolate or oblong-lanceolate leaf, closely allied to *T. fusco-lutea* and possibly a variety only. Sepals and petals narrow-lanceolate, acuminate, about 1 in. long. Column-hood at least 3 lines long and deeply fringed as in *T. fusco-lutea*, but the dorsal club-shaped appendage is densely callose-crested at the end, and the fringe of the hood rather more irregular, two of the long teeth sometimes much thicker than the others. Anther with a long terminal appendage.

W. Australia. Swan river, *Drummond, 1st coll.*

9. **T. fusco-lutea,** *R. Br. Prod.* 315. A stout glabrous species of 1 to 1½ ft., rarely smaller and more slender. Leaf ovate-lanceolate or oblong-lanceolate, shortly acuminate, 2 to 4 in. long. Raceme of few rather large flowers. Sepals and petals broadly oblong-lanceolate, acute or shortly acuminate, usually about ¾ in. long or rather more, yellow with dark brown spots. Column exceedingly short below the anther, the wing produced behind and beyond the anther into a broad hood nearly 3 lines long, deeply cut into a fringe of long linear lobes, shortly crested or bearded on the back in the centre, and with a dorsal club-shaped appendage proceeding from near the base, rather shorter than the fringe, entire slightly rugose or notched at the end but not crested. Anther ending in a thick obtuse fleshy appendage, long but shorter than the hood.—Lindl. Gen. and Sp. Orch. 519; Reichb. f. Beitr. 10.

S. Australia. Onkaparinga, *F. Mueller.*

W. Australia. King George's Sound, *R. Brown;* Swan river, *Drummond, 1st coll.*; Vasse river, *Pries.*

A drawing of Bauer's, of which there is a tracing in Herb. Lindley, represents the appendage of the hood as notched at the end, but as it is as broad in the side view as in the front view it must have been thick and club-shaped; in the finished coloured drawing in the British Museum it is represented as flatter more deeply 2-lobed and denticulate. Reichenbach f. therefore, thinking that Drummond's plant described by Lindley, in which the appendage is entire, must be specifically distinct, has, Beitr. p. 55, characterized it as new under the name of *T. Benthamiana.* Brown's own specimen has the flowers too far advanced for examination, but one from Capt. King, labelled by Brown as *T. fusco-lutea,* appears to me without doubt to be the same as Drummond's plant. Probably the dorsal appendage is variable and its shape exaggerated in Bauer's finished drawing.

Sect. 2. Macdonaldia.—Column-wing broadly produced behind the anther but much shorter than it, and not hood-shaped.

10. **T. carnea,** *R. Br. Prod.* 314. Stem slender, often flexuose, from under 6 in. to near 1 ft. high, with 1 to 3 pink flowers, the leaf narrow-

linear and 1 or 2 bracts sometimes leaf-like. Sepals and petals oval-elliptical or oblong, obtuse, usually about 4 lines long, less veined than in *T. ixioides* and its allies. Column nearly half as long as the perianth, the wings very broadly truncate and connected behind the anther but shorter than it, the 2 extreme lateral lobes short broad and denticulate, the intermediate 2 lobes very broad and scarcely prominent. Anther-connective produced beyond the cells, but broad and obtuse.—Lindl. Gen. and Sp. Orch. 519; Hook. f. Fl. Tasm. ii. 5, t. 102 B; Reichb. f. Beitr. 9.

N. S. Wales. Port Jackson, *R. Brown, Woolls.*
Victoria. Wendu Vale, *Robertson;* Port Phillip, *C. French.*
Tasmania. Hobarton, *Gunn;* Georgetown and Cheshunt, *Archer;* Southport, *C. Stuart.*
S. Australia. Lofty Range, *F. Mueller.*

11. **T. flexuosa,** *Endl. Nov. Stirp. Dec.* 23. Stems slender but usually wiry and flexuose, 6 to 9 in. high. Leaf narrow-linear, rather thick, and the empty bracts sometimes leaf-like. Flowers 1 or 2, yellow, smaller than in the other species, the sepals and petals obtuse and scarcely more than 3 lines long. Column-wing continued behind the anther but only about half its length, broadly truncate and very shortly and equally 3-lobed or rather 3-crenate, the lobes quite entire and rather thin, or the middle one broader and notched, the wings forming in front a loose cup nearly ½ line long, the sides broad and rounded almost into lateral lobes. Anther produced into a broad thick pubescent appendage.—*Macdonaldia Smithiana,* Gunn, and *M. concolor,* Lindl. Swan Riv. App. 50, t. 9 B, Gen. and Sp. Orch. 385; Endl. in Pl. Preiss. ii. 4; *Thelymitra Smithiana,* Hook. f. Fl. Tasm. ii. 4, t. 101 B.

Victoria. Port Phillip, *C. French;* Mount Abrupt and Bunip Creek, *F. Mueller.*
Tasmania. Circular Head, *Mrs. Smith;* Georgetown, *Archer;* Southport, *C. Stuart.*
W. Australia. *Drummond,* 1*st coll.*, 3*rd coll. n.* 309; Albany and Upper Hay river, *F. Mueller;* Thistle Cove, *Maxwell.*

SECT. 3. BIAURELLA, *Lindl.*—Column-wing not produced behind the anther, but with 2 lateral erect lobes or appendages longer than the anther and often connected by a short crest behind it.

12. **T. antennifera,** *Hook. f. Fl. Tasm.* ii. 4, t. 101 A. Stems erect, wiry, flexuose, 6 in. to 1 ft. high, with the narrow-linear rather thick leaves of *T. carnea* and *T. flexuosa.* Flowers 1, 2 or rarely 3, yellow, of the size of those of *T. carnea* or rather larger. Sepals and petals obtuse or acute, 4 to 6 lines long. Column-wings adnate to the anther as high as the cells but not produced behind it, with erect dark-coloured broadly spathulate lateral appendages, longer than the anther, very obtuse or emarginate. Anther produced into a broad thick concave appendage, bent forward, very obtuse, rugose-pubescent outside.—*Macdonaldia antennifera,* Lindl. Swan Riv. App. 50; Gen. and Sp. Orch. 385; Endl. in Pl. Preiss. ii. 4.

Victoria. Wendu Vale, *Robertson;* Portland, *Allitt;* Port Phillip, *Gunn;* ighton, *F. Mueller;* Grampians, *S. Fisher.*

Tasmania. Georgetown, *Archer.*
S. Australia. Encounter bay, *Whittaker;* Tamunda and Macclesfield, *F. Mueller.*
W. Australia. King George's Sound to Swan river, *Wakefield, F. Mueller, Drummond,* 1*st coll.*, 4*th coll. n.* 230, 5*th coll. n.* 116, *Preiss, n.* 2181, and many others; eastward to Bremer and Esperance bays, *Maxwell, Dempster.*

13. **T. Macmillani,** *F. Muell. Fragm.* v. 93. Habit and foliage of *T. antennifera,* but perhaps nearer allied to *T. variegata.* Sepals and petals about ½ in. long, apparently deep coloured (red ?). Column-wing produced into 2 lateral diverging lobes as long as the anther, curved and slightly denticulate at the end, separated by an acute sinus behind the anther, without any dorsal crest.

Victoria. Port Phillip, *M'Millan;* a single specimen in Herb. F. Mueller.—The species requires further investigation and may prove to be an abnormal form of one of the allied species or possibly a hybrid.

14. **T. variegata,** *Lindl. in Herb. Benth.* Stem not very stout, 1 ft. high or rather more, with 2 to 4 large flowers. Leaf with a villous sheath, the lamina usually glabrous, linear, much dilated at the base and often undulate. Sepals and petals lanceolate, shortly acuminate or acute, ¾ to 1 in. long, dark-coloured and variegated. Column nearly 3 lines long to the anthers, the broad wings deeply coloured, not produced behind the anthers, but with erect lateral lobes nearly 2 lines long, oblong, obtuse but not spathulate, connected by a crest behind the anther. Anther-cells short, the connective produced into a broad obtuse appendage as long as the lateral lobes of the column.—*Macdonaldia variegata* and *M. spiralis,* Lindl. Swan Riv. App. 50, Gen. and Sp. Orch. 385, 386; *Thelymitra porphyrosticta,* F. Muell. Fragm. v. 97.

W. Australia. Swan river, *Drummond,* 1*st coll.*; Kalgan river, *Maxwell.*—The form *spiralis* is scarcely a distinct variety, the undulate or twisted base of the leaf occurs in some of the larger as well as in the smaller and more slender specimens.

15. **T. venosa,** *R. Br. Prod.* 314. Habit of some of the rather larger forms of *T. ixioides.* Stem 1 to 2 ft. high, with a long narrow leaf and a raceme of 6 to 10 blue flowers. Sepals and petals ½ to ¾ in. long, veined as in *T. ixioides.* Column broadly winged, with 2 long linear obtuse erect lateral lobes, not connected behind the anther and more or less spirally involute. Anther scarcely acuminate, shorter than the column-lobes.—*Macdonaldia venosa,* Lindl. Swan Riv. App. 50, Gen. and Sp. Orch. 386; Reichb. f. Beitr. 9.

N. S. Wales. Port Jackson, *R. Brown, A. Cunningham, J. D. Hooker.*

16. **T. cyanea,** *Lindl. MS.* A smaller and more slender plant than *T. venosa,* of which it is by some considered as a variety. Leaves narrow-linear, channelled, not very long. Flowers 1 to 3, blue or white. Sepals and petals not above ½ in. long. Column-wings produced into erect lateral lobes as long as the anther, lanceolate or oblong, either laterally convolute or with one margin thickened, and more or less distinctly connected by a very short crest behind the anther. Anther acuminate and often 2-dentate.—*Macdonaldia cyanea,* Lindl. Swan Riv. App. 50; Gen. and Sp. Orch. 386; *Thelymitra venosa,*

Hook. f. Fl. Tasm. ii. 4, t. 102 A, as to the Tasmanian plant, not of R. Br.

Tasmania. Circular Head and Rocky Cape, *Gunn;* Cheshunt and Port Sorrell, *Archer;* Macquarrie Harbour, *Milligan;* Southport, *C. Stuart.*

Hooker reduces this to the Port Jackson *T. venosa*, but it appears to me sufficiently distinct in the smaller flowers, the acuminate anther, and differently shaped lateral lobes of the column.

29. EPIBLEMA, R. Br.

Sepals and petals all nearly equal and spreading. Labellum unguiculate, the claw with 2 erect thick lobes, the lamina ovate, concave, with a tuft of linear processes on the disk near the base. Column very short, with erect petal-like thin lateral lobes or appendages not connected behind the anther. Anther erect or slightly bent forward, the cells distinct, with a short recurved point. Pollen-masses granular. Rostellum short.—Terrestrial glabrous herb, with the habit of *Thelymitra.* Leaf narrow-linear. Flowers few in a terminal raceme.

The genus is limited to the single species endemic in Australia.

1. **E. grandiflorum,** *R. Br. Prod.* 315. Habit of the slender forms of *Thelymitra ixioides.* Stem erect, 1 to 1½ ft. high, with one long narrow linear leaf and 1 or 2 smaller erect leaves or sheathing scales. Flowers 2 to 5, pedicellate, in a short raceme, the bracts shorter than the ovary. Sepals and petals 7 to 8 lines long, broadly oblong-lanceolate, veined like those of *Thelymitra ixioides*, and all nearly similar, except that the dorsal one is more concave than the other, the lateral sepals shortly united under the labellum by a broad base, and the petals more ovate-lanceolate. Labellum as long as the sepals, the claw about 1½ lines long, the 2 appendages erect rounded and closely parallel, the processes of the disk of the labellum long, slightly clavate, ascending, or sometimes one or two of them deflexed and clasped by the appendages of the claw. Lateral appendages of the column broad, oblique, about 2 lines long.—Lindl. Gen. and Sp. Orch. 523; Endl. in Pl. Preiss. ii. 13; Reichb. f. Beitr. 10.

W. Australia. King George's Sound, *R. Brown, A. Cunningham, Baxter;* Swan river, *Drummond,* 1*st coll., Preiss, n.* 2219; Cape Le Grand, *Maxwell;* Lake Muir, *Muir.*

30. DIURIS, Sm.

Dorsal sepal erect, rather broad, closely embracing the column at the base, the upper part open; lateral sepals narrow-linear, almost herbaceous, parallel or sometimes crossed, spreading or deflexed; petals longer than the dorsal sepal, ovate-elliptical or oblong, on slender claws. Labellum usually as long as or rather longer than the dorsal sepal, deeply 3-lobed, the middle lobe much contracted at the base, with 1 or 2 longitudinal raised lines along the narrow part. Column very short, the wings produced into lateral erect lobes, but not continued behind the anther. Anther erect, often acuminate, the 2-valved cells occupying nearly the whole inner surface. Rostellum 2-fid, shorter than the anther. Pollen-masses granular or mealy, without any distinct caudicle.—Terrestrial glabrous herbs, with underground tubers. Leaves narrow, few at or near the base of the stem, with a few sheathing bracts higher up. Flowers 1, 2 or several in a terminal raceme, often rather large and conspicuous from the antenna-like green lateral sepals, the rest of the perianth yellow purple or white, in many species bright yellow with deep purple spots or blotches, the prominent petals often very spreading, whilst the shorter dorsal sepal remains close over the column, make it appear as if the petals were outside in æstivation, and they are sometimes so drawn, but in the bud the æstivation is quite normal with the sepals outside.

The genus is limited to Australia, and cannot be confounded with any other, although the species are very difficult to distinguish from dried specimens which do not show their real colours.

Labellum 3-partite (the lateral lobes divided to the base), with 2 longitudinal raised lines on the middle lobe.
  Lateral sepals usually much longer than the petals.
    Flowers white . . . . 1. *D. alba.*
    Flowers bluish purple . . . . 2. *D. punctata.*
  Lateral sepals usually scarcely longer than the petals. Flowers yellow, often spotted or blotched with purple . . . . . 3. *D. aurea.*
Labellum 3-fid (the lateral lobes separating from above the base) with 2 longitudinal raised lines on the middle lobe.
  Raised longitudinal lines on the labellum at some distance apart. Eastern species.
    Lateral lobes of the labellum as long as or more than half as long as the middle lobe. Dorsal sepal usually as long as the labellum.
      Lateral sepals much longer than the petals . . . . . 4. *D. palustris.*
      Lateral sepals scarcely so long as the petals . . . . . 5. *D. maculata.*
    Lateral lobes of the labellum less than half as long as the middle lobe. Dorsal sepal usually shorter than the labellum.
      Raised longitudinal lines of the labellum pubescent . . 6. *D. pedunculata.*
      Raised longitudinal lines crenulate or cristate . . . . 7. *D. pallens.*
      Raised longitudinal lines glabrous and smooth . . . . 8. *D. abbreviata.*
  Raised longitudinal lines closely contiguous so as to appear like one canaliculate line, and uniting in a single one about the middle of the lamina. Western species.
    Leaves filiform Flowers 1 to 3 . . . . 9. *D. setacea.*
    Leaves linear. Flowers several . . . . 10. *D. emarginata.*
Labellum 3-partite (the lateral lobes divided off from the base) with a single raised longitudinal centre not furrowed.
  Labellum lateral lobes much shorter than the middle lobe. Flowers yellow often with purple spots . . . . 11. *D. sulphurea.*
  Labellum lateral lobes nearly as long as the middle lobe. Flowers purple often mixed with buff-colour . . . . 12. *D. longifolia.*
  Labellum lateral lobes much shorter than the middle lobe. Leaves filiform and habit of *D. setacea.* Flowers yellow . 13. *D. pauciflora.*

1. **D. alba,** *R. Br. Prod.* 316. A rather slender species, under or above 1 ft. high, exceedingly difficult in the dried state to distinguish from the small flowered specimens of *D. punctata.* Leaves usually narrower and one of them often nearly as long as the stem below the

inflorescence, but sometimes all shorter. Flowers white with a few dark spots inside, the proportions and shape of the parts the same as in *D. punctata* but all smaller. Lateral sepals 1 to 1¼ in. long; petals about ½ in.; dorsal sepal and labellum nearly equal in length and shorter than the petals. Lateral lobes of the column more frequently dentate than in *D. punctata*, but variable.—Lindl. Gen. and Sp. Orch. 509; Reichb. f. Beitr. 13.

Queensland. Port Bowen and Shoalwater passage, *R. Brown; Brisbane* river, Moreton bay, *A. Cunningham;* Rockhampton and vicinity, *Bowman, Thozet, O'Shanesy;* Darling Downs, *Law;* Wide bay, *Bidwill;* Broad Sound, *Bowman;* Rockingham bay, *Dallachy.*

N. S. Wales. Port Macquarrie, *Backhouse;* Tweed river, *C. Moore;* New Zealand, *C. Stuart;* Clarence river, *Wilcox.*

A single specimen from Warwick in Herb. F. Mueller has very long filiform lateral sepals, and the rest of the flower very small, the colour not mentioned; others again from the northern districts of N. S. Wales have the petals longer and narrower, and the lateral sepals scarcely 1 in. long and rather broad. The lateral lobes of the labellum are very variable in their venation, sometimes broad at the apex and many-nerved, sometimes narrower with only 1 or 2 of the nerves conspicuous, and the 2 lateral lobes of the same labellum sometimes different in venation. So also with the lateral lobes of the column, one may in the bud overlap the labellum and the other be wholly inside, one with the nerve reaching almost to the apex, the other with it visible only halfway up.

2. **D. punctata,** *Sm. Exot. Bot.* i. 13, t. 8. Stems 1 to 2 ft. high, or even more. Leaves usually 2, linear, 3 to 6 in. long, with 2 empty sheathing bracts above them. Flowers 2 or 3, blue or purplish, often dotted but not blotched like several of the yellow species, the acuminate bracts often but not always exceeding the ovary. Dorsal sepal in the typical form broadly ovate-oblong, 7 to 8 lines long; lateral sepals deflexed, very narrow, nearly 2 in. long. Petals broadly elliptical-oblong, nearly 1 in. long including a claw of about 2 lines. Labellum about as long as the dorsal sepal, divided at the base into 3 lobes, the middle lobe obovate-oblong, the lateral ones about one-third as long, oblong-falcate, varying in breadth, entire or crenulate; disk of the middle lobe with 3 longitudinal raised lines or plates starting from the base; the 2 lateral ones ending somewhat abruptly and sometimes forming an acute tooth below the middle of the lobe, the central one not so prominent or obscure at the base, but continued further along the lamina. Lateral lobes of the column as long as the anther, lanceolate with undulate margins, more or less distinctly 1-nerved. Anther shortly acuminate, 2 lines long. Rostellum short, bifid. Wings of the column very shortly produced at the base in front, not meeting as in *Thelymitra*, but often continuous with the raised lines of the labellum. —*D. elongata*, Sw. in Schrad. Neu. Journ. i. 59; R. Br. Prod. 316; Lindl. Gen. and Sp. Orch. 509; Reichb. f. Beitr. 13; *D. lilacina*, F. Muell. in Linnæa, xxvi. 239.

N. S. Wales. Port Jackson, *R. Brown, Sieber, n.* 166, *and Fl. Mixt. n.* 627, and many others; Clarence river, *Wilcox;* Nangas, *M'Arthur;* Mudgee, *Woolls;* on the Murrumbidgee, *Nolan.*

Victoria. Port Phillip, *Gunn, F. Mueller;* Wendu Vale, *Robertson;* Yarra, Mount Alexander, Mount Abrupt, *F. Mueller;* East Gipps' Land, *Walter.*

Var. *minor.* Under 1 ft. high, with smaller flowers, the middle lobe of the labellum more rhomboidal.—New England, *C. Stuart;* Upper Clarence river, *Riley.*

Var. *longissima.* Lateral sepals at least 3 in. long.—Mudgee, *Taylor.*

I have been unable to ascertain for what reason Smith's name and figure have been ignored by all subsequent writers.

3. **D. aurea,** *Sm. Exot. Bot.* i. 15, t. 9. Stems 1 ft. high or more. Leaves narrow, not very long. Flowers 2 to 5, yellow or more or less blotched or tinged with brown. Petals obovate-oblong or elliptical, 6 to 8 lines long in the typical form including the claw; lateral sepals rather longer, more or less dilated above the middle; dorsal sepal shorter than the petals, broad and embracing the column at its base, the ovate upper portion more open. Labellum as long as the dorsal sepal, divided to the base, the middle lobe very broad, contracted at the base, the lateral lobes much shorter, broadly falcate, often undulate-toothed, the disk with 2 raised longitudinal plates ending usually in small teeth at or below the middle of the lamina. Lateral lobes of the column falcate, obtuse, sometimes irregularly toothed, the wings almost continuous at the base with the raised lines of the labellum.—R. Br. Prod. 315; Lindl Gen. and Sp. Orch. 509; F. Muell. Fragm. v. 172; Reichb. f. Beitr. 11; *D. spathulata*, Sw. in Schrad. Neu. Journ. i. 60; *D. oculata*, F. Muell. Fragm. v. 173, partly ?

Queensland. Gainsford, *Bowman?*

N. S. Wales. Port Jackson, *R. Brown*, *Woolls;* Port Stephens, *Lady Parry;* New England, *C. Stuart;* Clarence river, *Wilcox.*

Var. *obtusa.* Petals broadly ovate or obovate, very obtuse; dorsal sepal broad.—Hunter's river, *Herb. Lindley and Hooker;* Port Jackson, *Woolls.*

4. **D. palustris,** *Lindl. Gen. and Sp. Orch.* 507. Very near *D. maculata*, but usually a smaller plant with finer leaves, the stem rarely much above 6 in. high, bearing 1 to 4 flowers blotched with yellow and purple, but the dark colour prevailing. Lateral sepals longer and sometimes twice as long as the petals, usually 6 to 8 lines long; petals ovate or oblong tapering into the claw; dorsal sepal rather shorter than the petals, ovate, very obtuse. Labellum still shorter, 3-lobed at some distance from the base, the lateral lobes large, but not quite so long as the broad emarginate middle lobe; the longitudinal raised lines prominent between the lateral lobes, usually dying off gradually on the middle lobe near the base of its lamina.—Hook. f. Fl. Tasm. ii. 7.

Victoria. Wendu Vale, *Robertson;* Mount Alexander, *F. Mueller;* Burra-Burra, *Hinteracker;* Ararat, *Green.*

Tasmania. Marshy ground near Hobarton, Circular Head, &c., *Gunn*, and others.

S. Australia. Onkaparinga, *Whittaker;* Bugle Barossa and Lofty ranges, *F. Mueller;* York Peninsula, *Fowler.*

5. **D. maculata,** *Sm. Exot. Bot.* i. 57, t. 30. Rather a small slender species, usually under 1 ft. high. Leaves narrow. Flowers on long pedicels, yellow, much spotted or blotched with brown or purple and sometimes almost entirely dark-coloured except the yellow centre of the petals, under ½ in. long. Dorsal sepal erect rigid and embracing the column at the base, ovate-oblong and very open at the top; lateral

sepals at length recurved, narrow, rarely exceeding the petals; petals ovate, on a long rigid dark-coloured claw. Labellum shorter than the dorsal sepal 3-lobed from above the base, the lateral lobes large and usually as long or nearly as long as the broad middle lobe, the 2 raised lines of the disk ending usually in prominent angles or teeth a little above the base of the middle lobe. Lateral lobes of the column often toothed.—R. Br. Prod. 315; Bot. Mag. t. 3156; Lindl. Gen. and Sp. Orch. 507; Hook. f. Fl. Tasm. ii. 6, t. 104 B; Reichb. f. Beitr. 11; *D. pardina* and *D. curvifolia*, Lindl. Gen. and Sp. Orch. 507.

Queensland. Rockhampton, *Thozet.*

N. S. Wales. Port Jackson, *R. Brown*, *Sieber*, *n.* 165, and many others; New England, *C. Stuart;* southward to Twofold bay, *F. Mueller.*

Victoria. Wendu Vale, *Robertson;* Port Phillip, *Gunn;* Forest Creek, Sealer's Cove, *F. Mueller;* Grampians, *Fisher.*

Tasmania. Very abundant in pastures and loose forests throughout the colony, *J. D. Hooker.*

S. Australia. Bugle Barossa and Lofty ranges, *F. Mueller;* York Peninsula, *Fowler.*

Var. *concolor.* Flowers yellow, not at all or scarcely spotted.—*D. æqualis*, F. Muell. Herb.—Port Jackson, *Woolls*, *Fitzgerald;* Richlands, *M'Arthur.*—This form closely resembles the western *D. pauciflora*, except in the two raised lines of the labellum. It may possibly be Bauer's *D. pauciflora* from Port Jackson, cited by Reichb. f. Beitr. 13.

6. **D. pedunculata,** *R. Br. Prod.* 316. Stems 6 to 9 in. high. Leaves several at the base of the stem, usually about half its length, narrow-linear. Flowers 1 or 2, of a pale yellow, often with dark tinges at the base, from $\frac{1}{2}$ to $\frac{3}{4}$ in. long. Petals elliptical, stipitate, about the same length as the linear lateral sepals; dorsal sepal shorter, broad, embracing the column at the base, shortly open at the top. Labellum longer than the dorsal sepal, 3-lobed at about 1 line above the base, the lateral lobes curved, not broad, often somewhat toothed, about $\frac{1}{4}$ the length of the ovate-rhomboid middle lobe; the 2 raised longitudinal lines far apart, ending in pubescent calli at the base of the broad part of the middle lobe, the intervening pubescent centre continued more or less along the middle of the lobe.—Lindl. Gen. and Sp. Orch. 508; Hook. f. Fl. Tasm. ii. 8, t. 105 A; F. Muell. Fragm. v. 173; Reichb. f. Beitr. 12; *D. lanceolata*, Lindl. l.c.; *D. Behrii*, Schlecht, Linnæa, xx. 572.

Queensland. Armidale, *Perrott.*

N. S. Wales. Port Jackson, *R. Brown*, *Woolls;* in the interior, *M'Arthur;* Macquarrie river and vale of Clwyd, *A. Cunningham.*

Victoria. Wendu Vale, *Robertson;* Portland, Melbourne, and many other localities, *F. Mueller*, and others, to E. Gipps' Land, *Walter;* Grampians, *Fisher.*

Tasmania. Abundant in moist places in many parts of the island, *J. D. Hooker*, and others.

S. Australia. Mount Gambier to St. Vincent's Gulf, *F. Mueller*, *Behr*, *and* others.

This species, with the flower usually pale coloured and narrow, and easily known by the pubescence of the centre of the labellum, varies much nevertheless in the breadth of the several parts of the flower, and in the raised lines or plates of the labellum, which sometimes end in broad pubescent calli separated by the broad base of the central pubescence of the lamina, sometimes are much rounded, incurved at the end, almost meeting, the pubescent centre of the lamina very narrow. The latter form characterizes the *D. lanceolata*, Lindl.; but I have found many intermediates with slight differences in other characters variously combined.

7. **D. pallens,** *Benth.* A small plant, very nearly allied to *D. pedunculata*, but distinct as far as I am able to judge from dried specimens in several particulars. Flowers smaller, rarely $\frac{1}{2}$ in. long, the lateral sepals scarcely herbaceous or quite as petal-like and about as long as the petals; the dorsal sepal much shorter. Labellum as long as the petals, 3-lobed from above the base as in *D. pedunculata*, but the lateral lobes still smaller in proportion than in that species, the raised lines or plates of the disk converging and ending in a single line along the lamina, but fringed with small calli instead of being pubescent or ciliate.

N. S. Wales. New England, *C. Stuart.*

8. **D. abbreviata,** *F. Muell. Herb.* Habit rather more of *D. maculata* than of *D. pedunculata*, to both of which this species is allied. Leaves rather narrow. Flowers pale-coloured when dry, more or less blotched, usually several in a loose raceme, the pedicels long and the rhachis often remarkably flexuose. Petals oval-oblong, on very long slender claws; lateral sepals rather longer, narrow-linear and herbaceous; dorsal sepal scarcely so long as the labellum, erect and embracing the column at the base, oval-oblong and open in the upper part. Labellum 3-lobed from above the base, the lateral lobes small, triangular or lanceolate, falcate; the middle lobe much longer, broad but much contracted at the base, the disk with 2 very prominent raised lines or plates ending a little beyond the base of the broad part of the middle lobe, quite smooth and glabrous. Lateral lobes of the column acute, entire or denticulate, the wings continuous in front with the raised lines of the labellum.

Queensland. Armidale, *Perrott;* Darling Downs, *Law;* also a specimen from Port Bowen, marked *D. dubia*, in Herb. R. Brown, appears to be this species.

N. S. Wales. New England, *C. Stuart.*

9. **D. setacea,** *R. Br. Prod.* 316. Stems under 1 ft. high. Leaves usually very narrow-linear or filiform and rather short. Flowers 1 to 3, yellow and most frequently with a few purple spots or blotches at least at the base of the labellum and sometimes as much spotted as in *D. maculata*, but occasionally wholly yellow and very variable in size, from under $\frac{1}{2}$ in. to nearly $\frac{3}{4}$ in. long. Lateral sepals about as long as or rather longer than the elliptical petals; dorsal sepal and labellum about equal in length and from $\frac{2}{3}$ to $\frac{3}{4}$ that of the petals. Labellum 3-lobed from very little above the base, the lateral lobes broad, about half as long as the middle one, the plate or longitudinal line along the centre much raised and deeply furrowed, forming on the claw of the middle lobe a double keel merging into a single one on the broad part of the lobe, the labellum being thus characterized as bicarinate by Brown and unicarinate by Lindley. Lateral lobes of the column narrow acute, as long as the anther.—Lindl. Gen. and Sp. Orch. 508; Reichb. f. Beitr. 12; *D. filifolia*, Lindl. Swan Riv. App. 51, t. 8 B; Gen. and Sp.

Orch. 510, Endl. in Pl. Preiss. ii. 11; *D. carinata,* Lindl. Gen. and Sp. Orch. 510; Endl. l.c.

W. Australia. King George's Sound and adjoining districts, *R. Brown,* and many others; and thence to Swan river, *Drummond,* 1*st coll. n.* 842, 843, 4*th coll. n.* 323, *Oldfield,* and others; eastward to Esperance bay, Cape Le Grand and Cape Arid, *Maxwell.*—I have not seen Preiss's specimens, but there is very little doubt of their having been rightly referred to this species, readily recognised by its foliage.

10. **D. emarginata,** *R. Br. Prod.* 316. Allied to *D. setacea,* but a stouter and taller plant, usually 1 to 2 ft. high. Leaves narrow-linear but not subulate, the empty sheathing bracts long and broad. Flowers several, distant from each other in a loose raceme, but on erect pedicels, larger than in *D. setacea.* Lateral sepals $\frac{3}{4}$ to 1 in. long; petals rather shorter, elliptical, contracted into a short claw; dorsal sepal shorter than the sepals, firm at the base and embracing the column, open at the top. Labellum as long as the dorsal sepal, the lateral lobes broad, entire or toothed, from $\frac{1}{3}$ to $\frac{1}{2}$ as long as the middle lobe, and the double raised plate or keel merging into a single one on the lamina of the middle lobe as in *D. setacea.*—Lindl. Gen. and Sp. Orch. 508; Reichb. f. Beitr. 12; *D. Drummondii,* Lindl. Swan Riv. App. 51, Gen. and Sp. Orch. 511.

W. Australia. King George's Sound, *R. Brown;* Lake Muir, *Muir;* Gordon, Tone, Kalgan, Vasse rivers, *Oldfield;* Swan river, *Drummond,* 1*st coll.*, and perhaps also Murchison river, *Oldfield,* the specimen very imperfect.

D. laxiflora, Lindl. Swan Riv. App. 51, Gen. and Sp. Orch. 510, from Swan river, *Drummond,* appears to me to be a rather slender drawn-up state of *D. emarginata,* but the specimens are not perfect enough for examination. I have not seen those of Preiss named *D. laxiflora* by Endl. in Pl. Preiss. ii. 11.

R. Brown evidently derived his name from an emargination of the petals and sepals, which he fully describes in his notes, but of which I can find no trace in the specimens of his herbarium labelled as *D. emarginata,* of the Prodromus. Probably the emargination was accidental in the specimen described on the spot; the dried specimens agree in every other respect with his description, as well as with Drummond's described by Lindley, although here as in *D. setacea* Lindley regarded the labellum as unicarinate, Brown as bicarinate.

11. **D. sulphurea,** *R. Br. Prod.* 316. Stature and habit very much those of *D. aurea,* from which this species is difficult to distinguish without examination of the flowers. Leaves usually rather broader than in that species. Flowers 2 to 5, yellow, almost always blotched with purple at least at the base of the petals. Dorsal sepal about $\frac{1}{2}$ in. long, embracing the column at the base, narrow ovate and open upwards, lateral sepals longer but not very long, often reflexed, petals also longer than the dorsal sepal. Labellum 3-lobed from the base as in *D. aurea,* but shorter than the dorsal sepal; lateral lobes broad, several-nerved and more or less undulate-toothed; middle lobe at least twice as long, very broad but the sides closely reflexed, the disk with a single raised line along the centre gradually dying off above the middle of the lamina. Lateral lobes of the column acute, quite entire, as long as the anther; the wings joining at the base in the front of the column and continuous with the central keel of the labellum.—Lindl. Gen. and Sp. Orch. 509;

Hook. f. Fl. Tasm. ii. 7, t. 104 A; Reichb. f. Beitr. 12; *D. oculata*, F. Muell. in Linnæa, xxvi. 241.

N. S. Wales. Towards George's river, *R. Brown;* Port Jackson and Bathurst, *Woolls;* New England, *C. Stuart.*
Victoria. Glenelg river, *Robertson;* Port Phillip, *Gunn, F. Mueller;* Little river, *Fullager.*
Tasmania. Common in many parts of the colony, *J. D. Hooker*, and others.
S. Australia. Mount Gambier, *F. Mueller.*

The Victorian plant originally described by F. Mueller as *D. oculata*, is certainly the *D. sulphurea*, with the labellum shorter than the dorsal sepal, and with a single central ridge. The very similar *D. aurea* is, however, in several collections under the name of *D. oculata*, F. Muell., and is probably the one described under that name in the Fragmenta, v. 173, with the double keel or 2 longitudinal plates on the labellum.

12. **D. longifolia,** *R. Br. Prod.* 316. Stems from under 1 ft. to considerably above that height. Leaves linear, narrow or broad, one often but not always very long. Flowers usually 3 to 5, variable in size but usually large and dark-coloured when dry, described when fresh as purple and buff. Petals oval or oblong-elliptical often ¾ in. long including the claw; lateral sepals as long, linear or dilated above the middle; dorsal sepal much shorter and very broad. Labellum as long as the dorsal sepal, 3-lobed from the base; lateral lobes nearly as long as the middle one, very broadly cuneate and curved or obliquely obovate, entire; middle lobe contracted into a claw, with a faintly prominent raised line along the centre.—Lindl. Gen. and Sp. Orch. 509; Reichb. f. Beitr. 14; *D. porrifolia*, Lindl. Swan Riv. App. 51, Gen. and Sp. Orch. 511; Endl. in Pl. Preiss. ii. 12; *D. corymbosa*, Lindl. ll.cc.; Hook. f. Fl. Tasm. ii. 7, t. 105 B; F. Muell. Fragm. v. 172.

Victoria. Wendu Vale, *Robertson;* Queenscliff, *F. Mueller;* Portland, *Allitt;* Grampians, *Fisher;* E. Gipps' Land, *Walter.*

Tasmania. Port Dalrymple, *R. Brown;* common in the northern parts of the island, *J. D. Hooker*, and others.
S. Australia. Mount Gambier, and around St. Vincent's Gulf, *F. Mueller.*
W. Australia. King George's Sound, *Menzies, Harvey*, and others; thence to Vasse and Swan rivers, *Drummond*, 1*st coll. n.* 841; *Preiss, n.* 2195; and others; Murchison river, *Oldfield;* eastward to Bremer bay, *Maxwell.*

In some of the Western specimens (*D. corymbosa*), Lindl., the flowers are larger and the petals and labellum-lobes broader than in the Tasmanian and Victorian ones, in others quite as small or smaller.

13. **D. pauciflora,** *R. Br. Prod.* 316. A slender plant of 6 in. to 1 ft. resembling at first sight the var. *concolor* of *D. maculata.* Leaves very narrow, but not quite so fine as in *D. setacea.* Flowers 1 to 3, yellow or the petals and labellum with a purple base. Petals ovate, about ½ in. long, including the very short claws; lateral sepals not reflexed, linear but rather broader than in most species; dorsal sepal shorter. Labellum as long as or rather longer than the dorsal sepal, 3-lobed from the base, the lateral lobes not half as long as the middle lobe, with a single central raised line not reaching to half the length of the labellum.—Lindl. Gen. and Sp. Orch. 510; Reichb. f. Beitr. 13.

W. Australia. King George's Sound, *R. Brown*, who gathered several specimens, but I have not seen it in any other collection. Bauer's Port Jackson plant of the same name, referred to by Reichb. f., l.c., is probably the yellow variety of *D. maculata*.

31. ORTHOCERAS, R. Br.

Dorsal sepal erect, incurved, hood-shaped; lateral sepals narrow-linear, long and erect; petals short, erect, narrow. Labellum 3-lobed, the middle lobe larger and contracted at the base, a thick callus on the disk between the lateral lobes. Column very short, with lateral erect lobes not connected behind the anther. Anther erect or slightly incurved, tapering to the end, the 2-valved cells occupying the whole inner face. Pollen-masses granular or mealy. Rostellum very short.—Terrestrial glabrous herb, with an ovoid tuber, few narrow leaves, and several sessile flowers.

The genus is limited to the single Australian species, found also in New Zealand.

1. **O. strictum,** *R. Br. Prod.* 317. Stem rigid, erect, 1 to 1½ ft. high. Leaves several near the base, linear, 3 to 6 in. long, or one or two outer ones short and lanceolate, and 2 or 3 long sheaths with short erect laminæ above the leaves. Flowers distant, erect, in an interrupted spike, the subtending bracts sheathing, acute, sometimes scarcely exceeding the ovary, sometimes much longer than the dorsal sepal. Dorsal sepal broad and very concave, much incurved, acute or obtuse, about ½ in. long, greenish or white outside, brown purple or yellowish inside; lateral sepals antenna-like, slightly clavate, ¾ to nearly 1 in. long; petals thin, not 2 lines long, truncate notched or toothed at the end. Labellum 3 to 4 lines long, the lateral lobes broad and oblique, the middle lobe twice as long and ovate, the callus between the lateral lobes broad and prominent, but variable in shape. Lateral lobes of the column often nearly as long as the petals.—Lindl. Gen. and Sp. Orch. 512; Reichb. f. Beitr. 14; *O. Solandri*, Lindl. l.c.; *Diuris Novæ-Zelandiæ*, A. Rich. Fl. Novæ Zel. 163, t. 25.

N. S. Wales. Port Jackson to the Blue Mountains, *R. Brown*, *A. and R. Cunningham*, *Woolls*, and others, but said to be very rare; towards Illawarra, *A. Cunningham*.

Victoria. Dandenong, *F. Mueller;* Glenelg valley, *Robertson.*

S. Australia. St. Vincent's Gulf, *F. Muell*, *Behr*; Spencer's Gulf, *Wilhelmi*.

The New Zealand plant does not appear to me to differ in the slightest particular.

32. CRYPTOSTYLIS, R. Br.

Flowers reversed. Sepals and petals nearly similar, narrow linear-lanceolate, thin and membranous, convolute and appearing subulate when the flower opens, the petals usually smaller than the sepals. Labellum longer and thicker than the sepals, undivided, sessile with a broad base enclosing the column, more or less contracted above the column, extended into a narrow or broad convex or concave lamina. Column exceedingly short, the wings forming broad distinct auricles or connected into a membrane behind the anther, the margin toothed or jagged. Anther erect against the back of the stigma or bent forward over it, 2-celled, usually biconvex on the back, obtuse or shortly acuminate. Pollen-masses farinaceous.—Terrestrial glabrous herbs, with a short rhizome and thick fibres. Leaves few, radical, on rigid petioles, ovate to lanceolate. Flowering stems leafless, bearing 2 or more erect sheathing scales or empty bracts. Flowers rather large, green with a brown red or purple labellum, several in a terminal raceme. Bracts acute, membranous.

Besides the Australian species, which are all endemic, there are two or three in E. India and the Archipelago. The genus is nearly allied to *Calochilus*, differing in the rhizome and foliage, and in the labellum not fringed.

Labellum lamina convex with reflexed sides when fully out, the disk with a double raised line dilated into 2 prominent thick lobes near the end. Anther bent forward, with the column wings connected behind it.
  Leaves oblong or lanceolate. Eastern species . . . . 1. *C. longifolia.*
  Leaves ovate or broadly oblong. Western species . . . . 2. *C. ovata.*
Labellum lamina concave. Anther nearly erect, the column-wings or auricles not connected behind it.
  Labellum lamina very broad, with a broad membranous longitudinal vertical plate . . . . 3. *C. erecta.*
  Labellum lamina linear, without any longitudinal plate or raised line . . . . 4. *C. leptochila.*

1. **C. longifolia,** *R. Br. Prod.* 317. Leaves usually 2 or 3, on rather rigid petioles of 1 to 3 in. the lamina oblong or lanceolate, 2 to 4 in. long and erect. Scape 1 to 2 ft. high, bearing 2 or more distant sheathing scales of about $\frac{1}{2}$ in. Flowers usually 3 to 6, rather distant, nearly sessile within membranous acute bracts, the ovary narrow, longer than the bract, much recurved at the end after flowering. Sepals and petals very thin and membranous, their lanceolate shape only seen in the bud for the margins are rolled inwards the moment the flowers expand; dorsal sepal usually about $\frac{3}{4}$ in. long, the lateral sepals rather longer, the petals shorter, but all otherwise similar and acute. Labellum usually about 1 in. long, scarcely contracted above the short broad base which completely encloses the column and has no internal raised lines, the lamina broadly oblong or ovate-oblong, the sides convolute in the bud but reflexed when the flower opens exposing two raised lines forming a double keel along the centre, which commences immediately above the broad base and expands a little below the end into 2 thick prominent rounded auricles or lobes, and tapers beyond them almost to the end of the lamina which is obtuse or shortly acuminate; there are also in the middle of the labellum 1 or 2 additional short raised lines parallel to the two principal ones, the disk veined but not so strongly as in *C. erecta.* Anther bent forward, almost hood-shaped, with a small dorsal recurved point near the apex. Column-wing broad short and irregularly lobed toothed or jagged, shortly continuous behind the anther.—Lindl. Gen. and Sp. Orch. 445; Bauer, Ill. Orch. Gen. t. 17, 18; Endl. Iconogr. t. 17; Hook. f. Fl. Tasm. ii. 9, t. 108 A; *Malaxis subulata*, Labill. Pl. Nov. Holl. ii. 62, t. 212; *Cryptostylis subulata*, Reichb. f. Beitr. 15.

Queensland. Glasshouse Mountains, *F. Mueller.*
N. S. Wales. Port Jackson, *R. Brown, Harvey, Woolls;* Hastings river, *Beckler;* Tweed river, *C. Moore;* M'Leay river, *Fitzgerald;* Pennant Hills, *A. Cunningham;* Ashfield, *Ramsay;* Maneroo, *Mrs. Calvert.*
Victoria. Portland, Dandenong, Bunip Creek, Snowy river, &c., *F. Mueller.*
Tasmania. Port Dalrymple, *R. Brown;* Circular Head, *Gunn;* Port Sorell, Garretts Sugarloaf, Meander river, *Archer;* Oyster Cove, *Milligan;* Southport, *C. Stuart;* N. Huon river, *Oldfield.*

Labillardière having placed the species in a genus with which it has no connexion, there seems to be no sufficient reason for substituting his specific name founded upon the apparent not upon the real shape of the sepals and petals, for the one so generally adopted.

2. **C. ovata,** *R. Br. Prod.* 317. Habit inflorescence and flowers of *C. longifolia,* from which this species is scarcely to be distinguished except by the larger broader leaves more strongly ribbed, resembling those of *Alisma plantago,* varying from ovate to oblong and 3 to 6 in. long. Labellum the same as in *C. longifolia,* or the raised central lines rather closer together, and the lobes near the end perhaps more prominent and narrower, not different however from those represented in the figure of *C. longifolia* in the Fl. Tasm.—Lindl. Gen. and Sp. Orch. 445; Reichb. f. Beitr. 15; Endl. in Pl. Preiss. ii. 11.

W. Australia. King George's Sound, *R. Brown;* Stirling Range, *Maxwell, F. Mueller;* towards Swan river, *Drummond, A. C. Gregory;* Hampden, *Clarke.*

3. **C. erecta,** *R. Br. Prod.* 317. Closely resembling *C. longifolia* in habit leaves, inflorescence and perianth, but usually rather smaller in stature, the ovary less recurved at the end and the labellum and column very different. Labellum much contracted immediately above the broad base enclosing the column, the lamina very broadly ovate and obtuse, deeply concave, membranous with a few dark veins very conspicuous in the dried state, and a vertically broad longitudinal membranous and veined plate along the centre occupying a great part of the length of the lamina. Column-wing expanded into 2 broad denticulate or jagged lateral lobes, not connected behind the anther, which is more erect than in *C. longifolia* and shortly rostrate.—Lindl. Gen. and Sp. Orch. 446; Reichb. f. Beitr. 15; F. Muell. Fragm. vii. 115.

N. S. Wales. Port Jackson to the Blue Mountains, *R. Brown, Caley, A. Cunningham, Woolls, Daintree, Miss Atkinson, Fitzgerald.*

4. **C. leptochila,** *F. Muell. Herb.* More slender than the other species, our specimens 6 in. to 1 ft. high, with smaller flowers in a rather dense raceme. Leaf ovate on a short petiole. Perianth of *C. longifolia* but shorter. Labellum with the short broad base of the other species, abruptly contracted above it into an oblong-linear rather thick channelled lamina of ½ in.; a thick longitudinal raised line or plate along the centre of the broad base (which is veined only in the other species) ceases at the contraction, and the lamina has only two rows or interrupted lines of scarcely prominent calli or dark thick spots (at least in the specimen examined). Column-wing with 2 broad denticulate lobes interrupted behind the nearly erect anther as in *C. erecta.*

N. S. Wales. Springwood, *R. Cunningham*, a single specimen; Kurrajong, *Mrs. Calvert*, two small specimens.

33. PRASOPHYLLUM, R. Br.

Flowers reversed. Dorsal sepal lanceolate or broad, concave, usually arched over the column and sometimes adnate to it at the base; lateral sepals as long or longer, lanceolate or linear, free or more or less united; petals usually shorter but sometimes as long as the sepals, lanceolate or linear. Labellum sessile or on a short claw or claw-like basal appendage to the column, ovate oblong or lanceolate, undivided, the margins undulate-crisped or entire, usually erect and concave at the base, recurved towards the end, the disk with an adnate plate sometimes broad with free margins, sometimes reduced to a central longitudinal thickening. Column very short, not winged, but with 2 lateral erect appendages, usually adnate on one side to the stigma. Anther 2-celled, erect behind the rostellum, which is often produced beyond it, but sometimes shorter than the anther. Pollen-masses granular, attached to a linear caudicle.—Terrestrial glabrous herbs with globular or ovoid underground tubers. Leaf solitary, usually with a long sheath, the lamina terete and sometimes long, shortly opened near the stem, or the whole leaf reduced to a small sheath with a short erect point. Flowers variously coloured, often pale or greenish yellow, several or numerous in a terminal spike, usually abruptly bent above the ovary so as to appear very spreading or reflexed.

Besides the Australian species there are three in New Zealand, one of them apparently identical with an Australian one. The habit of the genus and many of its characters are those of *Microtis*, from which it differs in the reversed flowers, the more developed lateral appendages to the column (sometimes described as staminodia), and by the elongated caudicle of the pollen-masses which in *Microtis* is very small or obsolete.

The lateral sepals in two or three instances have been described as 2-dentate. I have never found them so, and believe the error to have arisen either from a slip of the pen referring to lateral sepals instead of the lateral appendages of the column, or the writer to have meant the lip composed of the two combined lateral sepals.

SECT. 1. **Euprasophyllum.**—*Labellum sessile at the base of the column.*

Flowers mostly above 3 lines long. Ovary elongated, narrow. Tall plants. Lateral sepals connate at least in the middle.
- Labellum with a broad gibbous thickish base, the inner plate broad, prominent, scarcely reaching beyond the bind. Leaf-lamina long . . . . 1. *P. australe.*
- Labellum with a rather narrow but obtuse base, the inner plate broad, but commencing only about the middle. Leaf-lamina very short and erect . . . . 2. *P. flavum.*
- Labellum slightly contracted at the base, the inner plate covering the greater part of the surface with its broad detached margins. Leaf-lamina long . . . . 3. *P. elatum.*

Flowers mostly above 3 lines long. Ovary obovoid or oblong.
- Lateral sepals connate, at least in the middle.
  - Labellum abruptly bent down in the middle. Eastern species . . . . 4. *P. brevilabre.*
  - Labellum gradually curved. Western species . . . 5. *P. hians.*

Lateral sepals free or very shortly connate at the base.
Labellum obtuse at the base or slightly contracted but not gibbous. Eastern species.
Labellum with the recurved end ovate or oblong, undulate, much broader than the thickened inner plate 6. *P. patens.*
Labellum with the recurved end linear-lanceolate, not much broader than the thick inner plate . . . . 7. *P. fuscum.*
Labellum with a gibbous base protruding between the lateral sepals. Western species . . . . 8. *P. cyphochilum.*
Flowers under 3 lines long. Ovary narrow-oblong. Slender Western species with numerous flowers.
Sepals and petals linear, all nearly equal . . . . 9. *P. ovale.*
Lateral sepals subulate, twice as long as the petals . . . 10. *P. macrostachyum.*

SECT. 2. **Podochilus.**—*Labellum obtuse at the base, on a short distinct horizontal claw, but continuous with it and the base of the column.*

Lateral sepals not saccate at the base, usually connate.
Inner plate of the labellum deeply and copiously fringed on the margin. Tall Western species . . . . 11. *P. Fimbria.*
Inner plate of the labellum entire. Small Eastern species . 12. *P. striatum.*
Lateral sepals saccate at the base. Western species.
Lateral sepals adnate at the base to the basal projection of the column, otherwise free . . . . 13. *P. parvifolium.*
Lateral sepals connate, the saccate base enclosing the basal projection of the column but free from it.
Flowers under 2½ lines. Sepals and petals of nearly uniform colour and mostly obtuse . . . . 14. *P. gibbosum.*
Flowers 3 lines long. Lateral sepals broad and white, contrasting with the deep-coloured acute dorsal sepal and petals . . . . 15. *P. cucullatum.*

SECT. 3. **Genoplesium.**—*Labellum obtuse or contracted into a claw at the base, articulate on a horizontal claw-like basal projection of the column, and usually moveable. Stem slender, the leaf almost or quite reduced to a sheathing bract. Flowers very small.*

Labellum neither fringed nor ciliate.
Lateral sepals broad, very gibbous at the base. Labellum broad. Rostellum as long as the anther . . . . . . 16. *P. nigricans.*
Lateral sepals narrow, scarcely gibbous. Labellum narrow.
Column shorter than the anther. Rostellum very short.
Flower dark coloured, about 1 line long . . . . . 17. *P. rufum.*
Column as long as or longer than the anthers. Rostellum half as long as the anthers.
Lateral appendages of the column 2-fid or 2-dentate. Flowers pale, nearly 2 lines long . . . . . . 18. *P. brachystachyum.*
Lateral appendages of the column entire and long. Flowers dark-coloured, 1½ lines long . . . . . 19. *P. despectans.*
Labellum ciliate or fringed.
Labellum oblong, fringed in the upper half with long hairs.
Lateral sepals gibbous at the base.
Lateral sepals 2 to 2½ lines. Labellum equal or dilated towards the end . . . . 20. *P. fimbriatum.*
Lateral sepals nearly 3 lines. Labellum tapering towards the end . . . . 21. *P. Archeri.*
Labellum broadly ovate, fringed or ciliate with long hairs.
Lateral sepals not gibbous at the base . . . . 22. *P. intricatum.*
Labellum ovate-oblong, bordered with very short cilia. Lateral sepals scarcely gibbous at the base . . . . . 23. *P. Woollsii.*

SECT. 1. EUPRASOPHYLLUM.—Labellum sessile at the base of the column.

1. **P. australe,** *R. Br. Prod.* 318. Stems often 2 to 3 ft. high, the leaf-sheath occupying about half its length or even more, the lamina much shorter than the spike. Flowers "striped with brown and yellowish green" in a spike of 3 to 6 in. or rather more. Ovary elongated. Sepals and petals all acutely acuminate, about 4 lines long, the dorsal sepal broad and concave, the lateral sepals united in the middle, sometimes free at the base and the points always free; petals broader than in *P. elatum.* Labellum sessile, with a broad gibbous somewhat fleshy erect base, abruptly recurved and reflexed in the middle of its length, the margins of the recurved part undulate, the inner plate very prominent, entire or minutely crenulate, scarcely reaching beyond the bend. Anther obtuse, shorter than the rostellum. Lateral appendages of the column adnate to the stigma at the base, lanceolate-falcate, obtuse, rather longer than the rostellum, dilated and sometimes thickened on the outer margin.—Reichb. f. Beitr. 17; *P. lutescens,* Lindl. Gen. and Sp. Orch. 514; Hook. f. Fl. Tasm. ii. 10, t. 110 B.

Victoria. Wet places or in water, Portland, *Robertson, Allitt.*

Tasmania. Adventure bay, *Nelson* (*Herb. R. Brown*); Rocky Cape, *Gunn;* Flinders island and Oyster Cove, *Milligan;* Southport, *C. Stuart.*

2. **P. flavum,** *R. Br. Prod.* 318. Stems stout, attaining 2 to 3 ft. the whole plant drying very dark. Leaf-sheath rather loose, the lamina rarely 1 in. long, erect, concave with a very short terete point. Bracts rather broad. Flower of a yellowish green. Ovary elongated. Sepals lanceolate, acute, nearly 4 lines long, the lateral ones connate the greater part of their length; petals narrower, but nearly as long. Labellum ovate-oblong, sessile, almost gibbous concave and erect at the base, recurved towards the end, broader in the middle; the inner plate commencing from about the middle, the oblique margins joining in the centre and dying away before the end of the lamina. Lateral appendages of the column adnate on one side at the base, short and broad, of a thicker texture than those of most species, 2-dentate or 2-lobed at the end.—Lindl. Gen. and Sp. Orch. 514; Hook. f. Fl. Tasm. ii. 11, t. 109 A; Reichb. f. Beitr. 17.

N. S. Wales. Port Jackson, *R. Brown;* Blue Mountains, *R. Cunningham;* Berrima, *Miss Calvert;* towards Durval, *Leichhardt;* New England, *C. Stuart.*

Tasmania. Cheshunt, *Archer;* Oyster Cove, *Milligan;* Huon river, *Oldfield;* Southport, *C. Stuart.*

3. **P. elatum,** *R. Br. Prod.* 318. Stems from under 2 ft. to above 3 ft. high, the long leaf-sheath covering a great part of it, the lamina often long but very variable. Flowers greenish, nearly sessile in a raceme or spike of 4 to 8 in. or even longer. Ovary elongated. Dorsal sepal and petals lanceolate, very acute, 4 to 5 lines long, lateral sepals often rather longer, connate at least from the middle upwards, the points sometimes also free. Labellum about as long as the petals, sessile but not gibbous at the base, ovate-oblong, the margins undulate, the inner

plate occupying the greater part of the surface, its free margins broad, converging and united a little below the apex." Lateral appendages of the very short column free, linear-falcate, obtuse, as long as the anther, the outer margin thickened near the base into an oblong gland-like prominence. Anther nearly 2 lines long, the acuminate rostellum about as long.—Lindl. Gen. and Sp. Orch. 515; Reichb. f. Beitr. 16; *P. australe*, Lindl. l.c. 514; Hook. f. Fl. Tasm. ii. 10, not of R. Br.

N. S. Wales. Port Jackson, *R. Brown*, *Sieber*, *n.* 167, *Woolls*, and others.
Victoria. Port Phillip, *F. Mueller;* E. Gipps' Land, *F. Mueller*, *Walter*.
Tasmania. Circular Head and Rocky Cape, in poor soil, *Gunn*.
S. Australia. Lofty Range and Guichen bay, *F. Mueller*.
W. Australia. King George's Sound, *R. Brown*, *F. Mueller*, and others; Swan river, *Drummond*, 1*st coll.*

These Western species vary much in the size and depth of colour of the flowers and in the precise shape of the labellum, which is sometimes very broad, sometimes much narrowed towards the end. They would include *P. giganteum* and *P. macrotys*, Lindl. Swan Riv. App. 54, Gen. and Sp. Orch. 515; *P. Brownii*, Reichb. f. Beitr. 16; *P. Drummondii*, Reichb. f. l.c. 58 and 60, and probably the *P. macrotys*, Endl. in Pl. Preiss. ii. 12, which I have not seen. The long almost transparent lateral appendages of the column with a thickened or glandular margin on the outer base, and the broad plate of the labellum are constant in all the forms.

4. **P. brevilabre,** *Hook. f. Fl. Tasm.* ii. 11, t. 110 A. Allied to *P. patens* in aspect, in the size of the flowers and in the ovary, but with the perianth more like that of *P. elatum*. Stems mostly under 1 ft. high, the leaf-sheath broad and loose, the lamina usually but not always rather short. Ovary obovoid or oblong, narrower than in *P. patens*, but much shorter than in *P. elatum*. Dorsal sepal under 4 lines long; petals rather shorter narrow and acuminate; lateral sepals longer and broader than the dorsal one, united almost to the end. Labellum sessile, the erect part rather narrow, the oblong lamina of the length of the erect part, but abruptly reflexed against it so as to give a very short appearance to the whole labellum; the margins much undulate; the inner plate commencing from the base, nearly as broad as the erect part, and continued very shortly on to the reflexed lamina. Lateral appendages of the column adnate to the stigma rather high up, entire, the outer margin without any appendage.

Queensland. Archer's station, Moreton bay, *Leichhardt*.
N. S. Wales. Hastings river, *Beckler*.
Victoria. Moe swamps, *F. Mueller;* E. Gipps' Land, *Walter*.
Tasmania. Rocky Cape, *Gunn;* Southport, *C. Stuart*.

5. **P. hians,** *Reichb. f. Beitr.* 59, 61. Stem 6 in. to 1 ft. high, the leaf-sheath loose. Flowers "whitish" or "of a reddish green" rather larger than in *P. patens*, in a rather dense spike. Sepals about 4 lines long, the lateral ones united almost to the apex, very thin and whitish near their line of junction; petals at least as long as the sepals and more dilated than in most other species. Labellum sessile, rather broad at the base but not gibbous, recurved above the middle, the margins undulate, the inner plate much narrower forming a longitudinal central thickening, ending at the bend or a little beyond it

in a thick papillose-fringed callus. Column rather long, the lateral appendages adnate on one side, falcate, acute, entire. Anther short.

W. Australia. King George's Sound, *F. Mueller*; Harvey river, *Oldfield;* also in *Drummond's collection*, but our specimens too much injured for positive identification.

6. **P. patens,** *R. Br. Prod.* 318. Stems usually tall but varying from 1 to 3 ft., the length of the leaf also very variable. Flowers usually smaller than in *P. elatum*, larger than in *P. fuscum*, of a yellowish green, the labellum bordered with white, but neither the size nor the shade of colour constant. Ovary obovoid or shortly oblong. Sepals lanceolate, acute or obtuse, scarcely 4 lines long, the lateral ones quite free, the petals of the same length but more obtuse and much more petal-like in consistence. Labellum as long as the petals, sessile but not gibbous at the base, the erect part not very broad, as long as the recurved or reflexed portion, which varies from ovate to oblong, or almost lanceolate, but always shows a considerable breadth of undulate margin, the inner plate not very prominent, much narrower than the labellum and scarcely extending beyond the bend or rarely reaching nearly to the end. Lateral appendages of the column adnate at the base on one side to the stigma, nearly as long as the rostellum, the outer margin without the glandular prominence of *P. elatum*, and rarely with a scarcely prominent tooth. Anther rather shorter than the rostellum.—Lindl. Gen and Sp. Orch. 513; Hook. f. Fl. Tasm. ii. 11, t. 111; Reichb. f. Beitr. 19.

Queensland. Armidale, *Perrott.*

N. S. Wales. Port Jackson, *R. Brown*, *Woolls*, and others; in the interior, Nangas, *M'Arthur;* Mudgee, *Taylor;* New England, *C. Stuart;* southward to Twofold bay and Murray river, *F. Mueller.*

Victoria. Wendu Vale, *Robertson;* mouth of the Glenelg, *Allitt;* Port Phillip, *Gunn;* Melbourne, Cobras mountains, Wimmera, *F. Mueller.*

Tasmania. Abundant in moist ground throughout the island, *J. D. Hooker* and others.

S. Australia. From Mount Gambier to St. Vincent's Gulf, *F. Mueller* and many others.

P. truncatum, Lindl. Gen. and Sp. Orch. 513, Hook. f. Fl. Tasm. ii. 12, t. 109, distinguished by the slightly prominent lateral tooth of the column appendages and by the inner plate of the labellum terminating abruptly a little beyond the bend and not dying off on the lamina, does not appear to me to be separable even as a marked variety, for differences in these respects may be observed almost from specimen to specimen.

7. **P. fuscum,** *R. Br. Prod.* 318. A very variable species nearly allied to some forms of *P. patens*, usually rather smaller with smaller flowers, and the sepals and petals narrow and darker coloured, but the extreme forms of the two species only to be distinguished by the labellum. Spike short or long, dense or interrupted. Ovary obovoid or shortly oblong as in *P. patens.* Flowers usually drying black or very dark, rarely pale-coloured. Dorsal sepal lanceolate, concave, acuminate, about 3 lines long in the typical form, but varying from 2½ to 4 lines; lateral sepals nearly as long, narrow, free or very shortly connate at the base; petals rather shorter and linear. Labellum nearly as long as the sepals, narrower than in *P. patens*, the erect part very concave, but neither gibbous at the base nor clawed, the spreading end very

narrow lanceolate; the inner plate not distinct at the base, showing slightly raised margins towards the end of the erect part and thence continued to near the end of the lamina in a thick often papillose centre, leaving a very narrow plain margin or occupying the whole breadth; the breadth however of the erect part and its contraction abrupt or gradual into the narrow lamina very variable. Lateral appendages of the column obliquely ovate-oblong, entire or irregularly 2-dentate, the outer margin often dilated at the base. Anther usually as long as the rostellum, obtuse or shortly and obtusely acuminate.—Lindl. Gen. and Sp. Orch. 516; Hook. f. Fl. Tasm. ii. 12, t. 112; Reichb. f. Beitr. 18; *P. alpinum*, R. Br. Prod. 318, Lindl. l.c. 515; Hook. f. l.c. ii. 12, t. 112; Reichb. f. Beitr. 19; *P. affine* and *P. rostratum*, Lindl. l.c. 516.

Queensland. Archer's station, Moreton bay, *Leichhardt.*

N. S. Wales. Port Jackson, *R. Brown;* Bathurst, *Woolls;* Emu plains, *A. Cunningham;* New England, *C. Stuart.*

Victoria. Portland, *Robertson;* Station Peak, Grampians, Munyong mountains, Gipps' Land, *F. Mueller;* Murray river, *Dallachy.*

Tasmania. Table mountain (Mount Wellington), *R. Brown;* Rocky Cape and Hampshire hills, *Milligan.*

S. Australia. Pine Forest, *Behr;* from various localities on St. Vincent's Gulf, *F. Mueller.*

Var. *grandiflorum*, flowers 4 to 5 lines long, but with the narrow petals and sepals and peculiar labellum end of *P. fuscum.*—Mudgee, *Woolls*; Pine Forest, *Behr.*

Botanists are generally agreed in distinguishing two species, but not as to the characters assigned to them, derived for the most part from the examination of specimens few in number or from few localities. With several hundred specimens before me from various stations I have been unable to sort them into marked varieties, although I would not deny that their study in a living state may point out more constant distinctions which have escaped me. Brown had only a rather large-flowered Port Jackson plant as *P. fuscum* and a small-flowered alpine Tasmanian plant as *P. alpinum*, without any of the very numerous intermediates. He describes the lateral sepals of *P. fuscum* as cohering at the base, those of *P. alpinum* as free; Archer (in Hook. f. l. c. ii. 13) reverses these characters. I have found them almost constantly free, though closely overlapping each other at the base, rarely shortly connate, and never connate in the middle as in *P. brevilabre.* Gunn and Hooker find that this partial connexion affords no constant character. Gunn observes that *P. alpinum* has a strong smell of Hyacinths, and *P. fuscum* is inodorous. But the Mediterranean *Orchis coriophora*, for instance, has three varieties, a sweet-smelling, a nauseous-smelling, and an inodorous one, which in the fresh state I was unable otherwise to distinguish. The colour of the flower of a dusky brown-green in *P. fuscum*, of a light green in *P. alpinum*, appears also to be very inconstant.

8. **P. cyphochilum,** *Benth.* Stems 1 to 2 ft. high. Leaf-lamina slender, sometimes short as in *P. brevifolium,* sometimes elongated. Flowers small, pedicellate, "white or pale-yellow." Ovary oblong-turbinate. Sepals about 2 lines long, the lateral ones quite free; petals rather shorter. Labellum sessile, gibbous at the base, forming a short pouch usually protruding between the lateral sepals, the erect part very concave, not broad, tapering upwards, the recurved portion about half its length and much undulate; the inner plate reduced to parallel not much raised lines along the centre. Lateral appendages of the column almost free, linear-falcate, quite entire, scarcely above half as long as the rather long rostellum. Anther shorter than the rostellum, shortly and obtusely acuminate.

W. Australia. Darling range, *Collie;* Swan river, *Oldfield;* King George's Sound and adjoining districts, *Oldfield, F. Mueller, Muir.*

9. **P. ovale,** *Lindl. Swan Riv. App.* 54, *Gen. and Sp. Orch.* 516. Stems rather slender, above 1 ft. high, the leaf-sheath and lamina long. Flowers small, in a spike of 3 to 6 in. not dense. Ovary narrow. Sepals about 2½ lines or sometimes nearly 3 lines long in the typical form, narrow, the lateral ones gibbous and very shortly united at the base, otherwise free, the petals scarcely shorter. Labellum sessile, the erect portion about 2 lines long, concave, of nearly equal breadth or slightly contracted at the base, the reflexed portion about half as long, rounded, undulate; inner plate much narrower than the labellum, ending at the bend in a thick papillose almost fringed callus. Lateral appendages to the column falcate, entire, as long as the rostellum. Anther shorter.

W. Australia. Swan river, *Drummond,* 1*st coll.*

Var. *triglochin,* Reichb. f. Beitr. 60. Leaf-lamina short. Flowers scarcely 2 lines long, the lateral sepals quite free from the base, the inner plate of the labellum ending less abruptly.—W. Australia, *Drummond.*

10. **P. macrostachyum,** *R. Br. Prod.* 318. A slender plant usually not so tall as *P. ovale,* but sometimes much resembling it or even taller, the flowers smaller, quite green, shortly pedicellate and usually rather distant in a long spike. Ovary narrow-turbinate or oblong. Lateral sepals about 2 lines long, lanceolate-subulate, acute, broad and shortly united at the base (or sometimes quite free ?); dorsal sepal as long or rather shorter; petals considerably shorter, lanceolate, acute. Labellum sessile, shorter than the sepals and narrower than in *P. ovale,* the erect part concave, the reflexed part as long, ovate, almost acuminate; the inner plate nearly as broad as the erect part, forming 2 calli at the bend and shortly continued along the centre of the reflexed part.—Reichb. f. Beitr. 17; *P. gracile,* Lindl. Swan Riv. App. 54, Gen. and Sp. Orch. 516; Endl. in Pl. Preiss. ii. 13; *P. nigricans,* Endl. in Pl. Preiss. ii. 12, not of R. Br. (*Reichb. f.*)

W. Australia. King George's Sound, *R. Brown, F. Mueller;* Swan river, *Drummond,* 1*st coll.*; Gordon river, *Oldfield.*

SECT. 2. PODOCHILUS.—Labellum obtuse at the base, on a short distinct horizontal claw, but continuous with it and with the base of the column.

11. **P. Fimbria,** *Reichb. f. Beitr.* 60. A tall species with the habit and size of flowers of *P. elatum,* but with a very different labellum. Ovary elongated. Flowers "pale violet, fragrant." Sepals about 5 lines long, the lateral ones free at the base but often connate above the middle; petals not nearly so long, linear. Labellum as long as the sepals, truncate at the base, on a distinct narrow horizontal claw of about ½ line, the lamina broadly oblong, slightly contracted at the bend in the middle, the upper part broad with fringed or crisped margins, the inner plate broad, commencing with 2 broad callosities at the base and ending above the middle in a broad free densely fringed margin, and within it in the centre of the labellum a more or less conspicuous

second plate with scarcely prominent entire margins. Lateral appendages of the column falcate, as long as the rostellum, with a small tooth on the outer margin.—*P. giganteum*, Endl. in Pl. Preiss. ii. 12, not of Lindl.

W. Australia. King George's Sound, *F. Mueller;* Kalgan river, *Harvey;* Swan river, *Drummond;* Gordon, Harvey, Swan, and Murchison rivers, *Oldfield;* Swan river, *Preiss, n.* 2215.

12. **P. striatum,** *R. Br. Prod.* 318. Stems scarcely exceeding 6 in., the leaf-sheath rather long, the lamina subulate. Flowers about the size of those of *P. patens,* in a raceme of 1 to 2 in. Ovary oblong-cylindrical. Lateral sepals united to near the end, narrow, acuminate, about 3 lines long; dorsal sepal rather shorter, narrow, concave; petals nearly as long as the sepals, lanceolate. Labellum fully 2 lines long, inserted on a narrow horizontal claw but continuous with it, oval-oblong, concave, the lower half erect, the upper half recurved, with undulate margins; the inner plate broad in the lower half, reduced upwards to a thick double raised line reaching almost to the end. Lateral appendages of the column narrow-linear, almost as long as the long slender-pointed rostellum. Anther very short, not acuminate; caudicle long and filiform.—Lindl. Gen. and Sp. Orch. 514; Reichb. f. Beitr. 18.

N. S. Wales. Port Jackson to the Blue Mountains, *R. Brown, A. Cunningham, F. Mueller,* and others.

13. **P. parvifolium,** *Lindl. Swan Riv. App.* 54, *Gen. and Sp. Orch.* 517. Stems slender, 9 in. to above 1 ft. long, the leaf above the middle of the stem, with a short slender lamina. Flowers resembling those of *P. striatum,* in a loose raceme of 2 to 3 in. Ovary narrow-oblong. Sepals about 3½ lines long, the lateral ones dilated at the base and adnate to the basal projection of the column, forming a short pouch, but otherwise free, the dorsal one lanceolate, concave; petals narrower and rather shorter. Labellum stipitate at the end of the basal projection, but apparently continuous with it, the total length of the projection and claw nearly 2 lines, the lamina lanceolate, concave, recurved, as long as the sepals; the inner plate nearly as broad as the lamina, and ending about the middle of its length. Lateral appendages of the column 1½ lines long, entire. Rostellum as long as the lateral appendages. Anther short. Caudicle of the pollen-masses long and linear.—Endl. in Pl. Preiss. ii. 13.

W. Australia. Swan river, *Drummond,* 1*st coll.*, *Preiss. n.* 2220; Cape Leuwin, *Cŏllie.*

14. **P. gibbosum,** *R. Br. Prod.* 318. Stems from under 6 in. to near 1 ft. long, the leaf-sheath long, with a narrow linear lamina. Spike rather dense. Lateral sepals united into a lip of about 2 lines, obtuse, very concave, projected at the base into a sack or short and very obtuse spur; dorsal sepal as long, obtuse, nearly similar in texture to the lateral ones; petals nearly as long, oblong-linear, tapering into a claw. Labellum nearly as long as the sepals, linear-oblong or linear-

cuneate, on a short horizontal claw, then erect to the middle, the upper part recurved, slightly undulate, truncate at the end; the inner plate not very conspicuous, not much narrower than the labellum at the lower end, and gradually disappearing long before the upper end. Lateral appendages of the column broadly falcate, obtuse, not half as long as the petals but longer than the rostellum and anthers, entire or with a small tooth on the outer margin.—Lindl. Gen. and Sp. Orch. 517; Reichb. f. Beitr. 18.

W. Australia. King George's Sound, *R. Brown*, also in *Drummond's collections, n.* 506.

15. **P. cucullatum,** *Reichb. f. Beitr.* 59. Very near *P. gibbosum* and probably a variety only. Apparently a smaller plant with a dense spike of 1 to 1½ in. Flowers about 3 lines long, similar in structure to those of *P. gibbosum*, but the broader white lateral sepals with their saccate base contrasting with the darker coloured narrow-lanceolate acuminate dorsal sepal; petals acute, almost as long as the dorsal sepal. Labellum like that of *P. gibbosum*, but the lamina ending in a much broader dilatation, and the inner plate or thickened centre scarcely prominent, ending before the dilatation. Lateral appendages of the column falcate, obtuse, about 1 line long. Rostellum with a long point. Anther short. Caudicle very long.—*P. gibbosum*, Endl. in Pl. Preiss. ii. 13.

W. Australia, *Drummond, n.* 443; *Preiss, n.* 2211; Gardner river, *Maxwell.*

I have great doubts whether this be really distinct from *P. gibbosum*, some of Brown's specimens look very much like it.

SECT. 3. GENOPLESIUM.—Labellum obtuse or contracted into a claw at the base, articulate on a horizontal claw-like basal projection of the column. Stem slender, the leaf almost or quite reduced to a sheathing bract. Flowers very small.

The species of this section have all very nearly the same habit, and are very closely allied to each other. They are distinguished by small differences in the shape and proportion of the parts of the flowers, some of which may prove to characterize varieties rather than species, or even to be individual, although mostly verified on several specimens. The study of living specimens, may therefore cause considerable modification in their circumscription.

16. **P. nigricans,** *R. Br. Prod.* 319. Stem very slender, under 6 in. high, with a single leaf or rather leafy sheathing bract very near the spike, ½ to ¾ in. long, the very short point or lamina erect. Flowers very small, in a spike of ½ to 1 in., usually dense, and drying of a dark purple or almost black. Ovary ovoid-globular, very oblique at the top. Sepals broader than in the following species, with minute points usually tipped with a small gland, the dorsal sepal about 1 line long, broadly hood-shaped, the lateral ones broadly lanceolate, 1½ to 1¾ lines long, often greener than the rest of the flower, their broad base either united under or adhering to the basal projection of the column forming a basal gibbosity but otherwise free; petals about 2 lines long, triangular-lanceolate, acute. Labellum articulate at the end of the claw-like basal projection of the column and said to be moveable, ovate or ovate-oblong, about 1 line long, almost acute, slightly contracted at the base; the

inner plate thick, occupying about ⅔ of the breadth and extending to the end. Lateral appendages of the column almost as long as the petals, acutely bifid at the end. Rostellum rather long. Anther shorter, with a very short fine inflexed point difficult to see and sometimes wanting.—Lindl. Gen. and Sp. Orch. 513; Reichb. f. Beitr. 19.

N. S. Wales. Port Jackson, *Woolls*.
Tasmania. Oyster Cove, *Milligan; Southport, C. Stuart.*
S. Australia. Port Lincoln, *R. Brown*.

17. **P. rufum,** *R. Br. Prod.* 319. Stem slender, 6 to 8 in. high, the leaf reduced to a sheathing bract near the spike ½ to ¾ in. long, the lamina short erect and subulate. Flowers the smallest in the genus, in a spike of ½ to ¾ in. Ovary oblong. Sepals tipped with a small point but (always?) without the gland of *P. nigricans*, the dorsal sepal ovate, concave, ¾ line long, the lateral ones lanceolate, quite free, 1 line long; petals lanceolate, the length of the dorsal sepal. Labellum articulate on a linear erect claw or claw-like projection of the column, said to be moveable, narrow-lanceolate or rarely broader, recurved, not ciliate; the inner plate with raised margins occupying the greater part of its breadth. Column very short below the anther, the lateral appendages about ¼ line long, 2-dentate or 2-fid. Anther mucronate, rather long. Stigma much shorter, with a minute rostellum.—Lindl. Gen. and Sp. Orch. 513; Reichb. f. Beitr. 20; *P. nudum*, Hook. f. Fl. N. Zel. i. 242, Fl. Tasm. ii. 14, t. 113 (partly).

Queensland. Rockingham bay *Dallachy*.
N. S. Wales. Port Jackson, *R. Brown*.
Victoria. Wilson's Promontory and Western Port (with the lateral appendages of the column almost entire), *F. Mueller;* Station Peak, *Fullagar*.
Tasmania. Cheshunt, *Archer*.

Var.? *intermedium*, with a broader labellum, but the small narrow sepals and petals seem to be rather those of *P. rufum* than of *P. nigricans*.—Port Jackson, *Woolls, Daintree.*

The species is also in New Zealand. *Genoplesium Baueri*, R. Br. Prod. 319, Reichb. f. Beitr. 21, of which no specimen appears to have been preserved, is founded on Bauer's drawing representing either an abnormal specimen or one in which there had been some confusion between the petals and lateral lobes of the column. It is evidently a *Prasophyllum*, and appears to me rather to belong to this species than to the following one, to which Hooker referred it. The plate of *P. nudum*, Hook. f., above quoted, represents the species correctly as to the general figure, but the analysis unfortunately must have been taken from a flower of the *P. intricatum*.

18. **P. brachystachyum,** *Lindl. Gen. and Sp. Orch.* 513. Nearly allied to *P. nigricans*, with the same habit, but readily known by its much narrower lighter coloured flowers, which are again distinguished from those of *P. rufum* chiefly by their size. Spikes usually short and dense. Lateral sepals nearly 2 lines long, shortly acuminate, but without the distinct gland of *P. nigricans*, scarcely dilated or slightly gibbous at the base; dorsal sepal shorter and broader, but not nearly so broad as in *P. nigricans*. Labellum articulate on a short basal projection of the column, lanceolate, neither ciliate nor fringed; the inner plate forming thick raised lines within the margin. Column about as

long below the anther as the anther, the lateral appendages of the column unequally 2-lobed, the inner lobe usually acuminate, the other short and broad. Rostellum shorter than the anther, but ovate and much more prominent than in *P. rufum.*—Hook. f. Fl. Tasm. ii. 13.

Tasmania. Circular Head, Rocky Cape, Hampshire hills, &c., *Gunn.*

P. nudiscapum, Hook, f. l.c., from Hobarton, seems to me to belong to this species rather than to *P. rufum*, although in some measure intermediate between the two; the flowers are however too far advanced to determine accurately the proportions of their parts.

19. **P. despectans,** *Hook. f. Fl. Tasm.* ii. 13, t. 113 A. Stems slender, leafless except a sheathing bract of about ½ in. below the spike. Flowers narrow and dark coloured as in *P. rufum*, but longer, the spike dense, ½ to 1 in. long. Lateral sepals lanceolate, acuminate, 1½ lines long, free and oblique at the base but not gibbous; dorsal sepal shorter, very concave, acuminate; petals broadly lanceolate, half as long as the lateral sepals. Labellum much shorter than the sepals, articulate on a short incurved basal projection of the column, moveable, lanceolate, recurved, channelled above. Column as long below the anther as the anther, the lateral appendages falcate, acuminate, longer than the anther, entire or with a small tooth halfway down. Rostellum shorter than the anther.

Tasmania. Sandy soil near Hobarton, *J. D. Hooker;* Cheshunt, *Archer;* Southport, *C. Stuart.*

20. **P. fimbriatum,** *R. Br. Prod.* 319. Stems very slender, about 6 to 8 in. high, with an erect leafy bract above the middle. Flowers small, drying dark, in a spike of ½ to 1 in. Ovary oblong-cylindrical. Lateral sepals nearly 2 lines long, with a very short claw and dorsal gibbosity or dilatation at the base, linear-falcate; dorsal sepal rather shorter, lanceolate, acute, concave, slightly ciliate; petals shorter, acuminate, striate. Labellum articulate at the end of a basal projection of the column of about ¼ line, linear-oblong, contracted into a short erect claw, recurved in the upper part, slightly dilated at the end and densely fringed with long fine cilia; the inner plate forming 2 raised thick lines extending along the claw and narrow part of the base of the dilatation. Lateral appendages of the column nearly as long as the anther, acutely bifid, the column below the anther very short. Anther with a fine point.—Lindl. Gen. and Sp. Orch. 517; Reichb. f. Beitr. 20.

N. S. Wales. Port Jackson, *R. Brown, A. Cunningham, Woolls.*

21. **P. Archeri,** *Hook. f. Fl. Tasm.* ii. 14, t. 113 B. Rather taller than *P. fimbriatum*, the leaf reduced to a sheathing bract with a short erect almost subulate lamina close under the inflorescence. Flowers rather larger than in the other species of the section, few together in a spike rarely 1 in. long. Ovary oblong, recurved. Lateral sepals nearly 3 lines long, of the shape and with the basal dilatation of those of *P. fimbriatum;* dorsal sepals and petals shorter, ciliate. Labellum articulate at the end of the basal projection of the column, oblong-linear and fringed with long hairs as in *P. fimbriatum*, but more tapering

towards the end. Column fully ½ line long below the anther, the lateral appendages longer than the anther, bifid, the outer lobe coloured like the petals, the inner lobe thin white and rather shorter. Rostellum shorter than the anther.

Tasmania. Cheshunt, *Archer;* Oyster Cove, *Milligan.*

22. **P. intricatum,** *C. Stuart in Herb. F. Muell.* A slender plant, with the habit and leafy bract of *P. fimbriatum.* Flowers brown or pale yellow, with the labellum purple. Lateral sepals rather more than 2 lines long, lanceolate, acute, neither oblique nor dilated at the base; dorsal sepal shorter and broader, concave, very acute; petals rather shorter and acute. Labellum articulate on the short basal projection of the column, broadly obovate, convex, recurved, fringed with shorter cilia than in *P. fimbriatum.* Lateral appendages of the column unequally bifid, the outer lobe ciliate. Rostellum shorter than the anther.

Tasmania. Southport, *C. Stuart.*—The analytical details given as those of *P. nudum* (*P. rufum*, Br.) in Hook. f. Fl. Tasm. ii. t. 113, appear to me to have been drawn from a flower of the present *P. intricatum*, but I have not met with any specimen from which it can have been taken.

23. **P. Woollsii,** *F. Muell. Fragm.* v. 100. Stem almost filiform, above 6 in. long, the leafy bract small and distant from the inflorescence. Spike ¾ in. long. Lateral sepals deeply coloured, lanceolate-subulate, slightly gibbous at the base, 1½ to 1¾ lines long; dorsal sepal paler coloured, lanceolate, acuminate and rather shorter; petals about half as long as the dorsal sepal, acutely acuminate, minutely ciliate. Labellum articulate on the short basal projection of the column, ovate-oblong, shorter than the sepals, obtuse, bordered with very short cilia. Column very short, the lateral appendages bifid, as long as the anther. Rostellum short.

N. S. Wales. Blue Mountains, *Miss Atkinson*, a single specimen in Herb. F. Mueller.

34. MICROTIS, R. Br.

Dorsal sepal erect, broad, incurved, concave; lateral sepals as long or shorter, lanceolate or oblong, spreading or recurved; petals usually narrower, incurved or spreading. Labellum sessile, oblong, obtuse truncate emarginate or 2-lobed, usually callous at the base and somewhat thickened along the centre. Column very short, nearly terete, with 2 small wings or auricles behind the stigma. Anther erect, 2-celled, the connective not produced; pollen-masses granular, without any or with a minute caudicle. Stigma obtuse or with a rostrum shorter than the anther.—Terrestrial glabrous herbs, with small globular underground tubers. Leaf solitary, the lamina elongated and terete, shortly opened out near the stem and continued in a closed sheath down the stem. Flowers small, green or whitish, usually numerous in a terminal spike, and owing to a bend immediately above the ovary the perianth is often horizontal or reflexed.

The genus extends to New Zealand, and in a single species to New Caledonia, the Indian Archipelago, and S. China. Of the six Australian species, one is common in New Zealand, another is the same as the Archipelago one, the remaining four are endemic. The habit of the genus is that of the small-flowered *Prasophylla*, the peculiar foliage is the same; it differs chiefly in the small rostellum and lateral appendages of the column, and the want of any long caudicle to the pollen-masses.

Dorsal sepal broad and very concave; lateral sepals recurved.
Labellum entire or emarginate.
Flowers above 1 line long. Labellum with a tubercle or callus on the disk near the end, the margin usually crisped . . . . 1. *M. porrifolia.*
Flowers scarcely 1 line. Labellum entire, without any callus on the disk except at the base. Eastern species . . . . 2. *M. parviflora.*
Flowers about 1½ lines. Dorsal sepal not so broad as in the foregoing. Labellum entire. Western species . . . . 3. *M. media.*
Lateral sepals revolute. Dorsal sepal acuminate, contracted at the base. Labellum broadly 2-lobed, crenate or fringed . . . . 4. *M. alba.*
Lateral sepals spreading, but not recurved.
Dorsal sepal very broad and obtuse. Labellum entire, broad, almost quadrate, not callous. Flowers about ½ line long . . 5. *M. atrata.*
Dorsal sepal not much broader than the lateral ones. Labellum contracted in the middle. Flowers about 1 line long . . . 6. *M. pulchella.*

1. **M. porrifolia,** *Spreng. Syst.* iii. 713. Usually tall and stout, often above 1 ft. high, with a long leaf and a dense spike of small green flowers, but sometimes slender with the flowers distant in a long spike. Pedicels short, subtended by small bracts. Dorsal sepal erect, broadly ovate, shortly acuminate, very concave, about 1⅓ lines long; lateral sepals spreading and recurved, about 1 line long; petals shorter, erect or spreading. Labellum sessile, as long as the lateral sepals, oblong, very obtuse retuse or shortly 2-lobed, the margin crisped or crenate, the disk with 2 oblong adnate calli or short longitudinal plates at the base, and above them the centre thickened and terminating in a tubercle or raised callus below the apex. Column very short, the auricles usually less prominent than in *M. parviflora.*—Lindl. Gen. and Sp. Orch. 395; *M. Banksii*, A. Cunn. in Bot. Mag. under n. 3377; *M. unifolia*, Reichb. f. Beitr. 62; *M. rara*, R. Br. Prod. 321; Hook. f. Fl. Tasm. ii. 24; Reichb. f. Beitr. 22; *M. pulchella*, Lindl. Gen. and Sp. Orch. 395, Hook. f. Fl. Tasm. ii. 24, t. 118, not of Br.; *M. arenaria*, Lindl. l.c. 396, Hook. f. l.c.; *M. frutetorum*, Schlecht. Linnæa, xx. 568.

Queensland. Burnett river, *F. Mueller;* Moreton bay, *C. Stuart.*
N. S. Wales. Port Jackson to the Blue Mountains, *R. Brown*, *A. Cunningham*, *Woolls*, and others; New England, *C. Stuart.*
Victoria. From Portland, *Allitt*, and Wendu Vale, *Robertson*, to Gipps' Land, *F. Mueller*, and in many other collections from various parts of the Colony.
Tasmania. Abundant throughout the island, *J. D. Hooker*, and others.
S. Australia. From the Murray to St. Vincent's Gulf, *F. Mueller*, *Behr*, and others.

The species is also in New Zealand, and appears to be the commonest form in South-Eastern Australia, not extending to the tropics. F. Mueller proposes to unite it with the two following ones under the name of *M. viridis.*

2. **M. parviflora,** *R. Br. Prod.* 321. A more slender species than *M. porrifolia*, the leaf and especially the sheath much narrower, the flowers smaller and less crowded in the majority of specimens although

sometimes this character is reversed, as in those described by Brown, both species having varieties or races with crowded and with attenuated inflorescences. Dorsal sepal broad, obtuse, concave, scarcely above 1 line long; lateral sepals shorter, and petals still smaller. Labellum as long as the lateral sepals, oblong, obtuse, entire, the transverse callus at the base not very prominent, the disk not thickened excepting near the apex where it usually forms a papillose protuberance rather than a callus. Column with distinct auricles between the stigma and the anther.—Lindl. Gen. and Sp. Orch. 395; Endl. Iconogr. t. 15; Bot. Mag. t. 2377; Hook. f. Fl. Tasm. ii. 25; Reichb. f. Beitr. 22.

Queensland. Port Bowen, *R. Brown;* Burnett river, *F. Mueller;* Rockhampton and neighbourhood, *Bowman, O'Shanesy;* Moreton bay, *C. Stuart.*

N. S. Wales. Port Jackson to the Blue Mountains, *R. Brown*, and many others; Macleay river, *Fitzgerald;* New England, *C. Stuart;* also in *Leichhardt's collection.*

Victoria. Gipps' Land, *F. Mueller.*

Tasmania. Circular Head, *Gunn.*

Var. *densiflora.* Flowers very numerous in a dense spike of about 2 in., the dorsal sepal very broad; perhaps a distinct species.

W. Australia. *Drummond, 4th or 5th coll. n.* 117, and perhaps the same from King George's Sound, *Maclean.*

The species extend to New Caledonia, the Indian Archipelago, and South China; it is the only tropical representative of the genus. F. Mueller unites it with the *M. porrifolia* as his *M. viridis;* and Woolls, who has supplied the Muellerian collection with very numerous specimens of both species, also suggests that they may not be really distinct. They appear, however, at any rate to be well-marked varieties. The minute differences in the form and the calli or papillose protuberances of the labellum may very often be individual only, and those above described in the two species must not be relied upon as constant.

M. Benthamiana, Reichb. f. Beitr. 24, from Sydney, *R. Brown*, does not appear to me to be distinct from *M. parviflora.*

3. **M. media,** *R. Br. Prod.* 321. A tall species, with the habit and the rather narrow leaves of the larger specimens of *M. parviflora*, but the flowers are considerably larger, much recurved. Dorsal sepal acute, not very broad and somewhat contracted at the base, coming nearer in shape to that of *M. alba*, about $1\frac{1}{4}$ lines long; lateral sepals short and revolute; petals still shorter. Labellum usually narrow, truncate or retuse, the margins entire or slightly crenulate. Column with prominent auricles.—Lindl. Gen. and Sp. Orch. 396; Bot. Mag. t. 3378; Reichb. f. Beitr. 23; *M. Brownii*, Reichb. f. Beitr. 24.

W. Australia. King George's Sound, *R. Brown;* Gordon and Blackwood rivers, *Oldfield;* Swan river, *Drummond*, 1*st coll.*; Murchison river, *Oldfield.*

This, the Western representative of the two preceding species, appears to have been included by Lindley in his *M. rara*, and forms part of F. Mueller's above-mentioned *M. viridis.*

4. **M. alba,** *R. Br. Prod.* 321. Stem usually tall, often above 1 ft. high. Leaf with a long sheath and the lamina often exceeding the spike. Flowers numerous, much incurved, nearly white when dry, said to be whitish green or cream-coloured by some collectors, in a spike sometimes very dense especially when young, sometimes long slender and interrupted. Dorsal sepal very prominent, lanceolate-

falcate, acute, concave but less hood-shaped than in *M. porrifolia*, and contracted at the base, 1½ to 2 lines long or in some specimens rather longer; lateral sepals nearly as long, oblong, at first erect or spreading but becoming revolute as the flower fades; petals shorter and narrower. Labellum as long as the sepals, narrow at the base, the upper half expanded into 2 lobes either large and broad or long narrow and divaricate, the margins always undulate and crisped crenate or fringed, the disk with an oblong callosity along the centre of the broad parts, and sometimes a pair of marginal calli below it. Column with very prominent narrow auricles.—Lindl. Gen. and Sp. Orch. 396; Reichb. f. Beitr. 23.

W. Australia. King George's Sound and adjoining districts, *R. Brown*, *Oldfield*, *F. Mueller*, and others; eastward to Esperance bay and Cape Le Grand, *Maxwell*; Swan river, *Drummond*, 1*st coll.*

Some specimens from various collectors have smaller flowers, with a narrower labellum and shorter capsules than the others, but I have been unable to sort them into distinct varieties; the larger-flowered ones have the capsules sometimes long sometimes short; the labellum is exceedingly variable as to the breadth and as to the shape of its lobes.

5. **M. atrata,** *Lindl. Swan Riv. App.* 54, *Gen. and Sp. Orch.* 395. The smallest of all the species, usually only 3 to 4 in. high and rarely exceeding 6 in., of a bright green when fresh, but usually drying black, especially the flowers. Leaf usually short, but the long sheath reaching almost to the inflorescence. Flowers minute, in a rather dense spike of ½ to 1½ in. Dorsal sepal concave, very obtuse, about ½ line long and broad; lateral sepals and petals nearly equal to it in length, oblong, very obtuse, spreading but not revolute. Labellum as long as the upper sepal, broadly oblong and very obtuse or almost square, convex, quite entire, without calli but marked by two longitudinal striæ.—Endl. in Pl. Preiss. ii. 6; *M. minutiflora*, F. Muell. Fragm. i. 90.

Victoria. Portland, *Allitt*; near Melbourne, *Adamson*; Grampians, *F. Mueller*.

W. Australia. King George's Sound, *R. Brown*; near Perongerup and Mount Clarence, *F. Mueller*; Swan river, *Drummond*, 1*st coll. n.* 852, *Preiss*, *n.* 2403.

6. **M. pulchella,** *R. Br. Prod.* 321. Stem slender, under 1 ft. high. Leaf narrow and short. Spike not dense, rarely 2 in. long, the bracts very small and the pedicels very short, the flowers drying of a yellowish tinge and the proportion of the parts different from those of any other species. Dorsal sepal about 1 line long, ovate, obtuse, slightly concave but much less so than in the other species; lateral sepals as long and almost as broad and not reflexed; petals also as long, but much narrower and incurved over the column. Labellum at least as long, oblong, truncate or retuse, slightly contracted in the middle, the margin entire, the disk thickened at the base into a broad callus, and bearing a small oblong thickening towards the end. Capsule ovoid.—Reichb. f. Beitr. 23.

W. Australia. King George's Sound, *R. Brown*, also *Drummond*, *n.* 307.

Lindley does not appear to have examined this very distinct species, of which there is no specimen in his herbarium, but only a tracing of Bauer's drawing of it; the Eastern specimens he mistook for it appear to me to be referrible *M. porrifolia.*

35. CORYSANTHES, R. Br.

Dorsal sepal erect, very much incurved and concave, hood-shaped or contracted into a stipes; lateral sepals and petals small, linear, sometimes minute. Labellum erect under the galea, broadly tubular, the margin of the oblique orifice either shortly recurved and denticulate, or produced into a large concave denticulate or fringed lamina closely reflexed. Column short, erect, variously thickened under the stigma or winged. Anther erect, 2-celled, the outer valves large, the inner small; pollen-masses granular, without any caudicle.—Dwarf terrestrial herbs, with small underground tubers, and a single ovate-cordate orbicular or reniform leaf, with a scarious sheathing bract below it. Flower solitary, sessile within the leaf or very shortly pedicellate, with a small subtending bract usually close to the leaf.

The genus is also in New Zealand and the Indian Archipelago. The Australian species are all endemic. I cannot agree with Reichenbach f. in reviving Salisbury's name of *Corybas* on the ground of priority of general publication. It has been universally rejected as having been surreptitiously described and figured, and falsely characterized from the inspection of a drawing of Bauer's with Brown's name attached to it, as was well known at the time, and was published on authority which could not be and was not denied. In the following descriptions I have been obliged to take from Bauer's finished drawings some details which it was impossible to verify from dried specimens.

Dorsal sepal with a narrow linear claw as long as the orbicular lamina. Labellum without basal spur, the tube broad, the lamina very short . . . . 1. *C. unguiculata.*
Dorsal sepal gradually contracted towards the base. Labellum slightly 2-gibbous at the base, the lamina large, reflexed, concave, denticulate, or fringed . . . . 2. *C. fimbriata.*
Dorsal sepal with a broad base. Labellum 2-spurred at the base, broad and very oblique upwards, with a slightly recurved convex margin . . . . 3. *C. bicalcarata.*

1. **C. unguiculata,** *R. Br. Prod.* 328. Leaf rather more ovate than in the two following species. Ovary rather long. Dorsal sepal abruptly contracted into a linear claw of 2½ to 3 lines erect at the base and then much incurved, the lamina nearly orbicular, concave, about 3 lines diameter; lateral sepals and petals narrow-linear, sometimes nearly as long as the dorsal one but variable. Labellum rather longer, the tube ovoid, oblong, incurved, somewhat inflated not unlike that of the corolla of some species of *Digitalis*, the orifice very shortly and obliquely expanded into a denticulate lamina, with a longitudinal hairy broad line inside. Column very short, 2-winged, the wings with a lower oblong reflexed lobe, as in several species of *Pterostylis.*—Endl. Iconogr. t. 18; *Corybas unguiculatus,* Reichb. f. Beitr. 43.

N. S. Wales. Port Jackson, *R. Brown, A. Cunningham,*

I have only seen three specimens of this species, all very small, one in Herb. R. Brown in the same sheet as one of *C. bicalcarata* (referred by Reichenbach f. by mistake to *C. pruinosa,* Cunn.), the two others in Herb. A. Cunningham, also mixed with the very distinct *C. bicalcarata*, a specimen of which was unfortunately sent to Lindley under the name of *C. unguiculata*, and represents it in his Herb. where the true plant is wanting.

2. **C. fimbriata,** *R. Br. Prod.* 328, *and App. Flind. Voy.* 610, t. 10. A small plant, usually drying black, rarely 2 in. high including the flower. Leaf orbicular-cordate, about 1 in. diameter, usually thicker and more opaque than in *C. bicalcarata*, the midrib and reticulate veins alone distinguishable or the latter sometimes united in a circular vein within the margin, the leaf rarely thinner with the veins more conspicuous. Flower or even the whole plant said to be of a violet purple, or by others as of a deep red, sessile or nearly so above the leaf with a small bract, the ovary short. Dorsal sepal $\frac{3}{4}$ to 1 in. long, varying from very much incurved to much straighter, probably at different periods of inflorescence, much contracted in the lower half, but not abruptly unguiculate as in *C. unguiculata*; lateral sepals and petals linear, small, but longer than in *C. bicalcarata.* Labellum-tube much narrower than in the other two species, erect against the dorsal sepal, 4 to 5 lines long, with 2 minute obtuse spurs or gibbosities at the base sometimes scarcely conspicuous, the lamina reflexed, very large, varying however longer or shorter than the tube, concave with inflexed fringed margins, the disk reticulate and hairy inside along the centre. Column very short, much thickened under the stigma, but not winged.—Lindl. Gen. and Sp. Orch. 393; Hook. f. Fl. Tasm. ii. 16, t. 117; *Corybas pruinosus* and *C. fimbriatus*, Reichb. f. Beitr. 42, 43.

N. S. Wales. Port Jackson, *R. Brown*, *A. Cunningham*, and many others.

A. Cunningham, as quoted by Lindley, Gen. and Sp. Orch. 393, distinguished two species, *C. fimbriata*, with a much incurved obtuse galea, and *C. pruinosa*, with a more erect mucronate galea. In the dried specimens those with an erect galea are quite as obtuse as the others, and in that state it is impossible to distinguish two distinct forms. The specimens of *Corysanthes* were all much mixed in Herb. A. Cunningham, two or even three species laid down on the same sheet, and none named by him *C. pruinosa.*

Var. *diemenica.* Labellum lamina rather shorter and denticulate only, not bordered by long cilia or fringe, but the teeth very variable. *C. diemenica*, Lindl. Gen. and Sp. Orch. 393.

Victoria. Port Phillip and Sealer's Cove, *F. Mueller;* Wendu Vale, *Robertson.*

Tasmania. Common in various parts of the island, *J. D. Hooker.*

S. Australia. St. Vincent's Gulf, *F. Mueller.*

W. Australia. Perongerup, *Mrs. Knight.*

It is possible that I may have included two or three species under *C. fimbriata*, but they cannot be separated in the dried state.

3. **C. bicalcarata,** *R. Br. Prod.* 328. Usually rather larger than the two preceding species, the stem often above 1 in. below the leaf, the leaf orbicular-cordate, larger and thinner than in *C. fimbriata*, and often almost transparent when dry showing besides the midrib 1 or 2 circular veins on each side connected by the transverse reticulations, but occasionally leaves of the two species scarcely distinguishable. Ovary long, cylindrical. Dorsal sepal very much incurved, very obtuse, not contracted at the base; lateral sepals and petals very small, linear-subulate, sometimes minute or almost obsolete. Labellum-tube broad in the upper part, incurved and concealed under the dorsal sepal, taper-

ing at the base with 2 short narrow-conical spurs, readily visible in the dried specimens, between which are the minute lateral sepals; the orifice oblique, with a recurved convex margin or lamina. Column much thicker and shorter than in *C. unguiculata*, the wings narrow, and a prominent gibbosity at the base between the column and labellum.—Lindl. Gen. and Sp. Orch. 394; *Corybas aconitiflorus*, Salisb. Parad. Lond. t. 83 incorrect as to details; Reichb. f. Beitr. 43.

Queensland. Brisbane river, Moreton bay, *W. Hill;* Rockhampton, *Thozet.*

N. S. Wales. Port Jackson, *R. Brown, A. Cunningham;* Paramatta and Currajong, *Woolls.*

Salisbury's above quoted plate contains rude copies of Bauer's three figures of the whole plant, with analytical details incorrectly borrowed. Whether Salisbury's story of the withered specimen from Lady Essex's garden, and the dried specimens of the two other species be a fiction or not cannot now be positively ascertained; but if they existed, they could never have been examined for his character and description. He was far too shrewd an observer to have overlooked the tubular nature of the labellum, and to have so grossly misdescribed other essential characters which he had misunderstood from a hasty inspection without study of Bauer's original drawings, as he had mistaken the colouring which was there only indicated by figures.

36. PTEROSTYLIS, R. Br.

Dorsal sepal broad, erect, incurved and very concave; petals lanceolate falcate, contracted at the base and attached to the basal projection of the column, falcate and curved under the dorsal sepal, nearly as long and forming with it an arched or almost hood-shaped upper lip or galea; lateral sepals more or less united in an erect or recurved 2-lobed lower lip, adnate at the base to the basal projection of the column, the lobes often terminating in long points. Labellum on a short claw at the end of the basal projection of the column, moveable, the lamina linear or oblong, channelled flat or convex, produced below its insertion on the claw into an appendage either very short and obtuse or longer linear incurved and forked or penicillate at the end with a tuft of three or more setæ or cilia. Column elongated within the galea and curved with it, with a pair of hatchet-shaped or quadrangular wings one on each side of the rostellum and sometimes narrowly winged lower down, the base produced into a short horizontal projection. Stigma oblong on the face of the column about the middle of its length below the wings. Anther erect, the cells distinct, 2-valved. Pollen-masses granular.—Terrestrial herbs, with small underground tubers. Radical leaves ovate, in a tuft at the base of the flowering stem or in a separate tuft or at a different time of year, the stem-leaves either developed and linear or lanceolate, or reduced to scarious sheathing scales. Flowers usually green often tinged or streaked with red or brown, large and solitary, or smaller and several in a raceme on short pedicels. The bend of the petals partaking always of that of the dorsal sepal, it has been thought useless to describe them separately for each species, they are comprised with the dorsal sepal under the name of galea.

The genus is chiefly Australian; one of the Australian species extends into New

Caledonia, another into New Zealand, where are also five or six species not Australian, the remainder of the genus is strictly endemic in Australia.

SECT. 1. **Antennæa.**—*Lower lip erect, the lobes or their points embracing the galea.*

SERIES 1. **Grandifloræ.**—*Flowers large (usually above $\frac{3}{4}$ in. and never under $\frac{1}{2}$ in.) solitary (or abnormally and very rarely 2). Labellum-appendage linear, penicillate or with 2 or 3 bristle-like lobes at the end (the cilia very rarely and abnormally deficient).*

Radical leaves rosulate at the base of the flowering-stems. No stem-leaves except sheathing scales.
  Labellum bifid at the end.
    Labellum-lobes narrow. Flowers above 1 in. long . . . 1. *P. ophioglossa.*
    Labellum-lobes short and broad. Flowers under 1 in. long 2. *P. concinna.*
  Labellum entire at the end.
    Flowers 1 in. long or more. Lobes of the lower lip lanceolate with an acute sinus between them.
      Flowers erect, curved only at or above the middle.
        Labellum very obtuse. Galea slightly curved, the point oblique . . . . 3. *P. curta.*
        Labellum acuminate. Galea much curved at the middle. The point horizontal . . . . 4. *P. acuminata.*
      Flowers much curved below the middle so as to appear nodding, the point reflexed . . . . 5. *P. nutans.*
    Flowers $\frac{1}{2}$ to $\frac{3}{4}$ in.
      Lobes of the lower lip lanceolate, with an acute sinus between them . . . . 6. *P. pedunculata.*
      Lobes of the lower lip separated by a broad truncate sinus with an inflexed tooth . . . . 7. *P. nana.*
Leaves crowded at the base of the flowering-stem, passing gradually into stem-leaves or scales.
  Flowers (of *P. nana*) $\frac{1}{2}$ to $\frac{3}{4}$ in. Lobes of the lower lip truncate, separated by a broad sinus, with an inflexed lobe or tooth . . . . 8. *P. pyramidalis.*
  Flowers 1 in or more. Lobes of the lower lip lanceolate, separated by an acute sinus.
    Flowers glandular-papillose . . . . 9. *P. cucullata.*
    Flowers quite glabrous outside . . . . 10. *P. furcata.*
Lower leaves reduced to scarious scales which pass into linear or lanceolate scales or leaves, the largest either subtending the pedicel or next to it.
  Points of the sepals straight.
    Labellum ending in a filiform point clavate at the end.
      Lower lip truncate between the lobes . . . . 11. *P. grandiflora.*
    Labellum tapering above the middle, acute or with a slender point. Lower lip notched or with an acute sinus between the lobes.
      Leaves usually lanceolate. Flower above 1 in. long . 12. *P. reflexa.*
      Leaves very narrow. Flower under 1 in. long . . . 13. *P. præcox.*
    Labellum very obtuse. Flower under 1 in. long. Lower lip truncate between the lobes . . . . 14. *P. obtusa.*
  Points of the galea and of the lower-lip lobes recurved in opposite directions. Flower large . . . . 15. *P. recurva.*

SERIES 2. **Parvifloræ.**—*Flowers* 2 *or more very rarely only* 1. *Labellum-appendage short entire or with* 2 *or* 3 *setæ. Stems leafless at the time of flowering, except empty sheathing bracts.*

Sepals with fine points. Labellum shortly 2-lobed at the base, with a small appendage between the lobes . . . . 16. *P. Daintreana.*

Sepals acute or very shortly pointed. Labellum-appendage short and narrow, with 2 or 3 setæ . . . . 17. *P. parviflora.*
Sepals obtuse. Labellum-appendage very short obtuse and entire . . . . 18. *P. aphylla.*

SECT. 2. **Catochilus.**—*Lower lip reflexed from the base or recurved from the middle, the lobes short or narrow. Labellum-appendage entire and obtuse, sometimes almost obsolete.*

Flower large, solitary Labellum linear-terete or filiform, bearded with long hairs, glabrous at the end.
Leaves broadly lanceolate, crowded at the base of the stem, diminishing upwards . . . . 19. *P. barbata.*
Leaves short, linear, acuminate, nearly equally distributed along the stem . . . . 20. *P. turfosa.*
Flowers several, under $\frac{1}{2}$ in. long.
Leaves in a radical rosette, persistent or fading away before flowering, those of the stem reduced to scarious sheaths.
Sepals all obtuse, the lower lip shortly 2-lobed . . . . 21. *P. mutica.*
Sepals with short or rarely long fine points, the lower lip deeply 2-lobed . . . . 22. *P. rufa.*
Stems leafy without a radical rosette.
Leaves linear or linear-lanceolate. Column-wings nearly square. Eastern species . . . . 23. *P. longifolia.*
Leaves lanceolate. Column-wings with an oblong lower lobe. Western species . . . . 24. *P. vittata.*

SECT. 1. ANTENNÆA.—Lower lip of the flower erect, concave, the lobes closing over the galea and embracing it by their points, which often extend far beyond it.

SERIES 1. GRANDIFLORÆ.—Flowers large and solitary at the end of the scape or stem, or very rarely 2 (only 2 biflorous specimens seen out of many hundreds). Labellum produced at the base into a linear appendage, curved up at the end and there terminating in 2 or 3 bristle-like lobes or in a tuft of cilia (penicillate).

1. **P. ophioglossa,** *R. Br. Prod.* 326. Leaves in a radical rosette, shortly petiolate, ovate or broadly oblong, obtuse or mucronulate, $\frac{3}{4}$ to $1\frac{1}{2}$ in. long, elegantly veined, the transverse veinlets usually uniting in 2 lateral nerves on each side of the midrib. Scape 1-flowered, rarely above 6 in. high, without any or with a single empty bract near the base, the terminal pedicel subtended by a rather broad very acute bract of 4 to 8 lines. Galea incurved, acuminate, fully $1\frac{1}{4}$ in. long, rather broad, striate; lower lip erect, broadly cuneate, deeply 2-lobed, at least $\frac{3}{4}$ in. long besides the long subulate points which embrace the galea. Labellum-claw flat and thin, about 1 line long; lamina oblong-linear, ending in 2 narrow lobes of about 1 line, the basal appendage linear-subulate, curved upwards, with a terminal tuft of setæ. Column nearly as long as the labellum, reflexed lobe of the wings oblong and obtuse, the erect lobe smaller lanceolate and acute.—Lindl. Gen. and Sp. Orch. 391; Reichb. f. Beitr. 35.

Queensland. Port Curtis, *R. Brown;* Brisbane river, Moreton bay, *F. Mueller, Fitzalan, Bailey.*

N. S. Wales. Port Jackson, *R. Brown, Harvey, Woolls.*

We have also what appears to be the same species from New Caledonia.

2. **P. concinna,** *R. Br. Prod.* 326. Nearly allied to *P. ophioglossa*, but a smaller plant. Leaves radical, under 1 in. long, ovate or broadly oblong, the petiole usually longer than in *P. ophioglossa*, the venation the same. Scape 1-flowered, rarely above 1 in. long and usually with an empty bract at or below the middle, besides the sheathing bract at the base of the terminal pedicel. Galea broader and more incurved than in *P. ophioglossa*, and under 1 in. long; lower lip also broader and shorter, the lobes more divaricate, the long points fine or slightly clavate. Labellum rather shorter than the column, broader than in *P. ophioglossa*, the terminal lobes or teeth very short and broad. Column-wings with the erect lobe acutely acuminate and rather long.—Lindl. Gen. and Sp. Orch. 391; Hook. Journ. Bot. i. 274, t. 136; Reichb. f. Beitr. 34; *P. acuminata*, Sieb. Pl. Exs. not of R. Br.

N. S. Wales. Port Jackson, *R. Brown, Caley, Sieber, n.* 157, and many others.

Victoria. Towards Brighton, *F. Mueller.*

S. Australia? Bugle Range, *F. Mueller;* the identification rather doubtful.

I do not find the tubercles at the base of the labellum mentioned by Reichenbach fil. Beitr. 34, nor are they represented in the excellent figure above mentioned. The plate in Bot. Mag. t. 3400 appears to me rather to represent one of the long-flowered forms of *P. curta*.

3. **P. curta,** *R. Br. Prod.* 326. Leaves in a radical rosette, usually on long petioles, ovate or broadly elliptical, 5- to 9-nerved, from under 1 in. to 1½ in. long. Scapes 1-flowered, usually about 6 in. high, with 1, 2 or 3 long loosely sheathing empty bracts besides the one subtending the terminal pedicel. Galea erect, about 1¼ in. long, acute but not acuminate. Lower lip cuneate, with 2 broadly lanceolate lobes, not so long as the galea and only shortly acuminate in the typical form. Labellum linear, obtuse and entire, rather longer than the column, the surface papillose, the basal appendage linear curved and penicillate. Column 7 to 8 lines long, with the basal projection rather long; wings with the lower lobe oblong and obtuse, the upper lobe short and broad with a narrow point at the front angle.—Lindl. Gen. and Sp. Orch. 390; Hook. f. Fl. Tasm. ii. 18; Bot. Mag. t. 3086; Reichb. f. Beitr. 35.

N. S. Wales. Port Jackson, *Caley, A. Cunningham, Woolls;* Liverpool, *Leichhardt;* Twofold bay, *F. Mueller.*

Victoria. Wendu Vale, *Robertson;* Melbourne, *Adamson;* Darebin Creek, Mount Disappointment, *F. Mueller.*

Tasmania. Port Dalrymple, *R. Brown;* common in shady places, *J. D. Hooker.*

S. Australia. Barossa, Lofty and Bugle ranges, *F. Mueller.*

Var.? *grandiflora.* Flowers above 2 in. long, the lobes of the lower lip ending in long points.

Queensland. Brisbane river, Moreton bay, *F. Mueller;* also in *Leichhardt's* collection.

N. S. Wales. Paramatta, *Woolls.*

4. **P. acuminata,** *R. Br. Prod.* 326. Leaves in a radical rosette, ovate or broadly elliptical and 5- or 7-nerved as in *P. curta*, some forms

of which this species closely resembles. Scape 1-flowered, 6 to 9 in. high with 1 or rarely 2 empty sheathing bracts besides the one embracing the terminal pedicel. Galea 1 to 1¼ in. long, erect but much incurved about the middle, and usually produced into a point; lower lip narrow cuneate, contracted into a claw, the lobes lanceolate and produced into long fine points embracing the galea. Labellum oblong-linear, tapering to a point. Column with a short basal projection, lower lobe of the wings broad and obtuse, upper lobe broad and scarcely prominent, with a linear point at the front angle.—Lindl. Gen. and Sp. Orch. 391; Bot. Mag. t. 3401; Reichb. f. Beitr. 36.

N. S. Wales. Port Jackson to the Blue Mountains, *R. Brown*, *A. Cunningham*, *Woolls*.

5. **P. nutans,** *R. Br. Prod.* 327. Leaves in a radical rosette, petiolate, ovate or elliptical, ½ to 1½ in. long. Scape 1-flowered, 6 in. to 1 ft. high, usually with a single long loosely sheathing empty bract besides the one under the terminal pedicel. Galea nearly 1 in. long, much curved near the base and again towards the end, so as to give the flower a nodding appearance, obtuse or acuminate in front; lower lip shortly and broadly cuneate, the lobes long and lanceolate, tapering into long points embracing the galea. Labellum oblong-linear, obtuse, 4 to 5 lines long, the surface smooth but sometimes minutely ciliate, the basal appendage narrow-linear, curved, penicillate. Column ¾ the length of the galea, the wings with a broadly oblong obtuse lower lobe and only a very small upper lobe or tooth, the stigma usually long and conspicuous.—Lindl. Gen. and Sp. Orch. 391; Hook. f. Fl. Tasm. ii. 18; Bot. Mag. t. 3085; Reichb. f. Beitr. 37.

Queensland. Brisbane river, Moreton bay, *Bailey*.
N. S. Wales. Port Jackson to the Blue Mountains, *R. Brown*, *Caley*, *Sieber*, n. 155, and many others.
Victoria. Portland bay, *F. Mueller*; E. Gipps' Land, *Walter*.
Tasmania. Port Dalrymple, *Paterson*; common in shady places in a poor soil, *J. D. Hooker*, and others.
S. Australia. Mount Gambier, *F. Mueller*.

6. **P. pedunculata,** *R. Br. Prod.* 327. Leaves in a radical rosette on rather long petioles, ovate or broadly oblong, ½ to 1 in. long, thin and usually 5-nerved. Scape 1-flowered, under or over 6 in. high, with 2 to 4 loosely sheathing empty bracts besides the one subtending the terminal pedicel. Galea about ½ in. long or rather more, erect but abruptly curved towards the end, acute or terminating in front in a short point; lower lip broadly cuneate, the entire part about 4 lines long, the lobes lanceolate with long points abruptly turned up and embracing the galea. Labellum oblong, very obtuse, 2 to 2½ lines long, thickened and pubescent on the surface along the centre towards the end, basal appendage rather long, linear, curved, dilated at the end and usually penicillate with few cilia. Column-wings broad, the upper front angle with a long almost hair-like tooth, the lower lobe recurved, broadly lanceolate, obtuse.—Lindl. Gen and Sp. Orch. 391; Hook. f. Fl. Tasm. ii. 19, t. 114 A; Reichb. f. Beitr. 36.

N. S. Wales. Cudgee, *R. Cunningham;* Grose river, *Miss Atkinson*, apparently the same species, although the basal appendage of the labellum has no tuft of cilia at the end.

Tasmania. Port Dalrymple, *R. Brown;* abundant in shady places, *J. D. Hooker*, and others.

7. **P. nana,** *R. Br. Prod.* 327. Near *P. pedunculata*, but a smaller and more slender plant. Leaves in a radical rosette, ovate, acute, usually only ¼ in. and rarely ½ in. long. Scape with a single empty sheathing bract, which as well as the one subtending the terminal pedicel is usually more acute and spreading than in *P. pedunculata.* Galea as in that species but little above ½ in. long, erect, abruptly curved towards the end, but obtuse or scarcely acute in front. Lower lip broadly cuneate, about 4 lines long without the lobes, which are linear-subulate, only shortly dilated at the base and separated by the broad truncate apex of the lip, with usually a small inflexed tooth in the middle, the long points of the lobes embracing the galea. Labellum linear, obtuse, about 2 lines long, the surface glabrous, the basal appendage linear, curved, with few setæ at the end usually 2 only of which are deeply divided into 3. Column scarcely more than half the length of the galea, the wings with a small lanceolate upper lobe or tooth, the lower lobe oblong and obtuse.—Lindl. Gen. and Sp. Orch. 391; Reichb. f. Beitr 37; Hook. f. Fl. Tasm. ii. 19, but not the plate 114 B, which may perhaps have been taken from *P. concinna.*

Victoria. Wendu Vale, *Robertson;* Port Phillip, *F. Mueller.*

Tasmania. Port Dalrymple, *R. Brown;* Woolnorth and Circular Head, *Gunn;* Bagdad, *Miss Forster;* Oyster Cove, *Milligan;* Southport, *C. Stuart.*

S. Australia. Mount Gambier, *Mrs Wehl.*

W. Australia. Swan river, *Drummond* (doubtful); Blackwood river, *Oldfield;* Monjerup, *F. Mueller;* Upper Hay river, *Miss Warburton.*

8. **P. pyramidalis,** *Lindl. Swan Riv. App.* 53, *Gen. and Sp. Orch.* 388. Very closely allied to *P. nana* and the flower almost identical in size and structure, but usually a rather taller and stouter plant and the leaves not strictly rosulate, but collected at or near the base of the stem and passing gradually into the smaller sessile stem-leaves or empty bracts which are nearly all spreading and leaf-like, thus placing the species in a different division of the genus as usually adopted. Lower lip of the perianth truncate, with an inflexed lobe or tooth between the antenna-like lobes as in *P. nana*, and thus readily distinguished from *P. pedunculata* in which the lanceolate bases of the lobes are separated by an acute sinus.—*P. barbata*, Endl. in Pl. Preiss. ii. 5, not of Lindl. (*Reichb. f.*)

W. Australia. Swan river, *Drummond;* Gordon river, *Oldfield;* Lake Muir, *Muir.*

9. **P. cucullata,** *R. Br. Prod.* 327. Usually a low plant, rarely much above 6 in. with a single large flower. Leaves crowded at the base of the stem and sometimes almost rosulate, often larger than in any other species, ovate or oblong-elliptical, 1 to 3 in. long, passing into 1 to 3 empty almost leaf-like bracts, the one subtending the terminal pedicel very loosely sheathing, ovate-lanceolate, above 1 in. long. Galea

erect, incurved, acute or shortly acuminate, $1\frac{1}{4}$ to $1\frac{1}{2}$ in. long, minutely glandular-scabrous or papillose-pubescent outside; lower lip rather narrowly cuneate, the entire part about $\frac{1}{2}$ in. long, the lobes lanceolate, tapering into fine points embracing the galea but not attaining its length. Labellum oblong-linear, equal in breadth or tapering towards the end, but always rounded at the end, about half as long as the gálea, the basal appendage linear, curved, dilated and penicillate at the end. Column as long as the labellum, the upper margin of the wings rounded with a short linear lobe or tooth at the front angle, the lower lobe oblong.—Lindl. Gen. and Sp. Orch. 390; Hook. f. Fl. Tasm. ii. 19, t. 115; Reichb. f. Beitr. 36; *P. dubia*, R. Br. Prod. 328, Lindl. l.c.; Reichb. f. Beitr. 42; *P. scabrida*, Lindl. l.c. 389.

Victoria. Port Phillip, *C. French;* Brighton Scrub, *Gulliver;* Gipps' Land, *F. Mueller.*

Tasmania. Port Dalrymple, *R. Brown;* common on poor soil in shaded places, *J. D. Hooker*, and others.

S. Australia. Mountains round St. Vincent's Gulf, *F. Mueller.*

Brown's specimen of *P. dubia*, from Derwent river, is a very unsatisfactory one, but appears to be a starved state of *P. cucullata.* In the Tasmanian specimens distinguished by Lindley as *P. scabrida*, the leaves are smaller, and the aspect somewhat different from the usual luxuriant habit of *P. cucullata*, but the two forms pass too much into each other to be distinguished as varieties.

10. **P. furcata,** *Lindl. Gen. and Sp. Orch.* 390. Very near *P. cucullata*, and perhaps a variety only, with a similar large erect solitary flower from within a large acuminate loosely sheathing bract, but the lower leaves are smaller, less crowded at the base of the stem or the lowest small and distant, and the bracts on the stem rather more leaf-like, the flower is perfectly glabrous outside, and the lobes of the lower lip end in longer fine points embracing the galea. The internal structure of the flower is the same as in *P. cucullata*.—Hook. f. Fl. Tasm. ii. 20.

Victoria? Some specimens from Plenty Range, *F. Mueller*, seem referrible rather to this species than to *P. cucullata.*

Tasmania. Shaded places, near Launceston and Deloraine, *Gunn;* Chudleigh and Cheshunt, *Archer;* Southport, *C. Stuart;* Hampshire Hills, *Milligan.*

P. dubia, Hook. f. Fl. Tasm. ii. 20, t. 115, seems to belong to this species rather than to *P. cucullata*, to which I would refer Brown's specimen so named. The two are, however, perhaps varieties only of one species.

11. **P. grandiflora,** *R. Br. Prod.* 327. Stems slender, 1-flowered, about 6 in. high, without any radical rosette of leaves at the time of flowering. Leaves along the stem, lanceolate, acuminate, not differing from the bract subtending the terminal pedicel. Galea above 1 in. long, abruptly curved forward about the middle, the petals as well as the dorsal sepals ending in front in short points; lower lip with the entire part broadly cuneate, fully $\frac{1}{2}$ in. long, truncate as in *P. nana* and *P. obtusa*, leaving a very broad straight and scarcely notched sinus between the lobes, which are very shortly dilated at the base, tapering into long filiform antenna-like points embracing the galea. Labellum oblong-linear at the base, tapering into a long filiform glabrous point somewhat clavate at the end; the basal appendage linear, curved, penicillate.

Column-wings with an erect linear acute lobe at the front angle, the lower lobe oblong, obtuse.—Lindl. Gen. and Sp. Orch. 387; Guillem. Ic. Pl. Austral. t. 6; Reichb. f. Beitr. 39.

N. S. Wales. Port Jackson to the Blue Mountains, *Caley*, *Woolls.*

12. **P. reflexa,** *R. Br. Prod.* 327. Stems slender, 6 to 9 in. high, glabrous or minutely scabrous-pubescent or papillose, without any rosette of radical leaves at the time of flowering. Leaves or empty scales lanceolate, erect or slightly spreading, acuminate, under 1 in. long in the typical form, and none usually so long as the bract subtending the terminal pedicel, more leaf-like and longer in some varieties. Galea $1\frac{1}{4}$ to above $1\frac{1}{2}$ in. long, curved but not abruptly so, the petals as well as the sepal tapering into fine points; lower lip cuneate at the base, the lobes lanceolate, separated by a sinus much narrower than in *P. grandiflora*, and almost acute, tapering into long filiform points embracing the galea. Labellum more or less lanceolate and tapering towards the end into a long or short point; the basal appendage linear, curved, penicillate at the end. Column-wings with a small erect acute lobe at the front angle, the lower lobe oblong and obtuse.—Lindl. Gen. and Sp. Orch 387; Reichb. f. Beitr. 38; *P. revoluta*, R. Br. Prod. 327; Lindl. l.c. 389; Reichb. f. l.c.; *P. scabra*, Lindl. Swan Riv. App. 53; Orch. Gen. and Sp. 388; *P. pyramidalis*, Endl. in Pl. Preiss. ii. 5, not of Lindl.

N. S. Wales. Port Jackson, *R. Brown*, *Woolls*, and others; New England, *C. Stuart;* Mudgee, *Taylor.*

Victoria. Grampians and Wimmera, *F. Mueller;* Little river, *Fullagar;* East Gipps' Land, *Walter.*

S. Australia. Mount Lofty ranges, *F. Mueller.*

W. Australia. Swan river, *Drummond*, *1st coll.*, *Preiss*, *n.* 2203, and others; Vasse river, *Oldfield;* Hampden, *Clarke;* Grenough Flats, *C. Gray.*

In the typical form the flowers are not very large, and the labellum has a long fine point. In Brown's *P. revoluta* the flowers are considerably larger, and the labellum tapers towards the end, but without the long point of *P. reflexa.* Somé of Woolls's specimens have the long flowers of *P. revoluta*, with the labellum of *P. reflexa.* Most of the Victorian, South Australian and Western specimens have shorter, more leafy stems, with the labellum of *P. revoluta*, many of the Western ones are more or less scabrous-pubescent. The Gipps' Land specimens are remarkably tall and pubescent. The long and short pointed labella and large and smaller flowers, however, pass so much one into another, that I have been unable to sort the specimens into distinct varieties. It is possible, however, that their study in a fresh state may point out more appreciable characters.

13. **P. præcox,** *Lindl. Gen. and Sp. Orch.* 388. Very nearly allied to *P. obtusa*, but the leaves usually more developed, green, narrow-linear or linear-lanceolate, the lower lip of the perianth less truncate between the lobes, and the labellum tapering above the middle into a rather obtuse or almost acute point, like that of the shorter-pointed forms of *P. reflexa*, from which species this one may be most readily distinguished by the narrow leaves smaller flowers and broader sinus between the lobes of the lower lip of the perianth.—Hook. f. Fl. Tasm. ii. 21; *Disperis alata*, Labill. Pl. Nov. Holl. ii. 59, t. 210; *P. alata*, Reichb. f. Beitr. 70.

Victoria. Wilson's Promontory, *F. Mueller;* Wendu Vale, *Robertson* (the specimens in fruit only, and therefore doubtful).

Tasmania. Circular Head, *Gunn;* Hobarton, *J. D. Hooker;* Flinders island, *Milligan.*

14. **P. obtusa,** *R. Br. Prod.* 327. Stems slender, 1-flowered, usually about 6 in. high, without any radical leaves at the time of flowering but often from a separate branch of the rhizome a rosette of ovate 5-nerved leaves like those of *P. concinna.* Stem-leaves or bracts lanceolate, acuminate, ½ to ¾ in. long, not different from the uppermost bract which subtends the terminal pedicel, the lower ones reduced to sheathing scales. Galea incurved, ¾ to near 1 in. long, besides the point which varies from 2 to 6 lines; lower lip with the entire part very broadly cuneate, almost truncate, 4 to 5 lines long, the lobes very divaricate, separated by a broad sinus notched in the centre, tapering into long subulate antenna-like points embracing the galea. Labellum the length of the column, oblong-linear, equally broad throughout and very obtuse, the basal appendage linear, curved, penicillate, the tuft consisting usually of 2 ciliate setæ. Column-wings with a prominent tooth or linear upper lobe at the front angle, the lower lobe oblong, the stigma very prominent.—Lindl. Gen. and Sp. Orch. 389; Hook. f. Fl. Tasm. ii. 19, t. 115 C; Reichb. f. Beitr. 38.

N. S. Wales. Port Jackson, *R. Brown, Woolls;* New England, *C. Stuart.*

Tasmania. Common in the northern parts of the island, *J. D. Hooker;* Southport, *C. Stuart.*

15. **P. recurva,** *Benth.* Stems 1 to 1½ ft. high, rigid, 1- or sometimes 2-flowered, without any radical rosette, the lower leaves reduced to small scales gradually increasing to linear or linear-lanceolate leaves of 1 to 2 in., the bract subtending the pedicel more lanceolate and sometimes shorter. Galea erect, above 1 in. long, not very broad and not much curved, the dorsal sepal as well as the petals ending in recurved points apparently variable in length; lower lip as long or longer than the galea, narrow-cuneate, divided to the middle into lanceolate lobes erect and embracing the galea, but recurved at the end and terminating in reflexed points of 2 to 6 lines. Labellum tapering towards the end but obtuse, the basal appendage linear, elongated, curved, bifid and penicillate at the end, but the tuft consisting of very few cilia.

W. Australia, *Drummond;* Upper Hay river, *Miss Warburton.*—A well-marked species, of which however I have seen but very few specimens.

Series 2. Parvifloræ.—Stems leafless at the time of flowering except empty sheathing scarious bracts, the leaves in radical rosettes at a different time of year, or if contemporaneous from a different branch of the rhizome. Flowers under ½ in. long, 2 or more in a raceme, very rarely reduced to 1. Basal appendage of the labellum short, entire or with 2 or 3 teeth or setæ.

16. **P. Daintreana,** *F. Muell. Herb.* Leaves (only seen in an imperfect state) like those of *P. parviflora,* small, ovate, in a radical rosette by the side of the scape or flowering stem. Scape slender, above 6 in.

high, with 3 or 4 empty sheathing bracts, the upper ones, like the bracts subtending the pedicels, produced into fine points. Flowers 4 or 5, distant, nearly the size of but much more slender than those of *P. parviflora*. Galea 3 to 3½ lines long, obtusely hood-shaped, produced in front into a long fine point; lower lip narrow, the entire part about 1 line long, the lobes narrow, produced into long fine points embracing the galea. Labellum narrow, obtuse and entire at the end, sagittate at the base with obtuse auricles and a small obtuse entire appendage between them. Column reaching to the end of the galea, the wings very broad with a small point at the upper front angle, the lower slender portion of the column bordered by narrow wings, the stigma scarcely prominent.

N. S. Wales. Near Sydney, *Daintree*, very few specimens in Herb. F. Mueller.

17. **P. parviflora,** *R. Br. Prod.* 327. Leaves in radical rosettes appearing at a different time of year from the flowering stem or if contemporaneous in a tuft by the side of it, ovate, under ½ in. and often only ¼ in. long, on a rather long petiole. Scape slender, 4 to 8 in. long, with 2 or 3 empty bracts or small sheathing leaves and a raceme of 2 to 5 small flowers. Galea much incurved, scarcely 5 lines long, very acute or shortly acuminate; lower lip cuneate, the entire part about 2 lines long, the lobes much incurved, the inner margin involute at the base, tapering into points variable in length but always shorter than the galea. Labellum very short, obtuse but entire, the basal appendage short and slender, terminating in the specimens examined in a tuft of 3 setæ. Column slender, the wing from a narrow base very prominent, with a narrow point at the upper outer angle, and a broad lower lobe.—Lindl. Gen. and Sp. Orch. 389; Hook. f. Fl. Tasm. ii. 22; Reichb. f. Beitr. 40.

Queensland. Brisbane river and Moreton island, *F. Mueller.*
N. S. Wales. Port Jackson, *R. Brown, Woolls;* Aitken Creek, *A. Cunningham.*
Victoria. Wilson's Promontory, *F. Mueller.*
Tasmania. Port Dalrymple, *R. Brown;* Huon river, *Oldfield;* Hobarton, *J. D. Hooker;* Cheshunt, *Archer.*

18. **P. aphylla,** *Lindl. Gen. and Sp. Orch.* 392. A smaller but rather stouter plant than *P. parviflora*, with the same foliage according to Archer and C. Stuart, but the radical tufts entirely gone at the time of flowering. Stems 3 to 5 in. high, with 1 to 3 flowers, which, when more than one, front each other in a peculiar way as described by C. Stuart and even apparent on dried specimens, but of which I see no trace in *P. parviflora.* Galea of the size of that of *P. parviflora* but rather broader and more obtuse, the lower lip with shorter points and the basal appendage of the labellum obtuse and undivided in the specimen examined, without even the 3 short points figured by Fitch.—Hook. f. Fl. Tasm. ii. 22, t. 116.

Tasmania. Huon river, *Oldfield;* Circular Head, *Gunn;* Cheshunt, *Archer;* Mersey river and plains near Southport, *C. Stuart.*

Sect. 2. Catochilus.—Lower lip of the perianth very spreading or reflexed from the base, or recurved from the middle, the lobes short or narrow. Basal appendage of the labellum-lamina entire and obtuse, or sometimes almost obsolete.

19. **P. barbata,** *Lindl. Swan Riv. App.* 53, *Gen. and Sp. Orch.* 388. Stems 1-flowered, from under 6 in. to nearly 1 ft. high. Leaves crowded at the base of the stem, ovate-lanceolate or lanceolate, acute or shortly acuminate, ½ to 1 in. long, sometimes extending halfway up the stem, sometimes almost rosulate at the base, passing more or less gradually into erect loosely sheathing bracts, the uppermost subtending the terminal pedicel. Galea erect, oblong, 1 in. long or rather more, the petals as well as the dorsal sepal ending in short subulate points; lower lip linear, very spreading recurved or reflexed, the lobes narrow, obtuse or ending in fine points. Labellum ½ to ¾ in. long, linear-terete or filiform, bearded with long yellow hairs except at the end, where it bears a broad glabrous nearly square complicate entire or toothed appendage, and at the base where it is glabrous, thickened and produced beyond the insertion on the claw into a short narrow appendage glabrous or shortly ciliate. Column slender, the wings broad, with the upper lobe from the front angle almost setiform, long and ciliate at the base, the lower lobe falcate-lanceolate, acute or obtuse, the lower part of the column very narrowly winged.—*P. squamata*, Lindl. Gen. and Sp. Orch. 388; Hook. f. Fl. Tasm. ii. 20, t. 116, not of R. Br.

Victoria. Wendu Vale, *Robertson.*

Tasmania. Common in sandy soil, *J. D. Hooker* and others.

S. Australia. Mount Lofty ranges, *F. Mueller.*

W. Australia. Swan river, *Drummond,* 1*st coll.*; King George's Sound, *Muir*; Mount Barker, *F. Mueller.*

The species is also in New Zealand.

20. **P. turfosa,** *Endl. in Pl. Preiss.* ii. 5. Stem short, slender, 1-flowered. Leaves short, linear, acuminate, all nearly equal and equally distributed along the stem, the upper one or bract subtending the terminal pedicel rather larger than the others. Galea erect, much like that of *P. barbata* but with a long filiform point, the linear lower lip also with long points to the lobes. Labellum linear-terete, bearded with long rigid hairs as in *P. barbata*, the end unknown, being broken off from the only specimen seen by Reichenbach, the basal appendage oblong-linear, incurved, obtuse at the end, glabrous. Column-wings with a long erect subulate upper lobe on the front angle, the lower lobe also long, oblong, ciliate at the end.

W. Australia. Stirling terrace, *Preiss, n.* 2632.—I have not seen any specimen; the above character is taken from a sketch and description sent to me by Reichenbach fil., and drawn up by him from the only known specimen now in the Lund Herbarium.

21. **P. mutica,** *R. Br. Prod.* 328. Leaves in a radical rosette at the base of the flowering stem sometimes but not usually withering away at the time of flowering, ovate, very shortly petiolate or almost sessile, mostly ½ to ¾ in. long. Stem 4 to 8 in. high, with 1 to 5 empty sheathing bracts, besides those subtending the pedicels, all obtuse or the

upper ones acute. Flowers 5 to 10, in a slightly spiral spike. Galea broad, much incurved, obtuse, about 3½ lines long; lower lip little more than 2 lines long and at least as broad, concave, reflexed, with 2 short broad obtuse lobes. Labellum on a rather long flat claw, broad, very obtuse, scarcely 1½ lines long, the basal lobe or appendage nearly as broad at the base, narrow, thick, obtuse and entire or emarginate at the end. Column reaching to the end of the galea; the wings broad, without any upper lobe or tooth, the lower lobe broad and obtuse.—Lindl. Gen. and Sp. Orch. 390; Hook. f. Fl. Tasm. ii. 21, t. 117; Reichb. f. Beitr. 42.

Queensland. Brisbane river, Moreton bay, *F. Mueller*, *Leichhardt.*

N. S. Wales. Port Jackson, *R. Brown*, *Woolls*, and others; Emu plains, *A. Cunningham;* New England, *C. Stuart;* southward to Illawarra, *Backhouse;* Gabo island, *F. Mueller.*

Victoria. Wendu Vale, *Robertson;* Melbourne, *Adamson;* Darebin Creek, Mount Disappointment, Grampians, &c., *F. Mueller.*

Tasmania. Common in rich pastures as well as in light sandy soil, *J. D. Hooker.*

S. Australia. Mount Gambier and Rivoli bay, *F. Mueller.*

22. **P. rufa,** *R. Br. Prod.* 327. Leaves in a radical rosette at the base of the stem, but most frequently withering away before the flowering, ovate, obtuse or acute, ½ to 1 in. long. Stem 6 to 10 in. high, with 2 to 4 loosely sheathing rather scarious empty bracts usually acute, besides the bracts subtending the pedicels. Flowers usually 3 or 4 in a short raceme. Galea about 5 lines long, hood-shaped, produced in front into a fine point 10 lines long in the typical form, very much longer in some varieties; lower lip on a rather long basal projection of the column, reflexed from it, broadly cuneate, 3 to 4 lines long without the points, divided to the middle into broadly lanceolate lobes ending in fine points, varying in length like that of the galea. Labellum on a short claw, ovate-oblong or narrow, concave or with involute margins, scarcely 1½ lines long, obtuse, bordered by few or many marginal cilia rarely entirely wanting, usually 1 long one on each side near the base, the basal appendage short, thick, entire, rugose, often ciliate. Column reaching to the end of the galea, the wings broad and nearly square, with a small point at the upper front angle, the lower angle or short broad lobe often ciliate, the middle part of the column narrowly winged.—Lindl. Gen. and Sp. Orch. 390; Hook. f. Fl. Tasm. ii. 21, t. 116; Reichb. f. Beitr. 41.

Queensland. Rockhampton, *Thozet.*

N. S. Wales. Port Jackson, common, *R. Brown*, *Woolls*, and others; New England, *C. Stuart;* Darling river, *Dallachy;* Upper Bogan and Lachlan rivers, *F. Marsh.*

Victoria. Murray river, *F. Mueller;* Wimmera, *Dallachy.*

Tasmania. Port Sorell, *Archer;* Meander river, *C. Stuart.*

S. Australia. Spencer's and St. Vincent's gulfs, *F. Mueller.*

W. Australia, *Drummond.*

The species varies much in stature, in the persistence of the radical leaves, in the size of the flowers, and especially in the length of the sepal-points. The following are the principal forms which have been distinguished mostly as species, but which pass very gradually into each other.

P. gibbosa, R. Br. Prod. 328; Lindl. Gen. and Sp. Orch. 390; Reichb. f. Beitr. 41, from Port Jackson, appears to me to be merely a tall-growing luxuriant state of the typical short-pointed form.

P. squamata, R. Br. Prod. 327, Reichb. f. Beitr. 41, not of Lindl., from Table Mountain (Mount Wellington), Tasmania, is a small-flowered variety with short points, in which the radical leaves are persistent, and the scarious empty scales more numerous. Almost similar small-flowered specimens, but with fewer empty scales, were gathered with large-flowered ones, in New England by *C. Stuart*.

P. Mitchelli, Lindl in Mitch. Trop. Austr. 365, from Mount Kennedy on the Maranoa, *Mitchell*, is a rather large-flowered form with long points to the sepals, the labellum narrow with usually numerous cilia. These cilia are marginal as in the other varieties, but owing to the involution of the margins appear to be along the upper surface as described. Some other Queensland specimens closely resemble *P. Mitchelli* in every respect except that the labellum is broader. Some of the Lachlan and Darling river specimens have the sepal-points fully ½ in. long, and in two specimens in Herb. F. Mueller, one from Queensland, *Bowman*, the other from Salt Creek, S. Australia, *F. Mueller*, these points vary from ¾ to above 1 in. in length.

23. **P. longifolia,** *R. Br. Prod.* 327. Stems rather slender, but often 1 ft. high or rather more, without any radical rosette, the lower leaves reduced to short sheathing scales, those at and above the middle of the stem linear or linear-lanceolate, acute or acuminate, from under 1 to above 2 in. long, very shortly sheathing at the base. Flowers 3 to 7, in a terminal raceme. Galea 5 to 7 lines, more or less incurved above the ovary and again abruptly curved towards the end, acute or with a short point in front; lower lip reflexed, 4 to 5 lines long, oblong, divided usually to about half its length into 2 narrow-lanceolate lobes. Labellum on a very short claw, oblong, about 3 lines long, more or less papillose on the surface, with a short obtuse or 2-toothed papillose process at the end, the basal appendage very short, obtuse and usually erect. Column-wings very broad, nearly square or slightly hatchet-shaped, the margins ciliolate or entire.—Lindl. Gen. and Sp. Orch. 388; Hook. f. Fl. Tasm. ii. 22, t. 117; Reichb. f. Beitr. 40.

N. S. Wales. Port Jackson to the Blue Mountains, *R. Brown*, *Sieber*, *n.* 160, *A. and R. Cunningham*, and many others; Illawarra, *Backhouse*.

Victoria. Forest Creek, Mount Disappointment, Wilson's Promontory, Naugatta range, &c., *F. Mueller;* Grampians, *Fisher;* E. Gipps' Land, *Walter*.

Tasmania. Common in dry soil in forest land, *J. D. Hooker*.

S. Australia. Mount Lofty range, between Mount Gambier and Rivoli bay, *F. Mueller*.

The species varies much in the length and breadth of the leaf, in the size of the flowers, the length of the lobes of the lower lip and the precise form of the labellum, especially of its terminal appendage, and it seems sometimes almost to pass into *P. vittata*.

24. **P. vittata,** *Lindl. Swan Riv. App.* 53, *Gen. and Sp. Orch.* 389. Allied to *P. longifolia* and like that species without any rosette of radical leaves to the flowering stem, but usually a stouter and much more leafy plant. Stems 8 in. to above 1 ft. high, often angular in the dried state. Leaves lanceolate, narrow or broad, acute, usually clasping the stem with rounded auricles, the lower one or two reduced to sheathing scales. Flowers in a more compact raceme than in *P. longifolia*, the bracts more leaf-like. Galea 5 to 6 lines long, broad, very much curved near the base and above the middle so as to be quite helmet-

shaped, with a short point in front directed downwards; lower lip rather broadly ovate, shorter than the galea, concave, recurved, with 2 short acuminate lobes. Labellum on a rather long claw, oblong, about 2 lines long, slightly contracted and emarginate or 2-lobed at the end, the margins ciliate, the basal appendage scarcely more than an obtuse thickening of the base of the lamina, and usually with a thick seta or linear tooth on one or both margins just above the base. Column-wings with an oblong lower lobe densely ciliate at the end, the middle of the column rather broadly winged.—Endl. in Pl. Preiss. ii. 5.

S. Australia? Some specimens from Bugle range, *F. Mueller*, mixed in his herbarium with others of *P longifolia* from Third Creek, under the name of *P. præcocissima* appear to belong rather to *P. vittata.*

W. Australia. Swan river, *Drummond;* King George's Sound and adjoining districts, *Maxwell, Preiss, n.* 2201, 2202, *Muir, Miss Warburton;* Vasse river, *Oldfield;* Hampden, *Clarke.*

I have been unable to ascertain whether the form of the column-wings be as constantly distinct from that prevailing in *P. longifolia* as it appeared in the few flowers examined.

37. CALEANA, R. Br.

Sepals and petals all linear, the dorsal sepal erect, the lateral sepals and petals spreading or reflexed (but the position apparently reversed by the resupination of the flower on the ovary). Labellum articulate at the base of the column or at the end of its basal projection and moveable, with a linear incurved claw, the lamina ovate or oblong, peltate, convex, entire, shorter below than above its insertion, the surface smooth or tuberculate. Column elongated, sometimes produced at the base into a linear projection, very broadly 2-winged in its whole length. Anther erect, not mucronate, the 2 cells distinct and nearly equally 2-valved. Pollen-masses granular.—Terrestrial glabrous herbs, with small underground tubers. Leaf linear lanceolate or oblong, solitary at the base of the stem which has also occasionally a small empty bract at or below the middle. Flowers 1 to 3 or rarely 4, shortly pedicellate, the subtending bracts acute. Ovary usually recurved, reversing the flower.

The genus is limited to Australia. Allied to *Drakæa*, it is readily known by the large petal-like wings of the column, forming a kind of pouch open or closed by the elastic motions of the lid-like labellum.

Claw of the labellum and lateral sepals inserted at the base of the column . . . . 1. *C. major.*
Claw of the labellum and lateral sepals inserted at the end of a basal projection of the column.
Leaf narrow linear. Eastern species . . . . 2. *C. minor.*
Leaf ovate or lanceolate. Western species . . . . 3. *C. nigrita.*

1. **C. major,** *R. Br. Prod.* 329. Leaf radical, linear or narrow-lanceolate, 2 to 4 in. long. Stem often above 1 ft. high, with a single closely appressed empty sheathing bract below the middle, and 1 to 4 red flowers on very short pedicels, the subtending bracts 2 to 4 lines long. Dorsal sepal narrow-linear, rather thick, channelled, erect or incurved

below the middle, often $\frac{3}{4}$ in. long; lateral sepals narrow-linear, acuminate reflexed (erect by the reversion of the flower) about 6 lines long; petals still narrower and shorter, erect (with reference to the floral axis). Labellum affixed to the base of the column, the claw linear, flat, incurved, about 3 lines long; lamina peltately attached, broadly ovate, fully 4 lines long and nearly as broad, shortly and broadly acuminate at each end, the upper surface smooth, the centre inflated and hollow, the cavity open on the under side. Column 4 to 5 lines long, bordered on each side from the base to the anther with a petal-like coloured wing about 3 lines broad. Stigma obscurely 2-pointed.—Lindl. Gen. and Sp. Orch. 429; Hook. f. Fl. Tasm. ii. 18, t. 107 A; Reichb. f. Beitr. 44; *Caleya major*, Endl. Iconogr. t. 8.

Queensland. Moreton bay, *F. Mueller.*
N. S. Wales. Port Jackson, *R. Brown*, *Backhouse*, *Woolls;* Blue Mountains, *Miss Atkinson;* New England, *C. Stuart.*
Victoria. Mount Sturgeon, Mount Abrupt, Latrobe river, *F. Mueller;* Mount William, *Sullivan;* Gipps' Land, *Walter.*
Tasmania. Rocky Cape, *Gunn;* Cheshunt, *Archer;* Southport, *C. Stuart;* South Huon, *Oldfield;* N. W. Bay, *Milligan.*

2. **C. minor,** *R. Br. Prod.* 329. Leaf radical, narrow-linear. Stem slender, about 6 in. high, without any or very rarely with a single small empty bract below the middle. Flowers 1 to 3, much smaller than in *C. major*, on longer pedicels. Sepals and petals linear, nearly equal, 4 to 5 lines long, the dorsal one often dilated above the middle and attached as well as the petals immediately above the ovary, the lateral sepals however attached to the extremity of the basal projection of the column on each side of the stipes of the labellum, which is linear and incurved. Lamina of the labellum peltate and convex, but narrower than in *C. major*, and tuberculate on the surface, the upper lobe obtuse or shortly 2-lobed, the lower lobe or appendage very short. Column about as long as the sepals, the broad wing adnate also to the basal projection, which is at least half as long as the column itself and nearly erect whilst the column is more spreading, the whole forming a broad sac or pouch.—Lindl. Gen. and Sp. Orch. 429; Reichb. f. Beitr. 44.

N. S. Wales. Port Jackson, *R. Brown;* New England, *C. Stuart* (a single specimen in Herb. F. Mueller differing in some slight particulars from Brown's and Gunn's).
Tasmania. Hobarton, *Gunn.*

3. **C. nigrita,** *Lindl. Swan Riv. App.* 54, *Gen. and Sp. Orch.* 429. Leaf radical, small, ovate or broadly lanceolate. Stem about 6 in. high, without any empty bract. Flowers 1 to 3, on pedicels of $\frac{1}{2}$ to 1 in. Sepals and petals linear, about 5 lines long, the dorsal sepal closely appressed to the column and apparently adnate to it at the base, the petals very narrow, the lateral sepals attached as in *C. minor* to the extremity of the basal projection of the column. Labellum with a claw of at least 3 lines, the lamina peltate, oblong, very convex, tuberculate on the surface, the upper end or lobe twice as long as the lower lobe or appendage, both ends obtuse or emarginate in the specimens examined.

Column nearly as long as the sepals, the broad wing adnate to the basal projection which is at least 2 lines long and forming a broad sac as in *C. minor*.—Endl. in Pl. Preiss. ii. 11.

W. Australia. Swan river, *Drummond*, 1*st coll. n.* 864.

38. DRAKÆA, Lindl.

(Spiculæa, *Lindl.;* Arthrochilus, *F. Muell.*)

Sepals and petals linear, the dorsal sepal erect, the lateral sepals and petals spreading or reflexed. Labellum articulate at the base of the column or at the end of its basal projection and moveable, with a linear claw; the lamina narrow, peltate, convex, shorter below than above its insertion. Column elongated, narrow, wingless except 1 or 2 pairs of narrow auricles variously placed. Anther erect, not mucronate, the 2 cells distinct and nearly equally 2-valved. Pollen-masses granular. Stigma large, orbicular, sometimes mucronate.—Terrestrial glabrous herbs, with small underground tubers. Leaf solitary at or near the base of the stem, usually broad, or none at the time of flowering. Scapes with 1 to 3 empty sheathing bracts. Flowers solitary or several in a raceme, the subtending bracts small and narrow. Labellum almost hammer-shaped and very irritable. Ovary straight or recurved, more or less reversing the flower.

The genus is limited to Australia, and is nearly allied to *Caleana*. The three species form one well-marked genus, the differences between *D.* (*Spiculæa*) *ciliata* and the original *D. elastica* correspond to those which distinguish *Caleana major* and *C. minor*.

Labellum articulate at the base of the column, without any intervening projection. Column with 2 pairs of auricles. Stem with 1 leaf. Flowers several . . . . 1. *D. ciliata.*
Labellum articulate at the end of a basal projection of the column.
Lateral sepals adnate to the basal projection of the column. Stem leafless. Flowers several . . . . 2. *D. irritabilis.*
Lateral sepals free from the linear basal projection. Leaf radical, ovate. Flowers solitary . . . . 3. *D. elastica.*

1. **D. ciliata,** *Reichb. f. Beitr.* 68. Stem under 6 in., the leaf a little above the base, lanceolate or ovate-lanceolate, complicate, shortly sheathing and erect at the base, recurved upwards; and a single small empty bract higher up either closely appressed or slightly spreading and leaf-like. Flowers 2 to 6, on short pedicels, the bracts 1 to 2 lines long. Sepals and petals narrow-linear, the dorsal sepal erect, about 6 lines long, the lateral sepals and petals rather shorter, spreading or reflexed. Labellum nearly as long as the sepals, articulate at the base of the column, the claw very slender, incurved; the lamina hammer-shaped, peltately attached, the upper lobe recurved, above 1 line long, very convex, papillose-pubescent on the surface of the lower half, terminating in a smooth obtuse point; the lower lobe or basal appendage apparently similar to the lower half of the upper lobe. Column erect, incurved, narrow, the very narrow wings dilated about halfway up into 2 long linear-falcate ascending auricles, and at the anther into 2 linear

recurved auricles, with a short triangular lobe at the base of each on the upper side. Stigma without any or only with a very short terminal point.—*Spiculæa ciliata*, Lindl. Swan Riv. App. 56, with a woodcut, Gen. and Sp. Orch. 428; Endl. in Pl. Preiss. ii. 10.

W. Australia. Swan river, *Drummond, 1st coll., 4th coll. n.* 325.

2. **D. irritabilis,** *Reichb. Beitr.* 68. Stems 6 in. to nearly 1 ft. high, leafless at the time of flowering except 1, 2 or 3 distant empty bracts sheathing at the base but sometimes spreading and almost leaf-like at the apex, and in one specimen the lowest is developed into a lanceolate complicate recurved leaf of nearly 1 in. Flowers green or whitish, often tinged with red, 3 to 8 on pedicels of 2 to 3 lines within small narrow bracts. Sepals and petals narrow-linear, slightly dilated above the middle, the dorsal sepal incurved, about 4 lines long, the lateral sepals and petals shorter, very spreading or reflexed, the former dilated at the base and adnate to the basal projection of the column. Labellum articulate at the end of the basal projection, the linear claw about 1 line long above the articulation, the lamina hammer-shaped and peltately attached, ciliate with long hairs on the upper surface, the upper lobe emarginate or terminating in a short smooth point, the lower lobe or appendage very obtuse and less hairy. Column incurved, with 2 narrow-linear acuminate auricles just under the stigma and 2 short points behind it.—*Arthrochilus irritabilis*, F. Muell. Fragm. i. 43.

Queensland. Brisbane river, Moreton bay, *W. Hill, F. Mueller, C. Prentice; Rockingham bay, Dallachy.*

C. Prentice collected in June, 1867, a specimen with the flowering stem proceeding from a tuft of lanceolate leaves 2 or 3 in. long. These may have been the ordinary leaves of the plant, which appear usually at a different time of year from the flowers; but in this instance the plant had probably flowered abnormally at the leafing time.

3. **D. elastica,** *Lindl. Swan Riv. App.* 55 *with a woodcut, Gen. and Sp. Orch.* 428. Leaf radical, broadly ovate-cordate or almost orbicular, apparently thick and fleshy, with recurved margins rarely above ½ in. diameter. Stem slender, 6 in. high or rather more, with a small empty sheathing bract below the middle, and a single flower on a pedicel much longer than the subtending bract. Sepals and petals linear, 5 to 6 lines long, the lateral sepals and petals reflexed, free from the basal projection, the dorsal sepal usually rather longer and erect. Labellum articulate at the end of the basal projection of the column, and "moving at the joint with every breeze," the claw narrow-linear, 3 lines long above the joint, the lamina hammer-shaped, peltately attached, broadly ovate but very convex and the sides completely folded back so as to conceal the under surface, the upper surface covered with short thick hairs or calli except the smooth tip, the lower lobe or appendage solid and fleshy, half as long as the upper one. Column nearly as long as the sepals, abruptly incurved in the middle, with very narrow wings produced into auricles at the base and somewhat dilated under the anther, the basal projection of the column supporting the labellum about 3 lines long and linear like the claw. Rostellum of

the stigma erect and sometimes nearly as long as the anther.—Endl. in Pl. Preiss. ii. 10.

W. Australia. Swan river, *Drummond*, 1*st coll.*; Cape Leschenault, *Oldfield;* King George's Sound, *F. Mueller*, *Muir;* known under the name of *Hammer Orchis.*

39. ACIANTHUS, R. Br.

Dorsal sepal erect or incurved over the column, concave, not very broad, and often produced into a fine point; lateral sepals narrow, erect or spreading; petals much shorter. Labellum about as long as the petals, sessile or nearly so, undivided, the margin entire, the disk smooth or papillose, with or without 2 adnate calli or tubercles at the base. Column erect or incurved, semiterete or 2-winged. Anther broad, erect, 2-celled, with broad outer valves, the connective sometimes produced into a short point; pollen granular, but less so than in *Caladenia* and more distinctly collected into 4 masses in each cell.—Terrestrial glabrous herbs, with small underground tubers. Leaf solitary, immediately above the basal scarious sheath or higher up the stem, broadly ovate-cordate. Flowers solitary or several in a terminal raceme, on a scape or stem without scales above the leaf, except the small bracts subtending the pedicels.

Besides the Australian species which are endemic, there is one from New Zealand. The genus is nearly allied to *Caladenia*, but without the calli and fringes to the labellum of that genus, a different foliage, and the pollen in two at least of its species appear to be of the more solid consistence of that of *Eriochilus*.

Column not winged. Sepals with fine points. Labellum oblong-lanceolate.
  Flowers 1 to 3. Dorsal sepal narrow, with a filiform point of ¾ to above 1 in. . . . . 1. *A. caudatus.*
  Flowers 3 to 10. Dorsal sepal with a point under 2 lines.
    Dorsal sepal ovate-lanceolate. Labellum with 2 papillose lines on the surface . . . . 2. *A. fornicatus.*
    Dorsal sepal lanceolate, contracted at the base. Labellum smooth on the surface . . . . 3. *A. exsertus.*
Column winged. Sepals without points. Labellum broader than long . . . . 4. *A. viridis.*

1. **A. caudatus,** *R. Br. Prod.* 321. Stems slender, sometimes filiform, 3 to 6 in. high. Leaf at or near the base, deeply cordate, ovate, rarely above 1 in. long. Flowers 1 to 3, of a dark colour, on short pedicels within small bracts. Dorsal sepal tapering into a filiform point varying from ½ to above 1 in. long, not very broad in the lower part, and contracted again at the base; lateral sepals shorter, filiform, shortly and slightly dilated at the base; petals lanceolate, reflexed, about 2 lines long, with short points. Labellum almost or quite sessile, oblong-lanceolate, shortly acuminate, as long as the petals, the 2 basal adnate calli very prominent, lanceolate or acuminate, the surface of the lamina with very few papillæ. Column semiterete, not winged. Pollen apparently more granular than in the two following species.—Lindl. Gen. and Sp. Orch. 397; Hook. f. Fl. Tasm. ii. 25, t. 119; Reichb. f. Beitr. 26.

N. S. Wales. Port Jackson, *R. Brown*, *A. Cunningham*, *Woolls*; Blue Mountains, *Miss Atkinson*.
Tasmania. Common in moist shaded woods, *J. D. Hooker*.

2. **A. fornicatus,** *R. Br. Prod.* 321. Stem slender, 6 to 8 in. high. Leaf at the base or below the middle, broadly ovate or orbicular, deeply cordate and stem-clasping with broad rounded auricles, usually 1 to 1½ in. long, sometimes sinuate or even rather deeply 3-lobed. Flowers 4 to 10, on short pedicels; bracts ovate or lanceolate, acute. Dorsal sepal ovate-lanceolate, 3 to 4 lines long, erect, incurved, concave, acute and the midrib produced into a fine point of 1 to 1½ lines; lateral sepals nearly as long but linear, with a long point and angular or toothed on each side of the point, close together or shortly united under the labellum; petals lanceolate, about half as long as the dorsal sepal, with a short point. Labellum much shorter than the dorsal sepal but variable in length, nearly sessile or on an exceedingly short claw, oblong-lanceolate, acuminate, concave at the base with 2 very short raised longitudinal plates or calli, smooth along the centre, with 2 very prominent broad raised papillose lines parallel to the reflexed margins, the point smooth. Column not 2 lines long, much incurved, semiterete, not winged, often concealed in the dorsal sepal, but sometimes bent forward as in *A. exsertus*. Anther with a very short point. Pollen-masses (in the flowers examined) 4 in each cell, of the somewhat solid consistence of *Eriochilus* but obtuse at the base, and perhaps becoming granular at a later stage. Stigma very prominent.—Lindl. Gen. and Sp. Orch. 397; Endl. Iconogr. t. 16; Reichb. f. Beitr. 25.

Queensland. Brisbane river, Moreton bay, *F. Mueller*; Rockingham bay, *Dallachy*; Mount Wheeler, *Thozet*.
N. S. Wales. Port Jackson, *R. Brown*, *Sieber*, *n.* 159, and others.

F. Mueller, Fragm. v. 96, unites this and the following species under the name of *A. Brunonis*.

3. **A. exsertus,** *R. Br. Prod.* 321. A slender delicate plant, much resembling *A. fornicatus*, but smaller. Leaf deeply cordate, ovate or orbicular and often sinuate as in that species. Flowers rather smaller, 3 to 6 in the raceme, the pedicels very short. Dorsal sepal slightly incurved, concave, but narrow and much contracted at the base, about 3 lines long, including the short point; lateral sepals almost subulate; petals lanceolate, about half as long as the sepals. Labellum nearly as long as the sepals, oblong-lanceolate as in *A. fornicatus* but on a more distinct claw, the raised plates at the base rather longer and the surface of the lamina smooth or with very few papillæ in 2 rows. Column slender, not winged, about half as long as the sepals, incurved and protruding forwards from the dorsal sepal. Pollen of *A. fornicatus*.—Lindl. Gen. and Sp. Orch. 397; Hook. f. Fl. Tasm. ii. 25, t. 119; Reichb. f. Beitr. 25.

N. S. Wales. Port Jackson, *Bauer*, *Woolls*, and others; Blue Mountains, *Caley*; New England, *C. Stuart*.
Victoria. Wendu river, *Robertson*; Seeler's Cove, *F. Mueller*; Portarlington, *Robertson*.

Tasmania. Circular Head, *Gunn;* Cheshunt, *Archer;* Southport, *C. Stuart;* Swanport, *Story.*
S. Australia. Lofty Range, *F. Mueller.*

Some specimens from Port Phillip, *F. Mueller*, have very much reduced flowers, either with much enlarged or with very small bracts, all probably abnormal states.

4. **A. viridis,** *Hook. f. Fl. Tasm.* ii. 372. A small plant, the cordate leaf scarcely above ½ in. diameter, above or rarely below the middle of the stem. Flowers 1 to 3, very nearly sessile, the bracts very small. Dorsal sepal erect, hood-shaped, much incurved, tapering at both ends as in *A. exsertus* but without any projecting point, 3 to 3½ lines long; lateral sepals about as long, linear, obtuse. Petals not 1 line long and very obtuse. Labellum 2 lines long, sessile, rhomboidal, concave, slightly undulate, possibly with 2 calli at the base but I could not clearly see them in the only flower I could examine. Column winged as in *Caladenia*. Pollen not seen.

Tasmania. Base of Mount Wellington, *Gunn, Oldfield.*

40. ERIOCHILUS, R. Br.

Dorsal sepal erect, slightly incurved and concave; petals nearly as long, usually narrower, erect or spreading; lateral sepals longer, spreading, oblong or elliptical, contracted into a distinct narrow stipes. Labellum much shorter, with a narrow concave erect claw, the margins often produced into small erect lateral lobes, the lamina or middle lobe recurved, very convex, entire, the surface villous, without calli. Column erect, the front angles sometimes ciliate or very narrowly winged. Anther erect, not mucronate, 2-celled, the outer valves large, folded over and concealing the small inner valves; pollen at length powdery or granular but much less so than in *Caladenia* and usually seen in 4 distinct and almost smooth masses in each cell, contracted at one end into points or short caudicles.—Terrestrial glandular-pubescent or hairy rarely glabrous herbs, with small underground tubers. Leaf solitary at the base of or higher up the stem, ovate or lanceolate. Flowers pink or white, 1 or more nearly sessile on a scape or peduncle, without empty bracts above the leaf, each flower subtended by a short loosely sheathing ovate bract.

The genus is limited to Australia. In all the flowers examined in which I have found the pollen-masses still in situ, their consistence has been much more solid and less granular than in any of the allied genera except *Acianthus*. F. Mueller, however, finds in the fresh state no difference between the pollen of *Eriochilus* and that of *Caladenia*.

Leaf at the base of the stem.
  Leaf broad. Stem glandular-pubescent. Labellum without lateral lobes, the middle lobe ovate-oblong. Eastern species . 1. *E. autumnalis.*
  Labellum with small erect lateral lobes, the middle lobe nearly orbicular. Western species.
    Leaf broad. Stem villous . . . . 2. *E scaber.*
    Leaf narrow. Stem glabrous . . . . 3. *E. tenuis.*
Leaf some way up the stem. Plant nearly glabrous. Western species.
  Labellum middle-lobe ovate-oblong. Flowers rarely above 3 . 4. *E. dilatatus.*

Labellum middle-lobe narrow-oblong. Flowers usually more than 4 . . . . 5. *E. multiflorus.*

1. **E. autumnalis,** *R. Br. Prod.* 323. A slender plant, rarely exceeding 6 in., more or less glandular-pubescent. Leaf radical, ovate, acute, usually dying away before the time of flowering, but occasionally still persisting at the base of some flowering specimens. Flowers pink, solitary or 2 or 3 rather distant, the subtending bracts loosely sheathing, 1 to 2 lines long, and no empty ones on the scape lower down. Dorsal sepal erect, slightly incurved, narrow-lanceolate, acute, scarcely contracted at the base, 3 to 3½ lines long; lateral sepals half as long again, very acute, elliptical-lanceolate, contracted into a distinct often slender stipes; petals rather shorter than the dorsal sepal, linear or linear-spathulate. Labellum about half as long as the lateral sepals, with an erect concave narrow claw, sometimes showing at the apex minute lateral lobes or angles, the lamina or middle lobe recurved, oval-oblong, convex and hairy but without prominent calli. Column shorter than the dorsal sepal, narrowly winged below the very broad concave stigma.—Lindl. Gen. and Sp. Orch. 427; Endl. Iconogr. t. 6; Hook. f. Fl. Tasm. ii. 26, t. 120 A; *Epipactis cucullata*, Labill. Pl. Nov. Holl. ii. 61, t. 211, f. 2; *Eriochilus cucullatus*, Reichb. f. Beitr. 27.

Queensland. Brisbane river, Moreton bay, *W. Hill* (with smaller flowers).

N. S. Wales. Port Jackson, *R. Brown* and others; frequent in the colony, *A. Cunningham;* New England, *Leichhardt, C. Stuart.*

Victoria. Wendu Vale, *Robertson;* Port Phillip, *R. Brown;* from the Yarra to Gipps' Land, *F. Mueller* and others.

Tasmania. Port Dalrymple, *R. Brown;* common in open and somewhat dry ground throughout the island, *J. D. Hooker.*

S. Australia. Near Mount Barker, *F. Mueller.*

2. **E. scaber,** *Lindl. Swan Riv. App.* 53, *Gen. and Sp. Orch.* 427. Closely allied to *E. autumnalis*, usually but not always shorter, not so slender, and hairy with articulate transparent hairs. Leaf radical, ovate or cordate, acute, usually persisting at the base of the flowering stem. Flowers 1 to 3, pink, the bracts broad and mostly acute. Sepals and petals rather shorter and broader than in *E. autumnalis*, but otherwise with the same proportions. Labellum-claw distinctly produced into small erect rounded lateral lobes, the lamina or middle lobe almost orbicular, very convex and densely hairy. Column not winged but the two angles ciliate as well as the outer valves of the anther. Pollen-masses distinct and almost contracted into caudicles as in *E. autumnalis.*—Endl. in Pl. Preiss. ii. 9.

W. Australia. Upper Hay river, *Miss Warburton;* Perongerup, *F. Mueller;* Swan river, *Drummond*, 1*st coll., Preiss, n.* 2207 (which I have not seen); Murchison river, *Oldfield.*

E. Lindleyi, Endl. in Pl. Preiss. ii. 10, from Swan river, *Preiss, n.* 2206, which I have not seen, is distinguished by the shortly pedunculate less erect flowers, but is probably only a slight variety of *E. scaber*, to which it is reduced by Reichb. f. Beitr. 62.

3. **E. tenuis,** *Lindl. Swan Riv. App.* 53, *Gen. and Sp. Orch.* 427. Stem glabrous, very slender, 3 to 6 in. high and single-flowered in all the

specimens seen. Leaf radical, but much narrower than in the two preceding species, lanceolate or oblong-lanceolate and almost obtuse. Bract small, acute. Flower of the size of those of *E. scaber*, but the sepals and petals not so broad, and the lateral lobes of the labellum scarcely prominent. Column ciliate on the angles as well as the valves of the anthers as in *E. scaber*.—Endl. in Pl. Preiss. ii. 10.

W. Australia. Swan river, *Drummond*, 1*st coll.*; King George's Sound, *Muir.*

4. **E. dilatatus,** *Lindl. Swan Riv. App.* 53, *Gen. and Sp. Orch.* 427. Stem glabrous or nearly so, usually rather above 6 in. Leaf at or below the middle but always far above the basal sheathing scale, linear-lanceolate in the typical form, sessile and stem-clasping. Flowers 1, 2 or rarely 3, resembling those of *E. autumnalis.* Dorsal sepal 3 to $3\frac{1}{2}$ lines long, oblong in the upper part, contracted below the middle; lateral sepals oblong-lanceolate, acute, 5 lines long, contracted into a slender claw; petals about as long as the dorsal sepal but narrower. Labellum much shorter, the claw erect, with slightly prominent rounded lateral lobes, the lamina or middle lobe ovate-oblong, longer than broad, very convex and recurved, pubescent above. Column neither winged nor ciliate. Anther-valves pubescent outside, minutely ciliate on the edges.

W. Australia. Swan river, *Mangles*; Cape Leeuwin, *Collie*; Kalgan river and Dillon bay, *Maxwell.*

Var. *latifolius.* Rather larger; leaves lanceolate. Flowers 2 to 4 and rather longer. *E. latifolius*, Lindl. ll. cc.—Swan river, *Mangles.*

Var. *brevifolius.* Leaves smaller, ovate-lanceolate.—Swan river, *Drummond*; Murchison river, *Oldfield.*

5. **E. multiflorus,** *Lindl. Swan Riv. App.* 53, *Gen. and Sp. Orch.* 427. Glabrous or nearly so, closely resembling *E. dilatatus* and perhaps a variety, the habit and foliage the same, but usually rather taller, often 1 ft. high or more, the flowers more numerous and rather smaller, sometimes above 10 in the spike or raceme. Labellum with very small lateral lobes or teeth at the end of the claw, the lamina or middle lobe oblong, much longer than broad, and usually as long as the claw.—Endl. in Pl. Preiss. ii. 10.

W. Australia. Swan river, *Drummond*, *Preiss*, *n.* 2190; Forest Hill, *Muir.*

41. LYPERANTHUS, R. Br.

Dorsal sepal broad, concave, erect or incurved over the column; lateral sepals and petals narrow, erect or spreading, all nearly equal in length. Labellum shorter than the sepals, with a broad erect claw sometimes dilated upwards into small erect lateral lobes, the lamina or middle lobe ovate or lanceolate, recurved, the claw or disk between the lateral lobes longitudinally thickened in the centre, the surface of the lamina or middle lobe papillose. Column erect or incurved, not winged. Anther terminal, erect, 2-celled; pollen-masses granular.—Terrestrial herbs often drying black, with small underground tubers. Scapes or stems 2- or more-flowered, either with one radical leaf and 2 or 3 almost leaf-like empty bracts, or with about 2 stem-leaves. Bracts usually rather large and leaf-like.

Reduced to the two following species, both of them endemic in Australia, the genus appears more naturally distinct from *Caladenia*, with which Reichenbach fil. proposes to unite it. It differs in the broader galeate upper sepal, the want of calli on the labellum, and the wingless column, and in habit the single empty bract of *Caladenia* is replaced by 2 or more leaves or somewhat leaf-like bracts. The *L. sauveolens* and *L. serrata* have however the narrow upper sepal, the calli on the labellum, and the winged column of *Caladenia*, and are correctly transferred to that genus. A New-Caledonian species and one from the Auckland islands require further examination to determine their affinities.

Leaf radical, ovate-cordate, with 2 or 3 empty sheathing-bracts. Flowers 2 to 4. Labellum lamina fringed . . . . 1. *L. nigricans.*
Leaves usually 2, ovate-elliptical or lanceolate. Flowers usually more than 4. Labellum lamina not fringed . . . . 2. *L. ellipticus.*

1. **L. nigricans,** *R. Br. Prod.* 325. Stems from a few inches to nearly 1 ft. high, rather stout, the whole plant drying black. Radical leaf rather broadly ovate-cordate, 1 to 2 in. long and often nearly as broad, thick and fleshy; empty sheathing bracts usually 2 or 3 below the middle of the stem, loose and leaf-like but erect, obtuse, often 1 in. long. Flowers "pale purple," 2 to 4, rather distant, nearly sessile within acute or acuminate bracts usually longer than the ovary. Dorsal sepal broad, concave, much incurved, acuminate, 7 to 8 lines long; lateral sepals and petals spreading or deflexed, narrow-linear, as long as the dorsal sepal or the petals shorter. Labellum shorter, the claw erect or slightly recurved, channelled, dilated upwards into small erect lateral lobes, the middle lobe recurved or revolute with a deeply fringed margin; disk thickened between the lateral lobes into a more or less prominent broad longitudinal line, the surface of the middle lobe glandular-papillose. Column incurved, nearly attaining the lateral lobes, not winged, the circular peltate stigma very prominent.—Lindl. Gen. and Sp. Orch. 392; Endl. Iconogr. t. 7, and in Pl. Preiss. ii. 5; Hook. f. Fl. Tasm. ii. 16, t. 106 B; Reichb. f. Beitr. 33; *Caladenia nigricans*, Reichb. f. Beitr. 67; *Leptoceras pectinata*, Endl. in Pl. Preiss. ii. 6, not of Lindl.

N. S. Wales. Port Jackson, *R. Brown*, *A. Cunningham*, and others.
Victoria. Between Melbourne and Brighton, *F. Mueller*; Genoa Creek, *Walter.*
Tasmania. Forest near George Town, *Archer*; Rocky Cape, *Milligan.*
W. Australia. King George's Sound and adjoining districts, *Menzies*, *F. Mueller*, *Preiss*, *n.* 2200; Esperance bay and Cape Le Grand, *Maxwell*; Swan river, *Drummond*, 1*st coll.*; Darling range, *Preiss*, *n.* 2186.

This appears to be one of those species which, like some of our European terrestrial Orchideæ, have a wide geographical range, but are often very rare in each locality.

2. **L. ellipticus,** *R. Br. Prod.* 325. Stems erect or decumbent, 6 to 9 in. high, without any radical leaf, but usually with 2 narrow-ovate elliptical or lanceolate acute or acuminate leaves higher up, both very variable in size, 1 to 1½ in. long when broad, longer when narrow. Flowers more numerous and nearer together than in *L. nigricans*, green

with the labellum variegated white and red. Sepals and petals acuminate with fine points; the dorsal sepal 6 to 7 lines long, broadly lanceolate, incurved and concave; lateral sepals and petals rather shorter, linear or narrowly linear-lanceolate, flat, slightly falcate, spreading or recurved. Labellum above half as long as the sepals, with a short broad erect concave claw, the lateral lobes obsolete, the middle lobe or lamina recurved, veined, the margin entire, the surface with a few raised papillæ at its base. Column incurved, not winged.—Lindl. Gen. and Sp. Orch. 392; Reichb. f. Beitr. 33; F. Muell. Fragm. vii. 133; *Caladenia elliptica*, Reichb. f. Beitr. 67.

N. S. Wales. Port Jackson to the Blue Mountains, apparently rare, *Caley, A. Cunningham, Miss Atkinson, Daintree, Fitzgerald.*

42. BURNETTIA, Lindl.

Sepals and petals nearly equal, erect or connivent, the dorsal sepal incurved and concave, the lateral sepals and petals falcate. Labellum shorter than the sepals, sessile, undivided, erect at the base, recurved towards the end, with 2 longitudinal raised plates along the centre broken up into calli above the middle. Column erect, incurved, winged. Anther erect, 2-celled, the outer valves broad; pollen-masses granular.—Terrestrial herbs with small underground tubers, leafless at the time of flowering except empty sheathing scales. Leaf solitary at a different time of year. Flowers few.

The genus is limited to a single species endemic in Tasmania. It is very near *Caladenia*, with which Reichenbach f. proposes to unite it, but from which it differs in habit and in the longitudinal plates on the labellum, as well as in the consistence of the perianth and its more connivent segments.

1. **B. cuneata,** *Lind. Gen. and Sp. Orch.* 518. Flowering stems 2 to 4 in. high, with several sheathing empty scales, the lower ones short and imbricate, the upper ones distant, loose, often $\frac{1}{2}$ in. long, obtuse or passing into the acute bracts subtending the pedicels. Leaf (which I have not seen) according to C. Stuart on a separate stem ovate-lanceolate, acute, $\frac{3}{4}$ in. long, disappearing some months before the flowering. Flowers 1 to 3, erect but much incurved. Sepals and petals about 5 lines long, of a thicker consistence than those of *Caladenia*, "reddish-brown outside pure white within" (C. Stuart). Labellum very broad, truncate or obscurely sinuate and sometimes slightly fringed, the longitudinal plates of the disk shortly lobed or broken up into a few calli above the middle and disappearing before the end of the lamina, and a few small calli sometimes scattered over the surface. Column-wings like those of *Caladenia*.—Hook. f. Fl. Tasm. ii. 17, t. 107; *Lyperanthus Burnettii*, F. Muell. Fragm. v. 96, vii. 134; *Caladenia cuneata*, Reichb. f. Beitr. 67.

Tasmania. Rocky Cape, *Gunn;* Woolnorth, Oyster Cove, Macquarrie Harbour; *Milligan;* Southport, *C. Stuart.*

43. CYRTOSTYLIS, R. Br.

Dorsal sepal linear or linear-lanceolate, erect and incurved, concave; lateral sepals and petals very narrow, spreading, nearly equal in length to the dorsal sepal, or the petals shorter. Labellum with a short claw, flat, undivided, entire, with 2 calli at the base produced into raised lines along the lamina. Column elongated, incurved, winged upwards. Anther terminal, erect, 2-celled; pollen-masses granular distinctly cohering in 4 masses.—Terrestrial glabrous herbs. Leaf solitary at the base of the stem, lamina broad and spreading. Scape without any empty bract, usually bearing a raceme of several flowers.

Besides the Australian species there are two in New Zealand. The genus is very near *Caladenia*, with which Reichenbach f. unites it. The habit is more that of *Acianthus*.

1. **C. reniformis,** *R. Br. Prod.* 322. A small delicate glabrous plant. Leaf orbicular-cordate or reniform, radical, sessile, the lamina varying from under ¾ to 1½ in. diameter. Scape from under 2 in. high and 1-flowered to 6 or 7 in. with 4 or 5 pale red flowers. Bracts subtending the pedicels short broad loose and truncate, or rarely acuminate. Dorsal sepal 4 to 5 lines long; lateral sepals and especially the petals often shorter and very narrow. Labellum nearly as long as the dorsal sepal, obtuse emarginate or with a short point, the medial raised lines terminating in the centre of the lamina or reaching nearly to the end. Column slender, ⅔ the length of the dorsal sepal, the wings sometimes very short under the anther, sometimes extending halfway down.—Lindl. Gen. and Sp. Orch. 398; Hook. f. Fl. Tasm. ii. 26, t. 119; Hook. Journ. Bot. i. t. 135; *Caladenia reniformis,* Reichb. f. Beitr. 67.

Queensland. Brisbane river, Moreton bay, *F. Mueller;* Rockhampton, *Thozet* (doubtful, the specimens too young to determine).

N. S. Wales. Port Jackson to the Blue Mountains, *R. Brown*, *Sieber*, *n.* 158, and many others; southward to Twofold bay, *F. Mueller.*

Victoria. Wendu Vale, *Robertson ;* Portland, *Allitt ;* Station Peak, *F. Mueller*.

Tasmania. Common in open and somewhat dry ground throughout the island, *J. D. Hooker*.

S. Australia. Encounter bay, Lofty Ranges, and other localities, *F. Mueller*, *Whittaker*.

Var. *Huegelii.* Bracts more acuminate and the labellum usually narrower. *C. Huegelii*, Endl. in Pl. Preiss. ii. 6.

W. Australia. King George's Sound, *Collie;* Vasse river, *Oldfield;* Upper Hay river, *Miss Clarke;* Swan river, *Drummond*, 1*st coll. n.* 862; Rottenest island, *Preiss*, *n.* 2204.

44. CALADENIA, R. Br.

Dorsal sepal erect or incurved over the column, usually narrow, lateral sepals nearly equal to it but flat and spreading, petals narrow, erect or spreading, or rarely sepals and petals all nearly equal and spreading. Labellum erect at the base, undivided or 3-lobed, the lateral lobes when present erect, the middle lobe or upper part of the undivided labellum recurved, the margins often fringed or toothed, the disk with sessile or stipitate oblong linear or clavate calli, in 2 or more longitudinal rows, or irregularly crowded or scattered; in one species no calli but the margin fringed. Column erect or incurved, more or less 2-winged in the upper part. Anther erect, 2-celled, the outer valves broad, the inner much smaller, the connective usually produced into a point. Pollen-masses granular.—Terrestrial herbs, usually hairy, with small underground tubers. Leaf solitary, linear-lanceolate or oblong, from within a scarious sheathing scale close to the ground. Flowers solitary or very few in a loose raceme, on an erect scape, leafless except a small narrow sheathing scale or empty bract about the middle, and a similar bract under each pedicel, and in one species the radical leaf and empty bract are deficient. Flowers usually erect, variously coloured.

Besides the Australian species, which are all endemic, there are three from New Zealand, or two only, if we refer the *C. bifolia* to *Chiloglottis*. The genus is a difficult one, not only as to the determination of the limits to be assigned to species, especially from dried specimens, in which the precise form of the labellum can often not be ascertained, but the limitation of the genus itself is very arbitrary. We might almost equally well separate generically some of the following sections, especially *Leptoceras* and *Pentisia*, or with Reichenbach fil. add to them by the incorporation of *Glossodia*, and some others.

Sect. 1. **Leptoceras.**—*Sepals acute or rather obtuse, the dorsal one erect or incurved and concave. Petals erect, linear-clavate, longer than the sepals (not exceeding them in the other sections).*

Petals much longer than the sepals. Labellum not fringed, the disk with 2 to 4 rows of calli . . . . 1. *C. Menziesii.*
Petals rather longer than the sepals. Labellum very broad, fringed at the end, the disk without calli . . . . 2. *C. fimbriata.*

Sect. 2. **Phlebochilus.**—*Sepals obscurely or distinctly acuminate, the dorsal one incurved and concave, erect behind the column or reflexed with it. Labellum broad, with deeply coloured diverging veins, undivided or with a very small and obscure middle lobe.*

Sepals scarcely acuminate. Labellum on a short claw, not fringed; calli in 2 rows, the lower ones linear-clavate, the upper depressed . . . . 3. *C. Cairnsiana.*
Sepals shortly acuminate. Labellum on a long claw, not fringed; calli few long and clavate, the lower ones connate . 4. *C. multiclavia.*
Sepals subulate-acuminate, all reflexed as well as the column. Labellum on a short claw, fringed; calli short, thick, and irregularly crowded . . . . 5. *C. discoidea.*

Sect. 3. **Calonema.**—*Sepals acuminate, with long or short points, the dorsal one erect and incurved. Labellum inconspicuously veined, the disk with* 2 *or more rows of calli.*

Sepal-points usually long. Lateral lobes of the labellum fringed or toothed, sometimes passing into the middle lobe.
Leaf narrow-linear. Calli of the labellum in 2 rows . . . 6. *C. filamentosa.*
Leaf linear-lanceolate. Calli of the labellum in 4 to 6 rows . 7. *C. Patersoni.*
Sepal-points usually long. Lateral lobes of the labellum very prominent and entire; calli in 4 to 6 rows.
Leaf narrow-linear. Eastern species . . . . 8. *C. clavigera.*

Leaf ovate-lanceolate. Western species . . . . 9. *C. Drummondii.*
Sepal-points usually short.
Leaf oblong or lanceolate. Labellum sessile, broad, undivided, fringed with 4 to 6 irregular rows of calli . . . . 10. *C. hirta.*
Leaf narrow linear. Labellum on a distinct claw, the lateral lobes broadly rounded; one long clavate callus at the top of the claw, and small crowded sessile ones along the centre of the disk . . . . 11. *C. Roei.*
Leaf narrow-oblong or lanceolate. Labellum on a long claw, the lateral lobes narrow falcate; one long clavate callus at the top of the claw, and 2 shorter linear ones, and 2 lines of small ones on the middle lobe . . . . 12. *C. Barbarossæ.*

SECT. 4. **Eucaladenia.**—*Sepals acute or obscurely acuminate, rarely obtuse, the dorsal one usually erect and concave. Labellum inconspicuously veined, the disk with 2 or more rows of calli (sometimes arranged or united at the base almost in a semicircle).*

Flowers large, yellow. Labellum broadly cordate, 3-lobed, the long calli almost in a semicircle . . . . 13. *C. flava.*
Flowers pink or white.
Leaf oblong or lanceolate. Labellum deeply 3-lobed, the long calli in 2 short rows, more or less converging in a semicircle.
Calli all free . . . . 14. *C. latifolia.*
Calli more or less united . . . . 15. *C. reptans.*
Leaf very long. Labellum with the lateral lobes small, the calli short and thick in 2 rows, the rest of the disk covered with papillæ or small calli.
Sepals and petals $\frac{3}{4}$ in. or more. Middle lobe of the labellum entire . . . . 16. *C. suaveolens.*
Sepals and petals little more than $\frac{1}{2}$ in. Middle lobe of the labellum usually toothed . . . . 17. *C. serrata.*
Leaf narrow-linear. Labellum with the lateral lobes broad and obtuse.
Sepals and petals lanceolate, usually 6 to 8 lines long. Calli of the labellum in 2 rarely in 4 rows . . . . 18. *C. carnea.*
Sepals and petals narrow, much contracted at the base, usually under 6 lines long. Calli of the labellum much crowded in irregular rows . . . . 19. *C. testacea.*
Leaf narrow-linear. Labellum with the lateral lobes falcate, almost acute; calli short, densely crowded . . . . 20. *C. congesta.*
Leaf none on the flowering stem. Labellum with the lateral lobes falcate, almost acute; calli linear-clavate, in 2 rows . 21. *C. aphylla.*
Flowers blue.
Leaf usually linear, glabrous or slightly hairy.
Labellum broadly sessile, lateral lobes broad and obtuse, middle lobe lanceolate; calli in 2 rows.
Leaf lanceolate-linear. Eastern species . . . . 22. *C. cærulea.*
Leaf narrow-linear. Western species . . . . 23. *C. saccharata.*
Labellum with a long linear-cuneate claw, lamina ovate, obscurely lobed; calli numerous and crowded . . . . . 24. *C. deformis.*
Leaf lanceolate, silky-pubescent. Flowers large. Labellum cuneate, equally 3-lobed; calli in about 4 rows . . . . 25. *C. sericea.*

SECT. 5. **Pentisia.**—*Sepals and petals obtuse and nearly equal and all spreading. Labellum and column very short; calli small and numerous in longitudinal rows.*

Flowers blue. Labellum broadly ovate, undivided . . . . 26. *C. gemmata.*
Flowers (yellow ?). Labellum broadly ovate, almost acuminate, obscurely 3-lobed . . . . 27. *C. ixioides.*

SECT. 1. LEPTOCERAS, *R. Br.*—Sepals acute or rather obtuse, the dorsal one erect, or incurved and concave, the lateral ones spreading. Petals erect, narrow, linear-clavate, longer than the sepals.

1. **C. Menziesii,** *R. Br. Prod.* 325. Stems slender, glabrous or slightly hairy, usually 6 to 9 in. high. Leaf ovate-lanceolate or oblong-lanceolate, 1 to 2 in. long. Flowers 1 or 2, on long pedicels. Sepals lanceolate, acute, 4 to 5 lines long, the dorsal one concave and incurved, the lateral ones falcate, slightly spreading. Petals much longer than the sepals, erect, narrow-linear, sometimes almost filiform but clavate towards the end, giving the flower a 2-horned aspect. Labellum shorter than the sepals, undivided, ovate-oblong broadly ovate obovate or almost orbicular, erect at the base, recurved towards the end, entire emarginate or produced into a broad obtuse point; calli shortly linear-clavate, more or less distinctly arranged in 2 or 4 rows. Column rather broadly winged.—Hook. f. Fl. Tasm. ii. 27, t. 121 A; *Leptoceras Menziesii*, Lindl. Gen. and Sp. Orch. 416; Endl. in Pl. Preiss. ii. 6; *Caladenia macrophylla*, R. Br. Prod. 325; *Leptoceras macrophylla*, Lindl. l.c.; *L. oblonga*, Lindl. Swan Riv. App. 53, Gen. and Sp. Orch. 416; Endl. in Pl. Preiss. ii. 6.

Victoria. Portland, *Allitt;* Wendu river, *Robertson;* Bunip Creek, *F. Mueller.*
S. Australia. Guichen bay, *F. Mueller.*
W. Australia. King George's Sound and adjoining districts, *Menzies, Oldfield, Preiss, n.*2213, and several others, Swan river, *Oldfield, Preiss, n.* 2212; also *Drummond, n.* 359.

Brown distinguished two species, *C. Menziesii*, with short leaves and two rows of calli to the labellum, and *C. macrophylla*, with long leaves and four rows of calli. Lindley also distinguished a second species with 4 rows of calli, but differently characterized as to leaves, and also as to the shape of the calli; but the 2 or 4 rows do not appear to me to be at all accompanied by any constant difference in the foliage, in the shape of the calli, or in any other respect.

2. **C. fimbriata,** *Reichb. f. Beitr.* 65. Glabrous or nearly so, 6 in. to 1 ft. high. Leaf sometimes very small and rarely above 1 in. long, ovate-lanceolate or oblong. Flowers usually 2 or 3, rather distant on erect pedicels of ½ to ¾ in. Dorsal sepal oblong-lanceolate, acuminate, concave, erect, about 4 lines long; lateral sepals about as long, lanceolate, acuminate, spreading or reflexed; petals erect, linear-clavate, longer than the sepals but not nearly so long as in *C. Menziesii.* Labellum half as long as the sepals and broader than long, truncate and fringed at the broad end, consisting of the very broad lateral lobes, with a very small short and broad middle lobe, either entire or slightly fringed or toothed, the disk without any calli.—*Leptoceras fimbriata* and *L. pectinata*, Lindl. Swan Riv. App. 53; Gen. and Sp. Orch. 416.

W. Australia. Swan river, *Drummond, 1st coll. n.* 856.—This is the only species without calli on the labellum, but it is evidently too closely allied to *C. Menziesii* to be generically separated from it.

SECT. 2. PHLEBOCHILUS.—Sepals obscurely or distinctly acuminate, the dorsal one incurved and concave, erect behind the column or reflexed with it. Labellum broad, with deeply coloured diverging simple or forked veins undivided (consisting entirely of the lateral lobes) or with a very small and obscure middle lobe.

3. **C. Cairnsiana,** *F. Muell. Fragm.* vii. 31. Stems hairy, rather slender, about 6 in. high in the specimens seen. Leaf linear. Flower solitary. Sepals and petals narrow linear, about 5 lines long, not produced into points, pink with dark purple lines. Labellum about as long as the sepals, on a very short claw, broadly ovate, consisting chiefly of the broad large lateral lobes elegantly marked with deep purple diverging simple or forked veins, the middle lobe very short, semi-orbicular, or sometimes obsolete, the margins entire, the calli crowded in 2 rows, those on the claw and at the base of the lobes linear-clavate, those between the lobes thick and depressed. Column narrow and much curved at the base, broadly winged upwards.

W. Australia. North of Stirling Range, *F. Mueller.*

4. **C. multiclavia,** *Reichb. f. Beitr.* 64. A hairy species of 6 in. to 1 ft. resembling some forms of *C. Patersoni.* Leaf linear-lanceolate. Sepals and petals lanceolate striate and spreading for about ½ in. then tapering into a long fine point, the dorsal sepal sometimes linear from the base and usually spreading or reflexed with the column. Labellum contracted into a long narrow claw almost appressed to and as it were lying over the column, the lamina spreading, broadly ovate-rhomboidal, marked with deep-coloured diverging veins, the margin entire; calli on the claw few, linear-clavate, the lower ones connate into a thick plate, multipapulose on the apex, and a number of small calli or papillæ crowded at the base of the lamina. Column about as long as the dilated part of the sepals, slender at the base, winged in the upper half, the wings produced, in the typical form, into oblong lobes.

W. Australia, *Drummond, n.* 441.—Reichenbach fil. refers here also Drummond's n. 440, which, however, in Herb. Lindl. represents a form of *C. Patersoni.*

Var. *brevicuspis.* Points of the sepals much shorter, the dorsal sepal very narrow, and the petals less dilated than in the typical form, the calli at the base of the labellum free or nearly so, long and clavate.—W. Australia, *Drummond.*

5. **C. discoidea,** *Lindl. Swan Riv. App.* 52; *Gen. and Sp. Orch.* 423. Usually very hairy and sometimes above 1 ft. high. Leaf broadly linear or lanceolate, usually long and sometimes above 6 in. Flowers often 2 or 3 on the stems. Sepals 6 to 7 lines long, shortly and acutely acuminate, the dorsal one narrow erect incurved and concave, the lateral ones lanceolate, somewhat falcate, spreading; petals rather longer and narrower. Labellum not much shorter than the sepals, broadly ovate or orbicular, undivided, marked with deep-coloured diverging forked veins and fringed with rather long cilia, very shortly contracted at the base; calli irregularly crowded along or near the centre, thick, obovoid or oblong, the lower ones often longer and clavate. Column narrow and incurved at the base, broadly winged in the upper half.—Endl. in Pl. Preiss. ii. 9 ?

W. Australia. Swan river, *Drummond*, 1*st coll.*—I have not seen Preiss's specimens.

SECT. 3. CALONEMA.—Sepals acuminate, with long or short points, the dorsal one erect incurved and concave. Labellum inconspicuously veined, the middle lobe developed, the lateral lobes sometimes broad and distinct sometimes obsolete, making the whole labellum to appear undivided; the disk with calli more or less distinctly arranged in 2 or more longitudinal rows.

6. **C. filamentosa,** *R. Br. Prod.* 324. Stature and inflorescence of the typical *C. Patersoni* and very nearly allied to it. Leaf narrow-linear. Sepals with the long points of *C. Patersoni*, but usually rather narrower and in the typical form apparently of a uniform red colour. Labellum as in that form tapering into the dark middle lobe and the margins more or less fringed but usually smaller and narrower, the calli varying in thickness but always in 2 rows only along the disk.—Hook. f. Fl. Tasm. ii. 27, t. 121 B; Reichb. f. Beitr. 31; *C. filifera*, Lindl. Swan Riv. App. 52, Gen. and Sp. Orch. 421; Endl. in Pl. Preiss. ii. 8; Field Sert. Pl. t. 73; *C. denticulata*, Lindl. ll.cc.; Endl. in Pl. Preiss. ii. 9.

N. S. Wales. Mudgee, *Woolls.*

Tasmania. Port Dalrymple, *Paterson, R. Brown;* George Town, *Archer;* Southport, *C. Stuart.*

S. Australia. Rivoli bay, *F. Mueller.*

W. Australia. King George's Sound and adjoining districts, *F. Mueller, Muir,* and others; Swan river, *Drummond, 1st coll.*

Var. *pallens.* Flowers apparently pink, the points of the sepals not so long nor so fine as in the ordinary *C. filamentosa.*—Swan river, *Drummond, n.* 442, *Mylne.*

The *C. filamentosa* may be, as suggested by F. Mueller, one of the numerous forms of *C. Patersoni*, chiefly prevalent in W. Australia, but the apparently constant reduction of the calli to 2 rows is accompanied by a difference in foliage which may justify its retention as a species, subject, however, to further investigation of living specimens.

7. **C. Patersoni,** *R. Br. Prod.* 324. More or less hairy and from under 1 ft. to near 2 ft. high, the upper portion and flowers often minutely glandular-pubescent. Leaf oblong-linear or lanceolate, 2 or more inches long. Flowers 1, 2 or very rarely 3, the bracts subtending the pedicels small and linear. Sepals more or less dilated in the lower part, tapering into a long point sometimes dilated again towards the end, the whole length varying from 1 in. to above 2 in., the points or the whole sepal usually glandular-pubescent. Petals shorter and not dilated at the base or rarely as long as the sepals. Labellum rarely half as long as the sepals, broadly ovate or ovate-lanceolate and undivided or dilated into broad lateral lobes, the margins more or less fringed ciliate or crenate, the end or middle lobe recurved; calli numerous, linear or clavate, more or less distinctly arranged in 4 or more longitudinal rows. Column as long as the erect portion of the labellum, shortly winged at the apex; anther-point usually prominent.—Lindl. Gen. and Sp. Orch. 422; Reichb. f. Beitr. 31; *C. pulcherrima*, F. Muell. Fragm. v. 93, 101.

Very variable in the length and proportions of the sepals, in the shape of the labellum and its fringes and calli, and in the colour of the flowers, which are usually of a pale greenish hue outside and yellowish or pink inside, the calli and end or middle lobe of the labellum usually and sometimes the whole disk of a rich purple. The following are the principal forms this species assumes, regarded by Lindley and others as distinct species, but passing too gradually into each other to be clearly marked out from dried specimens.

Var *dilatata*. Labellum with broad lateral usually pale-coloured deeply fringed lobes, the middle lobe ovate, dark-coloured; calli very numerous in several rows. Sepal points very long, especially in the Western specimens. *C. dilatata*, R. Br. Prod. 325; Lindl. Gen. and Sp. Orch. 422; Hook. f. Fl. Tasm. ii. 27, t. 122 B; Reichb. f., Beitr. 32; *C. filamentosa*, Lindl. l.c. 421, not of R. Br.; *C. longicauda*, Lindl. Swan Riv. App 52, t. 8 A; Gen. and Sp Orch. 422; Endl. in Pl. Preiss. ii. 9.

N. S. Wales. Port Jackson and Blue Mountains, *Woolls;* New England, *C. Stuart.*

Victoria. Melbourne and Yarra Yarra, *Adamson, Walter;* from the Grampians to Gipps' Land, *F. Mueller*, and others.

Tasmania. Circular Head, *Gunn;* forest lands near Cheshunt, *Archer.*

S. Australia. Bugle Range, *F. Mueller.*

W. Australia. King George's Sound to Vasse and Swan rivers, *Drummond, 1st coll.*, also *n.* 129, 439, 440, *Preiss, n.* 2217, *Oldfield*, and many others.—Apparently a common Western species, known by the name of "Spider Orchis," and usually with remarkably long points to the sepals, and the lateral lobes of the labellum large, broad, and much fringed.

Var. *typica*. Labellum usually ovate with fringed margins gradually tapering into the broad recurved crenulate undulate or shortly fringed apex or middle lobe, the lateral lobes but little prominent or quite obsolete. Calli of the disk very prominent or linear, distinctly arranged in 4 or 6 rows. Such at least are the usual characters; but the precise shape of the labellum, the prominence of the lateral lobes, the fringes of its margin, the number of calli, their shape usually longer more clavate or more slender towards the base of the disk, more depressed towards the end, the rows regular and distinct or crowded and confused towards the central line, vary often from specimen to specimen, and in a large proportion of the very numerous specimens I have had before me, the precise shape and arrangement of these calli cannot be well ascertained in their dried state. The whole species, and the allied *C. filamentosa* and *C. clavigera* included in it by F. Mueller, require critically working up in their native country, where alone it can be ascertained how far hybridism may have contributed to the confusion of different species or subspecies.—Hook. f. Fl. Tasm. ii. 28, t. 123 A.

N. S. Wales. Bathurst, *Woolls;* Murrumbidgee river, *F. Mueller.*

Victoria. Grampians, *F Mueller;* Burra Burra, *Hinteracker.*

Tasmania. Port Dalrymple, *Paterson* (*R. Brown*); abundant throughout the island, *J. D. Hooker.*

S. Australia. Several specimens from the neighbourhood of St. Vincent's Gulf, *F. Mueller*, and others, belong to this form.

C. pallida, Lindl. Gen. and Sp. Orch. 421, Hook. f. Fl. Tasm. ii. 28, from Circular Head, *Gunn*, is a slight variety of *C. Patersoni*, with a paler less fringed labellum. *C. Behrii* and *C. tentaculata*, Schlecht. in Linnæa, xx. 569 and 571, would appear, from the descriptions given, to be among the innumerable forms assumed by this species.

8. **C. clavigera,** *A. Cunn.; Lindl. Gen. and Sp. Orch.* 422. Stature hairiness and oblong-linear or lanceolate leaf of *C. Patersoni*, of which this is considered by F. Mueller to be a variety. Sepals usually about 1 in. long, lanceolate at the base with long fine points usually but not always clavate at the end; petals shorter and not clavate. Labellum under ½ in. long, the broad yellow lateral lobes quite entire, the purple middle lobe either entire or slightly crenate towards the base; calli in about 4 rows as in *C. Patersoni.*—Hook. f. Fl. Tasm. ii. 28, t. 222 A.

N. S. Wales. Near Bathurst, *A. Cunningham.*

Victoria. Ballarat, *Glendinning;* Malden, *Mrs. Nott.*

Tasmania. Circular Head, *Gunn;* Tamar river, *Archer;* Flinders island, *Milligan.*

9. **C. Drummondii,** *Benth.* Our specimens small, the leaf broader for its length than in any other species of *Caladenia*, ovate-lanceolate, ½ in. long and ¼ in. broad, but not yet fully developed. Flower solitary on a hairy scape of 2 to 3 in., without any empty bract but a long closely sheathing one subtending the terminal pedicel. Sepals and petals nearly of *C. hirta*, about ½ in. long, including the point which is shorter than or about as long as the dilated part, dark-coloured with the dilated margin whitish. Labellum nearly of *C. clavigera*, closely sessile, the lateral lobes broadly rounded, apparently white and entire, the middle lobe broadly oblong-lanceolate, recurved; calli acute, in about 4 rows. Column shortly and broadly winged under the anther, the connective apparently without any point.

W. Australia. Swan river, *Drummond.*—Without much of character in the flower, this species differs from the whole genus in the shape of its leaf. It is only known however from two specimens of Drummond's in Herb. Hooker.

10. **C. hirta,** *Lindl. Swan Riv. App.* 52; *Gen. and Sp. Orch.* 421. Very hairy and often above 1 ft. high, the root more creeping than in most species. Leaf oblong or lanceolate, 2 to 4 in. long. Flowers 2 or 3, more or less pink. Sepals and petals ¾ to 1 in. long, irregularly acuminate, but the points much shorter than in *C. Patersoni*, and always shorter than the dilated portion. Labellum at least half as long as the sepals, ovate-oblong or ovate-lanceolate, obtuse, undivided but more or less fringed from the middle upwards, contracted and erect at the base, recurved towards the end; calli linear, more or less regularly placed in 4 to 6 rows. Column winged upwards. Anther with a prominent point.—Endl. in Pl. Preiss. ii. 9; *C. mollis*, Endl. l.c. 8, according to Reichb. f.

W. Australia. Swan river, *Drummond*, 1*st coll*; Toodyay and Kalgan river, *Oldfield;* Stirling range, *F. Mueller;* West Mount Barren and Salt river, *Maxwell;* also *Preiss, n.* 2213 *and* 2218 (*Reichb. f.*).

11. **C. Roei,** *Benth.* Hairy, 6 to 8 in. high. Leaf narrow-linear, the empty bract on the stem 3 to 5 lines long and almost leaf-like. Flower solitary from a sheathing bract of 4 or 5 lines. Sepals and petals 6 to 7 lines long including the points which are much shorter than the dilated part. Labellum with a distinct unwinged claw of about 1 line, the lateral lobes very large and broad, light-coloured, not fringed, the whole labellum expanding to a breadth of nearly ½ in., the middle lobe much smaller, recurved dark-coloured and denticulate or fringed with short calli; one large long callus between the lateral lobes at the top of the claw, and small obtuse calli compactly crowded along the central line. Column long, incurved, broadly and shortly winged under the anther.

W. Australia, *Roe.*—Only seen in Herb. Hook. Allied to *C. Barbarossæ*, but differing in foliage and very much in the labellum.

12. **C. Barbarossæ,** *Reichb. f. Beitr.* 64. Stem hairy, 6 to 10 in. high. Leaf lanceolate or oblong-linear, attaining about 2 in. Flower solitary from a loosely sheathing bract. Sepals and petals 6 to 8 lines

long, linear or linear-lanceolate, produced into a short point, of a pale colour with a dark central line. Labellum with an unwinged channelled claw of about 2 lines, the lateral lobes linear-falcate and erect, the middle lobe twice as long, broadly oblong, recurved, the margin fringed, the disk with 2 curved lines fringed with small calli, a long erect clavate callus dilated or 2-horned at the end at the top of the claw between the lateral lobes, and on each side of it a smaller pubescent linear or oblong callus. Column winged from the middle upwards. Anther without any point.

W. Australia. Swan river, *Drummond*, *1st coll.* (n. 861, according to Reichb. f.). —Although the conspicuous callus on the labellum of this species and of *C. Roei* is much like the appendage of *Glossodia*, its position with regard to the labellum and column is very different, and does not necessarily require the union of the two genera.

SECT. 4. EUCALADENIA.—Sepals acute or obscurely acuminate, rarely obtuse, not produced into a distinct point, the dorsal one usually erect and concave but sometimes not very different from the others. Labellum inconspicuously veined; calli sometimes in 2 rows either parallel or short and almost joining in a semicircle, sometimes irregularly arranged in 2 to 4 rows or crowded along the centre.

13. **C. flava,** *R. Br. Prod.* 324. Hairy, more glandular than most species and usually low, rarely attaining 1 ft., the underground stems very woolly and knotty. Leaf lanceolate, rather large for the plant. Flowers large, yellow, usually 2 to 4 or even 5 on a flexuose rhachis. Sepals and petals broadly lanceolate, rather acute or almost acuminate, contracted at the base, the lateral sepals often above 1 in. long, with a somewhat darker middle line; dorsal sepal rather smaller, less yellow with a more or less distinct reddish line or red blotches along the centre. Petals still shorter, pale yellow or whitish and more red in the centre. Labellum 3 to 4 lines long and broad, with a very short broad concave claw, the broad lamina cordate at the base, deeply 3-lobed, the lateral lobes ovate, shortly acuminate, the middle lobe rather longer and lanceolate, bordered on each side by 2 or 3 long linear-clavate calli; calli of the disk linear-clavate, in 2 rows almost converging into a semicircle. Column winged from the base. Anther with a long point.—Lindl. Gen. and Sp. Orch. 418; Endl. in Pl. Preiss. ii. 7; F. Muell. Fragm. vi. 83; Reichb. f. Beitr. 29.

W. Australia. King George's Sound and adjoining districts, *Menzies*, *F. Mueller*, and many others, and thence to Swan river, *Drummond*, 1st *coll. n.* 827, *Preiss*, *n.* 2209.

14. **C. latifolia,** *R. Br. Prod.* 324. Hairy, from ½ to 1 ft. high. Leaf oblong-lanceolate, 1½ to 4 in. long. Flowers pink or rarely white, usually 2 or 3, rather distant, on short pedicels. Lateral sepals varying in different specimens from 6 to 11 lines long, oblong-lanceolate, obtuse or scarcely acute, the dorsal sepal rather shorter and more acute, the petals somewhat shorter and more lanceolate. Labellum not ⅓ the length of the sepals, shortly cuneate at the base, deeply 3-lobed, the lateral lobes oblong obtuse and entire, the middle lobe longer, ovate or

broadly lanceolate, fringed near the base with a few marginal calli, the calli of the disk linear-clavate, rather long, in 2 short converging rows, sometimes almost forming a semicircle, sometimes rather longer and more parallel. Column shortly and rather broadly winged at the apex. Anthers with a long point.—Lindl. Gen. and Sp. Orch. 419; Reichb. f. Beitr. 30, 64; Hook. f. Fl. Tasm. ii. 28; *C. mollis*, Lindl. Swan Riv. App. 51; Gen. and Sp. Orch. 419; *C. elongata*, Lindl. Swan Riv. App. 52; Gen. and Sp. Orch. 419.

N. S. Wales. Cape Howe, *F. Mueller.*

Victoria. Wendu Vale, *Robertson;* Melbourne, *Adamson;* Mount William, *F. Mueller;* Little river, *Fullagar.*

Tasmania. Port Dalrymple, *R. Brown;* Woolnorth, Circular Head, and Georgetown, *Gunn;* Hobarton, *Archer.*

S. Australia. St. Vincent's Gulf and neighbouring mountains, *F. Mueller* and others.

W. Australia (usually with rather larger flowers). King George's Sound, *F. Mueller* and others; Swan river, *Drummond, 1st coll. n.* 838; *Preiss, n.* 2184, *Oldfield.*

Var. *angustifolia.* Leaves almost linear.—Upper Hay river, *Miss Warburton;* Lake Muir, *Muir.*

C. marginata, Lindl. Swan Riv. App. 51, Gen. and Sp. Orch. 419, from King George's Sound, is, I think, a variety of *C. latifolia,* with the calli shorter, more numerous, in 2 longer rows. *C. ochreata,* Lindl. ll. cc., from Swan river, *Drummond,* 1st *coll.*, is a similar variety, with the middle lobe crenulate only, not prominently fringed. *C. marginata* and *C. ochreata* of Endl. Pl. Preiss. ii. 8, are referred by Reichenbach f. to the typical *C. latifolia.*

15. **C. reptans,** *Lindl. Swan Riv. App.* 52; *Gen. and Sp. Orch.* 419. A small 1-flowered species, with apparently a creeping underground stem, in other respects closely resembling *C. latifolia,* of which it is perhaps a variety. Leaf oblong or lanceolate. Sepals and petals of *C. latifolia* or rather more obtuse; labellum contracted into a longer claw, deeply 3-lobed, the middle lobe not fringed, the calli of the disk long and thick, more or less united at the base into 2 deeply-lobed laminæ, forming 2 short converging rows placed in a semicircle or almost transverse.

W. Australia. Swan river, *Drummond,* 1*st coll.*—Reichenbach fil. Beitr. 64, refers also to this species *C. Preissii* and *C. nana,* Endl. in Pl. Preiss. ii. 7.

16. **C. suaveolens,** *Reichb. f. Beitr.* 67. Glabrous and usually 1 ft. high or rather more. Leaf linear or linear-lanceolate, 6 to 8 in. long, and often 2 or 3 empty sheathing scales on the stem. Flowers 2 to 6, rather distant, almost sessile within sheathing bracts of $\frac{1}{2}$ to $\frac{3}{4}$ in. Dorsal sepal lanceolate, acuminate, incurved, concave, $\frac{3}{4}$ to nearly 1 in. long; lateral sepals and petals nearly as long, linear, spreading or recurved. Labellum not half so long as the sepals, the erect part broad with the erect lateral lobes rounded and not very prominent, the middle lobe ovate-oblong, obtuse, recurved; calli in 2 rows along the claw or erect part and between the lateral lobes, sometimes almost confluent, the remainder of the disk almost covered with smaller calli or papillæ more or less arranged in several rows. Column broadly winged. Anther

with a prominent point.—*Lyperanthus suaveolens*, R. Br. Prod. 325; Lindl. Gen. and Sp. Orch. 392; F. Muell. Fragm. v. 98; Reichb. f. Beitr. 32; *Caladenia sulphurea*, A. Cunn. in Field, N. S. Wales, 361. *Leptoceras sulphurea*, Lindl. Gen. and Sp. Orch. 416.

N. S. Wales. Port Jackson to the Blue Mountains, *R. Brown*, *Woolls*, *A. Cunningham*, and others; New England, *C. Stuart*.
Victoria. East Gipps' Land, *Walter*.
Tasmania. Dentrecastreaux Channel, *Milligan*.

17. **C. serrata,** *Reichb. f. Beitr.* 67. Nearly resembles *C. suaveolens* in habit and foliage, but usually a stouter and taller plant. Leaf broadly linear, often above 1 ft. long. Flowers 4 to 6, "greenish outside purplish pink inside, the labellum yellowish towards the tip." Bracts above 1 in. long, finely acuminate. Sepals shortly acuminate, the dorsal one lanceolate, incurved, concave, 6 to 7 lines long; the lateral sepals and petals narrower and rather longer. Labellum more than half as long, the lateral lobes erect as in *C. suaveolens*, but rather broader, the middle lobe more lanceolate, the margins usually undulate crisped or shortly fringed; calli linear, in 2 rows along the centre, and smaller ones in several rows on the remainder of the disk. Column winged. Anther acuminate.—*Lyperanthus serratus*, Lindl. Gen. and Sp. Orch. 393; Endl. in Pl. Preiss. ii. 6.

W. Australia. Swan river, *Drummond*, 1*st coll.*; Cape Leschenault, *Oldfield;* King George's Sound, *Muir;* near Wulgenup, *Preiss*, *n.* 2189, *F. Mueller*.

18. **C. carnea,** *R. Br. Prod.* 324. Slender, usually under 1 ft. high, sparingly hairy or nearly glabrous or glandular papillose. Leaf narrow-linear, often long. Flowers 1 to 3, pink in the typical form, often longitudinally veined and usually with transverse bands of a darker hue on the labellum and column. Sepals 6 to 8 lines long, or in some varieties rather smaller, lanceolate, acute, the dorsal sepal erect and as well as the petals rather smaller than the lateral ones. Labellum not above half as long as the lateral sepals, the lateral lobes broad obtuse and prominent, the middle lobe lanceolate, recurved, fringed with a few linear calli; calli of the disk linear in 2 rows in the typical form. Column narrowly winged.—Lindl. Gen. and Sp. Orch. 417; Endl. Iconogr. t. 51; Hook. f. Fl. Tasm. ii. 29, t. 124 A; Reichb. f. Beitr. 28; *Arethusa catenata*, Sm. Exot. Bot. t. 104; *C. alata*, R. Br. Prod. 324 (with smaller flowers); Lindl. Gen. and Sp. Orch. 418; Hook. f. Fl. Tasm. ii. 30, t. 125; Reichb. f. Beitr. 29; *C. angustata*, Hook. f. l.c.

Queensland. Keppel and Shoalwater bays, *R. Brown;* from Brisbane to Wide bay in great abundance, *Leichhardt*, *F. Mueller*, and others; Rockhampton, *O'Shanesy;* Nerkool Creek, *Bowman;* Darling Downs, *Law;* Mount Elliott, *Fitzalan*.
N. S. Wales. Port Jackson to the Blue Mountains, *R. Brown* and many others; northward to Richmond river, *Henderson;* New England, *C. Stuart;* southward to Twofold bay, *F. Mueller*.
Victoria. From the Glenelg to Gipps' Land, *F. Mueller*, *Robertson*, and others.
Tasmania. Abundant throughout the island, *J. D. Hooker* and others.
S. Australia. From the Glenelg to St. Vincent's Gulf, numerous localities, *F. Mueller* and others; Kangaroo island, *Waterhouse*.

Var. *alba*. Flowers white. I can see no other difference. *C. alba*, R. Br. Prod. 323; Lindl. Gen. and Sp. Orch. 417; Reichb f. Beitr. 28.—Port Jackson, *R. Brown, Woolls, A. Cunningham*, and a few of the southern specimens.

Var. *quadriseriata*. Labellum with 4 rows of calli; flowers pink.—Between Rivoli bay and Mount Gambier, *F. Mueller;* Southport, Tasmania, *C. Stuart.*

19. **C. testacea,** *R. Br. Prod.* 324. A slender plant with narrow linear leaves, very much resembling the smaller specimens of *C. carnea*, and united with that species by F. Mueller. Sepals and petals narrower and more contracted at the base. Labellum with the lateral lobes less prominent and more fringed, the calli of the disk numerous, much crowded in about 4 rows, the lower ones mostly stipitate or oblong-clavate, the upper ones very dense and depressed.—Lindl. Gen. and Sp. Orch. 420; Reichb. f. Beitr. 30; *C. gracilis*, R. Br. l.c.; Reichb. f. l.c.; Lindl. l.c. 423; *C. angustata*, Lindl. l.c. 420.

N. S. Wales. Port Jackson to the Blue Mountains, *R. Brown, A. Cunningham, Woolls*, and others.

Victoria. Port Phillip, *Gunn.*

Tasmania. Port Dalrymple, *R. Brown;* Southport, *C. Stuart.*

20. **C. congesta,** *R. Br. Prod.* 324. A slender glandular-pubescent or nearly glabrous species, allied to *C. carnea* but easily distinguished by the labellum. Stems 9 in. to 1 ft. high. Leaf narrow-linear. Flowers 1 or 2, pink. Sepals and petals narrow-lanceolate, acute, ½ to ¾ in. long, the dorsal sepal erect incurved and concave. Labellum fully half as long as the sepals, narrow, contracted into a claw, 3-lobed, the lateral lobes erect incurved and rather long, the middle lobe longer, narrow-lanceolate, recurved, densely covered with thick obtuse calli, either sessile or the lower ones somewhat contracted at the base.—Lindl. Gen. and Sp. Orch. 420; Hook. f. Fl. Tasm. ii. 30, t. 124; Reichb. f. Beitr. 31.

N. S. Wales. Near Bathurst, *A. Cunningham;* in the interior, *M'Arthur.*

Tasmania. Port Dalrymple, *R. Brown;* open forest land, Cheshunt, and Port Sorrel, *Archer.*

W. Australia? Swan river, *Mangles in Herb. Lindl.*—Perhaps not really belonging to this species.

21. **C. aphylla,** *Benth.* Stems slender almost filiform, glabrous, attaining 1 to 1½ ft., arising from a rather large ovoid tuber, without any leaf at all at the time of flowering, and with only 1 or 2 short scarious scales at the base, and no empty bract higher up except a small one at a short distance from the flower (subtending the pedicel). Flower solitary, slightly papillose or quite glabrous. Sepals and petals narrow-lanceolate, acutely but very shortly acuminate, tapering at the base, ¾ to nearly 1 in. long, the dorsal one erect and concave. Labellum more than half as long as the sepals, contracted into a claw, the lateral lobes erect incurved almost acute the middle lobe longer lanceolate and recurved, the margin entire; calli rather long, linear-clavate, numerous or few, in 2 rows sometimes not extending beyond the lateral lobes, sometimes reaching halfway along the middle lobe.

W. Australia. King George's Sound, *Harvey;* Hay and Kalgan rivers, *F. Mueller;* near Three-miles Plain, *Maxwell;* Forest Hill, *Muir.*

22. **C. cærulea,** *R. Br. Prod.* 324. A smaller plant than *C. deformis*, more glabrous, the empty bract on the stem smaller and more erect but with the solitary blue flower of that species. Leaf linear or linear-lanceolate. Stem rarely 6 in. high. Lateral sepals 4 to 5 lines long, oblong-lanceolate, glandular-dotted but scarcely veined; dorsal sepal as long but narrower; petals both narrower and shorter. Labellum more than half or nearly as long as the lateral sepals, broad almost from the base; lateral lobes broad, erect, obtuse, with transverse bands of a darker hue, middle lobe lanceolate, almost acute, entire or slightly fringed; calli linear-clavate, in 2 rows. Column rather narrowly winged almost from the base. Anther-point very short.—Lindl. Gen. and Sp. Orch. 417; Hook. f. Fl. Tasm. ii. 29; Reichb. f. Beitr. 28.

N. S. Wales. Port Jackson, *R. Brown*, *Sieber*, *n.* 163, and many others; New England, *C. Stuart;* southward to Twofold bay, *F. Mueller.*

Victoria. North of Wombayne river, *F. Mueller.*

Tasmania, *Bauer* (*Brown, Reichenbach fil.*).—I have seen no Tasmanian specimens; there are none in herb. R. Brown.

23. **C. saccharata,** *Reichb. Beitr.* 63. Very nearly allied to *C. cærulea* and perhaps only a local form of it, the structure of the flower apparently the same. The leaves are rather longer and narrower and the petals and sepals also longer and narrower, the colour of the flowers, probably blue, but not recorded. The station however is so very different from that of *C. cærulea*, which in the East has not the extended range of *C. deformis*, that a further examination of fresh flowers may prove it to be really distinct.

W. Australia, *Drummond, 4th coll. n.* 324.

24. **C. deformis,** *R. Br. Prod.* 324. A small species, usually only slightly hairy and rarely much above 6 in. high. Leaf linear, sometimes rather long. Scape with a single blue flower larger than in *C. cærulea.* Sepals and petals nearly equal, 7 to 8 lines long, linear-oblong or lanceolate, rather obtuse, slightly contracted towards the base, the dorsal sepal more erect than the others and concave. Labellum with an erect linear claw of 2 to 3 lines, the lamina ovate, recurved, more or less fringed on the margin, the lateral lobes scarcely prominent or quite obsolete; calli of the disk oblong linear or slightly clavate, very numerous and crowded, sometimes covering the whole of the reflexed portion of the lamina, fewer and smaller along the claw. Column narrowly winged the whole length. Point of the anther short.—Reichb. f. Beitr. 29; *C. barbata*, Lindl. l.c. 418; Hook. f. Fl. Tasm. ii. 29, t. 123 B; *C. unguiculata*, Lindl. Sw. Riv. App. 51; Gen. and Sp. Orch. 418; Endl. in Pl. Preiss. ii. 7.

N. S. Wales. Near Albury, *Beattie.*

Victoria. Portland, *Allitt;* Port Phillip and Melbourne, *Gunn, Adamson;* Nangatta Range and Genoa river, *F. Mueller;* Ararat, *Green;* Burra-Burra, *Hinteracker.*

Tasmania. Port Dalrymple, *R. Brown;* abundant throughout the island, *J. D. Hooker;* Flinders island, *Milligan.*

S. Australia. From the Murray to St. Vincent's Gulf, *F. Mueller;* York Peninsula, *Fowler;* Biscuit flat, *Schulzen.*

W. Australia. Point Henry, *F. Mueller;* Lake Muir, *Muir;* Swan river, *Drummond,* 1*st coll., Preiss, n.* 2191.

Var. *albiflora.* Flower white, with the lateral lobes of the labellum rather more prominent.—S. Australia, *F. Mueller.*

25. **C. sericea,** *Lindl. Sw. Riv. App.* 52, *and Gen. and Sp. Orch.* 418. Usually softly villous, the hairs especially on the leaves shorter more dense and silky than in any other species. Leaf oblong-lanceolate, often rather broad, 1 to 3 in. long. Stem from 6 in. to 1 ft. with 1 or 2 rather large blue flowers, much incurved in the bud. Sepals and petals nearly equal, $\frac{3}{4}$ to 1 in. long, oblong-lanceolate, obtuse or nearly acute, the dorsal one more erect than the others and concave. Labellum at least half as long as the sepals, contracted at the base, cuneate upwards, nearly equally 3-lobed at the end, the lateral lobes erect, shortly oblong, incurved and obtuse, the middle lobe recurved, shortly fringed with a few calli; calli of the disk short, linear or slightly clavate, in about 4 rows, with a few long linear-clavate ones at the base of the limb, the lowest ones sometimes united in linear or oblong plates.

W. Australia. Swan river, *Drummond,* 1*st coll., also n.* 119; King George's Sound, *Muir;* Upper Hay river, *Miss Warburton.*

SECT. 5. PENTISIA.—Sepals and petals nearly equal, all obtuse and spreading. Labellum and column very short. Calli of the disk small and numerous, in longitudinal rows.

As observed by Reichenbach fil., this section in its perianth connects *Caladenia* with *Glossodia,* but the calli of the labellum are entirely those of the former, without the remarkable basal appendages of the latter genus.

26. **C. gemmata,** *Lindl. Sw. Riv. App.* 52, *and Gen. and Sp. Orch.* 420. Loosely hairy and 6 to 8 in. high when 1-flowered, rather taller when 2-flowered. Leaf ovate or ovate-lanceolate, rarely above 1 in. long. Flower rather large, of a soft deep blue. Sepals and petals broadly elliptical-oblong, about $\frac{3}{4}$ in. long, obtuse, contracted at the base and almost clawed. Labellum broadly ovate, undivided, erect at the base but scarcely contracted into a claw, recurved at the end and obtuse; calli small and clavate in numerous longitudinal rows. Column the length of the labellum, narrowly winged the whole length. Anther with a prominent point.—Endl. in Pl. Preiss. ii. 8; *C. pellita,* Endl. l.c. (*Reichb. f.*).

W. Australia. King George's Sound to Swan river, *Drummond, Oldfield, F. Mueller,* and others, *Preiss, n.* 2193 (mixed with *Glossodia Brunonis,* according to Reichenbach f.); eastward to Salt river and South west bay, *Maxwell.*

27. **C. ixioides,** *Lindl. Sw. Riv. App.* 52, *and Gen. and Sp. Orch.* 420. Closely resembles *C. gemmata* in habit stature and size and general structure of the flower, and possibly a variety only, but the flowers (from the dried specimens) appear to have been yellow, the labellum is not so broad, more acuminate and sometimes obscurely 3-lobed, and the calli of its surface are more prominent, almost linear.

W. Australia. Swan river, *Drummond,* 1*st coll.*

45. CHILOGLOTTIS, R. Br.

Dorsal sepal erect, incurved, concave, contracted at the base; lateral sepals narrow-linear or terete; petals lanceolate falcate. Labellum on a very short claw, ovate or obovate, undivided, the disk with variously arranged calli. Column elongated, incurved, winged. Anther terminal, erect, 2-celled; pollen-masses granular.—Terrestrial herbs, with small underground tubers. Leaves 2, radical or nearly so. Scape 1-flowered, without any empty bract below the one subtending the terminal pedicel.

Besides the two Australian species which are endemic, there is one from the Auckland islands and another from New Zealand, if the *Caladenia bifolia*, Hook. f., be referred to it. It has the petals broader than the lateral sepals and the 2-leaved habit of *Chiloglottis*, but was referred to *Caladenia* because the wing of the column does not extend behind the anther as in the other species, a difference like that which separates the two sections of *Glossodia*. Reichenbach fil. proposes to unite the whole genus *Chiloglottis* with *Caladenia*.

Dorsal sepal 5 to 6 lines long, much contracted at the base; lateral sepals linear-terete. Calli of the labellum slender, scattered over the whole disk	1. *C. diphylla.*
Dorsal sepal 8 to 9 lines long, not much contracted at the base; lateral sepals linear. Calli of the labellum thick and crowded along the centre	2. *C. Gunnii.*

1. **C. diphylla,** *R. Br. Prod.* 323. Radical leaves 2, from ovate-elliptical to oblong-lanceolate, usually acute, $\frac{3}{4}$ to $1\frac{1}{2}$ in. long, contracted into a petiole of 2 to 3 lines. Scape from 3 to 6 in. high, bearing occasionally a single sheathing bract near the base besides the one subtending the terminal pedicel. Dorsal sepal cuneate, 5 to 6 lines long, shortly acuminate and much contracted in the lower half; lateral sepals linear-terete, spreading or reflexed, very slender or somewhat thickened in the upper half, as long as or longer than the dorsal one; petals lanceolate, attached by a broad base, acute, rather shorter than the sepals. Labellum more or less obovate, obtuse or acute, contracted at the base into a distinct long or short claw; calli covering the disk, mostly shortly linear, but some larger and thicker ones arranged more or less in 2 rows, and 1 or 2 at the base of the lamina, much longer rather thick and reflexed, all however very variable as to form and numbers. Column about as long as the petals.—Bauer, Ill. Pl. N. Holl. t. 8; Lindl. Gen. and Sp. Orch. 386; Hook. f. Fl. Tasm. ii. 23; Reichb. f. Beitr. 27; *Caladenia diphylla*, Reichb. f. Beitr. 67; *Epipactis reflexa*, Labill. Pl. Nov. Holl. ii. 60, t. 211, f. 1; *Acianthus? bifolius*, R. Br. Prod. 322; Reichb. f. Beitr. 26.

Queensland. Brisbane river, Moreton bay, *F. Mueller*.
N. S. Wales. Port Jackson, *R. Brown*, *Woolls*; Clarence river, *Wilcox*.
Victoria. Wilson's Promontory, *F. Mueller* (without flowers, but probably this species).
Tasmania. Shaded places, Woolnorth, Circular Head, &c., *Gunn*, *Archer*; Southport, *C. Stuart*.

The calli of the labellum vary much, the slender clavate ones are sometimes limited to a broad tuft or patch at the base of the lamina, sometimes extend over a great part of it and descend along the claw, the thick ones are usually few and the 1 or 2 long reflexed ones are often very prominent; sometimes also there are a pair of small ones near the base of the claw.

2. **C. Gunnii,** *Lindl. Gen. and Sp. Orch.* 387. Often a smaller plant than *C. diphylla*, but not so slender. Leaves rather larger and broader, sometimes nearly 2 in. long. Scape usually short, but in some specimens 7 to 8 in. long, the sheathing bract obtuse and loose, $\frac{1}{2}$ in. long. Dorsal sepal 8 to 9 lines long, obovate-oblong, much less contracted below the middle than in *C. diphylla*, acuminate acute or almost obtuse. Lateral sepals narrow-linear or slightly lanceolate at the base, acutely acuminate but not terete; petals rather broadly lanceolate-falcate, almost as long as the sepals. Labellum on a very short claw, broadly ovate, acute: calli all thick, either crowded along the centre of the disk or more or less arranged in 2 rows, without any of the slender linear-clavate ones of *C. diphylla*.—Hook. f. Fl. Tasm. ii. 23, t. 108; *Caladenia Gunnii*, Reichb. f. Beitr. 67.

Victoria. Dandenong range, higher mountains on the Avon river, summit of Mount William, at an elevation of 5000 ft., *F. Mueller.*

Tasmania. Shaded banks, &c., Circular Head, Cheshunt, Hobarton, *Gunn, Archer*, and others; Southport, *C. Stuart;* summit of Ben Lomond, at an elevation of 5000 ft., *Milligan.*

46. GLOSSODIA, R. Br.

Sepals and petals nearly equal, spreading. Labellum sessile, undivided, not fringed, without calli or plates on the disk, but, at its base, 2 (sometimes united into 1) linear clavate calli or appendages erect against the column and from half to nearly its whole length. Column erect, often incurved, 2-winged. Anther erect, 2-celled, the outer valves broad, the inner much smaller, the connective produced into a small point. Pollen-masses granular.—Terrestrial herbs usually hairy, with small underground tubers. Leaf solitary, oblong or lanceolate, from within a scarious sheath close to the ground; flowers 1 to 2 on an erect scape, leafless except an empty sheathing bract at or below the middle, and a similar bract under each pedicel. Flowers erect, blue or purple.

The genus is limited to Australia. It is closely allied to the section *Pentisia* of *Caladenia*, but the peculiar position and form of the calli, constant in all the species, and probably performing some special function in the fertilizing process, may justify its retention as a genus, rather than merging it into *Caladenia*, as proposed by Reichenbach fil. The two sections might almost be considered as distinct genera.

SECT. 1. **Euglossodia.**—*Labellum with a broad biconvex pubescent base. Column-wing not at all or scarcely extending above the base of the anthers. Eastern species.*

Sepals and petals 7 to 10 lines long. Labellum-appendage single, dilated and 2-dentate at the apex . . . . 1. *G. major.*
Sepals and petals 4 to 6 lines long. Labellum-appendages 2, linear-clavate, scarcely united at the base . . . . 2. *G. minor.*

SECT. 2. **Eleutheranthera.**—*Labellum narrow and glabrous. Column-wing extending beyond the anther and forming a hood over it. Western species.*

Flowers usually 2, deep blue. Labellum shorter than the column, lanceolate or linear . . . . 3. *G. Brunonis.*

Flowers usually solitary (reddish-purple?). Labellum as long as or longer than the column, oblong, emarginate . . . . 4. *G. emarginata.*

SECT. 1. EUGLOSSODIA.—Labellum with a broad biconvex pubescent base. Column-wing not at all or scarcely extending above the base of the anthers.

1. **G. major,** *R. Br. Prod.* 326. Hirsute with long spreading hairs with a few shorter ones sometimes glandular. Tuber ovoid, often ½ in. long. Leaf oblong or lanceolate, 1 to 2 in. long. Scape 6 in. to 1 ft. high with 1 or rarely 2 blue flowers. Sepals and petals oblong-lanceolate, obtuse, 6 to 8 lines long, not blotched. Labellum not half so long as the calyx, ovate broad biconvex and pubescent with white hairs in the lower half, the upper half lanceolate blue and glabrous, the basal callus or appendage single, linear, erect against the column, with a broad reflexed 2-dentate or 2-lobed head. Column as long as the labellum, broadly winged, but the wing not at all or scarcely extending along the anther.—Lindl. Gen. and Sp. Orch. 423; Endl. Iconogr. t. 41; Hook. f. Fl. Tasm. ii. 31, t. 120; Reichb. f. Beitr. 34; *Caladenia major*, Reichb. f. Beitr. 67.

Queensland. Moreton bay, *Leichhardt;* Armidale, *Parrott.*
N. S. Wales. Port Jackson to the Blue Mountains, *R. Brown, Sieber, n.* 162, *Fl. Mixt. n.* 519, and many others; in the N.W. interior, *Fraser;* New England, *C. Stuart*; and southward to Twofold bay, *F. Mueller.*
Victoria. Glenelg river, *Robertson;* Portland, *Allitt;* Port Phillip and Melbourne, very common, *F. Mueller*, and others; Ballarat, *Glendinning.*
Tasmania. Common in poor sandy soil throughout the colony, *J. D. Hooker.*
S. Australia. Encounter bay, *Whittaker;* Bugle and Lofty ranges, *F. Mueller.*

2. **G. minor,** *R. Br. Prod.* 326. Hirsute with long spreading hairs mixed with shorter sometimes glandular ones as in *G. major*, but a smaller plant. Stems rarely above 3 or 4 in. high and almost always 1-flowered. Leaf lanceolate, the small sheathing bract usually green. Flower blue. Sepals and petals oblong-lanceolate, 5 to 6 lines long. Labellum about ⅓ the length of the sepals, broad, biconvex and pubescent with white hairs or papillæ in the lower half, the spreading upper half triangular, acute, flat, glabrous, the basal calli or appendages 2, linear, flattened, clavate at the end, rather shorter than the column, very shortly united at the base. Column nearly as long as the labellum, broadly winged but the wing not produced on the anther.—Lindl. Gen. and Sp. Orch. 423; Reichb. f. Beitr. 34; *Caladenia minor*, Reichb. Beitr. 67.

Queensland. Archer's Station, Moreton bay, *Leichhardt.*
N. S. Wales. Port Jackson, *R. Brown, Sieber, n.* 161, and others; New England, *C. Stuart;* southward to Twofold bay, *F. Mueller.*
Victoria. Genoa river, towards Mount Imlay, *F. Mueller.*

Out of more than two hundred specimens from various localities I have only seen one, in herb. F. Muell., from Twofold bay, with two flowers.

SECT. 2. ELEUTHERANTHERA, *Endl.*—Labellum narrow and glabrous. Column-wing extending beyond the anther, adnate to it along the centre, and forming a hood over it.

3. **G. Brunonis,** *Endl. Nov. Stirp. Dec.* 16, *and in Pl. Preiss.* ii. 9. A pubescent or softly hairy plant of 6 in. to 1 ft., with 1 or 2 rather large blue flowers, much resembling *Caladenia gemmata.* Leaf narrow-lanceolate, 1 to 3 in. long, sometimes nearly glabrous. Sepals and petals ½ to ¾ in. long. Labellum reduced to an irregularly lanceolate or almost linear lamina, often shorter than the column, entire, without calli on its disk, but at its base are 2 long thick linear obtuse calli often as long as the lamina, sometimes united at the base, erect against the column. Column half as long as the sepals, with a broad wing produced beyond the anther into a concave hood. Anther-case pubescent, shortly acuminate.—Lindl. Gen. and Sp. Orch. 424; F. Muell. Fragm. vi. 83; *Caladenia Brunonis,* Reichb. Beitr. 67.

W. Australia. King George's Sound and adjoining districts, *Oldfield, F. Mueller,* and many others; Swan river, *Drummond, 1st coll.*; Bremer bay and Gales brook, *Maxwell.*

4. **G. emarginata,** *Lindl. Gen. and Sp. Orch.* 424. Nearly allied to *G. Brunonis,* with the same habit and foliage, the scape however more frequently 1-flowered, the flower usually larger, not so blue in the dried state, and described by Oldfield as rose-coloured. Column with a hood-shaped wing extending beyond the anther as in that species. Labellum more developed, often exceeding the column, broadly oblong-linear, very obtuse or truncate and usually emarginate, the basal calli or appendages linear, slightly clavate, about as long as the labellum.—*Caladenia emarginata,* Reichb. Beitr. 67.

W. Australia. Swan river, *Drummond, 1st coll.*; Vasse river, *Pries;* Tone, Tweed, Kalgan, and Vasse rivers, *Oldfield;* Greenough Flats, *C. Gray;* Lake Muir, *Muir.*

TRIBE 6. OPHRYDEÆ.—Anther adnate to the top of the column over the stigma, the cells usually forming 2 lobes. Pollen-masses 2, granular, attached by caudicles to one or two glands or pouches over the stigma. Terrestrial herbs; rhizomes with annually renewed tubers. Stems simple, leafy. Flowers spicate.

47. HABENARIA, R. Br.

Dorsal sepal erect, very concave; lateral sepals free, connivent or spreading; petals entire or bipartite, usually connivent under the dorsal sepal. Labellum 3-lobed or rarely entire, with a spur or pouch at the base. Column very short, with 2 anterior linear-clavate processes lying on or partially adnate to the base of the labellum. Anther erect, with a broad connective and marginal cells, but the connective usually so short that the erect diverging cells appear disconnected. Pollen granular in 2 masses in each cell, each pair with a caudicle attached to a gland or an appendage of the stigma opposite to the cell, and more or less confluent with it.—Terrestrial herbs, with underground usually ovoid tubers. Leaves alternate on the stem or collected near the base. Flowers several or many in a terminal spike.

A large genus dispersed over the warmer and temperate regions of both the New and the Old World. Of the five Australian species, two are East Indian, the other three, as far as known, endemic.

Stem leafy, with broad leaves. Petals 2-partite. Labellum with 3 narrow-linear lobes. Anther-connective as high as the cells . 1. *H. trinervis.*
Leaves narrow, near the base of the stem. Petals undivided. Anther-connective very much shorter than the cells.
Lateral lobes of the labellum long and very narrow-linear like the middle lobe.
Leaves narrow-oblong. Spur of the labellum above 1 in. long 2. *H. elongata.*
Leaves linear. Spur of the labellum under ½ in. long . . . 3. *H. graminea.*
Lateral lobes of the labellum lanceolate-falcate; middle lobe linear. Spur longer than the sepals . . . . 4. *H. ochroleuca.*
Lateral lobes of the labellum very short and broad or quite obsolete, rarely shortly acuminate; middle lobe oblong or broadly linear. Spur shorter than the sepals . . . . . 5. *H. xanthantha.*

1. **H. trinervis,** *Wight Ic. Pl. t.* 1701. Stems erect, leafy, 1 ft. high or rather more. Leaves ovate-oblong or ovate-lanceolate, 2 to 4 in. long, more or less prominently 3- or 5-nerved. Spike not long and rather dense, the broad bracts often as long as the ovary and very conspicuous. Sepals broad, about 3 lines long, all nearly equal, the dorsal one erect and concave, the lateral ones oblique; petals about as long, divided to the base into 2 linear segments, both almost erect and parallel. Labellum not exceeding the sepals, of 3 linear lobes; spur ½ in. long, clavate towards the end. Anther about 1 line long, the marginal cells not exceeding the connective and much longer than the processes of the stigma. Anterior processes of the column oblong, incurved, rather flat, free from the labellum.

N. Australia. Port Darwin, *Schultz, n.* 828.—The species is also in the Peninsula of India. It is very nearly allied to the *H. digitata*, Lindl., from the Silhet mountains and from Moulmeyn, which differs chiefly in the anterior lobes of the petals deflexed and curved like the lateral lobes of the labellum.

2. **H. elongata,** *R. Br. Prod.* 313. A rather stout species, above 1 ft. high. Leaves oblong, 2 to 4 in. long. Spike rather dense. Dorsal sepal ovate, obtuse, 2 to 2¼ lines long; petals about as long, broadly falcate; lateral sepals rather longer, adnate at the base to the sides of the spur. Labellum lobes very narrow-linear, the middle one above ½ in. long, the lateral ones rather shorter; the spur at least 1¼ in. long, thickened beyond the middle. Anterior processes of the column clavate and cristate.—Lindl. Gen. and Sp. Orch. 317; F. Muell. Fragm. vii. 15; Reichb. f. Beitr. 6.

N. Australia. Arnhem S. bay, and islands off the coast of Arnhem's Land, abundant, *R. Brown.*
Queensland. Rockhampton, *O'Shanesy.*

3. **H. graminea,** *Lindl. Gen. and Sp. Orch.* 318. Stems slender, under 1 ft. high. Leaves at the base of the stem linear or linear-lanceolate, acute, under 3 in. long. Spike slender, short or long, with short narrow bracts. Dorsal sepal in the typical form 2½ lines long, lanceolate, concave; petals as long, slightly falcate; lateral sepals rather longer, broadly falcate, all rather obtuse. Labellum deeply di-

vided into 3 narrow-linear lobes about as long as the sepals; the spur about 4 lines long, thickened beyond the middle. Anterior processes of the column curved, rather long, adnate at the base to the labellum. Anther-cells adnate to linear processes of the stigma quite separated from each other by the very short connective and rostellum.—F. Muell. Fragm. vii. 16.

Queensland. Rockingham bay, *Dallachy.*—The species is also in Khasia and Silhet.

Var. *arnhemica.* Habit and foliage of the typical form, but the flowers much smaller. Galea 1¼ to 1½ lines long, slightly recurved and acute with a broad base; lateral sepals rather longer, the spur not 3 lines long and scarcely thickened beyond the middle.—*H. arnhemica*, F. Muell. Herb.

N. Australia. Port Darwin, *Schultz, n.* 162, 188, and with still smaller flowers, Port Essington, *Armstrong.*

4. **H. ochroleuca,** *R. Br. Prod.* 313. Under 1 ft. high. Leaves few, distant or collected at the base of the stem, lanceolate, short. Spike dense or interrupted. Dorsal sepal broad, obtuse, about 2 lines long; lateral sepals rather longer; petals considerably shorter. Labellum with the lateral lobes lanceolate falcate, the middle lobe narrow-linear, as long as the sepals; spur longer than the sepals, much and shortly clavate at the end. Anterior processes of the column very shortly adnate.—Lindl. Gen. and Sp. Orch. 323; Reichb. f. Beitr. 6.

N. Australia. Islands of the North Coast, *R. Brown.*

5. **H. xanthantha,** *F. Muell. Fragm.* vii. 16. Stems slender, often above 1 ft. high. Leaves at the base of the stem, linear or linear-lanceolate and acutely acuminate as in *H. graminea.* Flowers yellow, in a rather dense spike. Dorsal sepal and petals erect, obtuse, about 3 lines long; lateral sepals nearly the same length, quite free from the spur. Labellum linear-oblong or lanceolate, obtuse, as long as the petals and more or less 3-nerved like them, sometimes quite entire, sometimes with lateral lobes short and broad or tapering into a linear point, and the two lateral lobes sometimes unlike each other; spur curved, shorter than the sepals. Anterior processes of the column quite free from the labellum, and a small lateral tooth on each side between the anther and the anterior processes. Anther-cells deeply separated as in *H. graminea.*—*H. propinquior*, Reichb. f. Beitr. 53.

Queensland. Rockingham bay, *Dallachy.*

TRIBE 7. APOSTASIEÆ.—Anthers 2, lateral, near the base of the style, with a dorsal rudimentary or rarely perfect anther. Stigma terminal. Pollen granular. Terrestrial herbs.

48. APOSTASIA, Blume.

(Niemeyera, *F. Muell.*).

Sepals and petals 3 each, nearly equal and similar, spreading or recurved, the labellum similar to the other petals. Anthers 2, oblong-linear, attached near the base of the style, erect and embracing the style, with occasionally a dorsal staminode or imperfect anther. Style linear or filiform, with a terminal somewhat dilated obtuse or 3-toothed stigma. Ovary 3-celled, the placentas affixed to the axis.—Erect herbs with leafy stems. Flowers small, yellow, in simple or branched spreading racemes in the upper axils.

Besides the Australian species, which appears to be endemic, there are two or three nearly allied to it in East India and the Archipelago.

1. **A. stylidioides,** *Reichb. f. in Herb. Kew.* An erect glabrous plant, with simple stems of 6 to 8 in. Leaves almost grass-like, linear, tapering into long points, their sheathing imbricate bases covering the stem, varying from 3 to 6 in. long, with 1, 3 or rarely more nerves prominent underneath. Racemes shorter than the leaves, slender but rigid. Bracts lanceolate, 1 to 2 lines long. Ovary nearly 3 lines long at the time of flowering, elongated but still very narrow when in fruit. Sepals and petals narrow-linear, slightly recurved, about 2 lines long, the sepals obtuse, the petals with a dorsal point just below the apex. Anthers 2, without any rudimentary one, the two cells very unequal at the base, the connective scarcely produced beyond the cells. Style rather thick, the stigma with 3 short unequal erect teeth or lobes.—*Niemeyera stylidioides*, F. Muell. Fragm. vi. 96.

Queensland. Rockingham bay, *Dallachy*.—The general habit inflorescence size of flowers, &c. are quite those of the Indian *A. nuda*, with the leaves perhaps narrower and the stature lower. The anthers are those of *A. Wallichiana*, but without the additional staminode of that species, except that in some flowers I find the style abortive or nearly so, and replaced as it were by a staminode. The dorsal points of the petals are very variable, and appear sometimes on the sepals also.

I have followed Brown in considering the group of *Apostasieæ* as a tribe of *Orchideæ* rather than as a distinct Order, notwithstanding the number of anomalies it unites. It has the exceptional inflorescence of *Corymbis*, the perianth of *Thelymitra*, two anthers as in *Cypripedium*, and the 3-celled ovary of *Cypripedium* (*Uropedium* and *Selenipodium*.)

Order CXXI. BURMANNIACEÆ.

Flowers hermaphrodite, regular. Perianth superior, persistent, tubular or campanulate, usually 6-lobed, the 3 inner lobes often smaller or sometimes wanting. Stamens 3 or 6, inserted in the tube and shorter than the perianth. Anthers 2-celled. Ovary inferior, 3-celled or with 3 parietal placentas, the ovules very numerous. Style single, with 3 short branches stigmatic at the clavate or dilated ends. Fruit a capsule opening in loculicidal slits or valves. Seeds minute, the embryo apparently homogeneous.—Herbs, often slender. Leaves entire, radical or nearly so, rarely alternate along the stem, sometimes all reduced to small scales. Flowers terminal, solitary or several along a 2-branched rarely 3-branched rhachis, centrifugally developed, each flower opposite to a small often minute bract.

A small tropical Order, usually frequenting swamps or wet places, or decaying vegetable soils, common to the New and the Old World. The only Australian genus has the general range of the Order.

1. BURMANNIA, Linn.

(Gonyanthes, *Miers.*)

Perianth tubular, 3-winged or 3-angled, the 3 inner lobes smaller or wanting. Anthers 3, sessile or nearly so, below the inner perianth-lobes, the cells short, separated by a broad connective, opening transversely, with a small crested appendage behind each cell. Ovary 3-celled. Capsule opening between the dissepiments.—Herbs with radical leaves or all the leaves rarely reduced to scales. Flowers sessile or pedicellate along the branches of a forked cyme, reduced sometimes to a single flower.

The genus is common to the New and the Old World. One and probably both the Australian species have a wide range over East India and the Archipelago.

Radical leaves lanceolate. Flowers several in a once-forked cyme. Perianth at least twice as long as broad . . . . 1. *B. disticha.*
Leaves linear-setaceous, very small. Flowers solitary or few. Perianth with the wings as broad as long . . . . 2. *B. juncea.*

1. **B. disticha,** *Linn. Spec.* 411. Stems simple or scarcely branched, erect, glabrous, attaining 1 to 2 ft. Leaves chiefly radical, sessile, sheathing at the base, lanceolate, acute, spreading, all under 1 in. in most specimens, 2 to 2½ in. long in luxuriant ones; a few along the stem sometimes similar to the radical ones but smaller and more erect, or more frequently reduced to sheathing scales. Flowers green more or less tinged with blue, or in the Australian specimens more frequently of a deep blue, in a once-forked cyme, sometimes very compact, sometimes each branch 1 to 2 in. long. Bracts shorter than the flower. Perianth including the wings about ½ in. long, and scarcely ¼ in. broad, the 3 outer lobes ovate, concave, not half so long as the tube, the dorsal wings commencing about the middle of the lobes, truncate or rounded at the top, and continued along the tube to the base of the ovary, tapering into the short pedicel; inner lobes of the perianth oblong-linear, from half as long to nearly as long as the outer ones. Anthers immediately under the inner lobes, the cells small, separated by a prominent connective, the dorsal appendages nearly as long as the cells. Capsule usually occupying about half the length of the perianth, but sometimes continued higher up, opening at the top between the ridges with a disposition to split transversely as observed by Thwaites.—Roxb. Corom. Pl. t. 242; *B. distachya*, R. Br. Prod. 265.

N. Australia. Liverpool river, *Gulliver.*—A single slender specimen with only 3 flowers, which however are quite those of *B. disticha.*

N. S. Wales. Port Jackson, *R. Brown, Woolls*; New England, *C. Stuart*; Clarence river, *Beckler*; edges of swamps, Tweed river, *C. Moore, Guilfoyle.*

2. **B. juncea,** *Soland. in R. Br. Prod.* 265. Stems very slender, almost filiform, 6 in. to near 1 ft. high. Leaves few at the base of the stem, linear-filiform, ¼ to ½ in. long, and sometimes one or two smaller ones higher up, but the stem usually with only a few small distant scales. Flowers sometimes only one at the end of the stem, sometimes

in a once-forked cyme with 2 to 4 on each branch, all on short slender pedicels. Perianth including the wings about 3 lines long and quite as broad when in fruit, the outer lobes broad and only ¾ line long, the inner lobes very minute in the flower examined, the dorsal wings rounded at both ends. Anthers at a little distance below the inner lobes. Ovary and capsule occupying about half the length of the tube, the capsule usually opening by a transversely oblique fissure.

N. Australia. Port Essington, *Armstrong*.
Queensland. Endeavour river, *Banks and Solander.*

This species is scarcely to be distinguished but by its longer stem and narrow filiform leaves from the *B. pusilla*, Thw. Enum. Pl. Ceyl. 325, or *Gonyanthes pusilla*, Miers in Trans. Linn. Soc. xviii. 537, t. 38. It is also very closely allied to the common Indian *B. cœlestis*, Don, with which Banks and Solander's specimens are laid down on the same sheet in the Banksian herbarium as one species, the leaves however of *B. cœlestis* are not quite so narrow, and the perianth with the wings is longer than broad, in the latter respect, however, the short broad shape is not so decided in the Banksian as in Armstrong's specimens. Further investigation may induce the reducing both *B. pusilla* and *B. cœlestis* to varieties of *B. juncea.*

ORDER CXXII. IRIDEÆ.

Flowers hermaphrodite, regular or irregular. Perianth superior, with a short or distinct tube, the limb of 6 petal-like segments, the 3 inner ones sometimes very small. Stamens 3, inserted at the orifice of the tube or base of the outer segments, or rarely (in *Campynema*) 6; all fertile or (in *Diplarrhena*) one reduced to a barren filament. Filaments free or united in a tube. Anther-cells 2, parallel, erect, opening outwards. Style more or less divided into 3 lobes or branches, usually stigmatic at the end and sometimes broad and petal-like. Ovary inferior, 3-celled, with several often numerous ovules in each cell. Capsule opening loculicidally in 3 valves. Seeds albuminous, with a small embryo, the radicle next the hilum.—Herbs with a perennial tuberous creeping bulbous or very short rhizome, rarely annuals. Leaves usually either radical or alternate and equitant, that is distichous, sheathing and laterally flattened at the base, produced into a linear lamina laterally or vertically not horizontally flattened so that the inner edge is towards the stem, the outer edge a continuation of the keel of the sheath. Flowers either solitary and terminal or in spikes or clusters within one or two bracts often called *spathas*, the bracts within the cluster usually imbricate, but each flower opposed to the bract of the same node, not in its axil. Perianths in the Australian genera mostly blue white or rarely yellow, in several South African genera showing a great variety of rich colours.

The Order is generally dispersed over the New and the Old World, more abundant in temperate than in tropical regions, and especially rich and diversified in South Africa. Of the seven genera here included, four are endemic, one, *Libertia*, extends over New Zealand and extratropical South America; the *Moræa* is a solitary very local representative of a South African genus, and the *Sisyrinchia* are introduced weeds from South America.

The peculiar inflorescence of the several-flowered *Irideæ* does not appear to have been generally noticed. It is a kind of cyme, each flower terminates an axis, which is continued by the development of an axillary bud between the subtending bract and the flower, which thus becomes opposed to the bract of the same node. As these subtending bracts are not superposed and unilateral as in the ordinary forked cyme, but alternate along the branch, the rhachis assumes a zigzag not a scorpioid character.

Stamens 3.
  Perianth with 3 large outer and 3 very small inner segments. Style longer than the anthers, with 3 broad-spreading laminæ.
    Spike or cluster simple and terminal or rarely lateral also. Only two anthers . . . . 1. DIPLARRHENA.
    Spike solitary and terminal, the outer bracts enclosing 2 sessile spikelets. Three anthers . . . . 2. PATERSONIA.
  Perianth with 6 nearly equal spreading segments. Style divided into 3 oblong petal-like segments opposite to and arching over the anthers . . . . 3. MORÆA.
  Perianth with 6 spreading segments. Style shorter than the stamens, with 3 linear or linear-subulate branches scarcely dilated at the end or acute.
    Perianth-segments nearly equal. Outer bracts erect and closely sheathing.
      Filaments free. Ovary and capsule oblong, sessile or nearly so . . . . 4. ORTHROSANTHUS.
      Filaments connate below the middle or to the top. Ovary and capsule obovoid or globular on long pedicels . . 5. SISYRINCHIUM.
    Perianth with 3 outer segments usually smaller or rarely nearly equal to the inner ones. Bracts all membranous and open. Filaments free . . . . 6. LIBERTIA.
Stamens 6.
  Perianth-segments nearly equal . . . . 7. CAMPYNEMA.

Besides the above, several South African *Irideæ* have occasionally been found to have escaped from gardens. Amongst them F. Mueller's collection includes *Trichonema ocholeucum*, Ker, *Watsonia angusta*, Sweet, *Sparaxis tricolor*, Bot. Mag., and *Iris spuria*, Linn., var. *halophila*.

1. DIPLARRHENA, Labill.

Perianth slightly irregular, divided to the ovary into 6 segments, the three outer erect at the base spreading upwards obovate, the upper one rather larger and more concave, the three inner ones much smaller narrow and less spreading. Filaments 3, free, the upper one without any anther and short, the other two unequal, each with an oblong-linear anther. Style slender, longer than the stamens, with 3 broad thin petal-like laminæ bearing the stigmas, one much larger than the others. Capsule oblong, acutely 3-angled, loculicidally 3-valved, the pericarp somewhat coriaceous. Seeds orbicular, flat.—Herbs with a very short rhizome or leafy base. Leaves mostly radical, long and flat. Stem erect, simple or branched. Flowers rather large, pedicellate, in a simple spike or cluster sessile within two rigidly herbaceous sheathing bracts, the spike solitary and terminal, or rarely one also in the axil of one or more leaves lower down.

The genus is limited to Victoria and Tasmania.

Stems 1 to 2 ft. high. Leaves mostly under ¼ in. broad. Flowers white . . . . 1. *D. Moræa.*

Stems above 2 ft. high. Leaves ½ to ¾ in. broad. Flowers variegated blue and yellow . . . . 2. *D. latifolia.*

1. **D. Moræa,** *Labill. Voy.* i. 157, t. 15. Perfectly glabrous in all its parts. Rhizome or leafy base of the stem short. Leaves chiefly radical, rigid, very flat, 1 to 2 ft. long, 2 to 3 lines broad. Stems 1 to 2 ft. high, with a few shorter leaves besides the radical ones. Spike single and terminal in all the specimens seen, resembling outwardly those of *Patersonia*, but simple, not composed of 2 spikelets, two outer bracts rigidly herbaceous, prominently striate, 2 to 2½ in. long, acutely acuminate, keeled, with narrow scarious margins, the inner ones membranous or scarious, each one as well as the second outer one opposed to a flower. Pedicels slender but shorter than the bracts, carrying usually the top of the ovary to the level of the bract, and somewhat longer when in fruit. Outer perianth-segments pure white, obovate, about 1½ in. long, the upper one less spreading broader and more concave than the other; inner segments oblong-cuneate, ½ to ⅔ as long as the outer, white or slightly tinged with violet and yellow.—R. Br. Prod. 304; Hook. f. Fl. Tasm. ii. 34; F. Muell. Fragm. vii. 94; *Moræa diandra*, Vahl, Enum. ii. 154.

N. S. Wales. Cape Howe, *C. Walter.*

Victoria. Wilson's promontory, and gregariously between Mount Barclay and Mount Lizar, *F. Mueller.*

Tasmania, *Labillardière;* Port Dalrymple, *R. Brown;* abundant in good soil throughout the island, *J. D. Hooker* and others.

Var. *alpina*, Hook. f. Leaves rather broader and shorter, but very different from those of *D. latifolia*, and flower white as in the typical form.—Western Mountains, *Gunn.*

2. **D. latifolia,** *Benth.* A much taller and stouter plant than *D. Moræa*, with the leaves from ½ to ¾ in. broad "flowering along the whole scape" according to Oldfield's notes; I have only seen a second spike in the axil of one of the leaves in a single specimen in Herb. F. Mueller. Flowers larger than in *D. Moræa* and "the outer segments more concave" (*Oldfield*); "variegated blue and yellow" (*C. Stuart*), the dried specimens showing a bluish tinge in the whole perianth which they never do in *D. Moræa.*

Tasmania. From one-third of the way up to the summit of Mount Lapeyrouse, *Oldfield, C. Stuart.*

2. PATERSONIA, R. Br.

(Genosiris, *Labill.*)

Perianth regular, with a filiform tube and 3 outer broad spreading segments, the 3 inner ones very small and erect or almost obsolete. Filaments united to the middle or almost to the top into a tube; anthers oblong or lanceolate, the cells usually separated by a narrow membranous connective. Style filiform, longer than the anthers, usually constricted or articulate either near the base of the anthers or near the top, and often bent down or breaking off at the constriction, the 3 stigmatic lobes obovate-orbicular or broadly oblong, contracted and united in a cup or narrow and free at the base, reflexed on the style in the bud, spreading horizontally when in flower. Capsule sessile within the bracts, linear or oblong, 3-angled, opening loculicidally in 3 valves.—Herbs with a perennial short rhizome. Leaves in radical distichous tufts or rarely on shortly elongated stems, long and grass-like or rigid. Scapes or peduncles long erect and leafless, bearing a single oblong or lanceolate terminal spike, with two outer bracts enclosing 2 sessile spikelets, each with 1, 2 or several flowers and as many membranous more or less scarious bracts, each bract of the spikelets opposed to a flower on the same node. Perianths blue or very rarely yellow or white.

The genus is limited to extra-tropical Australia, or scarcely crosses the tropics on the East coast, and is readily distinguished from all others by the inflorescence, as well as by other characters.

It has been proposed by F. Mueller to revert to Labillardière's generic name under the strict rules of priority. But Brown's has been so universally adopted with a full knowledge of the circumstances, and is so generally known by gardeners as well as botanists, that it would appear only to produce confusion now to substitute for *Patersonia* one so defective in composition as *Genosiris*.

The solitary terminal spike enclosing two spikelets appears to be constant in and peculiar to this genus; but I have sometimes found one spikelet subtended by the outermost bract, and the other by the second bract, but more frequently both spikelets are within the second bract, one of them subtended by a third; and I have observed sometimes the first bract only, sometimes the first and second, and once it appeared to me the third bract also without a flower opposed to it. I have not, however, had it in my power to dissect a sufficient number of spikes to show how far these differences may be specific.

Quite glabrous. Outer bracts of the spike prominently striate. Perianth-tube exserted. Staminal-tube short, trifid . . . 1. *P. glauca.*

Quite glabrous. Outer bracts smooth or obscurely striate. Perianth-tube not exserted. Staminal tube short, trifid.

  Scape usually much longer than the leaves. South-eastern species . . . . 2. *P. longiscapa.*

  Scape usually shorter than the leaves. Western species . . 3. *P. occidentalis.*

Glabrous or the leaves hairy on the edges. Outer bracts smooth or obscurely striate. Staminal-tube long, undivided.

  Quite glabrous. Leaves under 3 lines broad. Outer bracts 2 to $2\frac{1}{2}$ in. long, acute, prominently keeled and usually pale.

    Flowers blue . . . . 4. *P. umbrosa.*

    Flowers yellow . . . . 5. *P. xanthina.*

  Leaves often hairy on the edges. Outer bracts usually of a rich brown.

    Tall stout plant. Leaves 4 lines broad, with prominent usually red margins. Outer bracts of the spike broad . 6. *P. limbata.*

    Low or slender plants. Leaves 1 line broad or less. Spike narrow.

      Usually glabrous. Leaves rather thick, scarcely bordered. Scapes glabrous, $\frac{1}{2}$ to 1 ft. high . . . . . 7. *P. juncea.*

      Slightly hairy. Habit of *P. pygmæa.* Flowers white . 8. *P. Maxwelli.*

      Leaves hairy usually bordered. Scape woolly, under 6 in. 9. *P. pygmæa.*

Leaves woolly or hairy at the base. Scape woolly. Outer bracts dark and strongly striate, covered with deciduous wool. Staminal tube trifid, short.

  Leaves about 1 line broad. Eastern species . . . . . . 10. *P. longifolia.*

Leaves about 2 lines broad or more.
Leaves woolly on the edges only.
Wool of the spike close and silky. Eastern species . . 11. *P. sericea.*
Wool of the spike copious but loose. Western species . 12. *P. lanata.*
Leaves woolly all over near the base. Western species . . 13. *P. rudis.*
Leaves glabrous or hairy at the base. Outer bracts green, glaucous or pale, strongly striate, glabrous or slightly silky hairy. Staminal-tube (always?) long and undivided.
Outer bract 3 in. long, silky-hoary. Stem very short. Leaves glabrous, 3 lines broad . . . . 14. *P. macrantha.*
Outer bracts about 2 in. long, silky-hoary. Stem elongated. Leaves glabrous or hairy on the edges, 2 lines broad or less 15. *P. glabrata.*
Outer bracts about 1⅓ in. long, glabrous, glaucous. Stem very short. Leaves 1 line broad, the margins prominent, hairy near the base . . . . 16. *P. Drummondii.*
Outer bracts 1 to 1¼ in. long, the 2nd conspicuously higher attached than the 1st. Leaves very narrow.
Stem somewhat elongated. Leaves rigid, hairy at the base. Scape glabrous . . . . 17. *P. inæqualis.*
Stems very short, almost bulbous. Leaves grass-like, glabrous or finely ciliate. Scape cottony-woolly at the base 18. *P. graminea.*
Outer bracts 1 to 1¼ in. long, densely covered as well as the dwarf scape with very long soft hairs. Stems almost bulbous . . . . 19. *P. babianoides.*

1. **P. glauca,** *R. Br. Prod.* 304. Stems exceedingly short, clustered on the rhizome, with a few outer rigid sheathing scales, gradually passing into erect rigid leaves, the longest from ½ to 1½ ft. long and 1½ to nearly 2 lines broad, rather less flat than in *P. occidentalis* and the margins not thickened. Scapes usually much shorter than the leaves, but occasionally exceeding them, with 1 or 2 sheathing scales at their base. Spike with the two outer bracts 1½ to 1¾ in. long, acute, rigid, somewhat glaucous, finely but prominently striate, each spikelet containing 3 or 4 flowers, the bracts membranous and quite glabrous, as well as the flowers. Perianth-tube usually exceeding the bracts by 3 to 5 lines; outer lobes ovate or ovate-oblong, obtuse, above ½ in. long; inner ones oblong-lanceolate, scarcely more than 1 line long. Staminal column short, divided to about the middle. Style slightly thickened at the end, the laminæ ovate, but I do not find the fringes of long cilia figured by Endlicher from Bauer's drawing.—Hook. f. Fl. Tasm. ii. 34; Bot. Mag. t. 2677; Lodd. Bot. Cab. t. 1182; Endl. Iconogr. t. 50; *Genosiris fragilis,* Labill. Pl. Nov. Holl. i. 13, t. 9; F. Muell. Fragm. vii. 36.

N. S. Wales. Hunter's river, *R. Brown;* New England, *C. Stuart.*
Victoria. Glenelg river, *F. Mueller.*
Tasmania. Port Dalrymple, *R. Brown;* abundant in wet peaty soils in the northern parts of the island, *J. D. Hooker.*

2. **P. longiscapa,** *Sweet, Fl. Austral.* t. 39. A glabrous plant like *P. glauca* and *P. occidentalis*, but the leaves usually shorter, and whether short or long the scape almost always still longer. Leaves flat and the margins often nerve-like, as in *P. occidentalis,* of which this may be a variety. Scape usually thickened and striate close under the spike, but the outer bracts fully 1½ in. long, quite smooth or very obscurely

striate. Spikelets with 3 or more flowers each. Ovary with a few hairs towards the apex. Perianth-tube slightly hairy at the base, scarcely exceeding the bracts; outer segments ovate-elliptical, acute, about ¾ in. long, the inner about 1 line long, narrow-lanceolate but appearing subulate when shrivelled. Staminal column divided to about the middle. Style articulate above the middle.—*P. bicolor*, F. Muell. in several herbaria.

Victoria. Port Phillip, *R. Brown;* near Portland, *Robertson;* Latrobe and Glenelg rivers and Dandenong range, *F. Mueller;* Grampians, *Wilhelmi.*

Tasmania, *Archer.*

S. Australia. Onkaparinga and Lofty ranges, *F. Mueller.*

The name of *longiscapa* was originally proposed by Sims for some specimens of the true *P. glauca*, with a longer scape than usual, which, relying on that character alone, Sweet thought to be the same as the plant introduced from the South coast by Baxter. The latter, however, has the foliage, the smooth outer bracts, the pubescent apex of the ovary, and the shorter perianth tube of *P. occidentalis*, of which it may prove to be an Eastern variety as doubtfully suggested by R. Brown in his notes.

3. **P. occidentalis,** *R. Br. Prod.* 304. Stems very short, the whole plant glabrous except the ovary and perianth. Leaves rigid, the longest often above 1 ft. long and usually 2 to 3 lines broad. Scapes varying from rather shorter to rather longer than the leaves, dilated and striate under the spike. Outer bracts 1½ in. long or rather more, prominently or rather obscurely keeled, the striæ scarcely or not at all conspicuous, the inner membranous bracts often slightly exceeding the outer and sometimes pubescent on the keel. Flowers usually numerous, rarely only 3, in each spikelet. Perianth-tube and ovary more or less villous, the tube very shortly exceeding the bracts; outer segments often fully 1 in. long, broad and very obtuse, of a rich blue, the inner segments minute, ovate or lanceolate. Staminal column short, the filaments very shortly free; anthers almost lanceolate, spreading. Style articulate near the base of the anthers.—*P. sapphirina*, Lindl. Bot. Reg. 1839, t. 60 (passing frequently in gardens for *P. sericea*).—*Genosiris occidentalis*, F. Muell. Fragm. vii. 31.

W. Australia. King George's Sound and Lucky bay, *R. Brown;* King George's Sound and adjoining districts, *Oldfield*, *F. Mueller*, and others, and thence to Swan river, *Drummond*, *Collie.*

Var. *latifolia.* Leaves 3 to 4 lines broad. Flowers very numerous in the spike. Champion bay, *Oldfield.*

Var ? *angustifolia.* Leaves under 2 lines broad. Possibly a distinct species. *P. tenuispatha* and *P. turfosa*, Endl. in Pl. Preiss. ii. 31.—Swan river, *Preiss, n.* 2338; Murchison river, *Oldfield.*

Var. ? *eriostephana*, F. Muell. Fragm. vii. 32. Inner segments of the perianth subulate.—Cape Arid, *Maxwell.*

P. occidentalis is evidently a variable plant, and appears to be common in West Australia. *P. Diesingii*, Endl. in Pl. Preiss. ii. 30, from Swan river, n. 2356, has the flowers rather large and the staminal-tube rather longer than usual, but I can find no other character to distinguish it from the commonest form. I do not find the spatha articulate on the scape in Preiss's specimen. *P. nana*, *P. compar*, *P. flaccida*, *P. sylvestris*, and *P. montana*, Endl. in Preiss. ii. 30, 31, all from W. Australia, *Preiss*, are unknown to me: all are described as glabrous, and are only distinguished from each other and from *P. occidentalis*, as far as the diagnoses extend, by slight differences in stature, in the proportion of the scape to the foliage, in the prominence of the keel of the bracts, and in the equality or inequality of the ribs of the leaves, all of which are very variable in the true *P. occidentalis*. No mention is made of the flowers, stamens, or style of any one of them.

4. **P. umbrosa,** *Endl. in Pl. Preiss.* ii. 31. Stems short, with the short scales outside the leaves more numerous than in most species, the whole plant glabrous except the perianth-tube. Leaves usually very long, often above 2 ft., and under 3 lines broad. Spike 2½ in. long, much compressed, the outer bracts acuminate and acutely keeled, smooth or very obscurely striate. Perianth blue, the tube considerably longer than the bracts, slightly silky-pubescent, the outer segments above 1 in. long, the inner ones linear-subulate, often 2 lines long. Staminal-tube rather long and slender, the filaments scarcely free immediately under the anthers. Style showing no articulation in the specimen examined, the limb shortly campanulate below the division into spreading laminæ.—*Genosiris umbrosa*, F. Muell. Fragm. vii. 32.

W. Australia. King George's Sound, *Preiss, n.* 2348, *F. Mueller;* Middle Mount Barren, *Maxwell;* also in *Drummond's* collection.

5. **P. xanthina,** *F. Muell. Fragm.* i. 214. A glabrous plant, with the habit, the long leaves, the long acuminate much compressed acutely keeled and smooth outer bracts, and all the essential floral characters of *P. umbrosa*, but the flowers are said to be yellow, and the buds show none of the blue tint of all the allied species. The perianth-tube in the specimen examined was rather shorter and more hairy than in *P. umbrosa*, and the inner segments not quite so narrow, but I could detect no other difference, and the plant may be a variety only of that species.—*Genosiris xanthina*, F. Muell. Fragm. vii. 33.

W. Australia. Geographe bay, *Oldfield;* Busselton, *Pries* (mixed with *P. occidentalis*).—There are several Irideæ in which the same species appears to include blue-flowered and yellow-flowered varieties.

6. **P. limbata,** *Endl. in Pl. Preiss.* ii. 29. A stout species, with a very short or slightly elongated stem, allied to *P. occidentalis*, but the leaves usually broader and more rigid, ½ to 1½ ft. long, remarkable for their reddish nerve-like keel and inner margin, often woolly on the young leaf but soon becoming glabrous. Scapes rarely above 1 ft. high, much dilated under the spike. Spike rather broad full and many-flowered, about 2 in. long, the outer bracts obscurely striate, usually of a reddish brown as well as the scape, but in some specimens paler coloured. Perianth blue, the tube very silky-villous in the lower part, shortly exceeding the bracts; outer segments apparently nearly 1 in. long, the inner ones very small, linear-lanceolate. Staminal-tube slender, the filaments very shortly free at the apex. Style articulate nearly at the base of the anthers, the stigmatic laminæ stipitate oval-oblong.

W. Australia. King George's Sound and adjoining districts, *Baxter, Drummond, Preiss, n.* 2349, *F. Mueller;* Champion bay, *Oldfield.*

7. **P. juncea,** *Lindl. Swan Riv. App.* 58. Almost stemless and quite glabrous, or the young leaves slightly ciliate. Leaves in clustered

tufts, very narrow-linear, 4 to 8 in. long in the typical form and about ½ line broad, prominently striate, very shortly dilated and sheathing at the base. Scapes slender, as long as or longer than the leaves, quite glabrous. Spike about 1½ in. long, like that of *C. pygmæa* rather narrow, the bracts brown smooth shining and veinless. Perianth quite glabrous, the tube slender, exserted; the structure of the flower apparently like that of *P. pygmæa*, but not seen perfect.—Endl. in Pl. Preiss. ii. 31; *Genosiris juncea*, F. Muell. Fragm. vii. 33.

W. Australia. Swan river, *Drummond*, 1*st coll.*, *n.* 772, *Preiss*, *n.* 2352; Vasse river, *Oldfield*; Stirling range, *F. Mueller.*

Var. *elongata.* Leaves nearly 1 ft. long, and scape still longer.—Cape Naturaliste, *Oldfield.*

P. Roei, Endl. in Pl. Preiss. ii. 31, which I have not seen, is probably the same species.

8. **P. Maxwelli,** *F. Muell.* Stems slender but short as in *P. pygmæa*. Leaves strongly ribbed, with a prominent nerve-like margin, slightly hairy on the edges when young, not exceeding 6 in. in the specimens seen and about 1 line broad. Scapes glabrous, nearly as long as the leaves. Spikes about 1 line long, narrow, the outer bracts smooth with very fine scarcely prominent veins as in *P. pygmæa*. Flowers very few (1 or 2 in each spikelet), said to be white. Ovary glabrous or slightly ciliate on the angles. Perianth-tube quite glabrous, about as long as the bracts; outer segments broadly ovate, obtuse, about ½ in. long; the inner ones broadly obovate or orbicular, nearly 1 line long. Staminal tube slender, the filaments scarcely free at the top. Style articulate at the base of the anthers.—*Genosiris Maxwelli*, F. Muell. Fragm. vii. 34.

W. Australia. M'Callum's Inlet, *Maxwell.*

9. **P. pygmæa,** *Lindl. Swan Riv. App.* 58. Stems with a slender erect base as in *P. glabrata* but very short, rarely 2 to 3 in. long below the scape or peduncle, shortly rufous or villous between the leaves or nearly glabrous. Leaves erect, narrow, rigid, with the very fine veins of *P. glabrata*, but only 2 to 4 in. long, more or less silky-hairy especially on the margins when young, usually glabrous on the fruiting specimens. Scapes short or nearly as long as the leaves, clothed with a loose and copious but deciduous wool. Outer bracts 1½ to near 2 in. long, narrow, acute, brown, rather smooth and shining, with very fine scarcely perceptible veins, glabrous or silky-woolly on the keel. Ovary and perianth-tube quite glabrous in the flowering specimens examined, but in others the fruit sometimes slightly hairy; outer segments of the limb obovate, ¾ in. long, the inner ones minute. Staminal column rather long and slender. Style apparently articulate near the base of the anthers.—Endl. in Pl. Preiss. ii. 32; *Genosiris pygmæa*, F. Muell. Fragm. vii. 33.

W. Australia. Swan river, *Drummond*, 1*st coll. n.* 770; Gordon river, *Preiss*, *n.* 2352; King George's Sound and adjoining districts, *F. Mueller*, *Oldfield*, and others; Salt river, *Maxwell.*

10. **P. longifolia,** *R. Br. Prod.* 303. Very near to *P. sericea,* but the leaves scarcely above 1 line broad, sometimes very long, sometimes not much exceeding the spike. Scape 3 to 8 in. long. Spike smaller than in *P. sericea,* the outer bracts often not much above 1 in. long, but in other respects like those of that species drying black, covered when young with a silky wool, and prominently striate when the wool wears off.—*Genosiris longifolia,* F. Muell. Fragm. vii. 35.

N. S. Wales. Grose river, *R. Brown;* Hunter's river, *Oldfield;* Port Jackson or Blue Mountains, *C. Moore, Vicary,* and others; Cape Sturt, *Backhouse.*—Perhaps a variety of *P. sericea.*

11. **P. sericea,** *R. Br. in Bot. Mag.* t. 1041, *Prod.* 303. Stems scarcely any. Leaves radical, long, erect and rigid, rarely above 2 lines broad, the edges very woolly at the base when young. Scapes shorter than the longest leaves, usually about 1 ft. high, silky-woolly towards the end. Spike stout and usually many-flowered, the outer bracts 1½ to near 2 in. long, at first silky-woolly, but the wool often wearing off leaving the bracts prominently striate and usually dark-coloured or black in the dried specimens. Ovary very woolly; the perianth-tube less so or glabrous towards the end, not longer than the outer bracts; outer segments of the limb broadly ovate, almost truncate or emarginate, of a deep violet blue; inner segments small, ovate or lanceolate, sometimes very minute, filaments at first shortly free but at length often separate to the middle or even lower. Style jointed near the top, the laminæ broadly obovate-oblong or nearly orbicular, as in other species closely reflexed in the bud, spreading when the flower is open.—*Genosiris sericea,* F. Muell. Fragm. vii. 35.

Queensland. Port Bowen, *R. Brown;* Moreton island, *F. Mueller;* Wide bay, *Leichhardt;* Glasshouse Mountains, *Byerley;* Dawson river, *Woolls.*

N. S. Wales. Port Jackson, *R. Brown, Sieber, n.* 196, and others; New England, *C. Stuart;* Clarence river, *Wilcox;* Hastings river and Mount Mitchell, *Beckler;* southward to Twofold bay, *F. Mueller.*

Victoria. Genoa river and Mount Wellington, *F. Mueller.*

Var. ? *latifolia.* Leaves fully 3 lines broad.—Blue Mountains, *Fraser.*

P. glabrata, Edw. Bot. Reg. t. 51, seems to represent a glabrous state of *P. sericea,* it is certainly not the *P. glabrata,* R. Br.

12. **P. lanata,** *R. Br. Prod.* 303. Very closely allied to *P. sericea,* and perhaps its western representative. Leaves similarly silky-woolly on the margins only, but the silky-wool of the scapes and bracts is much looser and more copious although very deciduous. Outer bracts dark-coloured rigid and striate as in *P. sericea,* but rarely much above 1 in. long. Flowers usually numerous, the wool of the ovary and of the base of the tube loose and very copious. Perianth-tube not exceeding the bracts; outer segments of the limb very broad and rounded, of a deep blue-purple; inner segments small, obovate or oblong-cuneate, sometimes slightly notched.—Sweet, Fl. Austral. t. 15; *Genosiris lanata,* F. Muell. Fragm. vii. 35.

W. Australia. Lucky bay, *R. Brown,* and probably the same locality, *Baxter.*

Var. *latifolia.* Leaves 3 lines broad.—*P. pannosa,* Endl. in Pl. Preiss. ii. 29.

13. **P. rudis,** *Endl. in Pl. Preiss.* ii. 29. This species is again closely allied to *P. sericea*, but a coarser plant, usually larger in all its parts and the silky-wool not only abundant on the bracts and upper part of the scape and on the margins of the leaves near the base, but the sides also of the leaves are very silky-pubescent near the base, and are surrounded at the base by a dense tuft of long ferruginous hairs proceeding from the rhizome and from the outermost scales. The many-flowered spikes with dark-coloured prominently striate loosely woolly outer bracts the same as in the three preceding species.—*Genosiris rudis*, F. Muell. Fragm. vii. 35.

W. Australia. Swan river, *Drummond*, 1*st coll.*, *n.* 774, *Preiss*, *n.* 2347.

14. **P. macrantha,** *Benth.* Stem very short and thick. Leaves above 1 ft. long, about 3 lines broad, neither very rigid nor marginate like those of *P. limbata*, the keel often but not always woolly at the base. Scape flattened, above 1 ft. long in our specimens, silky-pubescent under the spike. Outer bracts about 3 in. long, very slightly hoary or silky, prominently veined, of a pale brown or green. Ovary slightly villous. Perianth-tube glabrous except near the base, shortly exceeding the bracts; outer segments of the limb above 1 in. long. Staminal tube long and slender.

W. Australia. Darling range, *Collie.*

15. **P. glabrata,** *R. Br. Prod.* 304. This species is at once recognised by its rather slender stem of 3 to 6 in. with distichous leaves not so close as in other species and the lower ones very short, the upper ones from 6 in. to above 1 ft. long, rarely above 2 lines broad, finely veined, the keel and sometimes also the inner edge or margins silky-woolly towards the base. Scapes or peduncles one or sometimes two on the stem, rather slender, rarely exceeding the leaves, either quite glabrous as well as the spike, or the upper part silky-hoary with a very short pubescence. Outer bracts 1½ to near 2 in. long, very acute, rather narrow, pale-coloured and often slightly silky-hoary, distinctly striate, with broad scarious margins. Perianth-tube shortly exceeding the bracts, thinly villous at the base as well as the ovary; outer segments of the limb about ¾ in. long in some specimens, above 1 in. in others; inner ones oblong-lanceolate or slightly cuneate, ¾ to 1½ lines long. Staminal column rather long and slender.—Lodd. Bot. Cab. t. 768; *Genosiris glabrata*, F. Muell. Fragm. vii. 35; *P. media*, R. Br. l.c.

Queensland. Shoalwater bay, *R. Brown;* Moreton island, *F. Mueller.*
N. S. Wales. Port Jackson, *R. Brown* and others; Hastings and Clarence rivers, *Beckler;* New England, *C. Stuart;* Newcastle, *Leichhardt.*
Victoria. Genoa Peak, *F. Mueller;* entrance to Snowy river, *C. Walter.*

The species varies much in the length of the stem, in that of the outer bracts, and in the glabrous or pubescent scapes and bracts.—R. Brown's *P. media* only differs from his *glabrata* in the latter respect.

16. **P. Drummondii,** *F. Muell. Herb.* Stems very short. Leaves narrow but very flat, glaucous, often twisted, the thickened nerve-like margins ciliate towards the base when young, the longest 6 to 8 in.

long, about 1 line broad. Scape shorter than the leaves, rigid, compressed. Spikes at least $1\frac{1}{2}$ in. long and rather broad, the outer bracts obtuse, more or less glaucous, rigid, with numerous prominent but fine nerves. Flowers blue, glabrous, but so much injured by worms in the specimens examined that I have been unable to ascertain the structure of the staminal column, the anthers and large reflexed stigmatic lobes quite those of the genus.

W. Australia. Probably Swan river, *Drummond.*

17. **P. inæqualis,** *Benth.* Stems lengthening to 1 or 2 in. and covered with the closely imbricate distichous bases of the leaves, which are rigid, erect, under 1 ft. long, 1 to $1\frac{1}{2}$ lines broad, very strongly striate, and loosely silky-woolly towards the base on both surfaces. Scapes usually nearly as long as the leaves. Outer bracts narrow, rigid, green, and strongly striate, with broad scarious margins, the outermost one about 1 in. long, the next one shorter, inserted higher up, with a prominent base so as to give the spike an unequally gibbous aspect. Flowers few. Ovary somewhat hairy. Perianth-tube glabrous, about as long as the bracts; outer segments of the limb obovate, $\frac{3}{4}$ in. long, the inner ones very short, broadly obovate or orbicular. Staminal column elongated, the filaments very shortly free at the apex. Style articulate at or about the middle.

W. Australia. Stokes Inlet, *Maxwell.*

18. **P. graminea,** *Benth.* Stems exceedingly short, clustered on the rhizomes, the outer brown scales and bases of the leaves thickened almost into bulbs. Leaves very narrow, flaccid, prominently veined, the longest 6 to 9 in. long, quite glabrous or hairy at the base when young. Scapes slender, terete, above 1 ft. long, covered at the base with a white cottony wool, glabrous from the middle upwards. Spikes 9 to 10 lines long, the second bract inserted above the lowest, but not so prominent at the base as in *P. inæqualis*, both of them finely ribbed. Flowers 1 or 2 only in each spikelet, quite glabrous, not seen fully open, but in the bud the staminal column is very short, the style articulate at the base of the anthers, the stigmatic lobes dilated but small.

W. Australia, *Drummond, n.* 196 *and* 5*th coll., n.* 326.

19. **P. babianoides,** *Benth.* Rhizome short with the stem and leaves slightly bulbous at the base as in *P. graminea.* Leaves 1 or 2 only, 3 to 6 in. long, 3 or 4 lines broad in the middle but tapering into a long point and at the base into a long petiole, the lower part of the leaves and the inflorescence densely woolly with very long soft hairs. Scapes mostly but little above 1 in., in one specimen nearly 2 in. long. Outer bract about 1 in. long, narrow, thinly herbaceous or slightly coloured, striate but the veins concealed by the long hairs. Spikelets both single-flowered in the specimen dissected. Ovary very short and densely covered with long hairs, the perianth otherwise glabrous, the tube about as long as the bracts, the three outer seg-

ments of the limb large and obovate, the inner ones small, with subulate points. Filaments very shortly united. Stigmatic laminæ of the style broad, but not reflexed.

W. Australia. Swan river, *Drummond*, 1st coll., n. 760; Hampden, *Clarke.*—The dried specimens have so much the aspect of some of the dwarf species of *Babiana*, that they have been sent as a supposed introduced species of that S. African genus. The structure of the spike and of the flowers is however totally different and in every respect that of *Patersonia*.

3. MORÆA, Linn.

Perianth regular, divided to the ovary into 6 segments spreading almost or quite from the base, all nearly equal or the inner ones rather narrower. Filaments short, free; anthers linear. Style deeply divided into 3 oblong or spathulate petal-like branches, opposite to and arching over the stamens. Capsule oblong ovoid or globular, opening in 3 valves, the pericarp coriaceous or thick.—Herbs with a thick or very short rhizome and fibrous roots. Leaves mostly radical. Stems erect, without any or with one or two leaves, besides the bracts subtending the branches of the inflorescences. Flowers usually rather large, pedicellate, in spikes or clusters, solitary within the subtending bract, each flower opposed to a bract on the same node, the outer bract of the spike usually longer than the subtending one.

The genus is chiefly South African, the only Australian species, as far as known, endemic in Lord Howe's island. The limits of the genus are perhaps somewhat uncertain, and F. Mueller proposes the uniting it with *Iris*. The differences are, however, as to most species, well marked especially as to the form of the perianth, and it is universally maintained by all those who have specially studied either the Order generally or the South African Flora, a judgment which it is unsafe to set aside, without a thorough revision of all the species of both genera as well as of their allies.

1. **M. Robinsoniana,** *F. Muell. Fragm.* vii. 153. Rhizome very short. Radical leaves attaining 5 or 6 ft. in length and 2 to 4 in. in breadth in the broadest part. Flowering stem 5 or 6 ft. high, with a few long leaves in the lower part, branching into a large panicle, the bracts leafy and 5 or 6 in. long under the lower branches, gradually diminishing, the upper ones broad thin and under 1 in. long. Spikes or clusters few-flowered, the outer bract, opposed to the first flower, usually 1½ in. long and always longer than the subtending one. Flowers shortly pedicellate, the perianth "measuring when fully developed 4 in. across," but smaller in the specimens seen, segments white, elliptical, the outer ones rather broader than the inner. Style-branches oblong-spathulate, obtuse, about ½ in. long. Capsule above 1 in. long, ¾ in. broad, coriaceous, the fruiting pedicel usually as long as the opposed bract.—*Iris Robinsoniana*, F. Muell. l.c.; G. Benn. in Gard. Chron. 1872, 393.

N. S. Wales. Lord Howe's island, *C. Moore*, *Fullagar.*—Known as the "Wedding Flower." It is the largest species of the genus, the habit is that of the nearly allied *Pardanthus Chinensis*, the flowers nearly those of *Moræa iridioides*.

4. ORTHROSANTHUS, Sweet.

Perianth regular, with a short tube and 6 nearly equal spreading segments. Filaments short, free; anthers oblong, the cells contiguous. Styles or style branches linear, almost filiform, minutely dilated and denticulate at the end, shorter than the anthers and alternate with them. Capsule sessile or nearly so, oblong, 3-angled, opening loculicidally in 3 valves.—Herbs with a perennial short rhizome. Leaves mostly radical, long and grass-like or rigid. Stems erect, bearing 1 or 2 short leaves. Spikes 1- or several-flowered, sessile or pedunculate, solitary or several together within the same sheathing bract, each with 2 outer bracts, the second as well as the inner membranous ones (when present) opposed each to a flower on the same node.

The genus is limited to Southern and Western Australia. It has been generally referred to *Sisyrinchium*, but besides the free stamens and the very different capsules, the inflorescence gives it a very distinct aspect.

Spikes with several (more than 3) flowers, the outer bracts usually brown-scarious at the end. Capsule obtuse.
  Plant glabrous, usually tall. Spikes all sessile or both sessile and pedunculate in the same bract . . . . 1. *O. multiflorus.*
  Plant low, the leaves woolly on the edge when young. Spikes few, all pedunculate or rarely one sessile . . . . 2. *O. Muelleri.*
Spikes with 2 or 3 flowers, few and pedunculate. Capsule more or less contracted at the end.
  Leaves 2 to 2½ lines broad. Outer bracts of the spikes white-scarious at the end, capsule shortly and obtusely acuminate 3. *O. laxus.*
  Leaves ½ to ¾ line broad. Outermost bract green to the end. Capsule distinctly beaked . . . . 4. *O. gramineus.*
Spikes 1-flowered, very numerous, all more or less pedunculate. Stem tall. Leaves long . . . . 5. *O. polystachyus.*

1. **O. multiflorus,** *Sweet, Fl. Austral.* t. 11. Rhizome very short and woody. Leaves chiefly radical or nearly so, flat and grass-like but rigid and striate at the base, ¾ to near 2 ft. long, 1½ to 2 lines broad. Stems 1 to 2 ft. high, bearing only 1 or 2 short leaves below the inflorescence. Inflorescence narrow, 4 to 8 in. long, with lanceolate acuminate bracts. Spikes several-flowered, oblong, about ¾ in. long, one usually sessile within each bract along the main rhachis and accompanied by 1 or 2 others within the same bract, but on rigid peduncles often longer than the subtending bract. Outermost bract of the spike 7 or 8 lines long, striate, with a broad scarious-brown margin; second bract opposed to the first flower, rather longer, brown-scarious at the top; inner ones entirely scarious. Perianth-tube about 1 line long above the ovary; segments of the limb ovate, about ¾ in. long, the outer ones rather narrower than the inner. Ovary contracted and empty at the base and at the top. Capsule ovoid-oblong, 3-angled, nearly as long as the bracts.—Lodd. Bot. Cob. t. 1474; *Sisyrinchium cyaneum*, Lindl. Bot. Reg. t. 1090.

S. Australia. Kangaroo island, *R. Brown*, *Baxter*, *F. Mueller;* Portland and Cape Nelson, *Allitt.*

W. Australia. Cape Naturaliste, *Oldfield;* Scott's brook, Thomas river, *Maxwell.*

Var. *hebecarpa*. Capsule villous.—W. Australia? *Herb. Hooker.*

Libertia stricta, Endl. in Pl. Preiss. ii. 32, from Cape Riche, is unknown to me, and the short diagnosis given is insufficient to distinguish it either from *O. multiflorus* or from *O. laxus*. *Wuerthia elegans*, Regel, Gartenfl. ii. t. 46, referred here by F. Mueller, Fragm. vii. 92 (evidently from conjecture only without comparing the plate) is a South African species of *Ixia*, stated by a gardener's mistake to be a native of Adelaide in South Australia.

2. **O. Muelleri,** *Benth.* A small slender species, the stems rarely 1 ft. high, glabrous or slightly woolly towards the base. Leaves shorter than the stems, 1 to 1½ lines broad, woolly on the keel and inner margin when young. Spikes 2 to 4 on the stem, all pedunculate or the lower one sessile, resembling those of *O. multiflorus* in the brown scarious apices of the bracts, but rather smaller. Flowers usually 3 or 4 in the spike. Capsule obtuse, not longer than the bracts.

W. Australia. Swan river, *Drummond*, 1st *coll.*, *n.* 767; Stirling range, *F. Mueller*.

3. **O. laxus,** *Benth.* Quite glabrous, the rhizome sometimes slightly elongated. Leaves ¾ to 1½ ft. long, 2 to 2½ lines broad. Inflorescence longer than the leaves, loose. Spikes 2- or 3-flowered, not numerous, all on peduncles much longer than the subtending bracts. Outer bracts of the spike rather broad, green, striate, shortly white-scarious and obtuse at the end, 6 lines long or rather more. Capsule shortly and obtusely acuminate, about as long as the bract.—*Libertia laxa*, Endl. in Pl. Preiss. ii. 32.

W. Australia. Swan river, *Drummond*, 1st *coll.*, *n.* 769, *Collie*, *Preiss*, *n.* 2230; Kalgan river, *Oldfield;* Stirling range, *F. Mueller;* Swan river and Bremer bay, *Maxwell.*

4. **O. gramineus,** *Benth.* Quite glabrous. Leaves 6 in. to near 1 ft. long, not 1 line and mostly not above ½ line broad. Stem scarcely exceeding the leaves, the leafy bracts at the base of the peduncles long and lanceolate. Spikes 1 to 3 on each stem, on long peduncles, each with 2 or 3 flowers. Outer bracts of the spike scarcely exceeding ½ in. or the outermost one longer and acuminate, rather broad, the outermost one rarely at all scarious, the second usually with a broad white scarious end. Capsule tapering into a distinct beak, rarely exceeding the bracts.—*Libertia graminea*, Endl. in Pl. Preiss. ii. 32.

W. Australia. Swan river, *Drummond*, 1st *coll.*, *n.* 768; near York, *Preiss*, *n.* 2229.

5. **O. polystachyus,** *Benth.* Quite glabrous. Leaves 1½ to 2 ft. long, 2 to 3 lines broad. Stems attaining 2 to 4 ft., the inflorescence 6 to 10 in. long and compound. Bracts of the main rhachis lanceolate, acuminate, striate, 1 to 2 in. long, those of the short branches short and broad. Spikes usually many within each bract, all shortly pedunculate and 1-flowered, each with 2 unequal bracts, the longest not ½ in. long. Flowers blue, apparently rather smaller than in *O. multiflorus*. Capsule glabrous, obtuse, longer than the bracts.

W. Australia. Probably Swan river, *Drummond*, *n.* 206 *and* 357; Cape Naturaliste, *Oldfield;* Warren river, *Walcot.*

*5. SISYRINCHIUM, Linn.

Perianth regular, the tube very short, the limb of 6 nearly equal spreading segments. Filaments united to above the middle or rarely at the base only; anthers oblong or lanceolate. Style shorter than the filaments, with 3 linear or subulate spreading branches, stigmatic at the end. Capsule ovoid or globular, 3-valved, the pericarp rather thin.—Herbs with fibrous roots, and a very short tufted stock, or rarely annuals. Stem erect, often branched, usually 2-edged. Leaves mostly or all radical, narrow. Flowers on slender pedicels, clustered within 2 sheathing herbaceous bracts, the outermost one subtending the cluster on the main axis, the second outer bract and the inner membranous ones each opposed to a pedicel within the cluster, fruiting pedicels exserted from the bracts.

The genus is widely spread over North and South America, one species also found in various districts of the Old World, but in many places introduced. The two Australian ones are both of modern introduction, though now well established.

Perennial, 6 in. to 1 ft. high. Stem very prominently 2-angled or 2-winged. Perianth ½ in. diameter or more . . . . . . 1. *S. Bermudiana.*
Annual under 6 in. Stem slightly 2-angled. Perianth ¼ in. diameter . . . . 2. *S. micranthum.*

*1. **S. Bermudiana,** *Linn.; Cav. Diss.* vi. 346, t. 192. A perennial with erect stems of 6 in. to 1 ft., prominently 2-angled or 2-edged, or with 2 narrow acute wings sometimes broader under the bracts. Flowers 2 to 6 together (usually 3 or 4) in a terminal cluster, the filiform pedicels almost concealed within the 2 outer sheathing bracts, of which the outermost one often ends in a leafy tip exceeding the flowers, but occasionally both are nearly equal. Perianth blue, the segments 3 to 4 lines long.

A very common North American species also indigenous perhaps in some parts of Western Europe and an introduced colonist in many other countries. It is said to be well established in some part of **N. S. Wales** and **Victoria.**

*2. **S. micranthum,** *Cav. Diss.* vi. 345, t. 191. A slender annual, not above 6 in. high, the branches flattened and 2-angled but not winged. Flower-clusters on peduncles longer than the subtending leaves or bracts, one or two at the end of the stem and often one or two from lower axils. Outer bracts of the cluster ¾ to 1 in. long. Pedicels very slender. Perianth-segments about 1½ lines long. Staminal tube about half that length.—Bot. Mag. t. 2116.

A South American species, introduced as a weed into several parts of **Queensland** and **N. S. Wales,** and now said to be exceedingly abundant about Brisbane and Port Jackson.

6. LIBERTIA, Spreng.

(Renealmia, *R. Br.;* Nematostigma, *Dietr.*).

Perianth regular, divided to the ovary into 6 nearly equal spreading segments. Filaments free; anthers linear-sagittate. Style shorter than the filaments, with 3 linear-subulate spreading branches, stigmatic and minutely toothed or fringed at the end. Capsule ovoid or globular, 3-valved, the pericarp thin.—Herbs with a short often very short leafy base, and grass-like flat leaves almost radical though distichous. Flowering stems erect, simple or branched, with 1 or 2 leaves below the inflorescence. Flowers clustered in the axils of sheathing bracts, each flower on a slender pedicel, opposed to a bract as in other *Irideæ*, but the shortness of the rhachis of the cluster, and the length of the pedicels give the cluster the appearance of an umbel, and occasion some difficulty in tracing its real structure on the dried specimens. The inspection of fresh specimens however prove it very clearly to be in conformity with the rest of the order.

The genus extends to New Zealand and extratropical South America, one of the Australian species apparently the same as a New Zealand one, the other endemic. The genus is reduced by F. Mueller to *Sisyrinchium*, in which Brown had in the first instance placed his three species; but besides the differences indicated in his Addenda and recognised by subsequent observers, the inflorescence gives it so distinct an aspect that I am unable to concur in the union.

Flower-clusters many, paniculate, the flowering-stem 1 to $1\frac{1}{2}$ ft. high.
Perianth-segments about 4 lines . . . . 1. *L. paniculata.*
Flower-clusters few or solitary, the flowering-stem under 1 ft.
Perianth-segments 2 to 3 lines long . . . . 2. *L. pulchella.*

1. **L. paniculata,** *Spreng. Syst.* i. 168. Rhizome and leafy base of the stem very short. Leaves almost radical, grass-like, flaccid, $\frac{3}{4}$ to $1\frac{1}{2}$ ft. long and mostly about 3 lines broad. Stem 1 to $1\frac{1}{2}$ ft. high, with sometimes a short leaf below the inflorescence. Panicle oblong loose and irregular, occupying often half the stem, glabrous as well as the whole plant or the pedicels slightly glandular-pubescent. Bracts membranous-scarious, the lower ones rather long and acuminate, those subtending the flowers 3 to 4 lines long, ovate or ovate-lanceolate, obtuse or acute. Flower-clusters umbel-like, the pedicels 3 to 4 lines long when in flower, at least twice as long under the fruits. Ovary obovoid, about $1\frac{1}{4}$ lines long under the flower. Perianth-segments about 4 lines long, ovate, spreading, the inner ones larger than the outer, but not so much so as in the non-Australian species. Filaments dilated at the base but not united. Style column very short, with 3 linear-subulate spreading branches, minutely fringed at the end. Capsule ovoid-globular, 3 to 4 lines diameter.—*Sisyrinchium paniculatum*, R. Br. Prod. 305; F. Muell. Fragm. vii. 91; *Renealmia paniculata*, R. Br. Prod. Addenda; *Nematostigma paniculatum*, Dietr. Sp. Pl. ii. 510.

N. S. Wales. Port Jackson and Hunter's river, *R. Brown;* Blue Mountains, *Woolls, Miss Atkinson;* northward to Hastings river, *Beckler;* New England, *C. Stuart;* southward to Illawarra, *A. Cunningham;* Twofold bay, *F. Mueller;* Cape Howe, *C. Walter.*
Victoria. Snowy and Genoa rivers, *F. Mueller.*

2. **L. pulchella,** *Spreng. Syst.* i. 169. A much smaller plant than *L. paniculata*, the leafy base of the stems more slender but often elongated to from 1 to 3 in. and slightly branched. Leaves rarely 6 in. long and 2 to 3 lines broad, often not above half that size. Scape or

peduncle from under 6 in. to nearly 1 ft. long, often with a single leaf below the inflorescence which is more simple than in *L. paniculata,* usually a single terminal cluster and one or two lower down on the stem. Rhachis of the cluster often somewhat elongated, and sometimes a pedunculate cluster from the axil of the same bract. Bracts membranous, striate, acuminate, the subtending ones $\frac{1}{2}$ to $\frac{3}{4}$ in. long, the inner ones smaller. Pedicels filiform, $\frac{3}{4}$ to 1 in. long, often glandular-pubescent as well as the ovaries. Ovary globular, about 1 line long at the time of flowering. Perianth-segments narrower than in *L. paniculata* and still more nearly equal, varying from 2 to 3 lines long. Filaments filiform, very shortly dilated at the base. Style of *L. paniculata.* Capsule about 2 lines diameter, opening to the base in three valves which often retain the seeds long after they open.—*Sisyrinchium pulchellum,* R. Br. Prod. 305; F. Muell. Fragm. vii. 92; *Renealmia pulchella,* R. Br. Prod. Addenda; *Nematostigma pulchellum,* Dietr. Spec. Pl. ii. 510; *Libertia Laurencii,* Hook. f. Fl. Tasm. ii. 34, t. 129; *L. micrantha,* A. Cunn. in Hook. f. Fl. N. Zeal. i. 252.

N. S. Wales. Grose river, *R. Brown;* Blue Mountains, *C. Moore;* Port Jackson, *Woolls.*

Victoria. Upper Targil and Upper Latrobe rivers, Mount Baw-Baw, sources of the Yarra, &c., *F. Mueller.*

Tasmania. Abundant in various parts of the island, ascending to 4000 ft., *J. D. Hooker,* and others.

The species is also in New Zealand. The common Tasmanian form described by Hooker as *L. Laurencii,* has rather large flowers, the perianth-segments about 3 lines long; but some Tasmanian specimens, especially from Gordon river, *Milligan,* Brown's own, and some of the New Zealand ones have them remarkably small, the perianth-segments scarcely 2 lines long. The majority of the New South Wales and New Zealand ones are more or less intermediate in size.

7. CAMPYNEMA, Labill.

Perianth regular, divided to the ovary into 6 nearly equal segments spreading from the base. Stamens 6, the filaments free; anthers opening outwards, but sagittate with short obtuse auricles at the base. Styles 3, distinct from the base, rather thick, obtuse and stigmatic at the end, at first connivent, at length spreading. Capsule oblong or turbinate, prominently and obtusely 3-angled. Seeds more or less flattened or angular, with a spongy testa.—Perennial herbs, with a short rhizome, and a single or rarely two leaves radical or nearly so. Stem single, with 1 to 4 pedunculate flowers, centrifugal in their development, at first terminal but becoming opposed to a linear bract by the development of its axillary shoot.

The genus is endemic in Tasmania, with exceptional characters, to whatever Order it may be ascribed. Brown, judging from Labillardière's figure and description, placed it amongst anomalous genera at the end of *Melanthaceæ,* from which it differs essentially in its strictly inferior ovary and capsule, a character to which less importance was then attached than has since been attributed to it. Lindley enumerated it amongst *Amaryllideæ,* from which Kunth appears to have advisedly expunged it, as it appears in his index, but not in the text of his fifth volume. F. Mueller refers it without hesitation to *Hypoxideæ*, although the seeds have not the crustaceous testa nor the projecting hooked hilum peculiar to that suborder or tribe. It differs from all the above groups in two important characters—the anthers opening outwards, and the centrifugal inflorescence, in both of which it agrees with *Irideæ*, amongst which it appears to me to constitute an exceptionally hexandrous genus.

Plant of ¼ to 1 ft., with a narrow linear or linear-lanceolate leaf . . . 1. *C. lineare.*
Dwarf plant, with usually 2 oblong lanceolate short leaves . . . . 2. *C. pygmæa.*

1. **C. lineare,** *Labill. Pl. Nov. Holl.* i. 93, t. 121. Rhizome very short, with rather thick fibrous roots. Stems slender, varying from 3 or 4 in. to at least 1 ft. high, with a single linear or linear-lanceolate leaf of 2 to 4 in. and sometimes a small one outside of it almost reduced to its scarious sheath, and often 1 or 2 small linear leaves or bracts higher up. Flowers either solitary and terminal or 2 to 4, the lateral ones opening later on short or long peduncles, the terminal one becoming opposed to a linear bract. Perianth-segments ovate or ovate-lanceolate, acute or mucronate, contracted at the base, about 3 lines long. Filaments inserted at the base of the segments and about half as long, at length recurved; anthers ovate-oblong. Styles quite distinct and rather broad at the base, the tips long cohering but at length recurved. Ovary oblong-turbinate or almost linear. Capsule when fully developed ½ in. long, tapering towards the base.—Hook. f. Fl. Tasm. ii. 48; *Campylonema lineare,* Schultz, Syst. vii. 1507.

Tasmania. Rocky Cape, Macquarrie Harbour, Recherche bay, &c., ascending to 4000 ft., *Gunn, Milligan;* Southport, *C. Stuart.*

2. **C. pygmæum,** *F. Muell.* A small stout plant of 1 to 1½ in. Leaves usually 2, oblong-lanceolate, obtuse, recurved spreading, ½ to ¾ in. long and 2 to 3 lines broad. Flowers 1 or rarely 2 on a stem not exceeding the leaves. Ovary shortly turbinate but the structure the same as in *C. lineare,* of which F. Mueller thinks it may be an alpine variety.

Tasmania. Summit of Mount Lapeyrouse in the tufts of *Donatia, Oldfield.*

Order CXXIII. AMARYLLIDEÆ.

Flowers hermaphrodite, regular or oblique. Perianth superior, with or without a distinct tube, the limb of 6 coloured or petal-like segments, all equal or rarely the 3 inner ones rather larger or smaller. Stamens 6, inserted at the orifice of the tube or base of the segments or rarely 3 only opposite the inner segments, or in a very few genera not Australian more than 6; filaments free or united at the base into a short tube or corona; anthers versatile or rarely attached at the base, with 2 parallel cells opening inwards or laterally. Style single, with 3 adnate stigmas, usually very small and confluent on the obtuse end, rarely oblong or linear and connate or diverging into stigmatic branches. Ovary inferior or rarely half-superior, 3-celled with several usually numerous rarely only 1 or 2 ovules in each cell, amphitropous anatropous or rarely orthotropous, attached to an axile placenta, or rarely the ovary reduced to 1 cell by the abortion of 2 carpels or by the obliteration of the dissepiment. Fruit usually a capsule, opening loculicidally either at the free apex only or to the base in 3 valves, rarely succulent and indehiscent, or bursting irregularly. Seeds albuminous, with a small or linear embryo, the radicle near to or more or less distant from the hilum.—Herbs with a perennial short or tuberous or creeping rhizome, or bulbous base. Leaves mostly radical, or nearly so, the sheathing base either distichous or imbricate or bulbous, the lamina entire, usually narrow with contiguous parallel veins, either laterally compressed (equitant), terete, channelled or flat, rarely broad with distant primary veins and transverse veinlets. Scapes or flowering stems terminal, leafless or with one or more sheathing bracts or leaves much smaller than the radical ones, with a single terminal flower or more frequently with several or many flowers either in a terminal umbel surrounded by 2 or more membranous or coloured bracts, or few in a terminal raceme or many in a terminal thyrsus or panicle, each branch and each flower subtended by a bract, the inflorescence usually centripetal. Perianths glabrous hairy or woolly in the Australian genera, most frequently white, yellow, or red, rarely purple, never blue.

The Order is generally distributed over the warmer and temperate regions of the globe, most abundant in dry, sunny countries. Of the thirteen Australian genera, three range over tropical Asia and tropical and Southern Africa, two of them sparingly represented also in America, a fourth (*Eurycles*) extends to the Archipelago, the remaining nine are endemic, and six of them limited to West Australia. The several tribes here distinguished are usually considered as so many independent Orders, or at any rate as referrible to three distinct Orders—*Hæmodoraceæ*, *Hypoxideæ*, and *Amaryllideæ*; but although these subordinate groups are in most respects distinct, it appears to me that it is only by their union in one general Order that we can obtain a well-defined group of the same grade as *Irideæ*, *Burmanniaceæ*, *Orchideæ*, *Scitamineæ*, and *Hydrocharideæ*, all of them clearly marked out by definite and important characters. It is generally admitted that the above suborders here united under *Amaryllideæ*, agree in the most important characters derived from the flower and seed, differing from *Hydrocharideæ*, *Orchideæ*, and *Burmanniaceæ* in their albuminous seeds, from *Scitamineæ* and *Orchideæ* in their regular (or only oblique) flowers, from *Irideæ* and *Burmanniaceæ* in their centripetal not centrifugal inflorescence and in their stamens, from *Taccaceæ* and the majority of *Orchideæ* and *Burmanniaceæ* in their axile placentum, from *Dioscorideæ* in their hermaphrodite flower, and in all cases there are other characters either less constant or of minor importance.

Amongst these Amaryllideous suborders, *Hæmodoraceæ* have been supposed to be distinctly characterized by equitant leaves and furfuraceous-tomentose flowers, but *Hæmodorum* and *Phlebocarya* are perfectly glabrous, the leaves are terete or channelled above in some species, and never equitant in *Vellozieæ*, now generally included in *Hæmodoraceæ*. Herbert, who unites the greater portion of *Hæmodoraceæ* with *Amaryllideæ*, would exclude *Hæmodorum* itself as being triandrous; but although *Irideæ* are almost universally triandrous and *Amaryllideæ* hexandrous, this number of stamens, single or double, if relied upon absolutely, separates the Orders much less naturally than the difference in inflorescence wherever it can be ascertained, accompanied as it is by an apparently constant difference in the anthers. When the scapes are uniflorous, the number of stamens or some other of the above mentioned characters may be called in aid. Taking therefore the *Amaryllideæ* as a whole as one Order, it would include besides the five tribes or suborders here enumerated, which are all common to the New as well as the Old World, the *Vellozieæ*, which are confined to America, except a single African species and the *Alstrœmerieæ*, all American, in which however the secondary inflorescence appears to be centrifugal.

TRIBE 1. **Hæmodoreæ.**—*Perianth glabrous, divided to the ovary into distinctly 2-seriate segments. Stigmas very small. Leaves laterally flattened or terete. Inflorescence compound or rarely simply racemose.*

Stamens 3. Ovules 2 in each cell. Capsule almost superior, 3-dymous . . . . 1. HÆMODORUM.
Stamens 6. Ovules 1 to each carpel, the ovary often 1-celled. Nut inferior, indehiscent, 1-seeded. Flowers small . . . . . . 2. PHLEBOCARYA.

TRIBE 2. **Conostyleæ.**—*Perianth plumose-woolly or tomentose outside, the tube usually continued above the ovary, the lobes apparently 1-seriate. Stigmas very small. Leaves laterally flattened or terete. Inflorescence compound simply racemose or capitate, rarely 1-florous. Stamens 6.*

Filaments broad, produced into 2 erect flat appendages above the insertion of the anthers. Ovules many. Flowers solitary or few in a head or small cyme . . . . 3. TRIBONANTHES.
Filaments inappendiculate.
Perianth campanulate or rarely tubular. Ovules several, usually many in each cell, not in rows. Capsule opening in 3 valves at the apex. Flowers in heads or cymes, rarely solitary . . 4. CONOSTYLIS.
Perianth tubular, the limb equal. Ovules several, in 2 rows in each cell, the ovary adnate at the angles only. Flowers in short unilateral racemes on a branching scape . . . . . 5. BLANCOA.
Perianth long, tubular, the limb very oblique, more split on the lower side. Flowers in unilateral racemes or spikes on a simple or dichotomous scape or peduncle.
Ovules many or rarely 2 to 4 in each cell. Capsule 3-valved at the apex. Wool of the perianth red green or yellow . . 6. ANIGOZANTHOS.
Ovules 1 in each cell. Seeds separating in 3 cocci, leaving the thick persistent septa. Wool of the perianth nearly black . . . . 7. MACROPODIA.

TRIBE 3. **Hypoxideæ.**—*Perianth hairy or rarely glabrous. Stigmas or stigmatic lobes large, free or connate. Ovules many, in 2 rows in each cell. Seeds with a crustaceous testa, the hilum produced into a hooked beak. Leaves horizontally flattened, channelled or terete. Flowers in a simple spike or raceme or solitary.*

Flowers sessile within sheathing or imbricate bracts in a dense spike. Perianth usually with a long tube . . . . 8. CURCULIGO.
Flowers solitary or few in a loose pedunculate raceme. Perianth divided to the ovary into spreading segments . . . . . . 9. HYPOXIS.

TRIBE 4. **Agaveæ.**—*Perianth glabrous. Stigmas small. Tall often woody plants, not bulbous. Leaves horizontally flat, channelled (or terete?). Inflorescence compound.*

Radical leaves very numerous. Flowers large, red, in a terminal compound head or thyrsus . . . . 10. DORYANTHES.

TRIBE 5. **Euamaryllideæ.**—*Perianth glabrous. Stigmas small. Bulbous plants. Leaves horizontally flat, channelled or terete. Flowers umbellate or rarely solitary on leafless scapes.*

No corona. Flowers large. Ovules several, in 2 rows in each cell 11. CRINUM.
Filaments connected below the middle by a corona.
Ovary 3-celled, with 2 ovules in each cell. Leaves broad, with distant primary veins . . . . 12. EURYCLES.
Ovary 1-celled, with 2 ovules. Leaves narrow with close veins, or broad with distant primary veins . . . . 13. CALOSTEMMA.

TRIBE 1. HÆMODOREÆ.—Rhizome or base of the stem short and thick, emitting fibrous roots and sometimes covered with the brown sheathing bases of old leaves so as to resemble bulbs. Leaves mostly radical, with distichous sheathing bases and long laterally flattened or terete laminæ. Stems often with a few short leaves. Flowers glabrous, in cymes or panicles, rarely reduced to a simple raceme. Perianth divided to the ovary into 6 segments in 2 rows. Ovules 1 or 2 to each carpel or cell.

1. HÆMODORUM, Sm.

Perianth persistent, divided to the ovary into 6 segments, all nearly equal or the outer ones shorter. Stamens 3, inserted at the base of the inner segments, which in the open flower are usually convolute round the filaments at the base. Ovary entirely or almost entirely inferior, the broad summit either flat or with 3 slight protuberances, 3-celled, with 2 ovules in each cell. Style simple, obtuse, entire or obscurely 3-furrowed at the stigmatic end. Capsule half or almost entirely superior, the free part 3-dymous and opening in 3 loculicidal slits. Seeds ovate, flat, with a wing-like margin, peltately attached to a prominent placenta.—Erect glabrous herbs, the base of the stem or rhizome sometimes thickened and enclosed in the persistent sheathing base of the leaves so as to resemble narrow bulbs, the fibrous roots sometimes very thick and spongy and often red. Leaves sheathing and equitant at the base, the lamina laterally flattened or terete, the lower ones sometimes very long, the upper ones few and short. Flowers black, red, of a livid green, or perhaps in some of the small flowered-species yellow, usually fragrant, in clusters compound heads cymes loose panicles or interrupted spikes, with a bract under each branch or pedicel, and usually 2 on each pedicel even when very short.

The genus is limited to Australia.

Flowers in dense globular or oblong heads on dwarf stems (under 6 in.). Leaves nearly terete.
  Leaves thick and short. Flower-heads solitary. Perianth-segments nearly equal. Tasmanian species . . . . . 1. *H. distichophyllum.*
  Leaves slender and long. Flower-heads oblong, usually two. Outer perianth-segments shorter. Tropical species . . 2. *H. brevicaule.*
Flowers (black), above 3 lines long, in clusters of 2 to 4 or singly pedicellate. Leaves very narrow or terete. Western species.
  Bracts under and on the pedicels narrow, not scarious.
    Flowers singly pedicellate in a dichotomous panicle . . 3. *H. sparsiflorum.*
    Flowers usually two together along the rhachis of a single interrupted spike.
      Stout plant of 2 to 3 ft. Outer perianth-segments more than $\frac{3}{4}$ the length of the inner . . . . . . 4. *H. spicatum.*
      Slender plant under 1 ft., oblong-bulbous at the base. Outer perianth-segments $\frac{1}{2}$ as long as the inner . . 4. *H. brevisepalum.*
  Bracts under and on the pedicels broad with scarious margins.
    Tall plants. Bracts at least $\frac{3}{4}$ as long as the flowers.
      Flower-clusters several along the branches of an erect panicle . . . . 6. *H. paniculatum.*

Flower-clusters singly terminating the branches of a spreading panicle . . . . 7. *H. laxum.*
Slender plant. Bracts less than ½ as long as the flowers.
Flower-clusters solitary or few on long peduncles . . 8. *H. simplex.*
Flowers (dark red, livid or black) above 3 lines long, very numerous, in dense terminal corymbose cymes. Leaves flat or terete.
Perianth-segments nearly equal.
Flowers (black) usually in 2 or more distinctly pedunculate compact cymes or heads. Western species . . 9. *H. simulans.*
Flowers (greenish purple) forming one compact compound terminal cyme. Eastern species.
Leaves flat, 2 to 3 lines broad . . . . 10. *H. planifolium.*
Leaves slender, terete . . . . 11. *H. teretifolium.*
Outer perianth segments considerably shorter than the inner. Flowers (red ?) in a compact compound terminal cyme. Tropical species . . . . 12. *H. coccineum.*
Flowers about 2 lines long, green or yellow, in panicles or spikes, the inner segments broader but not longer than the outer. Tropical or Eastern subtropical species.
Leaves long, flat, flaccid. Flowers densely cymose in a corymbose panicle . . . . 13. *H. subvirens.*
Leaves flat, rigid, and glaucous. Flowers loosely racemose along the branches of a spreading panicle . . . . . 14. *H. ensifolium.*
Leaves slender, terete.
Flowers very shortly pedicellate along the branches of a spreading panicle . . . . 15. *H. parviflorum.*
Flowers nearly sessile along a simple rhachis . . . . 16. *H. leptostachyum.*
Flowers singly pedicellate in a dichotomous panicle . . 17. *H. tenuifolium.*

1. **H. distichophyllum,** *Hook. Ic. Pl.* t. 866. A dwarf compact species, not exceeding 4 in., the lower part occupied by the broad distichous leaf-sheaths; laminæ of the leaves thick, not much compressed, rigid, 2 to 3 in. long and not above 2 lines broad. Stems with 3 or 4 loose sheathing bracts of ½ to ¾ in. Flowers not many in a compact terminal head scarcely ¾ in. diameter. Perianth shortly adnate at the base, the segments oblong or lanceolate, about 2 lines long and all nearly equal. Filaments as long as the perianth, with small anthers. Capsule almost entirely superior, deeply tridymous.—Hook. f. Fl. Tasm. ii. 35.

Tasmania. Heathy ground near Port Macquarrie, *Milligan.*

2. **H. brevicaule,** *F. Muell. Fragm.* i. 64. Stems not above 6 in. high, covered at the base with the broad sheathing bases of the leaves sometimes splitting into fibres. Radical leaves with a terete slender lamina, sometimes 1 ft. long or even more; leaves or bracts under the inflorescence 1 or 2, lanceolate, ½ to 1 in. long. Flowers drying black, in one two or three dense oblong heads of ¾ to 1 in., the bracts within the head small. Inner perianth-segments in some specimens a little above 2 lines, in others nearly 3 lines long, narrow and obtuse, the outer ones ¼ to ⅓ shorter. Filaments thick, nearly as long as the perianth; anthers rather short.

N. Australia. Sea range and dry plains towards M'Adam range, *F. Mueller;* Liverpool river, *Gulliver.*

3. **H. sparsiflorum,** *F. Muell. Fragm.* vii. 117. A rather tall species, with the habit and foliage of *H. laxum* and a similarly divaricately branched panicle, but the flowers smaller, on pedicels longer than the perianth, the bracts at the base of and on the pedicels linear or lanceolate, 2 to 3 lines long, without scarious margins. Perianth-segments about 3½ lines long, apparently black, very narrow, shortly dilated at the base, the outer ones rather shorter than the inner. Capsule apparently much smaller than in *H. laxum,* but not seen ripe.

W. Australia. *Drummond, n.* 58.

4. **H. spicatum,** *R. Br. Prod.* 300. Stems simple, attaining 2 to 3 ft., covered at the shortly thickened base by the short broad sheathing bases of the leaves, but loosely so and not so bulb-like (in the specimens seen) as in *H. brevisepalum* and *H. simplex.* Leaves from their short sheathing base tapering into a long very narrow linear-subulate lamina. Flowers black, mostly in pairs along a simple rhachis of 6 in. to 1 ft., each pair subtended by an ovate-lanceolate acuminate bract of 2 to 3 lines, with a small lanceolate bract on each pedicel, the pedicels always much shorter than the perianth. Inner perianth-segments linear or linear-lanceolate, about 5 lines long, the outer ones usually about 1 line shorter, more subulate, but dilated at the base. Filaments inserted near the base of the inner segments, enclosed in but free from them. Anthers oblong, rather short. Capsule about 4 lines broad.—Endl. Iconogr. t. 98, and in Pl. Preiss. ii. 15; *H. edule,* Endl. in Pl. Preiss. ii. 15.

W. Australia. Lucky bay, *R. Brown;* King George's Sound and adjoining districts, *Preiss, n.* 1423, *Oldfield,* and others; Swan river, *Preiss, n.* 1421; Murchison river, *Oldfield.*—The roots are said to be eaten by the natives.

5. **H. brevisepalum,** *Benth.* Stems simple, rather slender, under 1 ft. high and enclosed at the base in the closely-pressed membranous sheathing bases of old leaves, forming an oblong bulb as in *H. simplex.* Leaves tapering into linear-terete laminæ, usually longer than the stem. Flowers in distant pairs along a simple rhachis as in *H. spicatum,* but fewer, as large or rather larger, and the outer segments broadly ovate, very shortly acuminate and only half as long as the inner ones.

W. Australia. Swan river, *Drummond,* 1*st coll. n.* 743.

6. **H. paniculatum,** *Lindl. Swan Riv. App.* 44. Very near *H. laxum* and perhaps a variety only. It appears to be a still taller plant, the leaves rather flatter and not quite so rigid. Panicle with fewer more erect branches, the flowers in clusters or cymes of 2 to 4 along the branches and not all terminal. Bracts larger, more scarious on the margins. Perianths rather larger, the outer segments rather shorter and considerably broader than the inner ones.—Endl. in Pl. Preiss. ii. 15? partly.

W. Australia. Swan river, *Drummond,* 1*st. coll. n.* 742; Murchison river, *Oldfield.*

H. strictum, Endl. in Pl. Preiss. ii. 15, from Wellington district, probably belongs to *H. paniculatum,* as well as Preiss's n. 1425, from Victoria district.

7. **H. laxum,** *R. Br. Prod.* 300. Stems rigid, 1½ to 2 ft. high. Lower leaves 10 in. to above 1 ft. long, rigid, striate, thick or sometimes almost terete, 1 to 1½ lines broad or sometimes nearly 2 lines at the base before they open into a long sheath, those of the stem also gradually dilated into long sheaths. Flowers on short pedicels, usually 2 to 4 together at the ends of the dichotomous branches of a very loose corymbose panicle. Bracts at the base of the branches small, those under and upon the pedicels ovate or oblong, at least ⅔ as long as the flowers, with scarious margins. Perianth black, rather above 4 lines long, the outer segments rather shorter than the inner ones, with broad bases. Capsule nearly ½ in. broad.

W. Australia. King George's Sound, *R. Brown;* Perongerup, *F. Mueller;* Vasse and Blackwood rivers, *Oldfield.*

8. **H. simplex,** *Lindl. Swan Riv. App.* 44. Stem rather slender, 1 to 1½ ft. high, not much thickened at the base but covered with the broad membranous sheathing bases of the lower leaves, so as to form a kind of narrow oblong bulb. Leaves very narrow, linear-terete, the radical ones sometimes short sometimes 6 to 8 in. long; those of the stem few, abruptly dilated into short sheaths. Flowers "blackish," few together in compact heads or cymes which are either solitary or 2 together on short peduncles at the end of the stem, or sometimes another on a longer peduncle from the axil of a bract lower down. Bracts under and upon the very short pedicels broadly ovate or orbicular, not half so long as the flower, with scarious margins. Perianth about 4 lines long, the outer segments rather shorter than the inner, very broad at the base, the inner much convolute. Capsule 3 to 4 lines broad, but not quite ripe in the specimens seen.

W. Australia. Swan river, *Drummond,* 1*st. coll.*; Kalgan river, *Oldfield;* Lake Muir, *Muir.*

H. polycephalum, Endl. in Pl. in Preiss. ii. 16, from Swan river, is probably founded on specimens of this species with more than one head of flowers, those which Lindley described having had mostly only a single head.

9. **H. simulans,** *F. Muell. Fragm.* vii. 117. A stout rigid species, one of Drummond's specimens above 4 ft. high. Leaves flat, rigid, striate, the lower ones sometimes 1 ft. long and ¾ in. broad, but in other specimens not above 2 lines broad, the upper ones short and broad. Flowers "black and fragrant" rather numerous, in compact cymes at the end of the stem and of 1 to 3 long peduncles. Bracts linear or narrow-lanceolate, without scarious margins. Perianth about ½ in. long, the segments all narrow and nearly equal, shortly dilated at the base. Filaments not longer than the anthers, which sometimes do not reach to above half the length of the perianth, although sometimes nearly its length.

W. Australia. Swan river, *Drummond,* 1*st coll., also n.* 310; Murchison river, *Oldfield.*

This was considered by Lindley as the same as the Eastern *H. planifolium,* and indeed is scarcely to be distinguished from it except by the usually broader more prominently veined leaves, and by the flowers in more compact heads or cymes, but usually 2 or more such heads on unequal peduncles, instead of forming together only one terminal compact corymb.

10. **H. planifolium,** *R. Br. Prod.* 300. Stems from a thick base 2 to 3 ft. high, scarcely branched below the inflorescence. Lower or radical leaves long, grass-like, flat, from under 2 lines to nearly 3 lines broad, the upper ones few and short. Flowers numerous, of a livid purple or greenish at the base, in short forked racemes or cymes collected in a compact more or less corymbose panicle, but usually looser than in *H. coccineum.* Bracts narrow, subulate-acuminate, usually longer than the pedicel. Perianth-segments linear or linear-lanceolate, obtuse in the bud, but appearing acuminate when open the margins being involute, about 5 lines long, the outer ones scarcely shorter than the inner. Stamens much shorter than the perianth; anthers linear, about as long as the filaments. Ovary wholly inferior, the summit showing only 3 slight protuberances. Capsule more than half superior, tridymous, 4 to 5 lines broad.—Bot. Mag. t. 1610.

N. S. Wales. Port Jackson to the Blue Mountains, *R. Brown, Sieber, n.* 203, and many others; Hastings and Clarence rivers, *Beckler;* Clarence river, *Wilcox;* New England, *Leichhardt, C. Stuart.*

The species varies in some manner in inflorescence, the panicle compact or spreading, the ultimate racemes few and long or numerous and short.

11. **H. teretifolium,** *R. Br. Prod.* 300. Stature inflorescence and flowers precisely those of *H. planifolium,* but the leaves from a short sheathing base very long, slender, terete, or nearly so, about $\frac{1}{2}$ line broad, as in *H. tenuifolium.* Flowers very numerous in a compact but not very dense compound corymbose panicle, black when dry as in the allied species. Perianth-segments narrow and about 5 lines long, the outer ones scarcely shorter than the inner as in *H. planifolium.*

N. S. Wales. Port Jackson, *R. Brown;* Illawarra, *A. Cunningham,* an imperfect specimen, but apparently the same as Brown's plant.

12. **H. coccineum,** *R. Br. Prod.* 300. Stems from a thick base 2 to 3 ft. high, not branched below the inflorescence. Leaves at the base of the stem 1 to 2 ft. long, flat, 2 to 3 lines broad, very finely striate, the upper ones few and short. Flowers numerous "red" but drying black, in dense cymes forming a compact terminal compound corymbose panicle. Perianth-segments linear, the outer ones thickened and dilated at the base, 3 lines long, the inner ones rather narrower and $\frac{1}{3}$ longer. Stamens a little shorter or longer than the perianth, the anthers much shorter than the filaments. Ovary wholly inferior. Capsule half superior, about 4 or 5 lines broad, conspicuously 3-dymous or didymous by abortion of one cell.

N. Australia. Islands of the Gulf of Carpentaria, *R. Brown, Henne, Gulliver;* Upper Victoria river, *F. Mueller.*

Queensland. Cape York, *M'Gillivray, Daemel, Veitch;* Albany island, *F. Mueller;* Fitzroy island, and Mount Elliott, *Fitzalan;* Rockingham bay, very abundant, *Dallachy;* Cape river and Glenella creek, *Bowman.*

13. **H. subvirens,** *F. Muell. Fragm.* i. 63. A tall species. Lower or radicle leaves longer and more flaccid than in *H. coccineum*, 3 to 4 lines broad. Flowers numerous, in a rather looser and more spreading corymbose panicle than in that species. Bracts lanceolate, acuminate, nearly as long as the flowers. Perianth scarcely above 2 lines long, greenish according to F. Mueller's notes, somewhat yellowish when dry, the outer segments narrow-lanceolate, acute, the inner ones rather broader and more obtuse but not longer.

N. Australia. Rocky hills, Upper Victoria river, *F. Mueller*, a single specimen preserved in Herb. Hooker, represented in Herb. F. Mueller by one of *H. coccineum*, which, though resembling it in habit, has very different flowers.

14. **H. ensifolium,** *F. Muell. Fragm.* i. 64. A rigid glaucous species, 2 ft. high or more. Leaves rigid, with long open sheaths, the lamina flat, obtuse, 2 to 3 lines broad and about 1 ft. long in the radical leaves, those of the stem short. Panicle broad loose and divaricate, the ultimate branches loosely racemose, with very small bracts. Pedicels usually about as long as the flowers. Perianth-segments oblong-lanceolate, obtuse, scarcely 2 lines long, the outer ones quite as long as the inner. Stamens about as long as the perianth. Capsule nearly ½ in. broad.

N. Australia. M'Adam range, *F. Mueller;* Port Darwin, *Schultz, n.* 522.

16. **H. parviflorum,** *Benth.* Stems slender, 1½ to 2 ft. high. Leaves with rather long sheathing bases, the lamina slender and terete like that of *H. teretifolium*, the lower ones sometimes 1 ft. long, the upper ones few and short. Panicle consisting of few, sometimes only 2 or 3 slender spreading branches, along which the flowers are racemose, on pedicels usually shorter than the flower. Perianth-segments about 1½ lines long, narrow-oblong, obtuse, the inner ones not longer than the outer. Stamens shorter than the perianth. Capsule although nearly ripe not above 3 lines diameter.

N. Australia Brunswick bay, N.W. coast, *A. Cunningham;* Port Darwin, *Schultz, n.* 723.

16. **H. leptostachyum,** *Benth.* Base of the stem thickly covered in our specimens with the rigid fibrous remains of old sheaths. Stem quite simple or branching below the middle, slender, 1 to 1½ ft. high. Leaves short, the lamina very slender and terete but rigid. Flowers sessile or nearly so within very small obtuse bracts along a long slender simple rhachis. Perianth-segments broadly ovate, very obtuse, nearly 2 lines long, the inner ones not longer than the outer. Stamens about as long as the perianth-segments. Fruit apparently like that of *H. parviflorum.*

N. Australia. Port Darwin, *Schultz, n.* 659.

17. **H tenuifolium,** *A. Cunn. Herb.* Stems 1½ to 2 ft. high. Radical and lower leaves with a rather broad short sheathing base, very long, slender and almost terete, about ½ line broad. Panicle loosely divari-

cate, the ultimate 1-flowered branches much longer than the flower, with small distant bracts, and not at all assuming the racemose character. Perianth-segments broadly ovate, very obtuse, all nearly the same length, the outer almost membranous with a broad base, the inner ones of a thicker consistence, nearly orbicular, contracted at the base and of a deeper colour when dry. Filaments attaining nearly the length of the perianth; anthers ovate, usually slightly exserted.

Queensland. Peat and boggy ground, shores of Moreton bay and island, *A. Cunningham*, *F. Mueller*.
N. S. Wales. Duval, *Leichhardt*.

2. PHLEBOCARYA, R. Br.

Perianth persistent, divided to the ovary into 6 nearly equal segments. Stamens 6; anthers erect, on short filaments, inserted at the base of the segments. Ovary inferior, more or less 3-celled when very young, but often 1-celled at the time of flowering by the obliteration of the dissepiments, placenta central, with 3 ascending ovules (1 to each cell). Style simple, obtuse, entire or obscurely 3-furrowed at the stigmatic end. Fruit wholly inferior, nut-like and indehiscent, containing a single erect seed. Testa membranous; albumen fleshy.—Herbs usually more or less ciliate with long hairs. Leaves long narrow and grass-like. Flowering stems slender, shorter than the leaves, usually forked or dichotomously divided, with a compact or loose cyme of small flowers at the end of each branch.

The genus is limited to West Australia.

Leaves flat, ciliate on the margin only, and sometimes almost without cilia. Anther-connective not longer than the cells . . . 1. *P. ciliata.*
Leaves flat, hairy on the whole surface. Anther-connective shortly produced beyond the cells . . . . 2. *P. pilosissima.*
Leaves terete, filiform, sparingly ciliate. Anther-connective much produced beyond the cells . . . . 3. *P. filifolia.*

1. **P. ciliata,** *R. Br. Prod.* 301. Rhizome short and thick. Radical leaves with broad black rigid sheathing bases, narrow-linear, $\frac{3}{4}$ to $1\frac{1}{2}$ ft. long, 1 to 2 lines broad, rigid or rather flaccid, prominently striate, more or less bordered with long cilia usually distant and sometimes only to be seen on young leaves. Flowers mostly about 2 lines long, the cyme-like clusters usually rather dense at first consisting of about 6 to 12 flowers, each on a very short pedicel in the axil of a linear bract of 1 to 2 lines; as the flowering advances the branches of the cluster sometimes lengthen to near 1 in., the whole inflorescence forming a loose panicle always much shorter than the leaves. Perianth-segments lanceolate, evidently spreading when fresh although almost universally erect in the dried specimens. Filaments very short; anthers oblong, the connective not produced beyond the cells. Ovary at the time of flowering 1-celled, with a broad thick disk- or cup-shaped central placenta at the base of the cavity, upon which are seated 3 erect orthotropous ovules, but I have occasionally found a persistent axis in

the centre connected with the apex of the cell and generally along the walls of the cell the remains of three dissepiments, so that probably in a very young state the ovary is normally 3-celled. Fruit a small ovoid-globular nut, crowned by the persistent perianth, the pericarp thick and apparently indehiscent, containing a single nearly globular seed with a membranous testa and fleshy albumen, not quite ripe however in the specimens examined.—Endl. in Pl. Preiss. ii. 29.

W. Australia. King George's Sound, *R. Brown, F. Mueller;* Capel and Vasse rivers, *Oldfield;* Hampden, *Clarke.*

Var. *lævis.* Leaves rather long, broad, and rigid, with very few cilia only to be seen on the young leaves, which, however, I have never found to be absolutely without any. —*P. lævis,* Lindl. Swan Riv. App. 43; Endl. in. Pl. in Preiss. ii. 29.—Swan river, *Drummond,* 1*st coll.*, *Preiss, n.* 1558; Capel river, *Oldfield.*

2. **P. pilosissima,** *F. Muell. Fragm.* viii. 23 (as a var. of *P. ciliata*). Leaves erect, rigid, mostly under 1 ft. long, flattened but under 1 line broad, striate, with long hairs on the sides as well as on the margins. Stems short, the panicle rather loose and hairy. Bracts linear-subulate. Flowers about 2 lines long, the segments narrow but obtuse. Anther-connective produced beyond the cells but not so much so as in *P. filifolia.* Ovary in the flowers examined completely 3-celled, but the dissepiments very thin and readily disappearing as the flower withers.

W. Australia, *Drummond.*

3. **P. filifolia,** *F. Muell. Fragm.* viii. 23 (as a var. of *P. ciliata*). Leaves filiform, terete, not conspicuously striate, glabrous except a few long cilia near the base, the longest above 1 ft. long. Panicle loose and few-flowered, shorter than the leaves. Bracts small, lanceolate. Perianth about the size of that of the two preceding species, but the segments narrowly acuminate. Anther-connective conspicuously produced into a rather long appendage beyond the cells. Ovary more or less completely 3-celled, with one ascending acuminate ovule in each cell.

W. Australia, *Drummond, n.* 207, *and* 368.

Tribe 2. Conostyleæ.—Rhizome short with fibrous roots, the base of the stem short and sometimes covered with brown sheathing bases of old leaves so as to resemble bulbs, or shortly branched and densely tufted, or rarely elongated and proliferous-branched. Leaves radical or at the base of the flowering stems, with distichous sheathing bases or densely tufted, the lamina long, laterally flattened or terete. Scapes or flowering stems usually bearing 1 or 2 sheathing bracts or short leaves. Flowers plumose-woolly or tomentose outside, in heads, cymes racemes or dichotomous panicles, rarely solitary. Perianth-tube more or less continued above the ovary (except in *Conostylis breviscapa*) the limb continuous with the tube, regular or oblique, the lobes usually appearing uniseriate and almost induplicate-valvate. Stamens 6. Stigmas very small at the end of a filiform style.

The Australian genera are all endemic in West Australia, but some South African and American genera are referrible to the same tribe.

3. TRIBONANTHES, Endl.

Perianth persistent, more or less woolly, deeply divided into 6 segments nearly equal and similar, erect at the base and connivent or cohering into a short tube, then spreading. Stamens 6; filaments broad, lining the perianth-tube and usually adnate to it, produced beyond it into 2 short erect entire or toothed appendages; anthers sagittate, attached to the filaments between the appendages, the connective often produced beyond the cells. Ovary superior, 3-celled, with a conical summit produced into a short style. Ovules numerous in each cell, crowded into several rows on a placenta attached above the middle of the cell extending more or less towards the base. Capsule opening loculicidally at the conical apex in 3 coriaceous valves.—Rhizome producing tubers enveloped in loose membranous scales. Leaves few, with dilated sheathing bases, produced into a terete or channelled lamina. Flowers solitary or few, in a terminal cyme or head. Bracts lanceolate or ovate, usually acuminate.

The genus is limited to West Australia. The species appear to be very variable and difficult to distinguish by positive characters.

Filament-appendages bearing on the back several longitudinal prominent laminæ. Flowers 2 or more, sessile. Perianth-segments glabrous inside . . . . 1. *T. brachypetala.*
Filament-appendages flat on the back. Perianth-segments woolly on both sides.
Filament-appendages as long as or longer than the anthers.
Flowers solitary within a broadly-ovate bract; segments usually glabrous along the centre outside . . . . . 2. *T. uniflora.*
Flowers 2 or more, sessile or nearly so. Bracts ovate or lanceolate . . . . 3. *T. australis.*
Flowers 2 or more, distinctly pedicellate . . . . . . 4. *T. variabilis.*
Filament-appendages much shorter than the anthers. Flowers solitary or several, on pedicels longer than the ovary . . 5. *T. longipetala.*

1. **T. brachypetala,** *Lindl. Swan Riv. App.* 44. Stems thickened into bulbs at the base, often above 1 ft. high, more or less woolly in the upper part. Leaves 2 to 4, with broad sheathing bases, tapering into a very narrow almost terete but usually channelled lamina, the lowest sometimes 6 in. long, but usually very much shorter. Flowers 3 to 6 together, closely sessile in a dense terminal head. Bracts broadly ovate, glabrous or nearly so, 1 or 2 of the outer ones produced into a point longer than the flowers. Perianth-segments woolly outside, glabrous inside, shortly erect at the base, the spreading or reflexed laminæ 2 to 2½ lines long, always shorter than the ovary. Filaments broad, adnate to the erect part of the segments, produced above it into 2 obovate appendages longer than the anther, each with a dorsal appendage divided into 4 longitudinal laminæ. Protruding apex of the capsule 3-valved.—*T. odora*, Endl. in Pl. Preiss. ii. 28.

W. Australia. Swan river, *Drummond*, 1*st coll.*; Canning river, *Preiss, n.* 2394; between Swan river and King George's Sound, *Harvey*.

Endlicher does not explain upon what grounds he proposed to suppress Lindley's names for this and the *T. longipetala.*

2. **T. uniflora,** *Lindl. Swan Riv. App.* 44. A slender species, none of our specimens above 8 in. high and many not above 4 in., and less woolly than the others. Tubers of the rhizome with loose membranous coatings. Leaves slender, nearly terete. Flowers solitary in all our specimens. Bracts 1 or 2, broadly ovate, membranous almost scarious, the outer one with a short or long green point. Perianth-segments 3 to 4 lines long, more or less woolly on the upper surface and sometimes densely woolly throughout, but almost always green outside along the centre and sometimes the wool very thin all over. Filament-appendages as long or nearly as long as the anthers, smooth and flat on the back, often ending in points. Style variable in length. Placentas reaching to the base of the cells of the ovary.—*T. violacea*, Endl. in Pl. Preiss. ii. 28.

W. Australia. Swan river, *Drummond*, 1*st coll.*; Vasse river, *Mrs. Molloy;* between Mounts Elphinstone and Melville, *Preiss*, *n.* 1562; King George's Sound, *Muir;* Kalgan river, *Oldfield;* Cape Legrand, *Maxwell.*

The short appendages to the filaments described by Lindley, were probably injured in the specimen he examined. I find them about equal to the anthers in Drummond's as well as other specimens, but irregular.

3. **T. australis,** *Endl. Nov. Stirp. Dec.* 27, *Iconogr.* t. 109. Stems usually under 1 ft., woolly in the upper part. Leaves usually shorter and not so slender as in *T. brachypetala.* Flowers 2 to 6 together almost sessile in a terminal head or dense cyme. Bracts ovate or lanceolate, glabrous, 1 or 2 often produced into a point as long as or longer than the flowers. Perianth-segments ovate or elliptical, 3 to 5 lines long, woolly on both sides, the erect base connate with the filaments into a short tube. Filament-appendages erect, flat, as long as or longer than the anthers, sometimes almost petal-like, without dorsal appendages but sometimes a double keel along the centre of the filament before it divides produced into 1 or 2 small intermediate teeth or lobes.

W. Australia. King George's Sound and adjoining districts, *Huegel, Oldfield, F. Mueller, Muir;* eastward to Bremer and South-west bays, *Maxwell.*

4. **T. variabilis,** *Lindl. Swan Riv. App.* 44. Stems often 1 ft. high, woolly as in the allied species. Leaves tapering into a rather thick almost terete lamina as in *T. australis.* Flowers 2 to 6 together or sometimes more, in a terminal dense corymb, pedicellate, but the pedicels shorter than the ovary. Perianth-segments oblong, as long as or longer than the ovary, woolly on both sides. Filament-appendages longer than in *T. australis,* 2- or 3-toothed or entire and then sometimes quite petal-like.

W. Australia. Swan river, *Drummond*, 1*st coll. n.* 764, *Helmich;* Vasse river, *Mrs. Molloy;* Greenough flats, *C. Gray;* Busselton, *Preiss.*—Perhaps a variety only of *T. australis.*

5. **T. longipetala,** *Lindl. Swan Riv. App.* 44. Rhizome producing several oblong or globose tubers loosely enveloped in membranous coatings. Stems usually ¾ to 1 ft. high, woolly in the upper part. Leaves nearly terete, rather thick and not long. Flowers solitary or in

a loose cyme of 2 to 6, the pedicels as long as or longer than the ovary. Bracts lanceolate or linear. Perianth-segments 5 to 7 lines, usually about 6 lines long, narrower than in the other species, densely covered on both sides with a white cottony wool, the short erect bases distinct but connected by their wool. Filaments short and flat, produced into 2 or 4 erect lobes always much shorter than the anther. Style reaching only to the base of the anthers. Placentas pendulous, not half the length of the cells, densely covered with several rows of ovules.—*T. Lindleyana*, Endl. in Pl. Preiss. ii. 27.

W. Australia. Swan river, *Drummond*, 1*st coll. n.* 763, 764; *Preiss, n.* 1561; Upper Hay river, *F. Mueller, Miss Warburton;* King George's Sound, *Muir.*

4. CONOSTYLIS, R. Br.

Perianth persistent, shortly tubular or campanulate above the ovary (except in *C. breviscapa*), the limb of 6 lobes either all nearly equal and almost induplicate-valvate, or 3 inner ones rather smaller. Stamens 6, inserted at the base of the perianth-lobes; filaments short, erect, anthers oblong or linear, the cells free at the base or to the middle. Ovary inferior or partially superior, 3-celled, the summit conical, tapering into a filiform style, with 3 very small adnate terminal stigmas. Ovules several often numerous in each cell, crowded on a more or less stipitate placenta attached above the middle of the cell, and usually but not always in its inferior or adnate portion. Capsule opening loculicidally at the free conical apex in 3 coriaceous valves, the style itself often persistent and splitting almost to the end. Leaves in distichous or crowded tufts on a short rhizome or tufted or proliferous-branched stem, linear, sheathing at the base, the lamina laterally flattened or terete. Scapes from the centre of the leaf-tufts more or less tomentose or woolly, bearing 1 or more short sheathing bracts. Flowers usually of a dull yellow, more or less plumose-tomentose outside, in a terminal head rarely lengthening out into a shortly dichotomous cyme.

The genus is limited to West Australia.

SECT. 1. **Brachycaulon.**—*Perianth divided to the ovary into* 6 *spreading segments. Anther-cells distinct, pendulous from a short connective. Placentas small, with few reflexed ovules.*

Densely tufted branching plant. Flowers in dense heads almost sessile or on very short scapes within the leaves . . . . . . 1. *C. breviscapa,*

SECT. 2. **Catospora.**—*Perianth more or less tubular above the ovary, the lobes all equal or* 3 *inner ones smaller. Anther-cells adnate to the connective at least to the middle. Placentas recurved, dilated, with several ovules reflexed from the under surface.* (*Perianth usually with long hairs mixed with the plumose tomentum.*)

Perianth glabrous or loosely hairy inside, with equal lobes and stamens.
 Leaves terete, not striate. Flowers 2 or 3 together within several scarious bracts almost sessile within the leaves . . 2. *C. vaginata.*
 Leaves flat. Flowers in globular heads.
  Leaves with thick margins, scarcely ciliate. Scapes short, with 2 large concave keeled bracts under the head . . 3. *C. petrophiloides.*

Leaves with thin ciliate-setose margins. Scapes long. Bracts small. Placentas in the adnate part of the ovary . . . 4. *C. setosa.*
Leaves with ciliate margins. Scapes rather long. Bracts small. Placentas above the adnate part of the ovary . 5. *C. aurea.*
Perianth tomentose or woolly inside, 3 inner lobes smaller than the outer, with shorter stamens.
Leaves flat.
Leaves bordered by few or short setæ. Perianth-segments scarcely longer than the free part of the tube. . . . . 6. *C. melanopogon.*
Leaves bordered by long spreading setæ. Perianth-segments much longer than the free part of the tube.
Leaves mostly 6 in. long or more and 1 line broad, usually exceeding the scape . . . . 7. *C. setigera.*
Leaves under 2 in., very narrow or subulate-acuminate, shorter than the scape . . . . 8. *C. psyllium.*
Leaves terete or slightly flattened.
Leaves hirsute all over with upwardly appressed hairs . . 9. *C. villosa.*
Leaves white with a close tomentum . . . . 10. *C. Drummondii.*

SECT. 3. **Euconostylis.**—*Perianth more or less tubular above the ovary, the lobes all equal. Anther-cells adnate to the connective at least to the middle. Placentas more or less stipitate but scarcely recurved, covered all over in front with numerous ovules.*

SERIES 1. **Involucratæ.**—*Perianth with long scarcely denticulate hairs or setæ without any tomentum. Scapes short. Leaves long. Flowers capitate with lanceolate bracts.*

Leaves very narrow but flat, with prominent striæ . . . . . 11. *C. involucrata.*
Leaves terete, smooth . . . . 12. *C. juncea.*

SERIES 2. **Proliferæ.**—*Perianth shortly plumose-tomentose. Stems proliferous or stoloniferous. Leaves short (except in* C. candicans), *very densely tufted, usually white when young (except in* C. gladiata), *often becoming glabrous.*

Dwarf densely tufted plants, with rather large flowers, solitary or rarely 2 together, almost sessile within the leaves.
Leaves rigid, glabrous, 2 to 3 lines broad . . . . 13. *C. gladiata.*
Leaves flaccid, white when young, not above ½ line broad . 14. *C. seorsiflora.*
Leaves nearly terete, short, and rigid. Flowers in pedunculate heads.
Scapes not longer than the leaves . . . . 15. *C. teretiuscula.*
Scapes many times as long as the small leaves . . . . . . 16. *C. stylidioides.*
Leaves flaccid, very narrow but flat, green and grass-like or white only when very young, bordered with a few distant setæ, rarely above 3 or 4 in. long. Scapes long.
Flowers in a dense head. Perianth 4 to 5 lines long . . . 17. *C. prolifera.*
Flowers in a loose raceme. Perianth 6 lines long . . . . 18. *C. racemosa.*
Leaves flat, narrow, and rather rigid, mostly above 6 in. long. Flowers capitate on long scapes.
Leaves very white-tomentose when young, and scarcely losing the tomentum when old . . . . 19. *C. candicans.*
Leaves tomentose-pubescent when young, soon becoming glabrous . . . . 20. *C. dealbata.*
Leaves glabrous from the first.
Leaves rarely above 1 line broad. Scapes under 1 ft. long 21. *C. Preissii.*
Leaves 2 to 3 lines broad. Scapes 1 to 2 ft. . . . . . . 22. *C. bracteata.*

SERIES 3. **Normales.**—*Perianth shortly plumose-tomentose, rarely with longer hairs mixed. Stem short, rarely shortly proliferous. Leaves usually long, glabrous except marginal setæ. Scapes several- or many-flowered, much shorter than the leaves or very rarely nearly as long.*

Perianth-lobes scarcely longer than the free part of the tube.
Leaves subterete, rush-like, not setose . . . . 23. *C. filifolia.*
Leaves flat but very narrow, with long rigid distant setæ . . 24. *C. spinuligera.*
Leaves flat, rigid, with much thickened nerve-like margins and distant pungent setæ . . . . 25. *C. bromelioides.*
Leaves flat, rather rigid, the margins not prominent, with distant rather rigid setæ . . . . 26. *C. aculeata.*
Perianth-lobes twice or three times as long as the free part of the tube. Leaves flat, rarely almost terete, with few or short setæ.
Scapes nearly as long as the leaves. Flowers in a globular head.
Leaves rarely above 1 line broad. Stems usually proliferous 21. *C. Preissii.*
Leaves 2 to 3 lines broad. Stems very short. Scapes 1 to 2 ft. . . . . 22. *C. bracteata.*
Scapes not half so long as the leaves.
Flowers few in a loose oblong panicle. Perianth with long hairs intermixed with the tomentum . . . . 27. *C. laxiflora.*
Flowers numerous in a branching cyme . . . . 28. *C. cymosa.*
Flowers capitate. Perianth divided almost to the ovary.
Leaves mostly 2 to 3 lines broad . . . . 29. *C. serrulata.*
Leaves ½ to 1½ lines broad, or almost terete . . . . 30. *C. caricina.*

SECT. 3. **Androstemma.**—*Perianth tubular above the ovary, the lobes all equal and narrow. Filaments erect, filiform, much longer than the anthers. Ovules rather numerous, bordering a peltate placenta.*

Dwarf densely tufted plant, with large solitary flowers almost sessile within the leaves, surrounded by short scarious bracts . 31. *C. Androstemma.*

SECT. 1. BRACHYCAULON.—Perianth divided to the ovary into 6 spreading segments. Anther-cells distinct, pendulous from a short connective on short erect filaments. Placentas small, with few reflexed ovules.

The peculiar anthers and perianth of this plant might have afforded grounds for establishing it as a distinct genus of a value at least equal to that of *Androstemma.*

1. **C. breviscapa,** *R. Brown, Prod.* 301. Stems very short and branching, densely covered with tufts of distichous leaves, with brown sheathing bases, the lamina flat, rigid, 6 to 8 in. long and 1 to 2 lines broad, striate, glabrous, without marginal cilia. Scapes very short amongst the leaves, with a dense globose head of sessile flowers. Bracts lanceolate, the inner ones linear. Ovary narrow-turbinate, tomentose. Perianth divided to the ovary into lanceolate acute spreading segments, nearly 3 lines long, all nearly equal, densely and shortly plumose-tomentose outside, loosely so inside. Anthers scarcely 1 line long, appearing at first sight sessile and erect at the mouth of the tube, but the cells are really distinct from their insertion and pendulous from a small connective at the apex of a slender filament which they conceal. Style shortly protruding beyond the anther, tapering and splitting almost to the apex as in many other species. Placenta-bearing portion of the ovary inferior, with a free conical summit.

W. Australia. Lucky bay, *R. Brown*, and probably the same locality, *Baxter.* It is also in Cunningham's herbarium marked as from S. W. Australia, *Fraser*, but not in Cunningham's handwriting, and the specimen may in fact be of Baxter's collecting.

I have seen it from no other locality, but Schultz, Syst. vii. 294, evidently saw the true plant in Sieber's herbarium, and has correctly described it.

SECT. 2. CATOSPORA.—Perianth more or less tubular above the ovary, the lobes all equal or the 3 inner ones smaller with the stamens shorter or inserted lower down; the perianth usually with long hairs intermixed with the plumose tomentum outside, and often hairy or woolly inside. Anther-cells adnate to the connective at least to the middle. Placentas recurved, dilated, with several ovules reflexed from the under surface.

2. **C. vaginata,** *Endl. in Pl. Preiss.* ii. 23. Stems densely branched, forming tufts of 2 to 4 in. covered with the sheaths of old leaves. Leaves linear-terete, rush-like, channelled along the inner or upper side but not otherwise striate, 3 to 5 in. long. Flowers 2 to 4 together, sessile in little heads surrounded by imbricate scarious bracts and borne on very short hairy peduncles within the upper leaves. Perianth about ½ in. long, softly hairy outside, glabrous or slightly hairy within, the lobes narrow linear-lanceolate, about as long as the free part of the tube. Anthers longer than the filaments. Placentas in the adnate part of the ovary, projecting and dilated, with 3 or 4 ovules pendulous from the under side; the free summit of the ovary conical with a filiform style.—Hook. Ic. Pl. t. 853.

W. Australia. Upper Kalgan river, *Oldfield, F. Mueller;* towards Cape Riche, *Preiss, n.* 1383, and probably the same neighbourhood, *Baxter, Drummond, n.* 444; West Mount Barren, *Maxwell.*

3. **C. petrophiloides,** *F. Muell. Herb.* Leaves flat, glabrous and bordered with small appressed cilia as in some specimens of *C. aurea,* but much thicker, with few prominent veins, 2 to 6 in. long in the only specimen seen. Scapes shorter than the leaves, densely white-woolly. Flowers numerous in very dense globular heads above 1 in. diameter, subtended by 2 broad concave carinate and shortly acuminate coloured bracts, about as long as the flowers and glabrous. Perianth 7 to 8 lines long, densely woolly-villous outside with long hairs plumose at the base, sprinkled with a few hairs inside; the lobes linear, much longer than the free part of the tube. Anthers long linear, on very short filaments. Placentiferous portion of the ovary inferior but shorter than the free conical summit; placentas prominent and dilated, with several ovules reflexed from the under side.

W. Australia. Flats on the Phillips river, a single specimen in Herb. F. Mueller.

4. **C. setosa,** *Lindl. Swan Riv. App.* 44, t. 6. Stem very short. Leaves flat, rigid, finely veined, often nearly 1 ft. long, 1 to 2 lines broad, glabrous but bordered towards the base with long fine cilia. Scapes about as long as the leaves, loosely woolly. Flowers in a dense terminal head often 1½ to 2 in. diameter. Bracts narrow, acuminate, shorter than the flowers. Perianth 10 lines to 1 in. long, very densely silky-woolly outside with long hairs plumose at the base, more or less hairy or woolly inside; lobes about as long as the cylindrical tube, all

narrow-lanceolate and nearly equal or 3 inner ones rather smaller. Stamens all equal, the filaments slender, erect and at least as long as the small narrow anthers. Placentas in the adnate part of the ovary dilated, with the ovules reflexed from the under surface.—Endl. in Pl. Preiss. ii. 17.

W. Australia. Swan river, *Drummond*, 1*st coll.*, *Preiss*, *n.* 1408.

5. **C. aurea,** *Lindl. Swan Riv. App.* 44. Stem or rhizome short. Leaves flat, rigid, prominently veined, often 1 ft. long, 2 to 3 lines broad, glabrous but bordered by short rigid marginal cilia usually numerous. Scapes shorter than the leaves, densely covered with a loose plumose wool, with a linear-lanceolate silky-woolly bract usually about the middle. Flowers in a dense globular head of 1 to $1\frac{1}{4}$ in. diameter, the short bracts entirely concealed. Perianth 6 to 7 lines long, of a thicker consistence than in most species, densely covered outside with a plumose wool of a golden-yellow or rarely pale-coloured, quite glabrous inside, the adnate portion very short and turbinate; the lobes narrow and thick, rather longer than the free portion of the tube. Anthers linear, rather long, attached by the centre to very short filaments. Ovary more than half or almost entirely superior, the placentas attached in the free portion, stipitate and dilated, with many ovules reflexed from the under side. Styles slender.—Endl. in Pl. Preiss. ii. 17; *C. sulphurea*, Endl. l.c.

W. Australia. Swan river, *Drummond*, 1*st coll. n.* 750, 759, *Preiss*, *n.* 1381, 1382, and others; Toodyay and Cape Naturaliste, *Oldfield.*

6. **C. melanopogon,** *Endl. in Pl. Preiss.* ii. 18. A low species, but the leafy stem sometimes branching and elongated to 2 in. or rather more. Leaves rigid, flat, attaining 6 in. to 1 ft. in length and about 1 to $1\frac{1}{2}$ lines broad, striate, glabrous except the margins ciliate when young. Scape woolly, shorter than the leaves, bearing 1 or 2 bracts, the upper one produced into a long point. Flowers 6 to 10 in a terminal head, the subtending bracts narrow and short. Perianth plumose-woolly outside, with longer hairs often turning to a dark colour, more or less hairy or woolly inside, about $\frac{1}{2}$ in. long; lobes about as long as the free part of the tube, but irregularly separating, the three inner ones shorter and more petal-like than the outer. Anthers oblong, on rather thick filaments, the 3 inner ones much shorter than the outer. Placentas in the adnate part of the ovary, projecting and dilated, the ovules not numerous, reflexed from the under surface; conical apex of the ovary long and narrow.

W. Australia. Swan river, *Drummond*, 1*st coll. n.* 754, *Preiss*, *n.* 1387; Kalgan in sand near the sea, *Oldfield;* south of Stirling Range, *F. Mueller;* Perongerup, *Mrs. Knight.*—Very near *C. setigera*, but without the long setæ to the leaves of that species, and the perianth less deeply divided.

Var. *major.* Perianths 7 to 8 lines long. Vasse river, *Pries;* Swan river, *Helmich.*

7. **C. setigera,** *R. Br. Prod.* 300. Short leafy stems much branched in dense tufts. Leaves rather flaccid, flat but very narrow and grass-

like, striate, mostly about 6 in. long, fringed below the middle or nearly the whole length with long spreading setæ. Scapes shorter than the leaves, white-woolly or tomentose, bearing 1 or 2 bracts with broad sheathing bases below the inflorescence. Flowers 6 to 10 in a terminal head with very short subulate bracts, the outer ones very shortly dilated at the base. Perianth 5 to 6 lines long, plumose-tomentose and hirsute with longer hairs outside, shortly woolly inside; the tube campanulate, the free about equal to the adnate part; lobes narrow, much longer than the tube, the 3 inner ones rather smaller and more petal-like than the outer. Filaments shorter than the anthers, those opposite the inner perianth-lobes shorter and inserted lower down than the others; anther-cells free from the middle. Placentas in the adnate part of the ovary, dilated, with rather few ovules reflexed from the under side.—A. Rich. Sert. Astrol. t. 29 (not good); *C. æmula,* Lindl. Swan Riv. App. 45; Endl. in Pl. Preiss. ii. 20.

W. Australia. King George's Sound, *R. Brown, Baxter, A. Cunningham, F. Mueller;* thence to Swan river, *Drummond,* 1*st coll. n.* 757, *Oldfield, Preiss, n.* 1390; eastward to Cape Arid, *Maxwell.*

C. discolor, Endl. in Pl. Preiss. ii. 20, from Swan river, *Preiss, n.* 1392, only differs from the common *C. setigera* in the purplish tinge assumed by the external wool of the perianth. *C. assimilis,* Endl. l.c., which I have not seen, must also, from the character given, be but a slight variety of *C. setigera.*

8. **C. psyllium,** *Endl. in Pl. Preiss.* ii. 21. Very near *C. setigera* and considered by F. Mueller as a dwarf variety of that species. Leaves tufted, about 2 in. long, very narrow, often almost terete, usually falcate and tapering to a point, fringed with spreading setæ as in *C. setigera.* Scapes as long as the leaves or rather longer. Flowers rather smaller than in *C. setigera,* but their structure as well as their number in a terminal head the same as in that species.—*C. minima,* Endl. l.c.

W. Australia. York district, *Preiss, n.* 1391; sand plains south of Stirling range, *F. Mueller;* Box vale, *Miss Wells.*

C. pusilla, Endl. l.c. 20, only differs from *C. psyllium,* as *C. discolor* from *C. setigera,* in the purple tinge assumed by the external wool of the perianth.

9. **C. villosa,** *Benth.* Stems very short and tufted like those of *C. setigera.* Leaves 6 to 8 in. long, under 1 line broad, flat and thick or sometimes almost terete, striate, ciliate and hairy all over not on the margins only. Scapes densely white-woolly, shorter than the leaves. Flowers 10 to 20 in a terminal globular head. Bracts small and narrow. Perianth 5 or at length 6 lines long, plumose-woolly outside, woolly hairy inside; lobes narrow, about as long as the tube and ovary, the 3 inner ones rather smaller and the 3 inner stamens shorter than the others as in *C. setigera;* anther-cells adnate almost to the base. Placentas in the adnate part of the ovary, dilated, with several ovules reflexed from the under surface.

W. Australia, *Drummond, n.* 311.

10. **C. Drummondii,** *Benth.* Stems short, densely tufted. Leaves linear-terete or scarcely compressed, rigid, scarcely striate, 6 to 9 in.

long, covered with a close whitish tomentum and a few longer appressed hairs intermixed, especially on the younger leaves. Scapes much shorter than the leaves, loosely woolly. Flowers sessile in a terminal head, with a membranous bract terminating in a long leafy point either immediately under the head or lower down on the scape. Perianth scarcely 5 lines long, plumose-tomentose outside, pubescent inside, the lobes narrow, rather longer than the free part of the tube, the 3 inner ones rather smaller, with their stamens rather shorter than the others. Anthers oblong-linear; filaments short. Placentas prominent from the adnate part of the ovary, dilated, with reflexed ovules on the under side.

W. Australia. Probably to the eastward of King George's Sound, *Drummond.*

SECT. 3. EUCONOSTYLIS.—Perianth more or less tubular above the ovary, the lobes all equal. Anther-cells adnate to the connective at least to the middle. Placentas more or less stipitate but scarcely recurved, covered all over in front with numerous ovules, and always in the adnate part of the ovary.

11. **C. involucrata,** *Endl. in Pl. Preiss.* ii. 23. Stems very short. Leaves often above 1 ft. long, flat, varying from ½ line to 2 lines in breadth, rigid, striate, glabrous, the margins sometimes shortly and loosely ciliate near the base. Scapes 2 to 3 in. long, loosely woolly-villous. Flowers in heads or spikes not very compact and the rhachis sometimes forked, not very numerous, subtended by broadly lanceolate acuminate leafy bracts the outer ones often as long as or rather longer than the flower. Perianth varying from ½ to ¾ in. long, sprinkled outside with long rigid simple or minutely denticulate hairs, without the dense wool or tomentum of other species, glabrous inside; lobes narrow, at least twice as long as the free part of the tube. Placentas very densely covered with numerous ovules.

W. Australia. Swan river, *Drummond, 1st coll. n.* 756, *Preiss, n.* 1407, *Oldfield;* Cape Naturaliste, *Oldfield,* all with narrow but not terete leaves; Swan river, *Drummond, 1st coll. n.* 756; Hampden, *Clarke,* both with rather broader leaves.

12. **C. juncea,** *Endl. Nov. Stirp. Dec.* 19. This species, which I have not seen, is said to have the habit, long leaves, short scapes, capitate flowers with lanceolate bracts and perianths with simple rigid hairs of *C. involucrata,* and only to differ from that species in the leaves almost or quite terete, obscurely striate.

W. Australia. Raised in Huegel's garden from Swan river seeds (*Endlicher*).

13. **C. gladiata,** *Benth.* Stems short, densely branched and tufted like those of *C. breviscapa.* Leaves densely distichous, 3 to 4 in. long and 2½ to 3 lines broad in the middle, slightly falcate, tapering at both ends, rigid and striate, glabrous, the margins quite entire or with a few very small distant setæ. Scape only a few lines long, covered with 2 or 3 acuminate brown bracts, and bearing a single flower which although large is almost concealed amongst the foliage. Perianth 8 to 9 lines

long, rather narrow, shortly plumose-tomentose outside, more or less villous inside, the narrow lobes at least 3 times as long as the short free portion of the tube. Anthers long, on short filaments, all equal. Ovules very numerous, covering the stipitate placentas.

W. Australia, probably to the eastward of King George's Sound, *Drummond.*

14. **C. seorsiflora,** *F. Muell. Fragm.* i. 158, viii. 19. A dwarf species, stoloniferous or proliferously branched, forming dense tufts not exceeding 4 or 5 in. Leaves very densely tufted, very narrow linear, often cottony-white and ciliate with long soft hairs when young, glabrous or nearly so when full grown, rarely above 2 in. long, ending in fine points and much dilated at the base. Scapes shorter than the leaves, sometimes very short, tomentose, with 1 or 2 broad membranous sheathing bracts usually produced into long subulate points or laminæ. Flowers 1, 2 or rarely 3 on the scape, the perianth about 8 lines long, shortly plumose-tomentose outside with a few longer hairs, loosely pubescent or shortly villous inside; lobes lanceolate, narrow, rather longer than the campanulate free part of the tube. Anthers linear, on short filaments. Placentas densely covered with very numerous ovules.

W. Australia. King George's Sound or to the eastward, *Baxter;* W. end of Stirling Range, *F. Mueller;* Gardner and Oldfield rivers, *Maxwell;* Toodyay rivulets, *Oldfield.*

15. **C. teretiuscula,** *F. Muell. Fragm.* viii. 18. Stems probably proliferous. Leaves densely tufted, nearly terete and white-tomentose when young, like those of *C. stylidioides,* but rather longer, attaining 2 in. Scapes scarcely 1½ in. long, tomentose. Flowers few in the head, resembling those of *C. stylidioides,* but rather smaller, the perianth lobes narrower and more acute.

W. Australia, *Oldfield.* Described only from the fragmentary specimens in herb. F. Mueller, and requires further investigation.

16. **C. stylidioides,** *F. Muell. Fragm.* viii. 17. Stems hard, rigid, repeatedly branched, the branches clustered with the leaves at the nodes within 2 or 3 persistent sheathing bracts or scales. Leaves very dense in the clusters, linear-acuminate or linear-subulate, nearly terete, very acute, 3 to 9 lines long; the younger inner or longer ones densely white-tomentose, the outer or shorter ones often nearly glabrous and striate. Flowers closely sessile in globular heads, on a loosely tomentose peduncle of 3 or 4 in., arising from the centre of the principal clusters of leaves and branches, with usually a small broad bract at or below the middle, the outer bracts of the head very broad and shorter than the flowers. Perianth campanulate, about ½ in. long, villous and plumose-tomentose outside, glabrous inside, the adnate base very short, the lobes broadly lanceolate, about as long as the free portion of the tube. Anthers oblong, on very short filaments. Placentas small, covered with numerous ovules.

W. Australia. Near Oolingara, Murchison river, forming dense tufts on rocks and sand plains, *Oldfield.*

17. **C. prolifera,** *Benth.* Stems slender, proliferously branched. Leaves in dense tufts, narrow-linear, usually flaccid, tapering into fine points, varying in length from 1 to 3 or 4 in., glabrous, striate, bordered by fine distant rigid cilia. Scapes longer than the leaves, loosely villous with plumose hairs, bearing 1 or rarely 2 distant linear or acuminate bracts. Flowers sessile in a terminal globose head, the outer bract acuminate, usually but not always longer than the head. Perianth campanulate, 4 to 5 lines long, plumose-villous outside, glabrous or slightly hairy between the stamens inside, the lobes longer than the free part of the tube. Ovules rather numerous, covering the small ovate projecting placentas.

W. Australia. Swan river, *Drummond;* Murchison river, *Oldfield.*

18. **C. racemosa,** *Benth.* Stems slender and proliferously branched. Leaves densely tufted, narrow flaccid and grass-like, resembling those of *C. prolifera,* of which this may be a variety, but the flowers are considerably larger, the head extended into a loose simple or forked raceme. Perianth ½ in. long, plumose-villous outside, glabrous inside, the lobes scarcely longer than the campanulate free part of the tube. Ovules numerous, on short placentas.

W. Australia. White Peak, Champion bay, *Oldfield.*

19. **C. candicans,** *Endl. Nov. Stirp. Dec.* 20, *and in Pl. Preiss.* ii. 16. Stem proliferous, erect and short in some specimens, in others elongated and said to climb amongst bushes for several feet, loosely tomentose or glabrous, the branches and leaves densely clustered. Leaves linear, flat, the outer ones of the tuft or nearly all in the upper tufts short, but some much elongated, attaining occasionally in the lower tufts above 1 ft., varying in breadth from ½ line to near 2 lines, covered with a white tomentum concealing the veins and often ciliate on the margins, at length becoming sometimes nearly glabrous and showing a few striæ. Scapes or peduncles usually longer than the leaves, with a linear or linear-lanceolate bract at or above the middle. Flowers 10 to 20, almost sessile in a globular head, either dense or rather loose with a bifid rhachis. Bracts narrow, either all short, or one of them more leafy and as long as or longer than the flowers. Perianth campanulate, about ½ in. long, tomentose outside, hairy or nearly glabrous inside, the lobes glabrous inside, scarcely so long as the tube. Anthers linear, on very short filaments. Ovules numerous, covering the short placentas.—*C. albicans,* A. Cunn. Herb.; *C. propinqua,* Endl. in Pl. Preiss. ii. 17.

W. Australia. Swan river, *Drummond,* 1*st coll.*, *Oldfield;* abundant on the rocky and sandy banks of Rottenest island, *A. Cunningham, Drummond, Preiss, n.* 1400.

Var. *leptophylla.* Leaves very densely tufted, narrow, with long points, flaccid, 2 to 4 in. long. Scapes often bearing 2 or 3 pedunculate heads of flowers.—Swan river, *Drummond;* Champion bay, *Oldfield.*

20. **C. dealbata,** *Lindl. Swan Riv. App.* 45. Very nearly allied to *C. candicans* and perhaps a variety connecting it with *C. Preissii* and

C. bracteata, Endl. Stem less elongated than in *C. candicans*. Leaves, although still densely tufted, more normally distichous, longer and more rigid, tomentose pubescent and rather white all over when young, but soon losing the tomentum. Scapes about as long as the leaves. Flowers rather numerous in terminal heads. Perianth about ½ in. long, plumose-tomentose outside, glabrous inside, the lobes rather longer than the free part of the tube. Ovules numerous, covering the short placentas.

W. Australia. Swan river, *Drummond*, 1*st coll.*

C. bracteata, Lindl., Swan Riv. App. 45, from the same collection, appears to be established on vigorous specimens of *C. dealbata*, with the foliage nearly glabrous, some of the leaves 8 to 10 in. long, and the outer bracts under the flower-heads rather more leafy. A very gradual passage may be traced from the preceding five or six species, through the long-scaped less proliferous *C. Preissii* and *C. bracteata*, Endl., to the common *C. aculeata*, however remote may be the extreme species of the series.

21. **C. Preissii,** *Endl. in Pl. Preiss.* ii. 18. Stems usually shortly proliferous. Leaves from distichous sheathing bases, flat, ½ to 1 ft. long, rarely above 1 line broad, rather thick, glabrous, bordered in some specimens by a few very distant rigid cilia, in others almost or quite without. Scapes nearly as long as the leaves, bearing usually 1 or 2 ovate brown bracts more or less acuminate. Flowers sessile in a rather loose terminal head, branching out sometimes into a close cyme, the bracts short and narrow. Perianth about ½ in. long, plumose-tomentose outside with a mixture of longer hairs, glabrous or very slightly hairy inside; lobe narrow, nearly twice as long as the free part of the tube. Ovules numerous, covering the stipitate placentas.

W. Australia. Swan river, *Drummond*, 1*st coll. n.* 753, *Preiss, n.* 1384; Southern river, *Preiss, n.* 1386.

C. festucacea, Endl. in Pl. Preiss. ii. 18 (Preiss, n. 1386) has the leaves almost without cilia and the perianth more hairy, but the two forms are too closely connected by some of Drummond's specimens to justify their separation.

22. **C. bracteata,** *Endl. in Pl. Preiss.* ii. 16, *not of Lindl.* Stems short, perhaps sometimes slightly proliferous. Leaves glabrous, flat, rigid, 1 to 1½ ft. long, 2 to 3 lines broad, striate, the margins prominent and nerve-like, naked or bordered by a few small distant setæ. Scapes about as long as the leaves, loosely tomentose, with 1, 2, or 3 broad lanceolate bracts, and sometimes with a long peduncle or branch in the axil of one of them. Flowers numerous, in a dense globose terminal head, subtended usually by 1 or 2 lanceolate bracts, the other bracts small and linear, all shorter than the flowers. Perianth nearly ½ in. long, plumose-tomentose outside, glabrous or slightly hairy inside, the lobes nearly twice as long as the free part of the tube. Anthers linear, on short filaments. Ovules very numerous, covering the placentas, which are scarcely stipitate.

W. Australia. Swan river, *Drummond*, 1*st coll. n.* 751, *Preiss, n.* 1405.—The species seems to connect the preceding ones very closely with *C. aculeata*, from which it differs in the leaves scarcely ciliate, the longer scapes, larger heads of flowers, and the perianth more deeply lobed.

23. **C. filifolia,** *F. Muell. Fragm.* viii. 18. Stems short and tufted. Leaves above 1 ft. long, terete, rigid and rush-like, slightly striate, quite glabrous, 1 or 2 outer ones reduced to brown sheathing scales. Scapes very much shorter than the leaves, tomentose-woolly, bearing usually 2 distant ovate brown bracts. Flowers in a loose head often branching into a dense cyme, all nearly sessile within small bracts. Perianth about ½ in. long, plumose-tomentose outside, glabrous inside, the lobes narrow, as long as or rather longer than the free part of the tube. Anthers linear, on short filaments. Placentas covered all over with numerous ovules.

W. Australia, *Drummond.*

24. **C. spinuligera,** *F. Muell. Herb.* Stems densely tufted. Leaves very narrow but flat, 6 in. to nearly 1 ft. long, bordered by long rather rigid spreading distant cilia. Scapes nearly as long as the leaves. Flowers in a loose terminal head, growing out into a short raceme or once-forked cyme, the bracts short and narrow. Perianth about 4 lines long, broadly campanulate, plumose-tomentose outside, glabrous inside, the lobes rather broad, about as long as the free part of the tube. Anthers oblong-linear, as long as the filaments. Placentas covered with numerous ovules.

W. Australia, *Drummond.*

25. **C. bromelioides,** *Endl. in Pl. Preiss.* ii. 18. Stems short in our specimens, but perhaps stoloniferous or proliferous. Leaves glabrous, under 6 in. long, 1½ to 2 lines broad, thick and rigid, with remarkably thick nerve-like margins, and bordered by distant rigid almost pungent setæ. Scapes very short, loosely woolly, bearing 1 or 2 broadly ovate ovate-lanceolate or acuminate bracts. Flowers few, very shortly pedicellate in a rather loose terminal head, with small linear bracts. Perianth about 6 lines long, plumose-tomentose outside, slightly hairy inside, the lobes narrow-lanceolate, rather longer than the free part of the tube. Anthers linear, on short filaments. Placentas stipitate, covered with numerous ovules.

W. Australia. Swan river, *Drummond;* near Avondale, York district, *Preiss, n.* 1601. The specimens seen are very few, and they may prove to be a variety only of *C. aculeata.*

26. **C. aculeata,** *R. Br. Prod.* 300. Stems very short. Leaves rigid, flat, erect or recurved, 6 in. to 1 ft. long, 1 to 2½ lines broad, finely striate, glabrous but bordered by distant rigid more or less pungent setæ. Scapes usually much shorter than the leaves, rarely nearly as long, tomentose, often slightly branched, with an ovate or ovate-lanceolate brown bract under each branch. Flowers on very short pedicels in a loose head or dense cyme at the end of the scape or branches, with linear-subulate short bracts. Perianth campanulate, about 5 lines long, shortly plumose-tomentose outside, glabrous inside, the lobes lanceolate, obtuse, longer than the free part of the broad tube. Anthers linear, on very short flat filaments. Placentas covered all

over with numerous ovules.—Endl. in Pl. Preiss. ii. 18; Bot. Mag. t. 2989.

W. Australia. King George's Sound and adjoining districts, *R. Brown, A. Cunningham, Drummond, Preiss, n.* 1395, and many others; Greenough flats, *C. Gray,* and Murchison river, *Oldfield,* single specimens, but apparently the same species, which however in the Swan river district seems to be replaced generally by the *C. bracteata,* with longer scapes, larger flowers, the leaves with few or no marginal setæ, &c.

27. **C. laxiflora,** *Benth.* Stems tufted, very short. Leaves mostly above 1 ft. long, flat, striate, 1½ to 2½ lines broad, bordered when young by very short cilia as in *C. serrulata.* Scapes short, not above 6 in. high including the inflorescence, loosely tomentose-villous, branching from about the middle, each branch with a short loose raceme of 3 or 4 flowers, the whole forming a loose oblong somewhat one-sided panicle. Bracts subtending the branches lanceolate, membranous, villous, sometimes above ½ in. long, those under the pedicels small and narrow. Perianth broadly campanulate, about ½ in. long, tomentose outside and villous with long almost silky hairs, shortly plumose at the base, glabrous or sparingly hairy inside. Lobes lanceolate, very acute, twice as long as the free part of the tube. Anthers linear, on short filaments. Placentas covered with numerous ovules.

W. Australia. Vasse river, *Oldfield.*—Although allied in some respects to *C. serrulata,* this differs from the rest of the section in the indumentum of the perianth, and from the whole genus in its inflorescence.

28. **C. cymosa,** *F. Muell. Herb.* Leafy stem short, tufted or shortly branched. Leaves often above 1 ft. long, 1 to near 3 lines broad, finely veined, bordered by a few distant rigid cilia. Scapes much shorter than the leaves, woolly-tomentose, 4 to 6 in. high including the inflorescence. Flowers numerous in a loosely branched cyme, the pedicels 1 to 2 lines long subtended by small linear bracts. Perianth about 5 lines long, plumose-tomentose outside, glabrous inside, the lobes narrow, nearly twice as long as the free part of the tube. Anthers linear, longer than the filaments. Placentas stipitate, covered all over with numerous ovules.

W. Australia. Blackwood river and Champion bay, *Oldfield;* Greenough flats, *C. Gray;* Busselton, *Pries.*

29. **C. serrulata,** *R. Br. Prod.* 300. Rhizome often creeping; stems short, sometimes shortly proliferous. Leaves usually above 1 ft. long and 2 to 3 lines broad, more prominently striate than in *C. aculeata,* and bordered by very small scarcely spreading cilia. Scapes only 2 to 3 in. or very rarely nearly 6 in. long, tomentose-woolly, bearing several long lanceolate bracts. Flowers few, in looser heads or cymes than in *C. aculeata,* the pedicels often 1 to 2 lines long, subtended by linear bracts. Perianth 5 to 6 lines long, plumose-tomentose outside, glabrous inside, divided almost to the ovary into lanceolate acuminate segments. Anthers oblong, shorter than in *C. aculeata,* on very short filaments. Style rather short. Ovules numerous, covering the placentas. Capsules 3 to 4 lines diameter.

W. Australia. King George's Sound, *Menzies;* Kalgan river, *Oldfield,* probably from the same district, *Drummond, n.* 349; Swan river, *Drummond,* 1*st coll. n.* 758.

C. ensifolia, C. occulta, C. misera, and *C. longifolia,* Endl. in Pl. Preiss. ii. 21, 22, from the neighbourhood of Cape Riche, and *C. spathacea,* Endl. l.c. 22, from Darling range, do not appear to me to be distinguishable even as varieties from *C. serrulata. C. misera* is a starved specimen with short leaves and only 1 or 2 flowers in the head, such as we have also from Drummond. *C. longifolia* is said to have the perianth-lobes shorter than the tube. Preiss's specimen in Herb. F. Mueller is in old fruit only with the lobes worn short. In *C. occulta* the lobes are said to be equal to the tube, but in Preiss's specimen they are certainly longer than the fruit, and in all the above supposed species the lobes are free almost to the ovary, not forming a campanulate free tube as long as the ovary as in *C. aculeata.*

30. **C. caricina,** *Lindl. Swan Riv. App.* 45. Stems very short and densely tufted, the tufts sometimes almost bulbous. Leaves glabrous, 6 in. to nearly 1 ft. long, 1 to $1\frac{1}{2}$ lines broad, rigid, bordered by prominent nerve-like margins minutely serrulate-ciliate. Scapes short, tomentose-woolly, bearing 1 or 2 broad lanceolate acuminate bracts. Flowers few in the head, sessile or nearly so, the bracts linear. Perianth about $\frac{1}{2}$ in. long, plumose-tomentose outside glabrous inside, lobes narrow, 3 times as long as the very short free part of the tube. Anthers linear, on short thick filaments. Placentas prominently stipitate, the ovules numerous, covering the whole front as in other species of *Euconostylis* but reflexed almost as in *Catospora.*—Endl. in Pl. Preiss. ii. 19; *C. graminea,* Endl. l.c.

W. Australia. Swan river, *Drummond,* 1*st coll., Preiss, n.* 1380, 1385.

SECT. 4. ANDROSTEMMA.—Perianth tubular above the ovary, the lobes all equal and narrow. Filaments erect, filiform, much longer than the anthers. Ovules rather numerous, bordering the peltate placenta.

The long perianths buried amongst the leaves and surrounded by scarious bracts give this plant a peculiar aspect, which seemed to justify Lindley in establishing it as a separate genus. But subsequent discoveries have shown a nearly similar habit, foliage, and bracts in *C. vaginata,* solitary long flowers buried amongst the leaves in *C. gladiata* and *C. seorsiflora,* and nearly similar stamens in *C. setosa;* there remains as a distinct character only the placentation, which establishes it as a section of the same rank as the three preceding ones.

31. **C. Androstemma,** *F. Muell. Fragm.* viii. 19. Leafy stems short much branched, forming dense tufts of a few inches. Leaves with short sheathing bases, linear-subulate, rigid and rush-like, terete or slightly flattened, prominently striate, 4 to 8 in. long, glabrous in our specimens. Flowers solitary on hairy peduncles or scapes of $\frac{1}{4}$ to $\frac{1}{2}$ in. bearing several scarious bracts of which 3 or 4 usually close under the flower and rarely above 2 or 3 lines long. Perianth-tube from a very short turbinate adnate base broadly cylindrical, $\frac{3}{4}$ to 1 in. long, shortly plumose-tomentose outside, glabrous inside; lobes narrow-linear, about $\frac{1}{2}$ in. long, all equal and similar, spreading at the time of flowering. Filaments filiform and erect, nearly as long as the perianth-lobes; anthers several times shorter. Style as long as the stamens. Capsule half-superior, but little broader than the perianth at the time of flower-

ing.—*Androstemma junceum*, Lindl. Swan Riv. App. 46; Field. Sert. Pl. t. 33; Endl. in Pl. Preiss. ii. 24.

W. Australia. Swan river, *Drummond*, 1*st coll. n.* 762, *Preiss, n.* 1409; South Hutt river, *Oldfield;* Greenough Flats, *C. Gray.*

5. BLANCOA, Lindl.

Perianth persistent, tubular, the limb of 6 nearly equal almost conduplicate valvate short lobes. Stamens 6, with ovate-oblong anthers on very short filaments or nearly sessile at the orifice of the tube. Ovary with the angles wholly adnate to the perianth, the sides free almost from the base, 3-celled, the summit shortly conical. Style long, filiform, with 3 minute adnate terminal stigmas. Ovules several in each cell, in 2 rows on an adnate linear placenta. Capsule opening at the free apex in 3 coriaceous valves. Seeds few, oblong, striate.—Herb with very short branching stems. Leaves in distichous tufts, with sheathing bases and long laterally flattened linear laminæ. Flowers few together in unilateral racemes at the ends of the branches of the scape, the perianth densely plumose-woolly.

The genus is limited to the single species, endemic in West Australia. It has been united by F. Mueller with *Conostylis*, but is, in fact, much more nearly connected with *Anigozanthos*, but separated from both by characters which appear to be of full generic value, unless all the Australian *Conostyleæ* be treated as sections of one comprehensive genus.

1. **B. canescens,** *Lindl. Swan Riv. App.* 45. Leafy branched stems or rhizomes very short and shortly villous. Leaves rigid, 6 in. to 1 ft. long, about 2 lines broad, finely striate. Scape shorter than the leaves, loosely tomentose, bearing usually a narrow-linear leaf near the base and 2 or 3 short unilateral racemes, on short branches or peduncles, each with 2 or 3 large pendulous flowers subtended by small bracts. Perianth densely covered with a loose reddish plumose wool, the tube broadly cylindrical, about 1 in. long, the erect lobes 2 to 3 lines long, glabrous inside. Anthers much shorter than the lobes. Style usually shortly protruding from the perianth.—Endl. in Pl. Preiss. ii. 24; *Conostylis canescens*, F. Muell. Fragm. viii. 19.

W. Australia. Swan river, *Drummond*, 1*st coll. n.* 748, *Preiss, n.* 1410.

6. ANIGOZANTHOS, Labill.

(Schwægrichenia, *Spreng.*)

Perianth persistent, the tube much elongated above the ovary, often recurved at the end, the limb more or less oblique; lobes 6, lanceolate, almost induplicate valvate, equal or those on the lower side more deeply separated and the tube usually split open between them sometimes almost to the base. Anthers oblong or linear, on short filaments at the orifice of the tube, the cells free at the base only. Ovary wholly inferior, 3-celled, the summit flat or conical; style long, filiform, slightly clavate and stigmatic at the end. Ovules several, usually numerous in each cell, irregularly arranged or crowded on a projecting placenta. Capsule opening at the apex in 3 small valves. Seeds usually few, with a hard rugose or striate testa.—Herbs with a perennial usually horizontal thick rhizome. Leaves chiefly radical or nearly so, with a sheathing base, and linear lamina, either laterally flattened or nearly terete. Stems erect, usually bearing two or three smaller distant leaves. Flowers large, in close unilateral spikes or racemes, at the end of the simple stem or of the branches of a dichotomous spreading panicle. Perianth and inflorescence densely covered with a red green or yellow plumose wool; the stems sometimes, the leaves very rarely, bearing a shorter or looser tomentum, the leaves more frequently glabrous.

The genus is limited to West Australia. The derivation of the name has been frequently discussed, supposing it to have been taken from *ἀνίσχω* or *ἀνοίγω*, with meanings very inapplicable; it was, however, much more simple. Labillardière intending to express the unequal or oblique flower, *ἀνισος*, *ἀνθος*, merely changed the first s into a g, and the second into a z, for euphony sake.

SECT. 1. **Dianthesis.**—*Racemes or spikes several, in a divaricate dichotomous panicle. Anthers inappendiculate. Ovules 2 to 4 in each cell. Stems tomentose from the base.*

Leaves glabrous. Flowers red or of a pale or greenish yellow .	1. *A. rufa.*
Leaves mostly tomentose. Flowers of a rich yellow	2. *A. pulcherrima.*

SECT. 2. **Ceratandra.**—*Racemes or spikes several, in a divaricate panicle or 2 on a once-forked rhachis. Anthers tipped with a gland-like appendage. Ovules rather numerous in each cell. Stems glabrous at the base.*

Racemes or spikes several, in a divaricate dichotomous panicle. Perianth moderately curved	3. *A. flavida.*
Racemes 2 of 2 or 3 flowers each on a once-forked rhachis. Perianth very much curved, with a very oblique limb	4. *A. Preissii.*

SECT. 3. **Haplanthesis.**—*Racemes or spikes single or rarely 2, on a simple or rarely once-forked rhachis. Anthers inappendiculate. Ovules numerous and crowded in each cell.*

Perianth (under 2 in.) incurved, not contracted above the middle, redder upwards than at the base. Anthers shorter than the filaments. Leaves rather broad	5. *A. humilis.*
Perianth (nearly 3 in.) green, rarely yellowish throughout, not contracted above the middle. Anthers as long as the filaments. Leaves very narrow	6. *A. viridis.*
Perianth (about 3 in.) green with a red rarely yellow base, not contracted above the middle. Anthers much longer than the filaments. Leaves rather broad	7. *A. Manglesii.*
Perianth (2 to 3 in.) green with a red rarely yellow base, much contracted above the middle. Anthers as long as the filaments. Leaves very narrow	8. *A. bicolor.*

SECT. 1. DIANTHESIS.—Racemes or spikes several, in a divaricate dichotomous panicle. Anthers not appendiculate. Ovules 2, rarely 3 or 4 in each cell. Stems tomentose from the base.

1. **A. rufa,** *Labill. Voy.* i. 411, t. 22, *Pl. Nov. Holl.* ii. 119. Rhizome horizontal, thick and woody. Radical leaves above 1 ft. long, flat but

rather thick, about 2 lines broad, glabrous when full grown, the margins usually scabrous. Stem or leafy scape 3 to 5 ft. high, densely covered from the base with a short soft plumose tomentum, bearing a few short leaves, branching at the top into a broad dichotomous panicle, with a small lanceolate bract under each branch. Flowers on very short pedicels, in close unilateral racemes on the ultimate branches of the panicle, covered as well as the whole inflorescence with a plumose wool assuming a red or rich purple colour rarely varying to a brown or pale yellow. Perianth-tube including the adnate base about 1 in. long, with reflexed hairs or linear-ciliate scales inside below the middle; lobes lanceolate, 4 to 5 lines long, oblique but not so much so as in *A. Preissii.* Anthers oblong, without any terminal appendage, much shorter than the filiform filaments. Ovary short, with only 2 reflexed ovules to each placenta in all the flowers opened, the free summit very shortly conical.—R. Br. Prod. 301; Endl. in Pl. Preiss. ii. 25; *Schwægrichenia rufa*, Spreng. Syst. ii. 26; *A. tyrianthina*, Hook. Bot. Mag. t. 4507, copied into Lem. Jard. Fleur. t. 40 with an alteration of tint.

W. Australia. Lucky bay, *R. Brown;* King George's. Sound from the Kalgan to Cape Riche, *A. Cunningham, Drummond, n.* 327, *Preiss, n.* 1412, *Oldfield.*

2. **A. pulcherrima,** *Hook. Bot. Mag.* t. 4180 *copied into Fl. des Serres,* April, 1846. Very closely allied to *A. rufa,* the inflorescence and flowers the same in structure and indumentum, but the leaves are tomentose as well as the scape, and the wool of the flowers is of a bright yellow, sometimes slightly tinged with red. I find usually 2 perfect ovules to each placenta, but with the addition of 1 or 2 apparently abortive ones.

W. Australia, *Drummond, n.* 347.

SECT. 2. CERATANDRA.—Racemes or spikes several in a divaricate dichotomous panicle, or 2 on a forked rhachis. Anthers tipped with a gland-like appendage. Ovules rather numerous in each cell. Stems glabrous at the base.

3. **A. flavida,** *Red. Lil.* t. 176. Rhizome thick, with long radical leaves, and a stem of 3 or 4 ft., bearing a divaricately-branched dichotomous panicle as in *A. rufa,* but the leaves longer broader and thinner, attaining 3 to 4 lines in breadth and the stem at the time of flowering quite glabrous almost up to the panicle as well as the leaves, the panicle plumose-woolly as in that species. Flowers in one-sided racemes on the branches of the panicle, on pedicels of 1 to 2 lines, subtended by small narrow bracts, the wool of a dull yellowish green more or less red at the base of the perianth and sometimes a brown red nearly to the top. Perianth-tube about $1\frac{1}{4}$ in. long, glabrous and shining inside or minutely scabrous-dotted; lobes 4 to 5 lines long, pubescent inside, the lower ones more deeply divided than the upper. Anthers oblong-linear, almost as long as the filaments, the connective tipped with a small gland-like appendage. Ovules rather numerous in each

cell, crowded on oblong placentas. Capsule ovoid, 4 or 5 lines long, the conical summit within the perianth opening in 3 short valves.—Endl. in Pl. Pr. ii. 27; Bot. Mag. t. 1151; Bot. Reg. 1838, t. 37, and 64; *Schwægrichenia flavida*, Spreng. Syst. ii. 26; *A. grandiflora*, Salisb. Parad. Lond. t. 97; *A. coccinea*, Paxt. Mag. v. 271, with a plate; *A. Manglesii*, Maund, Botanist, t. 67, not of Don.

W. Australia. King George's Sound and adjoining districts, *R. Brown*, *Drummond*, *n.* 348, *Preiss*, *n.* 1411 *and* 1416, and many others; Blackwood river, *Oldfield*; Geographe bay, *Fraser*; Cape Naturaliste and Swan river, *Drummond*, *1st coll. n.* 746; Hampden, *Clarke*.

The species varies in the size of the flower, and very much in the colour of the wool, sometimes almost entirely red, sometimes green without any admixture of red, rarely with much of yellow.

4. **A. Preissii,** *Endl. in Pl. Preiss.* ii. 26. Stems 1 to 1½ ft. high, more or less clothed with a loose plumose reddish wool, more dense and redder towards the inflorescence. Leaves from a long sheathing base tapering into a narrow almost terete acuminate lamina, the lower ones 6 in. long or rather more, the upper ones smaller and distant, all glabrous. Flowers few, usually 3 to 6 on the short branches of the once-forked terminal peduncle, sometimes the whole inflorescence reduced to 2 or 3 flowers, and always appearing capitate when in young bud, although the pedicels ultimately attain 2 to 3 lines each subtended by a small bract. Perianth at least 2 in. long, very much curved in the bud, and the limb more oblique than in the other species, the adnate base very soon globular, the whole perianth densely plumose-woolly outside and usually more or less red, glabrous inside; lobes narrow-lanceolate, nearly ¾ in. long, the lower ones separated much lower down, and the tube often splitting between them. Anthers not very long, the connective tipped with a small gland-like appendage as in *A. flavida*. Ovules rather numerous, reflexed and irregularly crowded on the face of the placenta. Capsule globular, 5 lines diameter including the adnate perianth-base, opening at the top within the perianth in 3 very small rigid valves.—*A. minima*, Lehm. Pl. Preiss. ii. 274?

W. Australia. King George's Sound and adjoining districts, *Preiss*, *n.* 1413, *F. Mueller*, *Maxwell*.

F. Mueller, Fragm. viii. 23, places this species amongst those with simple inflorescence. I have, however, always found it once forked, except when reduced to 2 or 3 flowers, and then one of the pedicels seems to be rather a 1-flowered branch than a simple pedicel. The inflorescence in many specimens not yet fully developed appears capitate.

SECT. 3. HAPLANTHESIS.—Racemes or spikes single at the end of the stems with a simple rhachis, or rarely 2 the rhachis being once-forked, always unilateral several-flowered and rather dense. Anthers not appendiculate. Ovules numerous in each cell crowded on the placenta.

5. **A. humilis,** *Lindl. Swan Riv. App.* 46, t. 6 B. Rhizome thick, horizontal. Stems usually under 1 ft., rarely 1½ ft. high, loosely plu-

mose-woolly. Leaves chiefly radical, usually ciliate only with plumose hairs, but sometimes woolly-hairy all over or quite glabrous, flat, usually under 6 in. long and 2 to 3 lines broad but variable. Flowers nearly sessile in a simple terminal unilateral spike, or very rarely the rhachis once-forked forming a double spike, the pedicels rarely above 1 line long, the plumose wool red varying to a pale yellow and mixed with longer hairs. Perianth 1½ to 2 in. long, slightly curved, the limb very oblique, the lobes falcate-lanceolate, 4 to 5 lines long, the lower ones shorter but more deeply separated than the upper, the tube readily splitting on the lower side almost to the base; filaments filaform; anthers short, without terminal appendages. Ovules numerous, covering the placentas.—Endl. in Pl. Preiss. ii. 26.

W. Australia. King George's Sound and adjoining districts, *F. Mueller, Oldfield, Maxwell,* and others; Swan river, *Drummond, 1st coll. n.* 747, *Preiss, n.* 1418, *Oldfield.*

A. minima, Lehm. Pl. Preiss. ii. 274, which I have not seen, is referred by F. Mueller, Fragm. viii. 21, to *A. humilis,* but apparently only from the character given: the narrow leaves, almost capitate inflorescence, and curved perianth would rather indicate the *A. Preissii.*

6. **A. viridis,** *Endl. in Pl. Preiss.* ii. 25. Stems 1½ to 2 ft. high or more, glabrous or nearly so below the middle, loosely tomentose upwards. Leaves near the base of the stem from a broad sheathing base linear-subulate, 6 to 10 in. long. Flowers in a compact simple or unilateral terminal raceme, or rarely the rhachis once-forked forming a double raceme, the pedicels 1 to 3 lines long subtended by short linear-subulate bracts, the wool green throughout or yellowish towards the base of the flower. Perianth 2½ to near 3 in. long, the tube of equal breadth or very slightly contracted above the middle, splitting open on the lower side to near the base, with reflexed hairs or scales inside near the base, the limb not very oblique, the lobes 4 to 5 lines long, usually reflexed when open. Anthers linear, about as long as the filiform filaments. Ovules very numerous in each cell covering the placentas.

W. Australia. Swan river, *Drummond, 1st coll. n.* 745, *Preiss, n.* 1415; Vasse river, *Oldfield;* Pinjarrah, *J. S. Price;* Busselton, *Pries.*

7. **A. Manglesii,** *D. Don in Sweet, Brit. Fl. Gard.* ser. 2, t. 265. Stems 2 to 3 ft. high, slightly and loosely woolly towards the base, more densely so under the inflorescence. Leaves at the base of the stem flat, 6 in. to above 1 ft. long, usually 2 to 3 lines but sometimes ½ in. broad, quite glabrous. Flowers, the largest in the genus, in a simple terminal unilateral raceme, on pedicels usually of 2 to 3 lines, the plumose wool very dense, green except on the adnate base where it is usually red or rarely pale yellow. Perianth at least 3 in. long rather narrow and slightly incurved but not contracted above the middle, the lobes narrow, 4 to 5 lines long, the tube usually splitting open on the under side nearly to the base, glabrous inside except the long recurved hairs or ciliate scales near the base. Anthers linear much longer than

the short flat filaments, the connective without any appendage. Ovules numerous in each cell, covering the placentas.—Bot. Reg. t. 2102 (the hairs or scales inside the base of the tube represented as curved upwards instead of downwards) Bot. Mag. t. 3875.

W. Australia. Swan river, *Drummond*, 1*st coll. n.* 744, *Oldfield* (*Preiss, n.* 1420 ?); Blackwood river, *Oldfield; Busselton, Pries;* King George's Sound, *Baxter*, also *Drummond;* Lake Muir, *Muir* (the latter two with the base of the perianth of a dull yellowish white in the dried specimens); Gordon river, *Herb. F. Mueller* (with the raceme nearly 1 ft. long, the pedicels ½ in., the base of the perianth of a rich red).

8. **A. bicolor,** *Endl. in Pl. Preiss.* ii. 26. Resembles *A. Manglesii* and *A. humilis* in habit and in the red base of the otherwise green perianth, but readily distinguished from both by the shape of the flower. It is usually of the low stature of *A. humilis*, rarely much above 1 ft. high. Leaves chiefly at the base of the stems, glabrous, under 6 in. long, narrower and more tapering to the point than in *A. humilis.* Flowers 4 to 10, on pedicels of 2 to 4 lines in a close unilateral raceme, the rhachis simple in all the specimens seen, the wool of the adnate base of a rich red, the remainder green. Perianth 2 to near 2½ in. long, the tube at the base as broad as the adnate part but tapering towards the middle and much contracted in the upper half, much incurved at the end in the bud, the limb oblique, the lobes about 4 lines long, and the tube usually split open on the lower side to near the base. Anthers linear, but shorter than in *A. Manglesii*, on filiform very unequally inserted filaments. Ovules numerous in each cell, covering the placenta.

W. Australia. Kalgan, Perongerup, and other localities in the neighbourhood of King George's Sound, *Preiss, n.* 1417, *Oldfield, F. Mueller, Muir, Miss Warburton.*

Var. *minor.* Leaves under 3 in. long. Scape about 6 in., the perianth scarcely more than 1½ in. long.—M'Callum and Stokes inlets, *Maxwell.*

Var. *major.* Perianth fully 2½ in. long and rather less contracted above the middle. —Swan river, *Drummond*, 1*st coll*, *Fraser*, and an unknown collector who gathered it in 1839 and named it in Herb. Hooker *A. Mooreana*, mihi; Albany, *F. Mueller.*

7. MACROPODIA, Drumm.

Perianth persistent, the tube much elongated above the ovary, recurved and dilated at the end, the limb very oblique; lobes 6, lanceolate, almost induplicate-valvate, those on the lower side more deeply separated and the tube split open between them. Anthers oblong-linear, on slender filaments at the orifice of the tube, the cells free at the base only. Ovary superior, 3-celled, with thick double dissepiments, the summit scarcely prominent within the perianth. Style long, filiform, slightly clavate and stigmatic at the end. Ovules solitary in each cell, laterally attached. Fruit dry, not valvular, the seeds falling away separately with portions of the pericarp and adnate perianth-base, leaving the thick hardened dissepiments persistent with the axis. Testa somewhat crustaceous.—Herb with the habit and inflorescence of the paniculate *Anigozanthi.*

The genus is limited to a single species, endemic in West Australia. It is reunited with *Anigozanthos* by F. Mueller, notwithstanding the remarkable differences in the ovary and fruit.

1. **M. fumosa,** *Drumm. in Hook. Kew Journ.* vii. 57. Rhizome very short and thick. Leaves radical or nearly so, not above 1 ft. long, laterally flattened, often ½ in. broad, tapering to a fine point, of a rather thin consistence with acute edges. Stem stout, 3 to 4 ft. high, glabrous except the inflorescence, which is dichotomous but with few rather long branches, densely covered as well as the buds with a plumose wool very dark when fresh and black when dry; on the expanded flower the black wool is more scattered or entirely disappears leaving a closer but dense yellowish or whitish tomentum. Flowers almost sessile in dense unilateral spikes on the branches of the panicle. Perianth-tube about ¾ in. long, much incurved and expanded into the limb; lobes very oblique, nearly 1 in. long. Filaments almost as long as the lobes; anthers oblong-linear and tipped with a small gland-like appendage as in *Anigozanthos flavidus* and *A. Preissii.*—*Anigozanthos fuliginosus,* Hook. Bot. Mag. t. 4291.

W. Australia Moore river, *Drummond;* Hill river, *Oldfield;* Greenough flats, *C. Gray.*

TRIBE 3. HYPOXIDEÆ.—Rhizome or base of the stem short and thick, emitting thick fibrous roots and sometimes covered with the membranous or fibrous sheathing bases of old leaves so as to resemble bulbs. Leaves radical, horizontally flattened or rarely terete. Flowers solitary or in simple spikes or racemes. Perianth hairy or rarely glabrous. Seeds with a black crustaceous tuberculate or striate testa, the hilum produced into a hooked beak at the end of which the funicle is attached.

8. CURCULIGO, Gærtn.

Perianth persistent or at length withering away, the tube more or less elongated above the ovary (except in *C. recurvata*), the limb of 6 spreading nearly equal segments. Stamens 6, inserted at the mouth of the tube; anther-cells more or less free at the base. Ovary 3-celled, with numerous ovules in two rows in each cell. Style connate with the perianth-tube, shortly free above it with 3 erect or connate stigmatic lobes papillose outside. Fruit succulent, sessile within a sheathing bract. Seeds few, the testa striate, the funicle usually dilated.—Herbs with a thick rhizome and long flat or plicate-nerved radical leaves. Scapes very short or rarely longer than the spike. Flowers in short spikes or heads, each one subtended by a broad sheathing bract longer than the ovary and fruit.

The genus extends over tropical and Southern Africa and Asia, with one American species. Both the Australian species have a wide range over tropical Asia.

Leaves broad. Flowers in a dense nodding head on a scape of several inches. Perianth tube scarcely any . . . . 1. *C. recurvata.*
Leaves narrow. Flowers in an almost sessile spike. Perianth-tube filiform . . . . 2. *C. ensifolia.*

1. **C. recurvata,** *Ait. Hort. Kew.* ed. 2, ii. 253. Rhizome thick, with densely clustered fibrous roots. Leaves radical, glabrous or nearly so, the petioles 6 in. to near 1 ft. long with a broad sheathing base, the lamina oblong-lanceolate, 1 to 3 ft. long, strongly ribbed and plicate. Scapes densely woolly, from scarcely above ground to 6 to 8 in. high, recurved under the inflorescence. Spike or head of flowers nodding, very dense, ovoid or nearly globular, 1 to 2 in. diameter, with broadly lanceolate imbricate striate more or less woolly-hairy bracts, the outer ones usually empty and sometimes above 1 in. long, the upper ones subtending the flowers shorter. Perianth very woolly-villous on a short thick pedicel; segments of the limb ovate, spreading, 3 to 4 lines long, glabrous inside, separated almost to the ovary or united at the base in an exceedingly short ring. Filaments very short, the anthers oblong and erect, connivent into a cone round the style, which is slender slightly dilated and minutely 3-lobed at the stigmatic end. Capsule nearly globular, more or less succulent, softly hairy, about 3 lines diameter. Seeds globular, with a black rugose crustaceous testa.—Bot. Reg. t. 770.

Queensland. Rockingham bay, *Dallachy.*—The species extends over the eastern provinces of India and the Archipelago. Technically, the absence of any tube to the perianth above the ovary might place this species rather in *Hypoxis* than in *Curculigo;* but the inflorescence and other characters render it impossible to separate it generically from the closely allied *C. sumatrana,* which is in every respect a true *Curculigo.*

2. **C. ensifolia,** *R. Br. Prod.* 290. Stem short, produced into a descending rhizome with fibrous roots, and more or less covered with the scarious sheathing bases of old leaves. Leaves usually 6 to 9 in. long and $\frac{1}{4}$ to $\frac{1}{2}$ in. broad in the middle, but sometimes 1 to $1\frac{1}{2}$ ft. long and almost $\frac{3}{4}$ in. broad, tapering at both ends, with prominent nerves and more or less hairy especially towards the base. Spikes short and erect at the base of the leaves, the scarious sheathing bracts subulate-acuminate, often 1 in. long. Ovary almost sessile, elongated, enclosed in the bract. Perianth-tube filiform, hairy, $\frac{1}{2}$ to $\frac{3}{4}$ in. long above the ovary; segments of the limb usually 3 to 4 lines long, with lanceolate-pointed segments more or less hairy outside. Filaments short; anthers linear, the parallel cells shortly free at the base. Style column very short below the stigmas, which are as long as the anthers and connate or shortly free at the top. Capsule oblong, enclosed in the sheathing bract. Seeds several, the black testa elegantly striate but not tubercular.—*C. stans,* Labill. Sert. Austr. Caled. 18, t. 24; *C. orchioides,* Miq. Fl. Ind. Bat. iii. 585 and others, but not of Roxb.

Queensland. Prince of Wales and other islands off Cape York, *R. Brown;* Wide bay, *Leichhardt;* Rockingham bay, *Dallachy;* Fitzroy island, *C. Walter;* Broad Sound, *Bowman;* Keppel bay, *Thozet*; Moreton bay, *C. Stuart.*

N. S. Wales. Macleay river, *Beckler.*

The species has a wide range in eastern tropical Asia, for I can find no difference in the numerous specimens I have seen from Australia, New Caledonia, the Indian Archipelago, Bengal, China, and Japan. The Asiatic ones have been generally referred to the *C. orchioides,* Roxb. Corom. Pl. i. 14, t. 13, and I had myself considered them as a small variety of that species in the Hongkong Flora, p. 366 (where, however, I had by mistake described the perianth segments as 5 to 6 lines long instead of 3 to 4); but upon a more careful comparison with Roxburgh's plate and description, and with specimens probably authentic from Rottler's herbarium, it appears that the latter may be a distinct larger-flowered species of limited range in the Indian Peninsula, and possibly the same as Wight's *C. malabarica*. The South American (Guiana and W. Indian) *Hypoxis scorzonerifolia*, Lam., is scarcely distinguishable from the true *Curculigo ensifolia*, although for reasons unexplained it is still retained in *Hypoxis* by Seubert in the great Flora Braziliensis.

Var. *longifolia*. This may prove to be a distinct species if the characters are found constant. It is more slender and nearly glabrous. Leaves rigid, $1\frac{1}{2}$ ft long and only 3 to 4 lines broad in the broadest part, tapering into a long point and into a still longer petiole. Spike loose and elongated, each flower with its spatha or sheathing bract on a pedicel of $\frac{1}{4}$ in. or rather more. Perianth tube not so slender as in the typical form, but the segments of the limb the stamens style and fruit quite those of *C. ensifolia*.

N. Australia. Port Darwin, *Schultz, n.* 781.

9. HYPOXIS, Linn.

Perianth persistent, divided to the ovary into 6 rarely 4 nearly equal spreading segments. Stamens 6, rarely 4, inserted at the base of the segments; anthers oblong or linear, more or less lobed at the base. Ovary 3-celled rarely 2-celled, with many ovules in 2 rows in each cell. Style short, with 3, rarely 2, oblong or linear erect stigmas connate or free, papillose outside. Capsule globular oblong or linear, crowned by the persistent perianth, which (usually but not always in the Australian species) "at length falls off, carrying with it the top of the capsule, this then bursts into 3 valves and scatters the seeds." Seeds globular, with a crustaceous tubercular testa, the hilum prominent and hooked. —Herbs with bulbous or tuberous rhizomes, covered with sheathing membranous or fibrous scales. Leaves radical, flat or terete, usually hairy. Scape leafless or with a single sheathing leaf. Flowers white or yellow, solitary or few in a short raceme.

The genus is spread over tropical Asia and Africa, more abundant in South Africa, with two or three American species. Of the six Australian species, one is also in New Zealand, the others are all believed to be endemic.

Capsule globular or oblong, not above twice as long as broad.
- Anthers deeply divided at the base . . . . 1. *H. hygrometrica.*
- Anthers scarcely or very shortly lobed at the base.
  - Perianth-segments 3 to 5 lines long. Stamens nearly equal. Capsule ovoid or oblong . . . . 2. *H. glabella.*
  - Perianth-segments scarcely 2 lines long. Stamens alternately shorter. Capsule small, globular . . . . . 3. *H. pusilla.*

Capsule linear, 4 or 5 times as long as broad.
- Leaves subulate. Stamens alternately longer. Stigmas long and narrow . . . . 4. *H. leptantha.*
- Leaves linear terete. Stamens nearly equal. Stigmas short 5. *H. occidentalis.*
- Leaves narrow-linear but flat with prominent nerve-like margins . . . . 6. *H. marginata.*

1. **H. hygrometrica,** *Labill. Pl. Nov. Holl.* i. 82, t. 108. Rhizome thickened into a small tuber emitting thick clustered roots and covered at the top by the membranous leaf-sheaths not splitting into fibres. Leaves narrow-linear or almost filiform, from under 6 in. to nearly 1 ft.

long, usually sprinkled or ciliate with long slender hairs. Scape shorter than the leaves, almost filiform, from only 2 or 3 in. long with a single small flower, to near 6 in. with 2, 3, or very rarely more flowers. Ovary turbinate. Perianth yellow, the segments about 4 lines long in the common form but sometimes smaller, usually glabrous, the outer ones often darker coloured outside. Anthers deeply divided at the base into linear auricles. Style columnar, the stigmas ovate, erect, and connate. Capsule obovoid-globular, under 2 lines diameter. Seeds globular, elegantly tuberculate.—R. Br. Prod. 289; Hook. f. Fl. Tasm. ii. 36.

Queensland. Rockhampton, *O'Shanesy;* Armidale, *Perrott.*

N. S. Wales. Port Jackson to the Blue Mountains, *Sieber, n.* 153, *A. Cunningham*, and others; Gwydir river, *Leichhardt;* New England, *C. Stuart;* Clarence river, *Beckler.*

Victoria, *Harvey;* between Ballarat and Ballan, Loddon, *F. Mueller.*

Tasmania. Port Dalrymple, *R. Brown;* abundant throughout the island, ascending to 4000 ft., *J. D. Hooker.*

Var. *pratensis.* A small hairy slender variety, with 2 or 3 small flowers to the scape. —*H. pratensis*, R. Br. Prod. 289.—Hunter's river, *R. Brown;* New England, *C. Stuart, Leichhardt;* Rockhampton, *Thozet.*

Var. *elongata.* A larger hairy variety. Leaves flatter, often above 1 line broad. Flowers 2 to 5, each subtended by a subulate bract. Perianth segments fully 5 lines long.—Rockhampton, *Thozet, O'Shanesy;* Nerkool Creek, *Bowman;* Moreton bay and Condamine river, *Leichhardt;* Dawson river, *F. Mueller;* New England, *C. Stuart.*

2. **H. glabella,** *R. Br. Prod.* 289. The whole plant quite glabrous. Rhizome globular, bulb-like, covered with the fibrous remains of old leaf-sheaths. Leaves linear-subulate, terete or channelled above, expanded at the base into a narrow scarious sheath, the lamina varying from 2 to 3 in. long in some specimens, above 6 in. in others. Scape shorter than the leaves, with a long linear erect sheathing bract at or below the middle. Ovary oblong-turbinate. Perianth very variable in size, the segments from 3 to 5 lines long, but usually about 4 lines. Stamens nearly equal; anthers entire at the base or with exceedingly short obtuse auricles. Style-column shorter than the erect free stigmatic lobes. Capsule ovoid or oblong, but not above twice as long as broad when ripe.—Hook. f. Fl. Tasm. ii. 36, t. 130 A; *H. vaginata*, Schlecht. Linnæa, xx. 568.

Queensland? A specimen marked Warwick, *Beckler*, in Herb. F. Mueller appears to be this species.

Victoria. Port Phillip, *R. Brown;* Yarra-Yarra, Darebin Creek, *F. Mueller;* Werribee, *Fullagar.*

Tasmania. Abundant in pastures, &c., *J. D. Hooker.*

S. Australia. Lofty and Bugle ranges, Guichen bay, &c., *F. Mueller;* Yorke Peninsula, *Fowler.*

W. Australia? Specimens from Swan river, *Preiss, n.* 1601, referred here by Endl. in Pl. Preiss. ii. 14, have the longer stigmas of this species, but may yet perhaps belong to *H. occidentalis.* Specimens in flower only, without leaves or bulbs, from Greenough flats, *C. Gray*, have also the long stigmas of this species, but the anthers lobed at the base as in *H. hygrometrica.* The variety will require further investigation from complete specimens.

3. **H. pusilla,** *Hook. f. Fl. Tasm.* ii. 36, t. 130 B. A much smaller plant than *H. glabella*, with a similar globular bulb-like rhizome covered

with fibrous scales, but somewhat larger in proportion. Leaves filiform, rarely above 3 in. long, with short scarious sheathing bases. Scapes rarely 1 in. long, usually with 2 small setaceous bracts above the middle, and therefore perhaps sometimes 2-flowered, but only 1-flowered in all our specimens. Perianth-segments scarcely 2 lines long. Anthers shorter than in *H. glabella*, almost entire at the base, and 3 alternate stamens usually shorter than the others. Capsule globular, nearly $1\frac{1}{2}$ lines diameter, constricted under the perianth.

Victoria. Wendu Vale, *Robertson;* the capsule rather more obovoid, but distinctly contracted into a short neck.

Tasmania. Circular Head, *Gunn.*

The species is also in New Zealand.

4. **H. leptantha,** *Benth.* A small species, with the subulate leaves, short slender scapes with setiform bracts of *H. pusilla*, and the stamens as in that species alternately smaller with shorter anthers, but the ovary and capsule are long and narrow, the perianth-segments narrower and more acute, the inner ones considerably smaller than the outer, and the stigmatic lobes long and narrow.

W. Australia, *Drummond, Oldfield;* Greenough flats, *C. Gray;* Upper Hay river, *Miss Warburton.*

Should the characters derived from the inequality of the inner and outer perianth-segments and stamens, and the length of the stigmatic lobes prove to be inconstant, this may have to be reduced to a variety of *H. occidentalis.*

5. **H. occidentalis,** *Benth.* Nearly allied to *H. glabella*, equally glabrous, with a bulb-like rhizome and narrow leaves with scarious sheathing bases, and the scape with a sheathing almost leafy bract at or below the middle, but the ovary and capsule very differently shaped, the ripe capsule linear, often $\frac{1}{2}$ in. long and very narrow. Perianth-segments 3 to 4 lines long, the inner ones rather smaller than the outer; the parts of the flower frequently but not constantly reduced from 6 to 4. Anthers linear, very shortly lobed at the base. Styles very short, with lanceolate papillose almost plumose lobes, scarcely exceeding the filaments.

W. Australia. King George's Sound and adjoining districts, *F. Mueller, Muir, Miss Warburton.*

6. **H. marginata,** *R. Br. Prod.* 289. Leaves slightly hairy, flat though very narrow, the longest $1\frac{1}{2}$ ft. long, but some only 2 or 3 in., about 1 line broad, all with prominent nerve-like margins. Scapes filiform, 1-flowered, 2 to 4 in. long, with very long fine spreading hairs under the flower and upon the ovary. Capsule linear, nearly $\frac{1}{2}$ in. long and very narrow.

N. Australia. Islands of the Gulf of Carpentaria, *R. Brown.*

The specimens are very few, but quite distinct from any others known to me, whether from India or the other parts of Australia. An imperfect specimen, however, from Port Darwin, *Schultz, n.* 641, appears very nearly allied to it; it has only one imperfect small leaf, and I do not see the long hairs of *H. marginata.* The scape bears two flowers, one, already in fruit, has the long narrow capsule of *H. occidentalis* and *H. marginata.*

TRIBE 4. AGAVEÆ.—Tall often woody plants. Radical leaves usually numerous, either flat and thick or channelled or terete; stems also frequently leafy. Flowers usually large in terminal compound heads or thyrsoid panicles. Perianth glabrous, very petaloid, usually large.

10. DORYANTHES, Corr.

Perianth of 6 nearly equal deciduous segments spreading from near the base; the 3 inner ones slightly dilated at the base. Stamens 6; filaments linear-subulate or slightly dilated at the base; anthers elongated, the basal lobes closing over the filament. Style elongated, 3-furrowed, with an obtuse terminal 3-angled stigma. Capsule oblong-clavate or turbinate, with a woody endocarp opening loculicidally in 3 valves. Seeds flat, reniform, in 2 rows in each cell.—Very tall herbs, with clustered roots and very long and numerous radical leaves. Stems simple, with short leaves. Flowers large, red, in short spikes collected into a large terminal globose head or oblong thyrsus.

The genus is limited to Australia. Its nearest connexions are South African and American.

Flower-head nearly globular, 1 ft. diameter. Perianth-segments oblong-linear, 4 in. long . . . . 1. *D. excelsa.*
Flower-thyrsus oblong, 3 ft. long. Perianth-segments oblong-lanceolate, 2 in. long . . . . 2. *D. Palmeri.*

1. **D. excelsa,** *Correa in Trans. Linn. Soc.* vi. 213, t. 23, 24. Radical leaves "above 100," about 4 ft. long, "broadly sword-shaped." Stem attaining 10 to 18 ft., with numerous short linear-lanceolate erect leaves sheathing at the base. Flowers red, in a dense terminal globular head of 1 ft. diameter, surrounded by a few green acuminate leafy bracts or floral leaves, the spikes or clusters within the head of 3 or 4 flowers each, subtended by coloured lanceolate bracts, the rhachis 1 to 2 in. long, with an oblong or lanceolate coloured bract under each flower. Pedicels short and thick. Ovary or adnate perianth-tube about 1½ in. long; segments of the limb about 4 in. long, broadly oblong-linear, obtuse, thick, concave and undulate, nearly erect quite at the base, then spreading or recurved. Filaments about 3 in. long, adnate to the perianth at the base; anthers above 1 in. long.—R. Br. Prod. 298; Bauer Illustr. Pl. N. Holl. t. 12 to 14; Bot. Mag. t. 1685.

N. S. Wales. Port Jackson, *Bass;* George's river, *R. Brown;* Newcastle, *Leichhardt.*—W. Hill mentions a white-flowered variety which he found on Mount Lindsay.

2. **D. Palmeri,** *W. Hill.* Radical leaves "above 100," 5 to 6 ft. long and 2 to 2½ in. broad in the broadest part. Stem 6 to 8 ft. high, with linear-lanceolate acute leaves like those of *D. excelsa.* Flowers red, in an oblong terminal thyrsus about 3 ft. long, the rhachis and bracts of the same rich colour as the flowers, the spikes not close together, each with a thick short rhachis bearing 3 or 4 flowers. Bracts acuminate, the outer one of each spike as long as the flower, those subtending the flowers short. Perianth-segments oblong-lanceolate, pale

or white inside below the middle, about 2 in. long, spreading from near the base. Filaments thicker at the base than in *D excelsa*, and as well as the anthers much shorter than in that species.

Queensland. Mount Spicer near Cunningham Gap, *W. Hill;* Mackenzie river, *Hartmann.*—The above character is taken chiefly from W. Hill's description, and the drawings made by Miss Scott, from the specimen exhibited at Brisbane in 1870, confirmed by specimens of portions of the inflorescence.

TRIBE 5. EUAMARYLLIDEÆ.—Bulbous plants with horizontally flat channelled or terete radical leaves. Scapes leafless. Flowers in umbels or rarely solitary, surrounded by 2 or more membranous or coloured bracts. Perianth glabrous, often large. Stigma small.

11. CRINUM, Linn.

Perianth deciduous, slightly oblique, with a slender tube and 6 nearly equal lanceolate or oblong lobes. Stamens 6, inserted at the base of the lobes; filaments slightly dilated at the base; anthers linear or oblong, 2-lobed at the base. Style filiform, obtuse, with a terminal slightly 3-furrowed stigma. Ovary 3-celled, with several ovules (10 to 12 in the Australian species examined) in each cell, in 2 rows, bordering a narrow peltate placenta. Capsule often oblique and opening irregularly, with few rather large seeds.—Bulbous herbs usually tall. Leaves all radical, long, flat or channelled. Scape simple, leafless. Flowers large, mostly white, in a terminal umbel surrounded by a few membranous or slightly coloured bracts.

The genus extends over tropical and southern Africa and Asia. Of the five Australian species, one appears to be the same as a common Asiatic one, the four others are probably endemic.

Flowers sessile in the umbel or on pedicels shorter than the ovary.
  Filaments not ¼ as long as the perianth-lobes. Umbels few-flowered. Ovary usually beaked . . . . 1. *C. venosum.*
  Filaments more than ⅔ as long as the lobes.
    Umbels many-flowered. Ovary usually beaked . . . . 2. *C. asiaticum.*
    Umbels 1- or 2-flowered. Ovary not beaked . . . . . 3. *C. uniflorum.*
Flowers on pedicels usually longer than the ovary, which is not beaked.
  Perianth-lobes about 3 in. long and ¾ to 1 in. broad . . . 4. *C. flaccidum.*
  Perianth-lobes 2 to 2½ in. long and 3 to 5 lines broad . . . 5. *C. pedunculatum.*

1. **C. venosum,** *R. Br. Prod.* 297. Bulb and leaves not described, and none preserved to any of the specimens seen. Umbel of 6 to 8 flowers with 2 involucral bracts 2½ to 3½ in. long. Flowers sessile or here and there very shortly pedicellate, the ovary tapering at the top into a neck or beak, sometimes very short, sometimes longer than the ovary itself. Perianth-tube 3 to 4 in. long, the segments of the limb acuminate, 1½ to 2 in. long, 4 to 5 lines broad in the middle, not more conspicuously veined in the dried state than other species. Filaments 4 to 5 lines long; anthers nearly or quite as long.—Kunth, Enum. v. 567.

N. Australia. Coen river, Gulf of Carpentaria, *R. Brown;* Sweers island, *Henne.*
Queensland. Cumberland island, *R. Brown.*

2. **C. asiaticum,** *Linn.; Ker in Bot. Mag.* t. 1073. "Bulb produced into a column." Leaves long, attaining in the typical form a breadth of at least 2 in. Scape 1 to 2 ft. high. Flowers 8 to 20 or even more in the umbel, sessile or on pedicels shorter than the ovaries, the involucral bracts 2 to 4 in. long. Ovaries usually produced at the top into a beak varying from 2 or 3 lines to nearly 1 in. Perianth white, the tube above 3 in. long; segments of the limb 2 to near 3 in. long, acuminate, tapering at the base, 3 to 4 or rarely 5 lines broad in the middle. Filaments from $\frac{3}{4}$ the length to nearly the length of the segments, usually purple towards the base; anthers 3 to 4 lines long.—Kunth, Enum. v. 547; *C. arenarium*, Herb. in Bot. Mag. t. 2355; *C. australasicum*, Herb. Amaryll. 259, partly.

N. Australia. Victoria river, *F. Mueller;* Port Darwin, *Schultz, n.* 600.
Queensland. Rockingham bay, *Dallachy* (with shortly pedicellate flowers).

Var. *angustifolium.* Leaves under 1 in. broad.—*C. angustifolium*, R. Br. Prod. 297 partly; *C. confertum*, Herb in Bot. Mag. t. 2522.—King's Sound, N W. coast, *Chapman;* Adams bay, *Hulls;* Port Darwin, *Schultz, n.* 696; Port Denison, *Fitzalan;* Rockhampton, *O'Shanesy, Bowman.*

The specimen in Brown's herbarium, to which he has attached the label of *C. angustifolium*, appears to me certainly to belong to the narrow-leaved form of *C. asiaticum.* The marginal asperities described by him are distant, and appear occasionally but rarely on other specimens of *C. asiaticum.* In the same sheet are several specimens of *C. uniflorum*, to which the name of *C. angustifolium* would be more appropriate, but the leaves have no marginal asperities.

C. brachyandrum, Herb. in. Bot. Mag. under n. 2121, and Amaryll. 249, described from a single plant raised from tropical Australian seeds but since lost, should probably be reckoned amongst the varieties of *C. asiaticum.* The filaments, though said to be short, are described as very much longer than those of *C. venosum.*

3. **C. uniflorum,** *F. Muell. Fragm.* iii. 23. "Bulb ovate-globular." Leaves linear, long and flaccid, scarcely above 2 to 3 lines broad. Scape 6 in. to 1 ft. high, bearing only 1 or rarely 2 sessile or shortly pedicellate flowers enclosed in 2 bracts of about 2 in. Ovary not at all or scarcely contracted at the top. Perianth "white or slightly tinged with pink," the tube 4 to 5 in. long, recurved in the bud; segments of the limb $2\frac{1}{2}$ to 3 in. long, 4 to 6 lines broad in the middle. Filaments nearly as long as the segments; anthers long and narrow.

N. Australia. Coen river, Gulf of Carpentaria, *R. Brown;* Sweers island, *Henne.*
Queensland. Albany island, Cape York, *M'Gillivray;* sandy flats exposed to the sun, Somerset, Cape York, *Veitch.*

4. **C. flaccidum,** *Herb. in Bot. Mag. under n.* 2121 *and t.* 2133. Bulb not seen. Leaves as far as known elongated, varying in breadth from $\frac{1}{2}$ to 1 in. Scape $1\frac{1}{2}$ to 2 ft. high. Flowers white, usually 6 to 8 in the umbel, on pedicels varying from $\frac{1}{2}$ to 1 in., the bracts of the involucre much dilated at the base, 3 to 4 in. long. Ovary not beaked and scarcely contracted under the perianth. Perianth-tube 3 to 4 in.

long, the segments of the limb about $2\frac{1}{2}$ to 3 in. long, and much broader than in any other Australian species, $\frac{3}{4}$ in. in some specimens, nearly 1 in. in others, the whole size of the flower evidently variable. Filaments from $\frac{2}{3}$ to $\frac{3}{4}$ as long as the segments; anthers scarcely above 3 lines.—*Amaryllis australasica*, Ker in Bot. Reg. t. 426; *A. australis*, Spreng. Syst. ii. 54.

N. S. Wales. Darling river, *Mrs. Ford*, and thence to Cooper's Creek, *Victorian and other Expeditions;* Murray river, *C. Stuart;* "from the paddock of Mr. Bigge," *Leichhardt.*

S. Australia. Flinder's range, *F. Mueller;* in the interior, lat. 32° to 22°, *M'Douall Stuart.*

There is in the Banksian herbarium a specimen laid in as from New Holland, *R. Brown*, but no corresponding one in Brown's own herbarium. There may be therefore some mistake, for we have no evidence of the plant growing in those parts of Australia visited by R. Brown. *C. arenarium β*, Herb. in Bot. Mag. t. 2531, appears to me to represent the larger-flowered form of this species, which is that generally sent from the interior of N. S. Wales. F. Mueller's South Australian specimens belong to the smaller form figured Bot. Mag. t. 2133.

5. **C. pedunculatum,** *R. Br. Prod.* 297. Very near *C. asiaticum*, with apparently the same variation in the breadth of the leaf, and the flowers equally numerous in the umbel, but they are all on pedicels longer than the ovary, and the ovary is either not contracted at the apex or produced only into a very short beak. The perianth is also smaller, the tube rarely above 2 in. long, the segments about as long but usually very narrow.—*C. australe*, Herb. Amaryll. 246 (partly).

Queensland. Moreton bay, *Leichhardt?* in Herb. F. Mueller.

N. S. Wales. Port Jackson, *Backhouse;* Hastings river, *Beckler;* Glendon, *Leichhardt;* Springrove and Castlereagh river, *Herb. F. Mueller;* Lord Howe's island, *Fullagar.*

S. Australia. Murray river, *Behr;* Morunda to the eastern bend of the river, *F. Mueller.*

The figure of *C. pedunculatum*, Bot. Reg. t. 52, shows larger flowers and shorter pedicels than the dried specimens, and seems rather to represent the short-pedicellate form of *C. asiatica;* it was not drawn from any authentically Australian specimen. The *C. taitense*, Red Lil. t. 408, referred here by Ker, appears also to represent the same variety of *C. asiaticum.* Some fragmentary specimens from Burnet, *Haly*, and Curriwillighie, *Dalton*, seem almost intermediate between *C. pedunculatum* and *C. asiaticum.* The wild specimens in herbaria are, however, so unsatisfactory, and the cultivated ones in gardens so frequently uncertain as to their origin, that the distinction of species can only be established by studying them in their native country.

12. EURYCLES, Salisb.

Perianth deciduous, funnel-shaped, with a slender but usually short tube and 6 nearly equal broad lobes. Stamens 6, inserted at the mouth of the tube; filaments united at the base or to above the middle in a corona, produced between the free parts of the filaments into 2 lobes and sometimes separating between the filaments so as to leave them free but dilated into lateral appendages; anthers 2-lobed at the base. Ovary 3-celled, with 2 ovules in each cell collaterally attached to an axile placenta. Style filiform, obtuse, with a terminal stigma. Fruit more or less succulent, with 1 or few seeds.—Bulbous herbs. Leaves all radical, petiolate, the lamina broad with longitudinal rather distant veins and transverse veinlets between them. Scape leafless. Flowers usually white, in a terminal umbel surrounded by 2 or 3 membranous bracts.

The genus extends over some islands of the Indian Archipelago. Of the two Australian species, one has the general range of the genus, the other is endemic.

Leaves very broad and cordate. Corona not $\frac{1}{4}$ as long as the perianth-lobes . . . . 1. *E. amboinensis.*
Leaves ovate, not cordate. Corona not $\frac{2}{3}$ as long as the lobes 2. *E. Cunninghamii.*

1. **E. amboinensis,** *Loud. Encycl. Pl.* 242. "Bulb tunicate, brown." Leaves usually several, on rather long petioles, broadly ovate-cordate orbicular-cordate or almost reniform, often 8 to 10 in. long and as broad or broader, with numerous arcuate rather distant veins and transverse veinlets between them. Scapes 1 to 2 ft. high, bearing an umbel of numerous white flowers. Bracts of the involucre 2 or 3, shorter than the pedicels, which vary from $\frac{3}{4}$ to near 2 in. long, articulate immediately under the ovary. Perianth above 2 in. long, the lobes elliptical-oblong, rather longer than the tube. Corona 4 to 5 lines long, divided to below the middle into 6 lobes, each one divided at the apex into 2 lanceolate acuminate diverging lobes with the filament between them; stamens altogether shorter than the perianth.—Hook. Bot. Mag. under t. 3399; *Pancratium amboinense*, Linn., Red. Lil. t. 384; Bot. Mag. t. 1419; *Eurycles sylvestris*, Salisb. in Trans. Hort. Soc. i. 337; Kunth, Enum. v. 689, with several synonyms; *Pancratium australasicum*, Ker in Bot. Reg. t. 715; *Eurycles australasica*, Loud. Encycl. 242; *E. australis*, Schult. Syst. vii. 911; Kunth, Enum. v. 691.

Queensland. Island of Cairncross, *Veitch* (with flowers $2\frac{1}{2}$ in. long); Rockingham bay, *Dallachy;* Mount Elliot, *Fitzalan.*—The species is also in the Indian Archipelago.

2. **E. Cunninghamii,** *Ait. MS.; Lindl. Bot. Reg.* t. 1506. Bulbs about 1 in. diameter. Leaves on long petioles, ovate, not cordate, but veined as in *E. amboinensis*, varying from 4 to 10 in. long. Scapes about 1 ft. high, with an umbel of 6 to 10 flowers surrounded by 2 or 3 bracts. Perianth 1 to $1\frac{1}{4}$ in. long, the tube rather shorter than the obovate-oblong lobes. Corona about $\frac{2}{3}$ the length of the perianth, with 2 lanceolate lobes to each filament, and sometimes splitting lower down between them. Fruit apparently succulent, nearly globular, about $\frac{1}{2}$ in. diameter, but not seen quite ripe.—Bot. Mag. t. 3399.

Queensland. Brisbane river, Moreton bay, *A. Cunningham, F. Mueller, Leichhardt, C. Stuart,* and others; Rockhampton, *Bowman.*

13. CALOSTEMMA, R. Br.

Perianth at length deciduous, with a slender but usually short tube and a funnel-shaped or spreading nearly equally 6-parted limb. Stamens 6, inserted at the mouth of the tube; filaments united to about the middle in a tubular corona, truncate or shortly lobed between the free parts of the filaments; anthers 2-lobed at the base. Ovary 1-celled, with 2 or 3 ovules collaterally attached to one side of the cavity. Style filiform, obtuse, with a terminal stigma. Fruit globular, succulent, ripening usually only one rather large fleshy seed.—Bulbous herbs. Leaves all radical, narrow with close parallel veins or broad with more distant veins and transverse veinlets. Scapes leafless. Flowers variously coloured, smaller than in *Eurycles*, in a terminal umbel surrounded by 2 or 3 membranous bracts.

The genus is endemic in Australia. The remarkable reduction of the ovary to a single cell appears to be due to the early abortion of two of the carpels.

Leaves linear. Perianth tube dilated under the limb.
Flowers purple or pink (rarely white ?) . . . . 1. *C. purpureum.*
Flowers yellow (rarely white) . . . . 2. *C. luteum.*
Leaves ovate. Perianth-tube not dilated under the limb. Flowers white . . . . 3. *C. album.*

1. **C. purpureum,** *R. Br. Prod.* 298. Bulb truncate attaining $1\frac{1}{2}$ to 2 in. diameter. Leaves linear, usually developed after the flowering has commenced, sometimes very long, 2 to 4 lines broad, with close parallel veins. Scape attaining 1 to 2 ft., bearing an umbel of many flowers purple in the typical form but varying to pink or white. Bracts 2 or 3, acuminate, nearly equal or very unequal, the longest rarely exceeding the flowers and usually scarcely exceeding the pedicels. Pedicels $\frac{1}{2}$ to near 1 in. long, the outer ones at length often articulate above the middle. Perianth-limb usually about 5 lines long, the slender tube varying from half as long to nearly that length, and slightly dilated at the top gradually expanding into the spreading limb; segments obovate-spathulate, varying in breadth. Corona reaching to about half the length of the segments, very variable in its orifice, sometimes truncate between the filaments and entire 2-toothed or shortly divided, sometimes produced between the filaments into 1 or 2 obtuse lobes.—Bot. Mag. t. 2100; Bot. Reg. t. 422.

S. Australia. Head of Spencer's gulf, *R. Brown;* Barossa, *Behr;* Bugle range, *F. Mueller;* Burra-Burra, *Hinteracker.*

Var. *carnea.* Flowers rather larger with a longer tube, approaching those of *C. luteum*, but pink or white.—*C. carneum,* Lindl. in Mitch. Three Exped. ii. 39; Bot. Reg. 1840, t. 26.

N. S. Wales. Lachlan river, *Mitchell;* Murrumbidgee river, *Herb. F. Mueller;* Charleville, *Giles;* Lower Edward river, *Mein.*

2. **C. luteum,** *Sims, Bot. Mag.* t. 2101. Habit leaves and inflorescence of *C. purpureum,* of which it may be a variety. Flowers rather larger, the perianth-limb often above $\frac{1}{2}$ in. long, and yellow or rarely white. Corona usually truncate between the filaments or minutely toothed, but I have sometimes found it produced into short broad entire or notched lobes as in *C. purpureum.* Stamens often shortly exceeding the perianth.—Kunth, Enum. v. 688; Bot. Reg. t. 421 and 1840, t. 19; *C. candidum,* Lindl. in Mitch. Three Exped. i. 54; Kunth, l.c.

Queensland. Rockhampton, *Thozet*, *O'Shanesy;* Barcoo, *Schneider;* Darling Downs, *Law.*

N. S. Wales. Nammoy river, *Mitchell;* Liverpool Plains, *Leichhardt;* Castlereagh river, *Woolls;* Lower Darling river, *Mrs. Ford;* thence to Cooper's Creek, *Neilson;* Mount Murchison, *Bonney;* Mount Margaret, *M'Douall Stuart.*

The characters attempted to be drawn from the teeth of the corona and from the articulation or non-articulation of the pedicels vary in the same plant; the articulation appears only when the flowering is advanced; in some specimens I see no trace of it, in others it is very distinct upon some but not all of the pedicels.

3. **C. album,** *R. Br. Prod.* 298. Leaves resembling those of *Eurycles Cunninghamii*, but smaller, ovate, acute, tapering at the base, 3 to 5 in. long, 2 to 3 in. broad, with distant converging primary veins and transverse veinlets. Scapes 1 to 1½ ft. high, bearing an umbel of numerous white flowers on filiform pedicels of ¼ to ¾ in. surrounded by 3 or 4 bracts. Perianth-tube narrow, 4 to 5 lines long; segments of the limb narrow, not quite so long as the tube, spreading from the base so as to give the flower a more hypocrateriform shape instead of the more funnel-shape of *C. purpureum* and *C. luteum*. Corona produced between the filaments into lanceolate entire or bifid lobes as long as the filaments. Fruit globular, rather large.

N. Australia. Turtle island, Gulf of Carpentaria, *R. Brown.*—A remarkable species with the flowers of *Calostemma* and the leaves of *Eurycles.*

Order CXXIV. TACCACEÆ.

Flowers hermaphrodite, regular. Perianth superior, persistent, tubular or campanulate, 6-lobed. Stamens 6, inserted in the tube, opposite to the lobes; filaments broad, hood-shaped, shortly 2-lobed at the end; anther-cells 2, parallel, adnate to the inside of the hood and projecting between the lobes. Ovary inferior, 1-celled, with 3 parietal placentas and many anatropous or amphitropous ovules. Style short, with a broad umbrella-shaped stigma deeply divided into 3 bifid lobes. Fruit a berry, crowned by the withered perianth. Seeds ovoid, many-ribbed; albumen copious, embryo minute, near the hilum when basal, at a distance from it if lateral.—Herbs with a perennial tuberous creeping or fibrous rhizome. Leaves radical, on long petioles, large and much divided, or in species not Australian undivided. Flowers on a radical scape in a dense terminal simple umbel, usually intermixed with long filaments (barren pedicels), and surrounded by an involucre of a few large thin bracts.

The Order is limited to a single genus, represented in the tropical regions of the New as well as the Old World. The only Australian species has a wide range over the Indian Archipelago and the Islands of the South Pacific.

1. TACCA, Forst.

Characters and distribution those of the Order.

1. **T. pinnatifida,** *Forst.; Kunth, Enum.* v. 458. Rhizome tuberous, globular, attaining under cultivation a large size but not above

1 in. diameter in the few dried specimens where it has been preserved. Petioles erect, 1 to 3 ft. long below the ramification, divided always into 3 branches which are again often bifid or trifid or dichotomous or irregularly branched, each branch pinnate with remarkably variable distant or more or less confluent segments. In the larger varieties the three branches are often 2 to 3 ft. long, once bifid or trifid, the larger segments ovate-lanceolate acuminate and 4 to 6 in. long, but intermixed with smaller ones, some of which often very small, obovate or oblong and very obtuse; some, whether large or small, contracted into a short petiolule and distant, others decurrent along the petiole, or confluent with the next segments. In another Australian form (*T. Brownii*, Seem. Fl. Vit. 100) the leaf-branches are under 1 ft. long, the segments all more or less lanceolate and acuminate but very unequal in size. In a third rather distinct Australian form (var. *aconitifolia*, F. Muell., *T. maculata*, Seem. Fl. Vit. 103) the leaves are more regularly divided, the branches almost dichotomous with numerous linear-lanceolate segments regularly diverging so as to form a circle of about 6 to 8 in. diameter. Scapes the length of the petioles, arising from within its sheathing base, with sometimes a scarious sheathing scale outside the petiole. Involucres of few (about 4 to 6) oblong or lanceolate obtuse or acuminate bracts of a thin texture, 1 to 2 in. long, but those of the same involucre very unequal and often 2 or 3 additional very small ones. Flowers numerous, on pedicels of $\frac{1}{2}$ to 1 in. and usually mixed with numerous long filaments or capillary barren pedicels 3 to 4 in. long. Perianth-segments in some specimens ovate and 2 lines long, in others lanceolate and 3 lines, but the difference owing sometimes to different stages of development, the inner lobes scarcely larger than the outer. Berry ovoid-globular, $\frac{3}{4}$ to 1 in. diameter, contracted into a short neck crowned by the persistent withered perianth limb.—R. Br. Prod. 340.

N. Australia. Islands of the Gulf of Carpentaria, *R. Brown*, *Sweers;* King's Sound, N.W. coast, *Hughan;* Sea range and Victoria river, *F. Mueller;* Melville island, *Fraser;* Port Darwin, *Schultz*, *n.* 169, 195, 810, 832; Escape Cliffs, *Hulls.*

Queensland. Rockingham bay, *Dallachy;* Fitzroy island, *C. Walter.*—Both the large form considered by Seemann as typical, and the one named by him *C. Brownii* occur both on the N. coast and in Queensland, with some intermediates as to foliage; the var. *aconitifolia* was sent by F. Mueller with the *C. Brownii.* We have no means of matching any differences in the flower with these differences in foliage.

Order CXXV. DIOSCORIDEÆ.

Flowers unisexual, usually diœcious, regular. Perianth superior, of 6 lobes or segments. Stamens 6 or 3, inserted on the perianth or in the centre of the flower round a rudimentary ovary, shorter than the perianth-lobes; anthers with 2 parallel cells opening inwards. Ovary inferior, 3-celled, with 2 pendulous ovules in each cell. Stigmas 3, entire or 2-cleft, on a single or distinct styles, or sessile on the ovary. Fruit a 3-angled capsule, opening at the angles in 3 loculicidal valves, or an indehiscent berry. Seeds albuminous, with a minute embryo near the hilum.—Herbs often forming large tuberous rhizomes or woody half underground trunks, the stems usually slender long and twining. Leaves alternate or opposite, entire or divided, the nerves or primary veins parallel or digitate, the intervening veinlets reticulate. Flowers small, usually of a yellowish green, in axillary simple or paniculate spikes or racemes, the male inflorescence usually more branched than the female.

The Order comprises but few genera, dispersed over the tropical and temperate regions of the globe. The only Australian genus certainly belonging to it has the wide range of the Order. The second, imperfectly known and therefore doubtful genus subjoined, is endemic.

1. DIOSCOREA, Linn.

(Helmia, *Kunth.*)

Flowers diœcious. Stamens in the Australian species 6. Capsule 3-angled or 3-lobed, opening longitudinally at the angles, often leaving their nervelike margins free. Seeds winged.—Underground rhizomes often tuberous and known by the name of *Yams*. Stems twining.

A large tropical and subtropical genus, especially numerous in South America. Of the three Australian species, one is common in tropical Asia, the two others, as far as hitherto known, are endemic. The tubers of all three said to be eaten by the natives or colonists as native yams.

Perianth-segments broad. Capsule broader than long. Seeds winged all round.
  Leaves frequently opposite, ovate-cordate or triangular-hastate with a broad open basal sinus . . . . 1. *D. transversa.*
  Leaves all alternate, linear or linear-lanceolate, or the lower smaller ones triangular-hastate . . . . 2. *D. hastifolia.*
Perianth-segments narrow. Capsule longer than broad. Seeds winged at one end only. Leaves alternate, ovate-cordate . . 3. *D. sativa.*

D. lucida, R. Br. Prod. 295, described from a specimen without flowers or fruit, gathered on the Endeavour river by Banks and Solander, is evidently a species of *Roxburghia*, and probably *R. javanica*, Miq.

1. **D. transversa,** *R. Br. Prod.* 295. A slender glabrous twiner with a tuberous rhizome. Leaves opposite or alternate, triangular-hastate or ovate-cordate, the basal lobes always very spreading with a broad obtuse sinus, more or less acuminate, 5- or 7-nerved, usually 3 to 4 in. long and scarcely half as broad, but on barren shoots sometimes broader than long. Male flowers sessile and distant along the rhachis of interrupted slender spikes of 1½ to 2 in., the spikes clustered along the branches of axillary panicles. Perianth-segments orbicular, above 1 line diameter in some specimens, not half so large in others, but perhaps not fully developed, all much imbricate, the 3 inner ones rather smaller. Anthers on short filaments inserted on a disk, with a small rudiment of an ovary in the centre. Female racemes simple, 2 to 4 in. long, the flowers distant. Ovary oblong-linear. Perianth of the males but smaller. Styles short, distinct, the stigmas shortly 2-cleft. Capsule

shortly stipitate, the axis about $\frac{3}{4}$ in. long, the protruding angles or lobes about $\frac{1}{2}$ in. long.

N. Australia. Escape Cliffs, *Hulls.*

Queensland. Sent by numerous collectors as very common from Brisbane river, Moreton bay, *A. Cunningham*, and others, to Rockingham bay, *Dallachy.*

N. S. Wales. Hunter's and Paterson's rivers, *R. Brown;* Hastings and Clarence rivers, *Beckler*, and others; Richmond river, *C. Moore*, and others; New England, *C. Stuart.*

D. punctata, R. Br. Prod. 294, is referred by F. Mueller without hesitation to *D. transversa.* In Brown's herbarium there are two specimens in the same sheet, both in flower only, without the precise station for either. In both the leaves are similar to those of some specimens of *D transversa*, and one has the same obtuse broad perianth-lobes, but in the other the perianths are rather those of *D. sativa.* The *D. transversa* itself is evidently nearly allied to *D japonica*, Thunb., as described and figured by Decaisne in the Revue Horticole, 1855, p. 71, and confirmed by Japanese specimens in the Kew herbarium. The Australian plant has the leaves rather different, the basal sinus always very broad and open, and often the base quite straight and truncate. The fruit also appears to be rather larger, but that may be variable. The *D. batatas* Decsne., is certainly quite distinct from it, but probably the same as the common Indian *D. glabra*, Roxb.

2. **D. hastifolia,** *Endl. in Pl. Preiss.* ii. 33. A smaller species than *D. transversa*, with slender almost filiform twining stems. Lower leaves triangular-hastate, 3- or 5-nerved, under 1 in. long and sometimes as broad at the base as long, the upper ones all linear-lanceolate or linear, entire, $\frac{3}{4}$ to $1\frac{1}{2}$ in. long. Male spikes $\frac{1}{2}$ to $1\frac{1}{2}$ in. long, solitary in the axils of floral leaves, the upper ones appearing paniculate owing to the reduction of the subtending leaves to small bracts. Perianth-segments obtuse, nearly as large as in *D. transversa* but not so broad. Stamens similarly inserted in the centre of the flower round a minute rudimentary pistil. Female racemes very short, bearing usually only 2 or 3 flowers. Perianth small, with orbicular segments. Fruit and seeds of *D. transversa* or rather smaller, on a shorter stipes.—Kunth, Enum. v. 379.

W. Australia. Swan river, *Drummond, 1st coll. n.* 821, *Preiss, n.* 1954; Champion bay and Murchison river, *Oldfield;* Pinjarrah, *J. S. Pries.*

3. **D. sativa,** *Linn. Spec. Pl.* 1463, *Hort. Cliff.* t. 28, *but not of Kunth, Enum.* Stems from a tuberous rhizome elongated and twining, often bearing green globular bulbs in the axils of the leaves. Leaves alternate, broadly ovate, deeply cordate, usually 9-nerved, or when 7-nerved the extreme lateral nerves forked, very variable in size, in the Australian specimens usually 3 to 6 in. long and nearly as broad. Male spikes slender, 1 to 2 in. long when fully out, usually numerous, clustered along the branches of long narrow axillary panicles. Flowers very small, the perianth-segments very narrow, those of each series almost valvate in the bud, the inner ones linear. Stamens 6, the anthers almost sessile in the centre of the flower round a rudimentary pistil. Female flowers in longer slender single spikes. Capsule oblong, the axis about $\frac{3}{4}$ in. long. Seeds flattened, winged at the lower end only, assuming quite the form of a samara, the wing at least twice as long as the nucleus.—*D. bulbifera*, R. Br. Brod. 294; Wight, Ic.

t. 878 and of most modern authors but not of Linn.—*Helmia bulbifera* Kunth, Enum, v. 435.

N. Australia. Islands of the Gulf of Carpentaria, *R. Brown;* Goulburn island, *A. Cunningham;* Melville island, *Fraser;* Port Darwin, *Schultz, n.* 173; Port Essington, *Armstrong.*

Queensland. Rockingham bay, *Dallachy.*

The species is widely spread over East India and the Archipelago. I have already (Fl. Hongk. 368) expressed my belief that modern authors have transposed the Linnean names of the two species, *D. sativa* and *D. bulbifera*, both of which are apt to bear bulbs in the axils of the leaves.

2. ? PETERMANNIA, F. Muell.

Stamens 6; filaments capillary, with linear anthers. Ovary 3-celled with 4 or 5 ovules in each cell. Style filiform, undivided. Fruit a 1-celled berry, with several seeds.—Stems twining. Flowers in leaf-opposed panicles.

The genus is limited to the single imperfectly known species endemic in Australia.

1. **P. cirrosa,** *F. Muell. Fragm.* ii. 93. A glabrous climber with the aspect rather of the *Smilax* tribe than of the *Dioscorideæ*, the branches slightly angular and bearing a few minute prickles. Leaves alternate, rigid, oblong lanceolate or ovate-lanceolate, acuminate, 2 to 3 in. long, ¾ to 1 in. broad, tapering into a short petiole slightly twisted at the base as in *Smilax*, the longitudinal veins numerous, but mostly starting from the midrib below the middle and not reaching the base of the leaf, with reticulate veinlets between the veins. Panicles narrow and loose, leaf-opposed and rather longer than the leaves, the branchlets short and apparently few-flowered, but the specimens only retain 2 or 3 fruits the flowers having fallen away, and in the upper part the inflorescences are replaced by simple tendrils. Berries about 2 lines diameter, crowned by the scar of the fallen flower and containing several unripe seeds. There are no loose flowers with the specimen, but according to F. Mueller "Perianth-segments about 2½ lines long, only seen as well as the stamens in the female plant. Filaments glabrous, 1½ lines long, deciduous; anthers 1 line long. Styles 2 lines."

N. S. Wales. Cloud's creek, Clarence river, *Beckler.*—Of this I have only seen two very imperfect specimens in Herb. F. Mueller, one of which retains 2 berries in their place, and a few loose fragments of fruits. The "perfect stamens in female flowers," *i.e.*, hermaphrodite flowers, and several ovules in each cell of the ovary described by F. Mueller, are at variance with the ordinal character, and the venation of the leaves is different from that of any of the dictyogenous Monocotyledons known to me.

INDEX OF GENERA AND SPECIES.

The Synonyms and Species incidentally mentioned are printed in Italics.

END OF VOL. VI.